MICROBIAL TOXINS IN FOODS AND FEEDS

Cellular and Molecular Modes of Action

MICROBIAL TOXINS IN FOODS AND FEEDS
Cellular and Molecular Modes of Action

Edited by

Albert E. Pohland
Food and Drug Administration
Washington, D.C.

Vulus R. Dowell, Jr.
Late of Centers for Disease Control
Atlanta, Georgia

and

John L. Richard
Agricultural Research Service
Peoria, Illinois

Associate Editors

Richard J. Cole
Agricultural Research Service
Dawson, Georgia

Melvin W. Eklund
National Marine Fisheries Service
Seattle, Washington

Stanley S. Green
Food Safety and Inspection Service
Washington, D.C.

William P. Norred III
Agricultural Research Service
Athens, Georgia

Morris E. Potter
Centers for Disease Control
Atlanta, Georgia

PLENUM PRESS • NEW YORK AND LONDON

Library of Congress Cataloging-in-Publication Data

Symposium on Cellular and Molecular Mode of Action of Selected
   Microbial Toxins in Foods and Feeds (1988 : Chevy Chase, Md.)
     Microbial toxins in foods and feeds : cellular and molecular modes
of action / edited by Albert E. Pohland, Vulus R. Dowell, Jr., and
John L. Richard.
        p.   cm.
     "Proceedings of a Symposium on Cellular and Molecular Mode of
Action of Selected Microbial Toxins in Foods and Feeds, held October
31-November 2, 1988, in Chevy Chase, Maryland"--T.p. verso.
     Includes bibliographical references.
     ISBN-13:978-1-4612-7916-7   e-ISBN-13:978-1-4613-0663-4
     DOI:10.1007/978-1-4613-0663-4

     1. Microbial toxins--Mechanism of action--Congresses.
I. Pohland, Albert E.   II. Dowell, V. R.   III. Richard, J. L. (John
L.)   IV. Title.
     [DNLM: 1. Animal Feed--congresses.   2. Food Contamination-
-congresses.   3. Toxins--congresses.   QW 630 S988m 1988]
QP632.M52S96   1988
615.9'54--dc20
DNLM/DLC
for Library of Congress   90-14303
                                                                   CIP

Proceedings of a Symposium on Cellular and Molecular Mode of
Action of Selected Microbial Toxins in Foods and Feeds,
held October 31-November 2, 1988, in Chevy Chase, Maryland

ISBN-13:978-1-4612-7916-7

© 1990 Plenum Press, New York
Softcover reprint of the hardcover 1st edition 1990

A Division of Plenum Publishing Corporation
233 Spring Street, New York, N.Y. 10013

On behalf of the UJNR, we dedicate the
proceedings of this symposium
to the memory of V. R. Dowell, Jr.

UJNR TOXIC MICROORGANISMS PANEL MEMBERS

United States

Dr. John L. Richard, Chairman
Agricultural Research Service
Department of Agriculture
Peoria, IL   60604

Dr. Vulus R. Dowell[1]
Anaerobic Bacteria Branch
Centers for Disease Control
Atlanta, GA   30333

Dr. Albert E. Pohland
Food and Drug Administration
Department of Health and Human
  Services
Washington, DC   20204

Dr. Richard J. Cole
Agricultural Research Service
Department of Agriculture
Dawson, GA   31742

Dr. Melvin W. Eklund
National Marine Fisheries Service
Department of Commerce
Seattle, WA   98112

Dr. Stanley S. Green
Food Safety and Inspection Service
Department of Agriculture
Washington, D.C. 20250

Dr. William P. Norred III
Agricultural Research Service
Department of Agriculture
Athens, GA   30613

Dr. Morris E. Potter
Centers for Disease Control
Department of Health and Human
  Services
Atlanta, GA   30333

Japan

Dr. Ko Namba, Chairman
Environmental Health Bureau
Ministry of Health and Welfare
Tokyo, 100 Japan

Dr. Kageaki Aibara[2]
Department of Food and Environmental
  Safety
Hatano Research Institute F&DS Center
Kamagawa-Ken, 257 Japan

Dr. Masaru Manabe
Department of Applied Microbiology
National Food Research Institute
Ibaragi, 305 Japan

Dr. Susumu Kumagai
Department of Biomedical Research
  on Food
National Institute of Health
Tokyo, 141 Japan

Dr. Katsuro Ashida
Molecular Biology Section
National Research Institute of
  Fisheries Science
Tokyo, 104 Japan

Dr. Nobuyuki Terakado
2nd Laboratory of Bacteriology
National Institute of Animal Health
Ibaragi, 305 Japan

Dr. Masakatsu Ichinoe
Laboratory of Food Microbiology
National Institute of Hygienic
  Science
Tokyo, 158 Japan

Dr. Katsutosi Mise
Division of Microbiology
National Institute of Hygienic
  Science
Tokyo, 158 Japan

[1]Deceased
[2]Retired

PREFACE

Although toxigenic fungi have been known since ancient times, modern mycotoxinology probably began with the early work of Joseph Forgacs and his colleagues in the 1950s. This science grew tremendously with the discovery of aflatoxins and aflatoxicoses in the early 1960s, particularly following the finding of the carcinogenicity of aflatoxins in test animal species. The discovery and identification of new mycotoxins and mycotoxi-coses, development of analytical procedures, attempted measurement of human and animal exposure, evaluation of toxicological effects, estimation of risk due to human exposure, and development of regulatory control programs have been the major research goals over the past 30 years.

In recent years there has also been an explosive growth in our knowledge of the metabolites produced by the algae, especially the dinoflagellate-produced toxins, and the transmission of such toxins up the food chain to reef fish and shellfish. This knowledge has been invaluable in understanding and controlling human illness resulting from ingestion of seafood. We are now, for the first time, able to deal with such common seafood-related human diseases as paralytic, neurotoxic and amnesic shellfish and ciguatera poisoning.

The third, and perhaps most important area from a public health point of view, is the microbiological concern posed by the contamination of foods with bacterial pathogens and their toxic metabolites. Again, there has been impressive growth in our understanding of bacterial-related diseases in recent years. Also, the biotechnological advances in produc-tion, detection, and mode of action of microbial toxins was a major part of this area of research.

The U.S. Panel of the United States-Japan Cooperative Program on Development and Utilization of Natural Resources (UJNR) Joint Panels on Toxic Microorganisms decided that the area of microbial toxins had developed to a point that there was a need for a compilation of some of the latest information on the cellular and molecular modes of action of these toxins. It was felt that the best vehicle for exchange and dissemination of such information was an international symposium, with subsequent publication of the proceedings. Planning for the symposium began in early 1987, and with the cooperation of the Japanese panel in Tokyo in November 1987, the respective sections, topics and invited speakers were selected. This symposium and the resulting publication dealing with the cellular and molecular aspects of algal, bacterial and fungal toxins is the first of its kind. The speakers and contributors to the publication are to be commended for their willingness to participate and their timely submission of manuscripts.

The symposium was jointly sponsored by the U.S. and Japanese Panels on Toxic Microorganisms of the UJNR, the U.S. Department of Agriculture's

Agricultural Research Service, Food Safety and Inspection Service, and Office of International Cooperation and Development, and the Department of Health and Human Services, Public Health Service, Food and Drug Administration.

Albert E. Pohland
Vulvus R. Dowell, Jr.
John L. Richard

ACKNOWLEDGMENTS

The organizers appreciate the financial support of the U.S. Department of Agriculture's Agricultural Research Service and Food Safety and Inspection Service, and the U.S. Department of Health and Human Services, Food and Drug Administration. We also thank the U.S. Department of Agriculture, Office of International Cooperation and Development and the National 4-H Center for their help in organization and administration of this symposium. We thank the speakers, chairs and authors for their participation, as well as those who presented posters - yours was an important contribution to the overall success of the symposium. We especially thank Drs. Pohland and Green and their staffs who spent endless hours on local arrangements, travel and other organizational matters. Special thanks to S. Mewborn of the Agricultural Research Service and N. LaFon of the Food Safety and Inspection Service for their dedication in retyping all of the manuscripts. Finally, we thank all of the members of the U.S. Panel for serving as members of the editorial board for this publication.

CONTENTS

MYCOTOXINS

UJNR PANEL ON TOXIC MICROORGANISMS

C. W. Hesseltine

5407 Isabell
Peoria, IL   61614

INTRODUCTION

The UJNR (United States-Japan Cooperative Program on Natural Resources) was established in May, 1964, between Japan and the United States. The joint program of scientific and technical exchange on natural resources was a joint attack upon the environmental and resources problems that confronted both countries. As organized, each country had a coordinator, and areas of mutual interest between the two countries were established. Each area of interest was organized into a panel with a chairman and an alternate chairman on each side.

The original panels were:

1.	Desalting	5.	Forage seeds
2.	Air pollution	6.	Water evaporation
3.	Water pollution	7.	Toxic microorganisms
4.	Energy		

Initially, the Toxic Microorganism Panel had its emphasis on botulism with the title, Botulinum and other Toxic Microorganisms. At about this same time the fungal toxin, aflatoxin, was discovered; it and other fungal toxins shared equally with botulism.

In this paper Panel refers to the members of both the Japanese and United States Toxic Micororganisms Panels. In the beginning each side had seven members with C. R. Benjamin and Komei Miyaki as respective chairmen and Keishi Amano and C. W. Hesseltine as alternate chairmen. The Panel contained a mix of people who were experts on fungal mycotoxins and bacterial toxins.

Through the years the number of scientists has remained constant and with approximately an equal number of experts on bacterial and fungal toxins. The initial members of the Panel representing the United States, besides Benjamin and Hesseltine, were Keith Lewis, M. T. Bartram, John Graikoski, Carl Lamanna and Eugene Sporn. Representing the Japanese were the above named K. Miyaki and K. Amano along with K. Aibara, M. Kambayashi, H. Kurata, S. Matsuura, M. Nakano and S. Sakaguchi. Initially Sporn and Lamanna were consultants, and Matsuura was added at the first joint Panel meeting. Lamanna and Sporn had

Microbial Toxins in Foods and Feeds, Edited by A.E. Pohland *et al.,*
Plenum Press, New York, 1990

been appointed as consultants because they were from the Defense Department, and there was some questions as to whether they could serve on the Panel as full members.

In establishing the membership of the panels, different departments were given the responsibility for organizing certain panels. The Panel on Toxic Microorganisms was assigned to Agriculture; hence Benjamin, who was research leader of the mycological investigations at Beltsville, was selected by the Department to be the chairman. He in turn selected Panel members with the approval of the appropriate department and the acceptance by the United States Coordinator, Dr. John C. Calhoun, from the Department of the Interior. The original Panel members came from the Departments of Agriculture, Interior, Defense, and Health, Education and Welfare as they had to come from government laboratories.

After establishment of the panels, the first U. S. meeting progress report on the UJNR was held on January 30, 1965. At this time each U. S. chairman prepared a report of his panel's progress. There were informal meetings on September 18, October 19 and January 9 preceding the January 30 report for the purpose of allowing the United States chairmen to discuss ways of cooperating with the Japanese chairmen. In his report of January 12, 1965, Benjamin stated that no U. S. Panel meeting had been held but that one was planned in February. He had been in contact with Dr. K. Miyaki, the Japanese Chairman in December, 1964 and at that time they had agreed that Panel activities should be focused upon botulism and mycotoxins, especially mycotoxins in rice and the aflatoxins. They agreed there should be an exchange of research papers on these two subjects.

The first concrete accomplishment was the supplying to the Japanese of a 300-page proceedings of a symposium on botulism and a list of current research projects in the United States on both types of toxins. In exchange, the Japanese Panel, with considerable effort, prepared and translated into English, "References on Botulism in Japan," in June, 1965. This contained the titles and short abstracts of some 90 Japanese papers. A similar bibliography consisting of 260 Japanese references on toxins produced by fungi was also supplied.

In the earliest report, Benjamin suggested areas of cooperation because no guidelines existed. Besides literature exchanges, he suggested an exchange of scientists, visits of panel members to each other's country, joint conferences and symposia. Later the members of the Panel decided there should be a statement of the Panel's activities and the microorganisms to be considered. This was put into a formal charter which reads as follows:

> "The activities of the Toxic Micro-Organisms Panel are concerned with the pathogenic microorganisms and their toxins that contaminate foods and feeds. This contamination is a threat to the public health and causes unacceptable economic losses. The scope of Panel activites includes all aspects of these microorganisms--their distribution, the conditions of toxin production, their chemical characteristics, and the methods of control and prevention. Types of microorganisms specifically encompassed by this panel are toxic fungi, *Clostridium*, *Staphylococcus*, and *Vibrio parahaemolyticus*. This selection does not preclude future consideration of other microbial agents that produce food-borne diseases."

Joint Panel Meetings

The first joint meeting of the Panel was in June, 1966 in Tokyo with 4 American and all 6 of the Japanese members attending. The Japanese members came from the Ministry of Health and Welfare (3) and the Ministry of

Agriculture and Forestry (3). At this meeting the procedure of the joint Panel meetings was established. The first formal meeting was held between the two chairmen who discussed the week's program, agreed on a meeting procedure and location, arranged a study tour agenda (planned by the host panel), discussed the program for the scientific sessions, arranged special lectures and arranged for informal meetings. Many helpful suggestions were made by the U. S. Embassy staff. Besides the chairmen, all of the United States members as well as some of the Japanese members attended this first joint meeting.

The first day began with a formal meeting of the Panel with observers from the U. S. Embassy and the Japanese government in attendance. The coordinator for the Japanese UJNR panel, Dr. Y Shigihara, gave a formal speech that was translated into English. The U. S. Scientific Attache responded. The guest alternate chairman then served as moderator for the formal session with the host alternate chairman serving the second session. At these sessions members reported on pertinent research in Japan and the United States. At this time there was further exchange of reports, statistics and research papers. This included a list of 260 papers published in Japan on mycotoxins that had been translated, in part, into English.

The main business of this meeting was to engage in discussions on how best to effectively cooperate. Areas suggested were a symposium on toxic microorganisms to be held in 1968, exchange of young scientists between countries, joint projects, further literature exchange, and a second joint panel meeting in Washington, DC in 1967. The remainder of the week was devoted to site visits to laboratories of the host panel members and to giving special lectures, open to the public, on toxic microorganisms.

The 1966 joint Panel meeting was considered so successful by both countries that joint Panel meetings became a regular activity with meetings held in alternating countries. A typical joint Panel meeting began with a chairman to chairman meeting on Sunday afternoon or evening. Monday's session began with short talks by administrative people and by the two chairmen. Typically, moderators for the sessions were the alternative chairmen with the guest alternate chairman serving first. The remainder of the day and sometimes the next day was devoted to informal reports by each Panel member on the past year's development in his field, not only at his institute but information from other laboratories in his country. Each Panel member typically represented a special field. For example, on mycotoxins, one expert would be from his country's regulatory agency. Reprints of pertinent reports from both governments and private agencies as well as reports of poisoning outbreaks were exchanged among all panel members.

Also at these meetings, plans were made for the time, place and emphasis for the next meeting and included discussion of how to improve cooperation and future plans. Much valuable information was exchanged besides reports, such as, standards, samples and cultures. Considerable time was devoted to planning symposia and the publication of books on topics that had not been covered either by symposia or as books or reviews. The *Vibrio parahaemolyticus* symposium was a good sample because the Japanese were far ahead of the rest of the world in studies of this food poisoning bacterium.

After Dr. Okabe became chairman, the Joint Panel meetings spent less time on protocol items and more on scientific exchange. Since the time of his chairmanship the host panel, at considerable expense, had to furnish an interpreter which avoided many misunderstandings because none of the United States members understood or spoke Japanese, and some of the Japanese members were unfamiliar with spoken English. Typically the following day was a scientific session in which formal papers were presented. Usually outside experts from the host country in the field of toxic microorganisms were invited to participate, and the whole meeting was open to the public, although there were never any formal announcements made.

One day was devoted to a scientific session on mycotoxins, and another to bacterial toxins. Although at first botulism was the only bacterial toxin considered, the program was soon broadened to include many others (at these scientific meetings copies or abstracts of the talks were made available). Typically, on Thursday and Friday study tours were conducted, sometimes with the entire guest panel going together. At other times the members went to separate places which included industrial research laboratories, government institutes, academic institutions and factories. These tours were useful because the host members made it possible to visit places not open to the public.

One interesting and useful procedure was seen in a food handling factory where one could not leave a rest room until both hands were put in a sanitiz-ing liquid and a button pressed which unlocked a door to let you out.

The program was concluded either Friday or Saturday morning. Individual panel members usually made further arrangements to visit other institutions and specialists in their field. During these tours one or more of the panel members presented a lecture to the institute after the tour and discussions had ended. Table 1 is a listing of the types of places visited in the 1966 meeting in Japan.

This 1966 study tour was somewhat longer than usual because of the review of four projects supported by the United States involving the use of soybeans in foods.

Study tours in the United States were not as many, in part, because of the long distances between locations. They tended to be to university depart-ments of microbiology and plant pathology, government research institutes, and to a lesser extent, to private or trade association laboratories. Among the two Panels, over a hundred study tours to various locations have taken place. To compensate for the greater numbers of sites visited in Japan, the United States Panel held their meetings in many different regions of the United States; thus, meetings were held in Washington, DC, Atlanta, Madison, Seattle, Texas A & M, New Orleans, Peoria, Ames, Orlando, and Bear River.

The Panel also evolved in the production of lengthy annual reports of the meeting and study tours. Also we tried to include in the Annual Report the accomplishments over the past year so that a record could be made to show our accomplishments. Some 22 separate reports have been prepared including one for 1987.

Exchange of Mutual Interest Items

This was a continuous activity of the Panel and often this occurred directly between individual researchers. Initially 150 cultures of toxin producing fungi were exchanged. In one instance 60 pounds of *Fusarium* con-taminated grain containing vomitoxin was sent to Japan for evaluation. Myco-toxin standards of aflatoxin B_1 and M_1 were exchanged. Disease resistant stocks of peanuts including some which were rust resistant were compared. Standards of Type A botulinal toxin were sent to Japan and in exchange the United States received Type E botulinal toxin and Type E progenitor toxin. The Chiba Serum Institute has supplied the United States with a vial of Type E botulinal antitoxin. Type cultures of *Vibrio parahaemolyticus* were sup-plied to the United States. Another item exchanged was spectral data of some mycotoxins. Methods for mycotoxin detoxification of peanuts and corn were exchanged because the Japanese and the United States methods are different.

Table 1. Study Tours (1966).

Institution	Location	Subject of Interest
A. *Universities*		
Institute Applied Microbiology	University of Tokyo	Culture Collection
Botanical Institute	Tokyo Univ. Education	Mucorales & Cellulases
Lab. Industrial Microbiology	Tokyo Univ. Education	Fermentation in general
Dept. Agriculture Chem.	Univ. of Tokyo	Fermentations
Dept. Microbiology	Tokyo Univ. Science	Yellow rice toxin
Lab. Microbiol. Chem. Dept. Agri. Chem.	Nagoya University	Mycotoxins and soybean fermentations
Faculty Agriculture	Tohoku University	Miso fermentation
Res. Instit. Food Res. & Dept. of Agri. Chem.	Kyoto University	Fermented soybean foods
Faculty Agriculture	Tokyo Univ. Education	Soybean cheese
B. *Japanese Government Laboratories*		
Food Research Institute	Ministry of Agri. & Forestry, Tokyo	Mycotoxins
Dept. Food Research	National Instit. of Health, Tokyo	Mycotoxins
Institute for Fermentation	Osaka	Culture collection and fermentation in general.
Noda Institute of Research	Chiba	Shoyu fermentation
National Institute of Hygienic Science	Tokyo	Mycotoxins
C. *Trade Association Laboratories*		
Japanese Tofu Association	Tokyo	Tofu
Japanese Shoyu Institute	Tokyo	Shoyu fermentations
Central Miso Institute	Tokyo	Miso fermentation
D. *Companies*		
Sendai Miso Shoyu Co.	Sendai	Miso production
Koji Sanzaemon Roho	Kyoto	Tane Koji production
Takeda Chemical Industry	Osaka	Antibiotic & flavor fermentations
Kikkoman Shoyu Co.	Chiba	Shoyu production

Books, especially those dealing with methods of inspection, have been freely exchanged. For example, the United States received from the Japanese the book titled, *Training in Storage and Presenvation of Food Grains*, 307 pages; and the book *Official Methods for Hygienic Inspection of Foods*, 615 pages along with their Tables of Contents translated into English. The United States, in turn, has supplied such books as *Examination of Foods for Enteropathogenic and Indicator Bacteria*. This was translated into Japanese by one of the Japanese Panel members and 1,000 copies were distributed. There are more examples of book exchanges.

Publications

One of the major accomplishments of the Panel's activities was the publication of books. The following is a list of the books published to date.

1) Herzberg, M. Editor. 1970. Toxic Micro-organisms: Mycotoxins and Botulism. Pub. UJNR Joint Panels on Toxic Micro-organisms & U. S. Department Interior, 490 p.

2) Fujino, T. G. Sakaguchi, R. Sakazaki and Y. Takeda. Editors. 1974. International Symposium on *Vibrio parahaemolyticus*. Saikon Pub., Tokyo, Japan, 261 p.

3) Rodricks, J. V., C. W. Hesseltine, M. A. Mehlman. 1977. Mycotoxins in Human Health. Pathotox Pub., Forest Park South, IL, U.S.A., 807 p.

4) Kurata, H. and C. W. Hesseltine. 1982. Control of the Microbial Contamination of Foods and Feeds in International Trade: Microbial Standards and Specifications. Saikon Pub., Tokyo, Japan, 342 p.

5) Richard, J. L. and J. R. Thurston. 1986. Diagnosis of Mycotoxicoses, Martinus Nijhoff Pub., Dordrecht, Netherlands, 411 p.

6) Eklund, M. W. and V. R. Dowell, Jr. 1987. Avian Botulism. Charles Thomas Pub., Springfield, IL, U.S.A., 405 p.

Many of these books had wide circulation. For example, the first book listed was distributed to 21 countries and has been out of print for several years.

Each of these books, except the first and third, covered subjects that had never been summarized in a book or review. For example, the second book introduced the scientific world outside Japan, where all of the pioneering research was done, to the food poisoning bacterium, *Vibrio parahaemolyticus*. Two of the books dealt with bacterial poisoning, two with mycotoxins and one covered botulism and mycotoxins. One attempted to evaluate the problems and regulation of microorganisms in world trade. This book presented the need for better standards, information on the amounts of losses due to microorganisms and problems where no information is available. For example, one cannot establish the cargo losses due to microbial spoilage on ships, nor the amount of insurance paid for these losses. Spoilage covers a wide area from bacterial contamination of meats to losses of corn due to mold growth.

Another activity of the Panel was joint publications. For example, C. Lamanna and G. Sakaguchi published a paper on "Botulinal Toxins and Problems of Nomenclature of Simple Toxins" in Bacteriological Reviews 32:242-249, 1971.

Symposia

One of the more successful activities of the Panel was the holding of sym posia of several days duration for which special funds had to be secured to cover travel, printing, rental of space and editing. The first symposium was held at the University of Hawaii, Honolulu, with participants and attendees restricted to Japanese and American citizens. Not only were Panel members involved, but speakers were invited from non-government laboratories. Sixtynine papers about mycotoxins and botulism were presented from October 7-10, 1968. The detailed arrangements and funding came from four United States Government departments. By 1973, 2,174 copies of the papers had been sold; and it is now out of print.

The second symposium was funded and sponsored by the Japanese panel in Tokyo, Japan, September 17-18, 1973. At the suggestion of the Japanese members of the Panel, it was decided to invite participants from other countries besides the United States and Japan. Consequently scientists from 15 countries attended, and 43 papers dealing with *Vibrio parahaemolyticus* were presented. This bacterium is related to the cholera bacterium, *V. cholerae*.

V. parahaemolyticus was discovered by Japanese workers in 1950 as a cause of food poisoning from shell fish in an outbreak of illness in Osaka Prefecture where 273 persons were poisoned and 20 deaths occurred. The typical symptom was acute gastroenteritis. Professor T. Fujino of Osaka University studied the disease known only in Japan at that time. Later, the organism was found to be present in shell fish in many places in the world. After the United States Panel alerted U. S. Public health officials to the poisoning, eleven outbreaks had been traced to this *Vibrio* in the United States by 1973. The Panel felt that there should be a collection of all of our knowledge about this problem by bringing research people together who worked with *V. parahaemolyticus*.

This symposium set a precedent in organizing later symposia. Topics for symposia should be on a subject not previously covered by either international meetings or as a book. Secondly, these symposia should be truly international with experts invited from all over the world. Thirdly, the papers presented should be published. Fourthly, the symposia should be sponsored alternately by the members of the two countries, and the sponsoring panel should be responsible for financing. This second symposium was financed by the Japanese Ministry of Health and Welfare.

The third symposium, held October 4-8, 1976, at the University of Maryland, dealt with mycotoxins in human and animal health. It was funded by grants from Food and Drug Administration, National Cancer Institute, National Institute of Environmental Health Sciences and the United States Department of Agriculture. At this symposium 71 papers were presented by speakers from 17 countries including scientists from 3 communist block countries. By the time of this conference it was apparent to many that there were other equally important groups of mycotoxins besides the aflatoxins. Although aflatoxin received considerable attention, sessions were devoted to the trichothecenes and *Fusarium* toxins, zearalenone and other mycotoxins. Other topics included the biomedical assessment of mycotoxin risks and regulation in the market place.

The fourth symposium tackled a more difficult problem, namely, the losses and control of microbial contamination of foods and feeds in international trade and microbial standards and specifications. This was held in Tokyo from October 6-8, 1981 with 39 papers presented with speakers from 8 countries.

Unlike previous symposia, this conference was based on papers grouped by commodities (microbiological problems on meat and meat products and feed; problems in marine and freshwater fish and fish products; economic losses; mold caused losses; and microbial standards and specifications for food and feeds in international commerce). One accomplishment in this conference was the realization of the lack of information on microbial losses in international trade. Economic and scientific data were also identified as areas for which there was a great need for more information.

The problems are enormous because one economist specializing in international trade stated that in 1979 the total agriculture and fishery product exports in the world was $212 billion U. S. dollars with food products amounting to $119 billion, feed and farm products $26 billion, agriculture raw material $32 billion and residual $34 billion. At this same meeting there was panel discussion by various country representatives on their views of export and import countries' spoilage problems. Although specifications can be determined and agreed upon by scientists, this is overshadowed by economic and political considerations.

The fifth symposium held October 11-13, 1984, in Ames, Iowa, addressed specifically the problem of diagnosis of mycotoxicoses in animals. Probably the biggest problem with mycotoxins is their occurrence in animal feed with the consequent reduction of weight gain and death. This meeting had 39 presentations and was attended by scientists from three countries. The subjects covered included: (1) the occurrence and clinical manifestation of mycotoxicoses in the United States and Japan, (2) comparative biochemical changes associated with mycotoxicoses, (3) immunological changes associated with myco toxicoses and (5) methods of physical-chemical analysis in the diagnosis of mycotoxicoses.

The holding of international symposia has been a very successful activity of the Panel and has given the cooperation between Japan and the United States high visibility.

Regulations

Considerable time was spent in discussing regulations concerning counts and levels of toxins allowed in different food and feeds. Early in the Panel cooperation a 2 mg standard of aflatoxin B_1 was given to a Japanese scientist and a sample of a Japanese product was sent to the United States for comparison of purity. Because levels of only 20 ppb are allowed, the accurate measurement of contamination levels in exported peanuts and corn are extremely important both for the export and import countries.

Another important accomplishment was the change in the *Salmonella* regulations in Japan on imported meat. In 1981 the United States exported poul try to Japan with a value of $181 million. The following brief account was prepared by Dr. Johnston in 1982 showing this accomplishment.

"During the past decades, Japan has rejected shipments of uncooked meat and poultry from the United States due to the presence of *Salmonella* bacteria. Rejection has been based on a "Food Sanitation Law" enforced by the Japanese Ministry of Health and Welfare. These rejections have proven to be costly by U. S. exporters and discourage continued efforts. Approximately 2 years ago, the Office of International Cooperation and Development, USDA, assigned the *Salmonella* issue to the U. S. Toxic Microorganisms Panel of the United States - Japan Conference on the Development and Utilization of Natural Resources (UJNR). During the past 2 years, the U. S. Panel Chairman, Dr. Clifford Hesseltine, USDA, ARS, organized scientific discussions on the issue with the Japanese panel in an outstanding and skillful manner. In October of 1981, he co-chaired with Dr. Hiroshi Kurata, the Japanese Panel Chairman, an International symposium on the control of microbial contamination of foods and feeds in international trade. The symposium was held in Tokyo and was attended by scientific representatives from all Asian countries with existing or potential markets for U. S. meat and poultry. The U. S. panel explained the *Salmonella* problem as one which is unavoidable and one which exists in uncooked meat and poultry throughout the world. Resolution of the issue was extremely important since many Eastern countries follow Japanese leadership in food laws."

"In October 1982, the 17th annual meeting of the Toxic Microorganism Panel was held at Madison, Wisconsin, and the Japanese Panel reported on actions taken on the issue. In Japan as in the U. S., laws are difficult to change. In August of 1982, the Ministry of Health and Agriculture issued guidelines to all import stations in Japan dictating how imported foods will be inspected. Beginning last August, inspection procedures for uncooked meat and poultry from the U. S. will be subjected to *sensory examinations only*. This is the same type of inspection conducted on uncooked meat and poultry entering the U. S. No microbiological tests will be conducted. Since *Salmonella* will not be looked for, there will be no further *Salmonella* rejections. It is apparent that U. S./Japan scientific discussions within the UJNR coupled with the influence of the Japanese scientists on their government have terminated the *Salmonella* problem for uncooked meat and poultry."

Many other examples could be cited where there was much discussion on problems of methodology, sampling procedures, allowed levels of microorganisms and toxin levels. Japanese and U. S. members visited each other's laboratories to compare assay procedures.

Interaction with Other Societies

At the Fifth Joint Conference on Microorganisms at Brigham City, Utah, the Panel met with the Interagency Botulism Research Coordinating Committee for three days. This was the first time the Panel had met with another group with similar interests. This established a precedent for future joint Panel meetings with other scientific groups whose interest overlapped with the Panel's interests. Two days of scientific meetings were held on botulism with papers being presented by both groups followed by a joint field trip to see botulism in water fowl in the Bear River area. This encouraged the Japanese experts to look at this problem of botulism poisoning in their water fowl as well as the problem in domestic birds.

In 1971 the Panel met with the Japanese Association of Food Hygienic Sciences at Takamatsu, Japan. This was a two day meeting with 65 papers on food safety presented. Papers of interest to the U. S. Panel were summarized in English, and an English translation of all the titles was made available to the U. S. Panel members. These papers dealt with microbial and chemical contamination of foods as viewed from the Japanese perspective.

The 1974 Panel meeting in Washington, D. C. in October was arranged so that the Panel members could attend the annual meeting of the Association of Official Analytical Chemists which had papers devoted to mycotoxins and related microbial problems.

At the 1975 meeting in Tokyo members of the Panel on mycotoxins met with the newly formed Japanese Association of Mycotoxicology, and three U. S. Panel members presented papers. At the same time members of the Panel involved with bacterial toxins met with the ad hoc Working Group on Diarrheal Diseases Caused by Previously Unrecognized Etiological Agents at the National Institute of Health.

At the 1979 meeting we met with the Kanto area meeting of the Japanese Society of Bacteriology and the fall meeting of the Japanese Food Hygienic Society at Sendai, Japan. These meetings had papers dealing with topics of interest to some of our Panel members. Such meetings gave the Panel an idea of the Japanese research trends and also allowed United States Panel members to meet Japanese scientists working in the same field. Meetings of this type have contributed greatly to individual research. Most of such contacts and benefits are not documented because discussions and the exchange of ideas and things, such as samples, cultures, etc., are on a one to one basis.

Accomplishments

The Toxic Microorganisms Panel was one of the original panels of the

UJNR. Beginning in 1966 it has held 22 joint Panel meetings. It met continuously every year except 1985; however, it met twice in 1987. The accomplishments of the Panel are so numerous that only a partial list can be given. Many of the benefits of Panel cooperation are intangible and long range and, therefore, cannot be evaluated adequately.

1. U. S. Panel members first became aware of the importance of *Vibrio parahaemolyticus* as a food hazard during their first visit to Japan. Japanese panel members subsequently supplied the United States with cultures, isolation procedures and biochemical information useful for identification. This led to the discovery of outbreaks of seafood poisoning in the U. S. where the causal agent previously was unknown. The savings to the U. S. in research dollars is estimated to be $300,000.

2. Five international symposia were planned, arranged and financed; and the papers presented were edited and published. Participants came from a number of countries, and the topics selected for symposia were subjects not previously covered by conferences on a world-wide basis. One symposium was held jointly with the Third International Mycological Congress.

3. Literature exchange was a regular activity of the Panel. Material exchanged included original research papers (often in Japanese but with tables with English headings) and English abstracts. Other items included manuals, institute reports, and statistical data such as the number of food poisonings in the previous year and the amounts of contamination of domestic foods and feed and food imports with levels of contamination. Especially important were regulations in effect on levels of microbial contamination. Over the years it is estimated that 3000 papers have been exchanged among scientists. In one year alone 575 items were given to us by the Japanese. Each item, in most cases, was given to each host member.

4. Japanese Panel members supplied us in 1966 with an English translation of 260 papers on mycotoxins published from 1891 to 1966. These references represent a Japanese search of 23 journals. In 1970 this was updated with translation of abstracts of 59 papers on mycotoxins. A similar search was made on botulism in Japan in which 71 references were cited, and these were accompanied by English abstracts. Similar lists of publications were given the Japanese. The savings in library research and translation is estimated to be $50,000. These lists were distributed widely to researchers in the field in the United States.

5. Various members of the United States Panel presented special invited lectures. For example, Dr. K. Lewis gave a lecture at the Japanese NIH and C. Hesseltine gave one at the University of Tokyo which was published later by the Japanese in their journal *Science* in 1966.

6. Exchange of cultures of toxin producing bacteria and fungi occurred many times. Various types of mycotoxins and microbial toxins and antisera of botulinal toxin were exchanged. There is no exact number of such exchanges, but by 1968 at least 150 cultures had been exchanged. One particularly important culture was the type strain of *Vibrio parahaemolyticus* that causes serious food poisoning.

7. Exchange of toxin samples, including milligrams of pure aflatoxin, was made to check on purity so that the same results could be obtained in exported and imported oilseeds and cereals. Likewise there was an exchange of chemical structure data and official methods of toxin detection.

8. Scientist exchange was limited to a few months exchange. One Japanese scientist spent 4 1/2 months in the U. S. observing U. S. mycotoxin research. One U. S. Panel member made a month's tour of Japanese facilities studying the *Salmonella* problem in meat, especially beef. Another U. S. scientist spent a month looking at the problem of mold infestation resistance in Japanese strains of peanuts. Indirectly, arrangements were made for young Japanese scientists to work in U. S. laboratories on related subjects in fermentation and applied micro-biology. Panel members have had a few days tours of research facilities in Japan. During these visits they met many scientists in the host country.

9. U. S. Public Health publication No. 1142, "Examination of Foods for Enteropathogenic and Indicator Bacteria," was translated into Japanese and published by a Japanese company.

10. On numerous occasions throughout the Panel's existance, members of the Panel have served as intermediates to arrange contacts among non-Panel scientists in the two countries. The resulting conferences and exchange of technical information on subjects ancillary to the Panel's interests have been useful in facilitating international relationships and trade between Japan and the United States.

11. A joint research project compared the allergenic properties of botulinal antitoxin manufactured in the United States and Japan. The results showed the Japanese method of manufacture to result in a product with less potential for causing harmful allergenic reactivities than the United States product. This information was utilized in the U. S. and had an estimated savings of $100,000 in research.

12. Large losses among migratory waterfowl, especially in the Western flyway, occur annually due to Type C Botulism. This American experience was called to the attention of Japanese investigators. This led to the first-time discovery of Type C botulism as a cause for epidemic outbreaks of poisoning of wild waterfowl in Japan.

13. As a result of item 12 above and the large losses in domestic fowl, a Panel committee was formed to publish a book on avian botulism in wild and domestic birds. A publisher underwrote publication costs; and the book *Avian Botulism*, Charles Thomas Publisher, Springfield, IL, 405 pp., appeared in 1987. Authorship was international.

14. The Panel has developed an annual scientific program, in some cases in connection with other scientific meetings, both in the United States and Japan. This has allowed many contacts by Panel members with scientists in the other country.

15. At the Third International Mycological Congress at Tokyo in 1983, the Panel was one of the two sponsors of a number of sessions on mycotoxins. Papers presented at this meeting were published as a book, *Toxigenic Fungi - Their Toxins and Health Hazard*, Editors, H. Kurata and Y. Ueno, Kodansha Pub., Tokyo, 363 pp, 1984. Dr. Kurata was the chairman of the Japanese Panel at that time.

16. One of the United States Panel members worked closely with Japanese importers of peanuts and peanut producers on problems of preventing aflatoxin formation in peanuts, improving peanut quality, biological control of soilborne, toxin-producing, molds and exchange of peanut genetic stocks.

17. United States Panel members have had extensive contacts with Japanese trade associations importing U. S. soybeans and peanuts, and have given advice to the Japanese government on the regulations governing the import of these commodities. The Japanese have issued regulations that accept some of our ideas. These regulations favor commodities that have been handled properly to prevent mold growth, thus favoring U. S. products. Also pointed out to the Japanese was that soybeans, a very important item in the Japanese diet, are free of aflatoxin.

18. The Panel has published 6 books and contributed greatly to a seventh. Topics of the books were botulism and mycotoxins, *Vibrio parahaemolyticus*, international trade, mycotoxins in human health, mycotoxins in animals, avian botulism and ecology of mycotoxin producing fungi.

19. A rough estimate in dollars saved by the United States through projects of this Panel is in the range of 1 1/2 to 2 million dollars. Additionally there are a number of projects that have contributed greatly to the health and well-being of the people which cannot be given a monetary figure.

20. One of the most productive activities of the Panel is the scientific sessions of typically one-day duration for bacterial toxins and one day for mycotoxins. Abstracts, tables and papers are distributed. These usually are items that have not been published. Scientific meetings are open to the public and often papers are presented in the appropriate session by non-Panel members. These meetings are quite separate from our joint meetings with other scientific societies.

21. At the joint Panel meetings it has become customary to report outbreaks of food poisoning in each country. Besides botulism this was expanded to reporting on *Vibrio cholerae*, *V. parahaemolyticus*, *Legionella*, *Staphylococcus*, *Salmonella*, *Yersinia*, *Clostridium* and outbreaks where the causal agent is unknown. The mycotoxins receiving the most attention were aflatoxin and the *Fusarium* toxins.

22. One member of the U. S. Panel did an extensive tour of meat packing plants in Japan to look at sanitation conditions in slaughter houses and meat processing plants. The Panel was specifically assigned the responsibility of coordinating research on *Salmonella*, and this has become a major item of discussion at Panel meetings.

23. As a result of discussions among the two Panels, Japan's standards for accepting raw meat and poultry have been modified and are relaxed and now are similar to those of the United States. In 1982 inspection procedures for meat and poultry, to be cooked, from the United States consisted of sensory examination only.

24. Because of the information and cultures supplied by the Japanese of *Vibrio parahaemolyticus* on isolation and characterization, it was possible to identify outbreaks of food poisoning in the United States. By 1973, 11 such outbreaks from seafoods had been identified in the United States involving hundreds of people.

25. Study tours arranged in connection with Joint Panel meetings were helpful, especially to the U. S. Panel. We visited, separately and together, a number of government and academic institutions. The industries visited included chemical, meat packing, fermentation, food fermentation, aqua culture, rice harvesting and storage, box lunch preparation, mushroom production, peanut inspection, private basic research, chicken and dairy operations and vegetable oil. Many useful discussions and observations occurred which were made use of in our research. For instance a visit to koji fermentation plants opened our eyes to the use

of this technique to produce mycotoxins for experimental use. We were made aware of needed food research on soybeans to improve our export market.

Impact of Panel

The Panel's activities had a much broader impact than those described. Many long-time friendships were established between the members of the two Panels resulting in numerous frank, personal one-on-one contacts on research programs. These lasting contacts were not restricted to Panel members but extended to many academic and industrial people who are interested in fermentation and enzymology. As a result of international symposia sponsored by the Panel, many contacts were made between scientists from other countries, not only with the Panel members, but with scientists from all over the world. Dr. C. Benjamin, the first U. S. Panel chairman, in his report on the Panel, cited the intangible benefits of the Panel as follows:

"The intangible and long-range benefits of panel cooperation are hard to assess, but would seem of great importance as cooperation benefits. Input of the panels contributes to increased research efficiency, to increased interest and research by workers outside the program, increased understanding of toxic micro-organisms, to expanded international cooperation and improved international relations, and to improved quality of biological research in general.

This statement is as true today as in 1969.

On a personal note a couple of spin-offs occurred from trips to Japan at joint Panel meetings. At these times contacts were made with young Japanese scientists to study at the Northern Regional Research Center. At least five spent up to a year working on our problems which were not associated with mycotoxin work. None of these were supported by USDA funds and most were funded from Japan.

A second spin-off was the observation of how dry inoculum of molds is produced to start solid state fermentations. I brought back this technology and used it widely in our laboratory. Now this technique has been used widely in the United States as a useful method for production of enzymes and fungal metabolites including mycotoxins, and for biotransformations.

Problems

Two problems have existed from the first days of the Panel's formation. One of the hopes was to exchange young scientists between the two countries. This activity never was satisfactory because of lack of funds and the identification of suitable candidates who would be willing to study for up to one year in another country on a suitable research subject in the field of toxic microorganisms.

From the very beginning the financing of travel was a constant problem for both Panels. The U. S. Panel never had funds set aside for travel to meetings of the U. S. Panel or for travel to joint Panel meetings. The Japanese Panel did have limited amounts of funds but never enough for all its seven or eight members. All the travel to meetings of the U. S. Panel in the United States were financed by regular laboratory funds. The Japanese Panel was more fortunate in that its members were in Tokyo or the vicinity. At the first joint Panel meeting, part of the U. S. Panel traveled on funds from the PL 480 projects in microbiology and fermentation in India and Pakistan. However most of the travel funds came from specific grants for symposia or one-time travel funds from the various departments who had members on the Panel. The Panels are deeply grateful for the support of administrators in

13

making funds available and to the great help of the coordinators of the two Panels. We are also grateful to the scientific and agricultural attaches in the Japanese and United States embassies for support and guidance. Each year required a great effort to get approval and funds for travel; and in some instances Panel members felt so strongly about the importance of the meetings, they used personal funds.

Another expense item was for the hiring of an interpreter to translate at the joint Panel meetings. This was an essential cost because no American was able to understand Japanese, and some of the Japanese were unable to use English. This item was later covered by assessing a registration fee from the host Panel's members.

The publication of books and papers allowed for dissemination of information generated by the Panel. However a problem exists as to how to get other information of value to the scientific community. Although annual reports were prepared, these do not circulate to any degree. Information of a great deal of value would be the occurrence and level of aflatoxin contamination in peanuts and corn that is found in shipments in international channels. A second example might be large food poisoning outbreaks caused by unknown agents.

Reasons for Success of Panels

If one accepts the thesis that the Panel on Toxic Microorganisms was and is successful, the questions can be asked why did it succeed?

1) The Panels have been kept to no more than 8 members. Large numbers on a Panel make it unwieldy and individuals do not feel their personal responsibility. A good way to guarantee failure is to appoint a large number of poeple to a committee.

2) The Panels always kept a balance with one or more people with administrative ability, but with the majority of the Panel made up of bench scientists who are at the front of their fields.

3) Because there are different types of toxic microorganisms (algal, bacterial and fungal), a constant balance of experts in each field is necessary. Bacterial toxins are quite different from algal or mold toxins in the nature of their structure and in the method of detection and assay. At least for the U. S. Panel, the chairman and alternate chairman were fungal and bacterial toxin experts.

4) Every Panel member had specific responsibilities and actively contributed papers and reports for the Panel meetings. When there was a lack of interest and contributions or retirement, the member was replaced.

5) Each year the annual report of the Panel meeting continued a list of the accomplishments for the year. This was necessary to have books, reports, etc., available to justify the Panel's existance by showing solid accomplishments.

6) The Panel got exposure by holding international symposia, by meeting with other related societies, and by publication of books with world-wide contributors.

7) The meetings of the two Panels with a regular informal exchange of new work, recent outbreaks of food and feed poisoning, new regulatory guidelines and similar exchange of unpublished information

become very useful to government department programs.

8) The exchange of information and materials were about equal between the two countries.

9) The procedure of the Panel meetings became rather fixed with little time devoted to business matters. This took up less than two hours, while two days were devoted to each Panel member's reports on what had happened in his field since the last meeting. One day was devoted to more formal reports by those people involved with bacterial toxins and a second day devoted to mycotoxins. Study tours of one or more days followed.

10) It is essential that most of each Panel's members serve for several years so that there is continuity of activities and plans. Never were there more than two new panel members apointed in one year on each side.

This account briefly reviews the history and activities of the Toxic Microorganism Panel. This Panel is continuing to make outstanding progress as evidenced by this symposium, under the chairmanship of Dr. John Richard and Dr. K. Namba.

BACTERIAL TOXINS

OVERVIEW OF BACTERIAL TOXINS WITH A NONREDUCTIONIST APPROACH TO THE MODE OF ACTION OF BOTULINAL NEUROTOXIN

Carl Lamanna

3812 - 37 Street, N.
Arlington, VA 22207

INTRODUCTION

The word toxin was first used in 1886 by E. Ray Lankester in *Science* to name poisons for animals produced by pathogenic bacteria. Since the word was first applied to substances which later proved to be proteins, historical precedence would demand its restriction to bacterial proteins responsible for animal pathology. An implied characteristic of these proteins is that their poisonous character can be neutralized by specific antitoxins. While toxin is used in this sense, it more commonly designates any poison for plant or animal produced by a living organism while the generic term for any harmful substance of biological or nonbiological origin is poison. Poison the older term was first used in 1579. It is meaningful to record the first use of words concerning poisonous actions as a measure of the slow progress of knowledge of toxins until the end of the 19th century: toxicology, 1799; toxication, 1821; toxicosis, 1857; toxicity, 1881; toxicant, 1882. These dates are recorded in the Oxford Universal Dictionary on Historical Principles.

The Occurrence of Toxins and Food Poisoning

A recent estimate of the number of discovered bacterial toxins is 220 by Alouf.[4] Gram positive organisms produce 105 and gram negative organisms 115. Of these only 13 have been crystallized.

The known genera of bacteria causing food poisoning are *Salmonella*, *Shigella*, *Escherichia*, *Pseudomonas*, *Aeromonas*, *Vibrio*, *Pleiomonas*, *Listeria*, *Yersinia*, *Campylobacter*, *Bacillus* and *Clostridium*. While specific toxins have not been identified as the root cause of pathology in all of these cases, a reasonable ambition of bacteriologists is to be able to do so.

There are marked differences in the incidence of reported food poisoning caused by species of these genera, and these vary among nations. The nature of foods and their capacity to harbor and support the growth of pathogens and synthesis of their toxins, and the existence of a suitable physical environment are the determinants of the potential for food poisoning to occur. Obviously the opportunity for food to be in contact with the habitats of toxin-producing bacteria before and after marketing is a primary risk factor. Habits of community and family sanitation, personal hygiene, practices of food storage and techniques of cooking are associated risk factors.

Microbial Toxins in Foods and Feeds, Edited by A.E. Pohland *et al.,*
Plenum Press, New York, 1990

In light of the number of toxigenic bacterial species known, what opportunities remain for the discovery of new kinds of food poisoning? Certainly
with any change in food processing and packaging there is opportunity for new
factors to enter into the equation of food safety.

Historically the search for a food poisoning toxin is to wait for an
event of poisoning to occur, and then to seek for a causative toxin. Based
on the sophistication of present scientific kcnowledge is it not timely to
deduce from known properties of microbial growth and toxin production the
likelihood of situations apt to reveal as yet unknown bacterial food
poisons? Let me give an example of such a possibility. Theoretically any
household use of long term refrigeration without all the water in foods
turning to ice should result in sufficient growth of hitherto unrecognized
cryogenic bacteria to accumulate toxins harmful on ingestion. This possibility is sustained by knowledge that candidate bacteria for this situation
include *Pseudomonas* organisms. *Pseudomonas* organisms inhabit soils and surface waters making them frequent contaminants of foods, are producers of a
variety of toxins, and are prominent among cryogenic bacterial species.
Trial and error research on growing cryogenic bacteria in refrigerated foods
in a search for new sources of food poisoning may be intellectually low key.
But in human affairs the unexciting when mundane, that is, of this earth, is
not always without reward.

To stay a charge of simplistic reasoning in sponsoring a trial and error
approach, we recognize the deductive process in seeking for new food poisonings must be multifactorial analysis. Let me give an example. Mushrooms
have been traditionally grown on soils with horse manure. The horse is a
notorious shedder of intestional *Clostridium tetani*. One might deduce from
this fact that mushrooms are a source of tetanus by oral toxicity. Yet this
does not appear to be so. Why? Probably because the habits of preparation
for consumption of mushrooms do not permit sufficient growth of tetanus
organisms with accumulation of toxin. Additionally, it appears the tetanus
organisms do not compete successfully against other anaerobes for growth in
the human colon and possibly in foods. Yet we should not dismiss entirely
the possibility that in a rare concatenation of events the tetanus organism
is found to cause a food poisoning. In conclusion we recognize the mere
presence of a toxin producing organism is not of itself a sufficient condition of food poisoning.

In a lifetime probably everyone suffers events of short term gastrointestinal upset without fever, usually cramps with or without nausea or
diarrhea. These are not reported to physicians. They do not come to the
attention of public health authorities. Are any of these due to ingestion of
toxins, and if so, are any as yet unrecognized toxins? The lay public cannot
be depended upon to be concerned about one day stomach aches and diarrhea.
It responds to such events by neglect or the use of over the counter drugs.
Because of such a situation, Bacteriologists Arise! At the first sign of
intestinal upset collect samples of food and body effluvia and rush off to
the laboratory. In the name of scientific adventure investigate your individual tummy ache. Encourage family members, friends and students to report
such episodes to you. Study of this kind of event may well expand the list
of etiological agents of food poisoning.

In decay of flesh foods, numerous bacterial species, including anaerobes,
participate. Do human habits of consumption which treat such spoiled foods
as gustatorially and esthetically unattractive blind us from knowledge of new
toxins to be acknowledged? A significant scientific problem of toxicology
which remains almost ignored is the physiological and biochemical bases for
the apparent immunity which permits carrion feeding animals to subsist without apparent harm on what should be toxin loaded food sources. Research on
this subject might reveal as yet unknown mechanisms of resistance to toxins,

neutralization of toxins, and natural destruction of toxins by decay processes. A report of a vulture serum component neutralizing type C botulinal
toxin has not been followed up to definitive conclusion.[12,37] For populations regularly consuming contaminated foods resistance could be due to
acquired antitoxin immunity or evolutionary adaptation by anatomaic or physio
logical means not related to antitoxin neutralization.[7] In an uncompleted
study of a few hundred human sera from Viet Nam a few were found to have a
low level of botulinal antitoxin. Definitive studies along this line should
be supported by public health agencies.

In dealing with decaying foods, particularly with the participation of
anaerobes, we can consider reopening the ptomaine theory of food poisoning[43]
which had been unceremoniously discarded by the 1920's.[13] Ptomàines are
basic compounds produced by decarboxylation of amino acids. To justify
reconsideration of the theory there is now the example of scombroid poisoning which is due to microbial production of histamine by decarboxylation of
the amino acid histidine.[6] Ptomaines can be toxic and among the other pathologies can cause symptoms of intestinal upset. Scombroid poisoning opens
the question of whether other ptomaines than histamine can be produced in
certain foods under special circumstances where their production does not
involve repulsiveness of odor, taste or other unesthetic quality. In this
regard let us be reminded that one man's gourmet food, Limburger cheese for
example, is another's abomination. In certain primitive societies and under
circumstances of poverty putrefaction may not be an absolute barrier to food
consumption. In such social conditions products of putrefaction of microbial origin may well play roles in food poisoning that is unrecognized in
affluent societies inhabited by bacteriologists and tax supported public
health authorities.

In focusing on the dramatic actions of a specific food poisoning toxin we
should not be unaware that in the natural situation ingestion may include a
variety of other toxic substances. Particularly in the case of anaerobic
breakdown of foods, numerous complex metabolites are produced. Might such
substances complicate the clinical picture of food poisoning? Let me illustrate. In the old literature of human botulism poisoning, one third of the
victims are said to suffer nausea, vomiting, and diarrhea as preliminary
signs of poisoning.[13] These disturbances cannot be due to the botulinal
neurotoxin which is more likely to cause constipation as an intestinal event.
This contradiction must exist because the victim has ingested more than one
kind of toxic substance. These might well include ptomaines and other harmful products of protein breakdown described in the early literature of microbial toxins.[43] That multifactorial toxicity in food poisoning can be a
clinical reality is now well documented for avian type C botulism. Certain
strains of C. botulinum produce both the classic neurotoxin called C1, a protein, and an entirely different acting nonneurotoxic entertoxin called C2, a
binary protein.[36]

To conclude the subject of search for new enterotoxins two suggestions
for trial and error research are made. Both should include the use of the
ileal loop technique for revealing changes in intestinal physiology. First,
a food can simply be kept without refrigeration and growth of any naturally
present bacteria permitted. Such foods, or water or organic solvent extract,
could then be tested for toxicity. Second, extracts of normal fecal material
could be tested for enterotoxic activity. Such study might reveal sources of
endogenously produced enterotoxins acting poorly or not at all on the colon
but capable of acting on the small intestine. While Esherichia coli enterotoxins would no doubt often be present, the mixed flora could certainly
include as yet unrecognized producers of entertoxins particularly among anaerobes and spore formers. I recognize this is a return to a kind of now aban-
doned research popular in the nineteenth century. But a modest return to
such effort will be greatly enhanced by current day sophistication in methods

of separation of chemical components from mixtures of large numbers of diverse substances.

Classification and Nomenclature

The state and problems of classification and nomenclature of bacterial toxins has been adequately reviewed.[8,9,32] What remains to be done is for a scientific society to seize the initiative in organizing a movement to develop an internationally accepted nomenclature to meet required standards of clarity, thoroughness, and accuracy in communication.

On questions of nomenclature, historical precedence should be followed except for a countervailing logical reason of new knowledge. We have arrived at such a case for *Clostridium botulinium* toxins. The capital letter designation of the botulinal toxins should, by historical precedence, be limited to antigenically distinct but similarly acting neurotoxins. By this rule C2 toxin is misnamed and because it is confusing should not be so named. The label C2 implies the existence of two distinct neurotoxins by a particular strain of *C. botulinum*. The C2 toxin is not a neurotoxin nor is it always produced along with the type C neurotoxin, nor only by the type C organisms. Let us call the C2 toxin what it is, namely enterotoxin, and if other botulinal entertoxins are found they can be called entertoxin 1, 2, etc.

Role and Ecology of Toxins

Has a toxin evolved to be a toxin? In seeking the natural role of a toxin should the focus be on the fact that it is a poison? Bacteria have a long history of existence preceding the appearance of the multicellular hosts they learned to parasitize. Evolution is in response to existing conditions, it is present rather than future directed. To the extent the evolutionary history of a toxin producer came before the event of food poisoning it is not possible to recognize a role for the toxin as a toxin in the bacterial economy. As natural inhabitants of soils and the bottom muds of estuary waters, *Clostridium botulinum* must have synthesized neurotoxin long before man appeared on the scene as a victim of these toxins because of faulty measures of food preservation.

The useful place of a toxin in a producing organism can be sought within two categories, a place in the structure of the organism basic to or assisting in its life functions as for cell wall endotoxins or as a means for adaption of the organism within its ecological niche, for example, by conferring a competitive advantage in mixed populations of species. The staphylococcal delta toxin can inhibit the growth of the gonnococcus. If the staphylococcus and gonnococcus compete for the same ecological niche within the urinary tract this would be the rare known case of a bacterial toxin conferring a competitive advantage. The competitive advantage situation cannot apply to toxin producing pathogens that invade and occupy sterile intracellular space of host cells and the sterile internal body spaces of plants and animals.

Many toxins exist in the bacterium as protoxins, nontoxic molecules. They must be converted to the toxic entity either by the sloughing off of a piece of the molecule or by a change in molecular three dimensional shape before or after they leave the bacterium. In these cases it is the protoxin rather than the toxin that may have a role in the bacterial economy.

Toxins may have their actions far from the vicinity of producing bacteria and thus are in no position to affect the immediate environment of the bacteria. Such is certainly the reality of the food poisoning caused by *Staphylococcus* and *C. botulinum*. Toxins act on anatomophysiological systems of the host not present in the parent bacterium. This is the case for botulinal neurotoxins which interfere with acetylcholine releasing systems unique to nerve cells. These kinds of observations do not reveal any useful role for the toxicity of toxins in the economy of bacteria. The lack of a useful role of toxins in bacterial physiology or ecology is also supported by the mortality associated with infectious diseases. Crippling and killing the hand that feeds you, an unethical act, has no compensating economic value for the offender. These considerations lead one to the following conclusions. A toxin may have a anatomophysiologic role in a bacterium if it is produced during growth and incorporated into a functional structure or enzymatic system. If they appear in the declining growth phases they may be products of catabolism, excretions, with no role in anabolism or continued viability. Their toxicity for plants and animals is an adventitious property; depending on your theology, either an accident of circumstance or a curse visited on mankind.

In seeking a positive role for a toxin in bacterial ecology by focusing on the toxicity of clinical interest there is danger of saddling ourselves with an intellectual blinder. We must be attentive to the possibility that a toxin does unexpected things. There is the recent report of such a situation for *Pseudomonas syringae* p.v. *tabaci*, a plant pathogen for tobacco which produces a nonprotein inhibitor of glutamine synthetase catalyzed ammonia assimilation. This action results in inhibition of growth of the tobacco plant. Yet this same toxin is a stimulator of nitrogen fixation, and thus plant growth, when the bacterium infests the root nodules of a legume.[21] Which action, the harmful one or the useful one is the toxic molecule's historically first evolutionary role and which is the adventitious one?

There is a difference between bacterial toxins and venoms in their known function in nature. Venoms are a single toxic substance or a mixture of substances including toxins and enzymes. They act to protect against an individual's enemies, or immobilize and begin the process of digestion of prey. Thus venoms can have protection values and nutritional functions which provide the rationale for usefulness for their constituent toxins. This cannot be said for bacterial toxins associated with infectious diseases of plants and animals.

Reductionist Philosophy and Bacterial Toxins

In research two opposing principles operate. These are reductionism and serendipity. Reductionism is the philosophical commitment the scientist has to attempt to reduce the operations of nature to the fewest number of laws. Serendipity is its opposite, the accidental discovery, the occurrence of the unexpected which generally proves conventional scientific wisdom to have been too simplistic. Reductionism is generalization while serendipity is particularity challenging generalization. Philosophically, scientists live by reductionism yet pray for serendipity and the wisdom to recognize it.

It is possibly more realistic for mathematical physicists to hope to define creation at the subatomic level with a simple equation or two than biologists to hope for a few sweeping generalizations to encompass the multifaceted phenomena of viability and its pathologies. In spite of the greater commitment of fund givers to reductionism than unorthodox thoughts, researchers in toxicology are better advised to be adventurous challengers of reductionism. It has unexpected discovery as its reward. The history of the development of genetics which is so much a part of microbiology today proves this. The elucidation of the genetic code and the repressor-operator theory of gene regulation in some quarters induced a reductionist euphoria that denigrated the possibilities for further grand discoveries. Yet the proof of existence of jumping genes by McClintock and a positive arabinase operon by Ellisberg proved Mother Nature to be more versatile than was expected by pioneer experts. Another example is Beckwith's proof that sexuality in

Table 1. Ability of crystalline type A botulinal toxin to repeatedly induce bradycardia with recovery in an intact individual rat (male) exposed to crystalline toxin type A.

Heart rate as percent of normal

Time after toxin injection (min)	1st exposure		2nd exposure 7 days after 1st*		3rd exposure 21 days after 2nd*		4th exposure 10 days after 3rd exposure*	Solvent control 5 days after 1st exposure
	Toxin alone	Toxin plus antitoxin	Toxin alone	Toxin plus antitoxin	Toxin alone	Toxin plus antitoxin	Toxin plus antitoxin	
0	100 (525)**	100 (510)	100 (540)	100 (510)	100 (525)	100 (510)	100 (510)	100 (510)
5	91	94	90	94	91	100	100	98
15	80	90	83	94	86	94	100	98
30	74	82	78	94	86	94	100	92
45	74	90	78	90	80	90	100	92
60	74	94	78	100	80	100	108	92
75	80	90	78	100	91	100	100	92
90	74	94	83	100	86	100	100	98
105	91	100	83		awake			awake
120	91		83					
Rat's Weight		410 g	400 g		420 g			

The rat was given 12 mg Na pentobarbital/kg. For the first injection 50 mouse LD$_{50}$ were injected; for the 2nd, 3rd and 4th exposures 5 mouse LD$_{50}$ were injected.
After the toxin injection, the rat awakened and the heart rate stablized, the rat was anaesthetized and given another injection of toxin mixed with antitoxin sufficient to neutralize the toxin.
* The long interval between successive exposures was chosen to reduce the injected antitoxin before reexposure to toxin.
Botulinal toxin is an antigen protein readily inducing antitoxin production, therefore, the number of spaced injections of the toxin that can be given without interference with biological activity is limited. The lack of response to the 4th injection of toxin after the 3rd exposure to antitoxin is probably due to antitoxin remaining after the passive immunizations and any antitoxin induced by the repeated spaced injections of toxin.
** The values in parentheses are the heart rates (beats per minute) other numbers are percent of normal (beginning) beat.

bacteria involves a different mechanism than is true for higher organisms, i.e. multiplication independent of chromosome duplication. In toxinology the discovery of the diphtheria toxin plasmid failed to achieve universality when location on plasmids could not be found for other toxins. That differences exist in this regard even within what one might consider a single class of toxins, the botulinal toxins have been observed. Botulinal toxins also provide another example of the limitation of generalization. Though with the type A and B neurotoxins the molecule is a complex of toxin and hemagglutinin, in the type E organism a hemagglutinin is produced but it is not complexed with the toxin.[21]

Reductionism suggests it is unproductive to study toxins by routes of infection, species of animals, tissue reactions other than those associated with the lesions in natural disease.

But it is shortsighted to live by such a rule limiting the scope of research. The use of varied species, modes of exposure, and observations in systems seemingly remote from natural disease can increase the opportunity for observation of serendipitous events. This has been my experience with botulinal toxin. I give two such examples.

Based on clinical disease there was no reason to look for any effects of crystalline type A botulinal toxin on red blood corpuscles. Yet red blood cells are readily available, are an easy to study tissue element, and occur with or without a nucleus. Exposure of red cells to the toxin in happy surprise revealed the presence of an hitherto unknown hemagglutinin occurring as part of a molecular complex. The other example is as follows. The heart is a complex organ whose beat is subject to an endogenous muscle cell contraction (automaticity) and to a variety of nervous and hormonal influences. In clinical botulism effects on the heart have not been notable and play no apparent role in causing death. Nonetheless when the hearts of four laboratory animal species were exposed to crystalline type A toxin and hemagglutin-free toxin bradycardia occurred, and, most amazingly, was rapidly reversible.[26] Table 1 illustrates one kind of demonstration of this reversibility This reversibility, if confirmed by others, I believe has profound significance for hypotheses of the mode of action of the neurotoxin.

The Potential of Bacterial Toxins as Scientific Tools

The specificity of action of a bacterial toxin can make it a valuable tool for study of basic and applied research questions not immediately directed to medical and public health interest in its toxicity. When a secure knowledge has accumulated about the physiological or biochemical basis of a toxin's mode of action, and a high or absolute degree of purity has been achieved the toxin can be used to provide unique or supportive insight into phenomena outside clinical interest in the pathology caused by the toxin. The most imaginative, technically brilliant, and courageous of such applied research has been by Dr. Allen B. Scott.[38] Recognizing the medical value of long term specific muscle paralysis inducible by botulinal neurotoxin, he has pioneered the successful use of the crystalline type A toxin in the treatment and cure of strabismus and blepharospasm by immobilizing offending eye muscle. This breakthrough should father future efforts to find other clinical applications of bacterial toxins.

Now to relate my associates and personal experiences in the use of purified bacterial toxins as scientific tools, namely diphtherial and cholera toxins whose mechanisms of action at the biochemical level are well understood, and type A boutlinal toxin less well understood.

A classical question in mammalian reproductive physiology is the extent of time protein synthesis occurs during the process of ovulation. Protein synthesis is necessary to produce enzymes needed for the dissolution of the walls of ovary follicles so that the eggs can be released into the fallopian tubes for transport to the uterus. For study of the female reproductive cycle the hamster is an ideal experimental animal because of the regularity in lengths of times of segmented events from the beginning of ovulation to the time of implantation of the fertilized egg in the wall of the uterus. Ovulation takes place over an eleven hour period after the release of pituitary leutinizing hormone. Diphtheria toxin has a specific mechanism of action. It depletes cells of elongation factor-2 needed for growth of nacent peptide chains thus inhibiting protein synthesis. By the judiciously controlled injections of diphtheria toxin at different times during the ovulation process protein synthesis was found to be a continuous process during the length of time of ovulation.[1]

Clinical cholera does not involve pathology of the reproductive system. Yet when crystalline cholera enterotoxin was studied for possible effects on the estrous cycle of the hamster a new phenomenon was discovered.[2] Up to a six-fold increase by weight in the growth of the uterus was induced by the toxin when it was injected during the time of the 4 day estrous cycle when progesterone circulating hormone was high and estrogen low. This growth of the uterus called decidualization normally ensues when a fertilized ovum at the blastocyst stage of embryological development is implanted in the wall of the uterus. Thus the decidualization phenomenon can be studied in the absence of the fertilized egg by the use of cholera toxin. This finding, though unexpected, is not surprising. Cholera toxin acts on many cell types by irreversibly stimulating the adenylate cyclase-cyclic AMP system, and decidualization appears to be responsive to interactions of the cyclic-AMP system with progesterone and estrogen. Thus the toxin is a useful, if not unique, tool in probing this interrelationship. The finding with the toxin supports the hypothesis that induction of estrogen receptor by cyclic-AMP is the primary triggering event in decidualization. A side benefit of the study was the demonstration that the hamster uterine growth phenomenon was so regular, and could be well-controlled experimentally to be used as a new method for the specific detection and quantitative assay of cholera toxin.[3] Specificity of the assay was revealed by a lack of ability of the *Vibrio vulnificus* enterotoxin to stimulate hamster uterus growth even though it, like the cholera toxin, could increase vascular permeability in guinea pig skin. The latter phenomenon has been used to assay both these toxins and does not differentiate between them.

Much research has been done by neurophysiologists to learn the mechanism of action of botulinal neurotoxin. But underexploited is use of the toxin as a tool for exploration of other scientific objectives. An early such use was by Drachman[15] who showed embryological development of chick skeletal muscle to full maturity required nervous imputs which could be depleted by type A toxin.

Now for three examples of botulinal toxin as a probe for raising basic scientific physiological questions. The finding that botulinal toxin causes bradycardia must be telling us an instructive story about the heart's innervation by nerve cells using the acetycholine neurotransmitter.[26] The decrease in the heart beat is not 100 percent. Under our conditions of experimentation with massive doses of toxin the maximum decrease before spontaneous recovery was 70 percent for the mouse and rat and 75 percent for the dog. Apart from the automaticity of heart muscle cell pulsation what does this mean about the role of acetylcholine neurotransmitter relative to noncholinergic nerve cells in heart physiology? Obviously for the cardiologist the toxin's cardiac effect raises anatomic-physiological questions about innervation of the heart and provides a tool for exploration of answers. Among these questions are; is there a disparity in anatomic location of cholinergic and noncholinergic heart nerves; are some cholinergic nerves isolated structurally from contact with toxin in blood; what are the relative contributions cholinergic and noncholinergic nerves make to the normal heart beat, and under various hormonal influences?

The nervous system is remarkable for its plasticity. The degree and variety of experiences of life make for changes in reactivity of the nervous system to stimuli. Certainly the germ-free animal must be deprived of many stimuli experienced by the normal animal living in the microbe haunted world. Thus there are differences in the immune systems of these animals, but might there also be nervous system ones? To test this germ-free mice were exposed to type A botulinal toxin and found to be 2.3 times more sensitive to the toxin than conventional mice while specific pathogen-free mice were intermediate.[34] These results testify to the plasticity of the nervous system affected by the toxin. The next step is to identify the physiological and/or other bases for the plasticity.

Based on differences in sensitivity to a neurotoxin, inferences may be drawn as to variances in anatomy or physiology between young and old animals, and how these may be influenced by sexual development. Of itself a difference in responsiveness may not tell us much about the toxin if its mechanism of action is incompletely understood. Nonetheless such findings can reveal the toxin to be a means for better understanding, or revealing fruitful questions, about host biochemistry, anatomy and physiology at possibly clinically significant and, undoubtedly, a basic scientific level. A case in point is the finding that for mice the LD_{50} of botulinal neurotoxin is independent of the body weight over the whole range from weaning to adult maturity.[29] With the B and C toxins, female adult mice were slightly more resistant than males. In contrast, for the rat such independence of sensitivity from body weight does not apply.[27,28] Obviously these findings with the toxin are giving us biological messages about differences between the mouse and rat in nerve development after birth. These hints remain to be properly explored. The sex difference may or may not reside in the nervous system. While the fundamental cause of death in botulism is poisoning at cholinergic neuromuscular synapses it might be plausible to explain greater resistance by females by reason of a sexual difference in the capacity to compensate for interference with breathing.

The aforementioned three examples of use of bacterial toxins to raise questions outside the immediate interest of toxicology should be stimulus for bacteriologists to encourage collaborative research with colleagues in disciplines other than their own. If basic biochemical lesions of a toxin are monolithic then differences in response to a toxin by dissimilar tissues and species must have their origin at anatomical and physiological levels. Such differences may not be of interest to the clinician treating a patient but they may be revealing to the cytologist, histologist, physiologist, zoologist, botanist and evolutionist.

Oral Poisoning

The oral route is the mode of exposure to pathogens and toxins of specific interest for students of food poisoning. If a toxin causes pathology of the alimentary tract directly, it is properly referred to as an enterotoxin. The term oral poison is the broader one to include toxins acting on the alimentary tract directly or others whose primary mechanisms of action are at other sites. Botulinal and staphylococcal neurotoxins are examples of the latter. An effect on the alimentary tract can be due to actions on the sympathetic and central nervous systems. Cramp pain and emesis are prime examples.

How does an oral poison escape the confines of the alimentary tract to act at a distance? Four mechanisms and variations thereof have been considered for the passage of macromolecules such as whole proteins across cell membranes, namely, phagocytosis, pinocytosis, endocytosis and direct penetration of the membrane. Yet this list may not be complete for recognizing how macromolecules including bacterial toxins and larger particles such as microbes escape from the intestinal tract to enter the lymphatic and blood systems associated with the alimentary tract. There is the unresolved probability of particles not requiring crossing of cell membrane barriers but rather movement (slipping) between the surface of abuting cells. If this phenomenon occurs a theory is needed to explain it.

With botulinal toxin there is no question the toxin leaves the small intestine exclusively by way of the lymphatic route. What has not been definitively studied is the role of the lymphatic route for the toxin's ability to cross the buccal region, esophagus, stomach and colon.[31] Whether or not this occurs because of effects on the lymphatic route of absorption of toxin has not been studied. Foods could also act either directly on the toxin or on alimentary tract physiology. Prominent among these would be effects on the rate of peristalsis, enhancement or decrease in mechanisms for destroying toxin in the gut, competition for contact with the lumen mucosal surface or specific receptors, a detergent action on epithelial tissues, and complexing with the toxin.

It is probably of significance that the escape of microbes from the intestine by the lymphatic route is a normal event because nonpathogenic bacteria in small numbers do so, and how many escape is a rough function of their size.[44,45] Therefore it is reasonable to conclude that much smaller particles, such as bacterial toxins, are all potential poisons by the oral route. We have shown this to be true for the tetanal and diphtherial toxins not normally associated with clinical disease arising from natural intestinal exposure to these toxins.[24] The real world possibilities of any bacterial toxin to cause oral poisoning are functions or probabilities for exposure by ingestion, the quantity swallowed, length of time of residence in the intestine, and resistance to detoxification processes.

Escape of toxin from the intestine might be envisioned to be by protein coating the surface of fat particles which appear as the chyme entering lymph draining the intestine. But this is probably not an important event because toxin can cross the intestinal barrier without the presence of foods. Botulinal toxin appears in the liquid portion of the lymph draining from the intestine, and has the dimension of the intrinsic toxin molecule. While olive oil can increase oral toxicity of botulinal toxin so can egg albumen, an overall hydrophilic molecule.[31]

The lymphatic route of escape of toxin and nonpathogenic microbes from the intestine might argue against the existence of small holes or tears in the intestinal epithelium lining because in such cases macromolecules and the microbes should pass directly into the blood circulating in the alimentary tract. This assumes an ability of toxin to transfer across the capillary wall.

Infants and the elderly are said to be more susceptable to infectious disease presumably because of less efficient immune systems. If one is thinking only of classical antibody defense this may be too limited a view. Comparative studies of different aged animals might be productive of discovery of multifarious forces at work if more massive and earlier escape of toxins from the alimentary tract were observed in infants and the old. The existence of infant boutlism may be in part a reflection of easier escape of toxin from the colon than for older children and adults.

In the case of salmonellosis, it would be desirable to learn the details of how these bacteria make their way into eggs in a system where ovulation and egg laying is otherwise designed to produce eggs with sterile interiors.

A Nonreductionist Approach to the Mechanism of Action of Botulinal Neurotoxin

Understandably, success in explaining the biochemical mechanism of action of both diphtherial and cholera toxins as one consisting of specialization of separable pieces of the toxic protein molecule, one functioning to permit translocation across a cell membrane, and the other, the separated toxic portion, to act enzymatically in a deleterious manner has led to use of these findings as a model to explain the mechanism of other toxins.[39,40] Yet this reductionist effort has not lead to a commonly accepted explanation of the mode of action of botulinal neurotoxin. Perhaps bacterial toxins are testimony to the impish capability of nature to frustrate the human desire for simplicity by accomplishing the poisoning of multicellular organisms by protein molecules by fundamentally different means. In homage to this nonreductionist reality the mystery of the mechanism of action of botulinal neurotoxin will now be discussed. Essentially, the ideas that botulinal toxin must be transported across and leave the synaptic membrane of nerves before it can do harm, and that the damage it induces must be enzymatic will be questioned, examined and alternatives proposed.

There is an alternative to the concept that botulinal toxin must be transported across the synaptic membrane into the cytosol of the nerve cell and there express its toxicity. The other choice of explanation is that the toxin lodges at and in the membrane and by this location in some way prevents the escape of acetylcholine from vesicles and through the membrane into the synaptic gap between the nerve and the muscle cell it stimulates. The failure of acetylcholine neurotransmitter to appear in the synaptic gap after stimulation has been universally accepted as the explanation for the muscle paralysis of botulism.[11]

To concretize the thought that for botulinal toxin to be toxic it must be resident in the synaptic membrane and to suggest means for its testing is the purpose of the pipe and valve hypothesis of botulinal toxicity.[25] The first presentation of this hypothesis will now be modified and fleshed out in what follows.

The hypothesis recognizes that stimulation of a nerve permits the toxin molcule to enter the synaptic membrane where it acts in an as yet undefined manner to block the escape of acetylcholine from vesicles into the synaptic gap at the neuromuscular junction. In essence the stimulus acts as a valve which allows the toxin to enter preexisting closed membrane channels or channels appearing de novo in response to the nerve stimulation. The pipe is the analogy for the entered pathway, which convention calls a channel, through which movement of acetylcholine across the membrane normally takes place.

Two possibilities should be considered as to how the toxin in the membrane can prevent acetylcholine escape. The toxin might act as a mechanical blocker of the membrane channel through which acetylcholine moves. The other is that the toxin prevents exocytosis of acetylcholine from vesicles.

Observations of the normal situation show that nerve stimulation is followed by migration and attachment to the inner or intracellular surface of the synaptic membrane by acetylcholine containing vesicles. Following this attachment the vesicles exocytose and the released acetylcholine moves across the membrane in channels abutting the surface of the vesicles to be released into the synaptic gap. The location and number of membrane attachment sites for the vesicles are not random. Also known is that the stimulated toxin poisoned nerve cell does not accumulate acetylcholine in the cytosol. Accumulation may be expected to happen if exocytosis takes place since the acetylcholine released from the vesicle when it is not able to move across the toxin membrane blocked channels might spill out into intracellular space. Alternatively, the blockage of membrane passage could simply result in the acetylcholine remaining in the vesicle in spite of the appearance of a vesicle exit point.

Another choice is to think that the toxin inhibits exocytosis. A lack of change in the morphology of the vesicle is supportive of this idea. If at the point of attachment to the membrane the vesicle contacts toxin, rather than the part of its natural membrane receptor which triggers exocytosis, exocytosis is prevented. The membrane point of attachment then would have a dual function as a site of attachment for a vesicle and as an exocytosis initiator. The receptor could be a single entity with differentiated structure to accomplish each function separately or a combination of molecules (a complex?) one for each function. The pipe and valve hypothesis does not make a choice between these possibilities.

Whether toxin prevents exocytosis by masking a trigger or blocks channel translocation of acetylcholine either process can be nonenzymatic. Toxin may frustrate an enzymatic event such as exocytosis without itself acting as an enzyme. Enzyme inhibitors can be nonenzymatic. The nonenzymic aspect of the pipe and valve hypothesis satisfies the rule of Ockham's razor, namely, that entities should not be multiplied unnecessarily.

Vesicle exocytosis follows stimulation. Upon stimulation, to repeat, the vesicle attaches to a specific site on the synaptic membrane, and only then does exocytosis follow. If toxin does prevent exocytosis how can it do this? The membrane reception site for the vesicle may have a dual nature in acting both as a binding site and as a trigger for inducing exocytosis of the bound vesicles. This latter act may be the key to the role of toxin in inhibition of release of acetylcholine. Nerve stimulation not only results in vesicle activities but is also accompanied by other events, especially prominent among which are changes in the intracellular and membrane ionic environment in which calcium ions are major participants. The ionic environment is probably directly involved in exocytosis. Conceivably toxin in the synaptic membrane interferes with the access of ions to the trigger spot on the membrane, or masks the trigger spot and thus prevents normal functioning of ions at the vesicle membrane spot for exocytosis. Toxicity is explained if exocytosis can be proven to be a synaptic membrane-vesicle membrane interaction which is frustrated by toxin present at the membrane to membrane interface.

It is commonly accepted that toxin penetratiaon of the synaptic membrane requires nerve stimulation. This suggests that cytological study of nerve cells exposed to toxin but not first stimulated, for example, by severing axons, should show absence of toxin in the membrane. This kind of definitive study remains to be done.

If there are a limited number of specific channels in nerve membrane for transport of acetylcholine into the synaptic gap then a lethal dose of toxin must be consistent with the number of molecules needed to occupy these channels. Botulinal toxin is the most poisonous poison known, even being one and a half to three times more potent than tetanus toxin. That a strikingly small quantity of toxin can kill an animal has been father to the thought that the toxin must be an enzyme. Biologically only an enzyme is known to be capable to do an amount of work far beyond what its number of molecules suggests. Neurophysiologists agree on the finite range of numbers of muscle-nerve cell connections involved in muscle activity paralyzed in rodents by the toxin. These data are consistent with the number of molecules causing paralysis and death. Thus, the extraordinary toxicity of botulinal toxin is not of itself evidence for enzymatic activity.[18]

What evidence is there that toxin enters the synaptic membrane and stays there? It has not been an easy task to use a marker to label toxin without loss of toxicity and then find cytological evidence for the physical presence of the toxin in the membrane. Yet credible claims to this achievement exist.[14]

A noteworthy feature of botulism is the long term persistence (days, weeks, months) of paralysis. This persistence could be due to one of two possibilities. One, a biochemical botulinal lesion is of such a nature as to resist repair or rapid recovery. No knowledge exits either for or against this. Second, paralysis persists as long as the toxin remains in the synaptic membrane. It is quite probable that the half-life of the toxin in the membrane can be a long one since there is no clue to make one suspect destruction of the embedded toxin by proteolytic enzymes. Nor is there a compelling metabolic reason for believing the toxin in the membrane must be rapidly catabolized. Histological evidence of repair of poisoned nerve fibrils is consistent with persistence of toxin. Repair commences only after a long period of time and consists of synthesis and growth of new fibrils without the disappearance of the preexisting poisoned fibrils.[16] Poisoning does not interfere with regeneration following another kind of injury such as crush of motor nerves, and the new fibrils establish effective neuromuscular transmission after a few days.[42] Common sense suggests this to be a more likely event if the toxin were confined to the membrane rather than present in the cytosol where repair synthesis of new cellular components must take place. As yet undone are sequenced cytohistological searches for toxin in the membrane of the poisoned fibrils. Such studies should permit a definitive settlement of the question of the time scale of residence of toxin in the synaptic membrane.

If the toxin must enter the cytosol to do its harm, its presence in the membrane must be a transient phenomenon demonstrable by cytological means. Another way to test for the necessity for the toxin to cross over into the nerve cell cytosol is to bypass the membrane by direct injection of the toxin, and fractional parts of its molecule, into the nerve cell and its axon. I do not know of evidence of poisoning by this experimental approach.

Is there a specific membrane or other receptor for the toxin? An obvious candidate has been ganglioside, a quantitatively significant component of nerve tissue and with which toxin can combine. Unfortunately research results rule against this possibility.[46] The question of the existence and chemical nature of a specific toxin receptor remains unresolved. If such a receptor is found, our hypothesis would limit its cellular distribution to the synaptic membrane.

A so-called gating theory has been proposed wherein "toxins form large channels greater than 14 angstroms in diameter in planar lipid bilayers ..... these channels might act as tunnel proteins for translocation of the pharmacologically active fragments of these toxins from acidic (endocytic) vesicles into the cytosol".[17,19] This concept has a number of weaknesses. It assumes the necessity for botulinal toxin to pass beyond the nerve membrane and it assumes a fragment rather than the toxin molecule as a whole acts as the poisonous entity. Efforts to show toxicity by a fragment of toxin in analogy with diphtherial toxin have failed. This is in spite of the known intramolecular structural differences within the botulinal toxin molecule. While it may be that one part of the toxin molecule has a major, if not sole, role to play in attachment to synaptic membrane, its continued existence as an intergral part of the whole molecule may well be necessary to assure a three dimensional structure of the molecule required for expression of toxicity. This concept is not negated by the fact that monoclonal antibody against some fragments of the toxin antibody do not have the capacity to neutralize the toxicity of the whole molecule.[23]

The gate theory has not addressed the probability of an upper limit to the size of translocatable peptide chains. Such a limit could exceed the size of the toxin molecule that must abut or pass to its site of action. Other faults of the theory of the gate channel phenomenon in relation to botulism is that it does not account for reversibility of toxicity including spontaneously reversible cardiac toxicity. In casting these doubts we recognize and admire the alert candor of the chief student of the channel phenomenon who states "When we come to the channels formed by diphtherial, tetanus, and botulism toxins, however, we are confronted with a much more problematic and speculative situation" than may be the case for other proteins.[17]

If the toxin causes poisoning by its presence in the synaptic membrane, molecular kinetics and the intensity of intermolecular specific forces holding the molecule in its proper place would determine the efficiency and stability of poisoning. If the forces holding the toxin to the membrane were of a chemical nature, say, for example, covalent bonding, then strength of binding would tend to be invariable and only strong forces could be expected to displace the toxin and thus reverse toxicity. The fact is that there are indicators of variable and relatively weak binding of toxin, and this favors thinking in terms of physical rather than chemical bonding to the membrane.

The temperature coefficient (Q_{10}) of nerve cell poisoning is more than two and changes to become higher in the course of poisoning.[39] This is not definitive proof for enzymatic chemical forces at work. For example, overcoming high energy interfacial barriers and highly viscous systems would have high temperature coefficients.[30] Furthermore, if physical penetration, binding and alignment of toxin molecules in the membrane are separate and sequential events, each step can have a different energy barrier to satisfy which is reflected by dissimilar temperature coefficients appearing at different times.

Theoretically, for toxin acting as a physical barrier to membrane passage of acetylcholine or as an inhibitor of an enzymatic step in exocytosis there can be greater permissible variation in the chemical composition of botulinal toxin molecule than if it acted as an enzyme. Is this the significance of the fact that there are known to be at least seven similarly acting but different antigenic botulinal neurotoxic types but all are of the same order of molecular size? This situation is in sharp contrast to the existence of enzyme acting diphtherial toxin only as a single antigen type.

Indications of variable attractive intermolecular forces at work tying toxin to membrane are: large differences in potency of the different toxin types; differences among animal species in the sensitivity of their motor nerve terminals to a given toxin type;[41] relative to other types easy reversal of type E poisoning;[21] and rapid and spontaneous reversal of cardiac toxicity.

Intermolecular attractive forces can be expected to vary with a multitude of factors such as: differences in chemistry of toxin types; differences in the chemistry of membranes; aging, that is length of time of residence of the toxin in a membrane; and changes in the ionic environment of the membrane. The latter may explain the effects of ammonium ion on response of a stimulated muscle-nerve preparation to botulinal toxin.

In botulism continual stimulation of poisoned nerve preparations cause changes in contraction and relaxation times of skeletal muscle and results in supersensitivity of muscle to acetylcholine and variation in the miniature endplate potentials. These probably result from modifications in the ionic environment not normally seen. Modifications of forces of attraction of membrane to toxin might result in some if not all of these outcomes.

That a number of chemically unlike compounds reduce the paralysis of botulism argues for the existence of variable forces binding the toxin. Forces of intermolecular attraction, for examples, adhesion and interfacial tensions, are known to be influenced by similarities in physical configurations of molecules in spite of their differences in chemical composition.

An explanation of botulism must account for differences in sensitivity to toxin among the sympathetic, parasympathetic and central nervous systems in spite of their common possession of acetylcholine neurotransmitter producing neurons. The difference in response could rest on at least two factors affecting the capacity of the synaptic membrane to make contact with and hold the toxin. The first would be the ability of toxin to diffuse from lymph into the synaptic gap. As an example, if there is a limiting diameter of the gap that must be exceeded to permit entrance of the large-sized toxin molecule then sensitive nerve cells must have gaps exceeding the limit. The other factor affecting toxicity would be the strength of binding forces of synaptic membranes and these could differ among neurons associated with different parts of the nervous system and body.

An objection raised to the pipe and valve hypothesis is that it should predict blockage of acetylcholine transport to be an all or none phenomenon.[25] This necessity must be viewed critically when measurements are made on a nerve cell and not on an individual channel of movement of acetylcholine. The number of channels needed to be blocked for the all or none phenomenon to be observed may be more than the number of toxin molecules available. More important, if the forces holding the toxin to the membrane are physical in nature, then temporal variation in intensity of binding is possible as already discussed. The absence of the all or none phenomenon in botulism poisoning does not refute the pipe and valve hypothesis. In fact the hypothesis may help explain the phenomenon.

There is an ignored aspect of research on botulism poisoning requiring serious thought. When a catastrophe occurs, a primary causal event exists but its consequences may be many. Discovery of an outcome is not discovery of a cause. A consequence can be remote in time and place from the initiating act. In a railroad wreck discovery of misaligned rails does not tell us that the true cause of the wreck was the inefficiency of an alcohol besotted supervisory engineer in charge of the laying of the track. Thus discovery of a neurophysiological or neurochemical change is not necessarily discovery of primary cause, the initiating event. Nervous system activity takes place through catenary series, some single path distaxic, and other complex branching pathways. Blockage at one point can lead to pile up at one or more preceeding points near or far on a main branch or side branch. Such a pile up can then lead to accumulation of substances exceeding limits for normal behavior at places far removed from the initial blocking place. These facts are particularly pertinent in seeking explanations for the mechanism of action of botulinal toxin. Why? Because nerve stimulation is not inhibited by the toxin. After exposure to toxin the nerves involved with the musculature of the breathing apparatus continue to receive signals from active centers of the central nervous system. Each such natural stimulus, or repeated stimuli programmed by the experimenter, results in chemical messages with consequent ionic and other changes in the intracellular and membrane environments of peripheral nerves. Yet these events do not find normal expression in the release of acetylcholine into the synaptic gap. In this circumstance repeated chemical or electrical events must have outcomes which the neurophysiologist may well detect and legitimately report as new discovery. But such revelation is not of itself proof of discovery of an initial causal event. The burden of proof is on the discoverer to show he has observed the initial causal event and not a near or far removed consequence. A consequence may permit a logical mental construct back to a precipitating cause. But definitive evidence depends on direct proof of the initial event of toxicity. For botulism this remains to be accomplished.

   In summary, the pipe and valve hypothesis argues for a nerve stimulus to act like a valve which permits the toxin to enter the synaptic membrane where it acts to prevent acetylcholine from escaping into the synaptic gap. Two means for this are suggested. The toxin molecules can act as physical barriers of channels acetylcholine must traverse, or the toxin acts at the synaptic membrane-acetylcholine vesicle membrane where it prevents exocytosis. The exocytosis could be prevented by the toxin masking contact of the vesicle with a trigger in the synaptic membrane for exocytosis or by acting as a non-enzymatic inhibitor of a key enzyme of exocytosis. For these purposes the intact toxin molecule is required and this accounts for the reason that separated fractional parts of the molecule have not been observed to be toxic. The forces holding the toxin in the synaptic membrane are physical in nature rather than chemical bonds and this accounts for the observed variations in the efficiency of toxicity and instances of reversibility.

REFERENCES

1. Alleva JJ, Bonventre PF, Lamanna C, 1979, Inhibition of ovulation in hamsters by the protein synthesis inhibitors diphtheria toxin and cyclohexamide, *Proc Soc Exp Biol Med.*, 162:170-174.
2. Alleva JJ, Kenimer JG, Jordan AW, Lamanna C, 1983, Induction of estrogen and progesterone receptors and decidualization in the hamster uterus by cholera toxin, *Endocrin.*, 112:2095-2106.
3. Alleva JJ, Lamanna C, 1984, Characterization of uterine growth response to cholera toxin in hamsters and test of heat-labile enterotoxin from *Escherichia coli*, *J Clin Microbiol.*, 20:506-508.
4. Alouf JE, 1987, From "diphthcritic" poison to molecular toxicology, *ASM News*, 53:547-550.
5. Archer DL, Kvenberg JE, 1985, Incidence and cost of foodborne diarrheal disease in the United States, *J Food Prot.*, 48:887-894.
6. Arnold SH, Brown WD, 1978, Histamine (?) toxicity from fish products, *Adv Food Res.*, 24:113-154.
7. Blaker D, 1967, An outbreak of botulism among water birds, *Ostrich*, 38:144-147.
8. Bonventre PF, 1970, Nomenclature of microbial toxins: problems and recommendations, *In*: "Microbial Toxins" 1:29, Ajl SJ, Kadis S, Montie TC, eds., Academic Press, New York.
9. Bonventre PF, Lincoln RE, Lamanna C, 1967, Status of bacterial toxins and their nomenclature: need for discipline and clarity of expression, *Bacteriol Rev.*, 31:95-109.
10. Bryan FL, 1988, Risks associated with vehicles of foodborne pathogens and toxins, *J Food Prot.*, 51:498-508.
11. Burgen ASV, Dickens F, Zatman LS, 1965, The action of botulinum toxin on the neuromuscular junction, *J Physiol.*, 109:10-24.
12. Cohen GM, Pates AL, Easton, DM, Peterson MG, 1969, Vulture and rooster resistance to botulinum toxin, *Am Zool.*, 9:584.
13. Damon SR, 1928, "Food Infections and Food Intoxications," Williams & Wilkins Co., Baltimore.
14. Dolly JO, Tse CK, Black JD, Williams RS, Wray D, Givilt M, Hambleton P, Melling J, 1981, Botulinum neurotoxin type A as a probe for studying neurotransmitter release, *In*: "Biochemical Aspects of Botulism," Lewis GE, ed., Academic Press, New York.
15. Drachman DB, 1964, Atrophy of skeletal muscle in chick embryos treated with botulinum toxin, *Science*, 145:719-721.
16. Duchen LW, Strich SJ, 1968, The effects of botulinum toxin on the pattern of innervation of skeletal muscle in the mouse, *Quart J Exp Physiol Cogn Med Sci.*, 53:84-89.
17. Finkelstein A, 1985, The ubiquitous presence of channels with wide lumens

and their gating by voltage, *Ann NY Acad Sci.*, 456:26-32.

18. Hanig JP, Lamanna C, 1979, Toxicity of botulinum toxin: a stoichiometric model for the locus of its extraordinary potency and presistence at the neuromuscular junction, *J Theor Biol.*, 77:107-113.

19. Hoch D, Finkelstein A, 1985, Gating of large toxin channels by pH, *Ann NY Acad Sci.*, 456:33-35.

20. Kao I, Drachman DB, Price DL, 1976, Botulism toxin: mechanism of presynaptic blockage, *Science*, 193:1256-1258.

21. Kitamura M, Sakaguchi S, Sakaguchi G, 1968, Purification and properties of *Clostridium botulinum* type E. toxin, *Biochim Biophys Acta*, 168:207-217.

22. Knight TJ, Langston-Unkefer PJ, 1988, Enhancement of symbiotic dinitrogen fixation by a toxin-releasing plant pathogen, *Science*, 241:951-954.

23. Kozaki S, Kamata Y, Nagai J, Ogasawara J, Sakaguchi G, 1986, The use of monoclonal antibodies to analyze the structure of *Clostridium botulinum* type E derivative toxin, *Infect Immunol.*, 52:786-791.

24. Lamanna C, 1960, Oral poisoning by bacterial exotoxins exemplififed in botulism, *Ann NY Acad Sci.*, 88:1109-1114.

25. Lamanna C, 1976, The pipe and valve hypothesis of the mechanism of action of botulinal toxin, *Ann Acad Sci Arts Bosnia Herzegovinia*, 4:213-221.

26. Lamanna C, El-Hage AN, Vick JA, 1988, Cardiac effects of botulinal toxin, *Arch Internat Pharmacodyn Therap.*, 293:69-83.

27. Lamanna C, Hart ER, 1967, Potency of botulinal toxin as influence by body weight and mode of exposure, *In:* "Botulism 1966", Chapman and Hall, London.

28. Lamanna C, Hart ER, 1968, Relationship of lethal toxin dose to body weight of the mouse, *Toxicol Appl Pharmacol.*, 13:307-315.

29. Lamanna C, Jensen WI, Bross ID, 1955, Body weight as a factor in the response of mice to botulinal toxins, *Am J Hyg.*, 62:21-28.

30. Lamanna C, Mallette MF, Zimmerman LN, 1973, "Basic Bacteriology, Its Chemical and Biological Background," Williams & Wilkins Co., Baltimore.

31. Lamanna C, Meyers CE, 1960, Influence of ingested foods on the oral toxicity in mice of crystalline bootulinal type A toxin, *J Bacteriol.*, 79:406-410.

32. Lamanna C, Sakaguchi G, 1971, Botulinal toxins and the problem of nomenclature of simple toxins, *Bacteriol Rev.*, 32:242-249.

33. Leander S, Thesleff S, 1980, On the mode of action of botulinal toxin, *Acta Physiol Scand.*, 108:195-196.

34. Lincoln RE, Lamanna C, Foster WD, 1972, Enhanced sensitivity of germ-free mice to the botulinal neurotoxin, *Proc Soc Exp Biol Med.*, 139:1227-1230.

35. Ohishi I, 1983, Response of mouse intestinal loop to boutlinum C2 toxin: enterotoxin activity induced by cooperation of nonlinked protein components, *Inf Immunol.*, 40:691-695.

36. Ohishi I, DasGupta BR, 1987, Molecular structure and biological activites of *Clostridium botulinum* C2 toxin, *In:* "Avian Botulism," Edklund MW, Dowell JR, eds., Charles C. Thomas, Springfield.

37. Pates A8L, Davidson BL, Lueford K, 1970, The presence of a substance in vulture serum which neutralizes type C *Clostridium botulinum* toxin, *Proc SE Branch ASM* (abstract).

38. Scott AB, 1981, Botulinum toxin injection of eye muscles to correct strabismus, *Trans Am Ophth Soc.*, LXXIX:734-770.

39. Simpson LL, 1981, The origin, structure, and pharmacological activity of botulinum toxin, *Pharmacol Rev.*, 33:155-188.

40. Simpson LL, 1987, The pathophysiological actions of the binary toxin produced by *Clostridium botulinum*, *In:* "Avian Botulism," Eklund MW, Dowell VR, eds., Charles C. Thomas, Springfield.

41. Thesleff S, 1981, Neurophysiological aspects of botulinum poisoning, *In:* "Biochemical Aspects of Botulism,", Lewis GE, ed., Academic Press, New York.

42. Thesleff S, Zelena J, Hofmann WW, 1964, Restoration of function in botulinum paralysis by experimental nerve regeneration, *Proc Soc Exp Biol Med.*, 116:19-20.
43. Vaugh VC, Novy FG, 1902, "Cellular Toxins in the Causation of Disease," Lea Brothers & Co., Philadelphia.
44. Walker WA, 1981, Antigen uptake in the gut; Immunological implications, *Immunol Today*, Feb.:30-34.
45. Wolochow H, Hildebrand GJ, Lamanna C, 1966, Translocation of micro-organisms across the intestinal wall of the rat: Effect of microbial size and concentration, *J Infec Dis.*, 116:523-528.
46. Wonnecott S, 1980, Inhibition by botulinum toxin of acetylcholine release from synaptosomes: Latency of action and role of gangliosides, *J Neurochem.*, 34:1567-1573.

ADP-RIBOSYLATION OF ACTIN BY CLOSTRIDIAL TOXINS

Klaus Aktories, Udo Geipel, Monika Laux and Ingo Just

Rudolf-Buchheim-Institut fur Pharmakologie der
    Universitat Gießen
Frankfurter Str. 107
D-6300 Gießen, FRG

INTRODUCTION

ADP-ribosylation of regulatory proteins is an important mechanism by which bacterial toxins affect the eukaryotic organism.[9,27,32] Diphtheria toxin and *Pseudomonas aeruginosa* exotoxin A modify elongation factor 2 thereby inhibiting protein synthesis.[9,26,32] Cholera toxin, *Escherichia coli* heat-labile enterotoxin and pertussis toxin ADP-ribosylate regulatory G-proteins involved in a variety of transmembranous signaling processes in eukaryotic cells.[8,9,28,29,32] These G-proteins (G_s, G_i, G_o, transducin) comprise a family of highly homologous proteins of heterotrimeric structure,[25] which function by utilizing a guanine nucleotide binding and GTPase cycle. The toxins ADP-ribosylate the α-subunit of the G-proteins thereby either enhancing (cholera toxin) or inhibiting (pertussis toxin) signal transduction.[25]

Recently, it has been shown that various clostridial toxins possess ADP-ribosylating activity. Among these toxins are *Clostridium botulinum* C_2 toxin,[1,2,3,21,23] *C. perfringens* iota toxin,[20,22] *C. spiroforme* toxin[17] and a *C. difficile* ADP-ribosyltransferase.[18] Moreover, besides *C. botulinum* C_2 toxin, certain strains of *C. botulinum* type C produce a second ADP-ribosyltransferase, termed C_3.[5] With the exception of *C. botulinum* ADP-ribosyltransferase C_3, which modifies eukaryotic 21-24 kDa GTP-binding proteins,[4,6] all the clostridial toxins ADP-ribosylate actin.[1,2,3,17,18,22] Since actin is an ATP but not a GTP-binding protein, these clostridial toxins essentially differ in their substrate specificity from the above mentioned ADP-ribosyl-transferases and comprise a novel class of ADP-ribosylating toxins.[2]

Clostridium botulinum C_2 toxin

C. botulinum C_2 toxin is binary in structure and consists of components I and II with molecular weights of about 45,000 and 100,000.[14] Component I of the toxin possesses ADP-ribosyltransferase activity and component II is apparently involved in the binding of the toxin to the eukaryotic cell surface. The latter component has to be activated by trypsin treatment, which causes release of about an 80,000 Da peptide.[15] Component I of *C. botulinum* C_2 toxin ADP-ribosylates nonmuscle actin, whereas skeletal muscle actin is a poor substrate of the toxin.[1,3] Both β- and γ- isoforms of non-muscle actin are modified by the toxin.[31] The toxin's substrate is

Microbial Toxins in Foods and Feeds, Edited by A.E. Pohland *et al.,*
Plenum Press, New York, 1990

monomeric G-actin but not the polymerized F-actin.[1,3] Thus, phalloidin, which induces polymerization and stabilization of actin filaments completely blocks the ADP-ribosylation of actin by *C. botulinum* C_2 toxin.[1] Recently, it has been shown that modification of actin by *C. botulinum* C_2 toxin occurs in arginine-177.[31] This finding is surprising, since arginine-177 is identical both in nonmuscle and skeletal muscle actin.

Clostridium perfringens iota toxin

C. perfringens iota toxin is also a binary toxin with components i_a (M_r 48,000) and i_b (M_r 72,000).[24] Again, the low molecular weight component is the ADP-ribosylating enzyme, whereas the high molecular weight component is supposedly responsible for the toxin binding and transport into cells. In contrast to *C. botulinum* C_2 toxin, iota toxin accepts both nonmuscle actin and skeletal muscle actin as substrate[20] (Fig. 1). Figure 2 shows that pretreatment of chicken embryo cells with *C. botulinum* C_2 toxin reduced the subsequent ADP-ribosylation of actin by *C. perfringens* iota toxin in the lysates of pretreated cells indicating that the identical amino acid was modified by both toxins. That arginine-177 is also ADP-ribosylated by iota toxin was confirmed by protein chemical analysis of iota toxin-modified actin.[30]

Effects of ADP-ribosylation on binding and hydrolysis of ATP by actin

Since actin is an ATP-binding protein and possesses ATPase activity,[12,16] we studied the effects of ADP-ribosylation on binding of ATP and on ATP hydrolysis by actin.[10] For this purpose skeletal muscle actin was ADP-ribosylated by iota toxin. Thereafter, this modified actin was incubated in the presence of $[\gamma^{-32}P]ATP$ for up to 24 hours. After separation of the non-bound ATP by Dowex treatment,[13] the bound $[\gamma^{-32}P]ATP$ was released by addition of unlabeled ATP. Under these experimental conditions the k_{off} rate of ATP release by control actin was $3.2 \times 10^{-4} s^{-1}$ and increased to $5.9 \times 10^{-4} s^{-1}$ for the ADP-ribosylated actin. This effect was not simply caused by denaturation of actin since without addition of unlabeled nucleotide, $[\gamma^{-32}P]ATP$ remained bound to ADP-ribosylated actin for over 90 minutes.[10]

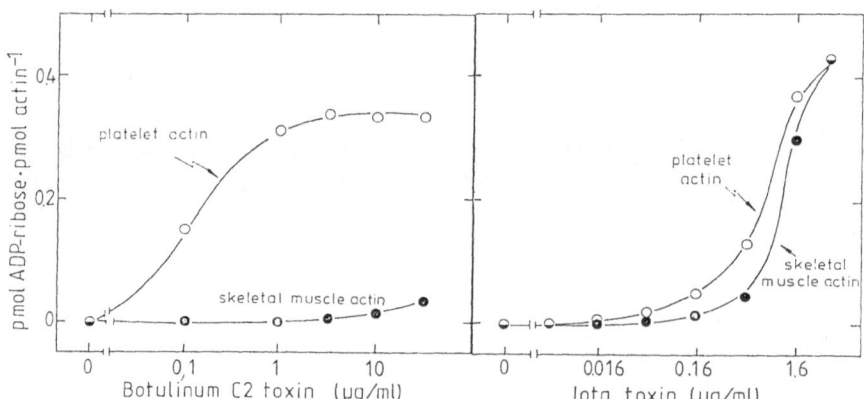

Fig. 1. ADP-ribosylation of skeletal muscle and platelet actin by *C. botulinum* C_2 toxin and *C. perfringens* iota toxin. Purified platelet actin (O) and skeletal muscle actin (●) were incubated for 1 hour with the indicated concentrations of *C. botulinum* C_2 toxin component I or *C. perfringens* iota toxin component i_a. The radioactivity of labeled actin was determined by filtration on nitrocellulose filters as described.[20]

ADP-ribosylation of actin greatly affected the ATP hydrolysis by actin.[10] Figure 3 shows the ATP hydrolysis catalyzed by rabbit skeletal muscle actin. Addition of Mg^{2+} caused a large increase in the ATPase activity. ADP-ribosylation of actin by iota toxin inhibited "basal" (without added Mg^{2+}) as well as Mg^{2+}-stimulated ATP hydrolysis by about 80%. Almost identical results were obtained with nonmuscle liver actin after ADP-ribosylation by *C. botulinum* C_2 toxin. Both "basal" and Mg-stimulated ATPase activities were decreased by the toxin in a dose dependent manner (Fig. 4). Moreover, inhibition of ATP hydrolysis correlated with the toxin-induced ADP-ribosylation of actin. One could argue that the decrease in ATP hydrolysis caused by the toxins was due to inhibition of polymerization since ATP hydrolysis occurs concomitantly with polymerization of actin. Several findings, however, indicate that the ADP-ribosylation of actin inhibits the ATPase activity of monomeric actin. First, ATPase activity was inhibited in the presence of 50 μM $MgCl_2$, a concentration of the divalent cation apparently too low to cause polymerization. Second, inhibition of ATP hydrolysis was observed even at an actin concentration (0.1 μM) below the critical concentration. Finally, ADP-ribosylation of actin inhibited ATPase activity of the DNase-Actin complex. Since it has been shown that DNase prevents actin polymerization,[11] the latter finding strongly supports the view that ADP-ribosylation inhibits ATPase activity of monomeric actin.

Recently, it has been reported that actin amino acid residues 114-118 are somehow involved in ATP-binding.[7] In the linear amino acid sequence the attachment site of ADP-ribose in actin (arginine-177) is too distant for direct interaction with this suggested nucleotide-binding site. Evidence exists, however, that the ATP-binding domain and the ADP-ribose attachment site of actin are located in functionally related areas. First, ADP-ribosylation of actin increases the rate of ATP exchange and inhibits ATP hydrolysis by actin.[10] Second, polymerization of actin increases ATP hydrolysis[16] and inhibits ADP-ribosylation of actin[1,20] and, finally, ADP-ribosylation inhibits the polymerization of actin.[1,3,20]

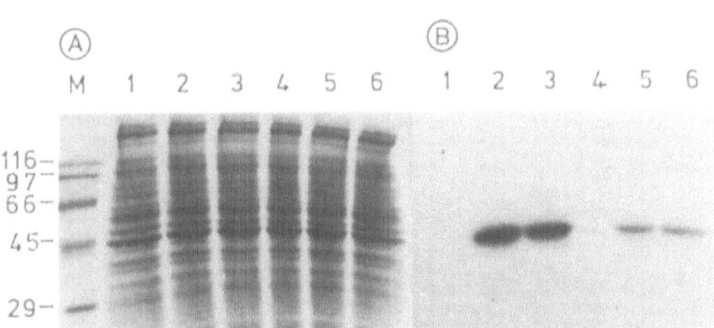

Fig. 2. Pretreatment of intact chicken embryo cells with *C. botulinum* C_2 toxin inhibits the subsequent ADP-ribosylation of actin by *C. perfringens* iota toxin. Intact chicken embryo cells were incubated 2 hours without (lanes 1, 2, 3) and with (lanes 4, 5, 6) *C. botulinum* C_2 toxin components I [50 ng/ml] and II [200 ng/ml]). Thereafter, the cell lysates were ADP-ribosylated 1 hour in the presence of [^{32}P]NAD with the enzymatic component of one or the other toxins: laes 2 and 5, C_2I (3 μg/ml); lanes 3 and 6, iota-a (1 μg/ml). Lanes 1 and 4 were incubated without the ezymatic component of either toxin.

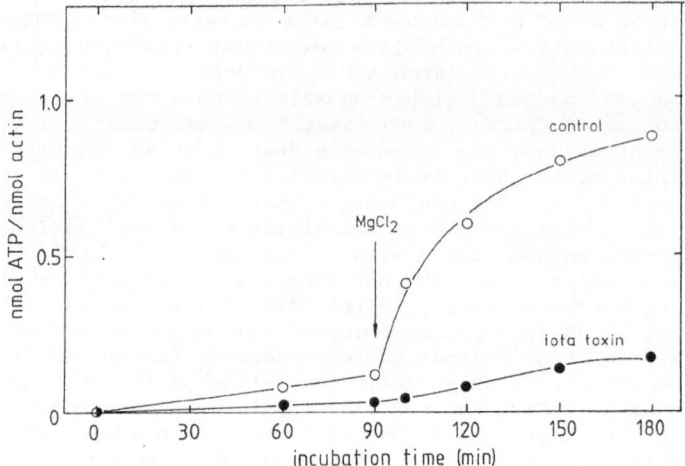

Fig. 3.   Influence of ADP-ribosylation of actin on ATP hydrolysis.
Skeletal muscle actin was ADP-ribosylated by *C. perfringens*
iota toxin component i_a.   The ADP-ribosylated actin was incu
bated with 50 μM [γ-^{32}P]ATP for 180 min. P_1 released was
determined at the indicated intervals.   ATPase activity was
stimulated by the addition of $MgCl_2$ (1 mM) after 90 min.
(Data and methods are from Ref. 10).

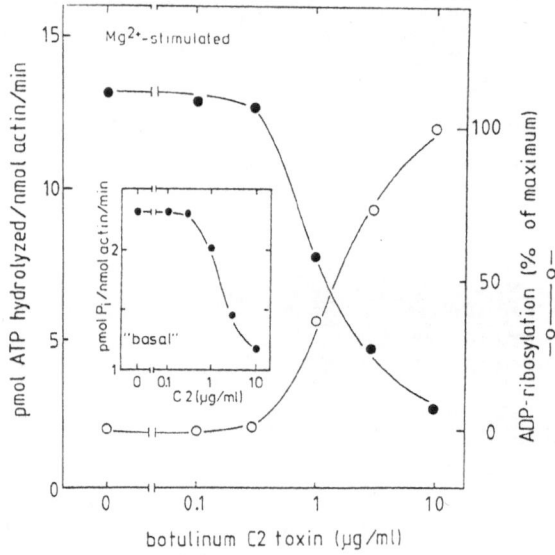

Fig. 4.   Inhibition of ATP hydrolysis and ADP-ribosylation of liver
actin by *C. botulinum* C_2 toxin.   Liver actin was ADP-ribosylated by the indicated concentrations of *C. botulinum* C_2
toxin.   Thereafter, actin was loaded with [γ-^{32}P]ATP and
P_1 release (●) was determined without (inset) or with added
$MgCl_2$ (1 mM).   ADP-ribosylation of actin (O) was determined
by the filter assay.   (Data and methods are from Ref. 10).

Actin ADP-ribosylated by either *C. botulinum* C_2 toxin or *C. perfringens* iota toxin loses its ability to polymerize.[1,3,20] Moreover, *C. botulinum* C_2 toxin was shown to destroy the microfilament network of intact cells and to increase the cellular content of G-actin.[19] To obtain insights into the molecular mechanisms by which the toxins cause these dramatic changes in the cytoskeleton organization, the effect of ADP-ribosylated actin was studied on polymerization of unmodified actin[33] (Fig. 5). These studies showed that ADP-ribosylated skeletal muscle actin inhibited the nucleated polymerization of actin at substoichiometric ratios. Furthermore, actin filaments were depolymerized by the addition of ADP-ribosylated actin. This depolymerization terminated when the concentration of monomeric actin reached the critical concentration of the pointed end of actin filaments. In contrast, ADP-ribosylated actin did not affect polymerization of non-modified actin, when the barbed end was capped by gelsolin. All these data indicate that ADP-ribosylated actin behaves like a capping protein at the barbed end of actin filaments.

Model of the molecular mechanism of the toxin's effects on the microfilament network of cells

Actin is one of the most important components in the architecture and motile functions of cells.[12,16] The protein is involved in processes like movement, cell streaming, phagocytosis and exocytosis. In nonmuscle cells these functions of actin largely depend on rapid changes in the polymeric state of actin.[12,16] The dynamic equilibrium between monomeric G- and polymeric F-actin is regulated by divalent cations, nucleotides and most importantly by actin binding proteins. *C. botulinum* C_2 toxin, *C. perfringens* iota toxin or related toxins disturb this balance between G- and F-actin by ADP-ribosylation of G-actin (Fig. 6). By capping the barbed ends of actin filaments ADP-ribosylated actin inhibits polymerization of unmodified actin and stimulates depolymerization. As the ADP-ribosylated actin is not incorporated into actin filaments, the modified actin is trapped in its monomer form and will accumulate. Both effects, "trapping" of monomeric actin and

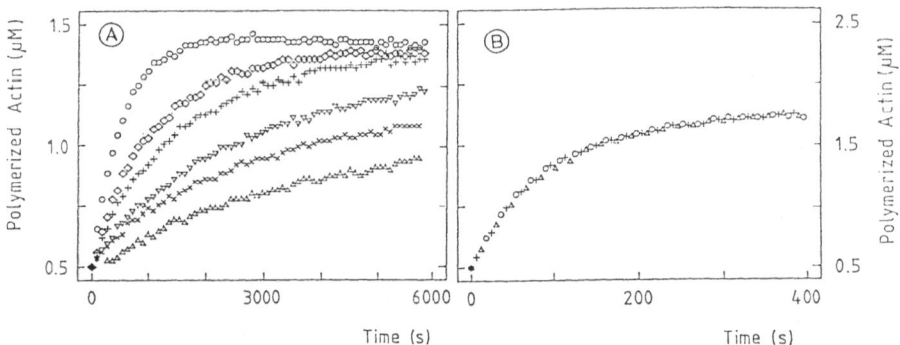

Fig. 5. Nucleated polmerization of actin in the presence of ADP-ribosylated actin. A. Polymerized F-actin (0.5 μM) was added to 1.0 μM monomeric G-actin in the presence of ADP-ribosylated actin (0 = 0 μM, □ = 0.1 μM, + = 0.2 μM, ▽ = 0.4 μM, X = 0.8 μM, △ = 1.6 μM). B. Gelsolin-capped actin filaments were added to 2 μM monomeric actin in the presence of the ADP-ribosylated actin (0 = 0 μM, △ = 1.0 μM, ● = 2.0 μM). The polymerization was measured by the increase in fluorescence intensity. (Data and methods are from Ref. 33).

"capping" of actin filaments, cause the depolymerization of polymeric actin and the destruction of the microfilament network. These processes finally result in the profound morphological changes of intact cells observed after treatment with actin-ADP-ribosylating toxins.

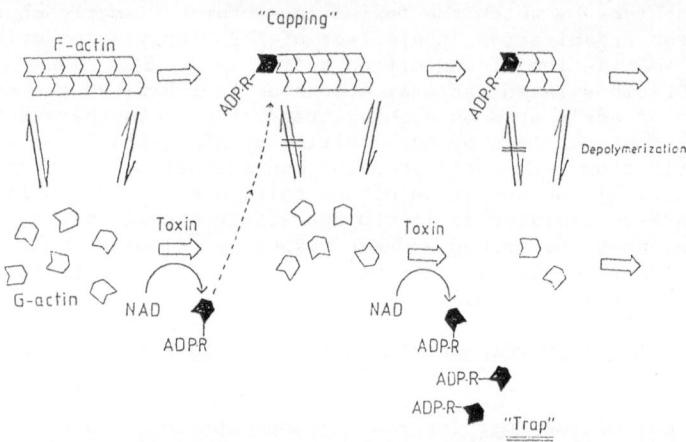

Fig. 6. Model of the molecular mechanism of the toxin's effects on the microfilament network of cells. In intact cells, ADP-ribosylation of actin by the toxins disturb the dynamic equilibrium between monomeric G- and polymeric F-actin. The ADP-ribosylated actin caps the barbed end of actin filaments, thereby preventing further actin polymerization. Because modified actin is not incorporated into actin filaments it is trapped in the monomeric form and accumulates. Both phenomena finally cause destruction of the actin filament network of intact cells.

REFERENCES

1. Aktories K, Ankenbauer T, Schering B, Jakobs KH, 1986, ADP-ribosylation of platelet actin by botulinum C_2 toxin, *Eur J Biochem.*, 161: 155-162.
2. Aktories K, Barmann M, Chhatwal GS, Presek P, 1986, New class of microbial toxins ADP-ribosylates actin, *Trends Pharmacol Sci.*, 8:158-160.
3. Aktories K, Barmann M, Ohishi I, Tsuyama S, Jakobs KH, and Habermann E, 1986, Botulinum C_2 toxin ADP-ribosylates actin, *Nature*, 322:390-392.
4. Aktories K, Frevert J, 1987, ADP-ribosylation of a 21-24 kDa eukaryotic protein(s) by C_3, a novel botulinum ADP-ribosyltransferase, is regulated by guanine nucleotide, *Biochem J.*, 247:363-368.
5. Aktories K, Weller U, Chhatwal GS, 1987, *Clostridium botulinum* type C produces a novel ADP-ribosyltransferase distinct from botulinum C_2 toxin, *FEBS Lett.*, 212:109-113.
6. Aktories K, Rosener S, Blaschke U, Chhatwal GS, 1988, Botulinum ADP-ribosyltransferase C_3, Purification of the enzyme and characterization of the ADP-ribosylation reaction in platelet membranes, *Eur J Biochem.*, 172:445-450.
7. Barden, JA, Miki M, Hambly BD, Dos Remedios C, 1987, Localization of the phalloidin and nucleotide-binding sites on actin, *Eur J Biochem.*, 162:583-588.
8. Cassel D, Pfeuffer T, 1978, Mechanism of cholera toxin action: Covalent modification of the guanyl nucleotide-binding protein of the

adenylate cyclase system, *Proc Natl Acad Sci USA.*, 75:2669-2673.

9. Foster JW, Kinney DW, 1985, ADP-ribosylating microbial toxins, *CRC Crit Rev Microbiol.*, 11:273-298.
10. Geipel U, Just I, Schering B, Haas D, Aktories K, 1989, ADP-ribosylation of actin causes increase in the rate of ATP exchange and inhibition of ATP hydrolysis, *Eur J Biochem.*, in press.
11. Hitchcock SE, Carlsson L, Lindberg U, 1976, Depolymerization of F-actin by Deoxyribonuclease, *Cell*, 7:531-542.
12. Korn ED, 1982, Actin polymerization and its regulation by proteins from nonmuscle cells, *Physiological Rev.*, 62:672-737.
13. Mockrin SC, Korn ED, 1980, *Acanthamoeba* profilin interacts with G-actin to increase the rate of the exchange of actin-bound adenosine 5'-triphosphate, *Biochemistry*, 19:5359-5369.
14. Ohishi I, Iwasaki M, Sakaguchi, G, 1980, Purification and characterization of two components of botulinum C_2 toxin, *Infect Immun.*, 30:668-673.
15. Ohishi I, 1987, Activation of botulinum C_2 toxin by trypsin, *Infect Immun.*, 55:1461-1465.
16. Pollard TD, Cooper JA, 1986, Actin and actin-binding proteins, A critical evaluation of mechanisms and functions, *Ann Rev Biochem.*, 55:987-1035.
17. Popoff MR, Boquet P, 1988, *Clostridium spiroforme* toxin is a binary toxin, which ADP-ribosylates cellular actin, *Biochem Biophys Res Com.*, 152:1361-1368.
18. Popoff MR, Rubin EJ, Gill DM, Boquet P, 1988, Actin-specific ADP-ribosyltransferase produced by a *Clostridium difficile* strain, *Infect Immun.*, 56:22992306.
19. Reuner KH, Presek P, Boschek CB, Aktories K, 1987, Botulinum C_2 toxin ADP-ribosylates actin and disorganizes the microfilament network in intact cells, *Eur J Cell Biol.*, 43:134-140.
20. Schering B, Barmann M, Chhatwal GS, Geipel U, Aktories K, 1988, ADP-ribosylation of skeletal muscle and non-muscle actin by *Clostridium perfringens* iota toxin, *Eur J Biochem.*, 171:225-229.
21. Simpson LL, 1984, Molecular basis for the pharmacological actions of *Clostridium botulinum* Type C_2 toxin, *J Pharmacol Exp Ther.*, 230:665-669.
22. Simpson LL, Stiles BG, Zepeda HH, Wilkins TD, 1987, Molecular basis for the pathological action of *Clostridium perfringens* iota toxin, *Infect Immun.*, 55:118-122.
23. Simpson LL, Zepeda H, Ohishi I, 1988, Partial characterization of the enzymatic activity associated with the binary toxin (type C_2) produced by *Clostridium botulinum*, *Infect Immun.*, 56:24-27.
24. Stiles BG, Wilkins TD, 1986, Purification and characterization of *Clostridium perfringens* iota toxin: Dependence on two nonlinked proteins for biological activity, *Infect Immun.*, 54:683-688.
25. Stryer L, Bourne H, 1986, G proteins: a family of signal transducers, *Ann Rev Cell Biol.*, 2:391-419.
26. Uchida T, 1983, Diphtheria toxin, *Pharmac Ther.*, 19:107-122.
27. Ueda K, Hayaishi O, 1985, ADP-Ribosylation, *Ann Rev Biochem.*, 54:73-100.
28. Ui M, 1984, Islet-activating protein, pertussis toxin: a probe for functions of the inhibitory regulatory component of adenylate cyclase, *Trends Pharmacol Sci.*, 5:277-279.
29. Ui M, Katada T, Murayama T, Kurose H, Yajima M, Tamura M, Nakamura T, Nogimori K, 1984, Islet-activating protein, pertussis toxin: A specific uncoupler of receptor-mediated inhibition of adenylate cyclase, *Adv Cycl Nucleotide Res.*, 17:145-151.
30. Vandekerckhove J, Schering B, Barmann M, Aktories K, 1987, *Clostridium perfringens* iota toxin ADP-ribosylates skeletal muscle actin in arg-177, *FEBS Lett.*, 225:48-52.

31. Vandekerckhove J, Schering B, Barmann M, Aktories K, 1988, Botulinum C_2 toxin ADP-ribosylates cytoplasmic β/v-actin in arginine 177, *J Biol Chem.*, 263:696-700.
32. Weggrett, KA, 1986, Bacterial toxins and the role of ADP-ribosylation, *J Rec Res.*, 6:95-126.
33. Wegner A, Aktories K, 1988, ADP-ribosylated actin caps the barbed ends of actin filaments, *J Biol Chem.*, 263:13739-13742.

MECHANISMS OF ACTION, DIAGNOSTIC AND RAPID METHODS OF ANALYSIS OF

STAPHYLOCOCCUS AUREUS ENTEROTOXINS

Merlin S. Bergdoll

Food Research Institute
University of Wisconsin-Madison
1925 Willow Drive
Madison, Wisconsin 53706

The staphylococci are involved in many human illnesses and infections, which include staphylococcal food poisoning, enterocolitis, diarrheas, and toxic shock syndrome (TSS). The staphylococcal enterotoxins are the causes of staphylococcal food poisoning,[11] enteritis,[38] some cases of TSS,[9] and are implicated in diarrheas.[29,44]

STAPHYLOCOCCAL DISEASES

Staphylococcal food poisoning

The most observable signs and symptoms of staphylococcal food poisoning are vomiting and/or diarrhea, with most victims experiencing both symptoms (Table 1). The more severe cases may result in prostration and hospitalization. An occasional death has occurred in children[42] or in older individuals. An illustration of a severe case is that of an individual purifying enterotoxin A (SEA) at the Food Research Institute. It is not certain how the toxin was ingested or how much, but it could have been only a very small amount because he is a very careful worker. The subject passed out from the severe vomiting and diarrhea and was taken to the emergency room by ambulance. Very little change in either temperature or blood pressure was

Table 1. Symptoms of Staphylococcal Food Poisoning from 122 Cases[a]

Symptom	Cases[b]	No reaction	Mild reaction	Severe reaction
Vomiting	122	15	12	95
Abdominal pain	122	6	40	76
Diarrhea	103	13	75	15
Headache	101	29	59	13
Muscular cramping	113	41	58	14
Sweating	100	33	67	0

[a] Includes 94 students, 8 lunchroom supervisors who took cream puffs home and 20 from cream puffs sold at 3 cafes and from the bakery truck.
[b] Cases from which this information was available.

Microbial Toxins in Foods and Feeds, Edited by A.E. Pohland *et al.,*
Plenum Press, New York, 1990

noted. About four hours after the attack began, the patient revived, was able to walk back to the laboratory, and drive the fourteen miles to his home. The only lasting effect was some weakness the following day. An example of a mild attack was experienced by the author while attending a meeting. Earlier a very small amount of toxin was ingested from eating a banana while pipetting an enterotoxin solution. This produced some discomfort in the stomach with one episode of vomiting and no after effects; the author returned to the meeting after the vomiting without anyone being aware of the illness. It is not unusual to find 10-50 ng of toxin/g in a food responsible for an outbreak.[24]

The minimal human dose is not known exactly because of the difficulty in estimating this from the food consumed. Estimations from food poisoning outbreaks indicated that as little as 100 ng may be sufficient to cause illness. Additional information from an outbreak in 1985 from chocolate milk consumed by school children (Table 2) indicated that 100 to 200 ng was sufficient to cause illness.[16] Analysis of twelve one-half pints (236 ml) of the chocolate milk showed the presence of an average of 150 ng/carton, the amount consumed by a majority of the young people who became ill.

Pseudomembraneous enterocolitis

In the early 1950's, Surgalla and Dack[38] associated enterotoxin production by staphylococcal strains isolated, many in pure culture, from the stools of surgical patients ill with pseudomembraneous enterocolitis. Initially, these strains were tested for enterotoxin production by the monkey feeding test[39] because none of the enterotoxins had been identified at the time. Examination of the strains for enterotoxin production after the enterotoxins were identified revealed that the majority of these strains (51 of 58) produced identified enterotoxins, with 80.9 percent of the enterotoxigenic ones producing SEA and enterotoxin B (SEB); a few produced only SEA. SEB was produced in relatively large amounts by many of the strains and may have been the major cause of the deaths that resulted from this illness before the cause was discovered. These authors undertook the investigation because of the similarity of the signs and symptoms observed in the surgical patients to those seen with severe staphylococcal food poisoning. Ordinarily, staphylococci cannot compete with the natural flora of the intestinal tract, but the administration of antibiotics to individuals undergoing surgery eliminated the natural flora and allowed the staphylococci, which had become antibiotic resistant, to grow uninhibited.

It was demonstrated that enterotoxigenic staphylococci administered intragastrically to monkeys, dogs, and chinchillas after antibiotic therapy produced enteritis in these animals.[40,41] Administration of a nonenterotoxigenic strain did not produce enteritis. Also, enteritis could be produced by the administration of SEB intragastrically, the severity depending on the amount of enterotoxin given. This type of staphylococcal illness seldom occurs today as a majority of the enteritis cases are accorded to *Clostridium difficile*. There was no question that these earlier enteritis cases were due to staphylococci because in many cases pure cultures of staphylococci were isolated. Admittedly, a specific set of circumstances must exist and this happens infrequently.

Toxic shock syndrome (TSS)

Although most cases of TSS are caused by toxic shock syndrome toxin-1 (TSST-1), evidence is available to show the staphylococcal enterotoxins, particularly SEB, can cause TSS, primarily in nonmenstrually related cases[8,9] (Table 3). The signs and symptoms observed in TSS are vomiting, diarrhea, fever, low blood pressure, decrease in urine output, increase in glutamic oxalacetic transaminase, thrombocytopenia, hyperfibrinogenemia,

Table 2. Enterotoxin Analysis of 2 Percent Chocolate Milk

Sample	SEA in concentrated extract (ng/ml)	SEA in milk[a] ng/ml	SEA in 1/2 pint ng	Average ng/1/2 pint
Group 1				
1	0.48	0.40	94	
2	0.53	0.45	106	
3	0.58	0.48	113	
4	0.75	0.63	149	
5	0.68	0.58	137	
6	0.87	0.73	172	129
Group 2				
1	1.20	0.75	177	
2	0.80	0.50	118	
3	1.00	0.62	146	
4	1.23	0.78	184	
5	1.20	0.75	177	
6	1.03	0.65	153	159
All samples		0.61	144	144

[a] ng/ml in concentrated extract was divided by concentration factor (3 for group 1 samples and 4 for group 2 samples) and then by recovery percentage (40 percent recovery was used).

Table 3. Toxin Production by Staphylococci Isolated from Patients with Toxic Shock Syndrome

Toxin	Menstrual patients No.	(%)	Nonmenstrual patients No.	(%)
TSST-1	115	(38.3)	22	(27.5)
TSST-1,SEA	142	(47.3)	33	(41.3)
TSST-1,SEC	12	(4.0)	13	(16.3)
TSST-1,SEA,SEC	8	(2.7)		
TSST-1,other	2	(0.7)		
SEB	10	(3.3)	9	(11.3)
SEC	1	(0.3)	1	(1.3)
Negative	9	(3.0)	2	(2.5)

initial leukopenia followed by a neutrophilic leucocytosis, pulmonary edema, increase in heart rate, pooling of blood in vascular beds with evidence of endothelial cell degeneration, gradual decrease in serum proteins, hypotension, shock, and death.[12] These symptoms are essentially identical to those observed when the enterotoxins are injected intravenously into monkeys.[3] Additional evidence for enterotoxin involvement is from patients with TSS from respiratory infections, such as pneumonia. The staphylococci from six of eight patients produced enterotoxin C (SEC) and TSST; all six died.[8] The other two patients survived even though the staphylococci from one produced SEA and TSST and the other produced TSST only. The SEC-TSST combination appears to be quite potent, as strains producing them have been implicated in peritonitis in calves resulting in death of several animals and in udder infections in cows that resulted in greatly reduced milk production. This combination of toxins is produced by a very high percentage of staphylococcal strains from mastitis in both sheep and goats.[13]

Diarrheas

Many cases of diarrhea in hospitals, particularly among babies, are related to staphylococcal infections. Little information is available about the staphylococcal product involved, although we have found some strains to produce identified enterotoxins. Other investigators have examined the enterotoxigenicity of staphylococci isolated from stools obtained from diarrheal patients and found a high percentage of these strains to be enterotoxigenic (Table 4).[29,44]

STAPHYLOCOCCAL ENTEROTOXINS

The staphylococcal enterotoxins are relatively low molecular weight proteins, 27,000 to 29,000 daltons.[2] They are produced by the staphylococci, primarily *Staphylococcus aureus*, although the newer species, *Staphylococcus intermedius*, primarily canine strains, is enterotoxigenic (Table 5)[14,20,21] and *Staphylococcus hyicus* has been reported to be enterotoxigenic.[1] The guidelines followed for many years in the Food Research Institute to determine whether staphylococci produce enterotoxin was to determine whether the particular strain produced either coagulase or thermonuclease. Formerly, both *S. intermedius* and *S. hyicus* would have been classified as *S. aureus*. Therefore, if we follow the current species classification, we cannot confine our testing of staphylococcal strains to *S. aureus* alone.

To my knowledge, no one is currently working on the identification of additional enterotoxins, although we do know that unidentified ones do exist. This is very difficult to accomplish because a specific biological test such as the monkey feeding test is necessary to identify the enterotoxin.[39] Apparently, only about five percent of staphylococcal food poisoning outbreaks are due to unidentified enterotoxins. The enterotoxins that have been identified so far are classified as SEA, SEB, enterotoxins C_1 (SEC$_1$), C_2 (SEC$_2$), C_3 (SEC$_3$), D (SED), and E (SEE).[2] The enterotoxin Cs are very closely related and can be identified by their cross-reactions with antibodies prepared against any one of the SECs. The other enterotoxins are identified by antibodies specific for each of the enterotoxins, although cross-reactions between SEB and the SECs[26], and SEA and SEE[27] do exist. Monoclonal antibodies have been prepared that are useful in the sandwich enzyme-linked immunosorbent assay (ELISA) for the detection of SEB and the SECs and SEA and SEE by one set of two antibodies for each group. Preparation of monoclonal antibodies that cross-reacted with SEA, SED, and SEE were unsuccessful, although an antibody was prepared that reacted strongly with SEA and SEE but only weakly with SED. Attempts to use this antibody in the sandwich ELISA, coupled with specific antibodies to each of the enterotoxins were unsuccessful.

Structure of the enterotoxins

Amino acid sequencing of SEA, SEB, and SEC$_1$ revealed that the enterotoxins have a similar structure, the cystine loop,[5] and a common amino acid sequence (Table 6).[4] It has been speculated that this common sequence may be the toxic site in the enterotoxin molecule as it is the only area of homology between the enterotoxins.[22] It was planned to synthesize a peptide containing this sequence and use it to prepare monoclonal antibodies; unfor-

Table 4. Enterotoxigenicity of Staphylococcal Isolates from
Diarrheal Patients

Enterotoxin	Poland No.*	%	Hungary No.	%	Czechoslovakia No.	%
All	276/365	75.6	39/54	72.2	59/205	28.8
SEA	101/276	36.6	32/39	82.1	37/59	62.7
SEB	209/276	75.7	7/39	17.9	4/59	6.8
SEC	130/276	37.1	8/39	20.5	14/59	23.7
SED			14/39	35.9	2/59	3.4
SEA + SEB	55/276	19.9	4/39	10.3	1/59	1.7

* No. enterotoxigenic/no. strains tested

Table 5. Enterotoxin Production by *Staphylococcus intermedius*

Enterotoxin	Number of strains		
	Diseased dogs Brazil	Spain	Healthy dogs Spain
A	0	3	6
B	1	1	1
C	1	17	17
D	4	4	0
E	5	1	0
A + C	1	0	0
C + D	1	2	0
D + E	0	1	1
D + C + E	0	1	1

Table 6. Common Amino Acid Sequence of the Enterotoxins

SE	Sequence
SEA	-Thr-Ala-*Cys-Met-Tyr-Gly-Gly*-Val-*Thr*-Leu-*His*-Asp-Asn-*Asn*-Arg-Leu-Thr-
SEB	-Lys-Thr-*Cys-Met-Tyr-Gly-Gly*-Val-*Thr*-Gln-*His*-Gly-Asn-*Asn*-Glu-Leu-Asp-
SEC$_1$	-Lys-Thr-*Cys-Met-Tyr-Gly-Gly*-Ile-*Thr*-Lys-*His*-Glu-Gly-*Asn*-His-Phe-Asp-
SEE	-Thr-Ala-*Cys-Met-Tyr-Gly-Gly*-Val-*Thr*-Leu-*His*-Asp-Asn-*Asn*-Arg-Leu-Thr-

tunately, this was not accomplished because the individual responsible for
the monoclonal work left the staphylococcal research group before this was
undertaken. It was hoped that such an antibody could be used to detect and
identify new enterotoxins. Attempts to detect unidentified enterotoxins
with the cross-reacting monoclonal antibodies to SEB and the SECs, and SEA,
SED, and SEE were unsuccessful.

DETECTION OF THE ENTEROTOXINS

All of the methods for detecting enterotoxins are based on the use of the antibodies prepared against the enterotoxins. Essentially all of the antibodies in use have been prepared in rabbits using the individual puri- fied enterotoxins.[33] These polyclonal antibodies react with the entero- toxins in gels to give precipitin reactions which makes them highly spe- cific. The monoclonal antibodies cannot be used in gels because their reactions with the enterotoxins do not result in the formation of precipitates.

Precipitin reaction in gels

Many types of gel reactions have been used to detect enterotoxins, the most common ones being some form of the Ouchterlony gel plate or some form of microslide.[10,30,31] These methods have been used widely to determine the enterotoxigenicity of staphylococcal strains. The modification of the Ouchterlony gel plate test that is used in the Food Research Institute and recommended to others is the optimum sensitivity plate (OSP) method.[34] It is easy to use and, in conjunction with production of the enterotoxins by the membrane-over-agar method[34] or the sac culture method,[15] is adequately sensitive to detect most enterotoxigenic staphylococci. The normal sensi- tivity is 0.5 μg/ml but can be increased to 0.1 μg/ml by a 5-fold concen- tration of the staphylococcal culture supernatant fluids. The microslide is used by some investigators, but it is a rather difficult method to set- up and make sure that good results will be obtained. Many things can go wrong with this method and experience is very important in using it successfully.[7]

Sensitivity of gel reactions

More recently a question has arisen regarding the sensitivity of the gel diffusion methods for determining the enterotoxigenicity of staphylo- coccal strains. Igarashi et al.[23] have reported that enterotoxin production of strains was observed by the reversed passive latex agglutination (RPLA) method that was not detectable by the OSP method (Table 7), in the neighbor- hood of 10 to 20 ng/ml. This was confirmed by concentrating the culture supernatant fluid from five strains that tested positive for SEA by ELISA about 100-fold for testing by the OSP method. Positive results were obtained. Importance of this low production may be questioned; however, we have found strains that were implicated in food poisoning outbreaks to be negative by OSP, but positive by ELISA. This resulted from the examination of strains by ELISA that were negative by OSP but positive by the monkey feeding test.[25] A number of these strains were positive for one or more of the identified enterotoxins by the ELISA method, particularly for SED (Table 8). Some of these strains had been isolated from food poisoning outbreaks which is significant because SED has been implicated as the second most important enterotoxin in food poisoning. It should be pointed out that, of the known enterotoxins, SED is produced in the smallest amounts.[34] It is interesting to note that only three of the strains produced low amounts of SEA, the enterotoxin implicated in 75 percent of staphylococcal food poi- soning outbreaks (Table 9).[43] The production of 10 to 20 ng of entero- toxin/ml is probably of significance because only 100 to 200 ng of SEA was shown to be necessary to produce food poisoning,[16] with the amount present in the vehicle, 2 percent chocolate milk, being 0.50 to 0.75 ng/ml. Admit- tedly, the amount of enterotoxin produced by the membrane-over-agar method is 5-10 times that produced in shake flasks or even possibly in foods, yet if growth is sufficient, 10^8 to 10^9, 1 to 2 ng of enterotoxin/g of food may be produced. This would be adequate to result in staphylococcal food poi- soning in sensitive individuals. It is important, therefore, that standard

Table 7. Comparison of Reversed Passive Latex Agglutination (RPLA) and Optimum Sensitivity Plate (OSP) Methods for Staphylococcal Enterotoxins

Strain	Enterotoxin Test	
	RPLA	OSP
188	SEA, SED	SEA, SED
228	SEA, SED	-
311	SEA	SEA
365	SEC	-
452	SEA	SEA
581	SEC	SEC
609	SEA, SED	-
754	SEA, SED	SEA, SED
802	SEB	SEB
887	SEA	-
896	SEA, SEB	SEA, -
965	SEA, SED	SEA, SED

Table 8. Detection of Enterotoxin from Staphylococcal Isolates Found Negative by the Optimum Sensitivity Plate (OSP) Method

Isolate Source	Enterotoxin Positive	
	Monkey Feeding	ELISA
Food poisoning	38	10
Foods	25	4
Fish (raw)	11	6
Nares (human)	26	4
Nares (horse)	4	1
Miscellaneous	6	1

procedures be developed for the examination of staphylococcal strains for enterotoxin production.

DETECTION IN FOODS

The detection of enterotoxin in foods requires much more sensitive methods than those required for the determination of the enterotoxigenicity of strains. The quantity of enterotoxin present in foods involved in food poisoning outbreaks may vary considerably, from less than 1 ng/g to greater than 1 μg/g, (10-20 μg). Usually, little difficulty is encountered in detecting the enterotoxin in foods involved in staphylococcal food poisoning outbreaks. Outbreaks, however, do occur in which the amount of enterotoxin is less than 1 ng/g, such as was the case with the 2 percent chocolate milk mentioned earlier. In such cases, the enterotoxin can be detected only by the most sensitive methods. Another situation in which it is essential to use a very sensitive method is to determine the safety of a food for con-

Table 9. Detection of Staphylococcal Enterotoxin (SET) in Foods from Outbreaks

Food	S. aureus count	SET by strain	SET detected in food			
			Gel-diff.	ELISA plate	ELISA kit	RPLA kit
Ham	1.5×10^9	A		A	A	A
Vanilla slice	1.3×10^9	A	A	A	A	A
Turkey & duck	3.0×10^8	A		A	A	A
Lasagne, dried	2.0×10^8	A	ND[a]	A	A	A
Salmon, canned	3.5×10^6	A	ND	A	A	A
Halloumi cheese	ND		ND	A	A	A
Sheep milk cheese	ND			A	A	NSA[b]
Corned beef	4.0×10^7	A,B	A	A,B	A,B	A,B
Ham	1.2×10^9	A,D	A,D,	A[c]	A,D	A[c]
Smokey bacon spd.	1.0×10^9	A,D	A,D,	A[c]	A,D	A,D
Salmon-mousse	9.0×10^8	A,D		A[c]	A,D	A[c]
Ham roll	5.0×10^8	A,D	A,D	A[c]	A,D	A[c]
Beef rolls	4.0×10^6	A,D		A[c]	A	ND[c]
Pork	6.0×10^9	B	B	B	B	B
Chicken	1.0×10^6	C			C	ND
Chicken chow mein	1.5×10^6	A		ND	ND	ND
Meat pies	1.0×10^6	A		ND	ND	ND
Corned beef	8.5×10^4	A		ND	ND	ND
Cold buffet	6.0×10^3	A,B		ND	ND	ND

[a] - ND, not detected
[b] - NSA, non-specific agglutination
[c] - Extract not tested for D

sumption. In such situations it is necessary to show that enterotoxin is absent by the most sensitive methods available.

The methods for the detection of enterotoxin in foods began with the concentration-extraction-microslide methods,[6,32] followed by the reversed passive hemagglutination method,[36] the radioimmunoassay (RIA),[28] and the ELISA method.[19,35,37] A major advantage of the highly sensitive methods is that it is not necessary to concentrate the food extracts, which makes it possible to do an analysis in one day.

Radioimmunoassay (RIA)

Considerable success was achieved with RIA by treating the unknown sample with the specific antibodies followed by the radioactive enterotoxin.[28] The enterotoxin-antibody complex was precipitated by use of protein A cells. The centrifuged precipitate which contained the antibodyenterotoxin complex was read in a scintillation counter to determine the amount of iodinated enterotoxin present. The amount of enterotoxin present in the unknown sample was indirectly related to the amount of radioactive enterotoxin present, the less radioactivity, the greater the amount of enterotoxin in the food sample. It was possible to detect enterotoxin at the level of 1 ng/g of food using a simple extraction procedure. The method was used in the Food Research Institute for several years before development of the ELISA methods. The basic requirements of handling radioactive materials along with the need for a scintillation counter and the purified enterotoxins limited the usage of this method to a few laboratories.

Enzyme-linked immunosorbent assay (ELISA)

The ELISA methods were applied to the detection of the enterotoxins in foods soon after they were originally developed for the detection of other proteins. Two types of ELISA procedures have been reported: (1) the competitive method in which the antibody reacts with the unknown sample before being treated with the enzyme-enterotoxin conjugate[37] and (2) the sandwich method in which the antibody is treated with the unknown sample before the antibody-enterotoxin complex is treated with the enzyme-antibody conjugate.[35] The latter procedure is preferred because the amount of enzyme and, thus, color developed, is directly proportional to the amount of enterotoxin present in the unknown sample. This eliminates the need for the highly purified enterotoxins, as crude or only partially purified enterotoxin is needed for preparation of a standard curve.

The majority of users of the ELISA method use microtiter plates to which the antibodies are attached. The large number of wells in a microtiter plate provide for doing several samples at one time, although there may not be uniformity in all of the wells, particularly those around the edge of the plate. The use of a plate reader for recording the results, which also adds expense to the method, is necessary.

An alternate procedure has been developed, and that is the use of polystyrene balls to which the antibodies are attached.[17,19,37] The ball method is more cumbersome because each ball must be handled separately. The main advantage is that a relatively large volume of the unknown sample can be used, and thus, increase the amount of enterotoxin absorbed per sample. This makes possible the use of 1 ml volumes of substrate so that the color developed can be read in a simple colorimeter, an instrument that most laboratories would have available. The sensitivity of the ELISA methods are between 0.5 to 1.0 ng/g of food (Table 10).

Polyclonal antibodies prepared in rabbits have been and are used in the enterotoxin detection methods. The development of monoclonal antibodies to the enterotoxins has made possible their use in enterotoxin analysis and currently they are being used in the development of an ELISA dip stick method for enterotoxin detection in foods. One kit that employs the ELISA ball method is available now commercially.[18] Those who have used it have found it to be a very good method for detecting enterotoxin in foods.[43]

Reversed passive latex agglutination (RPLA)

A RPLA kit is available commercially that is advertised for use in the detection of enterotoxins in foods. The method is adequately sensitive for the detection of enterotoxin in solution, but the extraction method recommended for foods (9 ml buffer/g food) is inadequate for detecting the small amounts of enterotoxin that can be present and cause food poisoning in sensitive individuals.[23] One paper has been published on the comparison of the various methods for the detection of enterotoxin in foods, including the RPLA method (Table 9).[43] In this work the same extraction procedure was used for all the methods, one that provides the sensitivity necessary for detection of enterotoxin in foods. In this comparison, the RPLA method is adequate in most situations and could be used if the extraction procedures were those used in making this comparison.

Conclusion

The current procedures are adequate for detection of enterotoxin in foods and the analysis can be done in one day. There is always the request that the time for doing an analysis be shortened, but in reality, there is no great need to obtain results in less than the time now required. We have

Table 10. Detection of Staphylococcal Enterotoxins in Foods by
ELISA AND RIA Procedures

Food	Enterotoxin	Amount of enterotoxin added (ng/g)	Amount of enterotoxin detected (ng/g) ELISA	RIA
Milk	SEA	0.63	0.63	0.54
		1.25		0.95
Ham	SEA	0.63	0.34	0.72
		1.25		1.14
Sausage	SEA	0.63	0.36	ND[a]
		1.25		0.56
Cheese	SEA	0.63	0.59	
	SED	0.63	0.15	
Potato salad	SEB	0.63	0.18	

[a] - ND, not detectable.

come a long way in improving the methods for enterotoxin detection in foods
and it is not improbable that further improvements will be made in the
future.

REFERENCES

1.   Adesiyun AA, Tatini SR, Hoover DG, 1984, Production of enterotoxin(s)
     by *Staphylococcus hyicus, Vet Microbiol.*, 9:487-495.
2.   Bergdoll MS, 1979, Staphylococcal intoxications, *In:* "Foodborne Infec-
     tions and Intoxications," 2nd edition, Riemann H, Bryan FL, eds.,
     Academic Press, New York.
3.   Bergdoll MS, 1983, Enterotoxins, *In:* "Staphylococci and Staphylococcal
     Infections," Vol. 2., Easmon CFS, Adlam C, eds., Academic Press, London,
     New York.
4.   Bergdoll MS, 1985, The staphylococcal enterotoxins - an update, *In:*
     "The Staphylococci," Jeljaszewicz J, ed., *Zbl Bakt Suppl.* 14, Gustav
     Fischer Verlag, Stuttgart, New York.
5.   Bergdoll MS, Robbins RN, 1973, Characterization of types of staphylo-
     coccal enterotoxins, *J Milk Food Technol.*, 36:610-612.
6.   Casman EP, Bennett RW, 1965, Detection of staphylococcal enterotoxin in
     food, *Appl Microbiol.*, 13:181-189.
7.   Casman EP, Bennett RW, Dorsey AE, Stone JE, 1969, The micro-slide gel
     double diffusion test for the detection and assay of staphylococcal
     enterotoxins, *Health Lab Sci.*, 6:185-198.
8.   Crass BA, Bergdoll MS, 1986, Toxin involvement in toxic shock syndrome,
     J Infect Dis., 153:918-926.
9.   Crass BA, Bergdoll MS, 1986, Involvement of staphylococcal enterotoxins
     in nonmenstrual toxic shock syndrome, *J Clin Microbiol.*, 23:1138-1139.
10. Crowle AJ, 1958, A simplified micro double-diffusion agar precipitation
     technique, *J Lab Clin Med.*, 52:784-787.
11. Dack GM, Cary WE, Woolperet O, Wiggers H, 1930, An outbreak of food
     poisoning proved to be due to a yellow hemolytic staphylococcus, *J
     Prevent Med.*, 4:167-175.
12. Davis JP, Chesney PJ, Wand PJ, LaVenture M, Investigation and Labora-
     tory Team, 1980, Toxic-shock syndrome. Epidemiological features, recur-
     rence, risk factors, and prevention, *New Engl J Med.*, 303:1429-1435.

13. De Buyser ML, Dilasser F, Hummel R, Bergdoll MS, 1987, Enterotoxin and toxic shock syndrome toxin-1 production by staphylococci isolated from goat's milk, *Int J Food Microbiol.*, 5:301-309.
14. De le Fuente R, Almazan J, Fuentes L, Valle J, Gomez-Lucia E, 1986, Enterotoxigenicity of *S. aureus* and *S. intermedius* strains isolated from dogs, *In:* "Proceedings, 2nd World Congress Foodborne Infections and Intoxications," Berlin (West), Vol. 1, Institute of Veterinary Medicine - Robert von Ostertag-Institute, Berlin (West).
15. Donnelly CB, Leslie JE, Black LA, 1968, Production of enterotoxin A in milk, *Appl Microbiol.*, 16:917-924.
16. Evenson ML, Hinds MW, Bernstein RS, Bergdoll MS, 1987, Estimation of human dose of staphylococcal enterotoxin A from a large outbreak of food poisoning involving chocolate milk, *Int J Food Microbiol.*, In press.
17. Fey H, Pfister H, 1983, A diagnostic kit for the detection of staphylococcal enterotoxins (SET) A, B, C and D (SEA, SEB, SEC, SED), *In:* "Immunoenzymatic techniques," Avrameas S, Druet, P, Masseyeff R, Feldmann, G, eds., Elsevier/North-Holland Publishing Co., Amsterdam.
18. Fey H, Pfister H, Ruegg O, 1984, Comparative evaluation of enzyme-linked immunosorbent assay systems for the detection of staphylococcal enterotoxins A, B, C and D, *J Clin Microbiol.*, 19:34-38.
19. Freed RC, Evenson ML, Reiser RF, Bergdoll MS, 1982, Enzyme-linked immunosorbent assay for detection of staphylococcal enterotoxins in foods, *Appl Environ Microbiol.*, 44:1349-1355.
20. Fukuda S, Tokuna H, Ogawa O, Sasaki M, Kishimoto T, Kawano J, Shimizu A, Kimura S, 1984, Enterotoxigenicity of *Staphylococcus intermedius* strains isolated from dogs, *Zbl Bakt Hyg.*, A 258:360-367.
21. Hirooka EY, Muller EE, Freitas JC, Vicente E, Yoshimoto Y, Bergdoll MS, 1988, Enterotoxigenicity of *Staphylococcus intermedius* of canine origin, *Int J Food Microbiol.*, In press.
22. Huang IY, Schantz EJ, Bergdoll MS, 1976, The amino acid sequence of the staphylococcal enterotoxins, *Jap J Med Sci Biol.*, 28:73-75.
23. Igarashi H, Fujikawa H, Shingaki M, Bergdoll MS, 1986, Latex agglutination test for staphylococcal toxic shock syndrome toxin 1, *J Clin Microbiol.*, 23:509-512.
24. Igarashi H, Shingaki M, Ushioda H, Terayama T, 1985, Detection of staphylococcal enterotoxins in food poisoning outbreaks by reversed passive latex agglutination, *Zbl Bakt Suppl.*, 14:255-257.
25. Kokan N, Bergdoll MS, 1987, Detection of low enterotoxin-producing *Staphylococcus aureus* strains, *Appl Environ Microbiol.*, 53:2675-2676.
26. Lee AC-M, Robbins RN, Bergdoll MS, 1978, Isolation of specific and common antibodies to staphylococcal enterotoxins A and E by affinity chromatography, *Infect Immun.*, 21:387-391.
27. Lee AC-M, Robbins RN, Bergdoll MS, 1980, Isolation of specific and common antibodies to staphylococcal enterotoxins B, C_1 and C_2, *Infect Immun.*, 27:432-434.
28. Miller BA, Reiser RF, Bergdoll MS, 1978, Detection of staphylococcal enterotoxins A, B, C, D, and E in foods by radioimmunoassay, using staphylococcal cells containing protein A as immunosorbent, *Appl Environ Microbiol.*, 36:421-426.
29. Osvath-Marton A, Ban E, Molnar L, 1976, Occurrence of enterotoxin-producing *Staphylococcus aureus* strains in enterocolitis in children, *In:* "Staphylococci and Staphylococcus Diseases," Jeljaszewicz J, ed., Gustav Fischer Verlag, Stuttgart, New York.
30. Ouchterlony O, 1949, Antigen-antibody reactions in gels, *Acta Pathol Microbiol Scand.*, 26:507-515.
31. Ouchterlony O, 1953, Antigen-antibody reactions in gels. IV. Types of reactions in coordinated systems of diffusion, *Acta Pathol Microbiol Scand.*, 29:231-240.
32. Reiser RF, Conaway D, Bergdoll MS, 1974, Detection of staphylococcal enterotoxin in foods, *Appl Microbiol.*, 27:83-85.

33. Robbins RN, Bergdoll MS, 1984, Production of rabbit antisera to the staphylococcal enterotoxins, *J Food Protect.*, 47:172-176.
34. Robbins R, Gould S, Bergdoll MS, 1974, Detecting the enterotoxigenicity of *Staphylococcus aureus* strains, *Appl Microbiol.*, 28:946-950.
35. Saunders GC, Bartlett ML, 1977, Double-antibody solid-phase enzyme immunoassay for the detection of staphylococcal enterotoxin A, *Appl Environ Microbiol.*, 34:518-522.
36. Silverman SJ, Knott AR, Howard M, 1968, Rapid, sensitive assay for staphylococcal enterotoxin and a comparison of serological methods, *Appl Microbiol.*, 16:1019-1023.
37. Stiffler-Rosenberg G, Fey H, 1978, Simple assay for staphylococcal enterotoxins A, B, and C: modification of enzyme-linked immunosorbent assay, *J Clin Microbiol.*, 8:473-479
38. Surgalla MJ, Dack GM, 1955, Enterotoxin produced by micrococci from cases of enteritis after antibiotic therapy, *J Amer Med Assoc.*, 158:649-650.
39. Surgalla MJ, Bergdoll MS, Dack GM, 1953, Some observations of the assay of staphylococcal enterotoxin by the monkey-feeding test, *J Lab Clin Med.*, 41:782-788.
40. Tan T-L, Drake CT, Jacobson MJ, Van Prohaska J, 1959, The experimental development of pseudomembraneous enterocolitis, *Surg Gynecol Obstet.*, 108:415-420.
41. Warren SE, Sugiyama H, Van Prohaska J, 1963, Correlation of staphylococcal enterotoxins with experimentally induced enterocolitis, *Surg Gynecol Obstet.*, 116:29-33.
42. Weed LA, Michael AC, Harger RN, 1943, Fatal staphylococcus intoxication from goat milk, *Amer J Pub Health*, 33:1314-1318.
43. Wieneke AA, Gilbert RJ, 1987, Comparison of four methods for the detection of staphylococcal enterotoxin in foods from outbreaks of food poisoning, *Int J Food Microbiol.*, 4:135-143.
44. Zak C, Jeljaszewicz J, Stochmal I, 1973, Serological types of enterotoxins produced by strains of *Staphylococcus aureus* isolated from feces, *In*: "Staphylococci and Staphylococcal Infections," Jeljaszewicz J, Hryniewicz W, eds, Polish Medical Publishers, Warsaw.

GENETIC ANALYSIS OF REGULATION AND STRUCTURE OF SHIGA-LIKE TOXIN I IN

ESCHERICHIA COLI

Stephen B. Calderwood[1,2] and John J. Mekalanos[2]

Infectious Disease Unit[1]
Massachusetts General Hospital
Boston, MA

Department of Microbiology and Molecular Genetics[2]
Harvard Medical School
Boston, MA

ABSTRACT

Shiga-like toxin I is an iron-regulated cytotoxin produced by certain strains of *Escherichia coli*. SLT-I expression is negatively regulated at the transcriptional level by the repressor protein Fur in the presence of iron. This regulation depends on the presence of a dyad symmetric operator binding site for Fur within the SLT-I promoter, a binding site that is shared by several other iron-regulated genes in *E. coli*. This binding site is sufficient to mediate iron regulation by Fur *in vivo* when inserted into a heterologous promoter. The A subunit of SLT-I (and other toxins in the Shiga family) are homologous with the A subunit of ricin and share an identical molecular mechanism of action. A glutamic acid residue at position 167 in the A subunit of SLT-I is conserved throughout the Shiga toxin and ricin families. Site-directed mutagenesis of this glutamic acid to aspartic acid reduces the specific activity of the A subunit to inhibit protein synthesis *in vitro* by 1,000-fold.

INTRODUCTION

In 1977, Konowalchuk et al., reported the first isolation of strains of *E. coli* that produced proteins cytotoxic for Vero cells in tissue culture.[17] Subsequent investigations by O'Brien et al. demonstrated that these proteins also killed HeLa cells in tissue culture and were neutralized by antibody raised against the classical Shiga toxin from *S. dysenteriae* 1.[22] Purification and characterization of one of these toxins showed it to have identical biological activities as well as subunit structure and size of classical Shiga toxin.[23] Because of these similarities, the name Shiga-like toxin (SLT) was proposed for Vero cell cytotoxin.[24] Strains of *E. coli* producing large amounts of SLT have been implicated in a variety of human illnesses, including diarrhea in adults and neonates, epidemics of hemorrhagic colitis, and the hemolytic-uremic syndrome.[4,14,16] Recently, Shiga-like toxins have been divided into two immunologically distinct types.[30] SLT-I toxins are neutralized by specific anti-Shiga toxin antibody, but SLT-II toxins are not. Despite this immunological difference, both toxins have similar biological activities and subunit size and the genes for the two toxins show weak DNA homology by cross-hybridization.[30] Shiga toxin, SLT-I, and SLT-II are related proteins that belong to a larger Shiga toxin family.

Members of the Shiga toxin family conform to an A-B structural model common to many bacterial toxins. A single enzymatically active A subunit is responsible for biological activity and five copies of a smaller B subunit are responsible for binding holotoxin to specific glycolipid receptors (containing ceramide trihexoside) on the surface of susceptible cells.[11,26] Following internalization of toxin, the A and B subunits dissociate and the A subunit irreversibly inhibits protein synthesis, by catalytic inactivation of 60S ribosomal subunits.[28]

Although the structural genes for Shiga toxin are chromosomally encoded in *S. dysenteriae* 1, the structural genes for both SLT-I and SLT-II in *E. coli* are carried on temperate bacteriophage that mediate conversion of the bacterial host to the toxinogenic state on lysogeny.[21,25,29,33] We have recently established the nucleotide and amino acid sequences for the SLT-I genes of *E. coli*;[1] identical results were independently established by two other groups.[5,13] Nucleotide and amino acid sequences have been determined as well for Shiga toxin from *S. dysenteriae* 1 and for SLT-II from *E. coli*.[12,31] The genes for the A and B subunits of all these toxins are transcribed as an operon from a promoter upstream of the gene for the A subunit. Translational controls for each subunit, however, are separate. SLT-I and Shiga toxin are virtually identical proteins, differing in only a single amino acid residue in their A subunit. SLT-II is more distantly related to the other two proteins, sharing 56% amino acid sequence homology.

We[1] and others[5] have demonstrated that the A subunits of these toxins share considerable amino acid sequence homology with the A subunit of ricin, a plant toxin that also inhibits protein synthesis by inactivation of 60S ribosomal subunits. Single chain, ribosome-inactivating proteins in plants (hemitoxins) are also homologous to the A chain of ricin and inhibit protein synthesis by a similar mechanism.[27] These hemitoxins are not toxic to intact cells, however, because they lack a B subunit (or domain) for binding to the cell surface.

Recent experiments have clarified the molecular mechanism of action of the ricin A chain. This protein catalyzes the cleavage of the N-glycosidic bond in adenosine 4324 of 28S ribosomal RNA; hydrolytic removal of adenine from this position leads to inactivation of the 60S ribosomal subunit.[7] Shiga toxin, SLT-I, and SLT-II have exactly the same molecular mechanism of action as ricin,[8,32] suggesting that the homology observed in amino acid sequence may identify residues important in catalytic function.

RESULTS

Regulation of SLT-I Expression in E. coli by Iron

The production of Shiga toxin is regulated by the concentration of iron in the growth medium, with more toxin made at low iron concentrations.[6] Other bacterial toxins, including diphtheria toxin and *Pseudomonas* exotoxin A, are similarly regulated. For diphtheria toxin, there is genetic evidence for a cis-acting operator binding site upstream of the bacteriophage-encoded toxin gene,[20] which interacts with a chromosomal locus in the host bacterium.[15] In *E. coli*, several genes involved in iron assimilation are repressed by a negative regulatory gene, the *fur* locus, in the presence of sufficient iron.[9] We hypothesized that this locus might also be involved in regulation of SLT-I expression.

The transposon vector Tn*phoA*[18] was used to construct a gene fusion between the amino terminal portion of the A chain of SLT-I and bacterial alkaline phosphatase. This gene fusion was under the transcriptional contr of the SLT-I promoter and provided a convenient assay for SLT-I expression We used a defined minimal medium (T medium) with or without iron supplementation to confirm that SLT-I expression is indeed tightly regulated by iron.[2] To examine the contribution of the *fur* gene to this regulation, we obtained a *fur* null mutant from Dr. J. B. Neilands that had been constructed as a Tn5 insertion in the *fur* structural gene. The *fur*^{o} mutation was moved by transduction with P1 bacteriophage into a bacterial host strain carrying a deletion in *phoA*. We confirmed transfer of the *fur*^{o} mutation to the recipient strain by demonstrating constitutive expression of iron-regulated outer membrane proteins in the presence of high iron. Assay of alkaline phosphatase activity from the plasmid carrying the gene fusion between SLT-I and *phoA* (pSC105), in the *fur*$^+$ and *fur*^{o} strains, confirmed that SLT-I expression is negatively regulated by Fur in the presence of iron (Table 1).[2]

We then constructed a series of BAL-31 deletion derivatives in the region upstream of the gene fusion and similarly assayed alkaline phosphatase activity determined by these derivative plasmids, in media containing low and high iron concentrations (Fig. 1). These experiments verified the location of the previously proposed promoter and confirmed that a site important in iron regulation is located between the -35 and -10 boxes of the promoter, corresponding to a previously noted 21 bp dyad repeat. We examined the promoter regions of three other Fur-regulated genes in *E. coli* and demonstrated homologous dyad repeats (Fig. 2). From this data, we proposed a dyad symmetric consensus sequence as a possible operator binding site for the Fur protein in the presence of iron.[2]

AP Activity of pSC105 Deletions in Three Bacterial Strains, Grown in Low Iron/High Iron T Medium

	CC118	DHB24	SBC24
Sna BI 175 -35 CTACGTACGTCAAGTAGTCGCATGAGATCTGACCAGATATGTTAAGGTTGCAGCTCTCTT			
Δ1	3,084/534	ND	ND
Δ2	2,797/281	457/54	595/369
Δ3	561/28	21/3	52/30
-10 250 TGAATATGATTATCATTTTCATTACGTTTATTGTTACGTTTATCCGGTGCGCCGTAAAACG			
Δ5	35/16	4/2	3/3
Δ6	46/26	ND	ND
Δ7	37/22	ND	ND
300 SD CCGTCCTTCAGGGCGTGGAGGATGTCAAGAATATAGTTATCGTATGGTGCTCAAGGAGTA Δ9	5/3	ND	ND
350 GTGTAAT ATG AAA ATA ATT ATT TTT AGA GTG CTA ACT TTT TTC MET Lys Ile Ile Ile Phe Arg Val Leu Thr Phe Phe			

Fig. 1. Deletion analysis of plasmid pSC105. (Left). Nucleotide sequence is shown from the *SnaBI* site in pSC105 to residue 12 in the signal sequence of slt-IA; nucleotide numbering above the sequence corresponds to that used by Calderwood et al.[1] Previously proposed -35 and -10 boxes, as well as the Shine-Dalgarno sequence (SD), are shown. The endpoints of individual deletions (1 to 9) are shown below the nucleotide sequence; each deletion was brought back to an identical upstream sequence. Horizontal arrows indicate a 21 bp interrupted dyad repeat overlapping the -10 box. (Right). Alkaline phosphatase (AP) activities of individual pSC105 deletions are indicated on the corresponding lines; activities were determined in low-iron T media/iron-supplemented T media. Strains CC118 and DHB24 are *fur*$^+$; strain SBC24 is *fur*^{o}; ND,

Table 1. Alkaline phosphatase and β-lactamase assays in low-iron and iron-supplemented T media.

Strain	Iron added (μM)	Alkaline phosphatase activity (U)	β-lactamase activity[a] (mU)	Induction ratio
DHB24 (pSC105) fur⁺10	None	547	65	12.7
		45	68	
SBC24 (pSC105) fur°	None	576	51	1.0
	10	487	44	

[a] Included as an internal control for plasmid copy number

To confirm that the proposed consensus sequence was sufficient to function in vivo as an operator binding site for Fur, we synthesized the consensus sequence and its complement as oligonucleotides. After annealing, this synthetic binding site was introduced at the start of transcription of an operon fusion between the *ompF* promoter and the *lacZ* structural gene.[3] β-galactosidase activities determined by the parental (pRT240) and engineered (pSC27.1) plasmids were compared, in the presence of high and low concentrations of iron in the media (Table 2). These experiments confirmed that the synthetic consensus sequence was sufficient to confer iron regulation to a heterologous promoter. β-galactosidase activities determined by both parental and engineered plasmids were constitutively high in the fur° mutant, regardless of iron concentration. We concluded that the synthetic consensus sequence was sufficient to mediate iron regulation by Fur in vivo and that

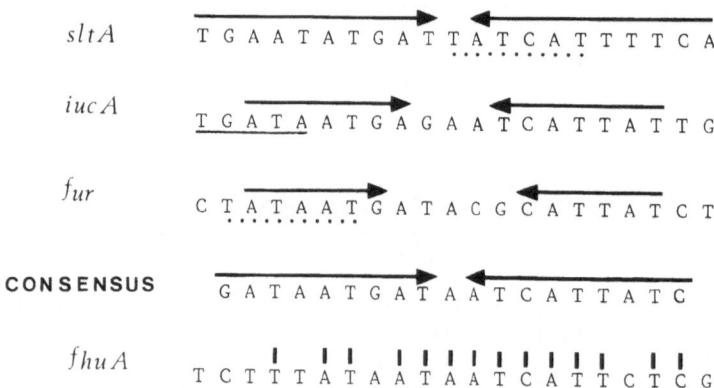

Fig. 2. Nucleotide homology between the promoter regions of several iron-regulated genes in *E. coli*. Arrows above the sequences refer to areas of dyad symmetry. A consensus sequence was derived as a two out of three (or better) match between *slt*-IA, *iucA* and *fur* that maintained dyad symmetry. Proposed -10 boxes are indicated by dotted lines under the nucleotides and -35 boxes by solid lines. The vertical dashes above the *fhuA* sequence indicate bases identical to the proposed consensus sequence.

Table 2. β-Galactosidase assays in low-iron and iron-supplemented T
media.

Strain	Plasmid carried	Iron added (μM)	β-Galactosidase activity (U)
SM796 (fur⁺)	pRT240	None	4,467
		10	3,634
	pSC27.1	None	2,575
		10	774
SBC796 (furᵒ)pRT240		None	4,147
		10	3,714
	pSC27.1	None	4,268
		10	3,185

iron regulation of SLT-I expression in *E. coli* is most likely dependent on
the 21 bp dyad repeat located in the promoter.[3]

Site-directed mutagenesis of the A subunit of SLT-I

A high resolution crystallographic structure for ricin demonstrated
that the ricin A chain was a globular protein with a prominent cleft hypothesized to contain the enzymatically active site.[19] When residues conserved
between the Shiga and ricin toxin families were plotted on the crystal structure of the ricin A chain, seven conserved amino acids were found to lie in
the proposed active site cleft. We hypothesized that one or more of these
residues were likely to be important in the identical catalytic activity of
these A chains. A glutamic acid residue at position 167 in the A subunit
of SLT-I (Slt-IA) was chosen for further study because carboxylate side
chains have been implicated in the N-glycosidase activity of other enzymes
and toxins. We constructed an expression vector for Slt-IA that brought
the structural gene and Shine-Dalgarno sequence under the control of the
lacZ promoter on pUC19.[10] Expression of Slt-IA from this plasmid was monitored in whole cell and periplasmic extracts by Western blotting, using
polyclonal rabbit anti-Shiga toxin antiserum. A 1,150 bp fragment from the
expression vector, containing the structural gene, was moved into M13mp19
to allow site-directed mutagenesis with an oligonucleotide-directed kit.
We designed a mutagenic oligonucleotide to replace the GAA codon for amino
acid 167 of Slt-IA, encoding glutamic acid, with GAT, encoding aspartic acid.
Aspartic acid was chosen as a highly conservative substitution that retained
the carboxyl function at residue 167 but altered its spatial position by
approximately one Angstrom (all other factors remaining equal). Fortui-
tously, mutations in the codon for glutamic acid-167 result in loss of the
unique HindIII restriction site within *slt*-IA, simplifying identification
of mutant DNA in subsequent experiments. After mutagenesis, the sequence
of the entire mutated fragment was confirmed to ensure that no second site
mutations had occurred and the fragment was moved back into the expression
vector. Both wild-type and mutant Slt-IA subunits were synthesized as full
length proteins that were processed correctly by signal peptidase during
transport to the periplasm. Both proteins were similarly susceptible to
cleavage by trypsin, providing evidence that there was no major change in
conformation induced by the mutation. The biological activity of the two
proteins was quantitated by inhibition of protein synthesis in an in vitro
reticulocyte lysate system.[10] Mutant Slt-IA had approximately 1,000-fold
less specific activity in this assay than wild-type protein (Fig. 3). We
concluded that glutamic acid-167 in Slt-IA is critical for the catalytic
activity of this protein and may be located in the active site.

DISCUSSION

Effective bacterial pathogenesis often requires the coordinate expression of a number of different virulence factors, including those encoded on accessory genetic elements such as plasmids or bacteriophage. There is very little free iron in mammalian tissues and pathogenic organisms may use the limited availability of iron as an important environmental trigger for the expression of virulence determinants. For enteric pathogens, the expression of an effective iron uptake system is essential for colonization and replication in the infected host. The negative transcriptional regulation by Fur of both iron uptake systems and SLT-I expression in *E. coli* ensures the coordinate regulation of these important virulence determinants.

The Shiga toxin and ricin families share an identical mechanism of toxic activity on ribosomes and show significant amino acid sequence homology in specific regions of their respective A subunits. For ricin, many of these conserved amino acids lie in a cleft in the three-dimensional structure that is suggestive of an active site. Site-directed mutagenesis of one of these conserved amino acids suggests it may play an important role in catalysis by Slt-IA. Additional site-directed mutants in this A subunit, as well as in the corresponding residues in ricin A, may provide additional insight into the mechanism of catalysis by these toxic proteins.

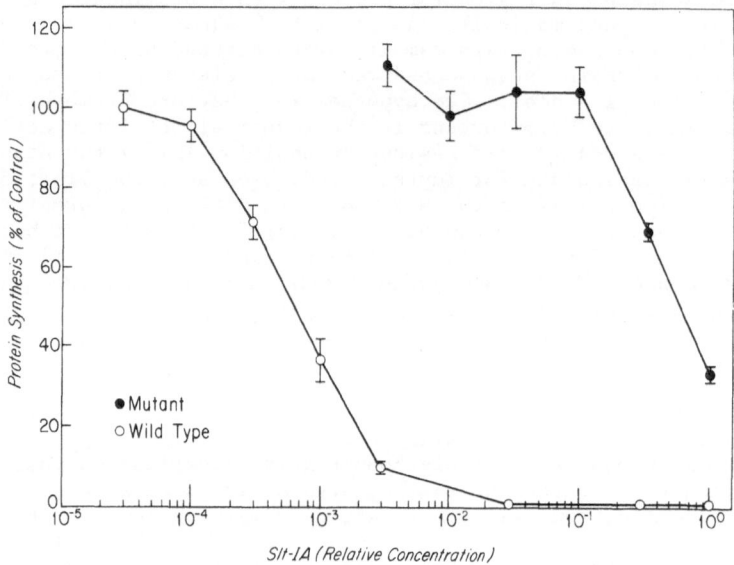

Fig 3.   Inhibition of protein synthesis by wild-type and mutant Slt-IA. Aliquots of rabbit reticulocyte lysate were preincubated for 30 minutes with various dilutions of periplasmic extracts containing wild-type (open circles) or mutant (closed circles) Slt-IA. These lysates were then assayed for protein synthesis and compared to a control lysate preincubated with an extract derived from the same bacterial strain carrying pUC19; background activity (without mRNA) was subtracted from all values. Concentrations of wild-type and mutant Slt-IA were normalized to the same amount of immunoreactive material and plotted as relative concentration.

REFERENCES

1. Calderwood SB, Auclair F, Donohue-Rolfe A, Keusch GT, Mekalanos JJ, 1987, Nucleotide sequence of the Shiga-like toxin genes of *Escherichia coli*, *Proc Natl Acad Sci USA*, 84:4364-4368.
2. Calderwood SB, Mekalanos JJ, 1987, Iron-regulation of Shiga-like toxin expression in *Escherichia coli* is mediated by the *fur* locus, *J Bacteriol.*, 169:4759-4764.
3. Calderwood SB, Mekalanos JJ, 1988, Confirmation of the Fur operator site by insertion of a synthetic oligonucleotide into an operon fusion plasmid, *J Bacteriol.*, 170:1015-1017.
4. Cleary TG, Mathewson JJ, Faris E, Pickering LK, 1985, Shiga-like cytotoxin production by enteropathogenic *Escherichia coli* serogroups, *Infect Immun.*, 47:335-337.
5. DeGrandis S, Ginsberg J, Toone M, Climie S, Friesen J, Brunton J, 1987, Nucleotide sequence and promoter mapping of the *Escherichia coli* Shiga-like toxin operon of bacteriophage H-19B, *J Bacteriol.*, 169:4313-4319.
6. Dubos RJ, Geiger, JW, 1946, Preparation and properties of Shiga toxin and toxoid, *J Exp Med.*, 84:143-156.
7. Endo Y, Mitsui K, Motizuki M, Tsurugi, K, 1987, The mechanism of action of ricin and related toxic lectins on eukaryotic ribosomes, *J Biol Chem.*, 262:5908-5912.
8. Endo Y, Tsurugi K, 1987, RNA N-glycosidase activity of ricin A-chain, *J Biol Chem.*, 262:8128-8130.
9. Hantke K, 1981, Regulation of ferric iron transport in *Escherichia coli* K12: Isolation of a constitutive mutant, *Mol Gen Genet.*, 182:288-292.
10. Hovde CJ, Calderwood SB, Mekalanos JJ, Collier RJ, 1988, Evidence that glutamic acid 167 is an active-site residue of Shiga-like toxin I. *Proc Natl Acad Sci USA*, 85:2568-2572.
11. Jacewicz M, Clausen H, Nudelman E, Donohue-Rolfe A, Keusch, GT, 1986, Pathogenesis of Shigella diarrhea, XI. Isolation of a Shigella toxin-binding glycolipid from rabbit jejunum and HeLa cells and its identification as globotriaosylceramide, *J Exp Med.*, 163:1391-1404.
12. Jackson MP, Neill RJ, O'Brien AD, Holmes RK, Newland JW, 1987, Nucleotide sequence analysis and comparison of the structural genes for Shiga-like toxin I and Shiga-like toxin II encoded by bacteriophages from *Escherichia coli* 933, *FEMS Lett.*, 44:109-114.
13. Jackson MP, Newland JW, Holmes RK, O'Brien AD, 1987, Nucleotide sequence analysis of the structural genes for Shiga-like toxin I encoded by bacteriophage 933J from *Escherichia coli*, *Microbial Pathogenesis* 2:147-153.
14. Johnson WM, Lior H, Bezanson GS, 1983, Cytotoxic *Escherichia coli* O157:H7 associated with haemorrhagic colitis in Canada, *Lancet* 1:76.
15. Kanei CT, Uchida T, Yoneda M, 1977, Isolation from *Corynebacterium diphtheriae* C7(beta) of bacterial mutants that produce toxin in medium with excess iron, *Infect Immun.*, 18:203-209.
16. Karmali MA, Petric M, Lim C, Fleming PC, Arbus GS, Lior H, 1985, The association between idiopathic hemolytic uremic syndrome and infection by Verotoxin-producing *Escherichia coli*, *J Infect Dis.*, 151:775-782.
17. Konowalchuk J, Speirs JI, Stavric S, 1977, Vero response to a cytotoxin of *Escherichia coli*, *Infect Immun.*, 18:775-779.
18. Manoil C, Beckwith J, 1985, TnphoA: A transposon probe for protein export signals, *Proc Natl Acad Sci USA*, 82:8129-8133.
19. Montfort W, Villafranca JE, Monzingo AF, Ernst SR, Katzin B, Rutenber E, Xuong NH, Hamlin R, Robertus JD, 1987, The three-dimensional structure of ricin at 2.8 Angstrom, *J Biol Chem.*, 262:5398-5403.
20. Murphy JR, Skiver J, McBride G, 1976, Isolation and partial characterization of a corynebacteriophage beta, *tox* operator constitutive-like mutant lysogen of *Corynebacterium diphtheriae*, *J Virol.*, 18:235-244.

21. Newland, JW, Strockbine NA, Miller SF, O'Brien AD, Holmes RK, 1985, Cloning of Shiga-like toxin structural genes from a toxin converting phage of *Escherichia coli*, *Science* 230:179-181.

22. O'Brien AD, LaVeck GD, Thompson MR, Formal SB, 1982, Production of *Shigella dysenteriae* Type 1-like cytotoxin by *Escherichia coli*, *J Infect Dis.*, 146:763-769.

23. O'Brien AD, LaVeck GD, 1983, Purification and characterization of a *Shigella dysenteriae* 1-like toxin produced by *Escherichia coli*, *Infect Immun.*, 40:675-683.

24. O'Brien AD, Lively TA, Chen ME, Rothman SW, Formal SB, 1983, *Escherichia coli* O157:H7 strains associated with haemorrhagic colitis in the United States produce a *Shigella dysenteriae* 1 (Shiga) like cytotoxin, *Lancet* 1:702.

25. O'Brien AD, Newland JW, Miller SF, Holmes RK, Smith HW, Formal SB, 1984, Shiga-like toxin-converting phages from *Escherichia coli* strains that cause hemorrhagic colitis or infantile diarrhea, *Science*, 226: 694-696.

26. Olsnes S, Reisbig R, Eiklid K, 1981, Subunit structure of *Shigella* cytotoxin, *J Biol Chem.*, 256:8732-8738.

27. Ready MP, Katzin, BJ, Robertus, JD, 1988, Ribosome inhibiting proteins, retroviral reverse transcriptases, and RNase H share common structural elements, *Proteins*, 3:53-59.

28. Reisbig R, Olsnes S, Eiklid, K, 1981, The cytotoxin activity of *Shigella* toxin, *J Biol Chem.*, 256:8739-8744.

29. Smith, HW, Green P, Parsell Z, 1983, Vero cell toxins in *Escherichia coli* and related bacteria: Transfer by phage and conjugation and toxic action in laboratory animals, chickens and pigs, *J Gen Microbiol.*, 129:3121-3137.

30. Strockbine NA, Marques LRM, Newland JW, Smith HW, Holmes RK, O'Brien, AD, 1986, Two toxin-converting phages from *Escherichia coli* O157:H7 strain 933 encode antigenically distinct toxins with similar biologic activities, *Infect Immun.*, 53:135-140.

31. Strockbine NA, Jackson MP, Sung LM, Holmes RK, O'Brien AD, 1988, Cloning and sequencing of the genes for Shiga toxin from *Shigella dysenteriae* type 1, *J Bacteriol.*, 170:1116-1122.

32. Takeda Y, Yutsudo T, Igarashi K, Endo Y, 1987, Mode of action of Verotoxins (VTI and VT2) from *Escherichia coli* and of Shiga toxin, Abstract, Twenty Third Joint Conference on Cholera, U.S.-Japan Cooperative Medical Science Program.

33. Willshaw GA, Smith HR, Scotland SM, Rowe B, 1985, Cloning of genes determining the production of Vero cytotoxin by *Escherichia coli*, *J Gen Microbiol.*, 131:3047-3053.

CHARACTERIZATION OF 50-KILOBASE PLASMID OF *SALMONELLA CHOLERAESUIS* STRAIN

AH1-S44

Hirofumi Danbara,[1] Kazuyoshi Kawahara,[1] Mayumi Tsuchimoto,[1]
Ryozo Moriguchi,[1] and Nobuyuki Terakado[2]

The Kitasato Institute[1]
5-9-1 Shirokane
Minato-ku, Tokyo 108
Japan

and

National Institute of Animal Health[2]
3-1-1 Kannonndai
Tsukuba City, Ibaraki 305
Japan

ABSTRACT

A plasmid of 50 kb was found in all six *Salmonella choleraesuis*
isolates originating from septicemic swine and humans. The lethality of
the parent strain by intraperitoneal injection into mice was found higher
than that of the plasmid-cured strain. The Tn*1*-tagged 50 kb plasmid was
introduced into the cured strain resulting, in a higher lethality than that
of the cured strain and an identical lethality to that of the parent. With
the injection of a sublethal number of cells of the parent and plasmid-
reintroduced strains, the spleen, liver, and lymph nodes in various sites
of mice became remarkably enlarged, and the bacteria were isolated from the
heart blood. In contrast to this, neither enlargement of the organs nor
recovery of organisms was observed with the plasmid-cured strain. A physical map with *Sal*I, *Eco*RI and *Hind*III restriction enzymes was established.
A mouse bacteremia test was applied for a total of 76 derivatives of the
parent strain with Tn*1* insertion on the 50 kb plasmid. Nine strains were
mutants incapable of causing mouse bacteremia. The Tn*1* insertions were
clustered within a 5.2 kb region of the 8 mutant strains, and one mutant
carried a smaller 50 kb plasmid with a deletion including the 5.2 kb region.
Tn*1* insertions of other derivatives, which caused mouse bacteremia, were
mapped outside the 5.2 kb region.

These results conclude that the 50 kb plasmid of *S. choleraesuis* is
closely related with mouse virulence, plays an important role in causing
mouse bacteremia, and the region responsible for the mouse bacteremia is
located within a 5.2 kb region of the plasmid.

Microbial Toxins in Foods and Feeds, Edited by A.E. Pohland *et al.*,
Plenum Press, New York, 1990

INTRODUCTION

The genera *Escherichia*, *Shigella*, *Salmonella*, and *Yersinia* all include species capable of causing invasive diarrhea in humans or animals. *Salmonella* and *Yersinia*, in contrast to *Escherichia* and *Shigella*, more likely invade underlying tissues and enter the blood stream. A characteristic feature of the *Salmonella* infection is the host specificity for the systemic infection (bacteremia or septicemia). *S. typhi*, *S. paratyphi A*, *S. paratyphi B* and, *S. sendai* are characterized by specificity for the human host. Other organisms are specific for animal hosts: *S. typhimurium* for mouse, *S. dublin* for cattle, *S. choleraesuis* for swine, *S. pullorum* and *S. gallinarum* for poultry, *S. abortusovis* for sheep, and *S. abortusequi* for horse. Virulence is multifactoral in *Salmonella* species and requires both chromosomal and plasmid-mediated traits for the systemic infections.

It has been reported that *Salmonella* strains carry cryptic plasmids that are specific in size for each serotype. Using parent, plasmid-cured or plasmid-reintroduced strains, the serotype-specific plasmids have been shown to be associated with the virulence of each serovar; i.e., *S. typhimurium*: 60 MDa,[11] *S. dublin*: 50 MDa or 80 kb,[4,16] *S. enteritidis*: 36 MDa,[10,14] and *S. gallinarum*: 85 kb.[2]

Helmuth et al.[9] have reported that all the *S. choleraesuis* isolates collected from a wide range of countries contained a 30 MDa serotype-specific plasmid. However, the function of the plasmid is not yet known. We[12] have demonstrated that a 50 kb plasmid, which was found in common among strains of *S. choleraesuis* isolated from swine and humans, was essential for the organism to cause bacteremia in mice.

In the present communication, we demonstrate the association of virulence phenotypes of *S. choleraesuis* with a 50 kb plasmid and the gene location on the plasmid required for mouse bacteremia.

RESULTS

Correlation between 50 kb plasmid of S. choleraesuis and its virulence

Six strains of *S. choleraesuis* isolated from swine and humans were analyzed for their plasmid content. All six strains carried a common plasmid in size. They also contained additional plasmids. As Fig. 1 shows, all 8 *EcoRI*-cleaved fragments of the common plasmids were found among the fragments generated from the plasmids of the 6 strains. The fragments that were found only among some of the isolates could be ascribed to the additional plasmids. The size of the common plasmid was calculated as 50 kb from the fragments generated by the single or double digestions with *EcoRI*, *HindIII*, and *PstI* endonucleases. The identical cleavage pattern suggested that molecular structure of the 50 kb plasmids was closely related to each other.

AH1-S44 (Fig. 1, lane 1) was chosen as a representative strain for further experimentation, and a rifampicin resistant mutant RF-1 was isolated from it. The transposon Tn*1* of a plasmid pTH10,[7] a maintenance-temperature-sensitive mutant of RP4, was introduced in the RF-1 strain to yield strain RF-2. The Tn*1* element was found to be located on the *EcoRI* fragment 3, E3 (Fig. 2), of the 50 kb plasmid. Strain 31N-1 cured of the 50 kb::Tn*1* plasmid was isolated after growth of RF-2 in the presence of novobiocin. The 50 kb::Tn*1* plasmid isolated from RF-2 was re-introduced to 31N-1 by transformation to yield 31N-1-T. The strains RF-1, 31N-1 and 31N-1-T had the same colony morphology, surface antigenicity (O7, Hc, H1, H5), and almost all the 24 biochemical properties tested, were identical among them. The doubling time of RF-1 at 37°C was slightly longer (30 min) than that of 31N-1 and 31N-1-T.

The virulence of the isogenic strains, RF-1 (the parent strain), 31N-(plasmid-cured strain), and 31N-1-T (plasmid-reintroduced strain) was compared by the two methods (Table 1). First, the lethality of the three strains was examined by the intraperitoneal injection into ICR or BALB/c mice. By the injection of 10^8 cells, eight or all of the 10 ICR mice were killed by RF-1 or 31N-1-T, respectively. Injection of the same number of 31N-1 cells did not kill any mice. BALB/c mice were then used to measure LD_{50} values of the three strains. The values of RF-1 and 31N-1-T were essentially the same and $10^{4.7}$ to $10^{5.8}$ fold higher than that of 31N-1. Second, enlargement of spleen, liver, and lymph nodes of mice was examined after intraperitoneal injection of a sublethal number of the three strains (10^6 cells) into ICR mice. As shown in Table 1, livers and spleens were remarkably enlarged and increased in weight after injection of RF-1 and 31N-1-T. The axillary, subiliac, mesenteric, and iliac lymph nodes were also enlarged following the injection of the two strains. In contrast with this, both the size and weight of organs of ICR mice injected with 31N-1 cells were similar to those of mice injected only with saline. These results strongly indicate that the 50 kb plasmid of *S. choleraesuis* is necessary for the virulence of the organism.

Physical and virulence mapping of 50 kb plasmid of S. choleraesuis

Bacteria were frequently isolated from the heart blood of ICR mice 1 to 14 days after injection of 10^6 cells of RF-1 and 31N-1-T (Table 2). On days

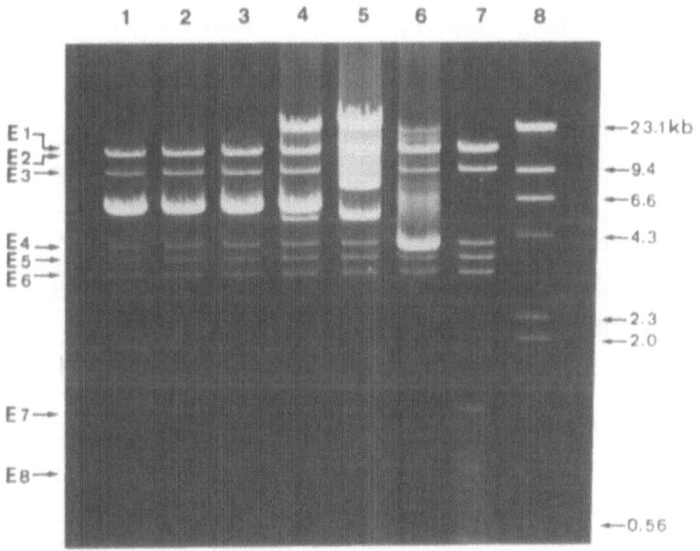

Fig. 1.   Cleavage patterns of plasmids of *S. choleraesuis* strains digested with *Eco*RI endonuclease.

Lane 1 (AH1-S44), lane 2 (AH1-S45), lane 3 (AH1-S90) and lane 4 (AH1-S91) are plasmid DNA of *S. choleraesuis* isolated from swine in Japan (1976), and lane 5 (TG-21) and lane 6 (TG-22) from children in Philippines (1987). Lane 7 is plasmid DNA of a SH1-S44 derived strain which lost an additional 6.7 kb plasmid, and lane 8 is Lambda DNA digested with *Hind*III as a molecular marker. E1 and E2 (ca. 14 kb), E3 (9.3 kb), E4 (4.0 kb), E5 (3.6 kb), E6 (3.1 kb), E7 (1.3 kb), and E8 (0.8 kb) are fragments generated from the 50 kb plasmid.

Table 1. Virulence of *S. choleraesuis* strains with or without 50 kb plasmid in mice.

Strain injected	Dead mice/ total injected[a]	LD_{50}[b]	Enlargement of organs[c]		
			liver (g)	spleen (g)	lymph nodes[d]
RF-1	8/10 (10^8) 0/ 5 (10^6)	$10^{1.3}$	0.31±0.09 (1.5)	2.2±0.5 (10.7)	+
31N-1	0/10 (10^8) 0/ 5 (10^6)	$10^{7.1}$	0.13±0.03 (0.5)	1.7±0.2 (7.1)	-
31N-1-T	10/10 (10^8) 1/ 5 (10^6)	$10^{2.4}$	0.45±0.13 (1.9)	2.5±0.5 (11.4)	+
Control[e]			0.11±0.02 (0.04)	1.6±0.2 (6.4)	-

[a] Cell suspensions in 0.2 ml saline injected intraperitoneally in 4-week-old female ICR mice. Lethality measured 14 days after injection. Figures in parentheses are the number of cells injected.

[b] Values (LD_{50}) were determined at 14 days after the injection in 4-week-old female BALB/c mice.

[c] 10^6 cells in 0.2 ml saline were injected intraperitoneally in 4-week-old ICR mice. Mice were humanely killed and necropsied after 14 days. Mean values in the three independent experiments are shown. Five mice for each bacterial strain were used in each experiment. Figures in parentheses represent the weight percentage to body weight.

[d] Possible enlargement of axillary, subiliac, mesenteric or iliac lymph nodes were recorded visually as + or -.

[e] 0.2 ml saline without bacteria was injected.

Table 2.   Isolation of bacteria from heart blood of mice injected
           with *S. choleraesuis* strains with or without 50 kb plasmid.

Strain	Days after injection[a]				
injected	1	3	5	9	14
RF-1	2/3[b]	3/3	3/3	8/8	8/13
31N-1	0/3	0/3	0/3	0/8	0/13
31N-1-T	0/3	3/3	3/3	8/8	11/13

[a] 10^6 cells in 0.2 ml saline were injected intraperitoneally into 4-week
     old female ICR mice.
[b]   Number of mice from which bacteria detected/number of mice tested.

3 to 9, bacteria were isolated from all mice.   By contrast, no bacteria
were detected in the heart blood of any mouse after injection of 31N-1 cells
These results suggest that the 50 kb virulence plasmid of *S. choleraesuis*
plays an important role in the mouse bacteremia caused by the organism.

    In order to localize a region responsible for mouse bacteremia on the
physical map of the 50 kb plasmid (Fig. 2), RF-1 derived strains with Tn*1*
insertion in the 50 kb plasmid (50 kb::Tn*1*) were isolated.   They were then
injected intraperitoneally into ICR mice.   On day 3, following injection,

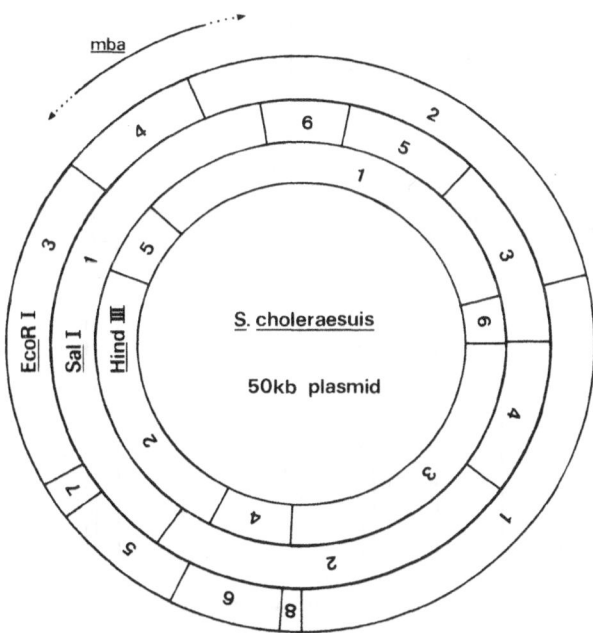

Fig. 2.   Endonuclease restriction map of 50 kb plasmid of *S.
           choleraesuis* AH1-S44 strain.

           8 *Eco*RI, 6 *Sal*I, and 6 *Hind*III sites are contained in the 50
           kb plasmid.   *mba* represents virulence region associated with
           mouse bacteremia.

69

mice were humanely killed and the heart blood spread on selective agar
plates for isolating RF-1 derived bacteria. This experiment was repeated 3
to 4 times for each strain. The 50 kb::Tn1 plasmid DNA from strains with
a diminished mouse bacteremia activity was introduced to 31N-1 by transformation. The mouse bacteremia assay was repeated for the 31N-1 derivatives. Nine out of the 76 strains tested were determined to be mutants
which lacked the ability to cause mouse bacteremia.

Tn1 insertion was localized on the 50 kb plasmid for the 76 strains
used for the mouse bacteremia test. As Fig. 3 shows, the insertions of the
mutant strains were found to be clustered on two restriction fragments; i.e.,
5 strains (F6-5, K16-2, L15-1, L15-4, L34-1) were within a 2.2 kb SalI/EcoRI
fragment, and 3 strains (A8-1, L30-3, B5-2) were within a 3.0 kb EcoRI/
HindIII fragment. The other mutant strain (K40-1) carried a smaller 50 kb::
Tn1 plasmid with a deletion. The deletion included the above mentioned 2.2
and 3.0 kb fragments. In addition, eight other strains with Tn1 insertion
within the 2.2 kb SalI/EcoRI fragment exhibited wild-type activity. These
results indicate that a part of the 5.2 kb region between SalI and HindIII
sites are essential for the mouse bacteremia. A detailed mapping of Tn1
insertion is required for the precise localization.

DISCUSSION

Virulence is multifactoral in Salmonella species, and survival within
phagocytic cells of the reticuloendothelial system is essential for virulence. Fields et al.[5] isolated a Tn10-labeled strain of S. typhimurium

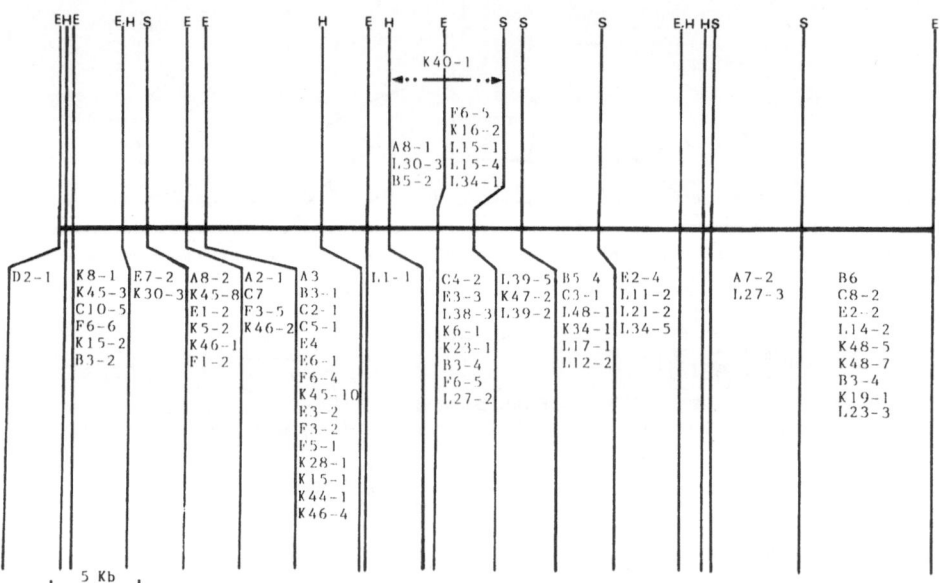

Fig. 3. Map of Tn1 insertions on 50 kb plasmid of S. choleraesuis
         AH1-S44 strain.

         A circular map of the 50 kb plasmid was linearized at an
         EcoRI site between E8 and E1 fragments (Fig. 2). Strains
         drawn above the map are Tn1 inserts exhibiting an abolished
         mouse bacteremia, and those drawn beneath the map are the
         inserts exhibiting a wild-type mouse bacteremia.

70

14028s with diminished capacity for survival within mouse macrophages. Phenotypic characters of those mutants were auxotroph, serum hypersensitive, altered response to oxidative stress, Tn10 in plasmid, lipopolysaccharide (LPS) alteration, nonmotile, and altered colony morphology. All the mutants were less virulent than the parent strain in vivo. Recently Heffernan et al.[8] have shown that S. dublin Lane and its pSDL2-cured strains colonized the intestine and invaded Peyer's patches with equivalent efficiency after the oral administration into mice. The parent strain multiplied in mesenteric lymph nodes and in the spleen, whereas, the plasmid-cured strain remained relatively static until the mice developed active immunity and cured themselves. Thus, the virulence plasmid of S. dublin allows multiplication within the reticuloendothelial system and does not have any effect on the organism's ability to colonize the intestine or invade Peyer's patches.

Serum resistance is also a virulence factor of Salmonella. The traT gene of a drug resistance plasmid R6-5 encodes for a 25,000-dalton outer membrane protein, which is responsible for serum resistance of the host E. coli strain.[13] Recently, Rhen et al.[15] identified a traT-like gene on the virulence plasmid pLT2 of S. typhimurium SH6749. Mutation of the traT-like gene of pLT2 produces a truncated protein, and results in increased outer membrane permeability to hydrophobic compounds, such as fusidic acid. Hackett et al.[6] cloned the gene from the virulence plasmid of S. typhimurium P9144 into a smaller plasmid containing 2.1 kb fragment. The cloned gene expressed an outer membrane protein of 11,000-daltons and mediated serum resistance in both E. coli K-12 and a virulence plasmid-cured strain of S. typhimurium. The virulence plasmid-cured S. typhimurium strain did not express normal LPS, but introduction of the 11,000 dalton-protein gene rendered the strain serum resistant without restoration of normal LPS synthesis. The 11,000 dalton-protein gene was not sufficient to restore either macrophage resistance or virulence to a virulence plasmid-cured S. typhimurium.

Terakado et al.[17] recently demonstrated that serum resistance of S. dublin 5240 strain was mediated by the 50 MDa virulence plasmid (pTE800). The increased serum susceptibility of plasmid-cured strains is related to the decreased neutral sugar content of the LPS. The reduced serum resistance and sugar content of the LPS of the cured strain were restored to a level of the wild-type strain by the introduction of the Tn1 tagged 50 MDa plasmid (pTE800). These results indicate that the 50 MDa virulence plasmid mediates serum resistance and alterations in the content of LPS in S. dublin.

Baird et al.[1] prepared transposon-insertion mutants of strains S. dublin 2229 and S. typhimurium 1275, and identified a region on their plasmids necessary for mouse virulence. The restriction maps for the regions of transposon insertion in the mutants of the S. dublin and S. typhimurium strains were identical. Beninger et al.[3] established a complete restriction endonuclease cleavage map of the 80 kb plasmid (pSDL2) of S. dublin Lane. They mapped a region encoding the virulence phenotype within a 6.4 kb portion of pSDL2 using transposon insertion mutagenesis with Tn5-oriT. Southern hybridization demonstrated that the virulence plasmids of S. dublin Vi[+], S. enteritidis, and S. choleraesuis shared a common 4 kb EcoRI fragment with the virulence region of pSLD2. Williamson et al.[18] showed that an 8 kb SalI/XhoI fragment, derived from the virulence plasmid harbored by S. typhimurium, hybridized with plasmids from 11 other serotypes of Salmonella.

Our present studies demonstrated that a 4 kb EcoRI fragment was, in fact, present in the 50 kb plasmid of S. cholerasuis AH1-S44 strain, and this fragment was a part of the 5.2 kb portion required for mouse bacteremia.

Interestingly, restriction cleavage sites neighboring the 4 kb *EcoRI* fragment are similar between the 50 kb plasmid of *S. cholerasuis* AH1-S44 strain
and pSLD2 of *S. dublin* Lane. This evidence, together with those obtained
by other groups, suggests that the virulence plasmids of *Salmonella* are
evolutionarily related, and the sequence for virulence has been conserved
during the evolutionary diversity of the *Salmonella* serotypes.

ACKNOWLEDGEMENTS

We thank N. Okamura for providing the *S. choleraesuis* strains from
children, and Y. Haraguchi for the valuable animal experiments. We also
thank Y. Nakase and I. Umezawa for their encouragement throughout the study.
We acknowledge stimulating discussions with A. Gohda and M. Ohsahi. This
study was financially supported by Grant-in-Aid (No. 62304036, No. 63570205)
provided by the Japanese Ministry of Education, Science and Culture.

REFERENCES

1.   Baird GD, Manning EJ, Jones PW, 1985, Evidence for related virulence
     sequences in plasmids of *Salmonella dublin* and *Salmonella typhimurium*,
     J Gen Microbiol., 131:1815-1823.
2.   Barrow PA, Simpson JM, Lovel MA, Binns MW, 1987, Contribution of
     Salmonella gallinarum large plasmid toward virulence of fowl typhoid,
     Infect Immun., 55:388-392.
3.   Beninger PR, Chikami G, Tanabe K, Roudier C, Fierer J, Guiney DG, 1988,
     Physical and genetic mapping of the *Salmonella dublin* virulence plasmid
     pSDL2, *J Clin Invest.*, 81:1341-1347.
4.   Chikami GK, Fierer J, Guiney DC, 1985, Plasmid-mediated virulence in
     Salmonella dublin demonstrated by use of a Tn5-*oriT* construct, *Infect
     Immun.*, 50:420-424.
5.   Fields PI, Swanson RV, Haidaris CG, Heffron F, 1986, Mutants of
     Salmonella typhimurium that cannot survive within the macrophage are
     avirulent, *Proc Natl Acad Sci USA*, 83:5189-5193
6.   Hackett J, Wyk P, Reeves P, Mathan V, 1987, Mediation of serum resist-
     ance in *Salmonella typhimurium* by an 11-kilodalton polypeptide encoded
     by the criptic plasmid, *J Infect Dis.*, 155:540-549.
7.   Harayama S, Tsuda M, Iiono M, 1980, High frequency mobilization of the
     chromosome of *Escherichia coli* by a mutant of plasmid RP4 temperature-
     sensitive for maintenance, *Mol Gen Genet.*, 180:47-56.
8.   Heffernan EJ, Fierer J, Chikami G, Guiney D, 1987, Natural history of
     oral *Salmonella dublin* infection in BALB/c mice: effect of an 80-kilo-
     base-pair plasmid on virulence, *J Infect Dis.*, 155:1254-1259.
9.   Helmuth R, Stephan R, Hoog CBB, Steibeck A, Bulling E, 1985, Epidemio-
     logy of virulence-associated plasmids and outer membrane protein pat-
     terns within seven common *Salmonella* serotypes, *Infect Immun.*,
     48:175-182.
10. Hoiv M, Sukupolvi S, Edward MF, Rhen M, 1988, Plasmid-mediated viru-
     lence of *Salmonella enteritidis*, *Microbial Pathogenesis*, 4:385-391.
11. Jones GW, Rabert DK, Svinarich DM, Witfield HJ, 1982, Association of
     adhesive, invasion, and virulent phenotypes of *Salmonella typhimurium*
     with autonomous 60-megadalton plasmids, *Infect Immun.*, 38:476-486.
12. Kawahara K, Haraguchi Y, Tsuchimoto M, Terakado N, Dannara H, 1988,
     Evidence of correlation between 50-kilobase plasmid of *Salmonella
     choleraesuis* and its virulence, *Microbial Pathogenesis*, 4:155-163.
13. Moll A, Manning PA, Timmis KN, 1980, Plasmid-determined resistance to
     serum bactericidal activity: a major outer membrane protein, the *traT*
     gene product, is responsible for plasmid-specified serum resistance
     in *Escherichia coli*, *Infect Immun.*, 28:359-367.

14. Nakamura N, Satoh S, Ohya S, Suzuki S, Ikeda S, 1985, Possible relationship of a 36-megadalton *Salmonella enteritidis* plasmid to virulence in mice, *Infect Immun.*, 47:831-833.
15. Rhen M, O'Connor CD, Sukupolvi S, 1988, The outer membrane permeability mutation of the virulence-associated plasmid of *Salmonella typhimurium* is located in a *tra*T-like gene, *FEMS Microbiological Letters*, 52:145-154.
16. Terakado N, Sekizaki T, Hashimoto K, Naitoh S, 1983, Correlation between the presence of a fifty-megadalton plasmid in *Salmonella dublin* and virulence for mice, *Infect Immun.*, 41:443-444.
17. Terakado N, Hamaoka T, Danbara H, 1988, Plasmid-mediated serum resistance and alterations in the composition of lipopolysaccharides in *Salmonella dublin*, *J Gen Microbiol.*, 134:2089-2093.
18. Williamson CM, Baird GD, Manning EJ, 1988, A common virulence region on plasmids from eleven serotypes of *Salmonella*, *J Gen Microbiol.*, 134:975-982.

PRIMARY STRUCTURE AND CONFORMATION OF *CLOSTRIDIUM BOTULINUM* NEUROTOXIN

Bibhuti R. DasGupta

Food Research Institute
University of Wisconsin
1925 Willow Drive
Madison, Wisconsin 53706

A set of neurotoxic proteins synthesized by certain strains of *Clostridium botulinum*, *C. butyricum*, *C. tetani* and presumably also by *C. barati*, are nearly identical in macro structure. Some aspects of their fine structures and pharmacological action are strikingly similar. These proteins, one called tetanus neurotoxin and the rest botulinum neurotoxins (NT), produce in their target nerve cells similar intracellular biochemical lesions; blockage of normal release of neurotransmitters, albeit inhibitory and excitatory. Diversity in the primary, secondary and tertiary structures among these proteins makes them antigenically distinguishable and endows them with the specificity to recognize and bind to the "acceptors/receptors" present on the susceptible nerve cells. This article considers the structural aspects of this generalized theme. The structural features of botulinum NT are presented first. The temptation to keep tetanus NT in the same focus as botulinum NT could not be overcome even twelve years after pointing out for the first time the similarities between botulinum and tetanus neurotoxins.[11]

Schematic structure of the neurotoxin

The generalized structure of botulinum NT is schematically shown (Fig. 1) in two ways; the protein stretched out and the protein with two folds. The single chain NT molecule undergoes a postsynthetic cleavage (nicking) by

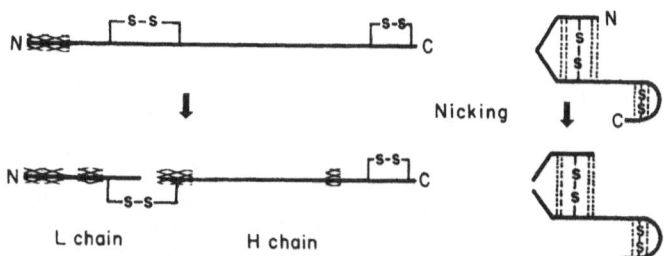

L chain   H chain

Nicking

Fig. 1   Structure of botulinum neurotoxin represented in the straight line and in folded configurations; the hatched segments have been sequenced. The single chain protein is nicked to the dichain protein. The dotted lines in the folded configuration represent noncovalent bonds.

Microbial Toxins in Foods and Feeds, Edited by A.E. Pohland *et al.*,
Plenum Press, New York, 1990

a protease endogenous or exogenous (e.g., trypsin) to the organism. The nicking site is one-third of the way from the N-terminal to the C-terminal end. The nicking does not change the molecular size (~150 kDa) of the single chain protein. The NT when isolated and purified from the bacterial culture is found as single chain, dichain, or as a mixture of single and dichain molecules. The light (L) chain of the nicked molecule is ~one-half the molecular size of the heavy (H) chain. Small regions of the protein, marked with hatched areas, have been sequenced by automatic Edman degradation. A half cystine residue located very near the N-terminal of the H chain (see amino acid sequences in Fig. 2) appears to be part of the -S-S- bond that links the H and L chains (see comment below). Another -S-S- on the H chain appears to be present near its C-terminal end. Size of this -S-S- loop is not known.

The diagram on the right in Fig. 1 depicts that i) the L chain is bound to the N-terminal half of the H chain by a -S-S- bond and noncovalent bonds (dotted lines between the L and H chains), ii) noncovalent interactions between the L chain and the C-terminal half of the H chain are virtually absent or extremely weak, and iii) the C-terminal end is folded (see comments below).

Chemical structures of the neurotoxins

One facet of the chemical structure of a NT is its empirical formula, the amino acid composition. "Amino acid analysis bears a relationship to the chemistry of proteins and peptides similar to that which elementry analysis bears to the chemistry of simpler organic molecules."[37] Amino acid compositions of botulinum NT types A, B, C_1, D, E and F and those of the H and L chains of types A, B, C_1, D and E are compared in Tables 1 and 2. These compositions, determined following HCl hydrolysis of the protein samples, are compared with the amino acid composition of tetanus NT computed from the complete amino acid sequence that was derived from its nucleotide sequence.[17,18] A point to remember is that while the composition of tetanus NT is error free, analysis of acid hydrolyzed protein is error prone; sources of error are the analytical methodology and the molecular weight of the protein used in the calculation of composition. Until the complete amino acid sequences of botulinum NTs are determined, the amino acid compositions will be the only basis to compare the NT serotypes in chemical terms.

Partial covalent structures of the neurotoxins

The partial amino acid sequences of botulinum type A, B, C_1 and E NT determined so far include the sequences of the NT produced by two different strains of type B (Okra and 657) and by *C. butyricum*. All of these sequences are compared with that of tetanus NT in Fig 2. Why include strain B-657 and *C. butyricum*? The NT produced by strain B-657 was reported to have serological specificity somewhat different than from the NT produced by the well characterized type B strain Okra.[19] The NTs from the two strains were found identical on the basis of partial amino acid sequences and serological reactions.[6] The type E NT produced by *C. butyricum* strain 5262[30] is a novelty for it is not a product of the *Botulinum species*. The NTs from *C. butyricum* and *C. botulinum* are similar in size and dichain structure (that results from tryptic cleavage of the single chain protein) but their serological reactions with anti-*C. botulinum* type E NT show some differences.[20] We compared partial amino acid sequences of the NTs produced by the two species; the H chains from the two NTs matched, but the two L chains differed at residues 2, 14 and 27.[21] The NT produced by *C. barati*[22] is yet to be biochemically characterized.

There are several interesting similarities among the botulinum NT serotypes and tetanus NT. Some of the examples are as follows. *In the L chain:* i) The N-terminal residue of *each* NT (or its L chain) is Pro. The first

Table 1.   Amino acid compositions of botulinum NT types A, B, C, D,
E, F and tetanus NT.

| | Botulinum NT types | | | | | | | Tetanus |
	A	B	C[a]	C[b]	D[c]	E	F	
Asx	200	212	229	225	210	195	240	218	88
Thr	75	54	65	78	67	69	75	80	73
Ser	79	83	103	88	78	101	98	105	99
Glx	114	130	149	116	112	119	118	128	70
Pro	44	46	45	41	50	55	45	47	53
Gly	64	61	70	59	54	58	58	69	63
Ala	53	44	54	44	51	40	40	47	52
Val	70 ·	54	64	65	69	73	62	72	63
1/2Cys	10	11	11	6	8	8	7	9	10
Met	22	23	27	16	19	17	17	14	25
Ile	111	144	147	114	138	111	123	128	134
Leu	104	107	99	91	103	103	107	104	114
Tyr	71	81	81	71	54	58	70	86	79
Phe	68	77	77	70	65	61	62	60	56
Lys	100	118	123	78	105	84	97	90	107
His	14	7	9	10	11	12	15	13	14
Arg	43	39	45	47	47	34	34	51	38
Trp	17	18	36	11	12	16	16	23	13
Asn	-	-	-	-	-	-	-	-	121
Gln	-	-	-	-	-	-	-	-	43
kDa	145	152	167	141	144	140	147	155	150
Reference	10	13	2	54	56	38	8	9	17

[a] = strain stockholm; [b] = strain 6813; [c] = strain 1873
In all cases (Table 1 and 2) except for tetanus Asx = (Asp + Asn) and
Glx = (Glu + Gln); for tetanus Asx represents Asp and Glx represents
Glu.

residue of C_1 NT appeared to be Pro but could not be identified unequivo-
cally (see explanation in ref. 57).   ii) A Ile.Trp (or I.W) pair is present
between residues 38 and 43 in *each* protein.   This includes the only Trp
residue of the L chain of tetanus NT (see Table 2 for amino acid composition).
The Trp residue in *each* protein is flanked on both sides by hydrophobic residues.   iii) Between botulinum NT types A, B and tetanus NT eight residues
are identical at positions 1, 7, 9, 11, 12, 18, 24 and 44.   iv) The segment
-D.P.V.N- (around residues 11-14) is identical between types A, E and tetanus.
v) The stretch -K.A.F.K.I- around residue 35 is present in tetanus and botulinum type A and B NTs but not in type E.   vi) Type B and tetanus NTs are 70%
homologous (31 out of 44 residues matched in identical positions); three segments of residues 3-7, 18-24 and 32-44 are identical.   vii) The 47 residues
sequenced from the 18 kDa fragment of type A (L chain) align with those of
tetanus NT beginning at its residue 259.   Between these 47 residues of each
of the two proteins 18 residues (i.e., 38%) are identical in positions.   *In
the H chain*:   i) An invariant pair - Cys.Ile- is present in *each* protein very
close to the N-terminal.   This Cys is part of the -S-S- bond between H and L
chain of tetanus NT.[17]   ii) Within the next nine residues (toward the C-terminal) Asn (N), Leu (L) and Phe (F) are conserved in *each* protein as
invariant residues.   iii) Between the type B and tetanus NT there are three

Table 2. Amino acid compositions of the H and L chains of botulinum NT types A, B, C, D, E and tetanus NT.

	H chain of botulinum NT type						Tetanus		L chain of botulinum NT type					
	A	B	Cᵃ	Cᵇ	Dᶜ	E	H	L	A	B	Cᵃ	Cᵇ	Dᶜ	E
Asx	149	153	148	135	125	177	55	33	65	68	76	71	70	77
Thr	39	38	43	43	38	50	49	24	33	17	28	26	31	22
Ser	55	59	60	55	58	65	68	31	28	25	29	29	43	32
Glx	79	90	80	77	76	86	46	24	39	43	39	35	43	38
Pro	21	18	19	22	26	16	29	24	20	22	25	27	29	21
Gly	37	34	40	36	32	34	39	24	30	25	20	21	26	28
Ala	35	32	25	27	26	30	37	15	20	14	18	25	14	13
Val	40	35	42	47	44	47	44	19	26	22	21	24	29	15
1/2Cys	6	7	4	6	6	6	5	5	4	5	2	2	2	4
Met	17	13	15	13	12	11	12	13	7	11	5	7	5	4
Ile	82	96	73	91	73	80	90	44	30	43	38	46	38	43
Leu	76	71	57	67	63	74	76	38	40	27	32	36	40	37
Tyr	49	60	46	39	40	47	51	28	25	24	22	16	18	23
Phe	37	52	42	37	32	40	33	23	34	28	26	25	29	25
Lys	63	79	53	66	54	72	74	33	39	41	28	38	30	32
His	6	2	5	5	5	8	8	6	6	4	5	4	7	6
Arg	30	25	28	25	18	21	24	14	15	18	20	23	16	14
Trp	16	18	7	10	12	16	12	1	4	3	2	2	4	3
Asn	-	-	-	-	-	-	79	42	-	-	-	-	-	-
Gln	-	-	-	-	-	-	27	16	-	-	-	-	-	-
kDa	97	104	91	92	85	102	98.3	52.4	53	51	50	52	55	50
Reference	44	44	54	56	38	44	17	17	44	44	54	56	38	44

ᵃ = strain stockholm; ᵇ = strain 6813; ᶜ = strain 1873

common stretches -N.E.D.L.-, -F.I.A- and -K.N.S.F-. The sequence of the 44 kDa fragment (the "C-terminal half") of the H chain from the botulinum type A NT[46] matches well with the comparable region of the tetanus NT (~52 kDa C-terminal fragment derived with papain, see ref. 17). The stretch of residues 7-10, -I.L.N.L- of botulinum NT is identical to the residues 885 to 888 of tetanus NT. The four residues blocked by the heavy line have skewed alignment; two pairs of residues are identical, and in two pair each amino acid differs by substitution of only one nucleotide. To the right of the skewed alignment, caused probably by one deletion and one insertion, are two positionally identical residues, Asp (D) and Ser (S), underlined and one Leu-Ile, i.e., -L.I- pairing (substitution of one base in code). Near the N-terminal, alignment of three pairs of residues, marked beneath with +, are conceivable because in each pair the amino acids differ by substitution of only one nucleotide in their code. Clearly these stretches of proteins are similar and highly homologous.

These proteins present many more cases of homology not cited here that can be considered on the basis of single base substitution, deletion and insertions as well as conservation of the charge, polarity, etc., of the side chains in the substituted amino acid residues e.g. Lys ▶ Arg, Ile ▶ Val.

C-terminal sequences of type A, B and E NTs although tentative do not appear identical.[58] The sequences were determined by digestion of these proteins with carboxypeptidase Y and using an amino acid analyzer. The probable sequential order of release from the C-terminals appeared as follows: for type A, -X-Gly-Leu (or -Gly-Leu-Leu); for type B, -Thr-Glu (or -Glu-Thr) and for type E, -Asn-Glu-Gln-Lys.

Location of -S-S- bonds

The exact location of the half cystine on the H chain of botulinum NT that forms the -S-S- bond with the L chain has not been experimentally determined, but three lines of other experimental evidence allow the model that the half cystine residue very near the N-terminal of the H chain is the one that forms disulfide bond with the L chain.

In botulinum type A NT, only the N-terminal-half region of the H chain is linked to the L chain by -S-S- bond(s); the C-terminal half is not linked to the L chain.[12,45,46,50] In type B NT, only one half of the H chain is linked to the L chain by -S-S- bond.[12] In tetanus NT, the N-terminal half of the H chain has only one Cys residue (466) which forms the only -S-S- link between L and H chains.[17] In this article, numbering of the residues of tetanus NT begins with Pro as 1, the amino terminal of the mature protein. Comparison between tetanus and the three botulinum NTs (Fig. 2) shows an invariant Cys very near the N-terminals of their H chains.

A note of caution: A dichain NT molecule without the -S-S- bond(s) linking the L and H chain is conceivable. The non-covalent interactions (ionic, hydrogen bonds, hydrophobic interactions) could keep the L and H chains together. Such a NT whether or not reduced with mercaptoethanol or dithiothreitol (DTT) would migrate in sodium dodecylsulfate-polyacrylamide gel electrophoresis (SDS-PAGE) as two bands of ~50 and ~100 kDa. In our laboratory, pure type A and type B NT preparations *not reduced* with merceptoethanol were found on many occasions to migrate in SDS-PAGE as three bands, most of which appeared as 150 kDa, with minor bands of 50 and 100 kDa. These samples when reduced yielded stronger bands of 50 and 100 kDa (unpublished). In the case of tetanus NT, a large proportion of the "intracellular" toxin, following trypsinization, did not have the -S-S- linking the two chains, whereas the entire population of the "extracellular" toxin molecule (nicked by endogenous protease) was made of the -S-S- linked H and L chains.[29] There are

Type   Light chain

```
                        1           10          20          30          40        49
A (Okra)   P.F.V.N.K.Q.F.N.Y.K.D.P.V.N.G.V.D.I.4.Y.I.K.I.P.N.A.G.Q.M.Q.P.V.K.A.F.K.I.H.N.K.I.W.V.I.P.E.R.D.T.
B (657)    P.V.T.I.N.N.F.N.Y.N.D.P.I.D.N.N.-
B (657)    P.V.T.I.N.N.F.N.Y.N.D.P.I.D.N.N.I.I.M.M.E.P.P.F.A.R.G.M.Q.R.Y.K.A.F.K.I.T.D.R.I.W.I.-
C          - I.T.I.N.N.F.N.Y.S.D.P.V.D.N.K.N.I.L.Y.L.D.T.H.L.-
E (botulinum) P.K.I.N.S.F.N.Y.N.D.P.V.N.D.R.T.I.L.Y.I.K.P.G.G.C.Q.E.F.Y.K.S.F.N.I.M.K.N.I.W.X.I.X.E.R.N.V.I.-
E (butyricum) P.T.I.I.N.S.F.N.Y.N.D.P.V.N.R.T.I.L.Y.I.K.P.G.G.X.Q.F.Y.K.S.F.N.I.M.K.N.I.W.I.I.P.E.R.N.V.I.G.-
Tetanus    P.I.T.I.N.N.F.R.Y.S.D.P.V.N.N.D.T.I.I.M.M.E.P.P.Y.C.K.G.L.D.I.Y.Y.K.A.F.K.I.T.D.R.I.W.I.V.P.E.R.Y.

  cont'd.           50          60          70
A          F.T.N.P.E.E.G.D.L.N.P.P.P.E.A.K.Q.V.P.X.S.Y.Y.D.-
Tetanus    E.F.G.T.K.P.E.D.F.N.P.P.S.S.L.I.E.G.A.S.E.Y.Y.D.-

                        1           10          20          30          40
A (18 KDa-piece)  -Y.E.M.S.G.L.E.V.S.F.E.E.L.R.T.F.G.G.H.D.A.K.F.I.D.S.L.Q.E.N.E.F.R.L.Y.Y.Y.N.K.F.K.D.I.A.S.T.L.-
E (" " )             -K.G.I.N.I.E.E.F.L.T.F.G.N.N.D.L.N.I.I.T.V.A.Q.Y.N.D.I.Y.T.N.L.L.N.D.Y.R.K.I.A.X.K.L.-
Tet. (#259-305)   -Y.M.Q.H.T.Y.P.I.S.A.E.E.L.F.T.F.G.G.Q.D.A.N.L.I.S.I.D.I.K.N.D.L.Y.E.K.T.L.N.D.Y.K.A.I.A.N.K.L.-
```

Heavy chain

```
                        1           10          20          30
A          A.L.N.D.L.C.I.K.V.N.N.W.D.L.F.F.S.P.S.E.D.N.F.T.N.D.L.N.K.G.E.P.I.T.S-
B (Okra)   A.P.G.I.C.I.D.V.D.N.E.D.L.F.F.I.A.D.-
B (657)    A.P.G.I.X.I.D.V.D.N.E.D.L.F.F.I.A.D.K.N.S.F.R.D.D.L.-
E (botulinum)  K.S.I.C.I.E.I.N.N.G.E.L.F.F.V.A.X.E.N.X.Y.-
E (butyricum)  K.S.I.C.I.E.I.N.N.G.E.L.F.F.V.A.S.E.N.X.Y.N.D.-
Tet. (#457→)  S.L.T.D.L.G.G.E.L.C.I.K.N.E.D.L.T.F.I.A.E.K.N.S.F.S.E.E.P.F.Q.D.E.I.V.S.Y.-

Type A (C-terminal       1           10          20
 half of H chain)    X.I.I.N.L.X.I.L.N.L.R.Y.E.X│N.H.L.I│D.L.K.X.Y.A.S.-
Tetanus
(residues #879-903)  I.L.K.K.S.T.I.L.N.L.D.I.N│N.D.I.I│S.D.I.S.G.F.N.S.-
                     + +                        -
```

Fig. 2 References:

Type		light chain		heavy chain
A	(Okra)	7,48		46,48,50
B	(Okra)	" 49	" "	49
B	(657)	" 6	" "	6
C		" 57	" "	-
E	(botulinum)	43,49	" "	43,49
E	(butyricum)	21	" "	21
E	(18 kDa piece):	DasGupta and Foley (manuscript in preparation)		
Tetanus		17,18	" "	17,18

at least two possible explanations for these observations: i) When trypsin cleaves the single chain to dichain form an additional cleavage occurs very close to the nicking site but outside the -S-S- bridge that links the two chains. (Nicking occurs inside the -S-S- loop.) Such a cleavage will generate a "heavy chain" that would *not* be linked to the L chain by -S-S- bonds. Its molecular size, slightly smaller than the H chain, would be essentially indistinguishable from the normal H chain. The N-terminal residue of this "heavy chain" would be different from the normal H chain. For such a scenario, the probable second cleavage sites (for trypsin or a trypsin-like enzyme) in botulinum type A NT are Lys-Val and/or Lys-Gly bonds (see Fig. 2); in type B NT Lys-Asn bond and in tetanus Lys-Ile and Lys-Asn (at two positions). Robinson and Hash[40] have found Asn as one of the N-termini of the H chain preparations of tetanus; Ser (residue 457) is the recognized N-terminal of the H chain of tetanus NT.[17] The other possibility, entertained by Matsuda and Yoneda,[29] is remote: The nascent protein may have the particular cysteine residues but the disulfide bridge may not form in some of the mature protein molecules during their folding. Pairing of cysteine (i.e. 1/2 cystine) residues to form correct disulfide bonds during folding is directed by non-covalent forces at appropriate oxidation-reduction potential.

Would a dichain molecule without the disulfide linking the subunit chains have neurotoxic activity similar to that of the -S-S- linked dichain NT? Available evidence and the following considerations suggest that a dichain NT without the -S-S- (that normally links L and H chains) need not be non-neurotoxic. The NT binds to the receptors on target cells via its H chain region (see references in 1). This binding also takes place between the isolated H chain and receptors,[1] therefore the -S-S- between the H and L chains does not appear to be critical for the binding activity, the first step in the mechanism of action of the NT. The isolated L chain in conjunction with the isolated H chain inhibits transmitter release.[1,39] No experimental evidence has demonstrated that specific reduction of the -S-S- that links the two chains (while other -S-S- bridges elsewhere in the molecule remain intact) detoxified the NT.

A disulfide loop at one end of the C-terminal half of the heavy chain of types A and B

SDS-PAGE examination of limited tryptic or chymotryptic digests of type A and B NT had demonstrated that each protease cleaved the H chains at about their mid-points to form: i) a ~50 kDa single chain fragment (SCF) representing about one-half of the H chain, ii) this fragment underwent an additional cleavage within an intrachain disulfide loop, and iii) the other half of the H chain was linked to the L chain hence 100 kDa in total.[12] The second conclusion could be developed because migration of the SCF was slightly faster under disulfide reducing conditions. Apparently, cleavage of the SCF near its C- or N-terminal end generated a small fragment (not identified by SDS-PAGE) that remained linked to the larger fragment (< ~50 kDa) by a -S-S- bond. A stronger case for this disulfide and the approximate span length of the -S-S- loop could not be made until two other half cystine residues, i.e., second and third cysteines, could be found in the H chain (one cysteine was already known linked to the L chain via -S-S-). Locations of cysteine residues in type A and B NTs were mapped employing the reagent 2-nitro-5-thiocyanobenzoate; two cysteines were found between the mid-point of the H chain and its C-terminal end (see Fig. 5, in ref. 5). Location of this -S-S- loop was (by inference) depicted at one end of the heavy chain (see Fig. 2 in ref. 5). This could be located near the N- or the C-terminal end of the C-terminal half of the H chain. The model was applicable for both types A and B. Schone et al.[50] came to the same conclusion from their tryptic cleavage study of type A NT and inferred that the small fragment cut off from one end of H chain or C-terminal half of the H chain was of 2-3 kDa. This is consistent with the reported cut at one end of the SCF.[12]

The secondary and tertiary structures of the NT and the state of tyrosine and tryptophan residues are being determined by different techniques, such as circular dichroism (CD), fluorescence spectroscopy, fluorogenic dye, and UV-difference spectroscopy (solvent perturbation and alkaline pH). Observations from CD studies:[14] The NTs have a high degree of ordered secondary structure (sum of α helix and β-sheet content) and the D-sheet content is clearly dominant. For example, the sum of secondary structure parameters found in type A at pH 6.0 (α helix 28%, β-sheet 42%, total = 70%) was similar to that found at pH 9.0 (α 22%, β 47%, total = 69%). Note that: the single chain type E at pH 6.0 (α 18%, β 37%, total = 55%) differed somewhat from these values at pH 9.0 (α 22%, β 43%, total = 65%) and the single chain type E and the dichain type A, both at pH 9.0, were closer in their ordered secondary structure (65% and 69%, respectively). Comparison of the single chain type E NT with dichain type A NT at the physiological pH 7.2 showed more similarities; both had ~21% α-helix and 43-44% β-sheet.[53] The above studies[14] indicate that the conformation of the single chain (unactivated) type E NT changes when exposed to an alkaline environment (pH 6 vs. 9) and at pH 6.0 the conformation of the single chain type E differs from that of the dichain (fully activated) type A NT. Conversion of the single chain type E to the dichain (brought about by mild trypsinization) produced changes in the secondary structure. The dichain type E at pH 6.0 assumed a structure (α 20%, β 47%, total 67%) that was different from its precursor the single chain at pH 6.0 (α 18%, β 37%, total = 55%) and more like the dichain type A (at pH 6.0, α 28%, β 42%, total = 70%). This similarity came mostly by a gain in the β-sheet content, the dominant feature of these NTs.

Similarity between tetanus and botulinum NTs is again apparent from the CD data: The ordered secondary structure (α + β contents) of botulinum NT, 70% for type A and 67% for dichain type E, agrees well with 65% of α + β contents of tetanus NT.[42]

Analysis by CD technique had parallel probing of the hydrophobic regions of types A and E (in the single and dichain forms) with the fluorogenic dye toluidine napthalene sulfonate.[14] The fluorescence intensity of the dye (emission maximum 428 nm) increases as more hydrophobic regions of a protein bind to it; this provides a means to titrate the hydrophobic regions. Hydrophobicity of the dichain type E NT was higher than the single chain protein and appeared similar to that of the dichain type A NT. The nicked type E NT thus appeared to have more hydrophobic sites than the unnicked single chain, and therefore these studies also suggest that conversion of the single chain type E NT to the dichain form (nicking by trypsin) induces conformational changes.

Examination of the ionization of phenolic groups of tyrosine residues of type A and E, based on UV difference spectra, has also detected changes resulting from nicking the single chain type E NT.[15] For dichain type A NT the plot of difference absorptivity (at 296 nm) vs. pH was a simple sigmoidal curve. The pK of the phenolic moieties of the tyrosine residues was 10.9. Nearly all tyrosine residues could be ionized. The single chain type E, unlike the dichain type A, yielded a two-step titration curve and pK values 11.3 and <7.5. Only ~60% of the total number of tyrosine residues were ionized. In contrast, all or nearly all tyrosine residues of the dichain (nicked) type E NT were ionized and thus the dichain type E NT appeared more like dichain type A NT.

The solvent perturbation difference spectra of type C_1 NT have revealed 3-5 tryptophan (out of a total of 11) and 20-32 tyrosine (out of a total of 71) residues exposed. In addition, 1-2 tryptophan and 12 tyrosine residues are probably partially buried.[55]

The C-terminal end of the NT seems folded (Fig. 1) because carboxypeptidase Y does not appreciably digest the C-terminal end of the NT types A, B and E unless they are unfolded with SDS.[58]

Structural domains

In the dichain NT the L and H chains appear to retain two quasi-independent structural domains although connected by -S-S- bond(s) and other weak noncovalent bonds. CD measurements of the isolated L and H chains and their parent dichain type A NT and other considerations lead us to this inference.[52] Note from Table 3 that the α helix, β pleated sheet, β-turn and random coil values of the type A NT (~150 kDa) are nearly equal to the sum of these secondary structure parameters of the L and H chain (~50 kDa and ~100 kDa, respectively) when calculated as weighted mean. This agreement indicates that the secondary structures of L and H chains do not significantly change when they are together as a dichain NT molecule and only weak interactions exist between the two subunit chains other than the -S-S- bond linking them.

The point on weak interaction can also be made on the basis of chromatographic behavior of the L and H chains[44,54] and two halves of the H chain.[46] To separate the L and H chains, the dichain NT (types A, B, C_1, D or E) is allowed to bind to an anion exchange column at pH 8.4; then the disulfide bond(s) is reduced with DTT in the presence of 2M urea. Washing the column with buffer containing 2M urea and DTT elutes the L chain out of the column while the H chain remains bound to the anion exchanger. Elution of the H chain requires higher ionic strength buffer. To obtain the two halves of the H chain, the dichain NT is digested with trypsin. This cleaves the H chain at about the mid-point. An anion exchange column (0.02 M Na phosphate buffer, pH 8.0) loaded with this digest does not retain the C-terminal half of the H chain; but the N-terminal half of the H chain linked to the L chain via -S-S- bond (~100 kDa) remains bound to the column. Elution with buffer at higher strength dissociates the ~100 kDa fragment (the N-terminal half of the H chain linked to L chain). These two polypeptides are then separated in the same way as L and H chains, as above. The association between the L chain and the C-terminal half of the H chain appears very weak if not insignificant because they are dissociated in 0.02M phosphate buffer, pH 8.0. The noncovalent association between the L chain and the intact H chain or N-terminal half of the H chain also appears weak; just 2M urea is enough to separate them following reduction of -S-S- bond(s).

Three other sets of observations also compliment the arguments for the "two quasi-independent domains" structural model. i) The ability of the isolated H chain to compete with its parent NT for the same acceptor ("receptors") site demonstrated that the NT's binding site for "receptors" was located on the H chain.[1] ii) The separated and purified L and H chains administered to susceptible tissue or cells *in vitro*, only in certain orders, induce paralysis or block acetylcholine release.[1,39] iii) When the isolated L and H chains are mixed together and carefully reoxidized they reconjugate and a -S-S- forms between the two chains. The reconjugated dichain molecule (~150 kDa) with regained toxicity is indistinguishable from the native dichain NT. This was first demonstrated with type B,[27] followed by C,[54] and then with A.[28] Clearly, the L and H chains do not undergo significant irreversible changes in their conformation when they are separated and isolated.

Structural features of NT and its binding with lipid membranes

To explain how the NT at 10^{-12}-10^{-14} molar concentration (a mouse lethal dose is ~1 ng/kg) is able to bind a high affinity receptor on nerve cells Montecucco[32] has proposed the following model. The NT first binds to the ganglioside rich lipid surface, then the lipid-NT complex moves laterally to

Table 3. Secondary structural parameters of the type A NT and its isolated L and H chains calculated from their CD spectra between 240 and 200 nm (at pH 8.1).

Proteins	α-helix %	β-sheet %	β-turn %	Random coil %
A NT	20.0	37.5	15.2	27.2
L chain	22.0	27.5	18.7	31.7
H chain	18.7	40.0	13.0	28.2
Weighted mean*	19.8	35.8	14.9	29.4

* Calculated as (1 x L chain + 2 x H chain)/3, because H chain is twice the size of the L chain; for example α-helix content is (22 + 2 x 18.7)/3 = 19.8. From Singh and DasGupta.[52]

reach and bind the NT specific receptor, which is trypsin sensitive. Accordingly, any docking of the NT molecule on the membrane results (following "catch and delivery effect") in a productive binding with the NT specific receptor protein. The lipid "binding step is actually equivalent to concentrating the NT and its protein receptor in a much smaller volume.... because the partners of the binding reaction are now restricted to the two dimensional plane of the plasma membrane rather than in the three dimensional water phase."

This idea seems to agree with our attempts to crystallize type A NT.[41] Two-dimensional, ordered arrays of the NT have been formed at the interface of a solution of the NT and phospholipid monolayer containing the trisialoganglioside G_{T1b}. The NT binds the hydrophilic moiety of the ganglioside and two-dimensional diffusions allow crystals to form in a couple of days.

The first two events of the intoxication process, binding of the NT to the membrane surface and internalization, are being studied in two ways: In collaboration with Dr. Cesare Montecucco, we have used the technique of hydrophobic labelling with photoreactive phosphatidylcholine analogues. These analogues,[34] stable in darkness, are interdispersed in trace amounts among the components of membranes. On illumination the photoreactive moiety is converted to a reactive intermediate able to cross-link and hence label neighboring protein molecules in the membrane. One analogue carries its photoreactive moiety near the phosphate group that probes the polar or surface region of the membrane. The other analogue bears its photosensitive group at the fatty acid methyl terminus to react with the regions of protein buried in the hydrophobic core of the membrane. Botulinum NT types A, B and E interacted, at neutral pH, with the polar head groups of phospholipids in membranes made of different lipid compositions. At acidic pH the NTs underwent conformational changes that were characterized by extensive hydrophobic interaction with lipids. Both the H and L subunit chains of the NTs were involved in the process.[35,36] In the other approach, formation of ion conducting channels are monitored. NT type A,[3,51] type B,[23] type C_1[16] and type E[4] have been tested. In each case, channel activity is stimulated by acidic pH on the cis side (the side to which protein is added) and neutral pH on the opposite (trans) side. The channel forming activity appears to depend on the dichain structure of the NTs. The single chain type E NT shows minimal activity, but addition of trypsin to the NT-containing side of the membrane dramatically enhances channel formation. This activation is blocked by the presence of soybean trypsin inhibitor. The tentative conclusion about the importance of dichain structure is consistent with the observations made with isolated L and H chains of types A and B NTs and the two halves of type A H chain. Only the H chain (~100 kDa) forms channels and the activity is confined to the N-terminal half (~50 kDa) of the H chain.[3,51] The role of these channels in the intoxication process, although not clearly understood, has been conjectured;[33] further discussion is outside the scope of this chapter. But how does a water soluble protein become partially embedded in the membrane lipid, and how does it cross the hydrophobic membrane barrier to reach the cytoplasm?

Potential amphiphilic secondary structures and their possible role

A cell surface membrane is not a rigid, nonfluctuating structure and its hydrocarbon domain is not completely devoid of polar groups. When a protein binds to such a surface the protein often adopts complementary secondary structures that are amphiphilic, i.e. one face of the polypeptide is rich in hydrophobic amino acid side chains, while the opposite face is preferentially occupied by lipophilic side chains.[25,26] An amphiphilic α-helix or β-sheet when oriented parallel to the plane of the membrane can immerse into the lipid membrane to different depths aided by a combination of hydrophobic and polar interactions.[31] Examples of membrane directed secondary structures with amphiphilic distribution of the individual amino acid side chains are the cytotoxic proteins melittin (principle toxic component of honey bee venom) and δ-hemolysin (from *Staphylococcus aureus*), hormones β-endorphin and calcitonin and also apolipoproteins.[26] Kaiser and Kezdy[25,26] developed the following guiding principles: "Polypeptides that bind to biological interfaces where the environments are amphiphilic will often contain important regions comprising amphiphilic secondary structures complementary to those of the surface. If such a membrane-directed secondary structure exists, it should be characterized by the amphiphilic distribution of the individual amino acid side chains.... Model building, to a first approximation, can be based largely upon secondary structural considerations, ignoring tertiary structural considerations."

I searched for the *possible occurrence* of amphiphilic secondary structures in botulinum and tetanus NTs from their amino acid sequences using described techniques.[25] Several α-helical and β-sheet regions were found (Fig. 3). Further examinations of the Schiffer-Edmundson wheel[47] patterns (Fig. 3a,b,c,d) on the basis of helical net diagram and π-helical conformation are needed.

Indirect experimental techniques are available to establish the existence of such conformations.[26] The ultimate confirmation of the role of active secondary amphiphilic structures requires the following considerations If an amphiphilic secondary structural feature, e.g. an α-helical segment, is critical for the biological activity of the NT, then this activity should be present in the NT made with a different amino acid sequence (the helical segment under study) but with essentially the same secondary structure. In other words, side chains of the amino acids in the amphiphilic region can be altered without significant loss of activity as long as the substitutions conserve the lipophilicity and hydrophicility of the residues. This issue can be approached by comparing NTs with certain amino acids replaced by mutations induced in laboratory or by nature. Many amino acid residues in botulinum NT have undergone replacement; differences in the amino acid compositions of the six serotypes (Tables 1, 2) and in the amino acid sequences of three serotypes (Fig. 2) are experimental confirmations of the premise. Progress in the sequence determination of the NT serotypes will offer further "first clues" in the possible existence of amphiphilic structures and opportunities to compare such segments among the "mutants" produced by nature.

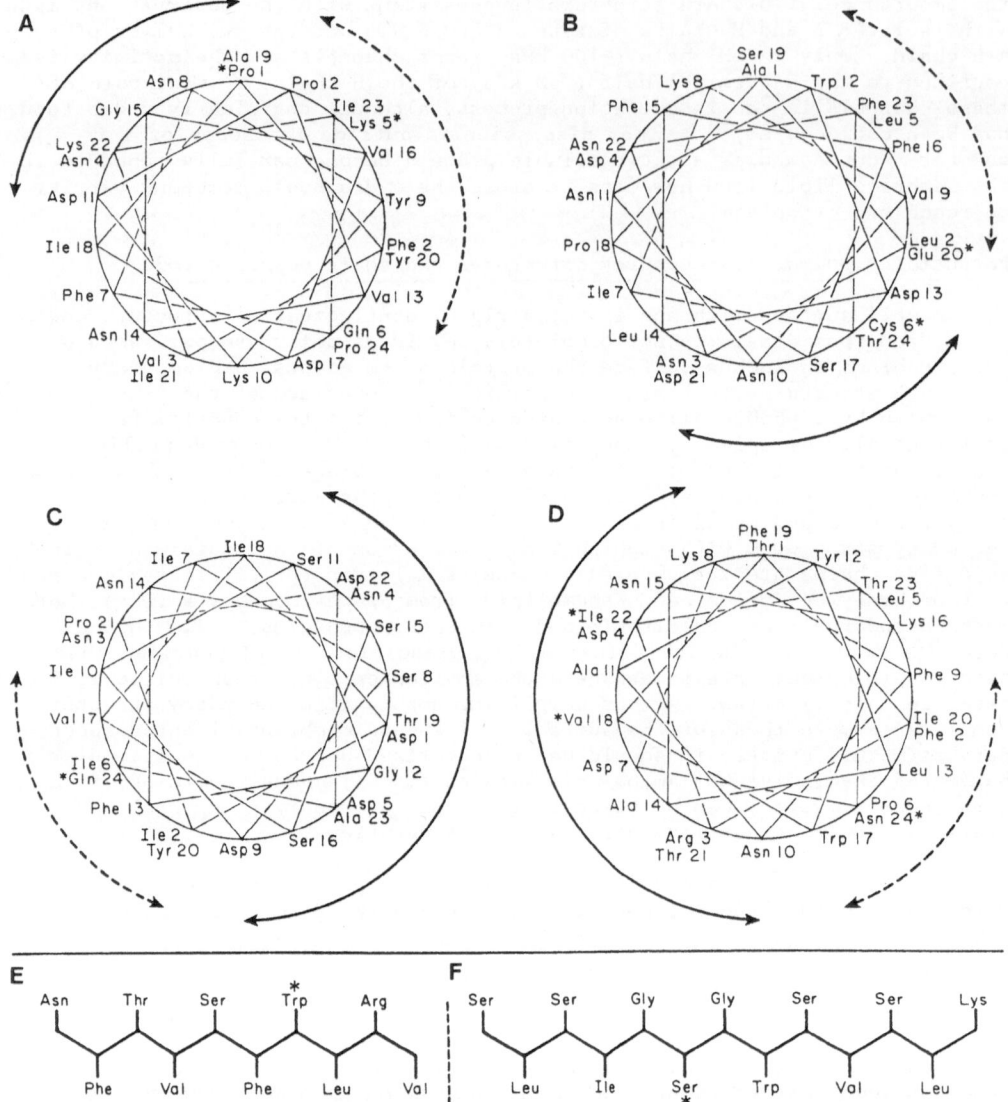

Fig. 3   Potential amphiphilic secondary structures in botulinum and
         tetanus NT. Circular panels: Axial projections of the amphi-
         philic helical (pitch 3.6 residues per turn) regions from
         botulinum (panels A, B) and tetanus (panels C, D) NT. Residues
         under a solid-line arch are hydrophilic, under a dashed-line
         arch are hydrophobic; exceptions are marked (*). Panel A is L
         chain, residue 1-24. Panel B is H chain, residue 1-24. Panel
         C is H chain, residue 889 (Asp)-912 (Gln). Panel D is H chain,
         residue 1014 (Thr)-1037 (Asn). Bottom linear panels: Amphi-
         philic β-pleated sheet conformation of two stretches from
         tetanus NT H chain, Panel E is residues 948 (Asn)-957 (Val),
         Panel F is residue 983 (Ser)-995(Lys). Hydrophilic and hydro-
         phobic residues are facing up and down; exceptions are marked
         (*).

Kaiser[24] pointed out one possible role of the amphiphilic secondary structure induced on the membrane: "in the vicinity of the receptor.... the peptide.... finds its receptor by absorption to the cell surface followed by diffusion in only two dimensions on that surface. This would eliminate the need for a three-dimensional search which would be required in the absence of the induction of an amphiphilic structure." This generalization is consistent with Montecusso's model for botulinum NT binding to neuronal membranes and our success in crystallizing the NT on lipid surface.[41]

ACKNOWLEDGEMENT

The studies in the author's laboratory were supported in part by NIH grant NS17742, U.S. Army Medical Research Development Contract DAMD 17-84-C-4245, Department of Defense-University Research Instrumentation Program award DAAG-29-83-Gr0063, the Food Research Institute and the College of Agricultural and Life Sciences of the University of Wisconsin-Madison.

REFERENCES

1. Bandyopadhyay S, Clark AW, DasGupta BR, Sathyamoorthy V, 1987, Role of the heavy and light chains of botulinum neurotoxin in neuromuscular paralysis, *J Biol Chem.*, 262:2660-2663.
2. Beers WH, Reich E, 1969, Isolation and characterization of *Clostridium botulinum* type B toxin, *J Biol Chem.*, 244:4473-4479.
3. Blaustein RO, Germann WJ, Finkelstein A, DasGupta BR, 1988, The N-terminal half of the heavy chain of botulinum type A neurotoxin forms channels in planar phospholipid bilayers, *FEBS Letters*, 226:115-120.
4. Blaustein RO, Hoch DH, DasGupta BR, 1988, Channels formed by botulinum type E neurotoxin in planar lipid bilayers, *FASEB J.*, A1750, (abstract).
5. DasGupta BR, 1981, Structure and structure function relation of botulinum neurotoxins, *In* "Biomedical Aspects of Botulism," Lewis, GE, ed. Academic Press, New York.
6. DasGupta BR, Datta A, 1988, Botulinum neurotoxin type B (strain 657); partial sequence and similarity with tetanus toxin, *Biochimie*, 70:811-817.
7. DasGupta BR, Foley J, Wadsworth C, 1988, Botulinum neurotoxin type A: partial sequence of L chain and its fragments, *FASEB J.*, A1750, (abstract).
8. DasGupta BR, Rasmussen S, 1983, Purification and amino acid composition of type E botulinum neurotoxin, *Toxicon.*, 21:535-545.
9. DasGupta BR, Rasmussen S, 1983, Amino acid composition of *Clostridium botulinum* type F neurotoxin, *Toxicon.*, 21:566-569.
10. DasGupta BR, Sathyamoorthy V, 1984, Purification and amino acid composition of type A botulinum neurotoxin, *Toxicon.*, 22:415-424.
11. DasGupta BR, Sugiyama H, 1977, Biochemistry and pharmacology of botulinum and tetanus neurotoxins, *In* "Perspectives in Toxinology", Bernheimer AW, ed., John Wiley & Sons, New York.
12. DasGupta BR, Sugiyama H, 1978, Limited proteolysis of *Clostridium botulinum* types A and B neurotoxin, *Am Soc Microbiol.*, (abstract), p. 25.
13. DasGupta BR, Woody MA, 1984, Amino acid composition of *Clostridium botulinum* type B neurotoxin, *Toxicon.*, 22:312-315.
14. Datta A, DasGupta BR, 1988, Circular dichroic and fluorescence spectroscopic study of the conformation of botulinum neurotoxin types A and E, *Mol Cell Biochem.*, 79:153-159.
15. Datta A, DasGupta BR, 1988, Botulinum neurotoxin types A, B and E; pH induced difference spectra, *Mol Cell Biochem.*, 81:187-194.
16. Donovan JJ, Middlebrook JL, 1986, Ion-conducting channels produced by botulinum toxin in planar lipid membranes, *Biochemistry*, 25:2872-2876.
17. Eisel U, Jarausch W, Goretzki K, Henschen A, Engels J, Weller U, Hudel M, Habermann E, Niemann H, 1986, Tetanus toxin: primary structure,

expression in *E. coli*, and homology with botulinum toxins, *EMBO Journal*, 5:2495-2502.

18. Fairweather NF, Lyness VA, 1986, The complete nucleotide sequence of tetanus toxin, *Nucleic Acids Res.*, 14:7809-7812.

19. Gimenez DF, 1984, *Clostridium botulinum* subtype Ba, *Zbl Bakt Hyg A*, 257: 68-72.

20. Gimenez J, Sugiyama H, 1988, Comparison of toxins of *Clostridium butyricum* and *Clostridium botulinum* type E, *Infect Immun.*, 56:926-929.

21. Gimenez J, Foley J, DasGupta BR, 1988, Neurotoxin type E from *Clostridium botulinum* and *C. butyricum*; partial sequence and comparison, *FASEB J.*, A1750, (abstract).

22. Hall JD, McCroskey LM, Pincomb BJ, Hatheway CL, 1985, Isolation of an organism resembling *Clostridium barati* which produces type F botulinal toxin from an infant with botulism, *J Clin Microbiol.*, 21:654-655.

23. Hoch DH, Romero-Mira M, Ehrlich BE, Finkelstein A, DasGupta BR, Simpson LL, 1985, Channels formed by botulinum, tetanus, and diphtheria toxins in planar lipid bilayers: relevance to translocation of proteins across membranes, *Proc Nat Acad Sci (USA)*, 82:1692-1696.

24. Kaiser ET, 1987, Design principles in the construction of the biologically active peptides, *Trends in Biochem Sci.*, 12:305-309.

25. Kaiser ET, Kezdy FJ, 1984, Amphiphilic secondary structure: design of peptide hormones, *Science*, 223:249-255.

26. Kaiser ET, Kezdy FJ, 1987, Peptides with affinity for membranes, *Ann Rev Biophys Chem.*, 16:561-581.

27. Kozaki S, Miyazaki S, Sakaguchi G, 1977, Development of antitoxin with each of two complementary fragments of *Clostridium botulinum* type B derivative toxin, *Infect Immun.*, 18:761-766.

28. Kozaki S, Togashi S, Sakaguchi G, 1981, Separation of *Clostridium botulinum* type A derivative toxin into two fragments, *Jpn J Med Sci.*, 34: 61-68.

29. Matsuda M, Yoneda M, 1974, Dissociation of tetanus neurotoxin into two polypeptide fragments, *Biochem Biophys Res Commun.*, 57:1257-1262.

30. McCroskey LM, Hatheway CL, Fenicia L, Pasolini B, Aurel P, 1986, Characterization of an organism that produces type E botulinal toxin but which resembles *Clostridium butyricum* from the feces of an infant with type E botulism, *J Clin Microbiol.*, 23:201-202.

31. Miller IR, 1987, Lipid bilayers and proteins: a fluctuating interaction?, *Trends in Biochem Sci.*, 12:461-462.

32. Montecucco C, 1986, How do tetanus and botulinum toxins bind to neuronal membranes?, *Trends in Biochem Sci.*, 11:314-317.

33. Montecucco C, 1987, Diphtheria toxin membrane translocation: an open question, *Trends in Biochem Sci.*, 12:181-182.

34. Montecucco C, Schiavo G, 1986, Tetanus toxin is labeled with photoactivatable phospholipids at low pH., *Biochemistry*, 25:919-924.

35. Montecucco C, Giao Z, Boquet P, Schiavo G, Bauerlein E, DasGupta BR, 1988, Interaction of botulinum and tetanus toxins with the lipid bilayer surface, *Biochem J.*, 251:379-383.

36. Montecucco C, Schiavo G, DasGupta BR, 1989, Effect of pH on the interaction of botulinum neurotoxins A, B and E with liposomes, *Biochem J.*, 259:47-53.

37. Moore S, 1972, The precision and sensitivity of amino acid analysis, *In* "Proceedings of the Third American Peptide Symposium," J. Meinhofer, ed., Ann Arbor Science Publishers, Ann Arbor, MI.

38. Murayama S, Syuto B, Oguma K, Iida H, Kubo S, 1984, Comparison of *Clostridium botulinum* toxins type D and C, in molecular property, antigenicity and binding ability to rat-brain synaptosomes, *Eur J Biochem.*, 142:487-492.

39. Poulain B, Tauc L, Maisey EA, Wadsworth JEF, Mohan PM, Dolly JO, 1988, Neurotransmitter release is blocked intracellularly by botulinum neurotoxin, and this requires uptake of both toxin polypeptides by a process mediated by the large chain, *Proc Natl Acad Sci (USA)*, 85:4090-4094.

40. Robinson JP, Hash JH, 1982, A review of the molecular structure of tetanus toxin, *Mol Cell Biochem.*, 48:33-44.

41. Robinson JP, Chiu W, DasGupta BR, 1988, Two-dimensional crystals of botulinum toxin type A, *FASEB J.*, A1750 (Abstract).

42. Robinson JP, Holladay LA, Hash JH, Puett D, 1982, Conformational and molecular weight studies of tetanus toxin and its major peptides, *J Biol Chem.*, 257:407-411.

43. Sathyamoorthy V, DasGupta BR, 1985, Partial amino acid sequences of the heavy and light chains of botulinum neurotoxin type E, *Biochem Biophys Res Commun.*, 127:768-772.

44. Sathyamoorthy V, DasGupta BR, 1985, Separation, purification, partial characterization and comparison of the heavy and light chains of botulinum neurotoxin types A, B and E, *J Biol Chem.*, 260:10461-10466.

45. Sathyamoorthy V, DasGupta BR, Niece RL, 1986, Botulinum neurotoxin type A, cleavage and partial sequence of the H chain, *Fed Proceed.*, 45:1793 (abstract).

46. Sathyamoorthy V, DasGupta BR, Foley J, Niece RL, 1988, Botulinum neurotoxin type A; cleavage of the heavy and light chain into two halves and their partial sequences, *Arch Biochem Biophys.*, 266:142-151.

47. Schiffer M, Edmundson AB, 1967, Use of helical wheels to represent the structures of proteins and to identify segments with helical potential, *Biophys J.*, 7:121-135.

48. Schmidt JJ, Sathyamoorthy V, DasGupta BR, 1984, Partial sequence of the heavy and light chains of botulinum neurotoxin type A, *Biochem Biophys Res Commun.*, 119:900-904.

49. Schmidt JJ, Sathyamoorthy V, DasGupta BR, 1985, Partial amino acid sequences of botulinum neurotoxins types B and E, *Arch Biochem Biophys.* 238:544-548.

50. Shone CC, Hambleton P, Melling J, 1985, Inactivation of *Clostridium botulinum* type A neurotoxin by trypsin and purification of two tryptic fragments, *Eur J Biochem.*, 151:75-82.

51. Shone CC, Hambleton P, Melling J, 1987, A 50-kDa fragment from the NH$_2$-terminus of the heavy subunit of *Clostridium botulinum* type A neurotoxin forms channels in lipid vesicles, *Eur J Biochem.*, 167:175-180.

52. Singh BR, DasGupta BR, 1989, Structure of heavy and light chain subunits of type A botulinum neurotoxin analyzed by circular dichroism and fluorescence measurements, *Mol Cell Biochem.*, 85:67-73.

53. Singh BR, DasGupta BR, 1989, Molecular topography and secondary structure comparisons of botulinum neurotoxin types A, B and E, *Mol Cell Biochem.*, 86:87-95.

54. Syuto B, Kubo S, 1981, Separation and characterization of heavy and light chains from *Clostridium botulinum* type C toxin and their reconstitution, *J Biol Chem.*, 256:3712-3717.

55. Syuto B, Kubo S, 1982, *Clostridium botulinum* type C toxin, *Mol Cell Biochem.*, 48:25-32.

56. Terajima J, Syuto B, Ochanda JP, Kubo S, 1985, Purification and Characterization of neurotoxin produced by *Clostridium botulinum* type C 6813, *Infect Immun.*, 48:312-317.

57. Tsuzuki K, Yokosawa N, Syuto B, Ohishi I, Fujii N, Kimura K, Oguma K, 1988, Establishment of a monoclonal antibody recognizing an antigenic site common to *Clostridium botulinum* Type B, C$_1$, D, and E toxins and tetanus toxin, *Infect Immun.*, 56:898-902.

58. Woody MA, DasGupta BR, 1988, C-terminal residues of botulinum neurotoxin types A, B and E, *Am Soc Microbiol.*, p. 40 (abstract).

THE *ESCHERICHIA COLI/VIBRIO CHOLERAE* FAMILY OF ENTEROTOXINS

Randall K. Holmes, Edda M. Twiddy, Carol L. Pickett,[a]
Hilda Marcus, Michael G. Jobling, and Francoise M. J.
Petitjean

Department of Microbiology
Uniformed Services University of the Health Sciences
Bethesda, Maryland 20814

ABSTRACT

The heat-labile enterotoxins (LT-I and LT-II) of *Escherichia coli* and
cholera enterotoxin (CT) from *Vibrio cholerae* belong to a family of related
protein toxins. Each toxin consists of an A subunit that activates adenylate cyclase in target cells by ADP ribosylation of the regulatory protein
Gs and an oligomeric B subunit that binds to plasma membrane receptors on
susceptible target cells. The *E. coli/V. cholerae* enterotoxin family is
divided into serogroups based on neutralization tests. LT-I and CT belong
to serogroup I, whereas LT-II belongs to serogroup II. Antigenic variants
of CT, LT-I and LT-II are produced by *V. cholerae* and *E. coli* strains from
natural sources. Genes encoding the A polypeptides for all of these toxins
are homologous, and both ADP-ribosyl transferase activity and stimulation of
the activity by ADP-ribosylation factor (ARF) are conserved functions of the
A subunits. Genes for the B polypeptides of the toxins in serogroup I are
also homologous, but they have no significant homology with the B subunit
genes of the toxins in serogroup II. Gangliosides to which toxins bind with
highest affinity are GM1 for LT-I and CT, GD1b for LT-IIa and GD1a for LT-
IIb. This paper presents an overview of the *E. coli/V. cholerae* enterotoxin
family and summarizes recent work from our laboratory on these toxins.

INTRODUCTION

Three different families of enterotoxins are now recognized among the
enteric, gram negative bacteria: i.e., the heat-labile enterotoxins (LT),
the heat-stable enterotoxins (ST), and the Shiga-like toxins (SLT), which
are also called Verotoxins (VT).[56] Evidence that each of these toxins
can function as virulence factors is reviewed elsewhere.[56,72] The *E. coli/
V. cholerae* heat-labile enterotoxin family is the subject of this paper.

Cholera enterotoxin (CT) was the first heat-labile enterotoxin from an
enteric gram negative bacterium to be purified and characterized.[20] In
experimental animals CT causes secretory responses after intraintestinal
injection into ligated ileal loops of adult rabbits, lethal diarrhea after

---

[a] Present address: Department of Microbiology and Immunology, University
of Kentucky, Lexington, Kentucky 40536 United States of America

Microbial Toxins in Foods and Feeds, Edited by A.E. Pohland *et al.,*
Plenum Press, New York, 1990

intraintestinal inoculation in infant rabbits, and many other biological effects; and in humans CT causes the massive secretory diarrhea which is the principal symptom of Asiatic cholera.[17] Nontoxinogenic mutants of *V. cholerae* are avirulent or dramatically attenuated both in experimental animals and in volunteers.[21,43,57] It is well established, therefore, that CT is an important virulence factor for *V. cholerae* and plays a central role in the pathogenesis of cholera.

Although enterotoxins are defined by their biologic activities in the gut of humans or animals, detection of putative enterotoxins is greatly facilitated by using tests that can be performed rapidly and inexpensively in cell cultures or *in vitro*. Particularly convenient tests for heat-labile enterotoxins use cultured mouse Y1 adrenal tumor cells[15] or Chinese hamster ovary (CHO) cells,[30] because the toxins cause changes in morphology of these cells that can be detected by direct microscopic observation. Purification of representative heat-labile enterotoxins and preparation of antisera or monoclonal antibodies against them led to the development of a wide variety of immunoassays,[5,6,47-50,90] and cloning of representative enterotoxin structural genes provided DNA probes to detect genes for related enterotoxins by DNA-DNA hybridization.[54,65,66,74,82] A broad arsenal of molecular methods is available, therefore, for analyzing the relationships among members of the *E. coli/V. cholerae* family of heat-labile enterotoxins.

Many bacteria associated with diarrheal diseases have been tested for production of enterotoxins. Antigenic relatedness between enterotoxins from different sources is usually established by neutralization tests involving homologous and heterologous combinations of toxins and antitoxins. Such tests provided the first evidence that enterotoxins related to CT were produced by other bacteria (Table 1) as well as *V. cholerae*.[17,18,34,35,91] Evidence that some enterotoxins are antigenically distinct from CT but, nevertheless, are related both structurally and genetically to CT emerged much later.[27,32,42,76,77] The *E. coli/V. cholerae* enterotoxin family is now divided into two serogroups, defined by cross-neutralization reactions.[32,42,76] CT and the type I heat-labile enterotoxins (LT-I) of *E. coli* belong to serogroup I, whereas type II heat-labile enterotoxins (LT-II) of *E. coli* belong to serogroup II. Antigenic variants of CT, LT-I and LT-II have also been recognized.[2,3,19,25,41,43,52,87] Representative bacteria that produce toxins belonging to the *E. coli/V. cholerae* enterotoxin family are summarized in Table 1, although not all of the toxins from these bacteria have been characterized in detail. The remaining sections of this paper will be limited to the heat-labile enterotoxins produced by *E. coli* and *V. cholerae*.

STRUCTURE AND IMMUNOCHEMISTRY

The heat-labile enterotoxins of *E. coli* and *V. cholerae* are oligomeric proteins consisting of one A polypeptide and five B polypeptides held together by noncovalent bonds.[26] Each polypeptide is synthesized as a precursor with a signal sequence that is removed during secretion and assembly of the holotoxin.[39] Table 2 compares some of the major properties of the A and B polypeptides of CT and representatives of the LT-I and LT-II toxin groups.

The CT structural gene is highly conserved among strains of *V. cholerae* representing the different biotypes and serotypes.[4,88] Minor differences in amino acid sequences between CT from various strains have been demonstrated, however, and some of them affect expression of specific epitopes on CT.[19] The B subunit of CT is immunodominant, and most of the neutralizing antibodies in polyclonal antiserum against CT are specific for the B subunit.[17,49] Most epitopes on CT-B appear to be dependent on conformation

Table 1. Representative Bacteria that Produce Toxins Belonging to
the *E. coli/V. cholerae* Family of Heat-labile Enterotoxins.

Bacterial Species	Reference
Vibrio cholerae	(20)
Noncholera vibrios	(80,91)
Escherichia coli	(27,35)
Salmonella typhimurium	(9)
Campylobacter jejuni	(53)
Aeromonas hydrophila	(1,51)
Plesiomonas shigelloides	(24)

the molecule, because reduction and carboxymethylation abolishes the ability
of CT-B to bind to most of the monoclonal antibodies directed against it.[61]
The minor differences in structure and antigenicity between CT from various
strains of *V. cholerae* have not yet been shown to have epidemiologic or
clinical significance.

All of the toxins assigned to serogroup I of the *E. coli/V. cholerae*
enterotoxin family crossreact immunologically. Each toxin in this group has
both unique and common epitopes. Antibodies against the common epitopes are
responsible for the cross-reactions seen in neutralization and immunobinding
assays. Polyclonal antisera contain antibodies against both unique and
common determinants of the immunizing antigen, and polyclonal antisera have
higher titers against their homologous enterotoxins than they do against
heterologous, crossreacting enterotoxins.[10,11,25,41,84] Monoclonal antibodies
(mAbs) against unique determinants recognize only the homologous enterotoxin,
but mAbs against common epitopes recognize all enterotoxins that express
these determinants.[2,3,19,40,58,78]

LT-I toxins produced by *E. coli* strains from humans (LTh-I) and from
pigs (LTp-I) are distinct antigenic variants of LT-I.[25,41,52,84] LTh-I and
LTp-I are much more highly homologous with each other than with CT.[88,89] In
epidemiologic studies the antigenic differences between LTh-1 and LTp-I that
had been defined with prototype strains were found to be consistently associated both with the animal species from which the enterotoxigenic *E. coli*
were isolated and with restriction fragment length polymorphisms in the Ent
plasmids that encoded the enterotoxins.[52,55]

The LT-II toxins of *E. coli* do not crossreact to any significant degree
with CT, LTh-I or LTp-I.[27,32,42,76,77] Two distinct antigenic variants of LT-
II, designated LT-IIa and LT-IIb have been identified.[32,42] LT-II producing
strains are rarely detected among isolates of *E. coli* from humans, and most
LT-II producing strains have been isolated from livestock or from food
intended for human consumption.[27,31,81] Among eleven mAbs raised against LT-
IIa (R. K. Holmes, E. M. Twiddy, and R. McNally, manuscript in preparation)
and fifteen mAbs raised against LT-IIb (F. M. J. Petitjean and R. K. Holmes,
manuscript in preparation), only one reacted both with LT-IIa and LT-IIb.[22]
The lower frequency of mAbs against common epitopes of LT-II, in comparison
with LT-1, is presumed to reflect the lower amino acid sequence homology
among the variants of LT-II in comparison with the variants of LT-I (Table
2).

The A polypeptides of all members of the *E. coli/V. cholerae* heat-labile
enterotoxin family are homologous (Table 2). Each mature A polypeptide
contains two cysteine residues (corresponding to residues 187 and 199 of the

mature CT-A polypeptide) that are joined to form an internal disulfide bond. The short loop subtended by this disulfide bond contains one or two arginine or lysine residues and is presumably exposed at the surface of the holotoxin. Treatment of each holotoxin with trypsin results in preferential cleavage within this loop and generates a nicked form of toxin in which fragments A1 and A2 remain joined by the disulfide bond. After reduction of the nicked toxin, the A1 and A2 fragments can be isolated by appropriate methods. The sequence within the A1 domain is much more highly conserved than is the sequence within A2,[77,88] (C. L. Pickett, E. M. Twiddy, C. Coker and R. K. Holmes, manuscript submitted). The amino terminal region of the A1 domain of these toxins also contains sequences that are homologous with the S1 polypeptide of pertussis toxin[59,71] and that presumably reflect the common function of these polypeptides as NAD-dependent ADP ribosyl transferase enzymes.

The B polypeptides are not homologous among all members of the *E. coli/ V. cholerae* enterotoxin family (Table 2). Although the B subunits of the enterotoxins within serogroup I[55,88,89] or within serogroup II[77] (C. L. Pickett, E. M. Twiddy, C. Coker, and R. K. Holmes, manuscript submitted) are homologous, the B subunits of the toxins from serogroup 1 do not have significant homology with those from serogroup II (Table 2). This lack of homology is correlated with differences in the binding specificities of the toxins in serogroup I and serogroup II.[22] In spite of these differences, however, the B subunits from the toxins in serogroups I and II share certain characteristics. They have similar hydropathy plots; they all have a single internal disulfide bond; and they all form pentameric B arrays that associate with their homologous A polypeptides to form the respective oligo- meric holotoxins[26,77,88] (C. L. Pickett, E. M. Twiddy, C. Coker, and R. K. Holmes, manuscript submitted).

MODE OF ACTION

All members of the family of *E. coli/V. cholerae* heat-labile entero- toxins are bifunctional molecules whose A and B subunits serve different functions. The B subunits mediate binding to specific receptors on target cells, and the A subunits catalyze the reactions that lead to activation of adenylate cyclase in the target cells.

Early studies with CT demonstrated that it binds with high affinity to the oligosaccharide moiety of ganglioside GM1,[12,13,44] which is a functional receptor for toxin in the plasma membrane of target cells.[67] LT-I also binds to ganglioside GM1,[16,29] but there is evidence that it also binds to a second receptor, which may be a glycoprotein, to which CT does not bind.[28,29,45,46] LT-IIa and LT-IIb also bind to gangliosides *in vitro*, but the specificity and affinity of their binding is different both from each other and from CT or LT-1.[22] LT-IIa binds with highest affinity to gang- lioside GD1b and also binds to ganglioside GM1, whereas LT-IIb binds with highest affinity to ganglioside GD1a and does not bind to ganglioside GM1.[22] It is not yet proven whether these gangliosides are functional receptors for LT-IIa or LT-IIb. It is tempting to speculate that the differ- ences in binding specificities among the members of the *E. coli/V. cholerae* enterotoxin family may be important as determinants of their relative activities against particular animals or tissues.

The A1 polypeptides of CT and LT-I have NAD-dependent ADP ribosyltrans- ferase activity that catalyzes ADP ribosylation of the alpha subunit of the GTP-binding protein Gs, the stimulatory regulatory protein of the plasma membrane adenylate cyclase complex of animal cells.[88] This results in activation of adenylate cyclase, accumulation of cyclic-3',5'- adenosine monophosphate (cAMP) in the intoxicated cell, and secondary effects that are

Table 2. Properties of the Mature A and B Polypeptides of Representative Members
of the *E. coli*/*V. cholerae* Heat-labile Enterotoxin Family.[1]

Toxin	A Polypeptide			B Polypeptide		
	Number of Amino Acid	Location of Cys Residues	Sequence Homology with LTh-I[2] (%)	Number of Amino Acid	Location of Cys Residues	Sequence Homology with LTh-I[2] (%)
CT	240	187;199	81	103	9;86	82
LTh-I	240	187;199	100	103	9;86	100
LTp-I	240	187;199	99	103	9;86	96
LT-IIa	241	185;197	57[3]	100	10;81	n.s.[3]
LT-IIb	243	185;197	55[3]	99	10;81	n.s.[3]

[1] Deduced from the published nucleotide sequences[77,88] (C. L. Pickett, E. M. Twiddy, C. Coker,
and R. K. Holmes, manuscript submitted.)

[2] Based on the deduced amino acid sequences.

[3] n.s. indicates not significant. The homologies between LT-IIa and LT-IIb were 85% for the
amino acid sequences corresponding to the mature A polypeptides and 58% for the amino acid
sequences corresponding to the mature B polypeptides.

mediated by cAMP and reflect the nature of the target cell. In the mucosa of the small intestine, the consequence is net secretion of fluid and electrolytes associated with inhibition of sodium chloride absorption and stimulation of secretion of chloride ions.[23] Diarrhea results when the rate of delivery of fluid and electrolytes from the small bowel to the colon exceeds the resorptive capacity of the colon.

The ADP ribosyltransferase activity of CT or LT-I requires reduction of the disulfide bond between A1 and A2, NAD, GTP, and the membrane-associated adenylate cyclase system, and activity is stimulated by soluble proteins designated ADP ribosylation factors, or ARF.[68] Recent experiments with purified LT-IIa and LT-Ilb showed that their A1 fragments have ADP ribosyltransferase activity with specificity similar to that of CT or LT-1 and demonstrated that stimulation of activity by ARF also occurs with enterotoxins that belong to serogroup II.[7,8]

GENETICS

The operons for CT of *V. cholerae*[60,63] and for LTp-I[14,55,83,88], LTh-I,[89] LT-IIa,[77] and LT-Ilb (C. L. Pickett, E. M. Twiddy, C. Coker, and R. K. Holmes, manuscript submitted) have been cloned and sequenced. All of these operons are organized in a similar manner. Each has a single promoter located just upstream from the structural gene for the A polypeptide, and the gene for the B polypeptide is located downstream from the A gene. The distal end of the A structural gene overlaps with the proximal end of the B structural gene; the overlap is four nucleotides in the case of the enterotoxins in serogroup I (CT, LTp-I and LTh-I) and eleven nucleotides in the case of those in serogroup II (LT-IIa and LT-Ilb). In each case, therefore, the coding sequences for the A and B polypeptides are translated in different reading frames from the polycistronic messenger RNA. Ribosome binding sites are located just upstream from the coding sequences for the A and B polypeptides. There is evidence that B coding sequence istranslated with higher efficiency, resulting in production of the B polypeptides in adequate amounts to fulfill the requirements for the 1A:5B stoichiometry of the holotoxin.[63]

The CT operon on the chromosome of *V. cholerae* is part of a larger genetic element that resembles a complex transposon.[4] Classical strains of *V. cholerae* contain two nontandem copies of the CT operon, and most El Tor strains contain a single copy; but tandem duplications of the CT operon can also occur.[62] The CT operon is positively regulated by the product of the *toxR* gene in *V. cholerae*.[64] The *toxR* gene product also regulates the toxin-coregulated pilus (*tcp*) operon and several other genes that may be important for the virulent phenotype of *V. cholerae*.[75,85] In contrast, the LT-I operon is located on plasmids in *E. coli*[33] and is not regulated by the *toxR* gene product of *V. cholerae*.[70] Chromosomal mutations in *E. coli* that affect expression of plasmid-borne LT-I genes have been described,[6] but their mode of action has not been characterized. In contrast to the LT-I operon, the LT-II operon is not located on plasmids and appears to be chromosomal.[27,76,77]

Comparison of the coding sequences for the A and B structural genes with the mature A and B polypeptides indicates that the primary translation products are precursor polypeptides with signal sequences. Analysis of synthesis and assembly of CT and LT-I reveals that the A and B polypeptides are secreted across the plasma membrane, and the holotoxins are assembled in the periplasm after removal of the signal sequences by signal peptidase.[36,39,70] In *V. cholerae*, but not in *E. coli*, CT or LT-I is excreted across the outer membrane and appears in the culture supernatant as an extracellular product.[38,70] Mutants that produce only the B subunit can assemble and excrete it, but mutants which produce only the A protein do not excrete it.[38] The B subunit is presumed, therefore, to be required both for assembly of the holotoxin and for interacting with the toxin excretory system of V. cholerae.[36-38] In E. coli LT-II, like LT-I, is formed as a cell-associated, and presumably periplasmic, product.[32,42] The LT-II operons have not yet been analyzed in V. cholerae. A mutant of V. cholerae, designated M14, has a specific defect that renders it unable to excrete either CT or LT-I from the periplasm to the extracellular milieu.[36,43,70] Recent work in our laboratory has led to the cloning from wild type V. cholerae strain 569B of a DNA fragment that appears to complement the defect in strain M14 and restore its ability to excrete CT (H. Marcus and R. K. Holmes, unpublished data). Experiments are in progress to characterize the cloned fragment, determine the structure of the corresponding gene product(s), and analyze its function in the excretory pathway for CT in V. cholerae.

Although the isolation and characterization of mutants is a powerful method to study the relationships between structure and function of toxins,[4,69] this technique is just beginning to be applied to the E. coli/ V. cholerae family of enterotoxins. A mutation in the B subunit gene of LT-I that resulted in substitution of aspartic acid for glycine at position 33 was associated with loss of receptor-binding activity,[86] and extensions at the carboxy terminal end of the B subunit affected the ability of the subunits to form oligomers and to be secreted.[79] A mutation in the A subunit gene of LT-I that resulted in substitution of glycine for arginine at position 146 abolished the acceptor site for auto-ADP ribosylation but had no effect on the ADP ribosylation activity required for activation of adenylate cyclase in target cells.[73] We have recently initiated a study of site-directed mutagenesis with the gene for the B subunit of CT (M. G. Jobling and R. K. Holmes, unpublished data). Preliminary results established that substitution of serine for cysteine at position 9 or position 86 or both resulted in loss of receptor-binding activity, and characterization of additional mutants is in progress.

ACKNOWLEDGMENTS

Research on enterotoxins in the authors' laboratory was supported in part by Public Health Service grant 5 R22 AI14107 from the National Institute of Allergy and Infectious Diseases and by protocol number R07301 from the Uniformed Services University of the Health Sciences. The opinions and assertions contained herein are the private views of the authors and should not be construed as official or as necessarily reflecting the views of the Uniformed Services University of the Health Sciences or Department of Defense.

REFERENCES

1. Annapurna E, Sanyal SC, 1977, Enterotoxicity of Aeromonas hydrophila, J Med Microbiol., 10:317-323.
2. Belisle BW, Twiddy EM, Holmes RK, 1984, Characterization of monoclonal antibodies to heat-labile enterotoxin encoded by a plasmid from a clinical isolate of Escherichia coli, Infect Immun., 43:1027-1032.
3. Belisle BW, Twiddy EM, Holmes RK, 1984, Monoclonal antibodies with an expanded repertoire of specificities and potent neutralizing activity for Escherichia coli heat-labile enterotoxin, Infect Immun., 46:759-764.
4. Betley MJ, Miller VL, Mekalanos JJ, 1986, Genetics of bacterial enterotoxins, Ann Rev Microbiol., 40:577-605.
5. Bramucci MG, Holmes RK, 1978, Radial passive immune hemolysis assay for detection of heat-labile enterotoxin produced by individual colonies of Escherichia coli or Vibrio cholerae, J Clin Microbiol., 8:252-255.

6.   Bramucci MG, Twiddy EM, Baine WB, Holmes RK, 1981, Isolation and
      characterization of hypertoxinogenic (*htx*) mutants of *Escherichia coli*
      KL320(pCG86), *Infect Immun.*, 32:1034-1044.
7.   Chang PP, Tsai SC, Adamik R, Moss J, Twiddy E, Holmes RK, 1988, Acti-
      vation of the ADP-ribosyltransferase activity of *Escherichia coli* heat-
      labile enterotoxins by 19 kDa guanine nucleotide-binding proteins.
      Clin Res., 36:578A (Abstract).
8.   Chang PP, Moss J, Twiddy EM, Holmes RK, 1987, Type II heat-labile
      enterotoxin of *Escherichia coli* activates adenylate cyclase in human
      fibroblasts by ADP ribosylation, *Infect Immun.*, 55:1854-1858.
9.   Chopra AK, Houston CW, Peterson JW, Prasad R, Mekalanos JJ, 1987,
      Cloning and expression of the *Salmonella* enterotoxin gene, *J Bacteriol.*,
      169:5095-5100.
10.   Clements JD, Finkelstein RA, 1978, Demonstration of shared and unique
      immunological determinants in enterotoxins from *Vibrio cholerae* and
      Escherichia coli, *Infect Immun.*, 22:709-713.
11.   Clements JD, Finkelstein RA, 1978, Immunological cross-reactivity
      between a heat-labile enterotoxin(s) of *Escherichia coli* and subunits
      of *Vibrio cholerae* enterotoxin, *Infect Immun.*, 21:1036-1039.
12.   Cuatrecasas P, 1973, Gangliosides and membrane receptors for cholera
      toxin, *Biochemistry*, 12:3558-3566.
13.   Cuatrecasas P, 1973, Interaction of *Vibrio cholerae* enterotoxin with
      cell membranes, *Biochemistry*, 12:3547-3558.
14.   Dallas WS, Falkow S, 1980, Amino acid sequence homology between cholera
      toxin and *Escherichia coli* heat-labile toxin, *Nature*, 288:499-501.
15.   Donta ST, King M, Sloper K, 1973, Induction of steroidogenesis in tissue
      culture by cholera enterotoxin, *Nature (New Biol)*, 243:246-247.
16.   Donta ST, Poindexter NJ, Ginsberg BH, 1982, Comparison of the binding of
      cholera and *Escherichia coli* enterotoxins to Y1 adrenal cells,
      Biochemistry, 21:660-664.
17.   Finkelstein RA, 1973, Cholera, *CRC Crit Rev Microbiol.*, 2:553-623.
18.   Finkelstein RA, 1976, Progress in the study of cholera and related
      enterotoxins, p 53-84, *In*: "Mechanisms in Bacterial Toxinology,"
      Bernheimer, AW (ed.), John Wiley and Sons, Inc., New York.
19.   Finkelstein RA, Burks MF, Zupan A, Dallas WS, Jacob CO, Ludwig DS, 1987,
      Epitopes of the cholera family of enterotoxins, *Rev Infect Dis.*,
      9:544-561.
20.   Finkelstein RA, LoSpalluto JJ, 1969, Pathogenesis of experimental
      cholera. Preparation and isolation of choleragen and choleragenoid, *J
      Exp Med.*, 130:185-202.
21.   Finkelstein RA, Vasil ML, Holmes RK, 1974, Studies on toxinogenesis in
      Vibrio cholerae. I. Isolation of mutants with altered toxinogenicity,
      J Infect Dis., 129:117-123.
22.   Fukuta S, Magnani JL, Twiddy EM, Holmes RK, Ginsburg V, 1988, Comparison
      of the carbohydrate-binding specificities of cholera toxin and
      Escherichia coli heat-labile enterotoxins LTh-I, LT-lIa, and LT-lIb,
      Infect Immun., 56:1748-1753.
23.   Gardner JD, 1980, Pathogenesis of secretory diarrhea, p 153-158, *In*:
      "Secretory Diarrhea," Field H, Fordtran JS, Schultz SG, (eds), American
      Physiological Society, Bethesda.
24.   Gardner SE, Fowlston SE, George WL, 1987, In vitro production of cholera
      toxin-like activity bv *Plesiomonas shigelloides*, *J Infect Dis.*,   156:
      720-722.
25.   Geary SJ, Marchlewicz BA, Finkelstein RA, 1982, Comparison of heat-
      labile enterotoxins from porcine and human strains of *Escherichia coli*,
      Infect Immun., 36:215-220.
26.   Gill DM, Clements JD, Robertson DC, Finkelstein RA, 1981, Subunit number
      and arrangement in *Escherichia coli* heat-labile enterotoxin, *Infect
      Immun.*, 33:677-682.
27.   Green BA, Neill RJ, Ruyechan WT, Holmes RK, 1983, Evidence that a new
      enterotoxin of *Escherichia coli* which activates adenylate cyclase in

eucaryotic target cells is not plasmid mediated, *Infect Immun.*, 41: 383-390.

28. Griffiths SL, Finkelstein RA, Critchley DR, 1986, Characterization of the receptor for cholera toxin and *Escherichia coli* heat-labile toxin in rabbit intestinal brush borders, *Biochem J.*, 238:313-322.

29. Guerrant RL, Brunton LL, 1977, Characterization of the Chinese hamster ovary cell assay for the enterotoxins of *Vibrio cholerae* and *Escherichia coli* and for specific antisera, and toxoid, *J Infect Dis.*, 135:720-728.

30. Guerrant RL, Brunton LL, Schnaitman TC, Rebhun LI, Gilman AG, 1974, Cyclic adenosine monophosphate and alteration of Chinese hamster ovary cell morphology: a rapid, sensitive in vitro assay for the enterotoxins of *Vibrio cholerae* and *Escherichia coli*, *Infect Immun.*, 10:320-327.

31. Guth BE, Pickett CL, Twiddy EM, Holmes RK, Gomes TA, Lima AA, Guerrant RL, Franco BD, Trabulsi LR, 1986, Production of type II heat-labile enterotoxin by *Escherichia coli* isolated from food and human feces, *Infect Immun.*, 54:587-589.

32. Guth BE, Twiddy EM, Trabulsi LR, Holmes RK, 1986, Variation in chemical properties and antigenic determinants among type II heat-labile enterotoxins of *Escherichia coli*, *Infect Immun.*, 54:529-536.

33. Gyles C, So M, Falkow S, 1974, The enterotoxin plasmids of *Escherichia coli*, *J Infect Dis.*, 130:40-48.

34. Gyles CL, 1974, Relationships among heat-labile enterotoxins of *Escherichia coli* and *Vibrio cholerae*, *J Infect Dis.*, 129:277-283.

35. Gyles CL, Barnum DA, 1969, A heat-labile enterotoxin from strains of *Escherichia coli* enteropathogenic for pigs, *J Infect Dis.*, 120:419-426.

36. Hirst TR, Holmgren J, 1987, Transient entry of enterotoxin subunits into the periplasm occurs during their secretion from *Vibrio cholerae*, *J Bacteriol.*, 169:1037-1045.

37. Hirst TR, Holmgren J, 1987, Conformation of protein secreted across bacterial outer membranes: a study of enterotoxin translocation from *Vibrio cholerae*, *Proc Natl Acad Sci.*, USA, 84:7418-7422.

38. Hirst TR, Sanchez J, Kaper JB, Hardy SJ, Holmgren J, 1984, Mechanism of toxin secretion by *Vibrio cholerae* investigated in strains harboring plasmids that encode heat-labile enterotoxins of *Escherichia coli*, *Proc Natl Acad Sci*, USA, 81:7752-7756.

39. Hirst TR, Welch RA, 1988, Mechanisms for secretion of extracellular proteins by gram-negative bacteria. *Trends Biochem Sci.*, 13:265-269.

40. Holmes RK, Twiddy EM, 1983, Characterization of monoclonal antibodies that react with unique and cross-reacting determinants of cholera enterotoxin and its subunits, *Infect Immun.*, 42:914-923.

41. Holmes RK, Twiddy EM, Bramucci MG, 1983, Antigenic heterogeneity among heat-labile enterotoxins from *Escherichia coli*, p 293-300, *In*: "Advances in Research on Cholera and Related Diarrheas," Kuwahara S, Pierce NF, (eds.), KTK Scientific Publishers, Tokyo.

42. Holmes RK, Twiddy EM, Pickett CL, 1986, Purification and characterization of type II heat-labile enterotoxin of *Escherichia coli*, *Infect Immun.*, 53:464-473.

43. Holmes RK, Vasil ML, Finkelstein RA, 1975, Studies on toxinogenesis in *Vibrio cholerae*. III. Characterization of nontoxinogenic mutants in vitro and in experimental animals, *J Clin Invest.*, 55:551-560.

44. Holmgren, J, 1973, Comparison of the tissue receptors for *Vibrio cholerae* and *Escherichia coli* enterotoxins by means of gangliosides and natural cholera toxoid, *Infect Immun.*, 8:851-859.

45. Holmgren J, Fredman P, Lindblad M, Svennerholm AM, Svennerholm L, 1982, Rabbit intestinal glycoprotein receptor for *Escherichia coli* heat-labile enterotoxin lacking affinity for cholera toxin, *Infect Immun.*, 38: 424-433.

46. Holmgren J, Lindlad M, Fredman P, Svennerholm L, Myrvold H, 1985, Comparison of receptors for cholera and *Escherichia coli* enterotoxins in human intestine, *Gastroenterol.*, 89:27-35.

47. Holmgren J, Soderlind O, Wadstrom T, 1973, Cross-reactivity between

heat-labile enterotoxins of *Vibrio cholerae* and *Escherichia coli* in neutralization tests in rabbit ileum and skin, *Acta Pathol Microbiol Scand.*, (B) 81:757-762.

48. Holmgren J, Svennerholm AM, 1973, Enzyme-linked immunosorbent assays for cholera serology, *Infect Immun.*, 7:759-763.
49. Holmgren J, Svennerholm AM, 1977, Mechanisms of disease and immunity in cholera: a review, *J Infect Dis.*, 136:S105-S112.
50. Honda T, Akhtar Q, Glass RI, Kibriya AK, 1981, A simple assay to detect *Escherichia coli* producing heat-labile enterotoxin: results of a field study of the Biken tests in Bangladesh, *Lancet*, 2:609-610.
51. Honda T, Sato M, Nishimura T, Higashitsutsumi M, Fukai K, Miwatani T, 1985, Demonstration of cholera toxin-related factor in cultures of *Aeromonas* species by enzyme-linked immunosorbent assay, *Infect Immun.*, 50:322-323.
52. Honda T, Tsuji T, Takeda Y, Miwatani T, 1981, Immunological nonidentity of heat-labile enterotoxins from human and porcine enterotoxigenic *Escherichia coli*, *Infect Immun.*, 34:337-340.
53. Johnson WM, Lior H, 1986, Cytotoxic and cytotonic factors produced by *Campylobacter jejuni*, *Campylobacter coli*, and *Campylobacter laridis*, *J Clin Microbiol.*, 24:275-281.
54. Kaper JB, Levine MM, 1981, Cloned cholera enterotoxin genes in study and prevention of cholera (letter), *Lancet*, 2:1162-1163.
55. Leong J, Vinal AC, Dallas WS, 1985, Nucleotide sequence comparison between heat-labile toxin B-subunit cistrons from *Escherichia coli* of human and porcine origin, *Infect Immun.*, 48:73-77.
56. Levine MM, 1987, *Escherichia coli* that cause diarrhea: enterotoxigenic enteropathogenic, enteroinvasive, enterohemorrhagic, and enteroadherent. *J Infect Dis.*, 155:377-389.
57. Levine MM, Kaper JB, Herrington D, Losonsky G, Morris JG, Clements ML, Black RE, Tall B, Hall R, 1988, Volunteer studies of deletion mutants of *Vibrio cholerae* O1 prepared by recombinant techniques, *Infect Immun.*, 56:161-167.
58. Lindholm L, Holmgren J, Wikstrom M, Karlsson U, Andersson K, Lycke N, 1983, Monoclonal antibodies to cholera toxin with special reference to cross-reactions with *Escherichia coli* heat-labile enterotoxin, *Infect Immun.*, 40:570-576.
59. Locht C, Keith JM, 1986, Pertussis toxin gene: nucleotide sequence and genetic organization, *Science*, 232:1258-1264.
60. Lockman HA, Galen JE, Kaper JB, 1984, *Vibrio cholerae* enterotoxin genes: nucleotide sequence analysis of DNA encoding ADP-ribosyltransferase, *J Bacteriol.*, 159:1086-1089.
61. Ludwig DS, Holmes RK, Schoolnik GK, 1985, Chemical and immunochemical studies on the receptor binding domain of cholera toxin B subunit, *J Biol Chem.*, 260:12528-12534.
62. Mekalanos JJ, 1983, Duplication and amplification of toxin genes in *Vibrio cholerae*, *Cell*, 35:253-263.
63. Mekalanos JJ, Swartz DJ, Pearson GD, Harford N, Groyne F, de Wilde M, 1983, Cholera toxin genes: nucleotide sequence, deletion analysis and vaccine development, *Nature*, 306:551-557.
64. Miller VL, Mekalanos JJ, 1984, Synthesis of cholera toxin is positively regulated at the transcriptional level by *toxR*, *Proc Natl Acad Sci.*, *USA*, 81:3471-3475.
65. *Moseley SL, Echeverria P, Seriwatana J, Tirapat C, Chaicumpa W, Sakuldaipeara T, Falkow S, 1982, Identification of enterotoxigenic Escherichia coli by colony hyhridization using three enterotoxin gene probes*, *J Infect Dis.*, 145:863-869.
66. Moseley SL, Huq I, Alim AR, So M, Samadpour-Motalebi M, Falkow S, 1980, Detection of enterotoxigenic *Escherichia coli* by DNA colony hybridization, *J Infect Dis.*, 142:892-898.
67. Moss J, Garrison S, Fishman PH, Richardson SH, 1979, Gangliosides

sensitize unresponsive fibroblasts to *Escherichia coli* heat-labile enterotoxin, *J Clin Invest.*, 64:381-384.

68. Moss J, Vaughn M, 1988, Cholera toxin and *E. coli* enterotoxins and their mechanisms of action, p 39-87, *In*: "Handbook of Natural Toxins, Volume 4, Bacterial Toxins, Hardegree M, Tu AT, (eds), Marcel Dekker, Inc., New York.

69. Neill R, Holmes RK, 1988, Genetics of toxinogenesis in bacteria, p 383-416, *In*: "Handbook of Natural Toxins, Volume 4, Bacterial Toxins," Hardegree MC, Tu AT, (eds), Marcel Dekker, Inc., New York.

70. Neill RJ, Ivins BE, Holmes RK, 1983, Synthesis and secretion of the plasmid-coded heat-labile enterotoxin of *Escherichia coli* in *Vibrio cholerae, Science*, 221:289-291.

71. Nicosia A, Perugini M, Franzini C, Casagli MC, Borri MG, Antoni G, Almoni M, Neri P, Ratti G, Rappuoli R, 1986, Cloning and sequencing of the pertussis toxin genes: operon structure and gene duplication, *Proc Natl Acad Sci.*, *USA*, 83:4631-4635.

72. O'Brien AD, Holmes RK, 1987, Shiga and Shiga-like toxins, *Microbiol Rev.*, 51:206-220.

73. Okamoto K, Miyama A, Tsuji T, Honda T, Miwatani T, 1988, Effect of substitution of glycine for arginine at position 146 of the A1 subunit on biological activity of *Escherichia coli* heat-labile enterotoxin, *J Bacteriol.*, 170:2208-2211.

74. Pearson GD, Mekalanos JJ, 1982, Molecular cloning of *Vibrio cholerae* enterotoxin genes in *Escherichia coli* K-12, *Proc Natl Acad Sci.*, *USA*, 79:2976-2980.

75. Peterson KM, Mekalanos JJ, 1988, Characterization of the *Vibrio cholerae* toxR regulon: identification of novel genes involved in intestinal colonization, *Infect Immun.*, 56:2822-2829.

76. Pickett CL, Twiddy EM, Belisle BW, Holmes RK, 1986, Cloning of genes that encode a new heat-labile enterotoxin of *Escherichia coli*, *J Bacteriol.*, 165:348-352.

77. Pickett CL, Weinstein DL, Holmes RK, 1987, Genetics of type IIa heat-labile enterotoxin of *Escherichia coli*: operon fusions, nucleotide sequence, and hybridization studies, *J Bacteriol.*, 169:5180-5187.

78. Remmers EF, Colwell RR, Goldsby RA, 1982, Production and characterization of monoclonal antibodies to cholera toxin, *Infect Immun.*, 37:70-76.

79. Sandkvist M, Hirst TR, Bagdasarian M, 1987, Alterations at the carboxyl terminus change assembly and secretion properties of the B subunit of *Escherichia coli* heat-labile enterotoxin, *J Bacteriol.*, 169:4570-4576.

80. Sanyal SC, 1983, NAG vibrio toxin, *Pharmacol Ther.*, 20:183-201.

81. Seriwatana J, Echeverria P, Taylor DN, Rasrinaul L, Brown JE, Peiris JS, Clayton CL, 1988, Type II heat-labile enterotoxin-producing *Escherichia coli* isolated from animals and humans, *Infect Immun.*, 56:1158-1161.

82. So M, Dallas WS, Falkow S, 1978, Characterization of an *Escherichia coli* plasmid encoding for synthesis of heat-labile toxin: molecular cloning of the toxin determinant, *Infect Immun.*, 21:405-411.

83. Spicer EK, Kavanaugh WM, Dallas WS, Falkow S, Konigsberg WH, Schafer DE, 1981, Sequence homologies between A subunits of *Escherichia coli* and *Vibrio cholerae* enterotoxins, *Proc Natl Acad Sci.*, *USA*, 78:50-54.

84. Takeda Y, Honda T, Sima H, Tsuji T, Miwatani T, 1983, Analysis of antigenic determinants in cholera enterotoxin and heat-labile enterotoxins from human and porcine enterotoxigenic *Escherichia coli*, *Infect Immun.*, 41:50-53.

85. Taylor RK, Miller VL, Furlong DB, Mekalanos JJ, 1986, Identification of a pilus colonization factor that is coordinately regulated with cholera toxin, *Ann Sclavo Collana Monogr.*, 3:51-61.

86. Tsuji T, Honda T, Miwatani T, Wakabayashi S, Matsubara H, 1985, Analysis of receptor-binding site in *Escherichia coli* enterotoxin, *J Biol Chem.*, 260:8552-8558.

87. Tsuji T, Taga S, Honda T, Takeda Y, Miwatani T, 1982, Molecular

heterogeneity of heat-labile enterotoxins from human and porcine enterotoxigenic *Escherichia coli, Infect Immun.*, 38:444-448.

88. Yamamoto T, Gojobori T, Yokota T, 1987, Evolutionary origin of pathogenic determinants in enterotoxigenic *Escherichia coli* and *Vibrio cholerae* 01, *J Bacteriol.*, 169:1352-1357.

89. Yamamoto T, Nakazawa T, Miyata T, Kaji A, Yokota T, 1984, Evolution and structure of two ADP-ribosylation enterotoxins, *Escherichia coli* heat-labile toxin and cholera toxin, *FEBS Letters* 169:241-246.

90. Yolken RH, Greenberg HB, Merson MH, Sack RB, Kapikian AZ, 1977, Enzyme-linked immunosorbent assay for detection of *Escherichia coli* heat-labile enterotoxin, *J Clin Microbiol.*, 6:439-444.

91. Zinnaka Y, Carpenter CC Jr, 1972, An enterotoxin produced by noncholera vibrios, *Johns Hopkins Med J.*, 131:403-411.

STRUCTURE AND FUNCTION OF *CLOSTRIDIUM BOTULINUM* DERIVATIVE TOXIN ANALYZED

WITH MONOCLONAL ANTIBODIES

Yoichi Kamata and Shunji Kozaki

University of Osaka Prefecture
College of Agriculture
804 Mozu-ume-machi 4-cho
Sakai-shi, Osaka 591, Japan

Botulinum toxin is a complex of two components; one is toxic and the other is nontoxic. The freed toxic component is called derivative toxin. The derivative toxin is produced as a single-chain polypeptide with a molecular weight of about 150,000. When treated mildly with trypsin or another lysine-specific endopeptidase, the polypeptide chain is nicked. When reduced with dithiothreitol, the nicked derivative toxin is separated into heavy and light chains with molecular weights of about 100,000 and 50,000, respectively.[1,12]

Botulinum toxin has been classified into seven types by the neutralization test with the type-specific antibodies. The antibodies are thus useful for studying the toxic action of the neurotoxins. Monoclonal antibody recognizes only a single antigenic determinant. This unique property makes monoclonal antibody a powerful research tool that is useful in various fields of investigation. It has been used for analyzing the relationship between the antigenic structure and the function of bacterial toxins such as cholera, tetanus, and diphtheria toxins.[4,9,11] We prepared monoclonal antibodies against the derivative toxin and examined for their reactivities to the toxin and its various fragments to correlate the molecular structure of the derivative toxin with the function.

Analysis of type B derivative toxin

Type B derivative toxin was treated for 18 hours with chymotrypsin at a toxin-to-enzyme ratio of 200:1 at 37°C in 0.05 M phosphate buffer, pH 7.5. The chymotrypsin-digested toxin was applied to a CM-Sephadex, C-50 column and eluted with a 0-0.5 M linear NaCl gradient. One major nontoxic protein peak was eluted. On SDS-PAGE, the protein in this fraction migrated in a single band in the presence or absence of dithiothreitol. The molecular weight of the protein was 115,000 daltons, which was a little higher than that of the heavy chain. Upon mild treatment of this fragment with a lysylendopeptidase and dithiothreitol, it dissociated further into two bands; one moved to the position identical to that of the light chain, while the other migrated as a band with a molecular weight of 57,000 daltons.

We prepared monoclonal antibodies against type B derivative toxin.[8] They were tested by enzyme-linked immunosorbent assay (ELISA) for their

Microbial Toxins in Foods and Feeds, Edited by A.E. Pohland *et al.,*
Plenum Press, New York, 1990

Table 1. Reactivities of monoclonal antibodies to type B derivative toxin and its fragments in ELISA.

Antibody	ELISA value (OD$_{450}$) in well coated with:[a]				
	Toxin	H-chain[b]	L-chain[c]	Chy-frag.[d]	Denatured Chy-frag.[e]
5	>2.000	>2.000	0	>2.000	> 2.000
6	>2.000	0	0.839	0.053	1.132
8	1.089	1.108	0	0.096	0.448
9	1.412	1.670	0	0	0
13	1.473	1.255	0	0.032	0.586
14	>2.000	>2.000	0.064	0	0
17	>2.000	>2.000	0.065	0	0

[a] The values were obtained with each antibody at 1 μg/ml. OD$_{450}$, optical density at 450 nm.
[b] Heavy chain.
[c] Light chain.
[d] Chymotrypsin-induced fragment.
[e] Chymotrypsin-induced fragment treated with urea and dithiothreitol.

reactivities to the derivative toxin, the heavy and light chains, and the chymotrypsin-induced fragment (Table 1). Of the seven monoclonal antibodies, six reacted to the heavy chain and the remaining one, B6, to the light chain. Four of the antibodies reacting to the heavy chain, namely B5, B6, B8 and B13, reacted to the chymotrypsin-induced fragment, but the reactivities of B6, B8 and B13 were considerably lower than those to the derivative toxin or to either chain. When the chymotrypsin-induced fragment was denatured with urea and dithiothreitol, the reactivities increased to levels similar to those to the derivative toxin and the heavy and light chains. It is conceivable that the chymotrypsin-induced fragment is composed of the light chain and a part of the heavy chain, which we named the H-1 fragment. The chymotrypsin-induced fragment was designated as L·H-1 fragment. Sathyamoorthy and DasGupta[13] reported that the light chain is situated at the amino-terminal end of the derivative toxin. The authors assume that the portion digested with chymotrypsin, named the H-2 fragment, contains the carboxyl-terminal end. The heavy chain-recognizing monoclonal antibodies, B8 and B13, bound not to the L·H-1 fragment but to the denatured one. The digestion of the H-2 fragment of the derivative toxin molecule may have caused a conformational change in the remaining L·H-1 fragment molecule. A probable location of the antigenic determinants on the type B derivative toxin molecule and its molecular structure are illustrated in Fig. 1.

We obtained strange monoclonal antibodies, B1 and B2. They reacted in ELISA to the intact derivative toxin but not to the one denatured with urea and dithiothreitol. They did not bind to the heavy nor the light chain nor the L·H-1 fragment in either ELISA or immunoblotting, however, they neutralized the derivative toxin (Table 2). The neutralizing potency of B2 was higher than those of the other monoclonal antibodies. B1 and B2 were examined for their reactivities to the derivative toxin by radioimmunoassay. The association constant of the derivative toxin to each monoclonal antibody was determined. Iodinated derivative toxin at various concentrations was incubated with a monoclonal antibody, the toxin-antibody complex was pre-

Table 2. Some properties of monoclonal antibodies against type B derivative toxin.

Antibody	Binding site[a]	ELISA value (OD_{450})[b]		Neutralizing activity(%)[d]	Association constant(M)
		Toxin	Denatured toxin[c]		
B1	ND[e]	>2.000	0.187	15	1.18×10^9
B2	ND	>2.000	0.530	1.2	1.45×10^9
B5	H-1	>2.000	>2.000	25	2.49×10^7
B6	L	>2.000	1.024	5.3	4.67×10^9
B9	H-2	1.412	1.385	76	$<10^7$
B10	H-1	1.308	1.529	0.2	1.62×10^{10}
B16	H-2	>2.000	>2.000	89	$<10^7$

[a] The binding site of each monoclonal antibody on the toxin molecule was determined by immunoblotting.
[b] The values were obtained with each antibody at 1 μg/ml.
[c] The toxin treated with urea and dithiothreitol.
[d] The figure shows the remaining toxicity.
[e] Not determined.

cipitated with anti-mouse IgG and Protein A, and the radioactivity of the precipitate was measured (Table 2). The binding of B1 and B2 to the derivative toxin was confirmed. Their association constants were higher than those of the monoclonal antibodies with no neutralizing activity. It has been reported that tetanus toxin contains conformational antigenic determinants.[9] From the results we obtained, it appears that botulinum toxin may also contain a determinant that relates to the steric conformation of the toxin molecule.

Analysis of type E derivative toxin

We prepared six monoclonal antibodies against type E derivative toxin.[7] The binding site of each monoclonal antibody on the derivative toxin molecule was determined by immunoblotting. Of the six monoclonal antibodies, five reecognized the heavy chain and the other one the light chain (Table 3). Three monoclonal antibodies reacting to the heavy chain (E14, E17, and E22) neutralized the toxin. E16 and E30 reacted also the type F derivative toxin.

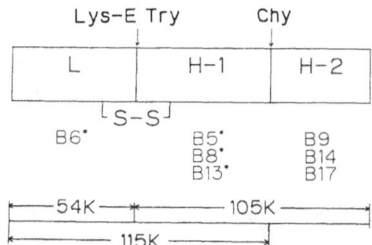

Fig. 1. Probable antigenic structure of type B derivative toxin. Arrows indicate the sites split by lysyl endopeptidase (Lys-E), trypsin (Try), and chymotrypsin (Chy). The monoclonal antibody with an asterisk neutralizes the derivative toxin. The molecular weights of the fragments are shown at the bottom.

Table 3. Some properties of monoclonal antibody against type E
derivative toxin.

Antibody	Subclass	Fragment	ELISA value(OD_{450})[a] in well coated with derivative toxin types		Neutralizing activity
			E	F	(%)[b]
14	G1	H	1.121	0	2.0
E16	G2b	H	0.798	0.720	100
E17	G1	H	1.140	0	31.4
E19	G1	H	0.682	0.020	92.7
E22	G1	H	0.964	0.005	36.3
E30	G1	L	1.567	1.641	100

[a] The values were obtained with each antibody at 1 μg/ml.
[b] The data express the remaining toxicity. When mixed with polyclonal
antibody, the remaining toxicity was less than 0.3%.

The derivative toxin was digested for 8 hours with trypsin (toxin-to-
enzyme ratio 40:1). The digested material was applied to a high-performance
liquid chromatography (HPLC) apparatus to separate the tryptic fragment. A
single major peak was eluted in HPLC-gel filtration. It contained about
of the original toxicity. In SDS-PAGE, it migrated in a single band with a
molecular weight of 105,000, the value being between those of the derivative toxin and the heavy chain and similar to that of chymotrypsin-induced
fragment of type B derivative toxin. When reduced with DTT, this fragment
dissociated into two bands with molecular weights of 56,000 and 49,000
daltons. The latter figure was identical to that of the light chain.
Chymotrypsin treatment of type E derivative toxin also induced a fragment
with a molecular weight of 110,000 daltons. Reduction of this fragment
with DTT did not yield any smaller fragment. When treated further with
trypsin and reduced, the fragment migrated as three bands; two had mobili-
ties identical to those of the heavy and light chains and the other had a
mobility similar to that of the slower band dissociated from the trypsin-
induced fragment with a molecular weight of 56,000 daltons.

In immunoblotting analysis of the trypsin-induced fragment, E14, E19
and E22 recognizing the heavy chain reacted to the larger fragment derived
from the trypsin-induced fragment. E30 recognizing the light chain bound
to the smaller fragment. Neither E16 nor E17, both recognizing the heavy
chain, reacted to any fragment (Fig. 2).

Neither E16 nor E17 bound to any chymotrypsin-induced fragment, either.
E14 bound to the 110,000-dalton fragment and also to the fragment induced
by further trypsinization and reduction of the chymotrypsin-induced fragment. E30 also reacted to the chymotrypsin-induced fragment and the light
chain contained by that fragment.

These results attested that the trypsin- and chymotrypsin-induced
fragments of type E derivative toxin consist of a light chain and a part of
the heavy chain, and that they resemble the L·H-1 fragment of type B derivative toxin induced by chymotrypsin treatment. The results also indicate
that the H-2 fragment or carboxyl-terminal portion of the heavy chain is
easily digested by trypsin or chymotrypsin.

Botulinum derivative toxin should consist of three fragments, L, H-1, and H-2. To validate this hypothesis, purification of the H-2 fragment was attempted. The heavy chain of type E derivative toxin was mildly trypsinized in the presence of dithiothreitol and urea. The material was applied to a QAE-Sephadex column. A protein peak was eluted with a buffer containing 0.1 M NaCl, 10 mM dithiothreitol, and 4 M urea. The eluted protein migrated in a single band with a molecular weight of 41,000 daltons. The sum of the molecular weights of the H-1 fragment, 56,000 daltons and of the isolated fragment, 41,000 daltons, is identical to that of the heavy chain, 97,000 daltons.

In immunoblotting analysis, monoclonal antibodies E16, E17, and E31 reacting to the heavy chain but not to L'H-1 fragment, reacted to the 41,000-dalton fragment. Thus it was demonstrated that the 41,000-dalton fragment was the H-2 fragment. The molecular structure of type E derivative toxin found by treatment with the proteases is illustrated in Fig. 3.

Several workers[2,6] have reported that the derivative toxin binds to brain synaptosomes. The neutralizing monoclonal antibodies were examined

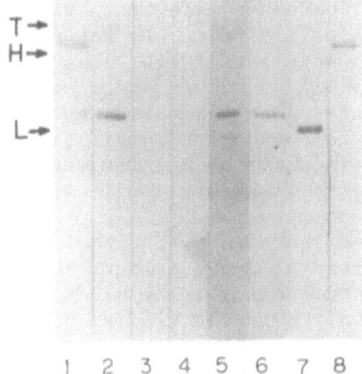

1 2 3 4 5 6 7 8

Fig. 2. Immunoblotting analyses of trypsin-induced fragment of type E derivative toxin bound to monoclonal antibodies. Before blotting, samples in lanes 2 to 7 were electrophoresed in the presence of dithiothreitol. Lanes 1 and 2, E14; lane 3, E16; lane 4, E17; lane 5, E19; lane 6, E22; lanes 7 and 8, E30.

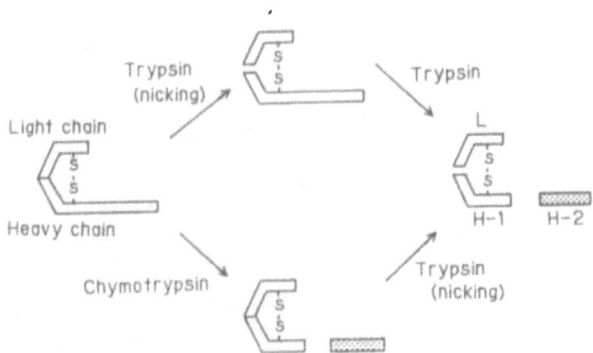

Fig. 3. Illustration of the molecular structure of type E derivative toxin scrutinized by with trypsin and chymotrypsin treatments.

for their inhibiting effects on the binding of iodinated type E derivative toxin to synaptosomes. E17, recognizing the H-2 fragment, completely inhibited the binding at 30 μg/mg of synaptosomes. Inhibition of the binding with E14 recognizing H-1 fragment was not complete even at 1,000 μg/mg of synaptosomes (Fig. 4). The iodinated derivative toxin, already bound to synaptosomes, was freed from them on incubation in the presence of E17. At a concentration of 15 μg of E17 per mg of synaptosomes, the count of the iodinated derivative toxin was close to the level of the nonspecific binding of the derivative toxin to synaptosomes. A slight displacement of the iodinated derivative toxin occurred at a high concentration of E14; about 20% of the bound derivative toxin was freed from the synaptosomes at the same concentration as that of E17. The association constants (Ka) of E14 and E17 were 1.24×10^9 and 1.48×10^9 M^{-1}, respectively. Since the binding affinities of the two antibodies were similar to each other, the sites on the toxin molecule recognized by the antibodies may cause the differences of efficacy of the binding inhibition and displacement.

In regard to the receptor of botulinum toxin, ganglioside, a sialic acid-containing glycolipid, has been one of the candidates.[3,10]. The interaction between type E derivative toxin and each of its fragments with gangliosides was examined by thin-layer chromatography (TLC)-immunostaining.[5] In brief, the derivative toxin was allowed to react with ganglioside on a TLC plate. After washing, the binding was detected with rabbit-antitoxin and peroxidase-conjugated anti-rabbit IgG. The derivative toxin bound to ganglioside G_{T1b}, G_{D1a}, and G_{Q1b}, but not to G_{M1} nor G_{D1b}. The heavy chain also bound to the same three gangliosides, but neither the light chain nor the L·H-1 fragment bound to any of the gangliosides. The results of the binding experiments suggest that the H-2 fragment is involved in the binding of the toxin molecule to the neural cell membranes and ganglioside.

The results described above show that the heavy chain of botulinum derivative toxin is composed of two fragments that are distinct from each other structurally and functionally. The H-2 fragment located in the carboxyl-terminal end is involved in the binding of the toxin (Fig. 5). The role the H-1 fragment and the light chain plays in the action of botulinum toxin is not clear. Monoclonal antibodies may also be useful tools for further study on the structure and function of botulinum toxin, particularly those of the H-1 fragment and the light chain.

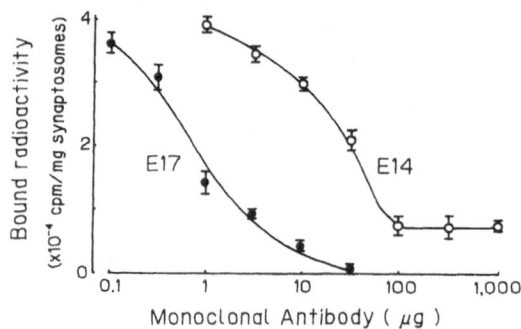

Fig. 4. Inhibition of binding of iodinated type E derivative toxin from the synaptosomes with the neutralizing monoclonal antibodies. E17 recognizes the H-2 fragment and E14 the H-1 fragment. Each point represents the mean + S.D. (n=6).

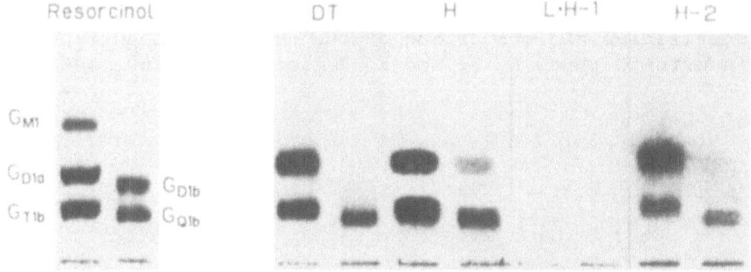

Fig. 5. Binding of type E derivative toxin and each of its fragments
to gangliosides as shown by TLC-immunostaining. The gangliosides developed on a TLC plate were colored by resorcinol
reagent. Abbreviations used are DT: type E derivative toxin;
H: the heavy chain of type E derivative toxin; L·H-1 fragment;
H-2: the H-2 fragment.

REFERENCES

1. Evans DM, Williams RS, Shone CC, Hambleton P, Melling J, Dolly JO,
    1986, Botulinum neurotoxin type B, Its purification, radioiodination
    and interaction with rat-brain synaptosomal membranes, Eur J Biochem.,
    154:409-416.
2. DasGupta BR, Sugiyama H, 1972, A common subunit structure in Clostri-
    dium botulinum type A, B, and E toxins, Biochem Biophys Res Commun.,
    48:108-112.
3. Habermann E, Dryer F, 1986, Clostridial neurotoxins: handling and
    action at the cellular and molecular level, Curr Topics Microbiol
    Immunol., 129:93-179.
4. Hayakawa S, Uchida T, Mekada E, Moynihan MR, Okada Y, 1982, Monoclonal
    antibody against diphtheria toxin: effect on toxin binding and entry
    into cells, J Biol Chem., 158:4311-4317.
5. Kamata Y, Kozaki S, Sakaguchi G, Iwamori M, Nagai Y, 1986, Evidence
    for direct binding of Clostridium botulinum type E derivative toxin
    and its fragments to gangliosides and free fatty acids, Biochem
    Biophys Res Commun., 140:1015-1019.
6. Kozaki S, 1979, Interaction of botulinum type A, B, and E derivative
    toxins with synaptosomes of rat brain, Naunyn-Schmiedeberg's Arch
    Pharmacol., 308:67-70.
7. Kozaki S, Kamata Y, Nagai T, Ogasawara J, Sakaguchi G, 1986, The use
    of monoclonal antibodies to analyze the structure of Clostridium
    botulinum type E derivative toxin, Infect Immun., 52:786-792.
8. Kozaki S, Ogasawara J, Shimote U, Kamata Y, Sakaguchi G, 1987, Anti-
    genic structure of Clostridium botulinum type B neurotoxin and its
    interaction with gangliosides, cerebrosides, and free fatty acids,
    Infect Immun., 55:3051-3056.
9. Mizuguchi J, Yoshida T, Sato Y, Nagaoka F, Kondo S, Matsuhashi T,
    1982, Requirement of at least two distinct monoclonal antibodies for
    efficient neutralization of tetanus toxin in vivo, Naturwissenschaften,
    69:5997-5998.
10. Ochanda JO, Syuto B, Ohishi I, Naiki M, Kubo S, 1985, Binding of
    Clostridium botulinum neurotoxin to gangliosides, J Biochem., 100:
    27-33.
11. Remmers EF, Colwell RR, Goldsby RA, 1982, Production and characteri-
    zation of monoclonal antibodies to cholera toxin, Infect Immun., 37:
    70-76.
12. Sakaguchi G, 1983, Clostridium botulinum toxins, Pharmacol
    Therap., 19:165-194.

13. Sathyamoorthy V, DasGupta BR, 1985, Separation, purification, partial characterization and comparison of the heavy and light chains of botulinum toxin types A, B, and E, *J Biol Chem.*, 260:10461-10466.

RAPID METHOD FOR DETECTION OF PATHOGENIC *YERSINIA ENTEROCOLITICA* AND

YERSINIA PSEUDOTUBERCULOSIS USING ENZYME IMMUNOASSAY

Seiji Kaneko and Tsutomu Maruyama

The Tokyo Metropolitan Research Laboratory of
    Public Health
Hyakunincho 3-24-1
Shinjuku-ku, Tokyo 160, Japan

ABSTRACT

In this paper we describe a convenient method for the detection of virulent *Yersinia* strains using an enzyme-immunoassay (EIA). An antiserum was prepared against virulent plasmid-coded specific proteins of Y. *enterocolitica* serotype 03. Then, EIA was performed with the antiserum and proteins of various *Yersinia* strains of different origin. Positive reaction in EIA was observed when the proteins of virulent plasmid-bearing strains were tested. These strains were Y. *enterocolitica* serotypes 05:27, 08 and 09, and Y. *pseudotuberculosis* serotypes 1b, 2a, 2b, 2c, 3, 4a, 4b, 5a, 5b, 6, 7 and 8. On the other hand, the reaction was negative when plasmid-cured strains of the pathogenic Y. *enterocolitica* and Y. *pseudotuberculosis* were tested, as well as non-pathogenic Y. *enterocolitica*, Y. *frederiksenii*, Y. *intermedia* and Y. *kristensenii*. The results clearly indicate the usefulness of the EIA method for the rapid detection of virulent *Yersinia* strains.

INTRODUCTION

Yersinia enterocolitica and *Yersinia pseudotuberculosis* are human pathogens and cause gastroenteritis, arthritis, erythema nodosum, septicemia and other diseases. The virulence of these two microorganisms was correlated with the occurrence of plasmids of 40 to 50 megadaltons (MDa).[3,14,17] When incubated at 37°C, the plasmid bearing strain expressed many temperature dependent phenotypes such as calcium dependent growth,[3] autoagglutination activity,[11] increase in cell surface hydrophobicity[10] and production of outer membrane proteins,[1,2,14] as well as production of V and W antigens.[13]

Isolation and identification of pathogenic Y. *enterocolitica* and Y. *pseudotuberculosis* is time-consuming and requires a lot of effort. Ordinarily, after isolation of these organisms, many tests are necessary for biochemical characterization and identification of the organism. Furthermore, tests for serotyping, as well as for virulence-associated properties such as autoagglutination activity, calcium dependent growth and the presence of a 40-50 MDa plasmid were needed for precise identification of the notorious food-poisoning bacteria. From the viewpoint of clinical and public health microbiology, it is very important and desirable to detect the virulent *Yersinia* strains rapidly.

Recently it was reported that virulent plasmid-bearing strains of Y. enterocolitica and Y. pseudotuberculosis produced the temperature-dependent proteins immunochemically identified.[4] If the proteins could be easily detected in all virulent strains in Yersinieae by means of enzyme immunoassay (EIA), the determination of virulent strains of Y. enterocolitica and Y. pseudotuberculosis could be done easily and rapidly.

In this paper we report a rapid method useful for the determination of virulent Yersinia strains using EIA.

MATERIALS AND METHODS

Bacterial strains

Virulence plasmid bearing strains employed were Y. enterocolitica serotype 03 biotypes 3 and 4, serotype 05:27 biotype 2, serotype 08 biotype 1 and serotype 09 biotype 2 and Y. pseudotuberculosis serotypes 1b, 2a, 2b, 2c, 3, 4a, 4b, 5a, 5b, 6, 7 and 8. Plasmid cured strains of these Y. enterocolitica and Y. pseudotuberculosis, as well as avirulent Y. enterocolitica serotypes 06, 07:8 and 012, Y. intermedia, Y. frederiksenii and Y. kristensenii, were also used.

Test for virulence of Yersinieae

Calcium dependent growth at 37°C of Yersinia strains was tested by using magnesium oxalate agar (MOX) as described by Higuchi and Smith.[5] Autoagglutination activity of Yersinieae at 37°C was tested by the method described by Laird and Cavanaugh.[11] The occurrence of plasmids was tested as described previously.[8,9]

Preparation of antiserum against plasmid-coded proteins

Yersinia enterocolitica serotype 03 carrying 44 MDa virulence plasmid was preincubated at 25° for 24 hours in brain heart infusion (BHI) broth. The culture was further incubated in growing medium[14] at 37°C for 24 hours. In this condition, plasmid-coded proteins were accumulating intracellularly in large amounts. Cells were then harvested by centrifugation, washed three times with phosphate buffered saline (PBS:pH 7.6) and suspended in PBS. The number of cells was adjusted to 10^9/ml in the suspension to which formaldehyde was added at a concentration of 3%. The formaldehyde-fixed cells were injected into the ear vein of a rabbit for immunization. Booster injections were repeated weekly. When the agglutination titer against Y. enterocolitica serotype 03 became higher than 1:2560, antiserum was collected from the immunized rabbit and stored at -80°C. To remove antibodies against chromosome-coded proteins of Yersinieae, the serum was absorbed with cells of a plasmid-cured stain of Y. enterocolitica serotype 03 grown at 37°C. The antiserum against plasmid-coded proteins thus obtained was used for the EIA reaction.

Detection of plasmid coded proteins by EIA

Each strain to be tested was preincubated in BHI broth at 25° for 24 hours. Ten microliters of the preincubated cultures were inoculated onto magnesium oxalate agar which was covered with a nitrate cellulose membrane filter and further incubated at 37°C for appropriate times. Then the filter was transferred to an empty petri dish and rinsed in TBS (20mM Tris, 500mM NaCl, pH 7.5). To avoid non-specific reactions, the filter was put in TBS containing gelatin at 3 percent and gently agitated for 30 minutes at room temperature. The filter was then transferred to the second petri dish containing 15 ml of 500-fold dilution of an antiserum against plasmid-coded

proteins and incubated for 3 hours with gentle agitation. To remove unbound
first antibody, the filter was washed with 1 percent gelatin in TBS. Then
the filter was transferred to the third petri dish containing 15 ml of 500-
fold dilution of goat anti-rabbit IgG-horseradish peroxidase conjugate and
incubated 1 hour with gentle agitation, followed by washing with 1 percent
gelatin in TBS. Finally the filter was transferred to color development
solution (30 mg of 4-chloro-1-naphthol, 10 ml ice cold methanol, 30 microliters ice cold 30 percent H_2O_2, 50 ml of TBS) and incubated for 30 minutes
at room temperature. In cases where *Yersinia* strains produce the plasmid-
coded proteins, purple spots will develop on the filter.

RESULTS AND DISCUSSION

Decision of incubation time of membrane filter in EIA test

As described in Materials and Methods, EIA tests using a filter containing bacteria were performed to evaluate the virulence of *Yersinia*
strains. The critical point of the EIA test is to determine the minimum
incubation time of the filter on the magnesium oxalate agar for detection
of virulence plasmid-coded proteins. For this purpose, *Y. enterocolitica*
serotype 03 strain with or without virulence plasmid was transferred to
membrane filters, with which EIA tests were carried out after 1-6 hours
incubation at 37°C. The results shown in Fig. 1 clearly indicate that the
plasmid-positive strain of *Y. enterocolitica* serotype 03 produced detectable
amount of plasmid-coded proteins in the EIA test after two hours incubation.
Plasmidless strain showed positive reaction, though very weakly, probably
due to non-specific reactions. However, the intensity of the latter reaction
was much weaker than that of the former. The difference between these two
reactions is apparent (Fig. 1). From these results and for safety, we
incubated the filter for three hours in the following EIA tests.

Temperature-dependent production of plasmid-coded proteins in virulent
Yersinia strains

To test the influence of incubation temperature upon production of
plasmid-coded proteins, five different *Y. enterocolitica* strains with or
without a virulent plasmid were established. These strains were spotted

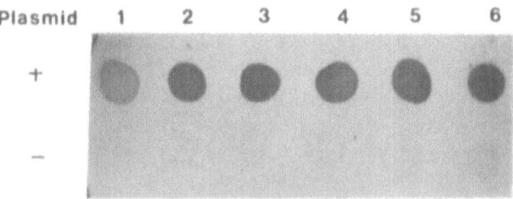

Fig. 1.   Detection of plasmid-coded proteins produced from *Yersinia*
          enterocolitica serotype 03 with EIA test.

          Ten microliters of BHIB cultured both plasmid bearing (the
          upper reaction) and curing (the lower reaction) strains of
          Yersinia enterocolitica serotype 03 were spotted on the
          membrane filter at intervals of one hour and incubated at
          37°C. Dark spots were positive and light spots were
          negative. Numbers in the figure indicate the incubation
          time (hours) after inoculation. Enzyme immunoassay
          procedure is described in the text.

on the membrane filter, incubated at 25°C and 37°C for 3 hours on magnesium oxalate agar, and the occurrence of plasmid-coded proteins on the filter was tested by the EIA test as described in Materials and Methods. The results shown in Table 1 clearly demonstrated that all strains carrying the virulence plasmid showed positive reactions in the EIA test only when incubated at 37°C; the reaction was negative when the plasmid positive strains were incubated at 25°C. The results in Table 1 also indicated that plasmid-coded proteins in serotype 03 are immunologically related to those in other serotypes.

Production of plasmid-coded proteins by the difference of incubation temperatures and the presence of virulence plasmid

Virulent *Y. enterocolitica* strains were examined to determine how the EIA reaction was influenced by different incubation temperatures and by the presence of virulence plasmid. Each strain was spotted on the membrane filter and incubated at 25°C and 37°C for 3 hours. The results of this EIA reaction were shown in Table 1 and the results clearly demonstrated that virulence plasmid-bearing strains alone showed positive reaction only when incubated at 37°C. But, the reaction was negative even when the plasmid-positive strains were incubated at 25°C. These results indicated that this anti-plasmid-coded protein serum reacted not only with plasmid-coded proteins produced from *Y. enterocolitica* serotype 03 but also the proteins from other virulent *Y. enterocolitica* strains when incubated at 37°C.

Table 1. Temperature-dependent Production of Plasmid-coded Proteins in Five Representative Strains of *Yersinia enterocolitica* with or without Virulent Plasmid.

Serotype	Biotype	Plasmid	Incubation Temperature	Production of Plasmid-coded Proteins
0:3	3	+	25°C	-
		-	25°C	-
		+	37°C	+
		-	37°C	-
0:3	4	+	25°C	-
		-	25°C	-
		+	37°C	+
		-	37°C	-
0:5,27	2	+	25°C	-
		-	25°C	-
		+	37°C	+
		-	37°C	-
0:8	1	+	25°C	-
		-	25°C	-
		+	37°C	+
		-	37°C	-
0:9	2	+	25°C	-
		-	25°C	-
		+	37°C	+
		-	37°C	-

Table 2. Correlation Between Virulence-associated Properties and EIA Test in Various *Yersinia* Strains.

Species	Biotype	Serotype	Plasmid	Autoagglutination	Calcium Dependent Growth at 37°C	Production of Plasmid Coded Proteins (EIA Test)
Y. enterocolitica	3	0:3	+	+	+	+
	3	0:3	-	-	-	-
	4	0:3	+	+	+	+
	4	0:3	-	-	-	-
	2	0:5, 27	+	+	+	+
	2	0:5, 27	-	-	-	-
	1	0:8	+	+	+	+
	1	0:8	-	-	-	-
	2	0:9	+	+	+	+
	2	0:9	-	-	-	-
	1	UT	-	-	-	-
	2	UT	-	-	-	-
Y. frederiksenii			-	-	-	-
Y. intermedia			-	-	-	-
Y. kristensenii			-	-	-	-
Y. pseudotuberculosis		1a	-	-	-	-
		1b	+	+	+	+
		1b	-	-	-	-
		2a	-	-	-	-
		2b	+	+	+	+
		2b	-	-	-	-
		2c	+	+	+	+
		2c	-	-	-	-
		3	+	+	+	+
		3	-	-	-	-
		4a	-	-	-	-
		4b	+	+	+	+
		4b	-	-	-	-

(Continued)

115

Table 2. Correlation Between Virulence-associated Properties and EIA Test in Various *Yersinia* Strains.

Species	Biotype	Serotype	Plasmid	Autoagglu-tination	Calcium Dependent Growth at 37°C	Production of Plasmid Coded Proteins (EIA Test)
		5a	+	+	+	+
		5a	-	-	-	-
		5b	+	+	+	+
		5b	-	-	-	-
		6	+	+	+	+
		6	-	-	-	-
		7	+	+	+	+
		7	-	-	-	-
		8	+	+	+	+
		8	-	-	-	-

Correlation between virulence-associated properties and EIA test in Y. enterocolitica and Y. pseudotuberculosis

The EIA test using an anti-03 serum was also applied to the representative *Yersinia* strains of different origin with or without a virulent plasmid (*Y. enterocolitica* serotypes 06, 07:8, *Y. frederiksenii*, *Y. intermedia*, *Y. kristensenii* and *Y. pseudotuberculosis* serotypes 1b, 2a, 2b, 2c, 3, 4a, 4b, 5a, 5b, 6, 7 and 8 in Table 2). There are several characteristics indicative of the virulence of *Yersinieae*. Some of these are the occurrence of 40-50 MDa virulence plasmids, autoagglutination activity and calcium dependent growth at 37°C. The correlation of these virulence associated properties and the result of the EIA test is summarized in Table 2. As clearly shown in this table, the plasmid-bearing strains of *Y. pseudotuberculosis*, as well as those of *Y. enterocolitica*, were all positive for EIA test, calcium dependent growth and autoagglutination activity at 37°C. These strains were shown to be pathogenic to mice.[9] In contrast, plasmid cured strains of these pathogenic strains and non pathogenic *Yersinia* strains of environment origin were all negative for the EIA test and virulence-associated properties.

We have not yet tested American strains of *Y. enterocolitica* serotypes 04:32, 013a:13b, 018, 020 or 021, or 40-50 MDa plasmid-bearing *Y. pseudotuberculosis* serotypes 1a or 4a. However, these strains are expected to be positive in the EIA test, because these strains have been reported to show the virulence-associated properties and carry the virulence plasmid.[8]

In *Y. enterocolitica* and *Y. pseudotuberculosis*, it is important to establish a rapid method for the differentiation of virulent and avirulent strains, because evaluation of the pathogenicity of these *Yersinia* strains generally consumes a lot of time. The methods for differentiation between virulent and avirulent strains have been reported by many workers, e.g., the difference in colony morphology,[12] congo red absorption,[15] and DNA colony hybridization,[6,7] and the presence or absence of specific proteins.[16] However, ambiguous results were occasionally obtained with these methods except for DNA colony hybridization.

As the pathogenicity of *Y. enterocolitica* and *Y. pseudotuberculosis* was dependent on the occurrence of the virulence plasmid, detection of the plasmid or plasmid-coded proteins results in the correct determination of the virulence of *Yersinia* strains tested. Our method reported in this paper is expected to be useful for rapid detection of virulent *Yersinia* strains. The applications of the reported method dramatically saves time and costs in the laboratory routine.

REFERENCES

1.   Bolin I, Norlander L, Wolf-Watz H, 1982, Temperature-inducible outer membrane protein of *Yersinia pseudotuberculosis* and *Yersinia enterocolitica* is associated with the virulence plasmid, *Infect Immun.*, 37:506-512.
2.   Bolin IL, Portnoy, DA, Wolf-Watz H, 1985, Expression of the temperature inducible outer membrane proteins of *Yersiniae*, *Infect Immun.*, 48: 234-240.
3.   Gemski P, Lazere JR, Casey T, 1980, Plasmid associated with pathogenicity and calcium dependency of *Yersinia enterocolitica*, *Infect Immun.*, 27:682-685.
4.   Heesemann J, Gross U, Schmidt N, Laufs R, 1986, Immunochemical analysis of plasmid-encoded proteins released by enteropathogenic *Yersinia* sp. grown in calcium-deficient media, *Infect Immun.*, 54:561-567.

5.   Higuchi K, Smith JL, 1961, Studies on the nutrition and physiology of *Pasteurella pestis*, VI. A differential plating medium for the estimation of the mutation rate to avirulent, *J Bacteriol.* 81:605-608.

6.   Hill WE, Payne WL, Aulisio, CCG, 1983, Detection and enumeration of virulent *Yersinia enterocolitica* in food by DNA colony hybridization, *Appl Environ Microbiol.*, 46:636-641.

7.   Jagow J, Hill W, 1986, Enumeration by DNA colony hybridization of virulent *Yersinia enterocolitica* colonies in artificially contaminated food, *Appl Environ Microbiol.*, 51:441-443.

8.   Kaneko S, Maruyama, T, 1986, Relationship between the presence of 44 megadalton plasmid and calcium dependency or autoagglutination to serotype 03 strains of *Yersinia enterocolitica*, *Jpn J Vet Sci.*, 48:205-210.

9.   Kaneko S, Maruyama T, 1987, Pathogenicity of *Yersinia enterocolitica* serotype C3 biotype 3 strains, *J Clin Microbiol.*, 25:454-455.

10. Lachica RV, Zink, DL, 1984, Determination of plasmid-associated hydrophobicity of *Yersinia enterocolitica* by latex particle agglutination test, *J C. in Microbiol.*, 19:660-663.

11. Laird WJ, Cavanaugh, DC, 1980, Correlation of autoagglutination and virulence of *Yersiniae*, *J Clin Microbiol.*, 11:430-432.

12. Mazigh D, Alonso JM, Mollaret HH, 1983, Simple method for demonstration of differ ntial colony morphology of plasmid-associated virulent clones of *Yersin a enterocolitica*, *J Clin Microbiol.*, 17:555-557.

13. Perry RD, Brubaker RR, 1983, Vwa$^+$ phenotype of *Yersinia enterocol tica*, *Infect Immun.*, 40:166-171.

14. Portnoy DA, Wolf-Watz H, Bolin I, Breeder AB, Falkow S, 1984, Characterization of common virulence plasmids in *Yersinia* species and their role in the expression of outer membrane proteins, *Infect Immun.*, 43:108-114.

15. Prpic JK, Robins-Browne RM, Davey RB, 1983, Differentiation between virulent and avirulent *Yersinia enterocolitica* isolates by using congo red agar, *J Clin Microbiol.*, 18:486-490.

16. Toyos J, Diaz R, Urra E, Moriyon I, 1986, Analysis by coagglutination of the distribution of a 24,000-dalton surface protein in *Yersinia enterocolitica*, *J Clin Microbiol.*, 23:804-805.

17. Zink DL, Feeley JC, Wells JG, Vanderzant C, Vickery JC, Roof WD, O'Donovan GA, 1980, Plasmid-mediated tissue invasiveness in *Yersinia enterocolitica*, *Nature* (London), 283:224-226.

OLIGODEOXYRIBONUCLEOTIDE PROBE SPECIFIC FOR THE 230 KILOBASE PAIR

VIRULENCE PLASMID IN ENTEROINVASIVE *ESCHERICHIA COLI* AND *SHIGELLA*

Keith A. Lampel, James A. Jagow and Megan L. Troxell

Molecular Biology Branch
Division of Microbiology
Food and Drug Administration
Washington, D.C. 20204

INTRODUCTION

Enteroinvasive *Escherichia coli* and the four *Shigella* species, *S. dysenteriae*, *S. flexneri*, *S. sonnei* and *S. boydii* are the causative agents of bacillary dysentery (shigellosis) in higher primates. *Shigella*-induced infections are due to the ingestion of either contaminated water or, as is now becoming increasingly apparent, foods. The infective dose for humans is 10^1 to 10^4 bacterial cells, with an incubation period of 12-50 hours. The illness is characterized by cramps, nausea, diarrhea and bloody stools and usually lasts about 4 days. According to Smith,[24] the primary reason for many outbreaks in the developed world is poor personal hygiene of food handlers who contaminate the food that they contact.

The methods for detecting foodborne pathogenic *Shigella* lags behind that used in clinical laboratories. As the number of outbreaks of shigellosis increases, identifying the suspect contaminated food(s) becomes a more significant matter of public safety. To improve the current methodology, a better understanding of the molecular biology of invasiveness by these pathogens is needed. As the basic mechanisms are clarified, more effective methods for identification of *Shigella* in foods and environmental specimens can be developed.

The target organ for invasive enteric bacteria is the human colon.[11] For *Shigella* to be fully virulent two critical events must occur. Initially, the organism must attach itself to and invade colonic epithelial cells.[9] Subsequently, the organism must multiply intracellularly and invade adjacent cells. Hale and Formal[6] have postulated that as enteroinvasive enteric bacteria pass through the duodenum and jejunum, proteotylic enzymes cleave species-specific bacterial outer membrane proteins which then activate the bacteria to adhere to and invade epithelial cells. After adherence to the epithelial cell membrane, the bacteria are engulfed by invagination without destruction of the plasma membrane. Yoshikawa (see Chapter) referred to this step as the primary invasion of colonic epithelial cells. Once inside the cell, the bacteria lyse the endocytic vacuole and multiply. The second step in invasion is the dissemination of the *Shigella* to adjacent epithelial cells.

Microbial Toxins in Foods and Feeds, Edited by A.E. Pohland *et al.*,
Plenum Press, New York, 1990

Several methods are used to assay the virulence of enteroinvasive enteric bacteria. An *in vitro* assay uses HeLa cells to monitor the invasiveness and multiplication of *Shigella*; bacteria are observed within these cells and destruction of the cell monolayer ensues.[13] In the Sereny test,[22] virulent strains inoculated into guinea pig cornea produce keratoconjunctivitis. Other methods involve oral feeding of starved, opiated guinea pigs or nonhuman primates.[24]

Four factors contribute to the pathogenicity of *Shigella* species: genetic loci in the chromosome, genetic determinants residing in a large virulence-associated plasmid, temperature regulation and toxins. Three regions of the *Shigella* chromosome are associated with virulence. One, the *kcpa* (keratoconjuntivitis provocation) locus, codes for a gene product necessary for *Shigella* to spread to adjacent cells. The *xyl-rha* region encodes for aerobactin[4] and possibly a Shiga-like toxin in *S. dysenteriae*[25] and a toxin in *S. flexneri*.[16] Loss of this region renders *Shigella* incapable of causing a fatal infection. The virulence-associated determinant encoded near the *his* locus is needed for the synthesis of the somatic antigen in *S. flexneri* and *E. coli*.[3]

It has been well established that a large (120-140 MDa) plasmid is necessary for invasion of colonic epithelial cells by enteroinvasive *E. coli* and *Shigella* spp.[15] Strains cured of this large plasmid were unable to manifest the virulent phenotype. Transfer of this plasmid into a cured strain restored its pathogenicity.

Transposon mutagenesis has been used to identify several loci that encode virulence-associated proteins in the large plasmid. Seven plasmid-associated proteins were identified by Hale et al.[7] Four of these proteins of 78, 57, 43 and 39 kDa were shown to cross react with convalescent-phase monkey sera. A 140 kDa plasmid-associated peptide was also found by Oaks et al.[14] to cross react by immunoblot analysis with human or monkey convalescent-phase sera. By Tn5 mutagenesis, Sasakawa[21] mapped seven genetic loci in the virulence plasmid that encoded for the virulence-associated proteins and regulatory factors (Fig. 1). Two loci, the *ipa* (invasion plasmid antigen) and *inv* locus, have been mapped to the invasion plasmid.[1,28] The *ipa* locus encodes for the four immunodominant antigens mentioned. The genes for these plasmid-associated proteins, *ipaA*, *ipaB*, *ipaC* and *ipaD*, have been mapped (Fig. 1) and sequenced.[1,26] The proteins synthesized from the *ipaB* and *ipaC* genes are thought to be responsible for the bacterial uptake into epithelial cells.[6] Observations by Hale[5] suggest that the *invA* locus is involved with transport of the *ipa* gene products to the outer membrane.

The 140 kDa protein is the product of the *virG* gene, which is necessary

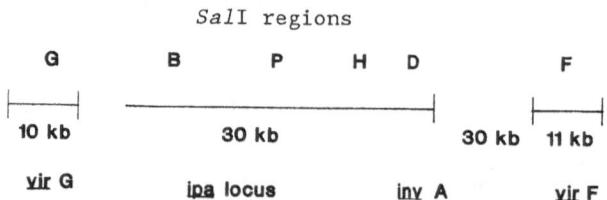

Fig. 1. Partial *SalI* restriction map of the 230 kb virulence plasmid involved with invasiveness. Data were modified from Hale.[5]

for the ability of enteroinvasive enterics to spread to adjacent cells but does not appear to be required for intracellular multiplication.[10] The *virF* gene is postulated to be a positive activator affecting the level of transcription of the *ipa* locus and the *virG* gene.[20]

Temperature plays a critical role in the expression of virulence-associated proteins in *Shigella*.[11] At 37°C, *S. flexneri* is invasive, able to elicit keratoconjunctivitis and binds Congo red dye. At 30°C, however, enteroinvasive enteric bacteria are avirulent but still retain the large plasmid. Maurelli and Sansonetti[12] have identified a genetic locus on the bacterial chromosome, *virR*, which is partially responsible for the expression of the virulence-associated genes of *Shigella*. The *virR* gene product acts as a transacting repressor molecule.

Several toxins are suspected of being involved in some aspect of virulence in enteroinvasive enteric bacteria. The Shiga toxin of *S. dysenteriae* is a well-characterized protein (reviewed by O'Brien[15]) but the exact function in shigellosis of this protein and Shiga-like toxin produced by *S. sonnei* and *S. flexneri* has not been elucidated. Sansonetti and coworkers[2,19] have suggested that a contact hemolysin interacts with the plasma membrane and is necessary for the *Shigella*-mediated lysis of vacuoles that facilitates the dissemination of the bacteria to other colonic cells.

There are three major differences among the virulence-associated plasmids in *Shigella*. The virulence plasmid isolated from *S. flexneri, S. dysenteriae* and *S. boydii* is 140 MDa, whereas in *S. sonnei* the plasmid is 120 MDa. The ability of each *Shigella* sp. to retain the large plasmid varies, with *S. flexneri* losing the plasmid at a frequency of 1 per 10^6 cells.[27] However, in *S. sonnei* the plasmid is lost at a faster rate, about 1 to 50 per 10^2 cells.[18] This high rate of loss for *S. sonnei* presents a problem for maintaining the large plasmid in clinical isolates. In the United States, *S. sonnei* has been implicated in the majority of shigellosis outbreaks in recent years and attempts to retain the pathogen with the large plasmid are often unsuccessful.

Synthesis of the O-antigen in some *Shigella* spp. is plasmid-dependent. In *S. sonnei*, the large plasmid encodes for a specific O-side chain of the lipopolysaccharide layer. Two forms of colonial morphology are seen on agar plates, depending upon whether or not the O-antigen is synthesized. Form I colonies are smooth; the bacteria are virulent and contain the large plasmid. Conversely, form II colonies are rough; the bacteria are avirulent and have not retained the large plasmid.[8] *S. dysenteriae* has two plasmids (140 MDa and 6 MDa); both plasmids are necessary for the complete synthesis of the O-antigen and both have a role in pathogenicity.[29] However, in *S. flexneri*, the large plasmid is not involved with antigenic determination; production of smooth colonies is solely dependent on chromosomal genes.[18]

CURRENT RESEARCH

Development of an oligodeoxyribonucleotide probe to detect Shigella in foods

Identification of the causative agent in a foodborne outbreak is hampered by difficulty in isolating the pathogen from foods. The methods to ensure recovery of *Shigella* from suspect foods has not been satisfactorily developed for several reasons. Dependent upon the food, *Shigella* may be easily outgrown by the resident microbial flora following direct inoculation onto semiselective agar plates or enrichment before plating. This selective culturing may not be specific enough to ensure detectable colonies of *Shigella* on agar plates. Our efforts to develop suitable methods are centered on two approaches. One is the application of molecular biology technology to develop more sensitive methods for isolating *Shigella* from foods. The other approach is to improve the standard methods of recovering these bacteria from foods by using more selective means.

We received a plasmid from S. Falkow's laboratory containing a 2.5 kilobase pair (kb) *Hind*III fragment that putatively carried genetic information for invasiveness in enteroinvasive *E. coli*.[23] Wood et al.[30] compared this fragment to a 17 kb *Eco*RI fragment for use as a probe to identify enteroinvasive enteric bacteria and found that these DNAs were able to discriminate between invasive and noninvasive enteric bacteria. The immediate goal was to use a short synthetic oligodeoxyribonucleotide derived from the 2.5 kb *Hind*III fragment as a probe to detect enteroinvasive *E. coli* and *Shigella* spp.

The nucleotide sequence of the 2.5 kb *Hind*III fragment was determined by using the dideoxynucleotide chain termination method of Sanger.[17] Specific restriction fragments were either cloned directly into the phage M13mp18 or the plasmid pUC19 and sequenced. To complete the entire sequence in both directions, deletions of the 2.5 kb *Hind*III fragment cloned into pUC19 were generated by using the Erase-a-base protocol of Promega Inc., Madison, Wisconsin. The complete sequence of the 2.5 kb *Hind*III fragment from the large plasmid of enteroinvasive *E. coli* is shown in Fig. 2.

From the sequence data, a 627 base pair (bp) open reading frame (108-735 in Fig. 2) was found. To confirm that this open reading frame was tran-

```
     HindIII
  1  AAGCTTATAGGTGTCTCAATATCTTTGCTACTTCTTTCTGGATGGTATGGTGAGGTTTTA  60

 61  TTGTCTTTTTGTCATGAAATAATGTTTTTAATTAAGAGTGGGGTTTGATGGACATTTCAA 120
                                        rbs            MetAspIleSerS

121  GCTGGTTCGAAAGTATTCATGTGTTTTTAATACTCCTGAACGGCGTTTTTTTTAGATTGG 180
     erTrpPheGluSerIleHisValPheLeuIleLeuLeuAsnGlyValPhePheArgLeuA

181  CTCCATTATTTTTCTTTCTTCCATTTTTAAATAACGGTATAAATTTCTCCATCTATTAGAA 240
     laProLeuPhePhePheLeuProPheLeuAsnAsnGlyIleIleIleSerProSerIleArgI

241  TACCTGTGATTTTCTTGTTGCATCAGGATTAATTACTTCTGGTAAGGTAGACATAGGTT 300
     leProValIlePheLeuValAlaSerGlyLeuIleThrSerGlyLysValAspIleGlyS

301  CTTCTGTTTTTGAACATGTTTATTTCCTTATGTTCAAGGAAATAATTGTTGGCCTCCTTC 360
     erSerValPheGluHisValTyrPheLeuMetPheLysGluIleIleValGlyLeuLeuL

361  TCTCTTTTTGCTTGTCTCTTCCCTTTTGGATATTTCATGCTGTTGGTAGCATTATTGACA 420
     euSerPheCysLeuSerLeuProPheTrpIlePheHisAlaValGlySerIleIleAspA

421  ACCAGCGTGGGGCAACGCTTAGTAGTTCAATTGATCCTGCCAATGGTGTTGATACGTCTG 480
     snGlnArgGlyAlaThrLeuSerSerSerIleAspProAlaAsnGlyValAspThrSerG

481  AGTTGGCAAAATTTTTCAATCTTTTTTTCTGCAGTTGTATTTCTATACAGTGGTGGAATGG 540
     luLeuAlaLysPhePheAsnLeuPheSerAlaValValPheLeuTyrSerGlyGlyMetV

541  TCTTTATTTTAGAATCCATACAATTGTCTTATAATATATGCCCGTTATTTTCTCAATGTT 600
     alPheIleLeuGluSerIleGlnLeuSerTyrAsnIleCysProLeuPheSerGlnCysS

601  CTTTCCGTGTCTCAAATATCTTAACATTTCTGACTTTATTGGCAAGTCAGGCTGTTATTT 660
     erPheArgValSerAsnIleLeuThrPheLeuThrLeuLeuAlaSerGlnAlaValIleL

661  TAGCCAGTCCTGTTATGATAGTATTGTTACTATCAGAAGTATTACTTGGTGTATTATCGA 720
     euAlaSerProValMetIleValLeuLeuLeuSerGluValLeuGlyValLeuSerA

721  GATTGCTCCGCAGATGAATGCTTTTTCCGTATCATTACTATTAAAAGTTTACTTGCAATA 780
     rgLeuLeuArgArg *
```

Fig. 2. Nucleotide sequence of the 627 bp open reading frame. The DNA and translated amino acid sequences of the putative gene in the 2.5 kb *Hind*III fragment from enteroinvasive *E. coli* are shown. The two base-pair differences in the nucleotide sequence between *E. coli* and *S. flexneri* are described in the text. The *Hind*III site and ribosome binding site (rbs) are shown.

scribed, total RNA was isolated from *S. flexneri* strains 270 and 354. The former is Sereny-positive and carries the large plasmid, whereas the latter is avirulent. RNAs were separated on agarose/formaldehyde gels and transferred to nitrocellulose filter paper. The RNAs were hybridized to a ^{32}P-labeled 2.5 kb *Hind*III fragment. Two transcripts were detected in strain 270; the sizes were estimated to be 700 and 500 bases. In contrast, RNA from strain 354, an avirulent strain lacking the large plasmid, did not hybridize to the 2.5 kb *Hind*III fragment.

The sequence data were used to synthesize several oligodeoxyribonucleotides from the 627 bp open reading frame in this region. These were used as probes against a number of gram-negative and gram-positive organisms. As shown in Table 1, the gene probe was specific for enteroinvasive enteric bacteria.

Table 1. Bacteria tested by hybridization with *Shigella* probe.[a]

Bacteria tested	No. of strains tested	Results of hybridization[b]	Presence of Plasmid[c]
Shigella flexneri	4	+	+
S. sonnei	1	+	+
S. dysenteriae	2	+	+
S. boydii	1	+	+
S. flexneri 354	1	-	-
S. sonnei	2	-	-
S. boydii	1	-	-
Escherichia coli	6	+	+
E. coli	3	-	-
Other gram-negative bacteria			
Enterobacter cloacae	1	-	NT[d]
Klebsiella pneumoniae	1	-	NT
Proteus vulgaris	1	-	NT
Salmonella typhimurium	1	-	NT
Serratia marcescens	1	-	NT
Yersinia enterocolitica	3	-	NT
Gram-positive bacteria			
Staphylococcus aureus	1	-	NT
S. epidermidis	1	-	NT
Streptococcus pyogenes	1	-	NT
Streptococcus faecalis	1	-	NT
Listeria monocytogenes	1	-	NT
L. ivanovii	1	-	NT
L. seeligeri	1	-	NT

[a] Bacteria were hybridized against the synthetic oligodeoxyribonucleotide derived from the 627 bp open reading frame.

[b] Colonies that hybridized with the *Shigella* probe are indicated with +; no hybridization is indicated with -.

[c] Plasmid presence determined by Southern hybridizations of isolated DNA separated on agarose gels.

[d] NT, not tested.

In order to assess the efficacy of these gene probes in detecting pathogenic invasive bacteria, six foods were seeded with *S. flexneri* and the homogenates spread on MacConkey agar plates. The bacterial growth was transferred to Whatman 541 filter paper, lysed, fixed and hybridized with one
oligodeoxyribonucleotide from within the open reading frame (5'-TAATACTCCTGA-
ACGGCG-3'). The amount of microbial flora varied in each food. In one
instance, the high microbial background in alfalfa sprouts interfered with
our ability to detect *S. flexneri* (Table 2).

The 140 MDa plasmid from *S. flexneri* was isolated, digested with *Hind*-
III, *Eco*RI and *Bam*HI, fragments were separated by agarose gel electrophoresis,
transferred to nitrocellulose filter paper and hybridized with the 2.5 kb
*Hind*III fragment. Three fragments (2.5 kb *Hind*III, 17 kb *Eco*RI and 20 kb
*Bam*HI) were seen from these Southern blots. The 2.5 kb *Hind*III fragment from
S. flexneri was cloned into pUC19 and sequenced. The nucleotide sequence of
the 627 bp open reading frame was nearly identical to that from enteroinvasive
E. coli; two base-pair differences were noted at positions 536 (A-T) and 609
(G-A).

CONCLUSIONS

Methods used to detect enteroinvasive *E. coli* and *Shigella* spp. in
foods must be refined and improved. Although we have designed a gene probe
to identify these organisms, it is quite apparent that a more selective
medium is required to reduce the background microbial flora of foods.

ACKNOWLEDGEMENTS

The authors thank Drs. Walter E. Hill and Thomas A. Cebula for their
critical reading of this manuscript. The oligodeoxyribonucleotides were
kindly provided by Dr. Mary Trucksess.

Table 2. Foods seeded with *Shigella flexneri*.

Food	Bacterial Background[a]	No. of *Shigella* seeded	Per cent recovered[b]
Scallops	1.1×10^4	97	53
Pasta	1.3×10^8	97	71
Alfalfa sprouts	1.4×10^8	97	0
Pecans	8.6×10^3	97	57
Chocolate	4.8×10^3	97	63
Crabmeat	7.0×10^6	82	49

[a] The background bacterial count was determined by plating dilutions
of the food onto Trypticase soy agar plates and incubating the
plates at 37°C.

[b] The percentage of *Shigella* cells recovered from each food was
determined by autoradiography of filters containing cultures from
each food hybridized with the *Shigella* probe as compared with the
total plate count.

REFERENCES

1. Buysee JM, Stover CK, Oaks EV, Venkatesan M, Kopecko DJ, 1987, Molecular cloning of invasion plasmid antigen (*ipa*) genes from *Shigella flexneri*: analysis of *ipa* gene products and genetic mapping, *J Bacteriol.*, 169:2561-2569.
2. Clerc P, Baudry B, Sansonetti PJ, 1986, Plasmid-mediated contact haemolytic activity in *Shigella* species: correlation with penetration into HeLa cells, *Ann Inst Pasteur Microbiol.*, 137A:267-278.
3. Gemski P, Sheahan DG, Washington O, Formal SB, 1972, Virulence of *Shigella flexneri* hybrids expressing *Escherichia coli* somatic antigens, *Infect Immun.*, 6:104-111.
4. Griffiths E, Stevenson P, Hale TL, Formal SB, 1985, Synthesis of aerobactin and a 76,000-dalton iron-regulated outer membrane protein by *Escherichia coli* K-12 *Shigella flexneri* hybrids and by entero-invasive strains of *Escherichia coli*, *Infect Immun.*, 49:67-71.
5. Hale TL, 1988, Organization and expression of plasmid virulence genes in *Shigella flexneri*, *In*: "UCLA Symposium on Molecular and Cellular Biology New Series, Bacteria-Host Cell Interaction," Horowitz MA, ed., Alan R. Liss, NY, 64:253-265.
6. Hale TL, Formal SB, 1988, Virulence Mechanisms of Enteroinvasive Pathogens, *In*: "Virulence Mechanisms of Bacterial Pathogens", Roth JA, ed., American Society for Microbiology, Washington, D.C.
7. Hale TL, Oaks EV, Formal SB, 1985, Identification and antigenic characterization of virulence-associated, plasmid-encoded proteins of *Shigella* spp. and enteroinvasive *Escherichia coli*, *Infect Immun.*, 50:620-629.
8. Kopecko DJ, Washington O, Formal SB, 1980, Genetic and physical evidence for plasmid control of *Shigella sonnei* Form I cell surface antigen, *Infect Immun.*, 29:207-214.
9. LaBrec EH, Schneider H, Magnani TJ, Formal SB, 1964, Epithelial cell penetration as an essential step in the pathogenesis of bacillary dysentery, *J Bacteriol.*, 88:1503-1518.
10. Lett M-C, Sasakawa C, Okada N, Sakai T, Makino S, Masatoshi Y, Komatsu K, Yoshikawa M, 1989, *virG*, a plasmid-coded virulence gene of *Shigella flexneri*: Identification of the *virG* protein and determina-tion of the complete coding sequence, *J Bacteriol.*, 171:353-359.
11. Maurelli AT, Blackman B, Curtiss R III, 1984, Temperature-dependent expression of virulence genes in *Shigella* species, *Infect Immun.*, 43:195-201.
12. Maurelli AT, Sansonetti PJ, 1988, Identification of a chromosomal gene controlling temperature-regulated expression of *Shigella* virulence, *Proc Natl Acad Sci USA.*, 85:2820-2824.
13. Oaks EV, Wingfield ME, Formal SB, 1985, Plaque formation by virulent *Shigella flexneri*, *Infect Immun.*, 48:124-129.
14. Oaks EV, Hale TL, Formal SB, 1986, Serum immune response to *Shigella* protein antigens in rhesus monkeys and humans infected with *Shigella* spp., *Infect Immun.*, 53:57-63.
15. O'Brien AD, Holmes RK, 1987, Shiga and Shiga-like toxins, *Microbiol Rev.*, 51:206-220.
16. O'Brien AD, Tompson MR, Gemski P, Doctor BP, Formal SB, 1977, Bio-logical properties of *Shigella flexneri* 2a toxin and its serological relationship to *Shigella dysenteriae* 1 toxin, *Infect Immun.*, 15:796-798.
17. Sanger F, Nicklen S, Coulson AR, 1977, DNA sequencing with chain terminating inhibitors, *Proc Natl Acad Sci USA.*, 74:5463-5467.
18. Sansonetti PJ, Kopecko DJ, Formal SB, 1982, *Shigella sonnei* plasmids: evidence that a large plasmid is necessary for virulence, *Infect Immun.*, 34:852-860.
19. Sansonetti PJ, Ryter A, Clerc P, Maurelli AT, Mounier J, 1986, Multi-plication of *Shigella flexneri* within HeLa cells: lysis of the phago-

cytic vacuole and plasmid-mediated contact hemolysis, *Infect Immun.*,
51:461-469.

20. Sasakawa C, Kamata K, Sakai T, Makino S, Yamada M, Okada N, Yoshikawa
M, 1988, Virulence-associated genetic regions comprising 31 Kilobases
of the 230-kilobase plasmid in *Shigella flexneri, J Bacteriol.*, 170:
2480-2484.

21. Sasakawa C, Makino S, Kamata K, Yoshikawa M, 1986, Isolation, character-
zation, and mapping of Tn5 insertions into the 140-megadalton plasmid
associated with loss of virulence and Congo red binding ability in
Shigella flexneri 2a, *Infect Immun.*, 51:470-475.

22. Sereny B, 1955, Experimental *Shigella* keratoconjunctivitis, *Acta Microbiol Acad Sci Hung.*, 2:293-296.

23. Small PC, Falkow S, 1986, Development of a DNA probe for the virulence
plasmid of *Shigella* spp. and enteroinvasive *Escherichia coli, In:*
"Microbiology," Leive L, Bonventre PF, Morello JA, Silver SD, Wu WC,
ed., American Society for Microbiology, Washington, D.C.

24. Smith JL, *Shigella* as a food pathogen, 1987, *J Food Prot.*, 50:788-801.

25. Timmis KN, Clayton CL, Sekizaki T, 1985, Localization of Shiga toxin
gene in the region of *Shigella dysenteriae* 1 specifying virulence
functions, *FEMS Microbiol Lett.*, 30:301-305.

26. Venkatesan MM, Buysee JM, Kopecko DJ, 1988, Characterization of
invasion plasmid antigen genes (ipaBCD) from *Shigella flexneri, Proc
Natl Acad Sci USA.*, 85:9317-9321.

27. Watanabe H, Nakamura A, 1985, Large plasmids associated with virulence
in *Shigella* species have a common function necessary for epithelial
cell penetration, *Infect Immun.*, 48:260-262.

28. Watanabe H, Nakamura A, 1986, Identification of *Shigella sonnei* Form I
plasmid genes necessary for cell invasion and their conservation among
Shigella species and enteroinvasive *Escherichia coli, Infect Immun.*,
53:352-358.

29. Watanabe H, Timmis K, 1984, A small plasmid in *Shigella dysenteriae* 1
specifies one or more functions essential for O antigen production and
bacterial virulence, *Infect Immun.*, 43:391-396.

30. Wood PK, Morris JG, Small PLC, Sethabutr O, Toledo MRF, Trabulsi L,
Kaper JB, 1986, Comparison of DNA probes and the Sereny test for
identification of invasive *Shigella* and *Escherichia coli* strains, *J
Clin Microbiol.*, 24:498-500.

USEFULNESS IN THE EPIDEMIOLOGY OF FOOD POISONING CASES OF DETECTION OF
SPECIFIC RESTRICTION ENDONUCLEASES IN SOME SEROTYPES OF *SALMONELLA* AND
YERSINIA

Katsutoshi Mise,[a] Michiko Miyahara,[a] Tsutomu
Maruyama,[b] Yasuo Kudoh[b] and Makoto Ohashi[b]

National Institute of Hygienic Sciences[a]
18-1, Kamiyoga 1-chome
Setagaya-ku, Tokyo 158, Japan

and

Tokyo Metropolitan Research Laboratory of Public Health[b]
24-1, Hyakunincho 3-chome
Shinjuku-ku, Tokyo 160, Japan

ABSTRACT

Restriction endonucleases have been employed as an extremely important tool in recombinant DNA technology. To date, the occurrence of more
than 100 restriction endonucleases with different specificities has been
reported.[2,8] Restriction endonucleases have been screened in this laboratory for pathogenic bacteria belonging to the *Enterobacteriaceae* in the hope
that: (i) the detection of specific restriction endonuclease is found at a
high frequency in the species; and (ii) new restriction endonucleases with
novel specificities might be found in pathogenic bacteria, since screening
of restriction endonucleases have been rarely carried out. Here, we report
that four clinically important food-poisoning bacteria produce specific
restriction endonucleases at high frequencies. Some of these are expected
to be useful for recombinant DNA technology after cloning of their gene into
Escherichia coli K-12.

Several satisfactory procedures are now available for the detection and
isolation of restriction endonucleases. In most cases, bacterial cells are
first disrupted with a sonicator or French press, and the resulting extract
is used as a source of restriction endonucleases.[7,9] However, in order to
avoid aerosol formation, a sonicator or French press should not be employed
when pathogenic bacteria are tested for the occurrence of restriction endonucleases. Fortunately, most restriction endonucleases are periplasmic,[1]
and can be obtained by gentle lysis of the cell wall with lysozyme. A safe
method using lysozyme for the detection of restriction endonucleases in
pathogenic bacteria has been devised in this laboratory.[3,5] A loopful
of bacteria grown on enrichment agar overnight was suspended in Tris-HCl
buffer (pH 7.5) containing 100 μg/ml lysozyme, 20% sucrose and 3 mM EDTA,
and stored at 0°C for 1-20 hours. Two drops of toluene were then added to

the suspension to kill residual bacteria. The mixture was centrifuged and the supernatant was tested for restriction endonuclease activity in TA buffer[6] supplemented with 0.08 μg lambda DNA as a substrate and 0.01 mM tRNA as an inhibitor of dnase I. The occurrence of restrictin fragments was tested by agarose gel electrophoresis.

Using the lysozyme lysis method, a cold active restriction endonuclease designated YenI was detected in a crude extract of *Yersinia enterocolitica* serotype 08 A2635.[4] Comparison of the cleavage patterns of lambda DNA produced by YenI and those of known endonucleases indicated that YenI is an isoschizomer of *PstI* endonuclease whose recognition sequence is CTGCAG.[10] The isoschizomer of YenI or *PstI* was found in 12 of 14 strains of *Y. enterocolitica* serotype 08 of different origins, but was not detected in any of 41 *Y. enterocolitica* strains belonging to other serotypes or in 33 *Y. pseudotuberculosis* or 3 *Y. pestis* strains tested (Table 1 and ref. 4). As the isoschizomer was not detected in any strain of serotype 07,08 or 08,019, the detection of the isoschizomer might result in more rapid determination of *Y. enterocolitica* 08, a highly pathogenic serotype.

The finding of the widespread occurrence of specific restriction endonuclease in *Y. enterocolitica* 08 led us to investigate the occurrence of restriction endonucleases in various strains of *Salmonella* belonging to clinically important serotypes. Specific restriction endonucleases with high activity were detected at a high frequency in *S. thompson*, *S. infantis* and *S. blockley* as described in Table 2. On the other hand, restriction endonuclease-positive strains were not detected at all, or they were detected at a low frequency in other clinically important serotypes (e.g., *S. typhi*, *S. typhimurium*). Among the three restriction endonucleases in Table 2, *S. thompson* endonuclease, an isoschizomer of *KpnI*, seems to be interesting and useful, since this endonuclease produces 5'-protruding ends unlike *KpnI*.[5,11] The recognition sequences of all *Salmonella* restriction endonucleases are similar to each other, suggesting that they might have originated from a common ancestor.

Table 1. The widespread occurrence of restriction endonuclease-producing strains in *Yersinia enterocolitica* 08.[a]

Yersinia strains (serotype)	Number of strains tested	Number of restriction endonuclease-producing strains
Yersinia enterocolitica 03	17	0
Yersinia enterocolitica 05 and 05,027	6	0
Yersinia enterocolitica 08	14	12 (isoschizomer of *PstI*)
Yersinia enterocolitica 09	8	0
Yersinia enterocolitica other serotypes (01,02,03; 06; 07,08; 08,019)	10	0
Yersinia pseudotuberculosis (13 different serotypes included)	33	0
Yersinia pestis	3	0

[a] Adapted from Miyahara et al.[4]

In summary, the detection of specific restriction endonucleases are useful in the epidemiology of some food poisoning episodes traced to *Y. enterocolitica* 08 and *S. infantis*. Moreover, new restriction endonucleases useful for recombinant DNA technology might be found in *Salmonella*, since stable restriction endonucleases with high activities have been detected at relatively high frequencies.

Table 2. The high incidence of restriction endonuclease-producing strains in *Salmonella thompson, Salmonella infantis* and *Salmonella blockley* isolated from humans in Japan, 1970 - 1986.

Serotypes	Number of strains tested	Number of restriction endonuclease-producing strains (%)	Specificity of restriction endonuclease (recognition sequence)[a]
Salmonella thompson	129	39 (30%)	Isoschizomer of *Kpn*I (GGTACC)
Salmonella infantis	88	77 (88%)	Isoschizomer of *Ava*II (GGWCC)
Salmonella blockley	8	3 (38%)	Isoschizomer of *Sty*I (CCWWGG)

[a] W, A or T

REFERENCES

1. Greene PJ, Betlach MC, Goodman HM, Boyer HW, 1974, The *Eco*RI restriction endonuclease, *Methods Mol Biol.*, 7:87-111.
2. Kessler C, Holtke HJ, 1986, Specificity of restriction endonucleases and methylases - a review (Edition 2), *Gene*, 47:1-153.
3. Mise K, Nakajima K, Terakado N, Ishidate M Jr, 1986, Production of restriction endonucleases using multicopy Hsd plasmids occurring naturally in pathogenic *Escherichia coli* and *Shigella boydii*, *Gene*, 44:165-169.
4. Miyahara M, Maruyama T, Wake A, Mise K, 1988, Widespread occurrence of the restriction endonuclease *Yen*I, an isoschizomer of *Pst*I, in *Yersinia enterocolitica* serotype 08, *Appl Environ Microbiol.*, 54:577-580.
5. Miyahara M, Mise K, 1988, Rapid method for detection of restriction endonuclease-producing strains in enteropathogenic bacteria, *Analytica Chimica Acta*, in press.
6. O'Farrell PH, Kutter E, Nakanishi M, 1980, A restriction map of the bacteriophage T4 genome, *Mol Gen Genet.*, 179:421-435.
7. Pirrotta V, Bickle TA, 1980, General purification schemes for restriction endonucleases, *Methods Enzymol.*, 65:89-95.
8. Roberts RJ, 1988, Restriction enzymes and their isoschizomers, *Nucl Acids Res.*, 16:r271-r313.
9. Schleif R, 1980, Assaying organisms for the presence of restriction endonucleases, *Methods Enzymol.*, 65:19-23.
10. Smith DL, Blattner FR, Davies J, 1976, The isolation and partial characterization of a new restriction endonuclease from *Providencia stuartii*, *Nucl Acids Res.*, 3:343-353.
11. Tomassini J, Roychoudhurry R, Wu R, Roberts RJ, 1978, Recognition sequence of restriction endonuclease *Kpn*I from *Klebsiella pneumoniae*, *Nucl Acids Res.*, 5:4055-4064.

GENETICS OF THERMOSTABLE DIRECT HEMOLYSIN OF *VIBRIO PARAHAEMOLYTICUS*

Mitsuaki Nishibuchi[1] and James B. Kaper[2]

Department of Microbiology
Faculty of Medicine
Kyoto University
Sakyo-ku, Kyoto 606, Japan[1]

and

Center for Vaccine Development
Division of Geographic Medicine
Department of Medicine
University of Maryland School of Medicine
10 South Pine Street
Baltimore, Maryland 21201[2]

INTRODUCTION

Vibrio parahaemolyticus causes seafood-borne gastroenteritis and is responsible for about half of bacterial food poisoning cases in Japan.[1] It also has been recognized as an important agent of gastroenteritis in other parts of the world.[2-4] Although many possible virulence factors have been suggested to explain the pathogenic mechanism of the organism, the thermostable direct hemolysin (TDH) has been considered the major virulence factor based on epidemiological findings.[5-6] TDH is toxic to various cultured cells and lethal to small experimental animals, and is a potent cardiotoxin, but its enterotoxic mechanism is not understood.[7]

Kanagawa phenomenon and the gene encoding TDH (tdh gene)

Almost all *V. parahaemolyticus* strains isolated from clinical sources produce TDH extracellularly and induce a clear zone of beta-type hemolysis around the colony when grown on a special blood agar, Wagatsuma agar. This phenomenon is called the Kanagawa phenomenon (KP[+]). A majority of isolates from the environment e.g., sea water and seafoods not implicated in gastroenteritis cases do not exhibit the Kanagawa phenomenon (KP[-]).[5,6]

The gene encoding TDH (*tdh* gene) was cloned from a KP[+] strain and TDH was produced in *Escherichia coli*[8] and the nucleotide sequence of the gene was determined.[9] The *tdh* gene encoded a polypeptide composed of 189 amino acid residues (Fig. 1). There existed a putative signal peptide composed of 24 amino acid residues. Molecular weight of TDH calculated from the deduced amino acid sequence of the mature protein was 18,496. These results indicated that the cloned gene encoded the subunit of TDH and supported the hypothesis that TDH is composed of two identical subunits.[10]

Microbial Toxins in Foods and Feeds, Edited by A.E. Pohland *et al.,*
Plenum Press, New York, 1990

A 415 bp DNA fragment derived from the internal portion of the *tdh* structural gene was employed to prepare the *tdh* gene probe.[11] The results of the DNA colony hybridization test with the *tdh* gene probe showed that all KP^+ strains carried the *tdh* gene while the majority of KP^- strains did not have the *tdh* gene (Table 1). Southern blot analysis of the representative *tdh*-gene positive strains revealed that KP^+ strains had two gene copies. KP^- strains that carry the *tdh* gene had only one gene copy. An exceptional KP^- strain had one chromosomal gene copy and another gene copy on a plasmid. The nucleotide sequence of the four representative gene copies were compared. The sequences of the coding regions were not completely identical and, thus, amino acid sequences of TDH encoded by the gene copies were slightly different (Table 2).

The two gene copies cloned from a KP^+ strain, WP1, (*tdh*1 and *tdh*2 in Table 2) encoded TDH with slightly different amino acid sequences. The amino acid sequence of TDH encoded by the *tdh*2 gene was more similar than the *tdh*1 gene to that determined by Edman degredation of TDH purified from the spent culture supernatant of strain WP1.[12] Out of 165 amino acid residues, TDH encoded by the *tdh*1 gene and *tdh*2 gene had 9 and 2 mismatches, respectively, with the TDH primary structure determined by protein sequencing. A *tdh*1-deficient mutant was derived from a KP^+ *V. parahaemolyticus* strain by *in vivo* recombination using the *tdh*1 gene mutated *in vitro*; the mutant strain remained KP^+. These results suggest that the *tdh*2 gene, rather than *tdh*1 gene, is contributing to the positive phenotype (KP^+). When the *tdh* genes of a KP^- strain, AQ3776, (*tdh*3 and *tdh*4 in Table 2) were cloned into *E. coli*, TDH produced in the *E. coli* cells was hemolytic to erythrocytes, indicating that the negative phenotype (KP^-) may not be simply due to mutations in the coding region of the *tdh* genes.

The gene encoding a hemolysin related to TDH

Few clinical isolates of *V. parahaemolyticus* are KP^-. We showed that 68% of such strains do not carry the *tdh* gene.[11] Recently, one such strain

```
  1  AA GCA TTT TTC ACA ATT ATT CAG TTT GCT TCT TTG GTT TTT TAG TTT TCA TAA CAT CCG TCA TTC TGG CAA   71
                                    -35                              -10
 72  AGT TAT TAA TCA ATT CAT GGG TTT TTT ATG AAA CAC CAA TAT TTT GCA AAA AAA TCA TTT TTA TTT ATA TCC  143
                          SD
  1                                  Met Lys His Gln Tyr Phe Ala Lys Lys Ser Phe Leu Phe Ile Ser   15

144  ATG TTG GCT GCA TTC AAA ACA TCT GCT TTT GAG CTT CCA TCT GTC CCT TTT CCT GCC CCC GGT TCT GAT GAG  215
 16  Met Leu Ala Ala Phe Lys Thr Ser Ala Phe Glu Leu Pro Ser Val Pro Phe Pro Ala Pro Gly Ser Asp Glu   39

216  ATA TTG TTT GTT GTT CGA GAT ACA ACT TTT AAT ACC CAA GCT CCG GTC AAT GTA AAG GTC TCT GAC TTT TGG  287
 40  Ile Leu Phe Val Val Arg Asp Thr Thr Phe Asn Thr Gln Ala Pro Val Asn Val Lys Val Ser Asp Phe Trp   63

288  ACA AAC CGT AAT GTA AAA AGA AAA CCG TAC GAA GAT GTT TAT GGT CAA TCA GTA TTC ACA ACG TCA GGT ACT  359
 64  Thr Asn Arg Asn Val Lys Arg Lys Pro Tyr Glu Asp Val Tyr Gly Gln Ser Val Phe Thr Thr Ser Gly Thr   87

360  AAA TGG TTG ACA TCC TAC ATG ACT GTG AAC ATT AAT GAT AAA GAC TAT ACA ATG GCA GCG GTG TCT GGC TAT  431
 88  Lys Trp Leu Thr Ser Tyr Met Thr Val Asn Ile Asn Asp Lys Asp Tyr Thr Met Ala Ala Val Ser Gly Tyr  111

432  AAG AGC GGT CAT TCT GCT GTG TTC GTA AAA TCA GGT CAA GTA CAA CTT CAA CAT TCC TAT AAT TCT GTA GCT  503
112  Lys Ser Gly His Ser Ala Val Phe Val Lys Ser Gly Gln Val Gln Leu Gln His Ser Tyr Asn Ser Val Ala  135

504  AAC TTT GTT GGT GAA GAT GAA GGT TCT ATT CCA AGT AAA ATG TAT TTG GAT GAA ACT CCA GAA TAT TTT GTT  575
136  Asn Phe Val Gly Glu Asp Glu Gly Ser Ile Pro Ser Lys Met Tyr Leu Asp Glu Thr Pro Glu Tyr Phe Val  159

576  AAT GTA GAA GCA TAT GAG AGT GGT AGT GGT AAT ATA TTG GTA ATG TGT ATA TCC AAC AAA GAA TCG TTT TTT  647
160  Asn Val Glu Ala Tyr Glu Ser Gly Ser Gly Asn Ile Leu Val Met Cys Ile Ser Asn Lys Glu Ser Phe Phe  183

648  GAA TGT AAA CAT CAA CAA TAA AAA TAA AAA GCC ACA GAA TAT TCT GTG GCT TTG AAA ATA ATT AAT AAT AAT  719
184  Glu Cys Lys His Gln Gln ***                                                                      190

720  TTT CTG ATT TA                                                                                    730
```

Fig. 1. Nucleotide sequence of the *tdh* gene cloned from strain WP1 of *Vibrio parahaemolyticus*. This *tdh* gene corresponds to *tdh*1 gene in Table 2. Amino acid sequence of TDH deduced from the nucleotide sequence is also shown. Underlined amino acid sequence indicates a presumed signal peptide. Nucleotide sequences homologous to the consensus sequences of *Escherichia coli* such as -35 region (-35), -10 region (-10) of the promoter, and Shine-Dalgarno sequence (SD) are designated by the dots. ***, termination code. <- ->, palindromic sequence.

Table 1.   Results of the DNA Colony Hybridization Test with the *tdh*
           Gene Probe.

tdh gene[a]	Number of *Vibrio parahaemolyticus* strains[b]		
	KP[+]	KP[+W]	KP[-]
+	66	12	10
-	0	2	51

[a] +, positive; -, negative
[b] , Kanagawa phenomenon. KP[+], KP positive; KP[+W], KP weakly positive;
    KP[-], KP negative.

Table 2.   Comparison of the Coding Regions of the *tdh* Genes Cloned
           from *Vibrio parahaemolyticus* Strains WP1 and AQ3776.

Gene designation	Origin	Percent homology	
		Nucleotide sequence (567 bp)	Amino acid sequence (189 residues)
*tdh*1	Chromosome of WP1	100	100
*tdh*2	Chromosome of WP1	97.2	94.7
*tdh*3	Chromosome of AQ3776	96.7	93.7
*tdh*4	Plasmid of AQ3776	96.7	94.2

Strain WP1 is Kanagawa-phenomenon positive and strain AQ3776 is Kanagawa-
phenomenon negative.

was found to produce a hemolysin related to TDH (TRH). The hemolysin
appeared to have a similar molecular structure to TDH and was immunologically
related to TDH, but it manifested significantly different physicochemical
characteristics than TDH.[13] Hybridization with the *tdh* gene probe under
reduced stringency allowed detection of the gene encoding TRH (*trh* gene).
Analysis of the nucleotide sequence of the *trh* gene indicated that the *tdh*
gene and the *trh* gene originated from a common ancestor and evolved by base
changes, retaining 68-69% homology in the coding region. Hydrophobicity
profiles of TDH and TRH predicted from the deduced amino-acid sequences were
very analogous.[14]

    A *trh* gene probe was derived from internal DNA sequences of the *trh*
gene and used as the hybridization probe in the colony hybridization test.
Sixty-seven percent of the *tdh*-gene negative, clinical strains of *V. parahaemolyticus* were *trh*-gene positive and 11% of clinical strains had both the
tdh and *trh* genes. The results suggest that TRH, like TDH, may be an
important virulence factor of *V. parahaemolyticus*.

tdh gene in diarrheagenic vibrios other than V. parahaemolyticus

    The DNA colony hybridization test with the *tdh* gene probe disclosed
that some clinical strains of *Vibrio mimicus*, *Vibrio cholerae* non-01, and
all strains of *Vibrio hollisae* carry the nucleotide sequences homologous to

the *tdh* gene. In fact, some strains of these species were shown to produce TDH or related hemolysins.[15-17] The genes encoding these hemolysins were cloned and analyzed. The genes had very similar nucleotide sequences to those of the *tdh* genes of *V. parahaemolyticus*, e.g., 93.1~96.7% DNA homology with *tdh*1 gene, and it was considered that these genes evolved from a common ancestor by base substitutions.

SUMMARY

Most clinical isolates of *V. parahaemolyticus* had the *tdh* gene and/or the *trh* gene. The *tdh* gene was also found in pathogenic *Vibrio* species other than *V. parahaemolyticus*. These genes, sharing more than 69% nucleotide sequence homology, appear to have evolved from a common ancestor by base substitutions. The hemolysins encoded by this group of the genes, although their physicochemical properties were not always identical, had similar molecular structures and appear to be closely related to pathogenicity.

ACKNOWLEDGEMENTS

This work was supported by a Grant-in-aid for Scientific Research from the Ministry of Education, Science and Culture of Japan and by grant AI19165 from the National Institute of Allergy and Infectious Diseases.

REFERENCES

1. Fujino T, Sakaguchi G, Sakazaki R, Takeda Y, (ed.), 1974, "International symposium on *Vibrio parahaemolyticus*," Saikon Publishing Co., Tokyo.
2. Blake PA, Weaver RE, Hollis DG, 1980, Disease of humans (other than cholera) caused by vibrios, *Ann Rev Microbiol.*, 34:341-367.
3. Morris JG Jr, Black RE, 1985, Cholera and other vibrioses in the United States, *N Engl J Med.*, 312:343-350.
4. Janda JM, Powers C, Bryant RG, Abbott SL, 1988, Current perspectives on the epidemiology and pathogenesis of clinically significant *Vibrio* spp., *Clin Microbiol Rev.*, 1:245-267.
5. Sakazaki R, Tamura K, Kato T, Obara Y, Yamai S, Hobo K, 1968, Studies on the enteropathogenic, facultatively halophilic bacteria, *Vibrio parahaemolyticus*. III. Enteropathogenicity, *Jpn J Med Sci Biol.*, 21: 325-331.
6. Miyamoto Y, Kato T, Obara Y, Akiyama S, Takizawa K, Yamai S, 1969, *In vitro* hemolytic characteristics of *Vibrio parahaemolyticus*: its close correlation with human pathogenicity, *J Bacteriol.*, 100:1147-1149.
7. Takeda Y, 1983, Thermostable direct hemolysin of *Vibrio parahaemolyticus*, *Pharmacol Ther.*, 19:123-146.
8. Kaper JB, Campen RK, Seidler RJ, Baldini MM, Falkow S, 1984, Cloning of the thermostable direct or Kanagawa phenomenon associated hemolysin of *Vibrio parahaemolyticus*, *Infect Immun.*, 45:290-292.
9. Nishibuchi M, Kaper JB, 1985, Nucleotide sequence of the thermostable direct hemolysin gene of *Vibrio parahaemolyticus*, *J Bacteriol.*, 162: 558-564.
10. Takeda Y, Taga S, Miwatani T, 1978, Evidence that thermostable direct hemolysin of *Vibrio parahaemolyticus* is composed of two subunits, *FEMS Microbiol Lett.*, 4:271-274.
11. Nishibuchi M, Ishibashi M, Takeda Y, Kaper JB, 1985, Detection of the thermostable direct hemolysin gene and related DNA sequences in *Vibrio parahaemolyticus* and other *Vibrio* species by the DNA colony hybridization test, *Infect Immun.*, 49:481-486.

12. Tsunasawa, S, Sugihara A, Masaki T, Sakiyama F, Takeda Y, Miwatani T, Narita K, 1987, Amino acid sequence of thermostable direct hemolysin produced by *Vibrio parahaemolyticus*, *J Biochem.*, 101:111-121.
13. Honda T, Ni Y, Miwatani T, 1988, Purification and characterization of a hemolysin producer by a clinical isolate of Kanagawa phenomenon-negative *Vibrio parahaemolyticus* and related to the thermostable direct hemolysin, *Infect Immun.*, 56:961-965.
14. Nishibuchi M, Taniguchi T, Misawa T, Khaeomanee-iam V, Honda T, Miwatani T, Cloning and the nucleotide sequence of the gene encoding the hemolysin related to the thermostable direct hemolysin (*trh* gene) of *Vibrio parahaemolyticus*, submitted for publication.
15. Yoh M, Honda T, Miwatani, T, 1985, Production by non-01 *Vibrio cholerae* of hemolysin related to thermostable direct hemolysin of *Vibrio parahaemolyticus*, *FEMS Microbiol Lett.*, 29:197-200.
16. Yoh M, Honda T, Miwatani T, 1986, Purification and partial characterization of a non-01 *Vibrio cholerae* hemolysin that cross-reacts with thermostable direct hemolysin of *Vibrio parahaemolyticus*, *Infect Immun.*, 52:319-322.
17. Yoh M, Honda T, Miwatani T, 1986, Purification and partial characterization of a *Vibrio hollisae* hemolysin that relates to the thermostable direct hemolysin of *Vibrio parahaemolyticus*, *Can J Microbiol.*, 32: 632-636.

GENETIC ASPECTS OF *CLOSTRIDIUM BOTULINUM*

K. Oguma,[1] M. Eklund,[2] N. Fujii,[1] F. Poysky,[2] K. Kimura,[1]
N. Yokosawa,[1] and K. Tsuzuki[1]

Department of Microbiology[1]
Sapporo Medical College
South 1, West 17
Sapporo, Japan

Northwest Fisheries Center[2]
Utilization Research Division
U.S. Department of Commerce
2725 Montlake Boulevard East
Seattle, Washington 98112

ABSTRACT

The relationship of bacteriophages and plasmids to the production of
neurotoxins was studied in strains of *Clostridium botulinum* types A through
G. Neurotoxins C_1 and D produced by types C and D, respectively, were shown
to be mediated by specific bacteriophages. Evidence is presented that
strongly suggests that both neurotoxin and bacteriocin production by type G
are in some manner related to a 81-MDa plasmid carried by toxigenic strains.

Antigenicity and host range of four type C and three type D converting
phages were studied. The phages were classified into three groups based on
their antigenicity and host range: group 1 consisted of c-st and c-468
phages; group 2 was c-203, c-d6f, and d-1873: and group 3 was d-sa and
d-4947.

Nucleic acids were extracted from groups 1 and 2 phages, and nonconverting mutant phage (c)-n71 which was obtained from C-Stockholm strain as
well as c-st phage. The susceptibility of phage DNAs to different types of
nucleases was observed. It was concluded that the nucleic acids of all six
phages were double-stranded DNA. The length of c-st, (c)-n71, c-468, and
c-d6f phage DNAs was about 110 kilobase pairs and that of c-203 and d-1873
was 150 kilobase pairs. PstI digested the DNAs from two group 1 phages and
(c)-n71 phage with very similar patterns, but did not digest the DNAs from
group 2 phages. On the contrary, Sau3A digested only the DNAs from group 2
phages though the similarity of digestion patterns was low.

The existence of the structure genes for the toxin in these five converting phages belonging to groups 1 and 2 and (c)n-71 was confirmed by the
hybridization test with phage DNAs and the oligonucleotide probe which
represented the DNA sequence predicted for the N-terminal amino acids (2 to
17) of *C. botulinum* type C toxin. The loss of the converting ability of

(c)-n71 phage may be caused not by the deletion of tox[+] gene but rather by the base mutation in c-st phage DNA.

INTRODUCTION

Clostridium botulinum is divided into seven types (A to G) based upon the toxins produced. The strains of these different types of *C. botulinum* can be classified further into four groups based on their biochemical and physiological characteristics, and their DNA and RNA homologies. Group 1 cultures are proteolytic and include A, B, and F; group 2 are nonproteolytic and include types B, E, and F; group 3 are nonproteolytic types C and D; and group 4 is weakly proteolytic type G. The properties of the third group of *C. botulinum* and *C. novyi* type A are very similar and differ only in the type of toxins produced.

In the current study, the relationship between plasmids and phages to toxigenicity of *C. botulinum* was investigated.

RESULTS AND DISCUSSION

Isolation of plasmids

Toxigenic *C. botulinum* and nontoxigenic clostridia resembling *C. botulinum* were screened for plasmids. Both nontoxigenic and toxigenic strains harbored plasmids ranging in mass from 2.1 to 81 MDa. All of the proteolytic type F strains tested were found to carry a single 11.5 MDa plasmid and all of the type G harbored an 81 MDa plasmid.[8]

Plasmids and the toxigenicity of C. botulinum type G

Both toxigenic and nontoxigenic derivatives were isolated from six different type G strains after sequential transfer of the cultures at elevated temperatures of incubation.

All of the 78 toxigenic isolates continued to harbor the 81-MDa plasmid and to produce type G neurotoxin. In contrast, the 81-MDa plasmid and the ability to produce type G neurotoxin were concomitantly lost in all of the nontoxigenic derivatives. In addition, all of the nontoxigenic derivatives ceased to produce a bacteriocin after they were cured of the 81-MDa plasmid.[9] This is the first evidence which suggests that the production of botulinal toxin and bacteriocin is in some manner related to the 81-MDa plasmid carried by the toxigenic strains. The data suggest that the structural gene for toxin production or a regulatory element that influences synthesis may be present on the plasmid.

Isolation of phages

The phages were induced from toxigenic cultures of *C. botulinum* types A to G by treatment with mitomycin C or ultraviolet irradiation. The induced lysates were filtered through a membrane filter with pore size of 450 nm, ultracentrifuged, and then the sediments were observed by electron microscopy. It was concluded that all types of toxigenic strains carried one or more phages.[1,11]

Isolation of nontoxigenic strains

Toxigenic cultures were treated with acridine orange, mitomycin C, nitrosoguanidine, or ultraviolet irradiation or allowed to sporulate. These cultures were plated on blood agar plates and incubated anaerobically for two or three days. Individual colony forming units were selected and inoculated into cooked meat medium, incubated, and checked for toxigenicity by

testing the culture supernatants using the mouse bioassay. Many nontoxigenic variants were obtained from *C. botulinum* types B, C, and D.[5,14]
Several spontaneous nontoxigenic variants were isolated from *C. botulinum*
types E and F.

Phage conversion to toxigenicity in C. botulinum types C and D

A portion, 0.5 ml, of the filtrates of mitomycin C-induced lysates were
mixed with 2.5 ml of actively growing nontoxigenic cultures. After incubation at 37°C for 3 to 4 hours, 0.2 ml of the mixtures were inoculated into
10 ml of cooked meat medium, incubated, and assayed for botulinal toxin.[12,13]
The conversion to toxigenicity was also studied by selecting plaques formed
on indicator nontoxigenic strains by bacteriophages.[2,3] The cells growing
in the plaques were transferred into cooked meat medium, incubated, and
assayed for lethal toxin. Conversion to toxigenicity occurred only in certain phage-culture pairings (Table 1).[16] The converting phages can be
classified into three groups on the basis of their conversion spectra: group
1 consists of c-st and c-468 phages; group 2 is c-203, c-d6f, and d-1873;
group 3 is d-sa and d-4947. Nontoxigenic strains (D)-134 and (D)-151 were
converted to produce type C or D toxin by the infection with type C or D
phage, respectively. Nontoxigenic strain HS15 from toxigenic strain of
C-153 and strain HS15 from the toxigenic strain of C-153 also showed the
same phenomenon.[5] C_2 toxin production was not related to any types of
phage.[4] Therefore, type C toxin described in this manuscript refers only
to C_1 toxin production.

Interspecies conversion to toxigenicity of C. botulinum type C to C. novyi type A

The lethal alpha toxin production of *C. novyi* types A and B was also
governed by specific phages.[7] Nontoxigenic strain HS37 from *C. botulinum*
type C strain 161 was found to be sensitive to three phages from types C
and 3C, types D and 1D, and *C. novyi* types A and NA1. HS37 was converted
to produce neurotoxin types C or D or alpha toxin by merely exchanging the
infecting phages.[6] These results showed that toxigenicity of *C. botulinum*
types C and D and *C. novyi* type A depends upon the continued participation
of specific tox$^+$ phages which are host-specific, and suggest that tox$^+$ gene
carried by each phage may be the structural gene for toxins.

Characterization of phage nucleic acids

Five converting phages belonging to groups 1 and 2 and nonconverting
phage (c)-n71[15] which was the mutant of c-st phage were purified. The phage
particles were precipitated from the lysates by polyethylene glycol 6000
(10%, wt/vol) in the presence of 1M NaCl. The sediments were resuspended
in 50 mM Tris hydrochloride buffer (pH 7.50), mixed with an equal volume of
chloroform, and vortexed for 30 sec. The aqueous phase was separated by
low speed centrifugation, and the phage particles were purified by cesium
chloride step gradients. Nucleic acids were purified from the phage
particles by phenol and chloroform extraction.[10]

Phage nucleic acids were treated with various restriction endonucleases
and nucleases, and the banding patterns were determined by 0.8% agarose horizontal slab gel electrophoresis. The nucleic acids were not digested by
RNase A, but were digested by DNase I and exonuclease II, indicating that
they were double-stranded DNA. The length of c-st, (c)-71, c-468, and c-d6f
phage DNAs was estimated to be about 110 kilobase pairs and that of c-203
and d-1873 was about 150 kilobase pairs. The digestion patterns of c-st,
(c)-71, and c-468 phage DNAs by PstI and Hind III were very similar. High
homology was observed in the dot hydridization test. A good similarity was
not observed with other phages and nucleases. Only a little similarity was

Table 1. Relationship between toxigenic and nontoxigenic strains and phages.

Toxic strain		C-ST		C-468	D-1873			D-SA
(Treatment)		(AO)	(NTG)	(UV)	(AO)			(SP)
Nontoxic strain[1]		(C)-A02	(C)-N71[3]	(C)-468U16	(D)-134	(D)-139	(D)-151	(D)-SA
Phage[2]	**Group**							
c-st	I	C	-	C	-	C	-	-
c-468		C	-	C	-	C	-	-
c-203	II	-	-	-	C	-	C	-
c-d6f		-	-	-	C	-	C	-
d-1873		-	-	-	D	-	D	-
d-sa	III	-	-	-	-	-	-	D
d-4947		-	-	-	-	-	-	D

[1] Nontoxigenic strains were obtained from toxigenic strains by the treatment with acridine orange (AO), nitrosoguanidine (NTG), ultraviolet irradiation (UV), or spontaneously (SP).

[2] Phages were also obtained from toxigenic strains.

[3] (C)-N71 strain was lysogenized by a nonconverting phage (c)-n71.

Table 2.　Susceptibility of phage DNA to various nucleases.

| Nucleases | Phages | | | | | |
| | Group 1 | | | | Group 2 | |
	c-st	c-n71	c-468	c-d6f	c-203	d-1873
DNase I	+	+	+	+	+	+
RNase	-	-	-	-	-	-
Exonuclease III	+	+	+	+	+	+
Hind III	+	+	+	+	+	+
Eco RI	+	+	+	+	+	+
Hpa I	+	+	+	+	+	+
Pst I	+	+	+	-	-	-
Sau3A	-	-	-	+	+	+
Xho I	-	-	-	-	-	-

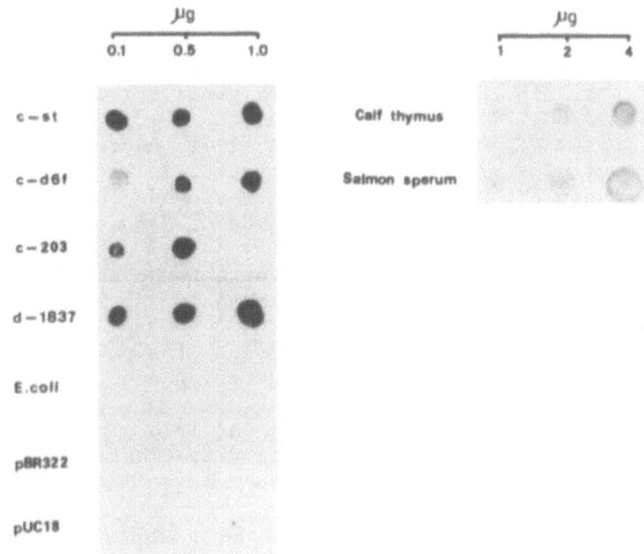

Amino acid sequence

X –Ile–Thr–Ile–Asn–Asn–Phe–Asn–Tyr–Ser–Asp–Pro–Val–Asp–Asn–Lys–Asn–Ile–Leu–Tyr
1　　　　　　5　　　　　　　　　　10　　　　　　　　　15　　　　　　　　　20

5'– ATC-ACT-ATC-AAC-AAC-TTC-AAC-TAC-TCC-GAC-CCG-GTT-GAC-AAC-AAA-AAC-AT – 3'

Probe

3'– TAI-TGI-TAI-TTI-TTI-AAI-TTI-ATI-AGI-CTI-GGI-CAI-CTI-TTI-TTI-TTI-TA – 5'

Fig. 1.　Hybridization of phage DNA and synthesized DNA probe.　The
DNA probe (50-mer) complementary to the amino acid sequence
of the c-st toxin N terminus was hybridized to five converting phage DNA preparations, which had been spotted on the
membrane in portions containing 0.1, 0.5, and 1.0 μg.　As
controls, plasmids and several DNA preparations were
employed.　The DNA preparation of nonconverting phage (c)-n71
also showed positive reaction (datum is not shown).

observed between c-203 and c-d6f phages, indicating that there are many differences in genes between the phage DNAs belonging to group 2. However, the susceptibility to PstI and Sau3A digestion agreed with the groupings proposed by infection spectrum and antigenicity. PstI digested the DNAs obtained from only group 1 phages, and Sau3A digested the DNAs from only group 2 phages (Table 2).

Structure genes for toxin

In an attempt to demonstrate the existence of a structural gene for toxin, the N-terminal amino acid sequence of type C toxin was first determined,[17] and then the oligonucleotide probe complementary to it was prepared. This probe hybridized all of six phage DNAs examined (Fig. 1). These data indicate that (c)-n71 phage, as well as c-st phage, carried all or some structural genes for type C toxin. Since (c)-n71 was obtained by nitrosoguanidine treatment, methylation of guanine may be induced and cause the loss of converting ability. Type D phages also showed a positive reaction. This is caused by the similarity of the N-terminal amino acid sequence of type C and D toxins.

REFERENCES

1. Eklund MW, Poysky FT, Boatman ES, 1969, Bacteriophages of *Clostridium botulinum* types A, B, E and F and nontoxigenic strains resembling type E, *J Virol.*, 3:270-274.
2. Eklund MW, Poysky FT, Reed SM, Smith CA, 1971, Bacteriophage and toxigenicity of *Clostridium botulinum* type C, *Science*, 172:480-482.
3. Eklund MW, Poysky FT, Reed SM, 1972, Bacteriophage and the toxigenicity of *Clostridium botulinum* type D, *Nature (London) New Biol.*, 235:16-17.
4. Eklund MW, Poysky FT, 1972, Activation of a toxic component of *Clostridium botulinum* types C and D by trypsin, *Appl Microbiol.*, 24:108-113.
5. Eklund MW, Poysky FT, 1974, Interconversion of types C and D strains of *Clostridium botulinum* by specific bacteriophages, *Appl Microbiol.*, 27:251-258.
6. Eklund MW, Poysky FT, Meyers JA, Pelroy GA, 1974, Interspecies conversion of *Clostridium botulinum* type C to *Clostridium novyi* type A by bacteriophage, *Science*, 186:456-458.
7. Eklund MW, Poysky FT, Peterson ME, Meyers JA, 1976, Relationship of bacteriophages to alpha toxin production in *Clostridium novyi* types A and B, *Infect Immun.*, 14:793-803.
8. Strom MS, Eklund MW, Poysky FT, 1984, Plasmids in *Clostridium botulinum* and related *Clostridium* species, *Appl Environ Microbiol.*, 48:956-963.
9. Eklund MW, Poysky FT, Miseitif LM, Strom MS, 1988, Evidence for plasmid-mediated toxin and bacteriocin production in *Clostridium botulinum* type G, *Appl Environ Microbiol.*, 54:1405-1408.
10. Fujii M, Oguma K, Yokosawa N, Kimura K, Tsuzuki K, 1988, Characterization of bacteriophage nucleic acids obtained from *Clostridium botulinum* types C and D, *Appl Environ Microbiol.*, 54:69-73.
11. Inoue K, Iida H, 1968, Bacteriophages of *Clostridium botulinum*, *J Virol.*, 2:537-540.
12. Inoue K, Iida H, 1970, Conversion of toxigenicity in *Clostridium botulinum* type C, *Jpn J Microbiol.*, 14:87-89.
13. Inoue K, Iida H, 1971, Phage conversion of toxigenicity in *Clostridium botulinum* types C and D, *Jpn J Med Sci Biol.*, 24:53-56.

14. Oguma K, Iida H, Inoue K, 1973, Bacteriophage and toxigenicity in *Clostridium botulinum*: an additional evidence for phage conversion, *Jpn J Microbiol.*, 17:425-426.
15. Oguma K, Iida H, Inoue K, 1975, Observations on nonconverting phage, c-n71, obtained from a nontoxigenic strain of *Clostridium botulinum* type C, *Jpn J Microbiol.*, 19:167-172.
16. Oguma K, Shiozaki M, Iida H, Inoue K, 1976, Antigenicity of converting phages obtained from *Clostridium botulinum* types C and D, *Infect Immun.*, 13:855-860.
17. Tsuzuki K, Yokosawa N, Syuto B, Ohishi I, Fujii N, Kimura K, Oguma K, 1988, Establishment of a monoclonal antibody recognizing an antigenic site common to *Clostridium botulinum* type B, C_1, D, and E toxins and tetanus toxin, *Infect Immun.*, 56:898-902.

MOLECULAR STRUCTURE AND BIOLOGICAL ACTIVITIES OF *CLOSTRIDIUM BOTULINUM* C_2 TOXIN

Iwao Ohishi and Genji Sakaguchi

University of Osaka Prefecture
College of Agriculture
Sakai, Osaka 591, Japan

Clostridium botulinum is known to produce an extremely potent neuro-toxin. In 1922, Bengtson isolated a nonproteolytic *C. botulinum* strain from *Lucilia caeser* larvae.[1] The toxin produced by this strain was not neutralized with anti-botulinum type A or type B toxin serum. Types A and B were the only *C. botulinum* types known at the time. For this reason, the new strain was designated *C. botulinum* type C. Since then, a number of *C. botulinum* type C and D strains have been isolated. However, the immuno-logical cross-reactions of toxins produced by type C and D strains have not been clearly elucidated. In 1935, Mason and Robinson reported that type C strains produced three toxins, C_1 and C_2 and a small amount of D, and that type D strains produced a small amount of C toxin in addition to a large amount of D toxin.[4] Jansen also reported that a type C_2 strain produced C_1, C_2 and D toxins, that a type C_β strain produced only C_2 toxin, and that a type D strain produced C_1 and D toxins.[3] It was also reported that non-toxigenic type C and D strains, either cured of their prophages or isolated from environmental sources, produced C_2 toxin.[2,6] Although it had been recognized that certain strains of *C. botulinum* types C and D produced C_2 toxin in addition to C_1 and D toxins, little was known about the molecular structure and the biological activity of C_2 toxin until the toxin was purified and characterized.

In 1979, we initiated the purification of botulinum C_2 toxin to allow characterization of its immunological properties and to compare its molec-ular structure with botulinum type A through F neurotoxins. The neurotoxin molecules have two polypeptide chains linked with disulfide bonds, desig-nated light and heavy chains respectively. We have found that C_2 toxin is entirely different in molecular structure and biological activity from botulinum neurotoxins.

Fig. 1 and Table 1 show the first evidence that C_2 toxin is constructed of two dissimilar proteins, which are linked with neither covalent nor non-covalent bonds. Chromatography of the culture supernatant of four strains of *C. botulinum* types C and D with a CM-Sephadex column yielded two frac-tions that were designated fractions I and II (Fig. 1). The toxicity of each fraction when tested by mouse assay was extremely low. However, full toxicity was restored when the percolated fraction (fraction I) and the eluted fraction (fraction II) were mixed together and trypsinized (Table 1). The same results were obtained with C_2 toxin produced by all four strains examined. The results indicated that C_2 toxin consists of two unlinked protein components, that can be resolved into two fractions by chromato-graphy and individually have extremely low toxicity.

Microbial Toxins in Foods and Feeds, Edited by A.E. Pohland *et al.,*
Plenum Press, New York, 1990

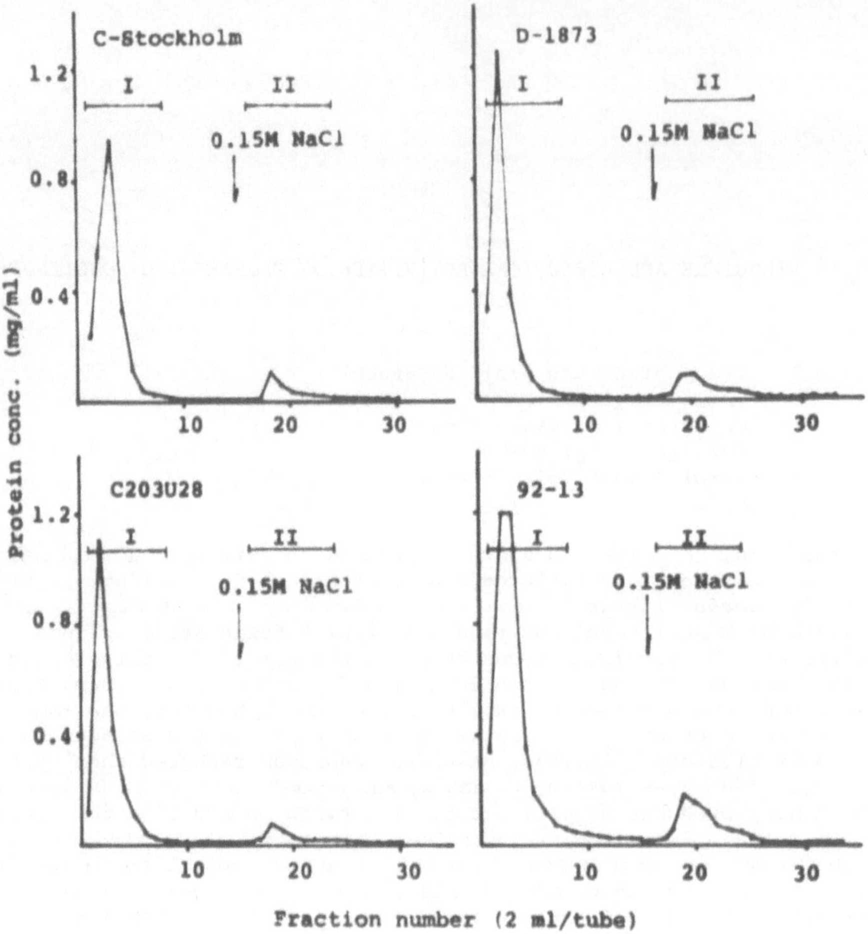

C-Stockholm
D-1873
C203U28
92-13
I
II
0.15M NaCl
1.2
0.8
0.4
10
20
30
Protein conc. (mg/ml)
Fraction number (2 ml/tube)

Fig. 1. Separation of C_2 toxin produced by *C. botulinum* type C and D strains into two fractions by CM-Sephadex chromatography. A 4-ml portion of the culture supernatant was applied onto the column, equilibrated with 10 mM acetate buffer, pH 6.0, and eluted with 0.15 M NaCl in the buffer.

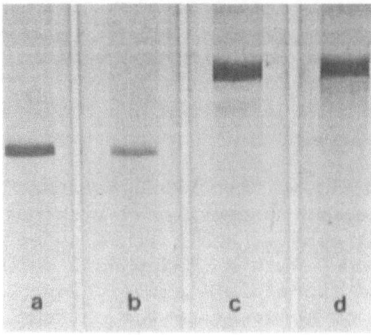

a
b
c
d

Fig. 2. Sodium dodecylsulfate-polyacrylamide gel electrophoresis of purified components I and II. (a) Component I before and (b) after reduction; (c) component II before and (d) after reduction.

Table 1. Toxicities of protein fractions of C_2 toxin separated by CM-Sephadex chromatography.

C. botulinum strains	Fraction	Toxicity (i.p.LD$_{50}$)[a]
Type C Stockholm	Applied	1,523
	I	44[b]
	II	D[c]
	I + II	1,645
Type D 1873	Applied	1,232
	I	23[b]
	II	D
	I + II	1,689
Type C 203U28	Applied	1,998
	I	D
	II	D
	I + II	1,678
Type C 92-13	Applied	2,879
	I	D
	II	D
	I + II	2,980

[a] Fractins were trypsinized and the toxicities were determined by the time-to-death method.[11]

[b] Most toxicities found in the fraction I of strains Stockholm and 1873 were possibly due to the presence of type C and D neurotoxins, respectively.

[c] D; mice lived for 2 hours, but died within 6 hours.

We have purified these two protein components of C_2 toxin from the culture of C. botulinum type C strain 92-13, which produces C_2 toxin but not C_1 or D neurotoxins.[6] Each of the two purified proteins, designated components I and II, gave a single band by sodium dodecylsulfate polyacrylamide gel electrophoresis and had molecular weights of 55,000 and 105,000, respectively (Fig. 2).[11] The reduction of these components with dithiothreitol or mercaptoethanol did not change the molecular weights, indicating that the components are not nicked as are type A through F neurotoxins. The lethality of the toxin was elicited by injecting a mixture of untrypsinized component I and trypsinized component II or trypsinized components I and II together, whereas each component alone even after being treated with trypsin and the combination of component I, either trypsinized or untrypsinized, and untrypsinized component II showed no or feeble toxicity (Table 2).[11] In agar gel double-immunodiffusion, purified components I and II against a mixture of rabbit anti-component I and anticomponent II sera gave a single precipitin line, that crossed each other (Fig. 3).[11] Neither anti-component I nor anti-component II serum formed a precipitin line with type C and D neurotoxins. The results indicates that components I and II of C_2 toxin are immunologically distinct protein molecules and neither component I nor II is antigenically related to botulinum type C or D neurotoxins.

In addition to lethal activity, C_2 toxin has vascular permeability

Table 2. Lethality of purified components I and II of C_2 toxin in mice.

Toxin component[a]	Lethality (i.p. LD_{50}/mg protein)
UT-I	-[b]
T-I	-
UT-II	-
T-II	D[c]
UT-I + UT-II	D
T-I + UT-II	D
UT-I + T-II	3,800
T-I + T-II	4,200

[a] UT-, untrypsinized; T-, trypsinized.
[b] -, Mice survived after intravenous injection.
[c] D, Mice survived for 100 minutes but not 6 hours.

activity when injected into guinea pig skin.[12] The activity of the toxin demonstrated as blueing response after intravenous injection of Evans blue dye is enhanced by trypsinization of a mixture of components I and II and is also attributable to the cooperation of two unassociated protein components, which individually have very low vascular permeability activity (Table 3).[12] When one component was injected intravenously and the other intradermally, an increase in vascular permeability was induced at the intradermal site of injection of component II but not at that of component I (Figs. 4-a and 4-b).[7] This indicated that the simultaneous injection of the two components of C_2 toxin is not always required to elicit the biological activity of the toxin. The vascular permeability response induced by separate injections of the two components suggests that the activity of C_2 toxin results from component II binding to the tissue around its injection site and component I recognizing the bound component II or the altered tissue.

C_2 toxin also induced fluid accumulation in mouse intestinal loops.[8] Neither component I nor component II alone induced the fluid accumulation in the intestinal loops, whereas a mixture of untrypsinized components I and II or untrypsinized component I mixed with trypsinized component II induced the response (Table 4). The activity of a mixture of untrypsinized components I and II is probably due to the activation of the toxin in the intestinal lumen with trypsin or trypsin-like protease(s). The fluid accumulation in intestinal loops inoculated with C_2 toxin was not diminished by removal of the toxin from the loops.[8] Moreover, the secretory response was positive when intestinal lumina were exposed to component II followed by the removal of the component and inoculation with component I, but it was negative when the intestinal lumina were exposed to component I followed by the removal of the component and inoculation with component II (Table 5). In addition, immunofluorescence studies showed that component II, either trypsinized or untrypsinized, bound to isolated epithelial cells of mouse intestine, whereas component I bound to the cells only in the presence of trypsinized component II (Fig. 5).[9] These results indicated that the secretory response of mouse intestinal loops to C_2 toxin is induced by the binding of component II to the epithelial cell surface of the intestine and the consequent binding or penetration of component I into the cells.

C_2 toxin, a mixture of untrypsinized component I and trypsinized component II, induced marked morphological changes in cultured monkey kidney cells.[5] The characteristic response of the cells to the toxin was round-

Table 3. Vascular permeability activity of two components of C_2 toxin.

Toxin component[a]	Blueing activity $(BU/\mu g)$[b]
UT-I	0.05
UT-II	0.03
UT-I + UT-II	0.13
T-I + UT-II	0.67
UT-I + T-II	92.4
T-I + T-II	103.1

[a] UT-, untrypsinized; T-, trypsinized.
[b] One blueing unit (BU) was defined as the quantity of toxin producing a blueing lesion of 10 mm in the mean cross diameter.

Table 4. Fluid accumulation in ligated mouse intestinal loops exposed to two components of C_2 toxin.

Toxin component[a]	Increase in wt of loop (mg)[b,c]
UT-I	39.1 ± 9.2
UT-II	32.8 ± 11.3
UT-I + UT-II	154.6 ± 62.9
UT-I + T-II	301.6 ± 63.7
T-I + T-II + anti-I serum	32.2 ± 23.0
T-I + T-II + anti-II serum	42.6 ± 5.9

[a] UT-, untrypsinized; T-, trypsinized.
[b] Due to the fluid accumulation in the intestinal loop.
[c] Each value represents the mean ± standard error of the mean of six mice.

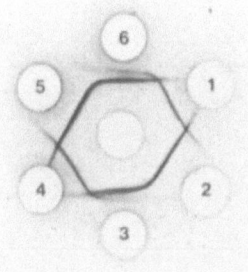

Fig. 3. Agar gel double-immunodiffusion with purified components I and II. Center well, a mixture of equal volumes of anti-component I and anti-component II sera; lateral wells (1) and (4), component I; (2) and (5), component II and (3) and (6), C_2 toxin, a mixture of components I and II.

Table 5. Effect of intestinal washing on fluid accumulation in mouse intestinal loops after exposure to two components of C_2 toxin.

Inoculation[a]	Time of exposure	Increase in wt of loop (mg)[b,c]
UT-I then T-II[d]	0	40.0 ± 12.1
	1	45.4 ± 12.5
T-II then UT-I[e]	0	261.6 ± 30.5
	1	244.4 ± 54.8
UT-I + T-II		244.8 ± 51.2

[a] UT-, untrypsinized; T-, trypsinized.
[b] Animals were killed 6 hours after the injection of the first component or the mixture of components I and II.
[c] Each value represents the mean ± standard error of the mean of six mice.
[d] Each loop was injected with 4 μg of component I, washed three times with 0.3 ml saline after the time of exposure indicated and injected with 8 μg of component II.
[e] Each loop was injected with 8 μg of component II and then with 4 μg of component I as described above.

ing, (Fig. 6). Again the components alone and a combination of untrypsinized components I and II showed little activity (Table 6).

The findings described above indicate that C_2 toxin elaborated by certain strains of *C. botulinum* types C and D has novel biological activity and molecular structure, which are not possessed by the neurotoxin of *C. botulinum* types A through F, and that the binding of trypsinized component II, but not of untrypsinized component II, to the cell surface membrane introduces the binding site for component I, which appears to be internalized into the cells.

Table 6. Rounding activity of two components of C_2 toxin to cultured cynomolgus monkey kidney cells.

Toxin component[a]	RD_{50} (ng/well)[b,c]
UT-I	940
UT-II	1,840
T-II	84
UT-I + UT-II	200
UT-I + T-II	0.5

[a] UT-, untrypsinized; T-, trypsinized.
[b] The tissue-cultured cells in 98-well plates were incubated with the toxin component for 18 hours and the percentage of rounded cells was determined by counting the cells with the aid of a phase contrast microscope.
[c] The toxin dose causing 50% rounding of the cells.

Recently we have found that component I of C_2 toxin preferentially ADP-ribosylates cytoskeletal actin.[13] The characteristic morphologic changes of cells in tissue cultures exposed to the toxin implied that the target molecule for component I, which has NAD-glycohydrolase activity,[10] may be tublin. Therefore, a crude tublin fraction prepared from rat brain was incubated with component I in the presence of $[^{32}P]$NAD. Fig. 7 shows that component I ADP-ribosylated actin but not tublin in the fraction. This was confirmed by affinity chromatography with a Sepharose column coupled with DNase I, which interacts with action. Fig. 8 shows that the ADP-ribosylated protein in the supernatant of mouse brain homogenate was specifically absorbed and eluted from the column as was purified hog brain actin. Moreover, the preferential modification of nonmuscle actin with component I was confirmed with the ADP-ribosylation of purified actins (Fig. 9).[13]

When the cultured cells were exposed to a mixture of component I and trypsinized component II and then the cell-lysate prepared from the cells was incubated with component I in the presence of $[^{32}P]$NAD, inhibition of radiolabeled cytoplasmic actin was observed (Fig. 10). This indicated that

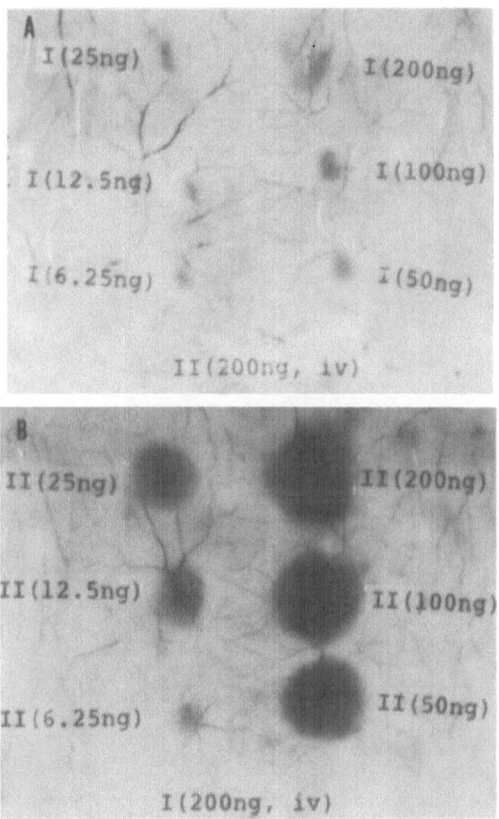

Fig. 4.   Vascular permeability response of the guinea pig skin by separate injection of two components of C_2 toxin. (A) The response by intravenous injection of component II after intradermal injection of component I and (B) that by intravenous injection of component I after intradermal injection of component II.

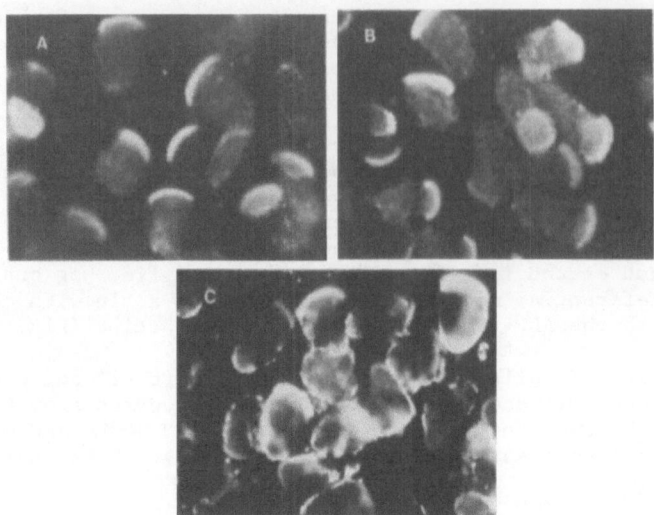

Fig. 5. Immunofluorescence labeling of isolated epithelial cells of mouse intestine with the two components of C_2 toxin. Isolated cells were incubated at room temperature for 30 minutes with (A) untrypsinized component II, (B) trypsinized component II and (C) component I in the presence of trypsinized component II. The cells were then reacted with rabbit anti-component II serum for (A) and (B) and rabbit anti-component I serum for (C) and stained with fluorescence-labeled antibody to rabbit IgG.

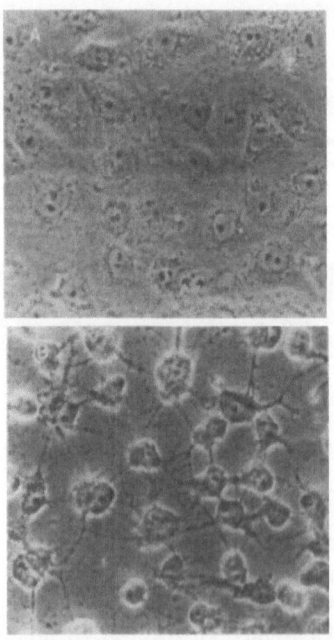

Fig. 6. Rounding response of tissue-cultured cynomolgus monkey kidney cells to C_2 toxin. The cells were (A) not exposed to the toxin and (B) exposed to 60 ng of the toxin for 2 hours at 37°C.

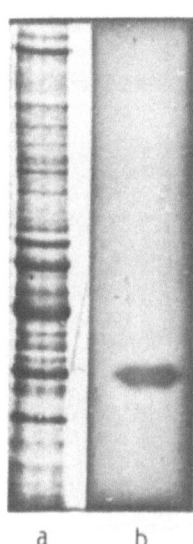

a   b

Fig. 7.   Sodium dodecylsulfate polyacrylamide gel electrophoresis of
crude tublin fraction of rat brain incubated with component
I of C_2 toxin in the presence of [^{32}P]NAD.   The tublin
fraction was prepared from rat brain.   (a) Coomasie blue stain
and (b) autoradiogram.

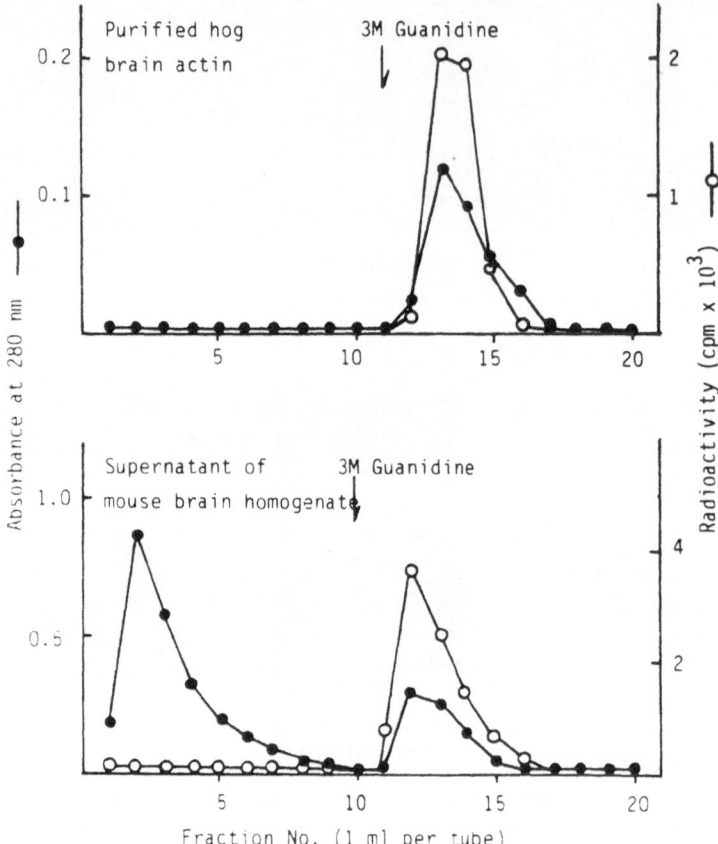

Fig. 8. Affinity chromatography of ADP-ribosylated actin on a column
of Sepharose 6B coupled with DNase I. The supernatant of
mouse brain homogenate (10% w/w) in 10 mM imidazole-HCl-1 mM
ATP-0.5 mM DTT-0.1 mM $CaCl_2$-1 mM NaN_3, pH 7.5, was prepared
by centrifugation at 10,000 g for 30 minutes. Nonmuscle actin
was purified from hog brain. The supernatant (1 ml) or the
purified actin (6 nmol) was incubated with component I (0.6
nmol) in the presence of [^{32}P]NAD (600 nmol) at 37°C for 30
minutes. After removing free NAD by gel filtration on a
Sephadex G-50 column equilibrated with the buffer, the
reaction mixture was applied onto a column of Sepharose 6B
coupled with DNase I and the actin absorbed was eluted with
3M guanidine.

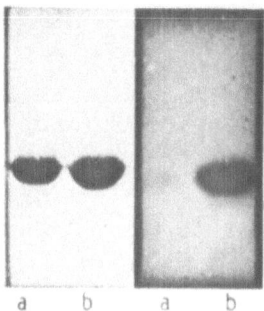

Coomasie blue   Autoradiogram
Staining

Fig. 9. Sodium dodecylsulfate polyacrylamide gel electrophoresis of
        nonmuscle and muscle actins. The nonmuscle actin was puri-
        fied from hog brain and the muscle actin from rabbit muscle.
        Actin was incubated with component I in the presence of
        [^{32}P]NAD at 35°C for 15 minutes and electrophoresed in 10%
        polyacrylamide gel.   (a) Rabbit muscle actin and (b) hog brain
        actin.

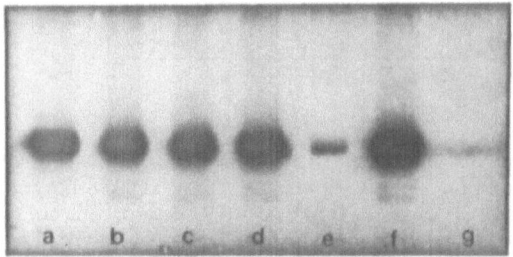

Fig. 10. ADP-ribosylation of intracellular actin with component I of
        C_2 toxin. Monolayer cultures of cynomolgus monkey kidney
        cells in 24-well plate were washed with phosphate buffered
        saline and incubated with two components of C_2 toxin for 60
        minutes at 37°C in incubator. The cells were washed again
        with phosphate buffered saline and lysed in 100 μl of 10 mM
        KCl-10 mM Hepes-0.1 mM $CaCl_2$-0.5 mM DTT-0.05 mM PMSF, pH 7.5
        (Buffer A). The lysate was transferred into test tubes and
        incubated again with 1 μl of component I and 200 nmol [^{32}P]
        NAD in a total volume of 150 μl of Buffer A for 15 minutes
        at 35°. The protein was precipitated, washed with 20% trich-
        loroacetic acid and dissolved in 40 μl of 2% sodium dodecyl-
        sulfate-0.2M Tris-HCl, pH 7.5. A 20-μl portion of the solu-
        tion was electrophoresed in 10% polyacrylamide gel, which
        was dried and exposed overnight to Kodak X-Omat AR film.
        Monolayer cells were incubated with (a) 1 μg of component I,
        (b) 2 μg of untrypsinized component II, (c) 2 μg of trypsi-
        nized component II, (d) a mixture of untrypsinized components
        I (1 μg) and II (2 μg), (e) a mixture of untrypsinized com-
        ponent I (1 μg) and trypsinized component II (2 μg), (f) no
        toxin component and (g) no component I in the reaction with
        cell lysate.

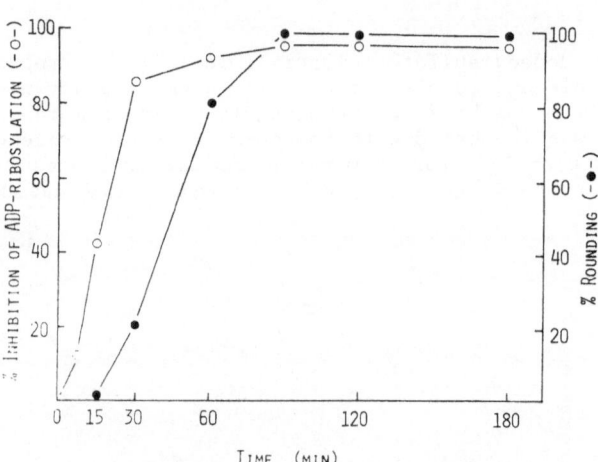

Fig. 11. Relationship between ADP-ribosylation of the intracellular
actin and the rounding of tissue-cultured cynomolgus monkey
kidney cells with C₂ toxin. Monolayer culture in 24-well
plate was washed with phosphate buffered saline and exposed
to 200 ng of C₂ toxin diluted in culture medium containing
2% bovine serum albumin instead of fetal calf serum at 37°C
for the time indicated in the figure. The inhibition of ADP
ribosylation was determined as described in the legend to
figure 10, except that the protein was precipitated with
trichloroacetic acid dissolved in sodium dodecylsulfate and
the ^{32}P content was counted in vials containing 2 ml of scin-
tillator. The rounded cells were counted with the aid of a
phase contrast microscopy.

the intracellular actin was ADP-ribosylated by C_2 toxin applied extracellularly. No inhibition of ADP-ribosylation of intracellular actin was observed with either components I and II alone or the mixutre of untrypsinized components I and II of C_2 toxin. Moreover, the inhibition by endogenous ADP-ribosylation of intracellular actin with C_2 toxin was correlated with the morphological changes of cells in tissue cultures induced by the toxin (Fig. 11). The results indicate that the biological activity of the toxin involves a novel enzymatic activity of component I, that catalyzes ADP-ribosylation of intracellular actin in target cells.

In conclusion, the two components of C_2 toxin together act as a molecule with dual functions. Component II recognizes the receptor site on the cell surface membrane and component I as the effector in the cytoplasm by ADP-ribosylating cytoskeletal actin. The diversity in the biological activity of C_2 toxin is probably dependent on specific differentiation of the target cells.

ACKNOWLEDGEMENT

A part of this work was supported by a grant-in-aid for scientific research from the Ministry of Education, Science and Culture of Japan.

REFERENCES

1. Bengtson IA, 1922, Preliminary note on a toxin-producing anaerobe isolated from the larvae of *Lucilia caesar*, *Publ Health Rep.*, 37:164-170.
2. Eklund MW, Poysky FT, 1972, Activation of toxin component of *Clostridium botulinum* types C and D by trypsin, *Appl Microbiol.*, 24:108-113.
3. Jansen BC, 1971, The toxic antigenic factors produced by *Clostridium botulinum* types C and D, *Onderstepoort J Vet Res.*, 38:93-98.
4. Mason JH, Robinson EM, 1935, The antigenic components of the toxins of *C. botulinum* types C and D, *Onderstepoort J Vet Sci Anim Ind.*, 5:65-75.
5. Miyake M, Ohishi I, 1987, Response of tissue-cultured cynomolgus monkey kidney cells to botulinum C_2 toxin, *Microbial Pathogenesis*, 3: 279-286.
6. Nakamura S, Serikawa T, Yamakawa K, Nishida S, Kozaki S, Sakaguchi G, 1978, Sporulation and C_2 toxin production by *Clostridium botulinum* type C strains producing no C_1 toxin, *Microbiol Immunol.*, 22:591-596.
7. Ohishi I, 1983, Lethal and vascular permeability activities of botulinum C_2 toxin induced by separate injections of the two components, *Infect Immun.*, 40:336-339.
8. Ohishi I, 1983, Response of mouse intestinal loop to botulinum C_2 toxin: enterotoxic activity induced by cooperation of nonlinked protein components, *Infect Immun.*, 40:691-695.
9. Ohishi I, 1985, Binding of the two components of C_2 toxin to epithelial cells and brush borders of mouse intestine, *Infect Immun.*, 48:769-775.
10. Ohishi I, 1986, NAD-glycohydrolase activity of botulinum C_2 toxin: a possible role of component I in the mode of action of the toxin, *J Biochem.*, 100:407-413.
11. Ohishi I, Iwasaki M, Sakaguchi G, 1980, Purification and characterization of two components of botulinum C_2 toxin, *Infect Immun.*, 30:668-673.
12. Ohishi I, Iwasaki M, Sakaguchi G, 1980, Vascular permeability activity of botulinum C_2 toxin elicited by cooperation of two dissimilar protein components, *Infect Immun.*, 31:890-895.
13. Ohishi I, Tsuyama S, 1986, ADP-ribosylation of nonmuscle actin with component I of C_2 toxin, *Biochem Biophys Res Commun.*, 136:802-806.

GENETIC STUDIES ON *VIBRIO FLUVIALIS* AND ITS ENTEROTOXINS

Stephen H. Richardson and John V. Kaspar

Department of Microbiology
Bowman Gray School of Medicine
Wake Forest University
Winston-Salem, North Carolina 27103

ABSTRACT

Vibrio fluvialis, an estuarine-based halophilic vibrio, causes opportunistic infections in humans resulting in bloody diarrhea. The organism
produces an enterotoxin (ET), a cytotoxin (CT), and a cytolysin (CL). We
created isogenic non-toxigenic mutants to assess the relative contribution
of each toxin to virulence. High level streptomycin (Sm) resistant mutants
of the prototype toxigenic strain 807-77 were used as conjugal recipients
for TnphoA, which creates gene fusions between target genes and *phoA*, which
encodes the alkaline phosphatase of *Escherichia coli*. Transposon insertions
into genes for secreted or periplasmic proteins were selected on indicator
plates. Putative CL- transconjugates (which appear as dark blue colonies)
were recovered at a rate of ca. 10^{-6}/recipient cell. Presumptive CL-
mutants were screened on blood agar, egg yolk agar, and with a sensitive
tube hemolysin assay. Twenty-two CL- mutants were tested vs. red cells from
two different species. CL- production was attenuated in 17 and negative in
5 of the mutants. The verified CL- mutants were tested for pathogenicity
in infant mice. The CL- mutants were unaltered in their ability to cause
fluid accumulation in this model. Various CL- mutants responded differently
to increased osmolarity in their growth media as measured by alkaline
phosphatase activity of permeabilized cells.

INTRODUCTION

Beginning from the mid-1970's, it became clear that there was a small
group of halophilic Vibrios that were serologically heterogeneous and were
often confused with members of the genus *Aeromonas*. The isolates were
originally called group F Vibrios and included strains that were both aerogenic and anaerogenic.[8] The aerogenic strains are now a separate species,
Vibrio furnissii. Group F Vibrios gained notoriety when the organisms were
isolated from patients in Bangladesh presenting with non-cholera diarrhea
during outbreaks in 1976-1977. Group F strains were epidemiologically
implicated as potential enteropathogens because no other bacterial pathogens
were recovered from the patient's stool specimens.[3] Some of these strains
were sent to the Centers for Disease Control where they were designated
group EF-6. It was concluded in 1980 from DNA-DNA hybridization data, that
group F and group EF-6 were identical and that the strains within these
groups should be classified as a new species which was named *V. fluvialis*.[7]

Microbial Toxins in Foods and Feeds, Edited by A.E. Pohland *et al*.,
Plenum Press, New York, 1990

Fluvialis, which means "river" in Latin, is a Gram negative halophilic, curved, rod-shaped bacterium whose natural habitat is stagnant rivers and estuarine waters. The organism causes opportunistic infections in humans resulting in diarrhea, vomiting, abdominal pain, moderate to severe dehydration, and fever. Direct examination of stool samples often reveals leukocytes and erythrocytes and some patients have overt bloody diarrhea.

Viable cells of V. fluvialis have been shown to stimulate fluid accumulation in rabbit ileal loops.[11] Different isolates also liberate various combinations of three physically and functionally distinct toxins: i) a cell-bound cholera-like enterotoxin (ET), eliciting diarrhea in animal models, and causing cell elongation in Chinese-Hamster-Ovary (CHO) cells; ii) a non-hemolytic cytotoxin (CT) which results in death and disintegration of CHO cells; iii) a separate hemolytic cytolysin (CL) which evokes morphological changes leading to loss of viability in CHO cells.[8] Oral challenge of infant mice with V. fluvialis is followed by fluid accumulation, diarrhea and death.[8] The fluid accumulation ratios recorded in these experiments were similar to those generated by V. cholerae.

Because these bacteria elaborate three distinct toxins and incite diarrhea in infant mice, we wanted to determine which toxin or combination of toxins was responsible for diarrhea through the creation of stable mutants defective in toxin production. To achieve this goal we decided to construct insertionally inactivated isogenic mutants using transposon TnphoA recently described by Manoil and Beckwith,[10] which is a fusion between an enzymatically active fragment of the alkaline phosphatase of E. coli and Tn5 which codes for kanamycin resistance. TnphoA is delivered by broad host range vectors by conjugation and has been used extensively for the analysis of secreted and periplasmic proteins of Gram negative bacteria.[13]

Here we describe the construction of V. fluvialis mutants that no longer produce a functional CL toxin. In addition, we used in vitro and in vivo techniques to verify that mutational events were restricted to the CL gene, tested the virulence of the organism in an animal model, and measured the expression of alkaline phosphatase in 2 of the CL- mutants under a variety of growth conditions.

MATERIALS AND METHODS

Bacterial and plasmid strains

Vibrio fluvialis strain 807-77 obtained from the Enterobacteriaceae Section of the Centers for Disease Control, Atlanta, Georgia, was a human isolate from an outbreak in Bangladesh. Of the 150 strains tested from our collection, 807-77 produced the highest titers of CL and was thus selected for further study.

Escherichia coli strain SM10pir containing plasmid pRT733 (oriR6K, tra , mob$^+$, Ap^r, Km^r) was used in conjugation experiments. pRT733 is a suicide vector and is a derivative of pJM703.1 which is a pBR325 derivative that has the ColE1 origin of replication replaced with a oriR6K. The plasmid is only functional in the presence of the pir gene which is supplied by a lambda-pir prophage.[11] This plasmid carries the mob site from RP4.[12] All E. coli plasmid bearing strains were kindly donated by Dr. John Mekalanos of Harvard Medical School (Boston, Massachusetts).

Frozen stock suspensions were prepared from all bacterial strains grown for 18 hours at 30°C on Columbia agar (CA; Oxoid, Ltd., England) or in Heart infusion broth (HIB; Difco Laboratories, Detroit, Michigan). Bacteria washed from agar plates with HIB or HIB broth cultures were adjusted to 40%

glycerol (v/v) and frozen at -70°C. Inocula for various investigations described were initiated from these frozen stocks.

Antibiotic resistant mutant construction

Wild type 807-77 was grown in HIB, sedimented by centrifugation, washed twice with sterile isotonic saline (0.85%) and resuspended in 1 ml of sterile saline. Samples of 200μl were then spread on CA plates containing 5μg/ml Sm and allowed to incubate at 30°C for 18 to 24 hours. Resultant colonies were picked and streaked on CA plates containing 1 mg/ml Sm.[5] Colonies were purified, amplified by growth in HIB and stored at -70°C.

Insertional mutation of V. fluvialis strain 807-77 Sm[r] by pRT733

Mobilization of pRT733 into the recipient was performed by growing 807-77 SmR in HIB or on CA containing 100 μg/ml Sm at 30°C and growing SM10 pir (pRT733) in HIB or on CA supplemented with 100 μg/ml Ap and 40 μg/ml XP. Aliquots of each strain were mixed (1:10; donor:recipient) on the surface of CA or in HIB without antibiotics. Conjugation was allowed to proceed for 2 to 24 hours (Fig. 1).

Isolation of transposon insertions

Transconjugants were plated from mating mixtures onto CA plates containing 100 μg Sm/ml, 45 μg Km/ml and 40 μg/ml 5-bromo-4-chloro-3-indoyl phosphate (XP; Sigma Chemical Company, St. Louis, MO) (40 mg XP/ml dissolved in a 1:1 ratio v/v formamide:methanol). After 2 to 5 days of incubation at 30°C, blue colonies (those with secreted alkaline phosphatase activity) were picked and purified by restreaking. In all cases transconjugates were ampicillin sensitive.

Screening putative CL- mutants

Potential CL- mutants were transferred to Columbia sheep blood agar plates without antibiotics. The plates were incubated 18 to 24 hours at 30°C. Frozen suspensions were prepared from the putative CL- colonies by growing the bacteria in HIB for 18 to 24 hours at 30°C in a shaker incubator. Samples were aliquoted, adjusted to 25% glycerol (v/v) and frozen at -70°C for use in further experiments.

Hemolytic activity

Hemolytic activity against mouse and sheep erythrocytes was determined by the tube method of Bernheimer and Schwartz.[2] One hemolytic unit was defined as the reciprocal of the dilution that results in the release of 50% of the hemoglobin in the standardized (0.7%, v/v) erythrocyte suspension.

807 SmR CL + SM10 (pRT 733) ApRKmR
HIB + Sm   HIB + Ap

30 C OVERNIGHT

1 SM10 : 10 807 on CA or in HIB

30 C 2 - 24 HRS

SPREAD ON CA + Sm and Km

30 C for 2 - 5 DAYS; PUTATIVE CL -

DARK BLUE KmR, SmR ApS

Fig. 1. TnphoA Mutation of V. fluvalis 807-7.

Culture supernatant fluids from CL- mutants were dialysed against ammonium bicarbonate (0.1 M NH$_4$HCO$_3$) for 20 hours at 4°C prior to performing the hemolytic assay to remove any interferring protease activity.

Infant mouse assay

Pregnant (17 to 18 days) white CFW mice were obtained from Charles River Breeding Laboratories, Inc. (Raleigh, North Carolina). Newborn mice weighing approximately 2.5 to 4.0 grams (4 to 6 days old) were starved for 10 hours at 30°C before being fed viable bacteria. Bacteria were grown for the infection studies by inoculating HIB (100 ml in 500 ml baffle flasks) with a loopful of culture from a CA plate and incubating the medium for 12 to 18 hours at 30°C with agitation at 250 rpm. The bacteria were collected by centrifugation, washed twice with PBS, and suspended in 10 ml of the same buffer. Viable counts (cfu/ml) were determined by spreading appropriate dilutions on Columbia agar plates, and incubating the plates 18 hours at 30°C.

The anorectal canals of mice were sealed with cyanoacrylate (Elmer's Wonder Bond; Borden, Inc., Columbus, Ohio), and a 0.25 ml glass tuberculin syringe fitted with a 23-gauge needle tipped with intramedic polyethylene tubing was used to inoculate mice intragastrically with 50 μl of each sample. After the desired time interval (3 to 4 hours), mice were humanely killed by cervical dislocation, and the fluid accumulation (FA) ratios were determined as described by Baselski et al.[1] FA ratios were expressed as 1,000 times the ratio of the weight of the stomach plus intestine to the remaining body weight (weight of stomach and intestines/(total body weight - weight of stomach and intestines)). The total of 40 mice in control groups which received 50 μl of PBS and were humanely killed at different times showed a mean FA ratio +/- standard deviation of 61.33 +/- 2.29. Based on this value, FA ratios above 65 were considered positive in these studies.

Activity against CHO cells

The ability to kill or cause a change in morphology of CHO cells was estimated by the general methodology of Lockwood et al.[9] Subconfluent monolayers were harvested by rinsing the cells with 1 ml of 0.05% EDTA-0.25% trypsin in Dulbecco phosphate buffered saline (PBS) at 37°C for 10 minutes, mixing the dislodged cells with 5 ml of supplemented EMEM containing 1% FCS, and collecting the cells by centrifugation at 200 x g for 5 minutes. Portions of the washed cell suspension (ca. 1000 cells in 0.1 ml of supplemented EMEM containing 1% FCS were added to wells of 96-well microtiter assay plates (Corning Glass Works, Corning, N.Y.). Serial two-fold dilutions of the preparations to be tested were made in suspension medium and 10 μl of each dilution was added to separate microtiter wells. The cells were examined (without staining) for necrosis, elongation and rounding after incubation in 5% CO$_2$ for a minimum of 18 hours at 37°C. One CHO cell unit was defined as the reciprocal of the highest dilution that caused >50% elongation of the cells or that killed >50% of the cells (Fig. 2).

Screening of parent and mutants of V. fluvialis 807-77 for activity against CHO cells

Cultures of 807-77 prototype and mutant strains to be examined for CT, ET, or CL activities were prepared as described above for infant mouse inocula. Culture supernatant fluids obtained by centrifugation were stored at -20°C until assayed. Crude CEF extracts of the washed (10 ml of 0.85% NaCl) bacteria were prepared by shaking the cells for 2.5 minutes in a 37°C water bath (150 rpm) in 10 ml of 0.005 M Tris-HCl buffered saline (pH 7.1) containing 1 mg of polymyxin B per ml, and removing debris and intact cells by centrifugation (27,000 x g, 15 min, 4°C). The extracts were stored at

-20°C until assayed. In some experiments, the crude extracts were freed of contaminating CT activity by acid precipitation and ammonium sulfate fractionation as described by Lockwood et al.[8]

Expression of alkaline phosphatase activity by CL- mutants

Ninety-six well microtiter plates containing 180 μl of HIB, supplemented HIB, or minimal medium/well were inoculated with 20 μl from overnight HIB cultures of the organism to be assayed. After 24-48 hours at 30°C, 100 μl aliquots of each mini culture were transferred to a new plate and the turbidity of each well was determined on an automated plate reader at 620 nm. Polymyxin B (20 μl of a 20 mg/ml solution in 5 mM Tris-maleate pH 7.0) was added and the plates were incubated at 37°C for 30 minutes to permeabilize the cells. Eighty μl of substrate (4 mg p-nitrophenylphosphate/ml in 1.5 M Tris-CL pH 8.5) was added and the plates were incubated at 37°C. The plates were read on an automated plate reader in the dual beam mode at 414 nm with the residual turbidity at 540 nm subtracted. Specific activity is expressed as [A (414 nm - 540 nm)/(A 620 nm per minute of incubation)] x 1000.

RESULTS AND DISCUSSION

Survey of clinical and environmental isolates of V. fluvialis for production of CT, ET, and CL

Culture supernatant fluids and cell extracts from 148 isolates were screened in the CHO cell assay for CT, ET and CL activities. Comparison of

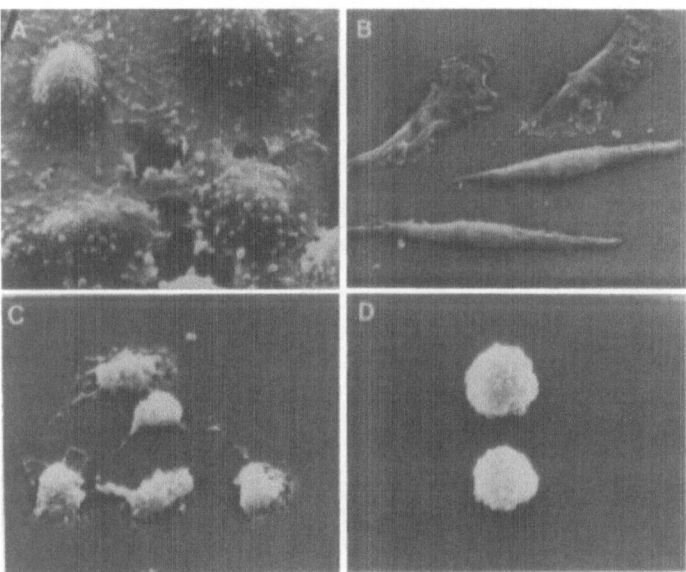

Fig. 2. Morphological changes elicited in CHO cells exposed to *V. fluvialis* products. Results are shown after exposure for about 18 hours at 37°C. (A) Control cells. (B) Cells treated with cell-free extracts containing enterotoxin activity (ET). (C) Cells treated with culture supernatant fluids containing cytotoxin (CT) and/or cytolysin (CL) activity. (D) Cells treated with culture supernatant fluids containing protease activity (CHO cell rounding).[8]

the activities found in 36 clinical specimens and 102 environmental isolates revealed that all except 2 of the environmental strains, which did not produce any detectable CHO toxin, elaborated one, two or all three of the putative virulence factors in various combinations with a wide range of titers. As shown in Fig. 3, strains producing titers of ET, CT and CL greater than 100 were more prevalent among the clinical isolates than in the environmental strains under the conditions of growth employed for screening. More of the environmental isolates were deficient in ET and CL production than clinical isolates but CT⁺ strains were more common in the former group. In order to choose the strain from our collection that produced the most cytolysin, several of the most active CL+ isolates were re-titrated. The results of this survey are shown in Table 1. Among the CL+ isolates tested, strain 807-77 consistently produced the highest CL titers and was used in all subsequent studies.

Construction of antibiotic resistant mutants and toxin negative mutants

Streptomycin resistant *V. fluvialis* 807-77 mutants were selected from concentrated overnight cultures for high or low resistance by their ability to grow in the presence of increasing concentrations of streptomycin as outlined in the methods section.

The broad host range transposon Tn*phoA* described by Manoil and Beckwith[6] was used to construct CL- mutants. Tn*phoA* is a derivative of *Tn5* that retains its broad host range and random insertion properties, and can create fusions between target genes and *phoA*, the alkaline phosphatase gene of *E. coli*. The gene fusions encode hybrid proteins composed of the carboxy terminal portion of alkaline phosphatase fused in frame to the amino terminal portion of the target gene product. In these experiments, plasmid pRT733 was used to transfer Tn*phoA* to the recipient 807-77 SmR as outlined in Fig. 1. Conjugation was permitted to proceed for 2 to 24 hours resulting in insertions, gene fusions and formation of hybrid proteins.

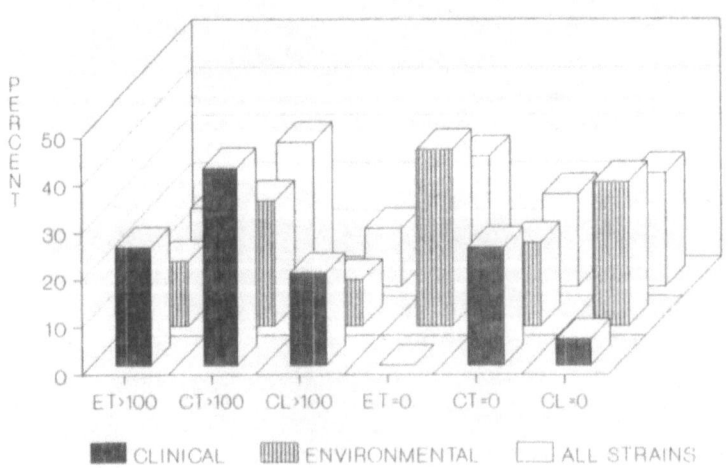

*TESTED IN CHO CELLS

Fig. 3.   Distribution of CT, ET, and CL in clinical and environmental isolates of *V. fluvialis*. Isolates were grown overnight with agitation at 30°C in HIB. Cells were removed by centrifugation and extracted with polymyxin B for ET. PB extracts and culture fluids were tested in CHO cells as described in the methods section.

164

Table 1. Screening of clinical and environmental isolates of *Vibrio fluvialis* for toxin production.[a]

Strain	Donor[b]	Source	CT	ET Titers	CL
807-77	1	Human	10	4800	300
5440	1	Human	640		100
8808	3	Human		15	180
SGM-547	4	Surface Water		320	50
10866	3	Human	800	320	100
C-1	5	Human	5120		
91-4	5	H_2O	640		40
79-10	5	H_2O	4	32	120

[a] Bacteria were grown as described in Materials and Methods.

[b] 1, Enterobacteriaceae section of the Centers for Disease Control (Atlanta, GA); 3, A. L. Furniss, Public Health Laboratory, Maidstone, United Kingdom; 4, Chesapeake Bay isolates received from R. R. Colwell, University of Maryland; 5, P. Echeverria, U.S. Army Research Laboratories, Bangkok, Thailand.

Determination of the temporal sequence of transconjugate formation by V. fluvialis strain 807-77 and pRT733

To determine the kinetics of transconjugate formation the time course of Tn*phoA*-positive colony appearance was followed using *V. fluvialis* strain 807-77 as the recipient and SM10 containing pRT733 as the donor as outlined above. As shown in Fig. 4, transconjugate formation was detected as early as 2 hours, steadily increased with time to reach a maximum at 6 hours, then rapidly decreased from that point. The average number of blue colonies

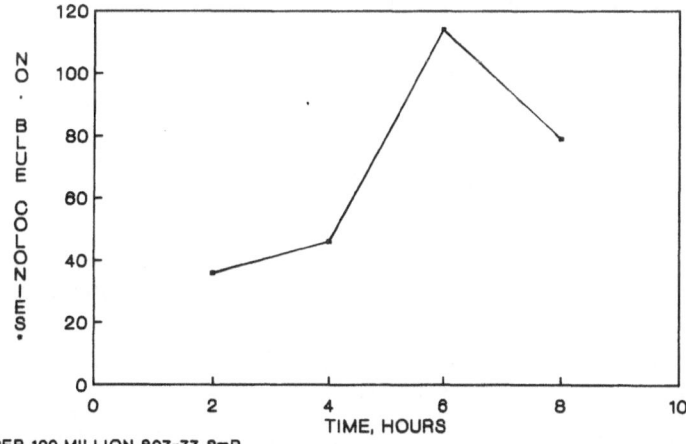

* PER 100 MILLION 807-77 SmR

Fig. 4. Temporal sequence of transconjugate formation. *Escherichia coli* strain SM 10 containing pRT733 was added at a 1:10 cell-cell ratio to 807-77 in HIB. Samples were removed at the indicated time, spread on CA plates containing Sm, Km and XP and the colonies scored as white, light blue or dark blue.

recovered for all experiments was approximately 1 per 10^6 recipients plated. All subsequent mating experiments were terminated at 6 hours.

Screening putative CL- mutants

Escherichia coli strain SM10 containing pRT733 was mated with *V. fluvialis* strain 807-77 SmR. Transconjugates were identified by the indicator plate assay for alkaline phosphatase as seen in Fig. 5a. This technique facilitates the isolation of large numbers of active gene fusions, many of which are targeted to genes encoding secreted or periplasmic proteins. The dark blue colonies represent fusions into genes coding for extracellularly secreted proteins such as cytolysin. The light blue colonies are inserts in periplasmic protein genes such as enterotoxin and the white colonies are insertions in nonsecreted protein genes.

Because we were primarily interested in the cytolysin which is known to be a secreted protein, dark blue colonies were selected as putative CL-mutants and were further screened. Several dark blue colonies from the XP plates were picked at random and gridded on blood agar plates. Halos around the colonies are CL+; colonies with no halos are potential CL- mutants (Fig. 5b). These experiments demonstrate the utility of the Tn*phoA* mutation technique for recovering numerous potentially useful mutants.

As shown in Fig. 5c, replicas of the initial blood agar plates showed that the potential mutant colonies were transferrable and stably maintained their non-hemolytic phenotype. Phospholipase A_2 is another known phenotypic expression associated with cytolysin activity that can also be used to directly select for CL- mutants. Egg yolk agar plates were replicated from the blood agar plates to test for concomitant loss of phospholipase A_2 activity. In all cases CL- colonies were also negative on egg yolk agar (data not shown). These results confirm our earlier biochemical studies which strongly suggested that the hemolysin molecule exhibited phospholipase A_2 activity.[14]

Verification of presumptive V. fluvialis CL- mutants

The putative CL- mutants identified on blood agar plates were confirmed by the highly sensitive tube hemolytic assay described in the methods section using sheep or mouse erythrocytes. The results summarized in Table 2 show that 807-77 and 807-77 SmR produced over 400 HU/ml. Of the twenty-two putative mutants screened, seventeen showed attenuated levels of hemolytic activity and five showed less than 2 HU/ml (below the limits of detection of the assay). The latter were considered confirmed CL- mutants. The other mutants are under study and may be useful in determining the mechanism(s) of action of the hemolysis and its location on the bacterial chromosome. All of the mutants were verified as to proper phenotype using the API rapid biochemical identification system.

Oral challenge of infant CFW mice with live V. fluvialis prototype strains and CL mutants

Groups of 10 or more mice were challenged with 5×10^8 viable organisms of *V. fluvialis* 807-77, 807-77 SmR and CL- 8, CL- 9, CL- 12, CL- 18, and CL- 20 mutants, each grown under conditions optimal for toxigenicity. The cells were harvested, washed free of preformed toxin and media constituents, resuspended to the required cell density and fed to the mice. Phosphate buffered saline (PBS) was fed as a negative control as described in Materials and Methods. Results, expressed as the means of FA ratios, are shown in Table 3. In some experiments, the intestines and stomachs were removed, homogenized in saline, and the viable bacteria present were quantitated. An average of 5×10^8 viable *V. fluvialis* were recovered from mice 4 hours post-feeding in these experimental groups. Four hours post feeding with 807-77 SmR and CL- mutants, there was an increased amount of fluid (FA = 75.14 +/- 6.85) within the stomach and intestines of the SmR mutants compared to the control animals. The mean FA ratios of the CL- mutant-challenged mice were significantly greater (P < 0.001) than the controls but were not significantly different from the prototype strain or the SmR mutant recipient (P = 0.186). These data strongly indicate that inactivation of CL has no detectable influence on FA ratios. Comparable mutants of appropriate CT+ and ET+ strains are under construction.

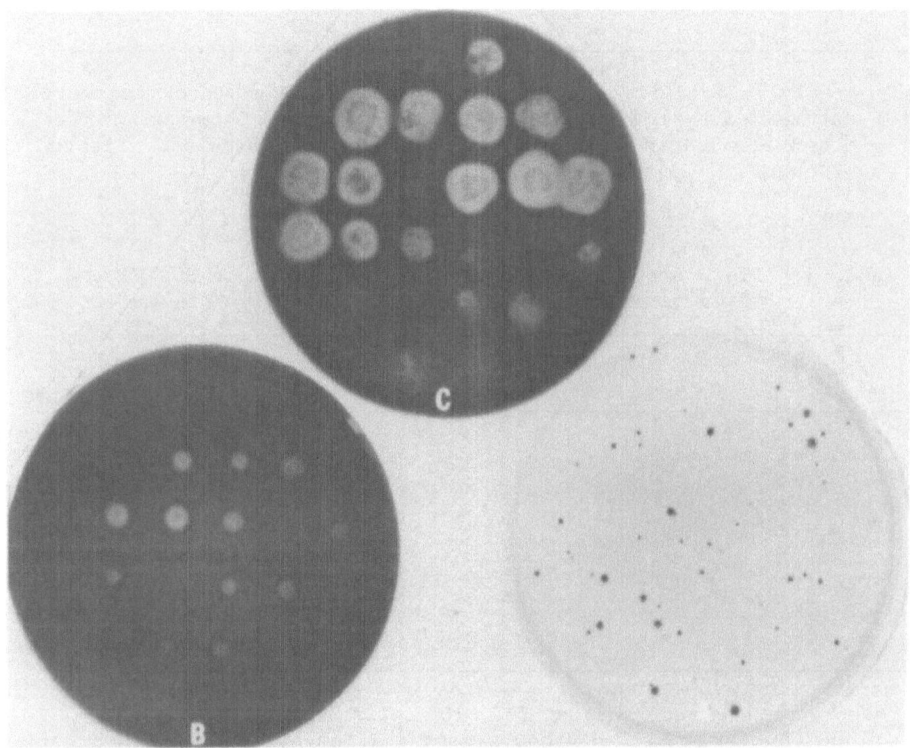

Fig. 5.   Selection and screening of CL- mutants. (A) 0.1 ml of a
          mating mixture from SM10 (pRT733) x 807-77 was spread on CA
          containing 100 µg/ml Sm, 45 µg/ml Km, and 40 µg/ml XP. Light
          blue colonies are inserts in periplasmic genes, and dark blue
          in genes coding for secreted proteins. The mixture was spread
          dilute to show more clearly the colony types. (B) Twenty-
          four dark blue colonies were picked at random from several
          plates prepared as described in Fig. 4a, stabbed into CA
          plates containing 5% sheep blood, and incubated at 30°C.
          Halos around colonies are CL+ colonies; no halos are potential
          CL- mutants. The plate shown is 24 hours old and contains 13
          CL+ and 11 CL- isolates. (C) Blood agar plate replicated from
          a plate similar to the one described in Fig. 4b. This plate
          is 72 hours old and contains 16 CL+ and 8 CL- isolates. The
          presumptive CL- clones were verified by tube hemolysin assays
          and phospholipase assays before being tested in infant mice
          for their potential to elicit fluid accumulation.

Table 2. Hemolytic activities of wild type and CL- *Vibrio fluvialis* mutants.

Sample	Cytolysin titers[a] HU/ml
807-77	400
807-77 SmR	400
CL- 8	<2
CL- 9	<2
CL- 12	<2
CL- 18	<2
CL- 20	<2

[a] Cytolytic activities against mouse and sheep erythrocytes were determined by the tube method (Bernheimer and Schwartz).[2] The other 17 mutants had titers between 10 and 100 units. Titers <2 cannot be detected with this assay.

Table 3. Fluid accumulation ratios of the prototype 807-77 and CL- mutants.

Sample	FA ratio[a]
Control (Buffer)	61.33 +/- 2.29
807-77	73.93 +/- 5.44
807-77 Sm[r]	75.14 +/- 6.85
CL- 8	78.80 +/- 10.24
CL- 9	78.50 +/- 5.34
CL- 12	77.75 +/- 8.44
CL- 18	78.87 +/- 11.29
CL- 20	76.40 +/- 4.88

[a] Mice were starved, sealed, fed ~10^9 cells in 50 μl PBS or PBS alone, humanely killed 4 hours post-challenge, and FA values were determined. FA ratios were determined in 10 mice per sample, values represent means +/- standard deviation.

Comparison of the Cytopathic effects of 807-77 wild type and CL- mutants on CHO cells

Since Tn*phoA* exhibited random transposition properties, a CHO cell assay was performed to determine whether the Tn*phoA* had inserted in sites in addition to the CL gene. CL- mutants were grown overnight at 30°C, washed with saline, polymyxin B extracted, and ammonium sulfate precipitated as described in Materials and Methods. The toxin levels were then determined as the reciprocal of the highest dilution of toxin production in CHO cells (Table 4).

Enterotoxin titers were masked in the polymyxin B extracts of CL- 9 and CL- 12 by the presence of low levels of cytotoxin which often contaminates crude extracts. After acid and ammonium sulfate precipitation

Table 4. Toxin titers[a] of wild type 807-77 and CL- mutants.

Sample[b]	Cytotoxin (CT)		Enterotoxin (ET)	
	Extract	AS	Extract	AS
807-77	40	8	1280	1280
807-77 Sm[r]	40	4	1280	1280
CL- 8	<10	4	160	64
CL- 9	40	10	Not detected	80
CL- 12	320	8	Not detected	640
CL- 18	160	40	1280	640
CL- 20	40	8	1280	640

[a] Toxin titers are the reciprocal of the highest dilution of toxin affecting CHO cells.
[b] Samples were grown overnight in HIB with appropriate antibiotics, washed twice in saline, polymyxin B extracted and acid and ammonium sulfate precipitated. Pellets were resuspended in 0.1M sodium phosphate buffer and assayed in CHO cells.

enterotoxin titers were close to normal for CL- 12, CL- 18, and CL- 20; however, CL- 8 and CL- 9 had clearly lower ET titers. Samples from the culture supernatants of mutants showed CHO cell killing and disintegration but the cells did not exhibit the compact and granular appearance typical of CL activity (data not shown). These data show that both cytotoxin and enterotoxin were still functional (attenuated activity or fully active) in these mutants when CL had been completely inactivated. It is possible that the attenuated mutants represent mutations in regulatory genes which often cause pleiotropic effects.

Alkaline phosphatase expression in Cl and V. hollisae TnphoA mutants

Representative CL mutants and an uncharacterized *V. hollisae* TnphoA mutant constructed by the procedure outlined in the methods section were assayed for alkaline phosphatase expression after growth in the standard HIB medium and in the medium supplemented with NaCl and lactose as osmotic stabilizers. The results of these experiments are summarized in Fig. 6. CL- 7 responded strongly to increasing concentrations of NaCl and lactose while CL-13 was less responsive. CL- 7 was also maximally stimulated by increased pH in the presence of 50 mM NaCl. The *V. hollisae* mutant (which is as yet uncharacterized) responded quite differently and was repressed in expression of alkaline phosphatase at increased NaCl concentrations. Lactose at 2 and 6% did not increase activity above that seen with unsupplemented HIB. These data indicate that the site of insertion in the target gene is important in determining gene-fusion activity and that these mutants will be useful in analysing the effects of ionicity and osmoregulators on protein secretion by halophiles.

The data presented here show that genes controlling hemolysin/cytolysin production by *V. fluvialis* growing at 30°C in aerated HIB are not necessary to elicit diarrhea in the infant mouse model. These results indicate that ET and/or CT is the major diarrheagenic factor. Experiments similar to those described here aimed at creating CT- and ET- mutants are underway. Another area of investigation is regulation of gene expression in these halophilic opportunists. Preliminary results (Fig. 6) show that TnphoA gene fusions in putative toxin genes will be useful monitors of which factors are expressed under conditions simulating the estaurine native habitat of these

microorganisms. Comparison of these results with those derived from organisms grown *in vivo* models such as the infant mouse or the sealed adult
mouse (SAM) may offer clues to the real functions of the putative factors.

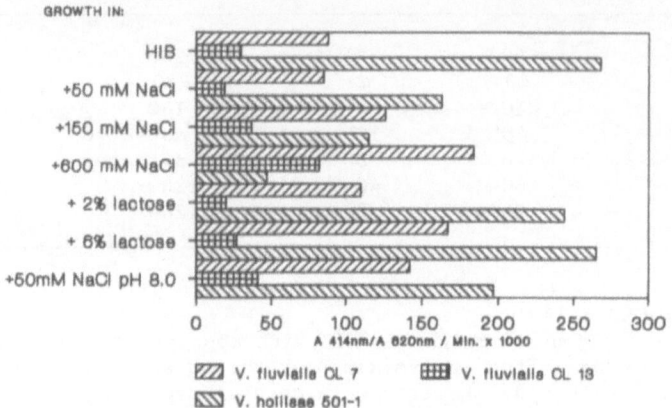

Fig. 6. Alkaline phosphatase activity of halophilic vibrio Tn*phoA*
mutants. Designated mutants were grown in media composed
as shown in the figure in 200 μl mini-cultures in microtiter
plates. Portions were removed and the turbidity measured on
an automated reader. The cells were then permeabilized with
polymyxin B, pnpp was added and the alkaline phosphatase
activity quantitated as indicated in the figure.

ACKNOWLEDGEMENTS

Portions of this work were supported by NIH grants AI 17840 and DK 38783.

REFERENCES

1.   Baselski V, Briggs R, Parker C, 1977, Intestinal fluid accumulation
      induced by oral challenge with *Vibrio cholera* or cholera toxin in
      infant mice, *Infect Immun.*, 15:704-712.
2.   Bernheimer AW, Schwartz LL, 1963, Isolation and composition of
      staphylococcal alphatoxin, *J Gen Microbiol.*, 30:455-468.
3.   Huq MI, Alam AKMJ, Brenner DJ, Morris GK, 1980, Isolation of Vibrio-
      like group EF-6, from patients with diarrhea, *J Clin Microbiol.*, 11:
      621-624.
4.   Kolter R, Inuzuka M, Helinski DR, 1978, Transcomplementation-dependent
      replication of a low molecular weight origin fragment from plasmid R6K,
      Cell, 15:1199-1208.
5.   Lederberg J, 1950, Isolation and characterization of biochemical mutants
      of bacteria, *Methods Med Res.*, 3:5-22.
6.   Lee JV, Donovan TJ, Furniss AL, 1978, Characterization, taxonomy, and
      emmended description of *Vibrio metschnikovii*, *Int J Syst Bacteriol.*, 28:
      99-111.
7.   Lee JV, Shread P, Furniss AL, 1981, Taxonomy and description of *Vibrio
      fluvialis* sp. nov. (synonym group F Vibrios, group EF-6), *J Appl Bact.*,
      50:73-94.
8.   Lockwood DE, Kreger AS, Richardson SH, 1982, Detection of toxins
      produced by *Vibrio fluvialis*, *Infect Immun.*, 35:702-708.
9.   Lockwood, DE, Richardson SH, Kreger AS, Aiken M, McCreedy B, 1983, *In*

vitro and *in vivo* biological activities of *Vibrio fluvialis* and its toxic products, *In*: "Advances In Research On Cholera And Related Diarrheas," Vol. 1, Kubuahara S, Pierce NF, eds., KTK Scientific Publishers, Tokyo, Japan.

10. Manoil C, Beckwith J, 1985, Tn*phoA*: A transposon probe for protein export signals, *Proc Natl Acad Sci USA.*, 82:8129-8133.

11. Seidler RJ, Allen DA, Colwell RR, Joseph SW, Daily OP, 1980, Biochemical characteristic and virulence of environmental group F bacteria isolated in the United States, *Appl Environ Microbiol.*, 40:715-720.

12. Simon R, Priefer U, Puhler A, 1983, A broad host range mobilization system for *in vivo* genetic engineering: Transposon mutagenesis in gram negative bacteria, *Biotechnol.*, 1:784-791.

13. Taylor RK, Miller VL, Furlong DB, Mekalanos JJ, 1987, Use of *phoA* gene fusions to identify a pilus colonization factor coordinately regulated with cholera toxin, *Proc Natl Acad Sci USA.*, 84:2833-2837.

14. Wall VW, Richardson SH, Kreger AS, 1984, Production and partial characterization of *Vibrio fluvialis* cytotoxin, *Infect Immun.*, 43:482-486.

MOLECULAR STRUCTURE OF *CLOSTRIDIUM BOTULINUM* PROGENITOR TOXINS

Genji Sakaguchi

University of Osaka Prefecture
College of Agriculture
804 Mozu-ume-machi 4-cho
Sakai-shi, Osaka 591, Japan

Progenitor toxin and its components

Clostridium botulinum "progenitor toxin" is defined as the ancestral toxin appearing first in foods and culture.[2] It is now known that the progenitor toxin is a complex of a toxic and a nontoxic component (Fig. 1). The molecule of the progenitor toxin dissociates when exposed to pH 7.2 or higher. The dissociated toxic component was named "derivative toxin."[2] The derivative toxin is the true "neurotoxin," but it does not seem entitled to be called "botulinum toxin," because it is highly toxic only when administered parenterally; ingestion of derivative toxin would not cause botulism. "Botulism" is the term used to designate a neuroparalytic disease such as that caused by oral ingestion of sausages containing the toxin.

Molecular structure of progenitor toxins

The molecular structure of the progenitor toxin is not uniform (Fig. 2). Until now, three forms have been distinguished based on the molecular size. They are "M toxin," "L toxin" and "LL toxin." M stands for "medium-sized," L for "large-sized," and LL for "extra-large sized." The molecular weight of M toxin is approximately 300 kDa with an $S_{20,w}$ of 12; that of L toxin approximately 500 kDa with an $S_{20,w}$ of 16; that of LL toxin approximately 900 kDa with an $S_{20,w}$ of 19.7[7]

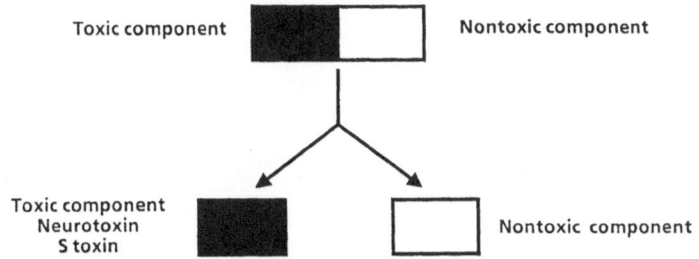

Fig. 1. Molecular structure of *C. botulinum* progenitor toxins.

Microbial Toxins in Foods and Feeds, Edited by A.E. Pohland *et al.,*
Plenum Press, New York, 1990

The toxic components of M, L and LL toxins are uniform in molecular size, with a molecular weight of approximately 150 kDa, whereas the nontoxic components of the three forms are different. The freed toxic component is called "derivative toxin," "neurotoxin," "small-sized" or "S toxin."

Any progenitor toxin is thus composed of one molecule each of the 7S toxic and the 7S or larger-molecular sized nontoxic components held together with at least one disulfide bond.

It is known that botulinum toxin contains "hemagglutinin," which is possessed by L and LL toxins only, or more exactly, the nontoxic components of L and LL toxins contain hemagglutinin, but that of M toxin does not.[7]

Progenitor toxins produced by C. botulinum types A-G

As shown in Table 1, type A strains produce the progenitor toxin in three different forms, LL, L, and M toxins; types B, C, and D in two different forms, L and M toxins; types E and F in a single form, M toxin; and type G also in a single form, L toxin. The quantities of LL, L and M toxins produced vary depending upon the strain, culture medium and cultural conditions

LL toxin is produced only by type A strains and has been crystallized.[3] We have some strains of C. botulinum type A that produce no hemagglutinin. Such strains should produce M toxin only and M toxin could not be crystallized.

We reported that in vegetables, such as string beans and mushrooms, a type A strain tended to produce a larger amount of LL and L toxins, while in meat, such as pork and tuna fish, the same strain tended to produce a larger amount of M toxin.[9]

Oral and parenteral toxicities

Using mice, we tried to correlate the molecular size of the toxin with the oral toxicity (Table 2). The oral toxicity of the progenitor toxin is

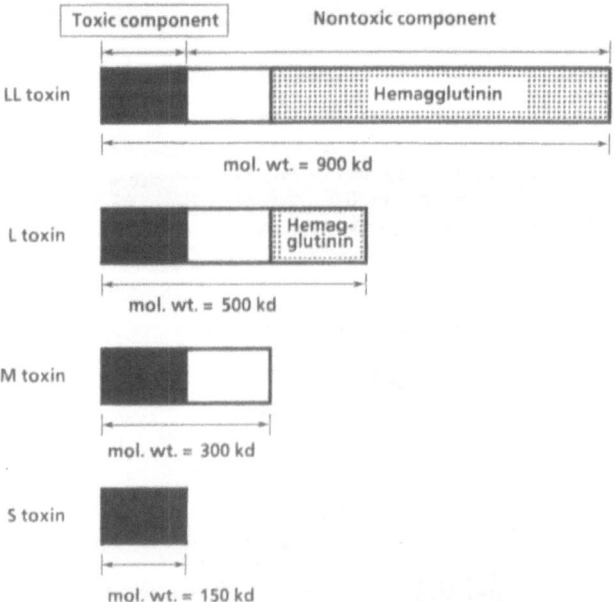

Fig. 2. Molecular forms of C. botulinum progenitor toxins.

Table 1. Production of progenitor toxins of different forms and of hemagglutinin by *C. botulinum* types A-G.

Type	Progenitor toxins produced			Hemagglutinin
	LL	L	M	
A	+	+	+	+
B	-	+	+	+
C	-	+	+	+
D	-	+	+	+
E	-	-	+	-
F	-	-	+	-
G	-	+	-	+

much higher than that of the derivative toxin and that of L toxin is apparently higher than that of M toxin of the same type. If the same toxicities in mouse ip LD_{50} were ingested, L toxin would be more toxic than M toxin, or if the same levels of toxin were produced in different kinds of foodstuffs and the same quantities of the foodstuffs were ingested, such foodstuffs that support production of L or LL toxin would be more fatal than those supporting production of M toxin.

Thus, within the same immunological type, the larger the molecular size of the toxin, the lower the specific parenteral toxicity, whereas the higher the specific oral toxicity.

Sensitivities of different forms of botulinum toxin to digestive enzymes

Next, attempts were made to correlate the molecular size of the toxin to the stability in the stomach.[10] Each toxin solution at a concentration of 50 μg/ml in 0.2 M sodium acetate-HCl buffer, pH 2.0, was incubated with pepsin at a concentration of 50 μg/ml at 35°C. Samples were taken periodically, diluted 10-fold, and injected intravenously into mice to titrate the remaining toxicity.

As shown in Fig. 3, types A and B derivative toxins lost their toxicities completely within 10 minutes. After incubation for 80 minutes, the remaining toxicities of type A crystalline and type B-L toxins were about 60% or more; those of type A-L, A-M, and B-M toxins were about 10%.

Thus, the larger the molecular size of the nontoxic component bound to the toxic component, the larger the protection afforded against destruction by gastric acid and pepsin. The nontoxic component, thus, plays an important role as a stabilizer of the unstable toxic component.

Intestinal absorption of botulinum toxin

We compared the intestinal absorption of the progenitor toxins of different molecular sizes (Fig. 4).[11] Wistar male rats were anesthetized and the abdomen opened along the midline. The thoracic duct was cannulated so that lymph could be collected continuously. After cannulation, a 5-cm segment of the duodenum of the rats was isolated with two ligatures. A 0.5-ml dose of type B-L, B-M, or B-S toxin containing 80,000,000 mouse ip LD_{50} was injected into a duodenal segment of a rat. Care was taken to prevent leakage of the toxin into the abdominal cavity. The incision was sutured

Table 2. Mouse intraperitoneal and oral LD_{50}/mg N of
botulinum toxins of different molecular sizes.

Toxin		Ip LD_{50}/mg N	Oral LD_{50}/mg N
Type	Size	$\times 10^{-3}$	$\times 10^{-3}$
A	LL	240,000	2,000
	L	300,000	140
	M	500,000	140
	S	1,000,000	23
B	L	300,000	190,000
	M	550,000	500
	S	1,100,000	46
C	L	57,000	12,000
	M	97,000	530
	S	180,000	
D	L	240,000	4,300
	M	500,000	1,000
	S	1,000,000	
E	M	50,000	230
	S	100,000	<130
F	M	120,000	45
	S	240,000	<40

and the rats were kept alive in restraining cages during collection of the
lymph.

After 6 hours, the lymph collected from the L-toxin recipient had a
toxicity of 25,000 mouse ip LD_{50}/ml and that collected from the M-toxin
recipient 200 mouse ip LD_{50}/ml. The lymph collected from the S-toxin
recipient failed to kill any mouse within 15 to 20 hours after injection.
The toxicity of lymph collected from the L-toxin recipient was about 0.01-
0.1% of the dose administered and that collected from the M-toxin recipient
only 1-2% of that collected from the L-toxin recipient.

Reverse passive hemagglutination tests detected both the toxic and nontoxic components to the same titers in lymph collected from the rat given any
of the toxins. The rates of appearance of the constituent components of each
toxin in lymph were nearly the same, reaching the highest titer in about 2
hours, in contrast to 6 hours required to attain the highest toxicity.
Reverse passive hemagglutination tests detected 0.6-1.5% of the toxic and
the same percentage of the nontoxic component in lymph collected from either
the L- or M-toxin recipient.

A 0.2-ml portion of the toxic lymph was centrifuged in 5 to 20% sucrose
density gradient at pH 6.0. The toxic component of the L-toxin recipient
showed the highest reverse passive hemagglutination titer at the position
of 7S, while the nontoxic component at the position of about 14S, indicating
that L dissociated but the two components did not re-associate. On the
other hand, both the toxic and nontoxic components in lymph collected from

the M-toxin recipient showed the highest reverse passive hemagglutination titers at the position of 12S, showing reconstruction of M toxin. The results were interpreted that both M and L toxins dissociated as soon as they were absorbed into the lymphatics and that the components of M toxin re-associated in a sucrose density gradient at pH 6.0, whereas those of L toxin did not.

No matter whether active or inactive, the progenitor toxin may not undergo molecular dissociation in the intestines before absorption. It is absorbed from the upper small intestines into the lymphatics, and the rate of absorption does not differ depending on the molecular size of the toxin. As soon as it is absorbed into the lymphatics, the molecular dissociation occurs and the toxic component reaches the susceptible nerve cells.

Persistence of both the components of botulinal toxin after absorption

To find the persistence of the lethal toxicity and the antigenicity of the toxin in the host body, a sublethal dose of M toxin of each of types B, C, D, E, and F was injected once intravenously into chickens and blood samples were withdrawn periodically from the chickens to determine the toxin remaining in the serum by the mouse injection test and by enzyme-linked immunosorbent assay for both the toxic and nontoxic components.[8] Type A toxin was not used for the experiments since the chicken is very highly susceptible to it. Inactivated type B and E toxins were used so that larger amounts could be injected.

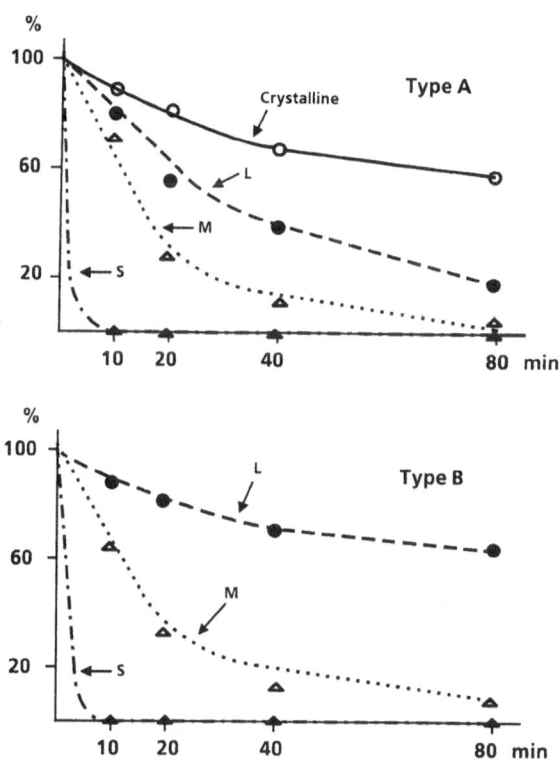

Fig. 3. Effects of pepsin at pH 2 on *C. botulinum* type A crystallin (mostly LL), L, M, an S toxins and type B L, M, and S toxins.

Figure 5 shows the results obtained with type B toxins. Twenty micrograms of inactivated type B-M toxin containing 20,000 and 1,800,000 mouse ip LD_{50} before and after tryptic activation, respectively, was injected intravenously into each of eight chickens. In 72 hours after injection, the mouse lethal activity was no longer detectable in the serum of the chicken, but considerable amounts of both toxic and nontoxic components were detected by ELISA. Both components were detected for at least a few days after the serum had become nontoxic to the mouse, indicating a higher stability of the antigenicities of both the components than the lethal toxicity in the chicken serum. For diagnosis of botulism, therefore, it seems justified to recommend detection of the antigen of the toxic component or nontoxic component or both in the serum by ELISA or any other sensitive immunological assay even if no toxin is detected by the mouse test. Such an immunological test would no doubt contribute to an increase in the rate of diagnosis of human and animal botulism cases.

In conclusion, the nontoxic components bound to the progenitor toxins play a very important role in pathogenesis of food-borne botulism and perhaps in diagnosis of botulism.

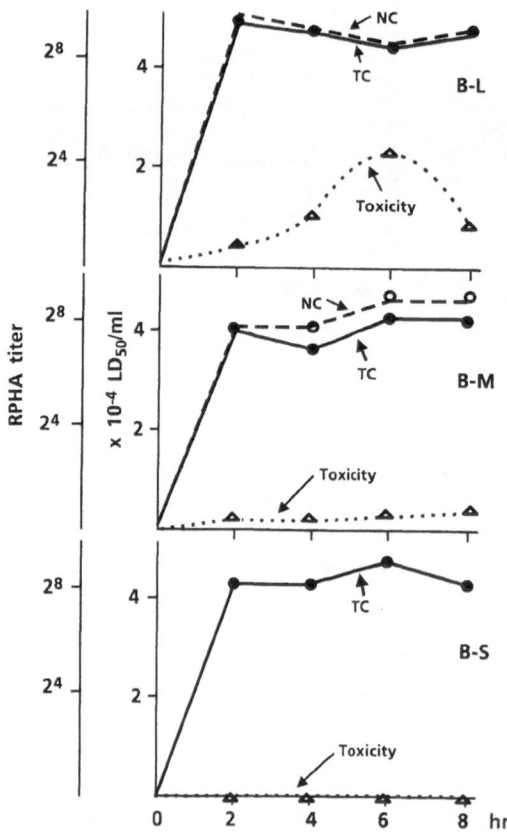

Fig. 4. Absorption of *C. botulinum* type B L, M, and S toxins from the ligated duodenum of the rat.

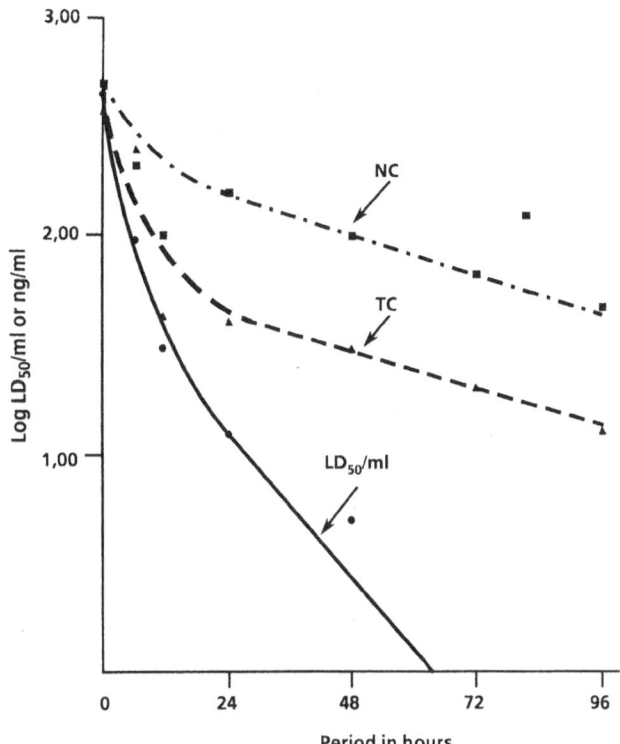

Fig. 5. Persistency of toxic (TC) and nontoxic components (NC) of
C. botulinum type B M toxin injected intravenously into
the chicken.

REFERENCES

1. Lamanna C, 1948, Haemagglutination by botulinal toxin, *Proc Soc Exp
    Biol Med.*, 69:332-336.
2. Lamanna C, Sakaguchi G, 1971, Botulinal toxin and the problem of
    nomenclature of simple toxins, *Bacteriol Rev.*, 32:242-249.
3. Lamanna C, MacElroy OE, Eklund MW, 1946, The purification and
    crystallization of *Clostridium botulinum* type A toxin, *Science*,
    103:613-614.
4. Nukina M, Mochida T, Sakaguchi S, Sakaguchi G, 1988, Purification of
    Clostridium botulinum type G progenitor toxin, *Zbl Bakt Hyg.*, A
    268:220-227.
5. Ohishi I, Sakaguchi G, 1980, Oral toxicities of *Clostridium botulinum*
    type C and D toxins of different molecular sizes, *Infect Immun.*,
    28:303-309.
6. Ohishi I, Sugii S, Sakaguchi G, 1977, Oral toxicities of *Clostridium
    botulinum* toxins in response to molecular size, *Infect Immun.*,
    16:107-109.
7. Sakaguchi G, 1983, *Clostridium botulinum* toxins, *Pharmac Therap.*,
    19:165-194.
8. Sakaguchi G, Sakaguchi S, Kurazono H, Kamata Y, Kozaki S, 1987,
    Persistence of specific antigenic protein in the serum of chickens
    given intravenously botulinum toxin type B, C, D, E or F, *FEMS
    Microbiol Let.*, 43:355-359.

9.   Sugii S, Sakaguchi G, 1978, Botulogenic properties of vegetables with special reference to the molecular size of the toxin in them, *J Food Safety*, 1:53-65.

10.   Sugii S, Ohishi I, Sakaguchi G, 1977, Correlation between oral toxicity and in vitro stability of *Clostridium botulinum* type A and B toxins of different molecular sizes, *Infect Immun.*, 16:910-914.

11.   Sugii S, Ohishi I, Sakaguchi G, 1977, Intestinal absorption of botulinum toxins of different molecular sizes in rats, *Infect Immun.*, 17:491-496.

PURIFICATION AND CHARACTERIZATION OF *BACILLUS CEREUS* ENTEROTOXIN AND ITS

APPLICATION TO DIAGNOSIS

Kunihiro Shinagawa

Department of Veterinary Medicine
Iwate University
3-18 Ueda, Morioka
Iwate, Japan 020

INTRODUCTION

Bacillus cereus is a large, gram-positive, spore-forming rod, and is aerobic or facultatively anaerobic. It is distributed widely in the environ ment including different kinds of foods and may cause food poisoning. Two types of B. cereus food poisoning are known;[2] the diarrheal type with an incubation period of 8 to 16 hours and the vomiting type with an incubation period of 1 to 5 hours. The diarrheal type characterized by diarrhea and abdominal pain resembles *Clostridium perfringens* food poisoning. The vomit- ing type is characterized by nausea and vomiting and resembles staphylo- coccal enterotoxin food poisoning. The agents of diarrheal type and vomit- ing type are toxins designated as "diarrheal toxin" (or "enterotoxin) and "emetic toxin," respectively. Diagnosis of *B. cereus* food poisoning by the detection of the toxin in clinical specimens or in incriminated foods has not been established.

Purification and characterization of *B. cereus* enterotoxin was attempted and an immune serum was produced against the purified material for establishing immunological methods for detection and quantitation of the enterotoxin. A reverse passive latex agglutination (RPLA) test was developed and enterotoxin produced by *B. cereus* strains isolated from outbreaks of food poisoning, various foods and soil was examined.

Incidence of B. cereus food poisoning and the biological activities of food poisoning strains

Incidence of B. cereus food poisoning. The first and most comprehen- sive description of the diarrheal syndrome was reported in 1950.[8] About 230 outbreaks of the diarrheal type were reported in Northern and Eastern European countries before 1976.[18] *Bacillus cereus* was the third most common cause of food poisoning in Hungary (117 outbreaks) between 1960 and 1968. Other countries that reported outbreaks are Finland (50 outbreaks), the Netherlands (11), and Canada(9).[18]

Another form of *B. cereus* food poisoning, now referred to as "vomiting syndrome," was found in the United Kingdom in 1972, when six cases were reported.[12] This type was most prevalent in the UK; there were 192 out- breaks with more than 1,000 cases from 1971 through 1984.[9] Other countries reporting outbreaks of this type include the Netherlands (50 outbreaks), Australia, Canada, and the United States.[18]

A total of 5,085 outbreaks of food poisoning with 175,133 cases were reported in Japan during 1983 through 1987. Bacteria were responsible for 3,720 outbreaks involving 143,861 cases. Eighty-seven percent of these were due to *Vibrio parahaemolyticus*, *Staphylococcus aureus*, and *Salmonella* (Table 1). Between 1977 and 1987, 130 outbreaks of *B. cereus* food poisoning were reported, 95 percent of which were the vomiting type, with 80 percent in the summer.

Biological activities of food poisoning strains

Some properties, biological activities, and assay methods for the diarrheal toxin have been reported.[1,3,4,15,16,19] The enterotoxigenicity was examined by biological assay.[14]

Brain Heart Infusion supplemented with 1 percent glucose was used for toxin production.[5,6] After incubation at 32°C for 6 hours, the culture was centrifuged and the supernatant fluid was filtered through a millipore filter (0.45 μm). The filtrate was assayed immediately for enterotoxin.

The vascular permeability (VP) test was used for assaying enterotoxin.[3,4] The culture supernatant fluid was injected intradermally into the back of a rabbit, followed in 3 hours with an injection of 2 percent Evans blue (2 ml/kg body weight) into the ear vein.

The animal was humanely killed 60 minutes post dye administration and the diameters of the blueing zones were measured. Culture filtrates of 15 diarrheal strains gave blueing zones of 15 to 24.9 mm followed by edema and hemorrhage. Seventy-one of 80 vomiting-type strains elicited no VP reaction (blueing zones less than 9.9 mm) (Table 2).

Bacillus cereus enterotoxin was also examined by the mouse ileal loop (MIL) test.[14] A 0.1-ml portion of the culture filtrate was injected into a 2.5 - 3.0 cm long section of ligated ileum. In 6 hours, the mice were killed and the abdominal cavities opened to measure the fluid accumulation in the ileal loop. Culture filtrates of five of 12 diarrheal strains induced fluid accumulation (0.03 to 0.1 ml/loop); two of them showed strong reactions (0.1 ml/loop). Culture filtrates of 52 vomiting-type strains failed to induce fluid accumulation.

The rabbit ileal loop (RIL)[15,19] test was also used for assaying the enterotoxins. A 1-ml portion of the fifty-fold concentrated culture filtrate was injected into a 7-8 cm long section of ileal loop in young rabbits, weighing 1.3 to 1.5 kg. After 8 hours, the rabbit was humanely killed and loops were examined for fluid accumulation. Six of 12 diarrheal strains induced fluid accumulation in the ileal loop (1.1 to 10 ml/loop). In comparison, only three of 30 vomiting-type strains induced fluid accumulation (1.1 to 5.0 ml/loop).

Purification and characterization of B. cereus enterotoxin

Several groups[1,16,19] reported purification and characterization of enterotoxin from culture filtrates, but the purification was apparently insufficient. Recently, enterotoxin produced by strain B-4ac of *B. cereus* was isolated by sequential chromatography on Amberlite CG-400, QAE-Sephadex, Sephadex G-75 and Hydroxylapatite was isolated. The toxin was associated with three antigenically distinct proteins. The activity was neutralized by antibody specific for two of the three proteins.

Table 1. Incidents of Bacterial Food Poisoning in Japan (1983 - 1987).

Agent	1983 Inci.*	Cases	1984 Inci.	Cases	1985 Inci.	Cases	1986 Inci.	Cases	1987 Inci.	Cases
Total	1,095	37,023(13)	1,074	33,084(21)	1,177	44,102(12)	899	35,556(7)	840	25,368(5)
Bacteria	769	31,125(3)	786	28,345(13)	877	36,566(3)	670	28,618	618	19,207
V. parahaemolyticus	305	11,235	384	8,222(1)	519	14,006(1)	343	12,138	321	8,149
Salmonella	109	3,612(1)	93	2,107(1)	82	2,412(1)	75	2,363	90	3,600
S. aureus	254	4,493(2)	205	4,813	163	4,968	155	3,885	139	3,616
E. coli, Ent. path.	30	3,355	27	6,151	34	3,899	28	2,141	15	1,616
C. botulinum	1	1	4	44(11)	1	1(1)	0	0	0	0
B. cereus	18	250	15	330	17	328	10	327	15	558
C. perfringens	16	4,571	12	1,733	9	971	22	3,258	9	288
Y. enterocolitica	0	0	0	0	0	0	0	0	0	0
C. jejuni/coli	31	3,405	39	4,652	50	9,497	34	4,368	28	1,378
Non-01 V. cholerae	2	16	3	156	1	8	2	38	0	0
Others	3	187	4	137	1	476	1	100	1	2
Unknown	202	5,136(1)	177	4,358	195	7,128	170	6,705	147	5,868

( ) Deaths, * Incidents

Table 2. Vascular Permeability Activities of *Bacillus cereus* Strains Isolated from Outbreaks of Food Poisoning.

Type of Food Poisoning	Starch Hydrolysis	Strains Tested	No. of positive strains: (mm^a)					Hemorrhage Positive
			0.0-4.9	5.0-9.9	10.0-14.9	15.0-19.9	20.0-24.9	
Diarrheal	+	15(2^b)	0	0	0	6	9(2)	13
Vomiting	−	80(1)	33	38(1)	9	0	0	0

[a] Average of diameters of the blueing zone.
[b] Reference strains, *B. cereus* B-4ac, 4433/73, and 4810/72.

We purified enterotoxin from a diarrheal-type strain possessing high VP and fluid accumulation activity.

Enterotoxin production

Bacillus cereus strain FM-1 isolated from a diarrheal-type food poisoning outbreak with high activities of VP and fluid accumulation in MIL and RIL tests was used. Brain Heart Infusion (Difco) was dialyzed against distilled water. The outer dialysate was supplemented with 1 percent glucose and used for enterotoxin production. After incubation at 32°C for 6 hours, the culture supernatant fluid was obtained by centrifugation.

Purification of enterotoxin

Step 1: The culture supernatant fluid contained enterotoxin at a titer of about 1,600 as determined by RPLA. The toxin was precipitated with ammonium sulfate at 56 percent saturation. The recovery was 85 to 90 percent and the ammonium sulfate precipitated preparation had a titer of 51,200 (Table 3).

Step 2: The precipitate was dissolved in distilled water and dialyzed against 0.005 M Tris-citric acid buffer, pH 4.8. The toxin was absorbed onto SP-Sephadex, C-25, equilibrated with the same buffer and eluted with 0.05 M sodium phosphate buffer, pH 7.0, containing 0.05 M NaCl. The eluate was concentrated to 10 ml with an Amicon hollow fiber. By this step, about 98 percent of the hemolysin was removed. The recovery of enterotoxin was 10.7 percent (Table 3).

Step 3: The concentrate was dialyzed against imidazole-HCl, pH 7.4. chromatographed on PBE 94 equilibrated with imidazole-HCl, pH 7.4, and eluted with 1/8th polybuffer 74, pH 4.0. The elution pattern of protein, enterotoxin, and hemolysin from chromatofocusing is shown in Fig. 1. Enterotoxin was eluted at a pH range between 5.5 and 5.7 with polybuffer 74. Hemolysin was eluted at a pH range between 4.1 and 4.3, being completely separated from the enterotoxin. The enterotoxin fractions were pooled and concentrated 30 times by lyophilization. The recovery was 8.0 percent (Table 3).

Step 4: The lyophilized toxin was dissolved in distilled water of about 1/30 the original volume, dialyzed against 0.075 M sodium phosphate buffer, pH 7.2 and passed through a column of Sephacryl S-300 equilibrated with the same buffer. The elution patterns of protein and enterotoxin of gel filtration on Sephacryl S-300 are shown in Fig. 2. The enterotoxin was eluted in the first protein peak. Purified enterotoxin (1.7 mg) was obtained (7.2 percent recovery) and the specific activity increased about 1,200 times from the first step (Table 3). The fractions containing the toxin were pooled and checked for purity. It was injected into rabbits for production of anti-enterotoxin sera.

The toxin was identified by SDS-PAGE and gel diffusion[11] test with antisera against the purified enterotoxin. The enterotoxin behaved as one component with a molecular weight of approximately 45,000 by SDS-PAGE in 7.5 percent polyacrylamide gel, pH 8.3 (Fig 3). The purified and crude enterotoxin formed a single precipitation line against the anti-enterotoxin serum, and showed a common line of identify with each other (Fig. 4).

Characterization of the enterotoxin

Biological activities. The biological activities and other properties of the purified enterotoxin are shown in Table 4.

Table 3. Purification of *Bacillus cereus* Enterotoxin.

Step	Volume (ml)	RPLA Titer	Protein (mg/ml)	RPLA Titer /mg Protein	Hemolysin Titer	Enterotoxin Recovery(%)
Culture supernatant	6,000	1,600	4.7	1	16	100
Precipitate at 56% saturated (NH$_4$)$_2$SO$_4$	164	51,200	83.6	1.8	256	87.5
Elution from SP-Sephadex (C-25)	10	102,400	23.8	12.7	64	10.7
Elution from Chromatofocusing on PBE94	1.5	512,000	5.2	289.3	8	8.0
Elution from Gel Filtration on Sephacryl S-300	2.7	256,000	0.62	1213.0	-	7.2

Table 4. Some Biological Activities and Other Properties of
         Bacillus cereus Enterotoxin

Nature	Antigenic protein
Molecular weight	*Ca.* 45,000
(Biological activities)	
Vascular permeability	+ (0.02 - 0.04 μg/rabbit)
Mouse lethality	+ (10 - 12 μg/mouse)
Mouse ileal loop	+ (2 - 4 μg/loop)
Cytotoxicity (Vero cell)	+ (0.1 - 0.2 μg)
Hemolysin	-
Phospholipase C	-
Neutralization of the	
above activities by	+
immune serum	
(Properties)	
Resistance to	
Trypsin (pH 8)	-
Pepsin (pH 5)	-
pH 6-9	+
<pH 3, >pH 11	-
Heating	
45°C, 10 min	+
56°C, 5 min	-
Storage	
-20°C, 60 days	+
4°C, 30 days	-

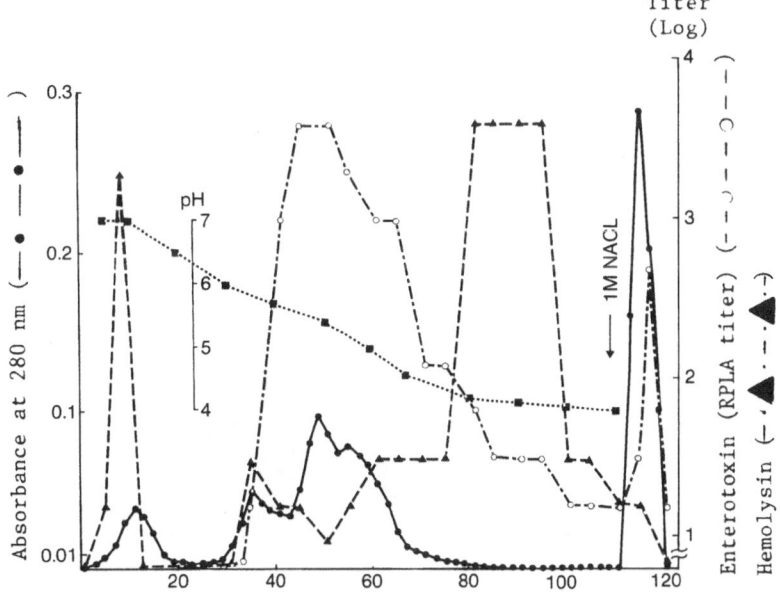

Fig. 1. Elution pattern from chromatofocusing column

Column	: Chromatofocusing PBE 94 (1 x 23 cm)
Eluting buffer	: Polybuffer 74, pH 4.0
Flow rate	: 33 ml/hr
Fraction volume	: (2 ml)

The enterotoxin increased VP (0.02-0.04 μg induced 7-mm or larger zones), induced fluid accumulation in MIL (2-4 μg/loop), was lethal to mice (10-12 pg/mouse) and showed cytotoxic activity (0.1-0.2 μg disrupted the Vero cell monolayer at 37°C for 18 to 20 hours). All the activities were neutralized with anti-enterotoxin serum at 10 to 100 fold dilution. It contained no hemolysin nor phospholipase C activity.

Properties. Purified enterotoxin was examined for stability under various conditions by assaying for VP activity.

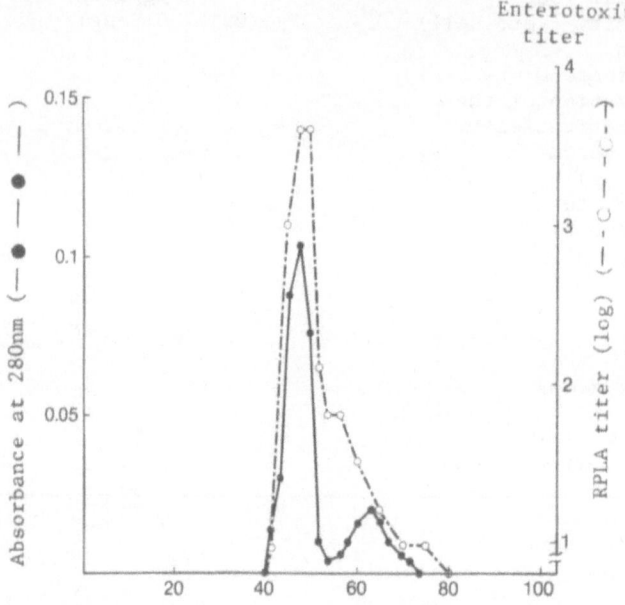

Fig. 2.   Elution pattern of gel filtration on Sephacryl S-300

        Column size   : 1.5 x 84 cm
        Eluting buffer   : 0.075M phosphate buffer, pH 7.2
        Flow rate   : 5.0 ml/hr.
        Fraction volume : (2 ml)

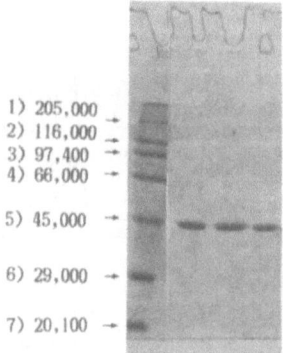

1) 205,000
2) 116,000
3) 97,400
4) 66,000
5) 45,000
6) 29,000
7) 20,100

Fig. 3.   Purity check and molecular weight determination of *B. cereus* enterotoxin by SDS-PAGE.

        1) Miosin,   2) Galactosidase,   3) Phosphorylase B
        4) Bovine albumin,   5) Egg albumin   6) Carbonic hydrase
        7) Trypsin inhibitor

The enterotoxin was labile. It was destroyed by treatment with trypsin (pH 7.8) or pepsin (pH 4.5), heat (56°C for 5 minutes), storage (4°C for 4 weeks), or exposure to pH values less than 3 or greater than 12. The biological activity of enterotoxin was neutralized with anti-enterotoxin serum.

Enterotoxin production by B. cereus strains

Methods for detection of enterotoxin. The symptoms of diarrheal-type *B. cereus* food poisoning may be caused by enterotoxin produced in the intestines. Detection of the toxin in feces, culture, and foods should be important for laboratory diagnosis, but a reliable method has not been developed. Biological assays such as VP test,[3,4] RIL[15,19] and MIL[14] test, and cytotoxicity test have been reported. A monkey feeding test for diarrheal activity was proposed[7,10] but the test proved to be nonspecific.[17] The biological assays are time consuming, costly and not satisfactorily specific. It seemed necessary to establish immunological assays that are simple and rapid for detection of enterotoxin.

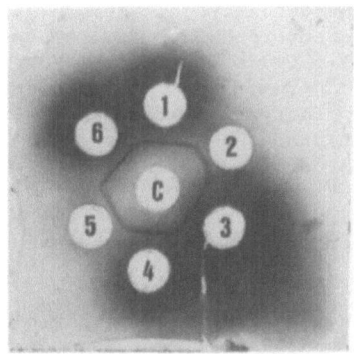

Fig. 4.   Agar gel diffusion with anti-enterotoxin serum and crude and purified enterotoxins.

Center well	: Anti-enterotoxin 1:2
Peripheral wells	: (1,4) purified enterotoxin (40µg/ml)
	: (2,5) purified enterotoxin (20µg/ml)
	(3,6) crude enterotoxin (20-40µg/ml)

RPLA test. We developed an RPLA test to simplify identification and quantification of enterotoxin.

Preparation of latex beads sensitized with anti-enterotoxin and with normal serum: 20 µg of anti-enterotoxin IgG and normal IgG purified by affinity chromatography were each mixed with an equal volume of 0.5 polystyrene latex beads at 37°C for 2 hours. The sensitized latex was blocked with 5 percent bovine serum albumin (BSA) and resuspended in phosphate buffered saline (PBS), pH 7.2.

Procedure of RPLA test: The culture supernatant was diluted with PBS containing 0.5 percent BSA and a 25-µl portion was mixed with an equal volume of control latex or sensitized latex in a well of a V-type microtiter plate and incubated in a humid chamber overnight at room temperature. Purified enterotoxin of a known concentration diluted with PBS containing 0.5 percent BSA was used as a controls.

The sensitivity and specificity of the RPLA test

The sensitivity and specificity of the RPLA test against purified enterotoxin, hemolysin and culture supernatants of various food poisoning bacteria and species from the genus *Bacillus* are shown in Table 5. Approximately 1 to 2 µg of enterotoxin per ml was detectable by the RPLA test. The specificity was very high reacting only with *B. cereus* enterotoxin and a culture supernatant fluid from *B. thuringiensis* var. *kurstaki*. Enterotoxins and hemolysins of other food poisoning bacteria, culture supernatant fluids of other members of the genus *Bacillus* and other unrelated bacteria failed to show positive reactions (Table 5).

Production of enterotoxin by B. cereus and B. thuringiensis strains as determined with the RPLA test

Starch hydrolysis[13] and enterotoxin production by *B. cereus* strains isolated from food poisoning outbreaks, normal foods, and soil are shown in Table 6. All 19 strains isolated from 13 outbreaks of diarrheal-type food poisoning were starch-hydrolysis positive and produced enterotoxin at levels ranging from 1.6 to 3.2 µg/ml. On the contrary, 105 strains isolated from 49 outbreaks of the vomiting-type were all starch-hydrolysis negative and 103 strains did not produce enterotoxin. The remaining two strains produced enterotoxin at a level of 1.6 µg/ml (Table 6). Forty of 50 starch-hydrolysis positive strains from normal foods produced enterotoxin at levels between 0.1 and 3.2 µg/ml. Forty-nine starch-hydrolysis negative strains isolated from foods did not produce enterotoxin. Of 98 strains isolated from soil, 70 were starch-hydrolysis positive and 28 were negative. Sixty-two (88.6 percent) of the starch hydrolysis-positive strains produced enterotoxin at levels between 0.1 and 3.2 µg/ml, and nine starch-hydrolysis negative strains produced enterotoxin at levels between 0.1 and 0.4 µg/ml (Table 6).

All of 51 *B. thuringiensis* strains of insecticide or soil origin were starch-hydrolysis positive and 49 strains (96 percent) produced enterotoxin at levels between 0.1 and 6.4 µg/ml. Twenty-two of *B. thuringiensis* strains produced a large amount of enterotoxin (1.6 to 3.2 µg/ml), the same level as *B. cereus* strains isolated from diarrheal-type food poisoning cases (Table 6).

Highly sensitive biological and immunological methods for detecting *B. cereus* enterotoxin, i.e., VP test (0.001-0.002 µg/ml in 4 hours), MIL test (2-4 µg/ml in 6 hours), cytotoxicity test (0.1-0.2 µg/ml in 24 hours), gel diffusion test (1.5-2.0 µg/ml in 48 hours) and RPLA test (1-2 ng/ml in 20-24 hours) in immunological assay have been developed. However, the biological assays are not necessarily specific. The immunological assays are simple and very specific, especially the RPLA test; it is the most sensitive method for detection of enterotoxin.

The RPLA test is the most practical method for detecting and quantitating *B. cereus* enterotoxin in cultures, extracts of stools from patients for diagnosis, and for assessing *B. cereus* in foods.

ACKNOWLEDGEMENTS

The author wishes to acknowledge the assistance of Professor Genji Sakaguchi, University of Osaka Prefecture, for his most helpful criticism and advice in preparing the manuscript and also extend my thanks to Mr. A. Agata, Nagoya City Health Research Institute, and Mr. H. Sugiya, Denka-Seiken Company for their cooperation with this research.

Table 5.   Sensitivity and Specificity of RPLA Test.

Toxin	Toxin (ng/ml)	RPLA Test
(Purified toxin)		
B. cereus enterotoxin	10	+++
B. cereus enterotoxin	1 - 2	+
B. cereus enterotoxin	0.1	-
E. coli enterotoxin		
Heat-labile	1,000	-
Heat-stable	1,000	-
C. perfringens		
enterotoxin	64,000	-
S. aureus		
enterotoxin A	128,000	-
enterotoxin B	25,000	-
enterotoxin C	6,400	-
enterotoxin D	3,200	-
V. cholerae		
enterotoxin	128,000	-
V. parahaemolyticus		
thermostable hemolysin	2,000	-
(Culture supernatant)		
B. thuringiensis var. *kurstaki*		+++
B. licheniformis IAM 11054		-
B. subtilis IAM 1069		-
B. megaterium IFO 3003		-
B. pumilis IFO 3813		-
S. aureus FRI-722(ET-A)		-
S. aureus 243(ET-B)		-
S. aureus FRI-361(ET-C)		-
E. coli 240-3(LT)		-
C. perfringens NCTC 8798		-
V. cholerae 569 B		-

Table 6. Enterotoxin Production by *Bacillus cereus* and *Bacillus thuringiensis* Strains Determined by RPLA Test.

Source	Starch Hydrolysis	No. of Strains Tested	No. of Strains Producing Enterotoxin (ug/ml)							
			<0.01	0.1	0.2	0.4	0.8	1.6	3.2	6.4
B. cereus										
Food poisoning										
Diarrheal-type	+	19	-	-	-	-	-	11	8	-
Vomiting-type	-	105	103	-	-	-	-	2	-	-
Untypeable	+	4	3	1	-	-	-	-	-	-
	-	2	2	-	-	-	-	-	-	-
Foods	+	50	10	1	8	9	10	10	2	-
	-	50	49	-	-	1	-	-	-	-
Soil	+	70	8	6	4	11	23	13	5	-
	-	28	19	2	3	4	-	-	-	-
Total		328	194	10	15	25	33	36	15	0
B. thuringiensis										
Insecticides, soil	+	51	2	1	6	9	11	16	3	3

REFERENCES

1. Ezepchuk YV, Bondarenko VM, Yakovleva EA, Koryagina IP, 1979, The
    Bacillus cereus toxin: isolation of permeability factor, *Zentralbl
    Bakteriol Hyg.*, I Abt Orig A, 244:275-284.
2. Gilbert RJ, 1979, *Bacillus cereus* gastroenteritis, *In*: Foodborne
    infections and intoxications, 2nd ed., Riemann H, Bryan FL, eds.,
    Academic Press, New York.
3. Glatz BA, Goepfert JM, 1973, Extracellular factor synthesized by
    Bacillus cereus which evokes a dermal reaction in guinea pigs, *Infect
    Immun.*, 8:25-29.
4. Glatz BA, Spira, WM, Goepfert, JM, 1974, Alteration of vascular permea-
    bility in rabbits by culture filtrates of *Bacillus cereus* and related
    species, *Infect Immun.*, 10:299-303.
5. Glatz BA, Goepfert JM, 1976, Defined conditions for synthesis of
    Bacillus cereus enterotoxin by fermenter-grown cultures, *Appl Environ
    Microbiol.*, 32:400-404.
6. Glatz BA, Goepfert, JM, 1977, Production of *Bacillus cereus* enterotoxin
    in defined media in fermenter-grown cultures, *J Food Prot.*, 40:472-474.
7. Goepfert JM, 1974, Monkey feeding trials in the investigation of the
    nature of *Bacillus cereus* food poisoning, *In*: Proc Fourth International
    Congress of Food Science and Technology, Vol 3, Institute Nacional de
    Ciencia Y Tecnoogia de Alimentos, Consejo Seuperior de Investigaciones
    Cientificas, Madrid.
8. Hauge S, 1955, Food poisoning caused by aerobic spore-forming bacilli.
    J Appl Bacteriol., 18:591-595.
9. Kramer JM, Gilbert RJ, 1988, *Bacillus cereus* and other *Bacillus*
    species, *In*: Foodborne bacterial pathogens, Doyle, MP, ed., Marcal
    Dekker, New York.
10. Melling J, Capel BJ, Turnbull PCB, Gilbert RJ, 1976, Identification
    of a novel enterotoxigenic activity associated with *Bacillus cereus*,
    J Clin Pathol., 29:938-940.
11. Ouchterlony O, 1949, Antigen-antibody reaction in gel, *Acta Pathol
    Microbiol Scand.*, 26:507-517.
12. Public Health Laboratory Service, 1972, Food poisoning associated with
    Bacillus cereus, *Brit Med J.*, 1:189.
13. Shinagawa K, Kunita N, Sasaki Y, Okamoto A, 1979, Biochemical charac-
    teristics and heat tolerance of strains of *Bacillus cereus* isolated
    from uncooked and cooked rice after food poisoning outbreaks, (In
    Japanese, English summary), *J Food Hyg Soc.*, Japan 20:431-436.
14. Shinagawa K, Matsusaka N, Konuma H, Kurata H, 1985, The relation
    between the diarrheal and other biological activities of *Bacillus
    cereus* involved in food poisoning outbreaks, *Jpn J Vet Sci.*, 47:
    557-565.
15. Spira WM, Goepfert JM, 1972, *Bacillus cereus*-induced fluid
    accumulation in rabbit ileal loops, *Appl Microbiol.*, 24:341-348.
16. Spira WM, Goepfert, JM, 1975, Biological characteristics of an
    enterotoxin produced by *Bacillus cereus*, *Can J Microbiol.*, 21:
    1236-1246.
17. Thompson NE, Ketterhagen MJ, Bergdoll MS, Schantz EJ, 1984, Isolation
    and some properties of an enterotoxin produced by *Bacillus cereus*,
    Infect Immun., 43:887-894.
18. Turnbull PCB, 1981, *Bacillus cereus* toxins, *Pharmacol Ther.*, 13:
    453-505.
19. Turnbull PCB, Kramer JM, Jorgensen K, Gilbert RJ, Melling J, 1979,
    Properties and production characteristics of vomiting, diarrheal, and
    necrotozing toxins of *Bacillus cereus*, *Amer J Clin Nutr.*, 32:219-228.

MODE OF ACTION OF SHIGA AND SHIGA-LIKE TOXINS

Yoshifumi Takeda[a]

The Institute of Medical Science
The University of Tokyo
4-6-1 Shirokanedai
Minato-ku, Tokyo 108, Japan

Enterohemorrhagic *Escherichia coli* 0157:H7 and other serotypes produce at least two distinct cytotoxins that affect Vero cells (Vero toxins).[9,14,15,17] Both Vero toxins have been purified to homogeneity and the properties of the purified toxins have been examined extensively.[9,17] One of them was found to be identical to Shiga toxin in molecular structure,[16] and thus its immuno-logical, physicochemical and biological properties are the same as those of Shiga toxin.[9] This toxin is called VT1, or Shiga-like toxin I.[15] The other Vero toxin, called VT2,[14] or Shiga-like toxin II,[15] is not related to Shiga toxin immunologically or physicochemically, although its biological activities are similar to those of Shiga toxin.[17] As the biological activities of VT2, such as its cytotoxicity to Vero cells, lethality to mice, and activity to cause fluid accumulation in ligated loops of rabbit ileum are similar to those of Shiga toxin and VT1,[17] it was of interest to study its mode of action and compare it with that of Shiga toxin and VT1.

Shiga toxin has long been known to inhibit protein synthesis in eukaryotic cells.[2,3,7] It inactivates eukaryotic 60S ribosomal subunits[3,13] and then inhibits polypeptide chain elongation by blocking the binding of amino-acyl-tRNA to ribosomes.[1] Obrig et al.[10] found that this inhibition of binding of aminoacy-tRNA to ribosomes by Shiga toxin was dependent on elongation factor-1(EF-1).

Like Shiga toxin, VT2 inhibited poly U-dependent polyphenylalanine synthesis in a rabbit reticulocyte cell-free system, but not in an *E. coli* system. Slight inhibition of polyphenylalanine synthesis was observed in a wheat germ system (Fig 1). Shiga toxin, VT1, and VT2 also inhibited globin synthesis in a rabbit reticulocyte system (data not shown). The inhibition of protein synthesis by VT1 and VT2, as by Shiga toxin, was due to inactiva-tion of 60S ribosomal subunits (Table 1).[8,12] In this experiment, ribosomes from rabbit reticulocytes were treated with either VT1 or VT2 and then hybrid ribosomes were prepared. As shown in Table 1, polyphenylalanine synthesis was greatly reduced when the 60S ribosomal subunits were derived from either VT1 or VT2 treated ribosomes. With 40 ribosomal subunits from toxin-treated ribosomes, no significant reduction of polyphenylalanine synthesis was observed.

To determine the step at which VT1 and VT2 inhibited protein synthesis,

---

[a] Present address: Department of Microbiology, Faculty of Medicine, Kyoto University, Sakyo-ku, Kyoto 606, Japan.

Microbial Toxins in Foods and Feeds, Edited by A.E. Pohland *et al.,*
Plenum Press, New York, 1990

Table 1. Abilities of VT1 and VT2 treated ribosomal subunits
of rabbit reticulocytes to support polyphenylalanine
synthesis.

Ribosomal subunits		[³H] Phe incorporated (p mol)	% Activity
40S	60S		
Normal	Normal	9.34	100
VT1-treated	Normal	9.22	98.7
Normal	VT1-treated	0.95	10.2
VT1-treated	VT1-treated	0.84	9.0
Normal	Normal	17.4	100
VT2-treated	Normal	15.2	87.4
Normal	VT2-treated	2.05	11.8
VT2-treated	VT2-treated	1.92	11.0

we examined their effects on various steps of protein synthesis. Results
showed that VT1 and VT2 did not inhibit the binding of Met-tRNA$_f$ to 40S
ribosomal subunits or 80S ribosomes, the formation of peptidyl puromycin,
or the formation of EF-2- or GTP-dependent phenylalanyl- or diphenylalanyl-
puromycin (data not shown). As shown in Fig. 2, VT2 inhibited EF-1-depen-
dent binding of aminoacyl-tRNA to ribosomes,[8,12] but did not inhibit non-
emzymatic binding of aminoacyl-tRNA to ribosomes. The same results were
obtained with Shiga toxin and VT1 (data not shown).

    The inhibitions of EF-1-dependent aminoacyl-tRNA binding to ribosomes
by Shiga toxin, VT1, and VT2 were assumed to be due to inactivation of 60S
ribosomal subunits by these toxins. Obrig et al.[11] reported that Shiga
toxin showed ribonuclease activity and suggested that this activity might
be responsible for ribosome inactivation. We examined the ribonuclease
activities of Shiga toxin and VT2, but failed to confirm the results of
Obrig et al.[11]

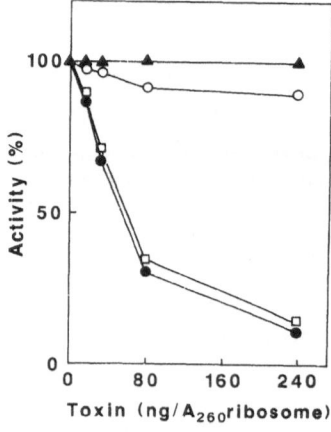

Fig. 1.   Inhibitions of protein synthesis by VT2.

        Poly U-dependent polyphenylalanine synthesis in *E. coli* system
        (▲), wheat germ system (○), rabbit reticulocyte system (●)
        and globin sythesis in rabbit reticulocyte system (□).

However, when 28S ribosomal RNA from 60S ribosomal subunits of rabbit reticulocytes was treated with aniline, the 28S rRNA from Shiga toxin- and VT2-treated ribosomes became smaller with concomitant appearance of a fragment of about 400 nucleotides (Fig. 3).[6] Endo and coworkers[4,5] have obtained with ricin and related lectins the same results as those described above and concluded that ricin and related lectins have RNA N-glycosidase activity. Thus we concluded that Shiga toxin, VT1, and VT2 also have RNA N-glycosidase activity.[6]

Several lines of evidence confirmed the conclusion that Shiga toxin and VT2 have RNA N-glycosidase activity. Table 2 shows that the production of a fragment of about 400 nucleotides by aniline treatment (aniline fragment) was correlated with inhibition of globin synthesis, and inhibition of EF-1-dependent phenylalanyl-tRNA binding to ribosomes in a rabbit reticulocyte system.[6] Table 2 also shows that the effects of Shiga toxin and VT2 on these reactions were prevented by homologous antibodies, but not by heterologous antibodies.

The nucleotide sequence of a fragment of 553 nucleotides from the 3' terminal of 28 SrRNA produced by Shiga toxin and VT2 was determined,[6] and results are shown in Fig. 4. This 553-nucleotide fragment of 28S rRNA was isolated from untreated, Shiga-toxin-treated and VT-2-treated ribosomes, and was labeled with [γ-^{32}P] ATP at the 5' terminus with polynucleotide kinase. The fragment was then partially digested with either RNAse phy M from *Physacium polycephalum*, RNAse T$_1$ or RNAse U$_2$. In some experiments, the fragment was treated with sodium carbonate buffer (pH 9.0) at 90°C for 1 minute (mild alkaline treatment) or 15 minutes (alkaline treatment). As shown in Fig. 4, the bands corresponding to the guanine base at position 4323 (G-4323) and adenine base at position 4324 (A-4324) were not detected on the gel of the 553-nucleotide fragments from Shiga toxin-treated and VT2-treated ribosomes. Treatment of the 553-nucleotide fragment with alkaline solution gave a band consisting of G-4323 and A-4324. This was clear after mild alkaline treatment.

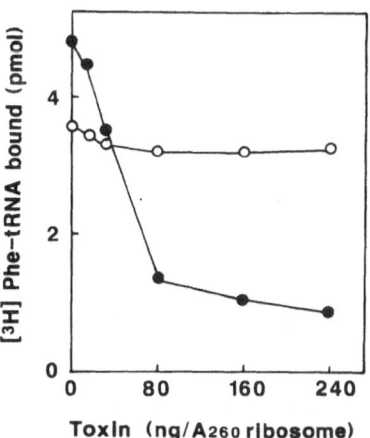

Fig. 2.   Inhibition of EF-1-dependent phenylalanyl-tRNA binding to ribosomes by VT2.

●, EF-1-dependent binding of Phe-tRNA; O, non-enzymatic binding of Phe-tRNA.

Table 2. Effects of antitoxins on inhibition of protein synthesis and degradation of 28S rRNA with aniline treatment by Shiga toxin and VT2.

Toxin	Antitoxin	Globin synthesis ([^{3}H] Leu incorporated (cpm)	EF-1-dependent Phe-tRNA binding (p mol)	Relative amount of aniline fragment
-	-	24,374	5.24	0
VT2	-	250	0.42	1.00
	anti-VT2	23,758	5.17	0.08
	anti-Shiga toxin	214	0.38	1.05
Shiga-toxin	-	225	0.53	0.98
	anti-VT2	324	0.51	0.94
	anti-Shiga toxin	23,158	5.09	0.07

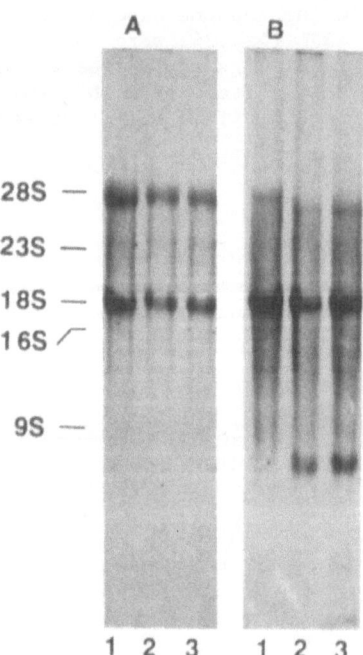

Fig. 3. Production of a fragment of 400-nucleotides by aniline treatment.

Ribosomal RNA from rabbit reticulocytes before (A) and after (B) aniline treatment. 1, rRNA from normal ribosomes; 2, rRNA from VT2-treated ribosomes; 3, rRNA from Shiga toxin-treated ribosomes.

The results in Fig. 5 demonstrate that Shiga toxin and VT2 cleaved the N-glycosidic bond of A-4324 of 28S rRNA in ribosomes.[6] In this experiment, the ethanol-soluble fraction of toxin-treated ribosomes was analyzed by thin-layer chromatography. As shown in Fig. 5, a marked increase of adenine, but not in other bases, was observed in the Shiga-toxin treated and VT2-treated samples. Quantitation of the released adenine showed that Shiga toxin and VT2 released 0.81 mol and 0.87 mol adenine per mol ribosomes, respectively. From these values, we concluded that only one of about 7000 N-glucosidic bonds in rRNA was cleaved by Shiga toxin and VT2. The RNA N-glycosidase activity of VT1 was not examined directly, but it seems reasonable to conclude that VT1 has the same enzymatic activity as Shiga toxin and VT2 because it is identical to Shiga toxin in primary structure.[16]

This study showed that Shiga toxin, VT1, and VT2 inactivate 60S ribosomal subunits by cleaving the N-glycosidic bond at A-3424 in 28S rRNA. By this inactivation of 60S ribosomal subunits, these toxins inhibit EF-1-

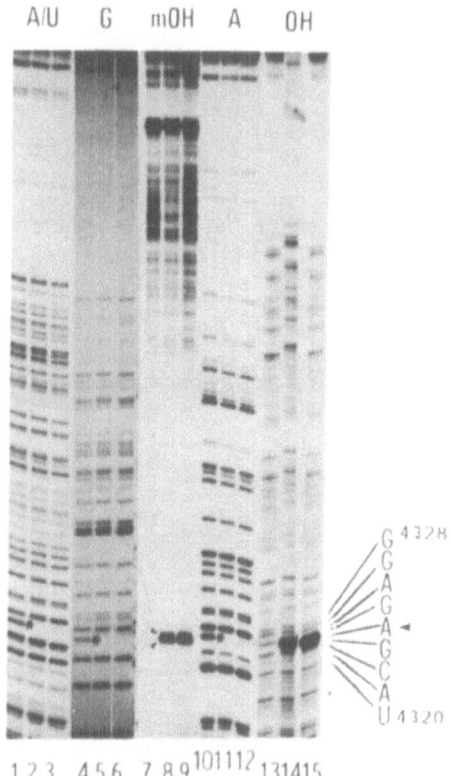

Fig. 4. Sequencing gel of the 553-nucleotide fragment of 28S rRNA.

A/U$_2$, treated with RNAse phy M; G, treated with RNAse T$_1$; A, treated with RNAse U$_2$; mOH, mild alkaline treatment; OH, alkaline treatment. Lanes 1, 4, 7, 10 and 13, fragment from untreated ribosomes; lanes 2, 5, 8, 11, and 14, fragment from VT2-treated ribosomes; lanes 3, 6, 9, 12 and 15, fragment from Shiga toxin-treated ribosomes. Asterisks denote bands missing on the gel. Arrow heads denote bands obtained by mild alkaline treatment that correspond to G-4323 and A-4324. The arrow denotes the site of action of Shiga toxin and VT2.

dependent binding of aminoacyl-tRNA to 60S ribosomal subunits with consequent inhibition of protein synthesis.   An important problem for future
study is how the inhibition of protein synthesis by these toxins is involved
in the pathogenesis of infections by *Shigella dysenteriae* and enterohemorrhagic *E. coli* producing Vero toxins.

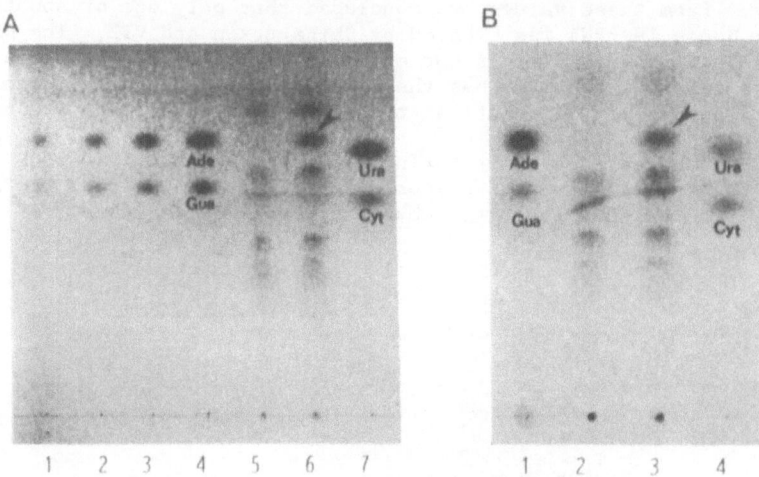

Fig. 5.   Thin-layer chromatogram of bases released from toxin-treated
ribosomes.

(A) Lane 5, sample from untreated ribosomes; lane 6, sample
from VT2-treated ribosomes; lanes 1-4 and 7, standard bases.
(B) Lane 2, sample from untreated ribosomes; lane 3, sample
from Shiga toxin-treated ribosomes; lanes 1 and 4, standard
bases.

REFERENCES

1.   Brown JE, Obrig TG, Ussery MA, Moran TP, 1986, Shiga toxin from
     Shigella dysenteriae 1 inhibits protein synthesis in reticulocyte
     lysates by inactivation of aminoacyl-tRNA binding, *Microbial
     Pathogenesis*, 1:325-334.
2.   Brown JE, Rothman SW, Doctor BP, 1980, Inhibition of protein synthesis
     in intact HeLa cells by *Shigella dysenteriae* 1 toxin, *Infect Immun.*,
     29:98-107.
3.   Brown JE, Ussery MA, Leppla SH, Rothman SW, 1980, Inhibition of protein
     synthesis by Shiga toxin: Activation of the toxin and inhibition of
     peptide elongation, *FEBS Lett.*, 117:84-88.
4.   Endo Y, Mitsui K, Motizuki M, Tsurugi K, 1987, The mechanism of action
     of ricin and related lectins on eukaryotic ribosomes, *J Biol Chem.*,
     262:5908-5912.
5.   Endo Y, Tsurugi K, 1987, RNA N-glycosidase activity of ricin A-chain,
     J Biol Chem., 262:8128-8130.
6.   Endo Y, Tsurugi K, Yutsudo T, Takeda Y, Ogasawara T, Igarashi K, 1988,
     Site of action of a Vero toxin (VT2) from *Escherichia coli* O157:H7 and
     of Shiga toxin on eukaryotic ribosomes: RNA N-glycosidase activity of
     the toxins, *Eur J Biochem.*, 171:45-50.
7.   Hale T, Formal SB, 1981, Protein synthesis in HeLa or Henle 407 cells
     infected with *Shigella dysenteriae* 1, *Shigella flexneri* 2a, or
     Salmonella typhimurium W118, *Infect Immun.*, 32:137-144.

8. Igarashi K, Ogasawara T, Ito K, Yutsudo T, Takeda Y, 1987, Inhibition of elongation factor 1-dependent aminoacyl-tRNA binding to ribosomes by Shiga-like toxin I (VT1) from *Escherichia coli* O157:H7 and by Shiga toxin, *FEMS Microb Lett.*, 44:91-94.

9. Noda M, Yutsudo T, Nakabayashi N, Hirayama T, Takeda Y, 1987, Purification and some properties of Shiga-like toxin from *Escherichia coli* O157:H7 that is immunologically identical to Shiga toxin, *Microbial Pathogenesis*, 2:339-349.

10. Obrig TG, Moran TP, Brown JE, 1987, The mode of action of Shiga toxin on peptide elongation of eukaryotic protein synthesis, *Biochem J.*, 244:287-294.

11. Obrig TG, Moran TP, Colinas RJ, 1985, Ribonuclease activity associated with the 60S ribosome-inactivating proteins ricin A, phytolaccin and Shiga toxin, *Biochem Biophys Res Commun.*, 130:879-884.

12. Ogasawara K, Ito K, Igarashi K, Yutsudo T, Nakabayashi N, Takeda Y, 1988, Inhibition of protein synthesis by a Vero toxin (VT2 or Shiga-like toxin II) produced by *Escherichia coli* O157:H7 at the level of elongation factor 1-dependent aminoacyl-tRNA binding to ribosomes, *Microbial Pathogenesis*, 4:127-135.

13. Reishig R, Olsnes S, Eiklid K, 1981, The cytotoxin activity of *Shigella* toxin: Evidence of catalytic inactivation of the 60S ribosomal subunits, *J Biol Chem.*, 256:8739-8744.

14. Scotland SM, Smith HR, Rowe B, 1985, Two distinct toxins active on Vero cells from *Escherichia coli*, *Lancet*, ii:885-886.

15. Strockbine NA, Marques LRM, Newland JW, Smith HW, Holmes RK, O'Brien AD, 1986, Two Toxin-Converting phages from *Escherichia coli* O157:H7 strain 933 encode antigenically distinct toxins with similar biological activities, *Infect Immun.*, 53:135-140.

16. Takao T, Tanabe T, Hong Y-M, Shimonishi Y, Kurazono H, Yutsudo T, Sasakawa C, Yoshikawa M, Takeda Y, 1988, Identity of molecular structure of Shiga-like toxin I (VT1) from *Escherichia coli* O157:H7 with that of Shiga toxin, *Microbial Pathogenesis*, 5:357-369.

17. Yutsudo T, Nakabayashi N, Hirayama T, Takeda Y, 1987, Purification and some properties of a Vero toxin from *Escherichia coli* O157:H7 that is immunologically unrelated to Shiga toxin, *Microbial Pathogenesis*, 3:21-30.

STRUCTURE AND FUNCTION OF *CLOSTRIDIUM PERFRINGENS* ENTEROTOXIN

T. Uemura, T. Akai and G. Sakaguchi

College of Agriculture
University of Osaka Prefecture
Mozu Ume-machi 4-cho
Sakai-shi, Osaka 591, Japan

INTRODUCTION

Food poisoning by *Clostridium perfringens* occurs after ingestion of food contaminated with large numbers of *C. perfringens* vegetative cells, and it is characterized by diarrhea and abdominal pain. The causative factor of the diarrhea is *C. perfringens* enterotoxin, which is produced in the digestive tract of humans by the organisms during sporulation. Studies made in the past 20 years have confirmed that *C. perfringens* enterotoxin is a single polypeptide with a M_r of 35,000 and manifests cytotoxic action to many mammalian cells by altering their membrane permeability.[3,7,11,21] In the present communication, we summarize our recent findings as well as a few of our unpublished ones involving the structure and functions of *C. perfringens* enterotoxin.

MICROHETEROGENEITY

Certain strains of *C. perfringens* type A of different Hobbs' serotypes produce immunologically identical enterotoxin.[12] Some *C. perfringens* type C and D strains also produce the same type of *C. perfringens* enterotoxin.[15,20] Multiple forms of the enterotoxin, however, have been reported. Two enterotoxin-like proteins were extracted with mercaptoethanol and dithiothreitol from mature spores of *C. perfringens* type A.[4] Also, two distinct proteins isolated from the sporangial extract of *C. perfringens* type A, appeared to be identical to the enterotoxin in the mechanism of action.[1] Other investigators showed that treatment of enterotoxin with trypsin caused a threefold increase in the biological activity.[6] Reports have not appeared on changes in biological activities of different forms of *C. perfringens* enterotoxin treated with trypsin.

Separation of derivatives of C. perfringens enterotoxin[16]

We fractionated the conventionally purified *C. perfringens* enterotoxin by high-pressure liquid chromatography (HPLC) and investigated the effect of trypsinization on the biological activity and physicochemical properties of each fraction. In this communication, the term "derivatives of the enterotoxin" will be used to designate the separated fractions.

Microbial Toxins in Foods and Feeds, Edited by A.E. Pohland *et al.*,
Plenum Press, New York, 1990

Clostridium perfringens enterotoxin was conventionally purified from the sporulating cells of the strain NCTC 8239 using methods previously published.[14] Several lots of the purified enterotoxin were freeze-dried and stored at -20°C. HPLC was run at room temperature with a DEAE-5PW 7.5 x 75 mm column equilibrated with Tris-HCl buffer (0.02M, pH 7.5). Elution was performed with a linear NaCl gradient (0 to 0.5 M) in the buffer at a flow rate of 1 ml/minute. Various preparations of purified enterotoxin were resolved into 5 peaks by HPLC. These peaks were named P1, P2, P3, P4 and P 5 in the order of the retention time (Fig. 1). Figure 1A shows the HPLC pattern of freshly purified enterotoxin; P3 was always the main peak. Figure 1B is the pattern with the purified enterotoxin stored at 4°C for less than 3 weeks. Other peaks appeared after storage of the purified preparation for longer periods. In HPLC of the purified enterotoxin stored for 7 years at -20°C, P3 became minor as seen in Figure 1C.

Trypsinization of derivatives of C. perfringens enterotoxin[18]

Three proteins, designated P1, P2 and P3, were separated from the purified *C. perfringens* enterotoxin by HPLC on a DEAE-5PW column with a shallow linear gradient of NaCl. Trypsinization of those derivatives was performed in Tris-HCl buffer by incubating an enterotoxin-trypsin mixture for 4 hours at 37°C with occasional stirring. The reaction was stopped by adding twice the amount of trypsin inhibitor. The weight ratio of trypsin to enterotoxin was 1:2. P1, when trypsinized and rechromatographed, revealed the same HPLC

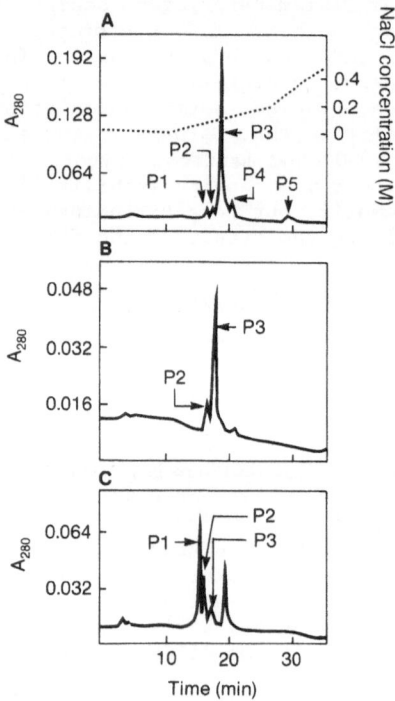

Fig. 1. HPLC of various preparations of conventionally purified enterotoxin on a DEAE-5PW column.[16]

The samples were: (1A) conventionally purified enterotoxin just after purification; (1B) the same lot depicted in 1A stored for 21 days at 4°C; (1C) purified enterotoxin after storage for 7 years at -20°C. Solid line, absorbancies at 280 nm; dotted line, NaCl concentration.

pattern as that of untrypsinized P1, but a new peak, P2', emerged on HPLC
from P2 and P3 upon trypsinization (Fig. 2). The biological activity was
determined by a Vero cell-staining assay.[17] The relative biological activity
of trypsinized P1 was on the same level as that of untreated P1, but those
of P2 and P3 upon trypsinization increased one and one half to three times
and finally reached the same level as those of P1 and P2' (Table 1).

From these results, the degradation process of *C. perfringens*
enterotoxin may be summarized as shown in Fig. 3.

FRAGMENTATION

Many bacterial toxins consist of at least two functionally distinct
domains, the binding and the biologically active domains. It is not known
whether *C. perfringens* enterotoxin contains these domains. A peptide of
16,000 daltons (16K peptide) was isolated by treating *C. perfringens* enterotoxin with trypsin in the presence of 0.05 percent SDS.[5] The 16K peptide
inhibited protein synthesis in a cell-free system but did not show any cytotoxic activity to Vero cells. No other report has appeared on structural
analyses of *C. perfringens* enterotoxin. In this communication we describe
the a) purification of a 15,000 dalton fragment (F15) from *C. perfringens*
enterotoxin by chromatography on a DEAE-Sephacel column after treating CPE

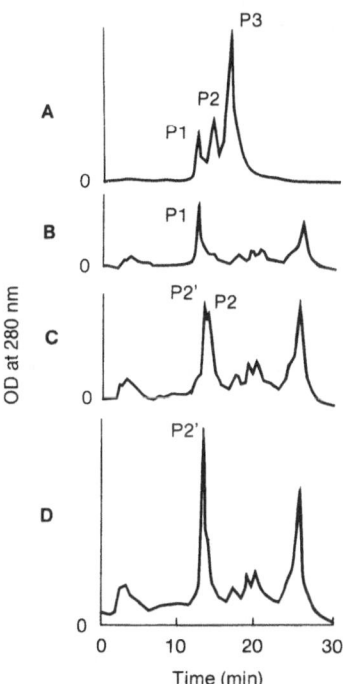

Fig. 2. HPLC of conventionally purified enterotoxin and its separated
peaks on a DEAE-5PW column.[18]

Three proteins, named P1, P2 and P3, were separated from the
purified *C. perfringens* enterotoxin by HPLC on a DEAE-5PW
column with a less steep linear gradient of NaCl than in Fig.
1. The samples were: (A) conventionally purified enterotoxin;
(B) trypsinized P1; (C) trypsinized P2; (D) trypsinized P3.
Absorbance at 280 nm.

Table 1. The Biological Activities of the Enterotoxin
Derivatives before and After Trypsin Treatment.[18]

| Derivative | TCD$_{50}$* in μg/ml (relative toxicity) | |
	Untreated	Trypsinized**
Peak 1	0.18 (3.0)	0.20 (2.7)
Peak 2	0.28 (1.9)	0.20 (2.7)
Peak 3	0.54 (1.0)	0.18 (3.0)

*   The dose of enterotoxin required to stain 50 percent of Vero cells.
** Trypsin treatment was performed for 4 hours at 37°C. The ratio of
trypsin to each derivative was 2.0 on protein basis.

with 2-nitro-5-thiocyanobenzoic acid (NTCB), a reagent which specifically
cleaves the amino-peptide bond of cysteine residues and b) discuss the
function of the F15 fragment.

Purification of the F15 fragment

     Conventionally purified *C. perfringens* enterotoxin was treated with
NTCB (Sigma Chemical Co., St. Louis, Missouri) by the method previously
reported.[8] A fragment was separated by chromatography on a DEAE-Sephacel
(Pharmacia, Uppsala) column equilibrated with 0.02 M Tris-HCl buffer, pH
8.0 by elution with a linear NaCl gradient (0-0.5 M). A major peak emerged
at a NaCl concentration of 0.045M (Figure 4). The fractions containing the
peak were further analyzed by slab gel electrophoresis[2] and SDS polyacrylamide gel electrophoresis (SDS-PAGE).[10] The peak fraction migrated as
a single band and showed neither agglutination nor complex formation on SDS-
PAGE as seen with conventionally purified enterotoxin (lane 2, Figure 5) or
NTCB-treated enterotoxin (lane 3, Figure 5). The molecular weight of the
protein in the fraction was determined at about 15,000 daltons and designated
the "F15" fragment. On slab gel electrophoresis at pH 9.4, F15 migrated to
a more basic position than did *C. perfringens* enterotoxin (Figure not shown).
Isoelectric focusing was performed as previously reported.[14] F15 had an
isoelectric point of 5.85 by isoelectric focusing (Figure 6), while that of
purified enterotoxin was 4.40.

     Richardson and Granum presented the amino acid sequence of *C. perfrin-
gens* enterotoxin consisting of 309 amino acid residues in which only one
cysteine residue was located at residue 177 from the N-terminal end of *C.
perfringens* enterotoxin.[13] The molecular size of the F15 fragment corresponds to that of the C-terminal part of the enterotoxin (133 amino acids).

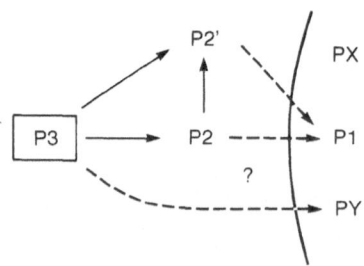

Fig. 3. Degradation of enterotoxin during storage.

Horiguchi et al.[8] showed that the amino acid composition of the protein corresponded to that of the C-terminal part of *C. perfringens* enterotoxin determined by Richardson and Granum.[13] These results confirm that the F15 fragment is located on the C-terminal part of the enterotoxin.

DETERMINATION OF THE BIOLOGICAL ACTIVITIES OF THE F15 FRAGMENT

The monoclonal antibodies (MCA) used in this study were those described in an earlier report.[9] ^{125}I-labeled enterotoxin or F15 was allowed to bind to Vero cells as previously reported.[8] Cytotoxicity of the enterotoxin and F15 was measured by ^{51}Cr-labeled Vero cells as reported previously.[9] The F15 fragment bound to each of the 4 MCAs prepared by immunizing mice with *C. perfringens* enterotoxin[8] as determined by immunoblot analyses.

Cytotoxic activity of the F15 fragment to Vero cells

The F15 fragment at concentrations up to 10 μg/ml did not reveal the cytotoxic activity, although it inhibited the *C. perfringens* enterotoxin-induced cytotoxicity.

Binding activity of the F15 fragment to Vero cells

[125]I-labeled F15 bound to Vero cells. The binding was saturable; 150 ng F15 against 10^5 Vero cells. Binding of ^{125}I-labeled F15 to Vero cells was inhibited by either enterotoxin or F15 (Figure 7A). F15 inhibited the binding of ^{125}I-enterotoxin to Vero cells (Figure 7B). Scatchard plot analysis revealed the K_a value of the binding of F15 with MCAs (Table 2). The K_a value of F15 with 2-B-4 was the largest among MCAs used. Those of 2-H-2 and 3-B-2 were almost the same and smaller than that of 2-B-4. The binding of ^{125}I-enterotoxin to Vero cells in the presence of MCA is different from that of ^{125}I-labeled F15. Figure 8A shows the binding of ^{125}I-enterotoxin to Vero cells in the presence of MCA; Figure 8B the binding of ^{125}I-labeled F15. One to two moles of 2-B-4 inhibited the binding of 1 mole of ^{125}I-enterotoxin or ^{125}I-F15 to Vero cells. MCA 3-B-2 partially inhibited the binding of both ^{125}I-enterotoxin and ^{125}I-F15. While 2-H-2 did not inhibit the binding of ^{125}I-enterotoxin to Vero cells, it did inhibit that of ^{125}I-F15.

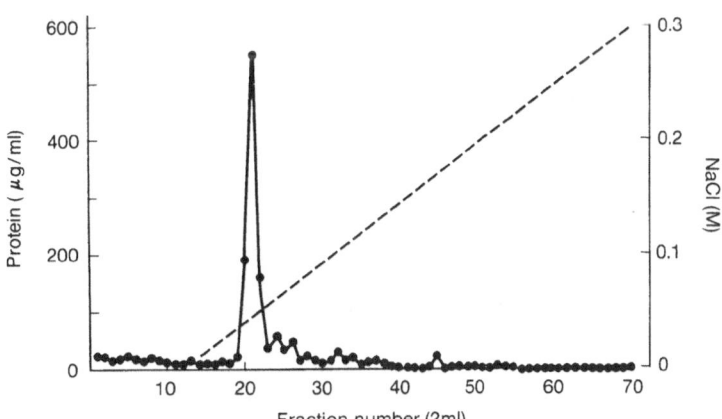

Fig. 4.   DEAE-Sephacel chromatography of enterotoxin treated with NTCB.

Solid line, absorbance at 280 nm; dotted line, NaCl concentration.

Table 2. Association Constants of MCAs* with F15 Fragment.

MCA	Association constant (M^{-1})
2-B-4	2.4×10^9
2-H-2	4.5×10^7
3-B-2	4.1×10^7

* Monoclonal antibody

During purification of the F15 fragment from *C. perfringens* enterotoxin, the N-terminal region, which interfered with the inhibition by 2-H-2
of enterotoxin-binding to Vero cells, may be simply removed and/or F15 may
stereochemically change the structure. Until now, we have not yet obtained
any MCA that binds to *C. perfringens* enterotoxin but not to F15.

In conclusion, we show a possible location of the binding sites on
the *C. perfringens* enterotoxin molecule binding with MCAs 2-B-4, 3-B-2 and
2-H-2, and Vero cells (Figure 9A), and illustrate the possible interpretation for inhibition of binding of enterotoxin and F15 to Vero cells with MCA
2-B-4 (Figure 9B) or 3-H-2 (Figure 9C). The cytotoxic sites of the enterotoxin molecule have still not been satisfactorily identified.

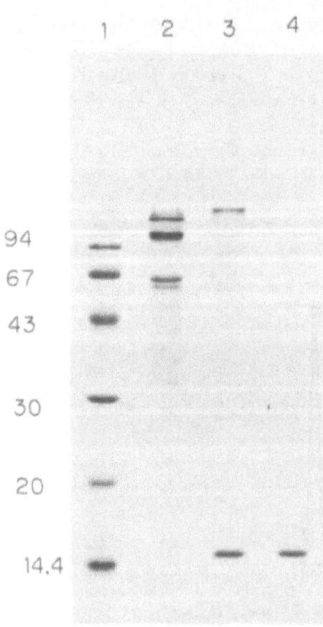

Fig. 5. SDS-PAGE of enterotoxin and F15 fragment.

The samples: (Lane 1) Low molecular weight markers; (Lane 2)
intact enterotoxin; (Lane 3) enterotoxin treated with NTCB;
(Lane 4) F15 fragment.

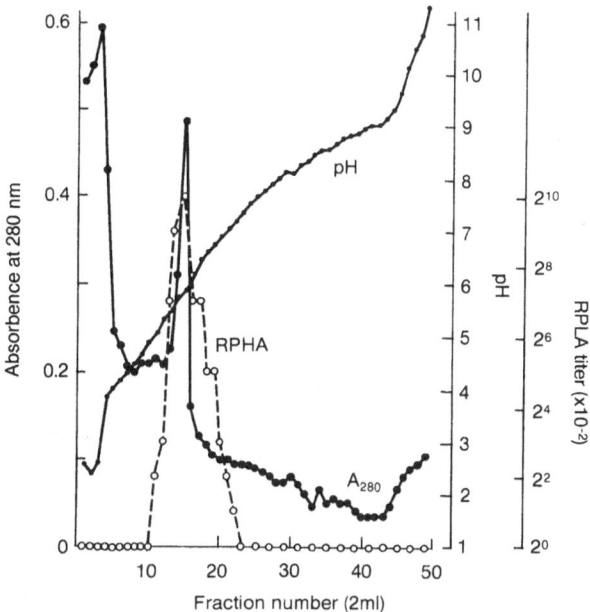

Fig. 6.   Isoelectric focusing of F15 fragment.

Protein,● ——● ; pH, •——• ; RPHA titer[21] ○ - - - - - ○

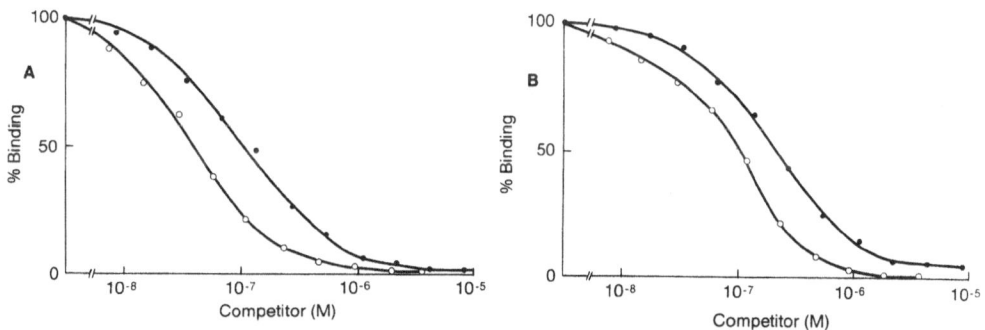

Fig. 7.   Inhibition of binding of ^{125}I-labeled enterotoxin or F15 to
Vero cells with unlabeled enterotoxin or F15.

(A) ^{125}I-labeled enterotoxin (150 ng) was added to Vero cells
(2.5×10^5) and the binding was inhibited with unlabeled enterotoxin or F15; (B) ^{125}I-labeled F15 (150 ng) was added to Vero
cells (2.5×10^5) and the binding was inhibited with unlabeled
enterotoxin or F15. Nonspecific binding was measured in the
presence of 50 μg/ml of unlabeled enterotoxin.

Enterotoxin, ○ ——○ ; F15,● ——●

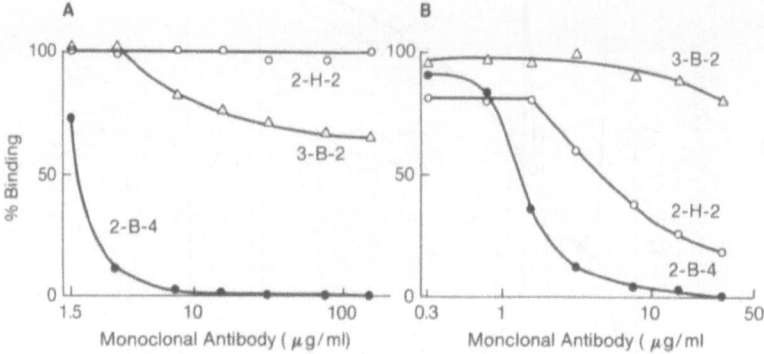

Fig. 8. Binding of ^{125}I-labeled enterotoxin or F15 to Vero cells in the presence of monoclonal antibodies.

(A) enterotoxin; (B) F15. The concentration of ^{125}I-labeled protein was 0.5 μg/ml.

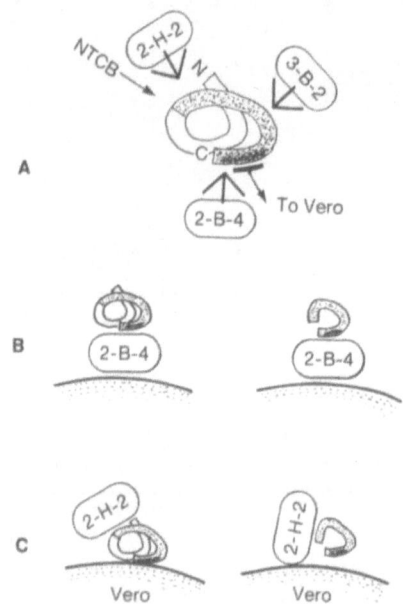

Fig. 9. (A) Possible illustrations for location of the sites on the enterotoxin molecule binding with MCA, 2-B-4, 3-B-2 and 2-H-2 and Vero cells; (B) inhibition of the binding of enterotoxin and F15 to Vero cells with MCA 2-B-4; (C) inhibition of the binding of enterotoxin and F15 to Vero cells with 2-H-2.

ACKNOWLEDGEMENTS

The authors acknowledge the efforts, cooperation and dedication of our co-workers, Y. Aoi, Y. Horiguchi, Y. Kamata, S. Kozaki, T. Maekawa and H. Nakahara, who assisted in this study. This research was supported in part by research grants (61304030, 62560296, 63560291) from the Ministry of Education, Science, and Culture of Japan.

REFERENCES

1.   DasGupta BR and Pariza MW, 1982, Purification of two *Clostridium perfringens* enterotoxin-like proteins and their effects on membrane permeability in primary cultures of adult rat hepatocytes, *Infect Immun.*, 38:592-597.
2.   Davis BJ, 1964, Disk electrophoresis II. Method and application to human serum proteins, *Ann N Y Acad Sci.*, 121:404-427.
3.   Duncan CL, 1970, *Clostridium perfringens* food poisoning, *J Milk Food Technol.*, 33:35-41.
4.   Frieben WR, Duncan CL, 1975, Heterogeneity of enterotoxin like protein extracted from spores of *Clostridium perfringens* type A, *Eur J Biochem.*, 55:455-463.
5.   Granum P, 1982, Inhibition of protein synthesis by a tryptic polypeptide of *Clostridium perfringens* type A enterotoxin, *Biochem Biophys Acta.*, 708:6-12.
6.   Granum PE, Whitaker JR, Skjelkvole R, 1981, Trypsin activation of enterotoxin from *Clostridium perfringens* type A, *Biochem Biophys Acta.*, 668:325-332.
7.   Hauschild AHW, 1971, *Clostridium perfringens* enterotoxin, *J Milk Food Technol.*, 34:596-599.
8.   Horiguchi Y, Akai T, Sakaguchi G 1987, Isolation and function of a *Clostridium perfringens* enterotoxin fragment, *Infect Immun.*, 55:2912-2915.
9.   Horiguchi Y, Uemura T, Kamata Y, Kozaki S, Sakaguchi G, 1986, Production and characterization of monoclonal antibodies to *Clostridium perfringens* enterotoxin, *Infect Immun.*, 52:31-35.
10. Laemmli UK, 1970, Cleavage of structural proteins during the assembly of the head of bacteriophage T4, *Nature*, 227:680-685.
11. McDonel JL, 1979, The molecular mode of action of *Clostridium perfringens* enterotoxin, *Am J Clin Nutr.*, 32:210-218.
12. Niilo L, 1973, Antigenic heterogeneity of enterotoxin from different agglutinating serotypes of *Clostridium perfringens*, *Can J Microbiol.*, 19:521-524.
13. Richardson, M, Granum PE, 1985, The amino acid sequence of the enterotoxin from *Clostridium perfringens* type A. *FEBS Let.*, 182:479-484.
14. Sakaguchi G, Uemura T, Riemann HP, 1973, Simplified purification of *Clostridium perfringens* type A enterotoxin, *Appl Microbiol.*, 26:762-767.
15. Skjelkvole R, Duncan CL, 1975, Enterotoxin formation by different toxigenic types of *Clostridium perfringens*, *Infect Immun.*, 11:563-575.
16. Uemura T, Aoi Y, Horiguchi Y, Wadano A, Goshima N, Sakaguchi G, 1985, Purification of *Clostridium perfringens* enterotoxin by high performance liquid chromatography, *FEMS Microbiol.*, Letters 29:293-297.
17. Uemura T, Maekawa T, Sakaguchi G, 1984, Biological assay for *Clostridium perfringens* enterotoxin with Vero cells, *Japan J Vet Sci.*, 46:715-720.

18. Uemura T, Nakahara H, Horiguchi Y, Sakaguchi G, 1986, Trypsinization of derivatives of *Clostridium perfringens* enterotoxin, *Lett Appl Microbiol.*, 3:31-33.
19. Uemura T, Sakaguchi G, Riemann HP, 1973, In vitro production of *Clostridium perfringens* type A enterotoxin and its detection by reversed passive hemagglutination, *Appl Microbiol.*, 26:381-385.
20. Uemura T, Skjelkvole R, 1976, An enterotoxin produced by *Clostridium perfringens* type D, *Acta Pathol Microbiol Scand.*, Sec B, 84:414-420.
21. Walker HW, 1975, Foodborne illness from *Clostridium perfringens*, *Critical Rev Food Sci Nutr.*, 7:71-104.

BIOCHEMICAL CHARACTERISTICS AND ACTIONS OF TOXINS FROM *CLOSTRIDIUM*

DIFFICILE ISOLATED FROM ANTIBIOTIC ASSOCIATED DIARRHEA IN JAPAN

Kazue Ueno, Hideki Kohno, Toyoko Kobayashi, and
Kunitomo Watanabe

Institute of Anaerobic Bacteriology
Gifu University School of Medicine
Gifu, Japan

INTRODUCTION

Clostridium difficile was isolated from the feces of patients with
pseudomembraneous colitis (PMC) and antibiotic-associated diarrhea
(AAD).[2,3,9,11] It was found that *C. difficile* produces at least two toxins,
enterotoxin (toxin D-1) and cytotoxin. These toxins are detected in the
stools of most patients suffering from pseudomembraneous colitis or antibiotic-associated diarrhea and in culture filtrates of the isolated *C.
difficile* strains.[12,15,16,17]

Studies on the partial purification of *C. difficile* have been reported
by many authors. The first report on the production of two biologically
and immunologically distinct toxins by *C. difficile* appeared in 1977.[2]
Taylor et al., (1979), and other investigators[1,5,16] reported on the partial
purification and characterization of these two toxins. These authors also
described the various and distinct biological effects in mouse, the rabbit
intestinal loop, and the skin (vascular permeability) of guinea pigs as well
as cytopathic effects (CPE) on various tissue culture cell lines.

Recently, we reported the purification and characterization of D-1
and D-2 toxins.[1] Rabbits were immunized with D-1 toxin to obtain anti-
D-1 serum. This antiserum was used to develop a latex agglutination test
to detect *C. difficile* D-1 toxin for clinical screening of *C. difficile*.[7]

Good correlation was obtained during a study comparing our later
agglutination test to the cytotoxigenic assay for D-2 toxin or toxin B
found in fecal samples.[6,8] There are a number of reports concerning the
detection of *C. difficile* toxin by cytopathogenic effect (CPE) and/or counterimmunoelectrophoresis (CIE).[13,19] In this report, we describe the partial purification of *C. difficile* enterotoxin from strain GAI 0280 and the
development of a latex agglutination immunoassay to detect *C. difficile*
enterotoxin in fecal specimens from patients with pseudomembraneous colitis
or antibiotic-associated diarrhea. In addition, we assayed fecal specimens
of subjects to determine the normal value of *C. difficile* enterotoxin.

MATERIALS AND METHODS

Culture Conditions. *Clostridium difficile* GAI 0280 was used for pro-

Microbial Toxins in Foods and Feeds, Edited by A.E. Pohland *et al.,*
Plenum Press, New York, 1990

duction of toxin. This strain displaying a high *in vivo* toxigenicity, was originally isolated at Gifu University Hospital from a patient suffering from pseudomembraneous enterocolitis. Ten colonies of *C. difficile* GAI 0280 were selected from a blood agar plate and each colony was inoculated into 10 ml of a medium consisting of 5 percent brain heart infusion (Difco) and 1 percent proteose peptone broth (Difco). After 48 hours of incubation at 37°C under anaerobic conditions, broth filtrates of each culture were prepared by centrifuging at 8,000 xg for 20 minutes and passing through a 0.45 μm membrane filter for toxin assays in mice.

Toxin production. The culture demonstrating the highest toxic activity was inoculated into 3.5 L of 3.0 percent brain heart infusion (Difco) and incubated in an anaerobic chamber (Hirasawa Works, Tokyo) for 5 days at 37°C. Toxin production increased gradually for 3 days and reached a plateau after 5 days. The culture broth was centrifuged at 8000 xg for 30 minutes at 10°C. The supernatant fluid was concentrated approximately 100 fold by ultrafiltration (YM-100). Materials less than 100,000 molecular weight were eliminated, AMICON.

Separation of C. difficile enterotoxin on Sephacryl S-300. The concentrated culture fluid (2.5 ml) was applied to a column (2.5 x 100 cm) of Sephacryl-S-300 (2.5 x 100 cm, Pharmacia). The column was eluded with 0.1M Tris-HCl buffer, pH 7.0, containing 0.9 percent NaCl at a flow rate of 15 ml/hour. Fractions were collected over a 72 hour period at 4°C. The eluate was monitored for optical absorbance at 280 nm and fractions 41 to 47 were pooled.

Partial purification of C. difficile enterotoxin on DEAE Sephadex. Selected fractions (50 ml) were applied to a column (1.6 x 20 cm) of DEAE Sephadex A-25 (Pharmacia). The DEAE column was eluded with a NaCl gradient (0-1.0M), pH 8.0, at a flow rate of 10 ml/hour. Each 3.0 ml fraction was monitored for optical absorbance at 280 nm. The fractions eluded with 0.17M NaCl and 0.38M NaCl, respectively were collected, pooled, centrifuged and concentrated to 2 ml volumes.

Mouse lethality. Groups of 10 mice, each weighing 20 g, were injected intraperitoneally with 0.2 ml of the sample. Lethality was recorded after 24 hours.

Fluid accumulation in rabbit ligated ileal loops. Samples (1 ml) were injected into ligated ileal loops (5 cm sections) prepared in New Zealand white rabbits weighing 1.5-2.0 kg.[4,5] The rabbits were humanely killed after 18 hours and the ratio of the volume of the accumulated fluid to the length of the loop was determined. A calculated ratio of 72 or greater was considered positive.

Cytotoxicity test cytopathic effect (CPE). The fecal specimen (0.5 g) was diluted with 0.5 ml of 0.1M Tris-HCl saline buffer (pH 8.0), mixed well, and centrifuged at 10,080 xg. The supernatant fluid was filtered through a 0.22 μm filter (millipore). The tests were performed as follows: CHO K-1 cells (Flow Lab) in 500 ml of Eagle minimal essential medium (GIBCO) containing 10 percent fetal calf serum (GIBCO) were incubated for 24 hours in a 96 well microtiter plate (Nunc). Fifty μl of supernatant fluid prepared from the fecal samples as described above were added to each well containing confluent CHO K-1 cells. The plates were incubated at 37°C with 5 percent CO_2 for 24 hours. A positive CPE test consisted of morphological changes in the CHO K-1 cells.

Preparation of antiserum. The purified enterotoxin was dissolved in 0.2 m phosphate buffered saline (PBS) and formaldehyde (33 percent) was added to a final concentration of 0.4 percent. The solution was kept at 4°C for 3 days and then dialyzed to remove residual formaldehyde.

214

Rabbit immunization was performed as follows: D-1 toxoid (0.3 mg) in 0.35 ml of saline was mixed with 0.5 ml of Freund's complete adjuvant (Difco) and the mixture was injected subcutaneously in multiple sites. After five weeks, 0.2 mg toxoid emulsified in incomplete adjuvant (Difco) was injected into the food pads of the rabbit. Two weeks later, the rabbit was exsanguinated. From this blood 50 ml of serum was obtained.

Preparation of antitoxin latex reagent from rabbit antiserum. Twenty ml of the rabbit antiserum was brought to 33 percent saturation with $(NH_4)_2SO_4$ at room temperature and centrifuged at 10,080 xg for 15 minutes. The precipitate was dissolved in distilled water and dialyzed against water to remove residual $(NH_4)_2SO_4$. The solution was lyophilized to obtain 400 mg of the immunoglobulin fraction. This fraction was dissolved in 5 ml of 0.1M Tris-saline buffer (pH 8.0) and vigorously stirred for 15 minutes. Five ml of 2 percent polystyrene latex particles (0.24 μm in diameter, Mitsubishi Chemical) suspended in 0.1M Tris-saline was added, stirred for 15 minutes, and centrifuged at 10/080 xg for 15 minutes. The pellet was resuspended in 10 ml of 0.1M Tris-saline containing 0.1 percent BSA, and stirred for 20 minutes at room temperature. The latex particle mixture was centrifuged at 10,080 xg for 15 minutes, the supernatant was removed, and the antitoxin D-1 latex particles were resuspended to a concentration of 2 percent with 0.2M Tris-buffer (pH 8.0). All of the preparations were stored at 4°C.

Collection and treatment of fecal samples. All samples were collected from patients at Gifu University Hospital and hospitals near Gifu City. The feces from 490 patients were each placed immediately into anaerobic containers (Clinical Supply, Tokyo) and transported directly to the laboratory.

Slide test for latex agglutination. Fecal samples (0.5 g) were resuspended in 0.5 ml of Tris-BSA buffer (0.2M Tris-HCl, 0.2 percent BSA, pH 8.0), vortexed for 1 minute, and centrifuged for 10 minutes (10,080 XG). The supernatant fluid samples were placed within the six elevated rings on a glass plate. Fifty μl of a 1.0 percent suspension of the prepared latex particles were added to each sample and mixed with an applicator stick. The glass plate was rotated slowly for 3 minutes and examined.

Latex photometric immunoassay. Agglutination was also measured by a turbidimetric rate assay using the automated latex photometric immunoassay system (LPIA Model L-1, Mitsubishi Chemical Industries, Japan) as follows: Fifty μl of *C. difficile* toxin D-1 diluted with 0.1M Tris-BSA buffer (pH 8.0) to concentrations of 15 ng - 4050 ng/ml were used to prepare calibration curves according to the manufacturer's recommendations. Fifty μl of fecal extract (mixture of 0.5 g of feces and 0.5 ml of 0.1M Tris-BSA buffer) were placed into a plastic cuvette and 400 μl of 0.1M Tris-BSA buffer and 50 μl of 1 percent latex reagent were added. The rate of the latex agglutination assay reaction was recorded at 1 second intervals by reading optical absorbance at 900 nm. The measurements were performed automatically.

SDS-polyacrylamide gel electrophoresis and transfer onto nitrocellulose. Prior to electrophoresis, *C. difficile* enterotoxin fractions were solubilized in buffer (40 mM Tris-HCl, pH 7.0, 2 percent SDS, 0.8 percent DTT, 5 percent glycerol and 0.002 percent bromophenol blue) and heated for 5 minutes at 100°C. A 3 percent stacking gel and 7.5 percent separating gel were used. The dimensions of the separating gel were 11 cm x 14 cm. A voltage of 75V was applied until the tracking dye reached the separating gel, and then increased to 150 volts until the dye was within 1 cm of the bottom of the gel. The proteins were immediately transferred onto a sheet of nitrocellulose.[18] Transfers were accomplished with a Bio-Rad Trans-Blot apparatus at 0.25A for 3 hours. The strips of nitrocellulose membrane were first blocked by incubation for 1 hour at 37°C, in 0.1M Tris-HCl, pH

7.3, containing 0.15M NaCl and 0.05 percent Tween-20. The buffer system consisted of 25 mM Tris-HCl (pH 8.3), 192 mM glycine and 20 percent methanol. The strips were incubated at 37°C for 1 hour with anti-*C. difficile* enterotoxin serum (diluted 500-fold). The strips were given two 10 minute washes in Tris-saline-Tween buffer and then incubated with HRP-conjugated goat antirabbit serum (diluted 1:2000 in Tris-saline-Tween buffer) for 2 hours at 37°C.

After two final 10 minute washings in Tris-saline-Tween buffer, the strips were rinsed briefly with deionized water and then placed in an HRP-colon development solution until the reaction product became visible (2-5 minutes). The strips were rinsed in distilled water and air dried.

RESULTS

The filtrate was tested for production of enterotoxin D-1 by the rabbit ileal loop test. Column chromatography was carried out after washing and equilibration of the columns with the appropriate buffers. Enterotoxin in culture broth was added to the column. The elution patterns of the enterotoxin on Sephacryl S-300 and DEAE-SEPHADEX are depicted in Fig. 1. The purified enterotoxin dissolved in 0.2M phosphate buffered saline (PBS) at pH 7.0 was subjected to SDS-polyacrylamide electrophoresis to assess its purity and approximate molecular weight (Fig. 2).

The sensitization of the latex beads with anti-*C. difficile* toxin was performed according to previous published methods.[13,14] Immunoglobulin was absorbed to the latex beads and BSA was added to stabilize the sensitized latex beads as depicted in the scheme of preparation (Fig. 3). The sensitivity of the coated latex beads to detect *C. difficile* toxin was determined as described in the Materials and Methods Section. Agglutination occurred with a minimum of 50 ng/ml toxin, while the latex photometric assay could detect 15 ng/ml toxin. Thus, the assay is sensitive enough to measure *C. difficile* toxin in patient's feces to determine *C. difficile* infection.

The latex slide test and the cytotoxin test utilizing CHO K-1 cells were used to compare fecal specimens from patients with antibiotic-associated diarrhea. The results of assaying 490 cases of antibiotic-associated diarrhea are presented in Table 1. Of 490 cases, 127 were positive for both toxins D-1 and cytotoxin while 339 of 490 cases were negative for both toxins. Eight cases were toxin D-1 positive, cytotoxin negative, and 16 cases were toxin D-1 negative but cytotoxin positive.

It was difficult to determine with certainty the eight cases of toxin D-1 positive and cytotoxin negative because of CHO K-1 cell deformation. We conclude from these results that the detection of *C. difficile* enterotoxin D-1 in the fecal specimens of patients by latex agglutination is useful and important for the diagnosis of antibiotic-associated diarrhea. Furthermore, it is a rapid and simple test compared with the conventional tissue culture assay of *C. difficile* cytotoxin. The latex agglutination test compared favorably with culturing *C. difficile* to determine colony forming units per gram feces (Table 2).

The latex photometric assay (LPIA) measures the latex agglutination reaction quantitatively at an optical density of 900 nm. The turbidity of the agglutinated latex beads is linearly related to the concentration of *C. difficile* toxin over a broad range of concentrations. The lowest concentration detected by the LPIA was 15 ng/ml. The latex reagent was able to detect *C. difficile* toxin by the agglutination test within 1-3 minutes (Fig. 4).

The fecal specimens of 18 of 39 cases of antibiotic-associated diarrhea were positive for *C. difficile* enterotoxin with the plate agglutination test. Fecal samples from healthy subjects contained 0-255 ng/ml toxin, however, samples from patients with antibiotic-associated diarrhea ranged between 800-5000 ng/ml toxin (Figs. 5 and 6).

DISCUSSION

Pseudomembraneous enterocolitis and antibiotic-associated diarrhea are thought to be caused by toxins produced by *C. difficile* in the intestines. These diseases are clinically important because they are in the category of opportunistic infections.

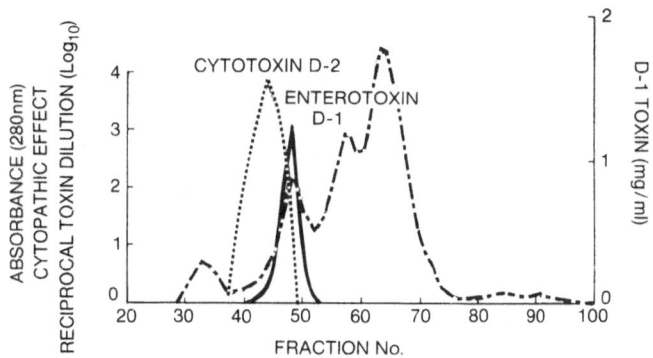

Chromatography of *C. difficile* toxins on the
column of Sephacryl S-300

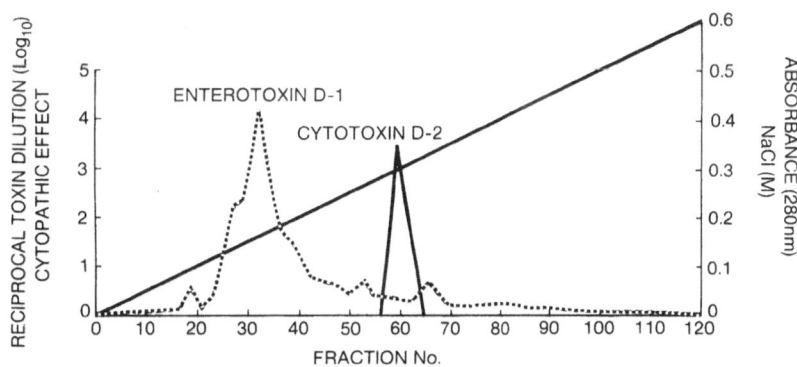

Chromatography of *C. difficile* toxins on the
column of DEAE-Sephadex A-25

Fig. 1. The column of Sephacryl S-300 was eluted with 0.1 M Tris-HCl, 0.9% NaCl and the DEAE Sephadex A-25 was eluted wth 0.1 M Tris-HCl by the gradient of NaCl (0-1.0 M). The fractions were monitored for optical desity at 280 nm. D-1 toxin (enterotoxin) was assayed by latex immunoassay and cytotoxin was assayed using the CHO-K1 cell line.

In order to study *C. difficile* infections, we have developed a latex agglutination test to detect *C. difficile* toxin. This immunoassay was compared to the cytopathic effect of fecal specimens collected from patients with antibiotic-associated diarrhea. Toxin-producing *C. difficile* GAI 0280 was isolated from a patient with pseudomembraneous colitis. This

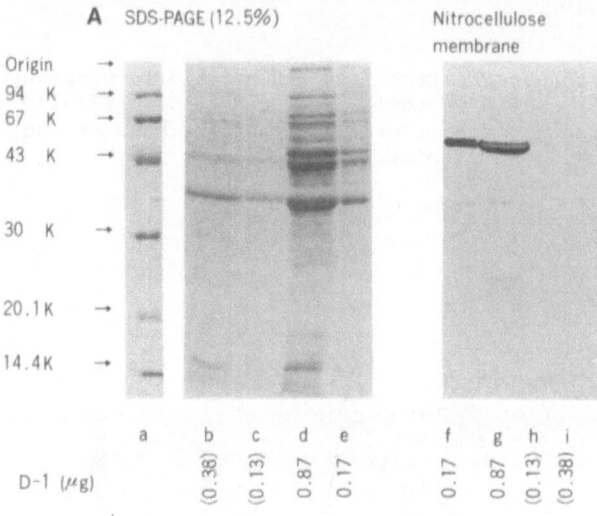

a : Molecular weight marker
d, e, f, g : *C. difficile* culture supernatant
b, c, h, i : Neutralized sample with anti D-1 latex

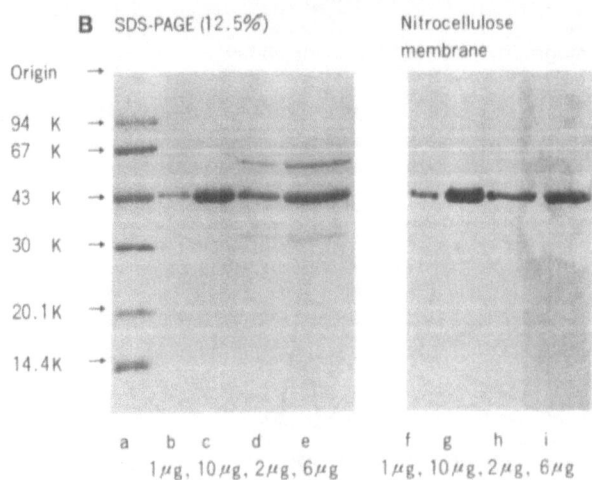

a : Molecular weight marker
b, c, f, g : Enterotoxin (DEAE-Sephadex A-25)
d, e, h, i : Enterotoxine (Sephacryl S-300)

Fig. 2-a. Immunoblotting of *C. difficile* culture supernatant with anti-enterotoxin (D-1) serum.

Fig. 2-b. Immunoblotting of purified enterotoxin (D-1) with anti-enterotoxin serum.

strain produced enterotoxin in high concentrations. The enterotoxin was used to immunize rabbits to produce anti-*C. difficile* sera and latex beads sensitized with the purified anti-*C. difficile* sera. Neutralization tests in mice using the developed latex reagent were conducted to confirm the efficiency of the reagent. The latex reagent absorbed *C. difficile* enterotoxin. Studies also were performed with clinical fecal specimens to correlate *C. difficile* enterotoxin and cytotoxin with numbers of *C. difficile* cultured from the fecal specimens. Our tests with the latex reagent correlated well with alternative methods, such as cytotoxicity or quantitation of *C. difficile* in specimens.

Our slide test utilizing the latex reagent is useful for clinical diagnosis of *C. difficile* infection because it is specific and sensitive. In addition, we measured *C. difficile* toxin quantitatively by a latex photometric immunoassay (LPIA system, Model L-1, Mitsubishi Chemical). The assay measures the increases in turbidity as it changes with the agglutination of the sensitized latex beads.

LPIA utilizing automated instrumentation enabled us to obtain more accurate measurements of the concentration of *C. difficile* enterotoxin in fecal specimens. Normal subjects showed less than 500 ng/ml toxin in their fecal specimens. However, patients with antibiotic-associated diarrhea demonstrated high levels of toxin in their fecal specimen greater than 1000 ng/ml. These results indicated that 700-800 ng/ml toxin concentration is the border line between normal *C. difficile* toxin levels in feces and the toxin levels in feces of patients with antibiotic-associated diarrhea.

In conclusion, our latex reagent prepared from anti-*C. difficile* enterotoxin sera showed a good correlation with alternative biological tests for *C. difficile* toxin or cultural methods for enumerating *C. difficile* in fecal specimens. Furthermore, the latex agglutination test to measure *C. difficile* enterotoxin D-1, discriminated between the levels of

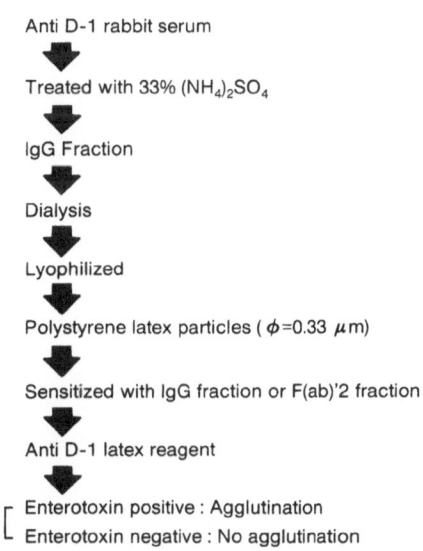

Fig. 3. Preparation of anti D-1 toxin latex reagent.

Anti *C. difficile* toxin latex reagent as prepared using immunoglobulin fraction of antisera. Latex beads (0.33 μm in diameter) were used to sensitize polystyrene particles.

Table 1. Comparison of latex agglutination test of toxin D-1
(enterotoxin) and cytopathic effect of toxin D-2.

Latex agglutination	Cytopathic effect	Number of cases
+	+	127 (25.9%)
+	-	8 (1.6%)
-	+	16 (3.3%)
-	-	339 (69.2%)
Total cases		490

For comparison, 490 fecal specimens of antibiotic-associated diarrhea were tested with the latex agglutination assay and for cytopathic effects on CHO-K1 cells.

Table 2. Detection of *Clostridium difficile* and toxin in feces of a healthy volunteer before and after administration of clindamycin.

Examination			Stool Analysis		
Administration of drug		No. of *C. difficile*	Detection of *C. difficile* toxin Ouchterlony[*]	Latex agg.[#]	Stool Character
Before		-	-	-	Normal stool
During	1	-	-	-	"
	2	-	-	-	"
	3	-	-	-	"
	4	2 cfu/g	-	-	"
	5	63 "	-	-	"
	6	4 "	-		"
	7	-	-		"
After	1	20 "	-	+	"
	2	20 "	-	+	"
	3	10^5 "	-	+	"
	5	10^5 "	++	+++	Soft stool
	6	10^2 "	-	+	Normal stool
	7	10^2 "	-	+	"
	10	10^2 "	-	+	"
	13	-	-	+	"
	18	-	-	+	"

[*] Ouchterlony double diffusion technique.
[#] Latex agglutination technique.

A group of healthy volunteers were administered clindamycin (450 mg/day for 7 days). As the fecal levels (CFU/g) of *C. difficile* increased, the detection of *C. difficile* enterotoxin by latex agglutination also increased.

toxin found in the feces of normal subjects and patients with antibiotic-
associated diarrhea. The latex agglutination test is rapid and simple to
perform. This test should be useful for the diagnosis of pseudomembraneous
colitis and antibiotic-associated diarrhea.

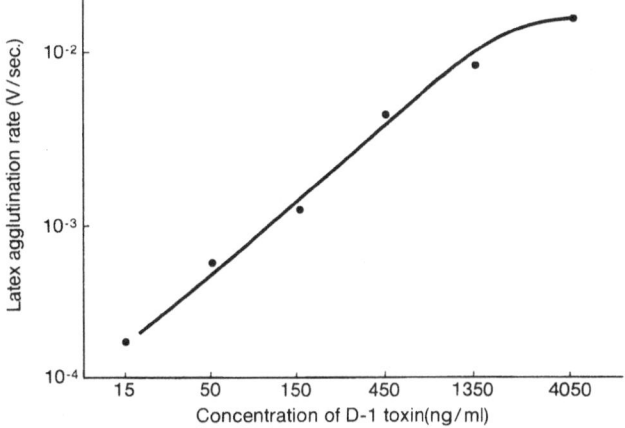

Fig. 4. Calibration curve of anti D-1 latex agglutination. The anti-
         D-1 latex was agglutinaed by D-1 toxin standards. The veloc-
         ity of agglutination was measured by latex photo immunoassay
         system, (LPIA system, Mitsubishi Chemical). Note linearity
         of the reaction from 15 ng/ml of D-1 toxin to 1,350 ng/ml
         D-1 toxin.

$$V = \frac{T1 - T2}{T1 \times t}$$

T1 = Initial transmittance (%)
T2 = % transmittance after t sec.
 t = Total time during which reaction is being measured.

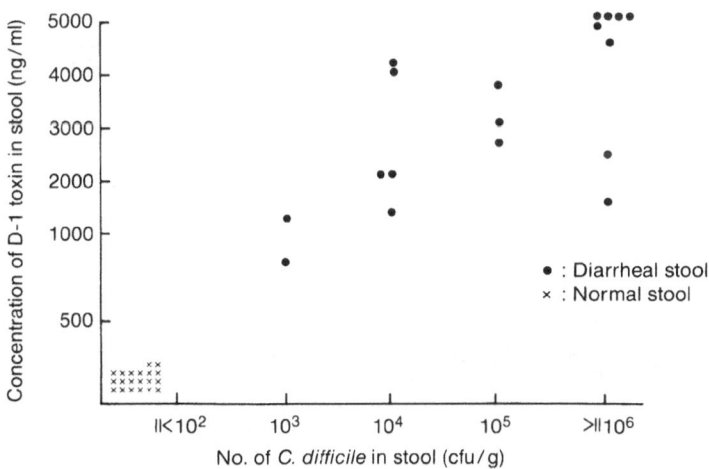

Fig. 5. Measurement of *C. difficile* toxin, in fecal specimens of
         patients with antibiotic-associated diarrhea and healthy
         subjects compared with the number of *C. difficile* (cfu/g)
         present. Healthy subjects showed toxin concentration
         under 500 ng/ml whereas, patients have more than 800 ng/ml.

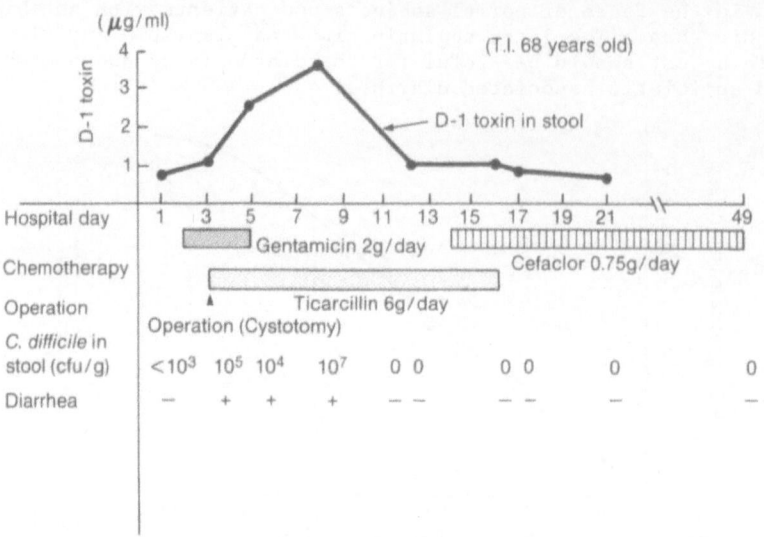

Fig. 6.   A case follow-up study comparing chemotherapy, number of *C. difficile* organisms and the amount of *C. difficile* D-1 toxin detected in stool.   The number of *C. difficile* recovered paralleled the amount of D-1 toxin in diarrheal stool.

REFERENCES

1.   Banno Y, Kobayashi T, Kohno H, Watanabe K, Ueno K, Nozawa Y, 1984, Biochemical characterization and biologic actions of two toxins (D-1 and D-2) from *Clostridium difficile*, *Rev Infec Dis.*, 6:Suppl 1, s11-s20.

2.   Bartlett JG, Onderdonk AB, Cisneros RL, Kasper DL, 1977, Clindamycin-associated colitis due to a toxin-producing species of *Clostridium* in hamsters, *J Infect Dis.*, 130:701-705.

3.   Bartlett JG, Moon N, Chang TW, Taylor N, Onderdonk AB, 1978, Role of *Clostridium difficile* in antibiotic-associated pseudomembraneous colitis, *Gastroenterology*, 75:778-782.

4.   Craig JP, 1965, A permeability factor (toxin) found in cholera stools and culture filtrates and its neutralization by convalescent cholera sera, *Nature*, 207:614-616.

5.   De SN, 1959, Enterotoxicity of bacteria-free culture filtrate of *Vibrio cholerae*, *Nature*, 183:1533-1534.

6.   Kohno H, Imada T, Kobayashi T, Watanabe K, Ueno K, 1985, Quantitative immunoassay of *Clostridium difficile* enterotoxin (D-1) by latex photo-metric immunoassay system, *Proc 14th Intersci Conf Antimicrob Agents Chemother*.

7.   Kohno H, Kobayashi T, Watanabe K, Ueno K, 1983, Latex immunoassay of *C. difficile* toxins (D-1, D-2) in human feces, *Proc 13th Intersci Conf Antimicrob Agents Chemother*.

8.   Kohno H, Kobayashi T, Watanabe K, Ueno K, 1985a, A rapid measurement of *C. difficile* enterotoxin in fecal samples from patients of anti-biotic-associated diarrhea and healthy subjects, *Amer Soc Microbiology*, Abstract C-247.

9. Larson HE, Prince AB, 1977, Pseudomembraneous colitis: presence of a clostridial toxin, *Lancet*, 25:191-201.

10. Masson PL, Gambiasco CL, Collet-Cassart D, Magnusson CGM, Richard, CB, Sindic, CJM, 1981, Particle counting immunoassay (PACIA), *Methods Enzymol.*, 74:106-139.

11. Rifkin GO, Fekety FR, Silva J Jr, Sack, RB, 1977, Antibiotic induced colitis: implication of a toxin neutralized by *Clostridium sordellii* antitoxin, *Lancet*, 2:1103-1106.

12. Rolfe RD, Finegold SM, 1979, Purification and characterization of *Clostridium difficile* toxin, *Infect Immun.*, 25:191-201.

13. Ryan RW, Kwasnik L, Tilton RC, 1980, Rapid detection of *Clostridium difficile* toxin in human feces, *J Clin Microbiol.*, 12:776-779.

14. Singer JM, Plotz CM, 1956, The latex fixation test. I. Application to the diagnosis of rheumatoid arthritis, *Am J Med.*, 21:888-896.

15. Sullivan NM, Pellett S, Wilkins TD, 1982, Purification and characterization of toxin A and B of *Clostridium difficile*, *Infect Immun.*, 35:1032-1040.

16. Taylor NS, Bartlett JG, 1979, Partial purification and characterization of a cytotoxin from *Clostridium difficile*, *Rev Infect Dis.*, 1:379-385.

17. Taylor NS, Thorne GM, Bartlett JG, 1981, Comparison of two toxins produced by *Clostridium difficile*, *Infect Immun.*, 34:1036-1043.

18. Towbin H, Staehelin T, Gordon J, 1979, Electrophoretic transfer of proteins from polyacrylamide gels to nitrocellulose sheets, *Proc Natl Acad Sci*, 76:4350-4354.

19. Welch DF, Menge SK, Matsen JM, 1980, Identification of toxigenic *Clostridium difficile* by counterimmunoelectrophoresis, *J Clin Microbiol.*, 11:470-473.

RESTRICTION ENZYME ANALYSIS IN THE EPIDEMIOLOGY OF *LISTERIA MONOCYTOGENES*

Irene V. Wesley,[1] Ronald D. Wesley,[1]
Judy Heisick,[2] Fannie Harrell,[2]
Dean Wagner[2] and John Bryner[1]

USDA, Agricultural Research Service,[1]
National Animal Disease Center, Ames, IA, and
Department of Health and Human Services[2]
Public Health Service
Food and Drug Administration
Minneapolis, Minnesota

ABSTRACT

Listeria monocytogenes is a foodborne bacterial pathogen which is transmitted to humans via consumption of dairy products and vegetables. The purpose of this study was to examine the genetic diversity of *L. monocytogenes* by restriction enzyme analysis and to localize the hemolysin gene by hybridization with a synthetic oligomer specific for the β-hemolysin (β-listeriolysin) toxin gene. Reference strains of serotypes 1A (1/2A), 1B (1/2B), 1C (1/2C), 3A, 3B, 3C, 4A, 4B, 4AB, 4C, 4D, and 4E and 39 isolates obtained from food and clinical specimens were serotyped and genomic DNA analyzed for restriction enzyme patterns. Three groups of isolates were examined: two groups were derived from environmental samples and food products, and unrelated sporadic clinical cases. The third set consisted of isolates recovered during the recent listeriosis outbreak in Los Angeles County, California. Isolates were identified as serogroup 1A (factor 1) or as serotype 4B (factor 6). Reference strains exhibited a distinct electrophoretic pattern after Hha 1 digestion. Isolates assigned to the same sero type exhibited different restriction patterns after Hha 1 digestion. However, when restriction fragments were blotted using the Southern method and probed with a synthetic oligonucleotide specific for the β-hemolysin gene, hybridization occurred with a limited number of restriction fragments. Ten strains recovered from Mexican-style soft cheese incriminated in the California listeriosis outbreak were antigenically identified as serogroup 4B (factor 6). All 10 strains exhibited identical restriction patterns after cleavage with Hha I. Only a single restriction fragment (0.9 kb) hybridized with the β-hemolysin gene probe. This suggests a common contaminating source of *Listeria* for the cheese samples examined.

INTRODUCTION

Listeria monocytogenes is a gram-positive, motile, facultative intracellular bacterium[22] which has recently emerged as an important foodborne pathogen. Consumption of milk,[11] coleslaw,[26] vegetables,[14] milk contaminated after pasteurization,[3,11] and Mexican-style cheese[1,18] have all been incriminated in transmission. Human listeriosis is characterized by flu-like symptoms, meningitis, and abortion, with mortality approaching 30 percent That immunocompromised individuals,[21] transplant recipients,[23,28] pregnant females,[32] their fetuses and neonates, and the elderly, are at risk suggests that impairment of cell-mediated immunity may predispose to listeriosis.

Bacterial restriction enzyme analysis (REA) is a sensitive technique in which the DNA is cleaved by restriction endonucleases. The restriction fragments, when size separated electrophoretically, yield the restriction enzyme pattern or DNA fingerprint of the bacterium. REA has been used in delineating the epidemiology of other bacterial pathogens, such as *Brucella*,[2] *Campylobacter*,[6,15,17] *Clostridium difficile*,[33] *Legionella pneumophila*,[30] *Neisseria gonorrhea*,[10] and enterotoxigenic *Escherichia coli*.

The β-hemolysin (listeriolysin O) toxin gene is only found in virulent *Listeria monocytogenes*.[4,5,16,24] Studies with the transposon Tn916-induced hemolysin negative mutants have shown that the loss of β-hemolysin toxin did not affect the initial entry of *Listeria* into mouse phagocytes, but did impair survival in phagolysosomes.[12] Studies with the human enterocyte-like cell line Caco-2 have suggested that *Listeria* penetrates the brush borders of enterocytes and is taken up by hagolysosomes. Here β-hemolysin toxin binds to the membrane, thus liberating *Listeria* into the cytoplasm where it replicates, shielded from the host macrophages.[13] β-Hemolysin stimulates the cell-mediated immune response which infers that a toxoid may be efficacious for immunization against *Listeria*.[4]

A Hind 1II-Hinc 1I DNA fragment of about 500 base pairs from a β-hemolysin gene of *Listeria* 10403S (cloned in pUC8) has been described which hybridizes only to DNA obtained from strains specifically elaborating the β-hemolysin toxin.[8,9] A 20-base oligonucleotide probe has been synthesized which hybridizes only to strains of *L. monocytogenes* which secrete β-hemolysin.

The purpose of this study was to examine the genetic diversity of *L. monocytogenes* by restriction enzyme analysis and to localize the β-hemolysin toxin locus by Southern blot hybridization with the synthetic oligomer specific for this gene.

MATERIALS AND METHODS

Isolates - Reference strains of *L. monocytogenes* were serotyped at the Center for Disease Control and were kindly provided by Dr. Robert Weaver (CDS, Atlanta, GA), with the exception of isolate 2155 which was part of our reference collection (Table 1). The clinical source and geographic geographic origin of the field isolates of *L. monocytogenes* (Table 2) examined in this study and *L. innocua* (Table 3) are shown. Live bacteria were antigenically characterized with respect to serotype and surface factors as described.[19] This information is also shown.

A total of 29 clinical isolates from sporadic cases of *L. monocytogenes* were antigenically characterized and genomic DNA subjected to Hha 1 digestion. The first set of isolates consisted of 14 strains, including strains recovered from ice cream products (2106, 2114, 2107, 2112), cheese (2115), bulk milk samples from the state of Vermont (2100, 2101), floor drains of a dairy plant (2108, 2110), clinical specimens from Massachusetts (2102, 2103, 2105, 2098), and potatoes purchased at a local supermarket in the Midwest (2163). The second group consisted of 11 clinical strains of *L. monocytogenes* (isolates 2075-2085, kindly provided by M. Scheier, M.D.) from Colorado. Isolates 2075-2080 were obtained from cerebrospinal fluid; iso-

Table 1. Serotype reference strains of *Listeria monocytogenes* examined in this study.

Reference Strain	Serotype
G845-888013731	1A
G848-88013665	1B
G1431-88027646	1C
G1127-88022440	3A
G1345-88027355	3B
G1407-88027622	3C
KC1710	4A
2155-1KA	4AB
G1310-88024944	4B
KC1705	4C
KC1706	4D
KC1717	4E

Table 3. Isolates of *Listeria innocua* examined in this study.

	Source	Serotype
2086-2411K1AH	Raw milk	4
2087-2412A	Raw milk	4
2088-2459B	Raw milk	1
2089-4846KE	Raw milk	Not Tested
2090-2420C	Raw milk	Not Tested
2091-2474A	Raw milk	4
2092-LA-1	Cheese	Not Tested
2093-C1-94	Cheese	6a
2094-BK-10	Cheese	Not Tested
2095-SE104	Crabmeat	Not Tested
2096-SE115	Crab legs	Not Tested
2097-SE123	Lobster	Not Tested

2126-C1P-8011[a]

[a] Dr. J. Rocourt, Institut Pasteur, Paris, France

lates 2081-2085 were recovered from blood cultures. Isolate 2075 was obtained from a female patient with hepatic cirrhosis admitted with fatal meningitis. Isolate 2076 was recovered from a male patient (69 years old) with meningitis. Isolate 2077 was obtained from a 59 year old male with alcoholic liver disease, who died despite therapy. Isolate 2078 was recovered from a 21 year old female with no predisposing factors (e.g., pregnancy, etc.,) evident. Isolate 2084 was recovered from a renal trans-plant patient (female, 48 years old) on immunosuppressive drugs. Isolate 2082 was obtained from a 77 year old man with *Listeria* bacteremia and under-lying diabetes and heart disease. No clinical histories were available for isolates 2079, 2080, 2081, 2083, and 2085.

Table 2. Field Isolates of *Listeria monocytogenes* Examined in this Study

NADC Isolate Number	Serotype	Factor
GROUP 1		
2098-Scott A	4B	6
2100-V7	1A	1
2101-V37CE	4B	6
2102-F5416	1A	1
2103-F5260	1A	1
2105-F5234	4B	6
2106-87-352-9006KE	4B	6
2107-87-479-752	1A	1
2108-86-445-642	4B	6
2110-86-448-196	1A	1
2111-87-496-958	1A	2
	(3A	4)
2112-87-340-892	4B	6
2114-87-340-897	4B	6
2115-85-354-9185E	4B	6
2163-4KC	1A	1
GROUP 2		
2075-DV-1	1A	1
2076-DV-2	1A	1
2077-DV-3	4B	6
2078-DV-4	1A	1
2079-DV-5	4B	6
2080-DV-6	1A	1
2081-DV-7	4B	6
2082-DV-8	1A	1
2083-DV-9	1A	1
2084-DV-10	1A	1
2085-DV-11	1A	1
GROUP 3		
2050-V97	Not typed	
2051-ATCC 7644	Not typed	
2047-RM 1	4B	6
2048-RM 11	4B	6
2099-Murray B	4B	6
2098-Scott A	4B	6
2100-V7	1A	1
2115-85-354-9185E-April 16	4B	6
2116-85-354-91840-September 3	4B	6
2117-85-354-91818E-June 9	4B	6
2118-85-354-91839E-August 30	4B	6
2119-85-354-91832A-August 8	4B	6
2120-85-354-91838A-August 29	4B	6
2121-85-354-91821C-June 29	4B	6
2122-85-354-91815B-May 28	4B	6
2123-85-354-91833B-August 16	4B	6
2124-85-354-91820D-June 28	4B	6

The third set of strains included human isolates obtained from a case of fatal meningitis (2051, ATCC 7644),[31] a human isolate recovered from the 1983 Massachusetts listeriosis outbreak (2099), two raw milk isolates recovered from the state of Massachusetts (2047, 2048), and two bulk milk isolates from Vermont (2100, 2050). Ten isolates (2115-2124) recovered from Mexican soft-style cheese incriminated in the listeriosis epidemic in Los Angeles County and bearing package code dates (1985) corresponding to that outbreak were also examined.

The 12 strains of *L. innocua* (kindly provided by Dr. J. Lovett, FDA, Cincinnati, Ohio) were recovered from dairy products (isolates 2086-2094) or seafoods (isolates 2095-2097).

Extraction of chromosomal DNA

All isolates were plated on brain-heart infusion agar containing 5 percent defibrinated bovine blood and incubated microaerophilically (37°C, 24 hours). Bacterial lawns were harvested in PBS (0.01 M, pH 7.2), pelleted (8000 x g, 30 minutes), resuspended in 0.4 ml of 25 percent sucrose in 1 mM EDTA, 10 mM Tris (pH 7.2), and frozen (-20°C) until the time of DNA extraction. DNA was extracted by a modification of two previously described techniques.[25,29] The bacterial suspension (300 μl) was transferred to ultracentrifuge tubes and lysozyme was added (130 μl at 50 mg/ml). After incubation (37°C, 1 hour), 20 μl of proteinase K at 50 mg/ml and 130 μl of 0.5 M EDTA (pH 8.0) was added and mixed gently. Cells were lysed by the addition of 130 μl of sarkosyl (25 percent) and incubated overnight at 65°c. DNA was recovered by equilibrium centrifugation (60,000 rpm, 3 hours, 15°C) of the lysate in CsCl (1.25 g/ml of 50 mM Tris, 5 mM EDTA, 5 mM NaCl) in a VTi 65.2 rotor (Beckman Instruments). The viscous DNA band was harvested from the side through a 16-gauge needle and dialyzed extensively against TE buffer (10 mM TRIS, 1 mM EDTA, pH 8.0). The final DNA concentration was determined spectrophotometrically in an Ultra-Spec 1I (Model 4050, LKB Instruments) with an OD_{260} of 1 = 50 μg DNA/ml.[20]

Restriction endonuclease digestion of DNA

Two micrograms of purified DNA were digested (3-4 hours, 37°C) with Bgl II, EcoRI or Hha I (Bethesda Research Laboratories, Gaithersburg, MD) in a 20-μl reaction mixture in buffer supplied by the manufacturer. Initial trials indicated that *Listeria* isolates were best differentiated with Hha I which was used thereafter. Following digestion, 5 μl of tracking dye (0.1 percent bromphenol blue, 20 percent Ficoll type 400) was added to each sample. DNA fragments were separated on 0.8 percent agarose gels (60V, 16 hours) in a horizontal gel bed (120 x 25 cm) with Tris-borate EDTA (0.089 M Tris, 0.089 M boric acid, 0.002 M EDTA) as the running buffer. At the completion of electrophoresis, gels were stained (1 hour) with ethidium bromide (0.125 μg/ml), visualized with shortwave UV light and photographed using a Kodak 23A red filter (Foto UV 300/mp-4 DNA Photographic Transilluminator System, Fotodyne, Inc., New Berlin, WI).

Southern transfer

Restriction fragments were transferred from agarose gels onto nylon membranes (GeneScreen, NEN Research Products, Boston, MA) as described.[27] UV cross-linking was used to covalently bind the DNA to the membrane filters.[7] Membranes were refrigerated until time of hybridization.

Hybridization

Prehybridization was carried out for 3 hours at 37°C in 6X SSC (SSC is 0.15 M NaCl, 0.015 M Na citrate. pH 7.0), 5X Denhart's solution (0.1

percent Ficoll, 0.1 percent polyvinyl pyrolidone, 0.1 percent bovine serum albumin), 0.5 percent SDS and 100 µg/ml of sonicated denatured salmon sperm DNA.[20] Hybridization was carried out at 37°C for 18 hours in fresh prehybridization solution containing 1-5 x 10^6 cpm/filter of the end-labelled ^{32}P synthetic oligomer. After incubation, filters were briefly washed once with 6X SSC preheated to 50°C, followed by two changes of 6X SSC at 50°C for 1 hour, each with gentle shaking. Dried filters were exposed to Kodak X-OMAT AR film at -80°C with two intensifying screens for 2-4 days.

RESULTS

The endonuclease Hha I provided the best resolution of the isolates, although genomic DNA of *Listeria* could be cleaved with EcoRI, Hind II, Bgl II, and Pst II.

Restriction enzyme analysis of serotypes

Genomic DNA of *L. monocytogenes* from reference strains of the serotypes 1A, 1B, 1C, 3A, 3B, 3C, 4A, 4B, 4C, 4D, and 4E and a field isolate serotyped as 4AB serotypes was cleaved with Hha I. Reference strains of serotypes 1 and 4, which are the predominant serogroups, exhibited a unique restriction enzyme pattern after Hha 1 digestion (Fig. 1). The presence of a restriction fragment of > 8 kb distinguished serovars 4B, 4C, and 4D. The DNA fingerprint of reference strains of serotypes 1A and 3A, however, were indistinguishable.

Restriction enzyme analysis of clinical isolates

The first set of isolates from humans, environmental samples, vegetables and dairy products were antigenically characterized as either sero-

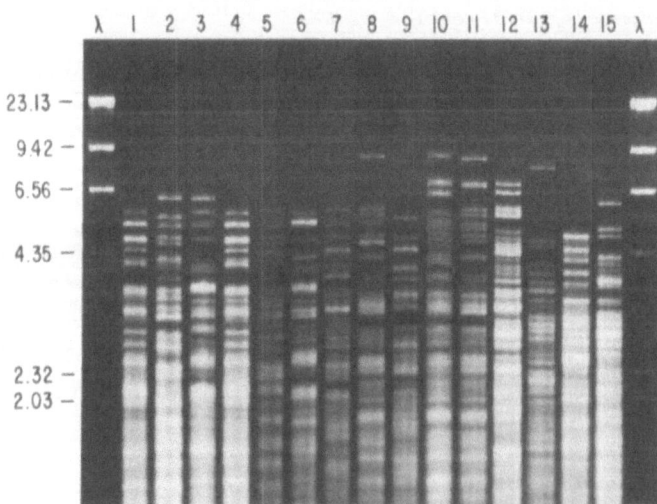

Fig. 1.   Hha I digests of 11 serotypes of *L. monocytogenes*,   Serotype 1A (lane 1), 1B (lane 2), 1C (lane 3), 3A (lane 4), 3B (lane 5), 3C (lane 6), 4A (lane 7), 4B (lane 8), 4AB (lane 9), 4C (lane 10), 4D (lane 11).   *L. welshimeri* (lane 12), *L. innocua* (lane 13), *L. ivanovi* (lane 14), *L. seeligeri* (lane 15). Hind III digest of bacteriophage lambda DNA (λ).   Sizes are indicated in kilobases.

type 1A (factor 1) (n = 6) or as serotype 4B (factor 6) (n = 8). Genomic
DNA was cleaved with the endonuclease Hha 1 (Fig. 2A). Five of the 8 isolates identified as serotype 4B, factor 6, exhibited remarkably similar DNA
fingerprints. These included isolates from ice cream (isolates 2106, 2112,
2114), a clinical isolate (2098) recovered from the 1983 Massachusetts
listeriosis outbreak, and a human isolate from Massachusetts unrelated
to this epidemic (2105). The genomic fingerprints of 5 of the 6 isolates
characterized as serotype 1A (factor 1) were different. An isolate
obtained from ice cream (2107) was remarkably similar to an environmental
sample obtained from an ice cream plant (2110). The DNA fingerprints of *L.
welshimeri*, *L. ivanovi*, *L. innocua*, and *L. seeligeri* are included for comparison. Restriction fragments were blotted using the Southern method and
probed with the oligomer specific for the β-hemolysin gene. Hybridization
occurred with the restriction fragments of approximately 0.6, 0.5, and 0.4

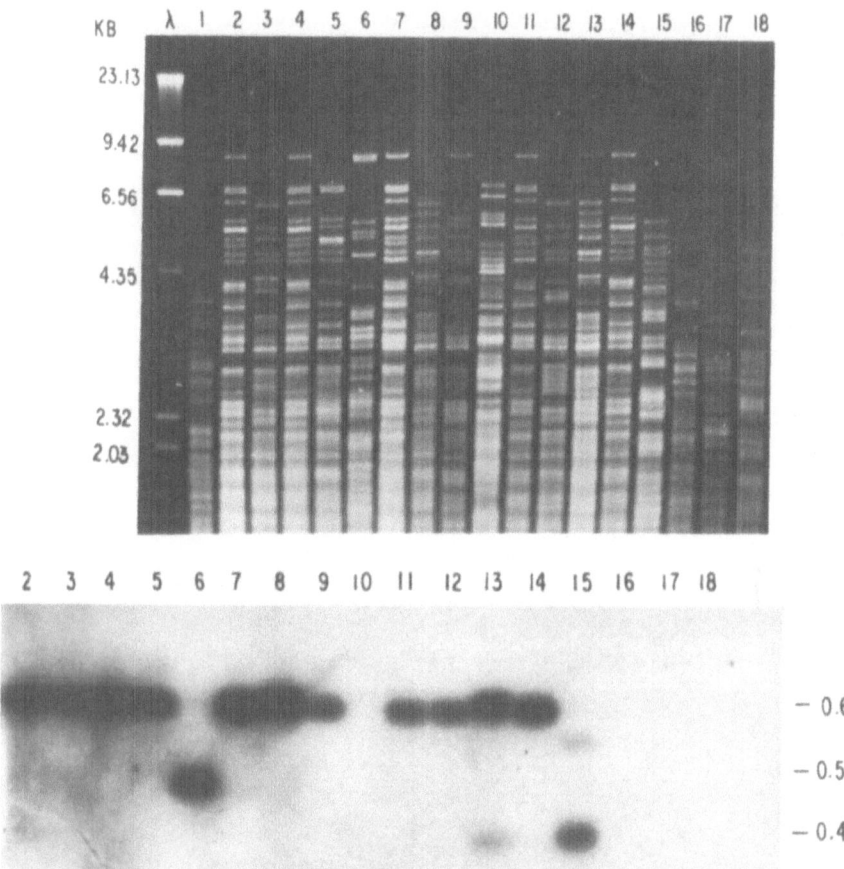

Fig. 2. (Upper panel) Hha 1 digests of isolates obtained from environmental samples and foods. Isolates 2108 (lane 1), 2114 (lane
2), 2110 (lane 3), 2106 (lane 4), 2101 (lane 5), 2163 (lane
6), 2098 (lane 7), 2102 (lane 8), 2115 (lane 9), *L. welshimeri*
(lane 10), 2105 (lane 11), 2107 (lane 12), 2103 (lane 13),
2112 (lane 14), 2100 (lane 15). *L. seeligeri* (lane 16), *L.
innocua* (lane 17), *L. ivanovi* (lane 18). Hind III digest of
bacteriophage lambda DNA (λ). Sizes are indicated in kilobases. (Lower panel) Localization of β-hemolysin gene within
0.6, 0.5, and 0.4 kb restriction fragments (arrows).

kb (Fig. 2B). The 0.6 kb fragment of the 8 isolates characterized as sero-type 4B (factor 6) hybridized with the β-hemolysin probe, which infers that restriction fragment length polymorphism was limited in this serogroup. Hybridization was limited in the serotype 1A (factor 1) isolates to restriction fragments of approximately 0.6, 0.5, and 0.4 kb. No hybridization occurred with *L. welshimeri*, *L. ivanovi*, *L. innocua*, and *L. seeligeri*, which do not elaborate the toxin. An environmental sample (isolate 2108, Fig. 2 lane 1) initially identified as *L. monocytogenes* exhibited a DNA fingerprint atypical of the species, and failed to hybridize with the β-hemolysin probe.

The second set of isolates included clinical samples from Colorado. Eight of the isolates were antigenically identified as serotype 1A (factor 1) and 3 isolates (2077, 2079, 2081) identified as serotype 4B (factor 6).

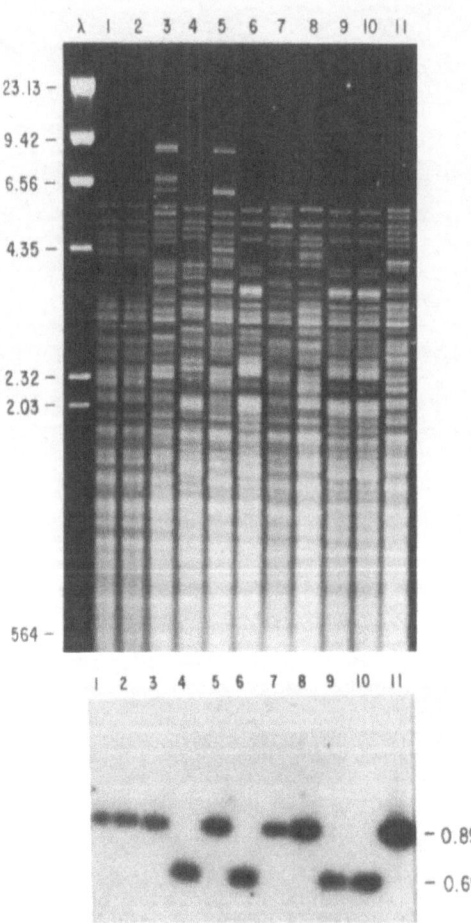

Fig. 3. (Upper panel) Hha I digests of clinical isolates (2075-2085). Isolate 2075 (lane 1), 2076 (lane 2), 2077 (lane 3), 2078 (lane 4), 2079 (lane 5), 2080 (lane 6), 2081 (lane 7), 2082 (lane 8), 2083 (lane 9), 2084 (lane 10), and 2085 (lane 11). Hind III digest of bacteriophage lambda DNA (λ). Sizes are indicated in kilobases. (Lower panel) Localization of β-hemolysin gene within .69 and .89 kb restriction fragments (arrows) of clinical isolates 2075-2085.

Cleavage with the endonuclease Hha I revealed six different restriction enzyme patterns in the 8 isolates identified as serotype 1A (factor 1) (Fig. 3, upper panel). Similar DNA electrophoretic patterns were seen in two clinically unrelated isolates (2075, 2076) characterized as serotype 1A (factor 1). The electrophoretic patterns of 3 isolates antigenically characterized as serotype 1A (factor 1) (2080, 2083, 2084) were also similar after Hha 1 digestion. Isolates 2077 and 2079, identified as serotype 4B (factor 6), were distinct from other isolates examined in this group because of the presence of two high molecular weight fragments (approximately 6 kb and 9 kb). Restriction fragments were blotted using the Southern method and probed with the oligomer specific for the β-hemolysin gene. Hybridization occurred with the 0.69 kb and 0.89 kb restriction fragments (Fig. 3, lower panel).

Restriction enzyme analysis of isolates recovered from clinical specimens and dairy products

A human isolate obtained from a case of fatal meningitis (2051),[31] a human isolate recovered from the 1983 Massachusetts listeriosis outbreak (2099), two raw milk isolates recovered from Massachusetts (2047, 2048), and two isolates obtained from a Vermont dairy (2100, 2050)[11] shortly after

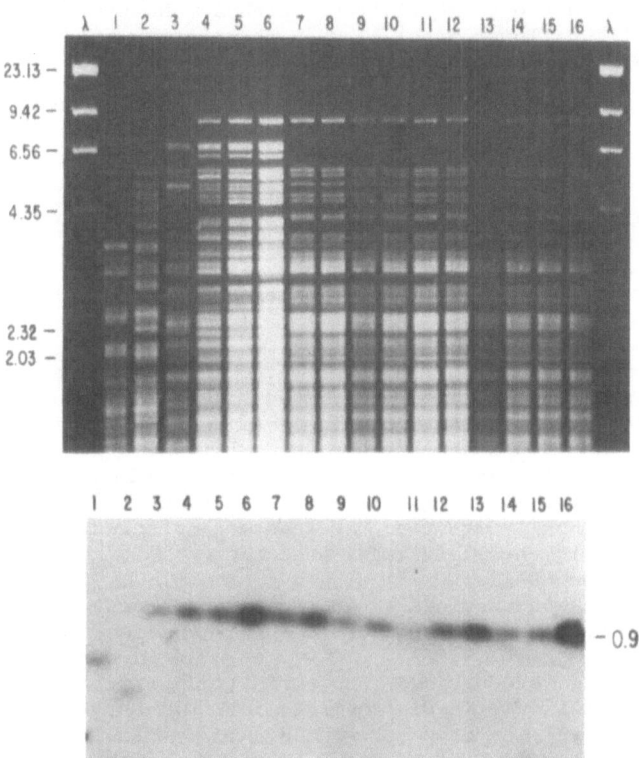

Fig. 4.   (Upper panel)   Hha I digests of L. monocytogenes isolates. 2051 (lane 1), 2100 (lane 2), 2050 (lane 3), 2047 (lane 4), 2048 (lane 5), and 2099 (lane 6); Mexican-style soft cheese isolates 2115-2124 (lanes 7-16). Hind III digest of bacteriophage lambda DNA (λ). Sizes are indicated in kilobases. (Lower panel)   Localization of β-hemolysin gene within 0.7, 0.6 and 0.89 kb restriction fragments.

the 1983 Massachusetts epidemic were cleaved with the endonuclease Hha 1 (Fig. 4; lanes 1-6). The three Massachusetts isolates (2047, 2048, 2099) displayed remarkably similar DNA fingerprints.

The 10 isolates recovered from the recent listeriosis epidemic in California, in which Mexican-style soft cheese was incriminated,[1,18] were antigenically characterized as serotype 4B (factor 6). The 10 isolates exhibited identical DNA patterns after Hha I digestion (Fig. 4, lanes 7-16). The identity of the DNA fingerprints, the serotype, and antigenic factors indicate a common contaminating source of L. monocytogenes of the cheese involved in the California epidemic. Hybridization with the synthetic probe localized the hemolysin gene in a 0.69 kb fragment for isolate ATCC 7644[31] and in a 0.50 kb fragment for the isolate 2100. Isolates 2047, 2048, and 2099 from Massachusetts and the 10 isolates associated with the California outbreak (2115-2124) hybridized with a single restriction band at approximately 0.9 kb (Fig. 4, lower panel).

Restriction enzyme analysis of L. innocua isolates

Genomic DNA of 12 strains of L. innocua after digestion with Hha 1 all exhibited restriction fragments greater than 0.56 kb, which may be characteristic of this species. Two epidemiologically unrelated strains (isolates 2086 and 2087) isolated from raw milk were identical (Fig. 5, upper panel), as were two isolated (2095, 2097) from seafoods. No hybridization signal was detected (Fig. 5, lower panel B) when restriction fragments were probed with the β-hemolysin gene probe.

DISCUSSION

We have developed a method for rapidly extracting genomic DNA from Listeria sp. which involves incubation (1 hour, 37°C) of bacterial suspensions with lysozyme,[25] followed by incubation (overnight, 65°C) with SDS and proteinase K.[29] Of the endonucleases examined during this study, Hha 1 provided the best resolution of electrophoretic patterns of genomic DNA.

Listeriosis occurs as sporadic unrelated clinical cases or, rarely, as epidemics. Twenty-nine isolates of L. monocytogenes, including epidemiologically unrelated strains from sporadic cases and two clinical isolates associated with the 1983 Massachusetts epidemic, which suggested milk as the vehicle of transmission[11] were examined. Fifteen isolates were identified as serotype 1A (factor 1) and 14 isolates were identified as serotype 4B (factor 6). After Hha I digestion, at least 11 different restriction enzyme patterns were seen in the 15 isolates of serotype 1A (factor 1). The 14 isolates of serotype 4B (factor 6) displayed a minimum of six distinct patterns. In addition, we examined 10 isolates from cheese recovered from the 1985 listeriosis outbreak which occurred in Los Angeles County, California.

The 39 isolates were divided into three groups for this study. The first set of isolates comprised epidemiologically unrelated isolates obtained from clinical, environmental, and food samples. Each of the six strains defined as serotype 1A (factor 1) exhibited a different DNA fingerprint. The 8 isolates assigned to serotype 4B (factor 6) displayed three distinct restriction enzyme patterns. A remarkably similar DNA fingerprint was evident in a clinical isolate (2098) associated with the 1983 Massachusetts epidemic, a clinical isolate (2102) not related to that epidemic, a strain derived from an environmental sample of a dairy plant (2100), and two strains recovered from ice cream (2106, 2112).

Group 2 isolates comprised of sporadic, clinically unrelated cases from Colorado. Eight of the 11 isolates were of serotype 1A (factor 1) and

234

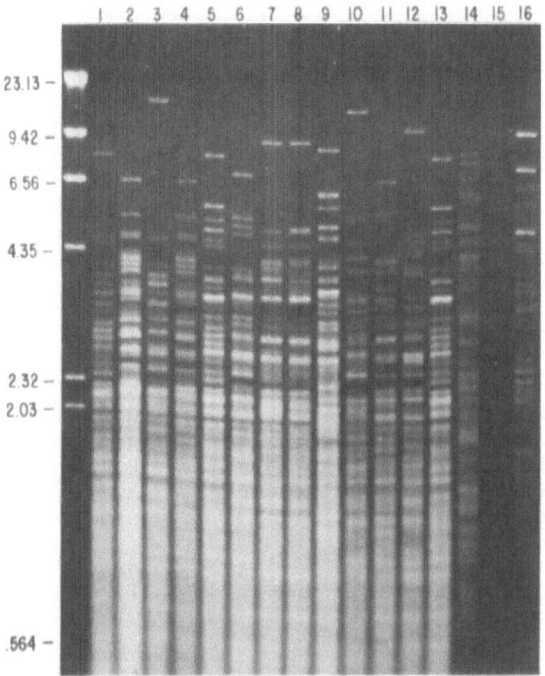

Fig. 5. (Upper panel) Hha I digests of *L. innocua* strains. 2086 (lane 1), 2087 (lane 2), 2088 (lane 3), 2089 (lane 4), 2090 (lane 5), 2091 (lane 6), 2092 (lane 7), 2093 (lane 8), 2094 (lane 9), 2095 (lane 10), 2096 (lane 11), 2097 (lane 12), 2126 (lane 13). Lanes 14-16: DNA from *Campylobacter hyointes-tinalis*, included as negative control. Hind III digest of bacteriophage lambda DNA (λ) included as molecular weight markers. (Lower panel) Failure of β-hemolysin probe to hybridize with *L. innocua*.

three were identified as serotype 4B (factor 6). Four different restriction enzyme patterns were evident in the 8 isolates of serotype 1A (factor 1). The three serotype 4B (factor 6) isolates (2077, 2079, 2081) each displayed unique DNA patterns. Despite the variety of restriction patterns, the β-hemolysin probe hybridized with either a 0.69 or 0.89 kb fragment. Multiple restriction enzyme patterns in isolates of the same serogroup and exhibiting identical surface antigens (factors) suggest the genetic diversity of the serogroup.

The third set included clinical and dairy product isolates. The DNA fingerprints of the two human isolates originating from the 1983 Massachusetts epidemic[11] (2089, 2099) were remarkably similar to two raw milk isolates from Massachusetts (2047, 2048) and to the five dairy-associated strains examined in the first group (2098, 2102, 2100, 2106, 2112). The epidemiological importance of the similarity of the DNA fingerprints of these dairy-associated isolates cannot be explained at this time. The Vermont bulk milk isolates (2100, 2101) recovered during a surveillance of dairy herds shortly after the 1983 outbreak and an isolate recovered from fatal meningitis (ATCC 7644)[31] were clearly unrelated. The multiple restriction enzyme patterns observed in isolates recovered from these sporadic cases confirm the heterogeneity of the isolates sharing the same antigenic profile. Despite the diverse origins of the isolates, hybridization with the β-hemolysin probe occurred with only three restriction fragments.

The largest single outbreak of human listeriosis occurred in Los Angeles County, California, from January to August 1985, and involved pregnant Hispanics, their offspring and consumption of Mexican-style soft cheese.[1,18] Recently, Linnan et al.[18] reported that the California outbreak consisted of epidemic cases (n = 93) superimposed on a background of sporadic cases (n - 49). A single phage type was isolated from 75 percent of the clinical cases and from environmental samplings of that factory. Although no conclusions could be reached regarding the initial source of *Listeria*, it was observed that more raw milk was delivered to the plant than could be accommodated by the pasteurizer. We examined *L. monocytogenes* isolates recovered from unopened cheese packages produced at the suspect cheese plant bearing package code dates (1985) corresponding to this interval. All of the 10 samples were identified as serotype 4B (factor 6). After Hha I digestion, all exhibited the identical restriction enzyme pattern. This is strongly indicative of a single contaminating source of *Listeria*. Hybridization of Southern blots with the synthetic oligomer specific for the β-hemolysin localized the gene encoding the toxin in a single low molecular weight restriction fragment (0.9 kb). We have subsequently examined a single human isolate from a diabetic patient in Arizona who died after eating the brand of cheese incriminated in this outbreak. Antigenically, the isolate was identified as serotype 4B (factor 6). The DNA fingerprint of this isolate and of the 10 isolates recovered from the cheese were identical after endonuclease digestion with Hha I.

In conclusion, this preliminary report on restriction enzyme analysis of *L. monocytogenes* demonstrates that DNA fingerprinting, when used in conjunction with serotyping schemes, may be extremely useful in deciphering the epidemiology of *L. monocytogenes*.

CONCLUSION

A technique for rapidly extracting genomic DNA for restriction enzyme analysis of *L. monocytogenes* was described. Genomic DNA was cleaved with the endonuclease Hha 1 and probed with a synthetic oligomer specific for the β-listeriolysin (β-hemolysin) toxin gene. Thirty-nine isolates from

236

clinical cases, food and environmental samplings were antigenically identified as either serotype 1A (factor 1) or serotype 4B (factor 6). At least
11 different DNA electrophoretic patterns were evident in the 15 isolates
assigned to serotype 1A (factor 1). The 24 strains characterized as serotype 4B (factor 6) exhibited a minimum of six different restriction enzyme
patterns. Despite variation in DNA fingerprints, hybridization occurred
with three DNA restriction fragments. In contrast, the 10 *L. monocytogenes*
isolates recovered from Mexican-style soft cheese incriminated in the recent
California epidemic were antigenically identical (serogroup 4B, factor 6)
and exhibited the same restriction enzyme pattern with Hha I. The β-
hemolysin toxin probe hybridized with a single restriction fragment (0.9
kb). This is strongly indicative of a single contaminating source of *L.
monocytogenes* in the cheese samples examined and demonstrates the usefulness
of restriction enzyme analysis in conjunction with other epidemiological
approaches in elucidating the transmission of *L. monocytogenes*.

REFERENCES

1. Anonymous, 1985, Listeriosis outbreak associated with Mexican-style
    cheese-California, *Morbid Mortal Weekly Report.*, 34:357-359.
2. Bailey KM, West DM, 1987, Restriction endonuclease (EcoRI) analysis of
    Brucella ovis DNA, *New Zealand Vet J.*, 35:161-162.
3. Barza M, 1985, Listeriosis and milk, *New Engl J Med.*, 312:438.440.
4. Berche P, Gaillard J, Alouf JE, 1987, T cell recognition of listerioly-
    sin O is induced during infection with *Listeria monocytogenes*, *J
    Immunol.*, 139:3813.3821.
5. Berche P, Gaillard JL, Sansonetti P, Geoffroy C, Alouf JE, 1987, Viru-
    lence of *Listeria monocytogenes*, *Ann Inst Pasteur Microbiol.*, 138:241-
    284.
6. Bradbury WC, Pearson AD, Marko MA, Congli RV, Penner JL, 1984, Investi-
    gation of *Campylobacter jejuni* outbreak by serotyping and chromosomal
    restriction endonuclease analysis, *J Clin Microbiol.*, 19:342-346.
7. Church GM, Gilbert W, 1984, Genomic sequencing, *Proc Natl Acad Sci.*,
    81:1991-1995.
8. Datta AR, Wentz BA, Hill WE, 1987, Detection of hemolytic *Listeria
    monocytogenes* by using DNA colony hybridization, *Appl Environ
    Microbiol.*, 53:2256-2259.
9. Datta AR, Wentz BA, Shook D, Trucksees M, 1988, Gene probes for detec-
    tion of *Listeria monocytogenes*, *Amer Soc Microbiol.*, p 32 (Abstract).
10. Falk ES, Danielsson D, Bjorvatn B, Melby K, Sorensen B, Kristiansen B,
    1985, Genomic fingerprinting in the epidemiology of gonorrhoea, *Acta
    Derm Venereol.*, (Stockholm) 65:235-239.
11. Fleming DW, Cochi SL, MacDonald KL, Brondum J, Hayes PS, Plikaytis NF,
    Holmes MB, Audurier A, Broome CV, Reingold AL, 1985, Pasteurized milk
    as a vehicle of infection in an outbreak of listeriosis, *New Engl J
    Med.*, 312:404-407.
12. Gaillard JL, Berche P, Sansonetti P, 1986, Transposon mutagenesis as a
    tool to study the role of hemolysin in the virulence of *Listeria
    monocytogenes*, *Infect Immun.*, 52:50-53.
13. Gaillard JL, Berche P, Mounier JS, Richard S, Sansonetti P, 1987, In
    vitro model of penetration and intracellular growth of *Listeria
    monocytogenes* in the human enterocyte-like cell line CaCo-2, *Infect
    Immun.*, 55:2822-2829.
14. Ho JL, Shands KN, Friedland G, Eckind P, Fraser DW, 1986, An outbreak
    of type 4B *Listeria monocytogenes* infection involving patients from
    eight Boston hospitals, *Arch Intern Med.*, 146:520-524.
15. Kakoyiannis CK, Winter PJ, Marshall R, 1984, Identification of
    Campylobacter coli isolates from animals and humans by bacterial
    restriction endonuclease DNA analysis, *J Appl Environ Microbiol.*,
    48:545-549.

16. Kuhn M, Kathariou S, Goebel W, 1988, Hemolysin supports survival but
     not entry of the intracellular bacterium *Listeria monocytogenes*,
     Immunol., 56:79-82.

17. Langenberg W, Rauws EA, Widjojokusumo A, Tytgat GNJ, Zanen HC, 1986,
     Identification of *Campylobacter pyloridis* isolates by restriction
     endonuclease DNA analysis, *J Clin Microbiol.*, 24:414-417.

18. Linnan MJ, Mascola L, Lou XD, Goulet V, May S, Salminen C, Hird DW,
     Yonekura L, Hayes P, Weaver R, Audurier A, Plikaytis BD, Fannin SL,
     Kleeks A, Broome CV, 1988, Epidemic listeriosis associated with
     Mexican-style cheese, *New Engl J Med.*, 319:823-828.

19. Lovett J, 1987, *Listeria* isolation, *In:* "FDA Bacteriological Analytical
     Manual," suppl 6th ed., Association of Official Analytical Chemists,
     Arlington, VA.

20. Maniatis T, Fritsch EF, Sambrook J, 1982, "Molecular cloning: A
     laboratory manual," Cold Spring Harbor Laboratory, New York.

21. McLauchlin J, 1987, *Listeria monocytogenes*, recent advances in the
     taxonomy and epidemiology of listeriosis in humans, *J Appl Bacteriol.*,
     63:1-11.

22. Murray EGD, Webb RA, Swann MBR, 1926, A disease of rabbits characterized
     by large mononuclear leukocytosis, caused by a hitherto undescribed
     bacillus, Bacterium monocytogenes (n.sp.), *J Pathol Bacteriol.*, 29:
     407-439.

23. Nieman RE, Lorber B, 1980, Listeriosis in adults: A changing pattern.
     Report of eight cases and review of the literature, 1968-1978, *Rev
     Infec Dis.*, 2:207-277.

24. Portnoy D, Jacks PS, Hinrichs DJ, 1988, Role of hemolysin for the
     intracellular growth of *Listeria monocytogenes*, *J Exp Med.*, 167:
     1459-1471.

25. Rocourt J, Grimont F. Grimont PAD, Seeliger HPR, 1982, DNA relatedness
     among serovars of *Listeria monocytogenes sensu lato*, *Curr Microbiol.*,
     7:383-388.

26. Schlech WF, Lavigne PM, Bortolussi RA, Allen AC, Haldane EF, Wort AJ,
     Hightower AW, Johnson SE, King SH, Nicholls ES, Broome CV, 1983,
     Epidemic listeriosis: Evidence for transmission by food, *New Engl J
     Med.*, 309:203-206.

27. Southern E, 1975, Detection of specific sequences among DNA fragments
     separated by gel electrophoresis, *J Mol Biol.*, 98:503-517.

28. Stamm AM, Dismukes WE, Simmons BP, Cobbs CG, Elliot A, Budrich P,
     Harmon J, 1982, Listeriosis in renal transplant recipients: Report of
     an outbreak and review of 102 cases, *Rev Infect Dis.*, 4:665-682.

29. Thiermann AB, Handsaker AL, Moseley SL, Kingscote B, 1985, New method
     for classification of leptospiral isolates belonging to serogroup
     Pomona by restriction endonuclease analysis: Serovar *kinnewicki*, *J
     Clin Microbiol.*, 21:585-587.

30. van Ketal R, Schegget J, Zanen HC, 1984, Molecular epidemiology of
     Legionella pneumophila serogroup I, *J Clin Microbiol.*, 20:362-364.

31. Webb RA, Barber M, 1927, *Listerella* in human meningitis, *J Pathol
     Bacteriol.*, 45:523-539.

32. Weinberg ED, 1987, Pregnancy-associated immune suppression risks and
     mechanisms, *Microbiol Pathol.*, 3:393-397.

33. Wren BW, Tabaqchale S, 1987, Restriction endonuclease DNA analysis of
     Clostridium difficile, *J Clin Microbiol.*, 25:2402-2404.

MOLECULAR GENETIC ANALYSIS OF VIRULENCE FACTORS OF *SHIGELLA FLEXNERI*

Masanosuke Yoshikawa, Chihiro Sasakawa, Takashi Sakai
Masatoshi Yamada, Nobuhiko Okada, Sou-ichi Makino

Department of Bacteriology
The Institute of Medical Science
The University of Tokyo
4-6-1 Shiroganedai
Minato-ku, Tokyo 108, Japan

INTRODUCTION

LaBrec et al.[7] established in 1964 that invasion into epithelial cells was the essential early step in the pathogenesis of bacillary dysentery. The virulent translucent strain of *Shigella flexneri* invaded the intestinal epithelium, while the avirulent opaque variant did not, in spite of the finding that both had the same LD_{50} upon intraperitoneal injection. Ogawa et al.[15] performed similar experiments and obtained essentially the same results. However, they made one important additional observation; within 24 hours of infection, the bacteria existed exclusively in the epithelial cells without direct penetration into the lamina propria and bacterial multiplication therein. Thus, virulent shigellae invade the colonic epithelium and spread successively to adjacent epithelial cells resulting in a severe inflammatory reaction, further spread of the bacteria to deeper tissue and, ultimately, to ulceration and bloody diarrhea.

Typical bacillary dysentery is confined to humans and monkeys. To overcome this obstacle experimentally, several other models have been devised. The Sereny test[26] (Ser is used hereafter as the phenotype designation) that measures the capacity of the bacteria to invade the corneal epithelium of guinea pigs, resulting in keratoconjunctivitis, is a model which has long been believed to best reflect the pathogenicity of the bacteria although it is unrelated directly to diarrhea. The invasion of tissue culture cells, followed by bacterial multiplication (phenotype Inv), the measurement of fluid accumulation (phenotype Flu) and the examination of histopathological changes of the intestinal mucosa following the rabbit ileal loop test are apparently related more directly to the pathogenic mechanism leading to diarrhea. Pigmentation of colonies on nutrient agar containing Congo red[12,23] (phenotype Pcr), bacterial growth inhibition in an artificial medium[23] (phenotype Igr), and hemolysis of sheep erythrocytes after mechanical packing with the bacteria followed by incubation at 37°C[21] (phenotype Chl) are attributes frequently seen only in virulent bacteria and, presumably, are related to the possession of some virulence factor(s) by the bacteria tested.

The discovery by Luria and Burrous[9] of an intergeneric conjugation between *Escherichia coli* K-12 Hfr and *S. flexneri* has made feasible the genetic analysis of the *Shigella* chromosome. The virulence-associated chromosomal loci were no exception to this. Thus, genetic loci for the synthesis of the group antigen 3, 4 of *S. flexneri* 2a linked to *his*, a locus *kcpA* after keratoconjunctivitis provocation responsible for Ser+ phenotype near *purE*, and the *arg-mtl* region necessary both for Flu+ and Ser+ phenotypes were proven to be essential virulence-associated genetic determinants,[4,5,6,19] although their biochemical implication for pathogenesis leading to bacillary dysentery has not been elucidated. Furthermore, a 100 to 140 MDa plasmid was found to be responsible for the early step of colonic invasion in four species of shigellae and in enteroinvasive *E. coli*.[10,16,19,20,24,27] Recent application of recombinant DNA technology has made it possible to characterize these genetic determinants in terms of molecular biology and molecular genetics.

To characterize these virulence-associated genetic determinants at the molecular level we chose a strain of *S. flexneri* 2a, YSH6000. It contains the 140 MDa (230 kb) virulence plasmid pMYSH6000 which consists of 23 *Sal*I fragments.[23] (Recent study has revealed the existence of an additional small fragment between *Sal*I fragments B and P). A *Sal*I restriction map of pMYSH6000 was made by analyzing 69 spontaneous deletion mutants and an IS1 insertion mutant isolated independently and further confirmed on the basis of 359 *Sal*I-generated partial digests of pMYSH6000 cloned into pBR322.[23] Then, more than 300 independent Tn5 insertion mutants[25] into pMYSH6000 were isolated.[24] Half of them were found to be Ser- as judged by mouse Sereny test.[13] These Ser- insertions were assigned to two individual *Sal*I fragments, F and G and to four contiguous *Sal*I fragments, B-P-H-D.[24] The fifth fragment between *Sal*I-B and *Sal*I-P was found recently but in this communication we do not refer to this.

An approximately 1 kb *virF* region essential for virulence and located on the 230 kb plasmid pMYSH6000 has been cloned in *E. coli* K-12 by selecting for Pcr+ phenotype.[16] The DNA sequence of the *virF* region has been determined and a minicell analysis has shown that three proteins of 30, 27 and 21 kd are produced from this region[17] which corresponds to the virulence region identified by separate experiments on the *Sal*I fragment F by analysis of Tn5 insertion mutants.[24] An approximately 4 kb *virG* region essential to provoke positive Sereny tests has been located on the *Sal*I fragment G of pMYSH6000.[10] Mutants with Tn5 insertions in this region invade cultured cells and multiply therein but do not spread to adjacent cells.[10] More precisely, the DNA region necessary for the *virG* function including the *virG* gene and its own regulatory sequence has been localized to a 3.6 kb fragment.[8] The nucleotide sequence of 4472 bp covering this region has been determined and a large open reading frame encoding 1102 amino acid residues has been identified.[8] A protein product corresponding to this open reading frame has been identified by the minicell method[8] and by Western immunoblotting using serum from a dysentery patient.[18] The molecular weight experimentally estimated by these methods was 130 kDa,[8,18] but the molecular weight predicted from the nucleotide sequence is 117 kDa.[8] External labeling of bacteria with radioactive iodine has indicated that the protein is exposed on the bacterial surface.[8] By analyzing more precisely the sites of 134 independent Tn5 insertions in the four contiguous *Sal*I fragments, B-P-H-D, five virulence-associated regions have been identified; four (regions 1, 3, 4 and 5) were associated with Ser, Inv, Pcr and Igr phenotypes and one (region 2) was associated with the Ser and Inv but not with the Pcr phenotype.[22] These five regions presumably correspond to a 37 kb minimum sequence necessary for invasion which was shotgun cloned into a high copy cosmid by Maurelli et al.,[11] although the 37 kb sequence presumably does not contain the *virF* region. The reason why the 37 kb fragment has apparently expressed Inv+ phenotype without the *virF* region seems to be that it was cloned into a high copy vector. The *EcoRI*, *HindIII*, and *BglII* cleavage patterns of region 2 of the B-P-H-D cluster of pMYSH6000 correspond to those of a 9 kb fragment cloned and analyzed by Buysse et al.[1] which encode 3 immunodominant proteins of 57, 43, and 39 kDa. The *HindIII*, *EcoRI*, *BglII*, *SalI* and *XhoI* cleavage patterns of region 5 of pMYSH6000 were similar to a virulence-associated insert of 20 kb cloned and analyzed by Watanabe and Nakamura[28] from a 120 MDa plasmid, pSS120, of *S. sonnei*. All of the 7 separable regions, *virF*, *virG* and region 1 through 5 were highly conserved among 4 species of shigellae and enteroinvasive *E. coli*.[10,18,22]

As the further extension of these previous studies, we describe in this communication our recent results on functional characterization of the *virF* gene[18] and the molecular genetic analysis of a chromosomal cistron *kcpA*.[29]

The 30 kDa virF protein is a transcriptional activator for the expression of the four virulence-associated antigens coded by the virG cistron and Region 2.

The production of virulence-associated antigens was analyzed by Western blotting with the serum of a bacillary dysentery patient in various YSH6000 derivatives containing Tn5 insertions in pMYSH6000 and/or carrying a plasmid with a cloned fragment from pMYSH6000. Several antigens including those of 130, 57, 43, and 39 kDa were expressed by the virulent strain but not by the plasmid-cured avirulent strain. This indicates that these antigens are shi gella-specific and associated with the presence of the 230 kb virulence plasmid. A Tn5 insertion in the *virG* gene resulted in the production of either very decreased amounts of the 130 kDa antigen or none at all, simultaneously with the loss of virulence. Two different Tn5 insertion mutants into region 2, which were Ser-, produced decreased amounts, or none at all of either or all of the 57, 43 and/or 39 kDa proteins. In these cases, both the production of antigens and virulence were simultaneously restored by the cloned *virG* plasmid or by the cloned region 2 plasmid. In contrast, the pro duction of all four antigens of 130, 57, 43, and 39 kDa was markedly reduced in an avirulent Tn5 insertion in the *virF* gene. All the changes, both in antigen production and virulence, were restored by adding the cloned *virF* gene which did not encode these antigens by itself. This suggests that *virF* is required for the full expression of the four antigens.

To confirm this quantitatively, a *virG-lacZ* fusion was made using a fragment containing *lacZ* with a deletion of the transcriptional regulator sequences and first 7 codons,[2] and β-galactosidase activity was measured. Although the β-galactosidase activity produced by the *virG-lacZ* fusion plasmid alone was 471 units, 2560 units of β-galactosidase activity was produced in the presence of the cloned *virF* fragment. Similar results were obtained with the *lacZ*-fusion with Region 2.

Then the same amount of RNA extracted from the strain with a cloned *virG* plasmid alone and the same strain containing both the cloned *virG* plasmid and the cloned *virF* plasmid was electrophoresed and hybridized with a 1.2 kb *virG* specific probe. An RNA transcript was found to be transcribed from the *virG* region of the cloned *virG* plasmid far more in the presence than in the absence of the cloned *virF* plasmid. Thus, it was concluded that the *virF* region regulates the expression of the 130 kDa antigen positively at a tran- scriptional level.

Because the four antigens were positively regulated by the *virF* region *in trans*, the proteins encoded by the *virF* region were likely to be respon- sible for this regulation. From the nucleotide sequence of the *virF* region previously determined[17] an open reading frame coding for 262 amino acids was predicted, and corresponded to a protein of 30 kDa. Within this open reading frame seven methionine codons were found and protein analysis by the minicell method revealed that the *virF* region produced not only the 30 kDa protein but two others of about 27 and 21 kDa.[17] The latter proteins were likely to be translated within the region coding for the 30 kDa protein.

A plasmid, pMYSH6521 was made by ligating a 1.3 kb DNA fragment containing the cloned *virF* region, from which pMYSH6522 was constructed by an 81 bp deletion around the initiation codon of the 30 kDa protein and by the insertion of an 8-mer *Bam*HI linker into the deleted region. Minicell analysis revealed that the 27 and 21 kDa proteins were produced both by pMYSH6521 and by pMYSH6522, but that only the former produced the 30 kDa protein. Western blot analysis of the whole cell extracts with the serum of a bacillary dysentery patient clearly demonstrated that pMYSH6521 but not pMYSH6522 had the *virF* activity required for restoring the production of the four antigens as well as virulence. These observations led us to conclude that the 30 kDa protein encoded by the *virF* region is responsible for the positive regulation.

Molecular cloning and characterization of chromosomal virulence region KcpA

The chromosomal region KcpA had been mapped near *purE* by intergeneric conjugation between *E. coli* Hfr and *S. flexneri* by Formal et al.[5] some 20 years ago. Since no virulence-associated loci other than KcpA have thus far been localized near *purE*, we expected to isolate a KcpA- intergeneric transconjugant from the virulent *S. flexneri* 2a YSH6000 by a conjugational cross with *E. coli* HfrH by selecting for Lac+ phenotype without alteration of any other virulence-associated genetic determinants. Thus, we obtained a KcpA- intergeneric transconjugant, YSH6700.

In addition, chromosomal DNA was isolated from YSH6000, partially digested with *Sau*3A, size-fractionated to approximately 4 to 30 kb by NaCl density gradient, ligated to the *Bam*HI site of a pBR322-derived vector and then transformed into *E. coli* MC1061, thus making a library consisting of more than 1300 *E. coli* transformants containing the *Shigella* genome. Plasmid DNA was again prepared from each *E. coli* transformant and transformed into the KcpA- intergeneric transconjugant YSH6700 carrying the 230 kB virulence plasmid, pMYSH6000, which was derived from YSH6000. Each transformant was examined for the Ser phenotype by the simplified focus-plaque assay.[22] This method, using a 96-well microtiter plate, was based on the plaque method devised by Oaks et al.,[14] but was technically much simpler.

Thus, a plasmid pMYSH6700 was found to convert the intergeneric transconjugant YSH6700 to KcpA+ and later to contain a cloned *Shigella* chromosomal fragment of approximately 27 kB derived from the DNA genome linked to the *purE* locus. By the use of various techniques to minimize the size of fragments necessary and sufficient to restore the KcpA+ phenotype in YSH6700 the *kcpA* region was finally localized within 1.0 kB between the *Hinc*II site proximal to the *Hpa*I site, and the *Stu*I site. The direction of transcription was determined by the use of a chloramphenicol resistance cartridge.[3]

Then the nucleotide sequence of 1014 bp between *Hinc*II and *Stu*I sites was determined. An open reading frame translatable in the direction determined by the chloramphenicol resistance cartridge was found between positions 457 and 784. It coded for 109 amino acid residues and the calculated molecular weight of the protein was 12.3 kDa. Upstream of the open reading frame, a -35 region, a -10 sequence and a ribosome binding sequence were found. A typical terminator was found downstream of the coding region.

The protein products were examined by the minicell method. Proteins of approximately 15 kDa and 8 kDa were detected. Analysis of various insertion mutants revealed that the former was the main product and the latter presumably a degradation product from the former.

The KcpA region originally identified genetically as an essential genetic region for provocation of keratoconjunctivitis or for the positive Sereny test in guinea pigs has thus been demonstrated to consist of a single cistron coding for a single protein of 12.3 kDa. Furthermore, the KcpA-intergeneric transconjugant YSH6700 has been found to lack only a single genetic element with respect to virulence. Under such circumstances comparative morphological studies were made by the use of the virulent wild type YSH6000 and the kcpA mutant YSH6700.

The kcpA mutant was positive in conventional invasion tests using tissue-culture monkey kidney (abbreviated as MK) cells, and almost no difference was found between YSH6000 and YSH6700. A kinetic study, however, revealed an interesting difference between them. Although all the MK cells were infected by YSH6000 at 6 hours after infection, the percentage of the MK cells infected by YSH6700 remained almost unchanged during this time and then drastically decreased. At 4 hours postinfection, YSH6000 spread within the cytoplasm and then to adjacent cells and the intracellular bacteria were apparently morphologically normal. On the other hand, YSH6700 multiplied in the cells more rapidly than YSH6000 and the number of YSH6700 in the infected cells was far greater than that of YSH6000 at 4 hours after infection. YSH-6700 bacteria apparently spread within the cytoplasm but never spread intercellularly and took on a so-called parade arrangement as if their intracytoplasmic movement was mechanically hindered.

To confirm these observations the infected cells were examined by electron microscopy. YSH6000 showed membrane lysis as reported by Sansonetti et al.,[21] whereas YSH6700 never showed membrane lysis even at 5 hours postinfection, when the bacterial number of YSH6700 per infected cells was far greater than that observed in cells infected with YSH6000.

When guinea pigs' eyes were infected with YSH6000, a typical inflammatory reaction characteristic of keratoconjunctivitis shigellosa was observed within 24 hours. Indirect immunofluorescence using anti-Shigella antisera revealed that YSH6000 invaded the full layer of corneal epithelium at 24 hours after infection, several surface layers of the corneal epithelium were peeled off, and a small ulceration reaching to the lamina propria was observed. In contrast, similar infection with YSH6700 never provoked any changes characteristic of keratoconjunctivitis. In immunofluorescence staining, the mutant invaded not at all, or at most only sporadically, the single surface layer of the corneal epithelium.

These observations suggest that the KcpA mutant is in some way hindered from free movement within the cytoplasm and, eventually, from spreading into adjacent cells. The membrane of the phagocytic vacuole surrounding mutant bacteria may hinder their free movement. Thus, intercellular spread, an important step in the pathogenesis of bacillary dysentery, is apparently defective in the KcpA mutant similar to the virG mutant.[8,10] However, the steps affected in these mutants are probably different from each other.

SUMMARY

The 130 kDa virG product and 3 antigens of 57, 43 and 39 kDa encoded by region 2 were found to be positively regulated at the transcriptional level by the 30 kDa protein encoded by the virF gene. In another series of experiments, the KcpA region, one of the virulence-associated chromosomal regions, was cloned into a KcpA- intergeneric transconjugant using a new simple method to assay the Ser+ phenotype. The nucleotide sequence of, and the minicell product from, the KcpA clone was determined, and demonstrated that the KcpA region contained only a single cistron. The KcpA product, like the virG product, was found to be required for the invading bacteria to spread into adjacent cells.

ACKNOWLEDGEMENT

The studies from the authors' laboratory were supported by grants provided by the Ministry of Education, Science and Culture, Japanese Government (No. 61440035, 62304036, and 63480151).

REFERENCES

1. Buysse JM, Stover CK, Oaks, EV, Venkatesan M, Kopecko DJ, 1987, Molecular cloning of invasion plasmid antigen (*ipa*) genes from *Shigella flexneri*: analysis of *ipa* gene products and genetic mapping, *J Bacteriol.*, 169:2561-2569.

2. Casadaban MJ, Chou J, Cohen SN, 1980, In vitro gene fusions that join an enzymatically active *β*-galactosidase segment to amino-terminal fragments of exogenous proteins: *Escherichia coli* plasmid vectors for the detection and cloning of translational initiation signals, *J Bacteriol.*, 143:971-980.

3. Close TJ, Rodriguez RL, 1982, Construction and characterization of the chloramphenicol-resistance gene cartridge: a new approach to the transcriptional mapping of extrachromosomal element, *Gene*, 20:305-316.

4. Formal SB, Gemski P Jr, Baron LS, LaBrec EH, 1970, Genetic transfer of *Shigella flexneri* antigens to *Escherichia coli* K-12, *Infect Immun.*, 1: 279-287.

5. Formal SB, Gemski P Jr, Baron LS, LaBrec EH, 1971, A chromosomal locus which controls the ability of *Shigella flexneri* to evoke keratoconjunctivitis, *Infect Immun.*, 3:73-79.

6. Formal SB, LaBrec EH, Kent TH, Falkow S, 1965, Abortive intestinal infection with an *Escherichia coli-Shigella flexneri* hybrid strain, *J Bacteriol.*, 89:1374-1382.

7. LaBrec EH, Schneider H, Magnani TJ, Formal SB, 1964, Epithelial cell penetration as an essential step in the pathogenesis of bacillary dysentery, *J Bacteriol.*, 88:1503-1518.

8. Lett MC, Sasakawa C, Okada N, Sakai T, Makino S, Yamada M, Komatsu K, Yoshikawa M, Identification of the *virG* protein and determination of the complete coding sequence: A plasmid-coded virulence gene of *Shigella flexneri*, *J Bacteriol*, in press.

9. Luria SE, Burrous JW, 1957, Hybridization between *Escherichia coli* and *Shigella*, *J Bacteriol.*, 74:461-476.

10. Makino S, Sasakawa C, Kamata K, Kurata T, Yoshikawa M, 1986, A genetic determinant required for continuous reinfection of adjacent cells on large plasmid in *S. flexneri* 2a, *Cell*, 46:551-555.

11. Maurelli AT, Baudry B, d'Hauteville H, Hale TL, Sansonetti PJ, 1985, Cloning of plasmid DNA sequences involved in invasion of HeLa Cells by *Shigella flexneri*, *Infect Immun.*, 49:164-171.

12. Maurelli AT, Blackmon B, Curtiss R III, 1984, Loss of pigmentation in *Shigella flexneri* 2a is correlated with loss of virulence and virulence-associated plasmid, *Infect Immun.*, 43:397-401.

13. Murayama SY, Sakai T, Makino S, Kurata T, Sasakawa C, Yoshikawa M, 1986, The use of mice in the Sereny test as a virulence assay of shigellae and enteroinvasive *Escherichia coli*, *Infect Immun.*, 51:696-698.

14. Oaks EV, Wingfield ME, Formal SB, 1985, Plaque formation by virulent *Shigella flexneri*, *Infect Immun.*, 48:124-129.

15. Ogawa H, Nakamura A, Nakaya R, 1968, Cinemicrographic study of tissue cell cultures infected with *Shigella flexneri*, *Japan J Med Sci Biol.*, 21:259-273.

16. Sakai T, Sasakawa C, Makino S, Kamata K, Yoshikawa M, 1986, Molecular cloning of a genetic determinant for Congo red binding ability which is essential for the virulence of *Shigella flexneri*, *Infect Immun.*, 54: 395-402.

17. Sakai T, Sasakawa C, Yoshikawa M, 1986, DNA sequence and product analysis of the *virF* locus responsible for Congo red binding and cell invasion in *Shigella flexneri* 2a, *Infect Immun.*, 54:395-402.
18. Sakai T, Sasakawa C, Yoshikawa M, 1988, Expression of four virulence antigens of *Shigella flexneri* is positively regulated at the transcriptional level by the 30 kilodalton *virF* protein, *Molec Microbiol.*, 2: 589-597.
19. Sansonetti PJ, Hale TL, Dammin GJ, Kapfer C, Collins HH Jr, Formal SB, 1983, Alterations in the pathogenicity of *Escherichia coli* K-12 after transfer of plasmid and chromosomal genes from *Shigella flexneri, Infect Immun.*, 39:1392-1402.
20. Sansonetti PJ, Kopecko DJ, Formal SB, 1982, Involvement of a plasmid in the invasive ability of *Shigella flexneri, Infect Immun.*, 35:852-860.
21. Sansonetti PJ, Ryter A, Clerc P, Maurelli AT, Mounier J, 1986, Multiplication of *Shigella flexneri* within HeLa cells: lysis of the phagocytic vacuole and plasmid-mediated contact hemolysis, *Infect Immun*, 51:461-469.
22. Sasakawa C, Kamata K, Sakai T, Makino S, Yamada M, Okada N, Yoshikawa M, 1988, Virulence-associated genetic regions comprising 31 kilobases of the 230-kilobase plasmid in *Shigella flexneri* 2a, *J Bacteriol.*, 170: 2480-2484.
23. Sasakawa C, Kamata K, Sakai T, Murayama SY, Makino S, Yoshikawa M, 1986, Molecular alteration of the 140-megadalton plasmid associated with loss of virulence and Congo red binding activity in *Shigella flexneri, Infect Immun.*, 51:470-475.
24. Sasakawa C, Makino S, Kamata K, Yoshikawa M, 1986, Isolation, characterization, and mapping of Tn5 insertions into the 140-megadalton invasion plasmid defective in the mouse Sereny test in *Shigella flexneri* 2a, *Infect Immun.*, 54:32-36.
25. Sasakawa C, Yoshikawa M, 1987, A series of Tn5 variants with various drug-resistance markers and suicide vector for transposon mutagenesis, *Gene*, 56:283-288.
26. Sereny B, 1955, Experimental shigella keratoconjunctivitis, *Acta Microbiol Acad Sci Hung.*, 2:293-296.
27. Watanabe H, Nakamura A, 1985, Large plasmids associated with virulence in *Shigella* species have a common function necessary for epithelial penetration, *Infect Immun.*, 48:260-262.
28. Watanabe H, Nakamura A, 1986, Identification of *Shigella sonnei* form I plasmid genes necessary for cell invasion and their conservation among *Shigella* species and enteroinvasive *Escherichia coli, Infect Immun.*, 53:352-358.
29. Yamada M, Sasakawa C, Okada N, Makino S, Yoshikawa M, Molecular cloning and characterization of chromosomal virulence region *KcpA* of *Shigella flexneri, Molec Microbiol.*, in press.

MYCOTOXINS

MYCOTOXINS AND MYCOTOXICOSES: OVERVIEW

Hiroshi Kurata

Tokyo Kenibkyoin, Institute of Food Hygiene
4-8-32 Kudan-Minami
Chiyoda-ku, Tokyo 102, Japan

INTRODUCTION

At this symposium of the UJNR, Toxic-Microorganisms Panels which has a
brilliant history of more than 20 years under the auspices of the United
States-Japan Cooperative Program on Development and Utilization of Natural
Resources, the discussion will be conducted on the main subject of cellular
and molecular mode of action of selected microbial toxins in foods and
feeds. The scheme of the symposium encompasses somewhat expanded fields as
compared with several panel meetings held in the past. Such a symposium,
certainly, is a more valuable meeting to promote international research
activity on mycotoxins.

It is a great honor and privilege to have been given this opportunity of
presenting a special lecture at the first session of this symposium.

Mycotoxins and Mycotoxicoses

The toxic substances produced by the fungi are generally called mycotoxins, i.e. fungal products encompass numerous classes of chemical compounds
that have complex and diversified structures and biological properties ranging from antibiosis to mammalian toxicity. Mycotoxins have been defined as
secondary metabolites of fungi that alter physiological processes in higher
animals regardless of the route of administration. This broad definition has
caused some confusion in the literature, because many fungal metabolites are
toxic only if a particular test system, dose, or route of administration is
chosen. Few fungal metabolites are proven mycotoxins as the etiological
agents of mycotoxicoses. Others are suspected of toxic activity in higher
animals. Additionally, there are diseases suspected of being mycotoxicoses
for which no toxin has been found. In contrast to those secondary metabolites, which usually have no apparent biochemical role, primary metabolites
are functional in metabolic pathways. As a practical means, the secondary
fungal metabolites can be grouped into 3 categories: (1) those that are
economically beneficial, (2) those that are detrimental, and (3) those that
are neither beneficial nor detrimental. Mycotoxins belong to category (2),
but antibiotics are included in category (1) although some of them reveal
toxic effects on human and animals.

Mycotoxins, therefore, include toxic compounds active in certain toxicological assay systems. Those compounds may well be called bioactive

Microbial Toxins in Foods and Feeds, Edited by A.E. Pohland *et al.,*
Plenum Press, New York, 1990

substances produced by fungi and should be separated from the group of mycotoxins, particularly in the field of food and feed hygiene.

From 1960 to 1970, extensive mycotoxin screening trials were carried
out by many mycotoxin research groups in the world. As the result of these
experiments, numerous toxic substances have been obtained such as ochratoxins, citrinin, penicillic acid, cytochalasins, chaetoglobosin, PR toxin
and so on. Presently more than five hundred toxic compounds of fungal origin
have been found. However, only few compounds are true mycotoxins implicated
in human and animal mycotoxicoses.

Numerous reviews, proceedings of symposia, reports of research and surveys have been published on the various aspects of the mycotoxin problem
involving humans and domestic animals.[1,5,7,8,10,11,13-25,27,29-36,39,40,42]

Today I do not have enough time to introduce mycotoxins in detail, so I
will speak about only several important mycotoxins and mycotoxicoses, their
present status and some future problems, mainly in Japan and in other Asian
countries.

Brief History of Mycotoxin and Mycotoxicoses

Prior to entering into the main subject, I would like to describe a brief
history of the mycotoxins and mycotoxicoses. Although the history of the
fungal poisoning is actually very old, the term mycotoxin was first used in
1960. It is said that the record of the occurrence of ergotism was found in
the Biblical times. In the beginning it was called toxic wheat. At that
time, of course, a causative agent was not known. Around 1875, it was clarified that the causative agent was the sclerotium of the fungus *Claviceps
purpurea*. Around 1970, many people in Europe died following ingestion of
cereal grains infested with the sclerotium of this fungus. However, since
1900 this mycotoxicosis (ergotism) has diminished remarkably by eradicating
toxic sclerotia (ergots) from harvested cereals, but in India, Canada and
Egypt outbreaks of ergotism in domestic animals still occurs.

Feed poisonings caused by a mycotoxin occurred in horses in the Soviet
Union from 1931 to 1940. This mycotoxicosis was quite endemic and resulted
in massive illnesses. The etiological survey indicated that this disease was
associated with the ingestion of hay infested with *Stachybotrys atra* (Table
1) Subsequently, this seasonal toxicosis was named stachybotryotoxicosis.
It is of interest that this toxicosis occurred in farmers due to possible
inhalation of a volatile compound emitted from the contaminated hay. At that
time the causal mycotoxin was not chemically described.

From 1940 to 1950, Forgacs and his associates[42] reported several deaths
of poultry resulting from feed poisoning in the United States, and they
pointed out that the causal agents were moldy feeds. They confirmed myco-
toxigenicity by feeding of artificial mold-inoculated feeds to poultry and
recognized *Penicillium rubrum*, *Aspergillus flavus* and other species of fungi
as toxin-producing fungi. Forgacs was a pioneer researcher in the world, and
we recognize his great contribution of work on so-called moldy corn toxicoses
in domestic animals. After the discovery of aflatoxins from *A. flavus* a
fungal toxin was chemically characterized from *P. rubrum* and named rubratoxin
(Table 1).

Before Forgacs' study, on the other hand, in the Orient-Japan, Sakaki
reported a suspected mycotoxicosis in man caused by moldy rice and demonstrated toxicity of moldy rice in animal experiments. He did not directly
detect a causal mycotoxin from the deteriorated rice. The disease was often
fatal within 3 days. Cardiac symptoms included irritability, palpitation,
dilatation of the right ventricle, and other symptoms of cardiac failure.

Table 1. Mycotoxins probably involved in mycotoxicoses

Mycotoxin	Mycotoxicosis	Producing fungi	Susceptible host	Biological effects
Ergot alkaloids	Ergotism	*Claviceps purpurea* *C. paspali*	human, cattle sheep, horse	Nervous or gangrenous forms (Ataxia)
Aflatoxins	Aflatoxicosis	*Aspergillus flavus* *A. paraciticus*	mammals, birds, fishes	Acute hepatotoxicity & cardiotoxicity Chronic liver toxicity Indian childhood Cirrhosis Reye's Syndrome Primary liver cancer Esophageal cancer Kwashiorkor in Sudanese children Death
Ochratoxin A	Ochratoxicosis Balkan nephropathy Renal porcine nephropathy	*Aspergillus flavus* *Penicillium viridicatum*	human? swine turkey, chickens horse	Nephropathy Urinary tract tumors Balkan nephropathy Renal porcine nephropathy
Citrinin		*Penicillium citrinum* *P. viridicatum*	swine	Nephropathy
Zearalenone (F-2)	Estrogenic myco- toxicosis	*Fusarium graminearum* (*Gibberella zeae*) *F. sporotrichioides*	swine	Estrogenic syndrome Vulvovaginitis Vaginal & rectal prolapse
Trichothecene toxins: Deoxynivalneol (Vomi- toxin, DON) Nivalenol (NIV) Fusarenon-X (4-acetyl-NIV)	Fusariotoxicosis (Akakabi-byo)	*Fusarium graminearum* *F. sporotrichioides*, *F. tricinctum, F. poae*	human livestock birds	Feed refusal nausea, vomiting, diarrhea, leukopenia, hemorrhage, reproduc- tion disorder, immuno- suppression (continued)

251

Table 1 (continued)

Mycotoxin	Mycotoxicosis	Producing fungi	Susceptible host	Biological effects
T-2 toxin Diacetoxyscirpenol (DAS)	Alimentary toxic aleukia (ATA)	F. sporotrichioides F. poae	human livestock	Leukopenia, sepsis, hemorrhage diathesis, necrotic angina Death
Satratoxin H Verrucarin Roriden E	Stachybotrio-toxicosis	Stachybotrys atra Fusarium spp. (?)	horse cattle sheep swine human	Rhinitis, fever, chest pain, leukopenia Gastrointestinal and pulmonary hemorrhages Death
Salframine Sporidesmins	slobber photosensitizing disease	Rhizoctonia leguminicola Pithomyces chartarum	cattle cattle, sheep	Excess salivation Hepatotoxic, facial eczema
Psoralens Maltoryzin Phomopsin Diplodiatoxin Secalonic acid (D.F.) Sterigmatocystin	"	Sclerotinia sclerotiorum Aspergillus oryzae Phomopsis leptostromiformis Diplodia maydis Penicillium oxalicum Aspergillus versicolor Chaetomium thieloavioideum	human cattle sheep cattle, sheep human rat chicken	Dermotoxic Death Nephropathy Mucoenteritis Death Hepatocarcinoma Acute oral toxicity
Penitrem A	Tremorgen intoxication	Penicillium palitans P. crustosum	cattle, sheep sheep	Tremorgenic
Yellow rice toxins: Luteoskyrin Cyclochlorotine Citreoviridin Rugulosin	Yellow rice toxicoses	Penicillium islandicum " P. citreo-viridie P. ruglosum	rat rat human rat	Hepatotoxic, cirrhosis Carcinogenic Neurotoxic Hepatotoxic
Rubratoxin Cytochalasin E PR toxin Patulin Penicillic acid		P. rubrum A. clavatus P. roqueforti P. patulum Penicillium spp.	cattle human (?) cattle cattle livestock	Hepatotoxic Death Abortion Death ?

The symptoms are similar to cardiac beriberi. The causative mycotoxin is now recognized as citreoviridin, one of the neurotoxic mycotoxins, produced by *Penicillium citreo-viride*. Following this, Japanese mycotoxicologists obtained two mycotoxins from moldy rice in Japan: citrinin and cyclochloro-tine from *Penicillium citrinum* and P. islandicum, respectively. They demon-strated that citrinin is nephrotoxic and cyclochlorotine is hepatotoxic to the rat. Several years later, cyclochlorotine and luteoskyrin, produced by several strains of *P. islandicum*, were found to cause cirrhosis and hepatoma. Presently, we have not experienced these two mycotoxins causing mycotoxicoses in Japan.

In 1960, a dramatic mycotoxicosis of turkeys occurred in England and sub-sequently, aflatoxin was discovered as the etiological agent of this disease. Related compounds, aflatoxins B and G were found from the culture of *A. flavus* which had contaminated Brazilian peanut meal mixed with other feeds for turkeys. Since the aflatoxins were proved to be highly potent hepatocarcino-gens, world-wide concern by microbiologists caused concentrated research on this matter. More information of aflatoxins will be given later.

Following the discovery of aflatoxins, mycotoxicologists of many coun-tries in the world have accelerated study of all mycotoxins as well as of aflatoxins. Mycotoxin research is still continuing and it seems to be end-less. Prior to 1960, besides stachybotryotoxicosis which was already described, there were reports such as alimentary toxic aleukia (ATA) occur-ring in man in the Soviet Union, and facial eczema in sheep in New Zealand (sporidesmins), diplodiosis, salivation syndrome of domestic animals, and celery photodermatitis of man.

In the last 27 years, ochratoxin A, produced by *A. ochraceus*; sterigmato-cystin, produced by *A. versicolor*; trichothecenes, deoxynivalenol and nivale-nol, produced by several different species of *Fusarium*; and zearalenone pro-duced by *Fusarium graminearum* were shown to be etiological agents for myco-toxicoses of man and livestock. These mycotoxins may well be called impor-tant mycotoxins, because they are highly toxic, some are carcinogenic, and they are heat stable. Except for sterigmatocystin, they have been detected chemically and biologically from various kinds of food and feedstuffs. Citrinin, a nephrotoxic mycotoxin, is likely a minor mycotoxin, even though it has promotion activity in renal carcinoma, since natural contamination of it in an agricultural commodity has not been found.

Under such a background of mycotoxin research, I will attempt to review the recent advances of research confining my remarks to two or three impor-tant mycotoxins which have played an etiological role in human and animal mycotoxicoses.

For more detailed information of mycotoxins and mycotoxicoses, this author would recommend three books edited by Busby and Worgan,[5] Shank[30,31] and Smith and Moss.[33]

Important Mycotoxins and Mycotoxicoses

Aflatoxins. Although 20 compounds, all designated aflatoxins, have been isolated, the term aflatoxin usually refers to four compounds of the group of bis-furano-coumarin metabolites produced by *A. parasiticus*, named B_1, B_2, G_1 and G_2, which occur naturally in plant products. Cows fed feeds contain-ing aflatoxin B_1 and B_2 excrete metabolites in the milk called aflatoxin M_1 and M_2. Extensive studies of aflatoxins have been carried out by myco-toxicologists in the world and have established their chemical characteris-tics, chemical and biological analyses, distribution, and natural occurrence in food and feedstuffs, toxicology, mode of action of toxigenicity, metabolic pathways, epidemiology, detoxication, and control.

Since 1975, an epidemiological survey has been conducted to obtain clear evidence concerning the relationship between aflatoxin (AF) ingestion and incidence of hepatoma in populations who are living in areas with a high incidence of hepatoma such as in Uganda, Thailand, Kenya, Mozambique, Swaziland, Indonesia,[30,31,42] and Philippines.[3,4,5] These results showed that surveys in Kenya and Thailand showed a positive correlation. However, most of the population living in the areas surveyed were infected with hepatitis B virus, so the hepatoma incidence caused only by AF is difficult to estimate from these results. Moreover, only primitive analysis methods could be used for the detection of aflatoxins from samples of foods and serums. We would recommend that such epidemiological study be re-conducted to confirm the exact correlation by newer analytical methods such as radio-immunoassay (RIA), enzyme-linked immunoabsorbent assay (ELISA) and monoclonal antibody of DNA-probe techniques, if available.

Regarding human aflatoxicosis, it should be noted that mass-death cases occurred in India in 1979. Reye's syndrome in Thailand children named Udoron syndrome, and recently Kwashiorkor disease of Sudanese children[35] which may be caused by protein energy malnutrition, were regarded as subacute aflatoxicosis because aflatoxins were detected from blood and urine samples of the patients.

Quite recently, an extensive workshop concerning monitoring AF in human fluids organized by IARC, Lyon, reported that recently developed immunological assay techniques such as RIA, ELISA, and HPLC-RIA are recognized as acceptable and available techniques. They are highly sensitive in detecting AF from clinical materials in patients with primary hepatocarcinoma.[12]

The biological effects and susceptible hosts of AF are shown in Table 1, and a list of agricultural commodities and foodstuffs in which AF has been detected are shown in Table 2. Corn is the grain most highly contaminated by AF, and tree nuts and cereal grains are also heavily contaminated. No domestic AF contaminated grains and peanuts have been observed in Japan.

In most countries, the standards for AF contamination levels in foods and feedstuffs and their products have been established. These limitatiaons are different among the countries. For instance, the US standard is 20 ppb of total AF for finished products, but the Japanese standard is 10 ppb of AFB for peanuts and tree nuts. Now the most fearful problem concerning AF is the AFM contaminated milk resulting from feeding highly contaminated feeds to domestic dairy animals. It goes without saying that in every country where weather conditions are appropriate during the harvest season, we should be on the alert.

Sterigmatocystin. This mycotoxin is the secondary metabolite produced by *A. versicolor* group, and is a precursor of AF. The chemical structure is a bis-dihydro-ring of AF with a xanthone nucleus. As compared with AF, the toxicity is fairly low. In 1978, the carcinogenicity was discovered in rats by a Japanese pathologist.[17] So it might be ranked in the group of important mycotoxins, but, as shown in Table 2, the natural occurrence of this toxin occurs in quite a small number of food and feedstuffs and level of the contamination is of extremely low incidence.

Ochratoxin A. Ochratoxins are mycotoxins produced primarily by *Aspergillus ochraceus* and *Penicillium viridicatum*, and it has renal toxicity to experimental animals and possibly to man. The chemical structure is a group of seven isocoumarin amides of L-β-phenylalanine.

Generally the natural contamination in agricultural commodities has been

254

Table 2. Natural occurrence of mycotoxins in foods and feeds

Mycotoxin	Commodity and Product	Range of Mycotoxin microgram/Kg
Aflatoxins	corn, sorghum, peanuts, chili pepper, copra, wheat, millet, tree-nuts, milk, cheese, barley, flour, beans, sunflower seed, rice, noodle, soybean, green coffee bean, oats, spaghetti, coconut, buckwheat, meats, rye, cotton seed, cassava, dried fish	Corn 10-400 Peanuts 20-5200 Tree-nuts 3-4950 Cotton seed 2000 Milk 0.002-6.5 ng/ml
Ochratoxin A	corn, wheat, barley, flour, rice, oats rye, bean, pea, green coffee bean, heavily molded cereals, pancake mix, mixed feeds	5-27500
Sterigmatocystin	heavily molded rice, green coffee bean, beans, salami, home cured ham, cheese wheat	Rice 0.05-0.45 Cheese 0.005-0.6 Wheat 0.3-?
Trichothecene toxins	barley, wheat, corn, roasted barley flour cookies, biscuits, pop corn, barley-tea, buckwheat, sunflower seed, Job's tears, mixed feeds	DON 0.001-204 NIV 0.004-522 T-2 0.001-2.0 DAS 0.004-1.8
Zearalenone	corn, barley, feeds, pig feeds, silage, sorghum, hay	feeds 0.1-2900
Citrinin	barley, mixed barley and oats, corn, corn flour, yellow peanut kernels	?
Patulin	apple juice	1-45
Penicillic acid	corn (blue eye)	5-230

255

found mainly in cereal grains, green coffee bean and molded feeds, but these concentrations were not high. We have never seen ochratoxicoses in domestic animals in Japan or Korea[20] and other oriental countries. In other countries, the presence of ochratoxin A in feedstuffs and foodstuffs was associated with a number of outbreaks: porcine nephropathy was associated with ochratoxin in Danish barley, indemic human nephropathy was related to ingestion of grains in Yugoslavia, and avian nephropathy was caused by ochra-toxin contaminated feeds in Denmark and the US.[11,36] It is sometimes present in the tissues of swine and poultry consuming contaminated feed. In Denmark, a swine carcass is condemned if the concentration of ochratoxin A in the kidney exceeds 25 ppb. This toxin is thus of major importance to agri-systems.[33]

Trichothecene Toxins. These are sesquiterpenoid mycotoxins produced mainly by several species of the genus Fusarium, Myrothecium, Trichoderma and Stachybotrys. The non-macrocyclics, e.g. diacetoxyscirpenol (DAS), Fusa-renon-x, HT-2 toxin, nivalenol (NIV), neosolaniole, T-2 toxin and deoxyni-valenol (DON, vomitoxin), and the macrocyclic, e.g. verrucarin A, satratoxins and roridin A, are members of the trichothecene family. Clinical signs depend upon a number of variables, including the toxins present, their levels and the time period of ingestion, as well as on the species and nutri-tional health status of the domestic or laboratory animal involved. Fre-quently associated with the trichothecenes are nausea, emesis, decreased food consumption, food refusal, skin irritation, diarrhea, hematological changes (leukemia), hemorrhaging, and reproductive problems in man and domestic animals (Table 1). Since the fungi involved are widely dispersed, the toxins have been found in several cereal grains and feeds (Table 2). There have been reports of illness in humans; Akakabi-byo in Japan and Korea and alimen-tary toxic aleukia (ATA) in the Soviet Union. Disease also occurs in farm animals associated with ingestion of or exposure to trichothecene toxins. Thus the toxins are ranked as important mycotoxins involved in human and animal mycotoxicoses. So human and animal health regulatory officials are concerned about trichothecene toxins and the government of Canada is now intending to set up regulations for some of these toxins in cereal foods. The long-term feeding test with NIV in mice for one year was carried out by a Japanese task group, but no carcinogenic change was observed in 1988.[29] The chemical analysis technique using HPLC-FD was developed recently. In a recent report, Tanaka[37] indicated that 40% of 500 samples of barley from 20 countries averaged 255, 284 and 46 ng/g of NIV, DON and zearalenone (ZEN), respectively, and the highest contamination indicated was NIV, 75%, 390 ng/g; DON, 75%, 145 ng/g; and ZEN, 74%, 33 ng/g.

Zearalenone (F-2 Toxin). ZEN is a mycotoxin produced by F. graminearum on plants such as corn and wheat, in the field and during storage. The chemical structure is ((E, S)-2, 4-dihydroxy-6-(6'-(6'-oxo-10'-hydroxy-1-undecenyl)-benzoic acid un-lactone. ZEN is an estrogenic compound that causes reproductive problems in farm animals, especially swine. Signs may include vaginal swelling and in severe cases, vaginal and rectal prolapses, especially in mature gilts, as shown in Table 1. Despite the fact that ZEN is quite often detected in wheat and barley in Japan, its toxicosis has never been observed in Asian countries including Japan. In the U.S., porcine hyperestrogenism occurred from contaminated feed containing 2909 ppm of ZEN.

Less Important Mycotoxins

The outline of the important mycotoxicoses and their causative mycotoxins are illustrated in the foregoing. Finally, several less important mycotoxins should be mentioned as shown in Table 1, they are as follows: rubratoxin, cytochalasin, PR toxin, patulin, penicillic acid and so on. Most of them have been obtained from culturing of fungi from contaminated food and feed-stuffs which caused animal toxicoses or resulted from the mycotoxicological

256

screening test using mice or rats. Details of this knowledge refer to several reviews which are mentioned in the references cited in this paper.

Control of Mycotoxicosis

For the most part, the mycotoxicoses classified as important in the field of food hygiene are the result of ingestion of rather badly abused feeds and foods, or of the failure to guard against consumption of crops which had been invaded in the field by a toxigenic fungus. It must now be recognized that moldy food or feed can frequently be far more than a problem of aesthetics. Moldy feed can cause a wide variety of diseases in livestock, ranging from depressed growth rate to mortality. Further, as mentioned above, consumption by livestock of feeds containing mycotoxins may result in human exposure through contamination of edible tissues such as meat, milk, eggs, etc. The number of known toxigenic fungi and mycotoxins is high. Furthermore, since most fungi have not been investigated for their toxigenicity, the likelihood is that there are many more toxigenic fungi than are now recognized. The practical solutions to the types of acute or sub-chronic toxicoses listed in Table 1 are based on the assumption that moldy food and feed must be avoided as much as possible.

Future Problems for Mycotoxicological Study

Finally the future problems for the study of mycotoxicology are suggested in the order of priority. There are many problems still remaining in every part of the mycotoxicological field of study. The following items were based upon questionnaires from world-wide mycotoxin specialists disseminated and reported by Hesseltine[36] in the IUPAC meeting in South Africa in 1980, and some of the authors comments are supplemented.

1. Rapid and improved analytical methods in not only chemistry, but also in biotechnology available for mycotoxin monitoring systems.
2. Effect of mycotoxins on humans and domestic animals with the cooperation of a veterinary or human pathologist.
3. Detoxification.
4. Etiological and mycological investigations.
5. Mycotoxicology.
6. Establishment of international standards and specifications.
7. Epidemiological survey supported by multi-disciplinary assay teams having a high level of facility for the detection of mycotoxins.
8. Biosynthetic pathways and pharmacokinetics.

In addition, I would like to say that we want to have more active cooperation from plant pathologists and veterinarians, especially in the East Asian countries, including Japan, Korea, Taiwan, China, etc.

Finally, I would like to thank the Toxic Microorganisms Panel of the U.S., UJNR for inviting me to give this lecture and for their financial support. Also, I would like to thank the related organization of the UJNR for the help extended to me in many ways.

REFERENCES

1. Berry CR, 1988, The pathology of mycotoxins, *J Pathol.*, 154:301-311.
2. Bhat RC, Nagarajan R, Tulpule PG, 1978, "Health hazards of mycotoxins in India," ICMR Offset Press, New Delhi.
3. Bulatao-Jayme J, Almero EM, Salmmat LA, 1976, Eppidemiology of primary liver cancer in the Philippines with special consideration of possible aflatoxin factor, *J Phil Med Assoc.*, 52:129-150.
4. Bulatao-Jayme J, Almero EM, Casino CA, Valandria FV, 1981, Dietary

aflatoxin and hepatocellular carcinoma in the Philippines, *Phil J Int Med.*, 19:95-101.

5. Busby WF, Wogan GN, 1978, Food-borne mycotoxins and alimentary mycotoxicoses, *In*: "Food Borne Infections and Intoxications", Riemann H, Bryan FL, eds., Academic Press, New York.

6. Carlborg FW, 1979, Cancer Mathematical models and aflatoxin, *Fd Cosmet Toxicol.*, 17:159.

7. Cole RJ, 1980, Tremorgenic Mycotoxins: An Update, *In*: "Antinutrients and Natural Toxicants in Foods," Ory RF, ed., Food and Nutrition Press, Inc., Westport, CT.

8. Diener UL, 1979, "Aflatoxin and Other Mycotoxins: An Agricultural Perspective," CAST Task Force Report No. 80, Council of Agricultural Science and Technology, Ames, IA.

9. Dovorackova I, Storac I, Ayraud N, 1981, Evidence for aflatoxin B_1 in two cases of lung cancer in man, *J Cancer Res Clin Oncol.*, 100:221-224.

10. Enomoto M, Saito M, 1972, Carcinogens produced by fungi, *Ann Rev Microbiol.*, 26:279-312.

11. Environmental Health Criteria 11:1979, Mycotoxins, United Nations Environment Programme and WHO, Geneva.

12. Garner C, Ryder R, Montesano R, 1985, Monitoring of aflatoxins in human body fluid and application to field studies, *Cancer Res.*, 45:922-928.

13. Goldblatt LA, 1969, "Aflatoxin, Scientific Background, Control, and Implications," Academic Press, New York.

14. Hadidane R, Roger-Rognault H, Bouattour H, Ellouze F, Bacha H, Creppy EE, Dirheimer G, 1985, Correlation between alimentary mycotoxin contamination and specific diseases, *Human Toxicol.*, 4:491-501.

15. Heraberg M, 1970, "Toxic Micro-Organisms, Mycotoxins-Botulism," UJNR Joint Panels on Toxic Micro-Organisms and the US Department of Interior. Washington, DC.

16. Joffe AZ, 1986, "*Fusarium* Species: Their Biology and Toxicology," John Wiley & Sons, New York.

17. Kurata H, Hesseltine CW, 1982, Control of the Microbial Contamination of Foods and Feeds in International Trade: Microbial Standards and Specifications, *In*: "Proc. 14th Int. Symp. on Toxic Micro-Organisms, UJNR, Toxic Micro-Organisms Panel", Saikkon Publishing Co. Ltd., Tokyo.

18. Kurata H, Ueno Y, 1984, Toxigenic Fungi - Their Toxins and Health Hazard, *In*: "Proc. of the Mycotoxin Symposium held in the Third International Mycological Congress, Development in Food Science 7," Kodansha Ltd., Tokyo.

19. Lacey J, 1985, "Trichothecenes and Other Mycotoxins, Proc. of Int. Mycotoxin Symposium held in Sydney, Australia," John Wiley & Sons, New York.

20. Lee Su-Rae, 1985, Present status of mycotoxins studies in Korea, *Korean J Toxicol.*, 1:17-30.

21. Mateles RI, Wogan GN, 1967, "Biochemistry of Some Foodborne Microbial Toxins", The MIT Press, New York.

22. Marasas WFO, van Rensburg SJ, 1979, Mycotoxins and their medical and veterinary effects, *In*: "PLant Disease, Vol. IV," Horsfall JF, Cowling EB, eds., Academic Press, Cambridge.

23. Marasas WFO, Nelson PE, Toussoun TA, 1984, "Toxigenic Fusarium Species, Identity and Mycotoxicology," The Pennsylvania State University Press, University Park.

24. Oltjen RR, 1979, "Interaction of Mycotoxin in Animal Production," National Academy of Sciences, Washington, DC.

25. Purchase IFH, 1974, "Mycotoxins," Elsevier, Amsterdam.

26. Richard JL, Cysewski SF, Pier AC, Booth GD, 1978, Comparison of effects of dietary T-2 toxin on growth immunogenic organs, antibody formation and pathogenic changes in turkeys and chickens, *Am J Vet Res.*, 39:1674-1679.

27. Rodricks JV, Hesseltine CW, Melhman MA, 1977, "Mycotoxins in Human and

Animal Health," Pathotox Publishers, Inc., Park Forest South, IL.

28. Ryu Jae-Chun, Ohtsubo K, Izumiya N, Nakamura K, Tanaka T, Yamamura H, Ueno Y, 1988, The acute and chronic toxicities of nivalenol in mice, *Fund Appl Toxicol.*, 11:38-47.

29. Scott PM, Trenholm HL, Sutton MD, 1985, "Mycotoxins: A Canadian Perspective, " NRCC No. 22848, NRCC/CNRC, Ottawa.

30. Shank RC, 1981, "Mycotoxins and N-Nitroso Compounds: Environmental Risks, Vol. I," CRC Press, Inc. Boca Raton.

31. Shank RC, 1981, "Mycotoxins and N-Nitroso Compounds: Environmental Risks, Vol. II," CRC Press, Inc. Boca Raton.

32. Sieber SM, Correa P, Dalgard DW, Adamson RH, 1979, Induction of osteogenic sarcomas and tumors of the hepatobiliary system in nonhuman primates with aflatoxin B_1, *Cancer Res.*, 39:4545-4554.

33. Smith JE, Moss MO, 1985, "Mycotoxins, Formation, Analysis, and Significance," John Wiley & Sons, New York.

34. Steyn PS, Vlegaar, 1985, Tremorgenic Mycotoxins, *Progress in the Chemistry of Organic Natural Products*, 48:1-80.

35. Steyn PS, Vlegaar R, 1986, "Mycotoxins and Phycotoxins, Vol. 1," Elsevier, New York.

36. Stich HF, 1983, "Carcinogens and Mutagens in the Environment, Vol. I," CRC Press, Inc., Boca Raton.

37. Tanaka T, 1988, Worldwide natural occurrence of *Fusarium* mycotoxins, nivalenol, deoxynivalenol, and zearalenone, 7th International Symposium in Mycotoxins and Phycotoxins (Abstract P. 19), Tokyo.

38. Tsuboi S, Nakagawa S, Tomita M, Seo H, Ono H, Kawamura N, 1984, Detection of aflatoxin in serum samples of male Japanese subjects by radioimmuno-assay and high-performance liquid chromatography, *Cancer Res.*, 44:1231-1234.

39. Ueno Y, 1983, "Trichothecenes - Chemical, Biological and Toxicological Aspects," Kodansha Ltd., Tokyo.

40. Uraguchi K, Yamazaki M, 1978, "Toxicology, Biochemistry and Pathology of Mycotoxins," Kodansha Ltd., Tokyo.

41. Watoson DH, 1984, An assessment of food contamination by toxic products of alternaria, *J Food Prot.*, 47:485-488.

42. Wogan GN, 1965, "Mycotoxins in Foodstuffs," The MIT Press, Cambridge.

INTERACTION OF T-2 TOXIN WITH CELL MEMBRANES

John R. DeLoach[a] and George G. Khachatourians[b]

US Department of Agriculture, Agricultural Research Service
Veterinary Toxicology and Entomology Research Laboratory
P. O. Drawer GE
College Station, TX 77841[a]

Department of Applied Microbiology and Food Science
University of Saskatchewan
Saskatoon, Saskatchewan, Canada[b]

The trichothecene mycotoxins are well studied and known to cause toxicological problems in man and animals.[25] The most potent of this family of
toxins is T-2 toxin. T-2 toxin is cytotoxic, having a radiomimetic effect on
rapidly dividing cells. Immune function impairment has been reported for T-2
toxin both at the level of humoral immunity and cell mediated immunity.[2,17,27]

In vitro experiments have shown that T-2 toxin inhibits protein synthesis
in both cellular and cell free systems, however, much higher toxin concentrations are required in cell free systems.[37-39] Ribosomal binding is generally
accepted as the site of action of T-2 toxin. But definitive experiments
still leave some doubt as to the extent of mechanism of action of T-2 toxin.
Because T-2 toxin has been reported as a weak protein synthesis inhibitor in
cell free-systems,[38] there is reason to suspect other sites of action.

To have an effect on cellular function T-2 must cross the cell membrane.
Thus, membrane solubility and membrane interaction are important parameters
in T-2 toxicology. Due to its amphipathic nature, T-2 toxin may be predisposed to mediate some of its toxic effects at the level of the membrane[19]
Studies of the effect of T-2 toxin on rat liver mitochondrial[32] and on yeast
mitochondrial electron transport systems[28] which indicate the effect of T-2
toxin on mitochondrial oxygen consumption, carbon dioxide reduction, and
activities of succinic dehydrogenase could also be explained by the effect of
T-2 on subcellualr organellic membranes. Additionally, studies on the effect
of T-2 mycotoxin on the uptake of amino acids in L-6 myoblasts[3] would indicate effects of T-2 toxin on the aspects of transport of amino acids and
hence a reduction of both protein synthesis and amino acid uptake. This
effect may be specific to some cells in that a study of the effect of T-2
toxin on amino acid uptake, release of 51-Cr was found to be unaffected by
T-2 toxin, but there was evidence of significant leakage after 18 hrs of
treatment.[18] Ueno makes a reference to the lipophilic aspect of T-2 toxin
and its higher affinity to cellular membranes than fusarenon-X and that the
phenomenon of this aggregation of the polysomal structure could be some of
the early effects of T-2 toxin on animal cells.[41] Thus, in general one
can suggest that the inhibition of protein synthesis or some of the other

Microbial Toxins in Foods and Feeds, Edited by A.E. Pohland *et al.,*
Plenum Press, New York, 1990

cellular functions such as DNA, RNA synthesis are cause effect relationships mediated through the cell membrane. Other studies which depict the inhibition of platelet aggregation[42] and inhibition of phagocytosis by polymorphal nuclear cells[43] suggest that T-2 acts at the cytoplasmic membrane level.

Segal et al.[37] first reported a hemolytic activity of T-2 toxin on rat erythrocytes. In fact, their hemolysis results were so clear cut that the authors raised the question of why it had gone unreported before. A point worth noting is that the T-2 concentration required for hemolysis ($2.5-7.5 \times 10^{-4}$ M) is much greater than that required to inhibit protein synthesis (6×10^{-6} M). In later studies, Gyongyossy-Issa et al. used 6.7×10^{-4} M T-2 toxin to hemolyze guinea pig erythrocytes.[22,23] Later studies by DeLoach et al. suggested that all species of erythrocytes were not susceptible to T-2 toxin hemolysis.[13-15]

Because membrane interaction with T-2 toxin may be an important aspect of its toxicological effects, we undertook experiments with several species of erythrocytes to describe properties of the system. Several questions that we attempted to answer were 1) effect of time, temperature, and concentration on hemolysis; 2) effect of T-2 on cell morphology; and 4) effect of membrane lipid composition on susceptibility to T-2 toxin hemolysis.

Hemolysis conditions. T-2 toxin is only sparingly soluble in water and must first be dissolved in an organic solvent such as isopropyl or ethyl alcohol or dimethyl sulfoxide. Stock solutions of T-2 toxins (6 mg/ml) were prepared daily in isopropyl alcohol. Addition of T-2 was made directly to buffer solutions containing erythrocytes. To avoid solvent effects on the cells, the final alcohol concentration did not exceed 8%.[13] Afterward, the erythrocyte, toxin, buffer mixture was incubated at the desired temperature for different times up to 24 hr. For most experiments, 20 μl of erythrocytes and 1.9 ml buffer and 100 μl of solvent were used.

Thirteen species of erythrocytes have been assayed for susceptibility to T-2 induced hemolysis.[13-15] Pig, human, rabbit, chicken, guinea pig, horse, dog, rat, and mouse erythrocytes are all lysed to some degree by T-2 toxin; yet, sheep, cow, goat, buffalo, and deer erythrocytes are not lysed by T-2 toxin (Table 1). Susceptibility to lysis seems to be dependent upon the presence of phosphatidyl choline in the membrane. All five erythrocyte species from animals with ruminant physiology contain little or no phosphatidyl choline and are not susceptible to T-2 induced lysis.

A ususal characteristic of T-2 induced lysis is a time lag of 5-15 min (Fig. 1c,d).[14,16,22] Doubling T-2 concentration decreases the time lag somewhat, but not completely. Both goat and bison erythrocytes are shown to be resistant to T-2 induced lysis (Fig. 1a,b). Twenty μl of human erythrocytes with T-2 toxin is equivalent to 1.2×10^8 cells and 1.2×10^{18} molecules of T-2 toxin (approximately 10^{10} molecules T-2 per cell). Mouse erythrocytes appear more susceptible to T-2 induced lysis (Fig. 1c). A 5×10^{-4} M T-2 concentration produces maximum hemolysis on 1.2×10^8 cells of 5×10^9 molecules T-2 per cell. The characacteristic time lag is supported by data which indicates that 20 min is required for T-2 to reach equilibrium into erythrocyte membrane.[24]

The total number of T-2 molecules required to reach maximum hemolysis varies from 0.38 to 1.1×10^{10} per cell (Table 2). A calculation of the average total phosphatidyl choline per cell[30,31] and the total T-2 required for maximum hemolysis reveals 40 to 277 molecules of T-2 per molecule of phosphatidyl choline (Table 3). For example, dog erythrocytes required 0.38×10^{10} molecules T-2 per cell and contain 9.6×10^7 molecules phosphatidyl choline per cell. For simplicity in these calculations, the molecular weight of phosphatidyl choline was taken as 750. The partition coefficient

Table 1.   Comparison of erythrocyte phosphatidyl choline and
susceptibility of hemolysis of T-2 toxin.

Erythrocytes	Ruminant physiology	Phosphatidyl choline in membrane (%)	Hemolysis by T-2 toxin (%)
Cow	Yes	0	0
Sheep	Yes	0	0
Goat	Yes	0	0
Buffalo	Yes	0	0
Deer	Yes	0	0
Mouse	No	47	95[a]
Dog	No	47	72[b]
Man	No	28	68[b]
Guinea pig	No	41	66[b]
Pig	No	23	60[b]
Rabbit	No	34	48[c]
Rat	No	48	30[d]
Horse	No	42	20[e]

Data on phosphatidyl choline from Nelson.[30,31]   Percentage hemolysis
is from a 4 hr incubation of 20 μl of erythrocytes in 2 ml of T-2
toxin (150 μl/ml) in isotonic saline.   Data in column followed by a
common letter are not significantly different (P < 0.05).

of T-2 toxin in erythrocyte/water is 0.8;[24] thus, the number of T-2 molecules actually in the membrane is approximately one-half of the total mentioned above.

An interesting characteristic of T-2 interaction with erythroyctes is the
osmotic protection afforded by a short duration incubation.[23]   For example,
a 3 min incubation of T-2 with guinea pig erythrocytes renders the cells more
resistant to osmotic shock.   Further incubation beyond 20 min causes hemolysis.   The threshold amount of T-2 needed to provide initial osmotic protection and later hemolysis is 1 x 10^{10} molecules per cell which translates to
approximately 1 x 10^{-4} M.[23]   This kind of interaction is reminiscent of
some of the anesthetics which function in the range 10^{-6} to 10^{-4} M to interrupt
anion transport in cells.   The lysis reaction is also temperature dependent
having an energy of activation of approximately 4.5 K cal.[22]

Encapsulation of T-2 toxin.   Since T-2 toxin causes no apparent hemolysis
of erythrocytes containing little or no phosphatidyl choline[12,13,15] studies
were conducted to determine if T-2 toxin could be encapsulated in erythrocytes.
Encapsulation in erythrocytes is achieved by a variety of techniques.[11]   Hypo-
osmotic dialysis of erythrocytes allows for entrapment of exogenous molecules
as small as sucrose and as large as 600,000 dalton proteins.[8]   Erythrocytes prepared to encapsulate molecules for slow release in circulation are
called carrier-erythrocytes.[11]   Carrier-erythrocytes are normal in all
respects with control erythrocytes with a tendency toward microcytosis (for
a review, see reference 11).

The rationale for using entrapment technology to discern membrane-inter-
active effects is based on 1) the fact that phospholipids are asymmetrically
distributed in the lipid bilayer; 2) some molecules diffuse very slowly
across the membrane; and 3) entrapment of molecules to the inside of erythrocytes allows for interaction with inner leaflet of bilayer without diffusion
through the bilayer.   Studies on the effects of drugs on entrapment and leakage of ^{14}C-sucrose reveal whether the drug causes stomatocytes or echinocytes.[9,10]

Cattle and sheep erythrocytes are both resistant to T-2 hemolysis.[13] Thus, we encapsulated T-2 toxin in these cells. Up to 13% of the added T-2 toxin (20 μg) could be encapsulated in 0.8×10^{10} cattle erythrocytes.[13] Thus, 2.6 μg of toxin (5×10^{15} molecules of toxin) were encapsulated. Very little T-2 toxin binds to bovine erythrocytes. In fact, the maximum amount bound to cells or adsorbed onto their membrane was 9×10^4 molecules per cell. With sheep erythrocytes up to 16 μg of T-2 can be entrapped which equates to 3×10^7 molecules per cell.[13] T-2 toxin at 10^{10} molecules per cell lyses at least 8 different erythrocyte species. While only 9×10^4 molecules could be bound to bovine and ovine erythrocytes, at least 3×10^7 molecules per cell could be entrapped.

T-2 toxin induced lysis of susceptible erythrocytes can be reduced by addition of 0.3 M sucrose or raffinose but not glucose. It was suggested that the membrane pertubation caused by T-2 was smaller than a 5.5 A lesion.[22] A further confirmation that bovine and ovine erythrocytes are resistant to T-2 induced membrane pertubation is that ^{14}C-sucrose encapsulation is unaffected by the toxin.[13,16]

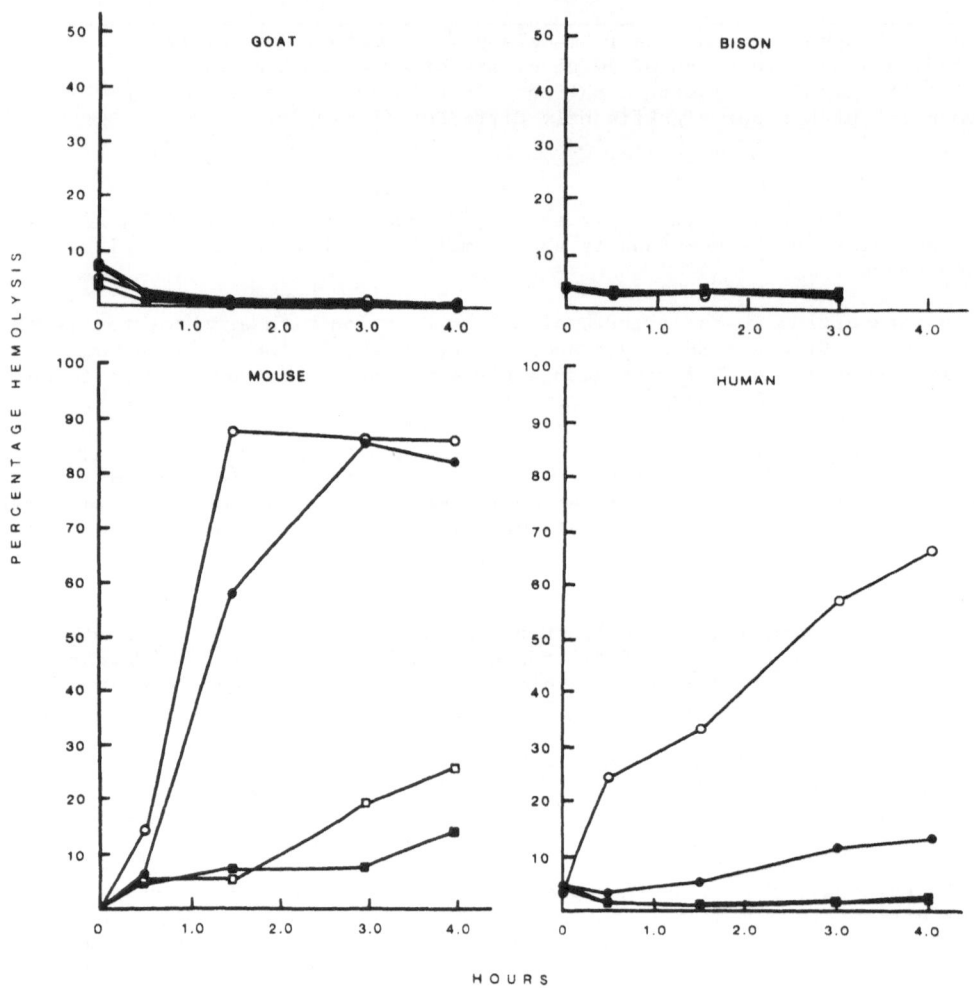

Fig. 1.   Time course of T-2 induced lysis of erythrocytes.
■, 0 μg T-2;   □, 30 μg T-2;   ●, 150 μg
O, 300 μg/ml T-2.

Table 2. Molecules of T-2 required for maximum hemolysis of susceptible erythrocytes.

Animal	No. of erythrocytes (10^8)	T-2 molecules/cell (10^{10})
Equine	1.6	0.75
Murine	1.2	0.50
Human	1.3	1.10
Rabbit	1.2	1.00
Chicken	0.6	1.00
Pig	1.6	0.75
Dog	1.6	0.38
Guinea Pig	1.1	1.10
Rat	1.3	0.46

Table 3. Ratio of T-2 Toxin to phosphatidyl choline in erythrocyte membrane.

Erythrocyte	Molecular ratio
Horse	167
Murine	62
Human	160
Rabbit	152
Pig	277
Dog	40
Guinea pig	138
Rat	58

Morphological Characteristics of Membrane Interaction. Scanning electron microscopy (SEM) has been used as a tool to discern the morphology of erythrocytes from several species after exposure to T-2 toxin. Five examples are given below and are characteristic of the action of T-2 action.[12-16,23] Since solvents, pH, and other treatments can cause artifacts, careful attention was paid to conditions and the use of solvent controls.[13-15]

Guinea Pig Erythrocytes. Solvent control cells are slightly ruffled with a tendency toward echinocytes I (Fig. 2a). T-2 exposed cells (exposure less than 3 min) are completely echinocytic (Fig. 2b). Within 30 min, T-2 treated cells are prehemolytic sphero-echinocytes and by 1 1/2 hr cells are disintegrating (Fig. 2d). However, solvent control treated cells are not appreciably changed after 6 hr exposure (Fig. 2c). Thus, T-2 causes immediate echinocytosis by inserting in the outer leaflet of the lipid bilayer increasing the surface to volume ratio. Later movement of T-2 to inner leaflet results in a spheroid prehemolytic cell.

Chicken Erythrocytes. Solvent control treated cells (zero time) appear elliptical (Fig. 3a). T-2 treated cells (zero time, <3 min exposure) are elliptical with small undulations in the membrane (Fig. 3c); however, T-2 treated cells are almost completely hemolyzed and appear mainly as amorphous stroma (Fig. 3d).

Sheep Erythrocytes. Sheep erythrocytes contain only minute quantities of phosphatidyl choline and are not susceptible to T-2 induced lysis. Solvent treated control cells appear to lose some of their normal shape after 6 hr exposure. T-2 treated cells at 6 hr appear sphero-echinocytic but interestingly do not lyse (Fig. 4a-d).

Human Erythrocytes. Human cells are slightly echinocytic after only a 3 min exposure to T-2 toxin (Fig. 5a-d). Solvent control cells are unaffected by solvent after 3 min or 6 1/2 hr exposure. A longer exposure of human cells to T-2 toxin causes some stomatocytosis.

Bison Erythrocytes. Erythrocytes exposed to T-2 toxin for 6 hr are morphologically indistinguishable from solvent control treated cells (Fig. 6a-b). The lack of effect of T-2 toxin on morphology of bovine cells has been reported.[13]

Yeast Cells. Yeast cells were exposed to T-2 toxin at 10 μg/ml. When

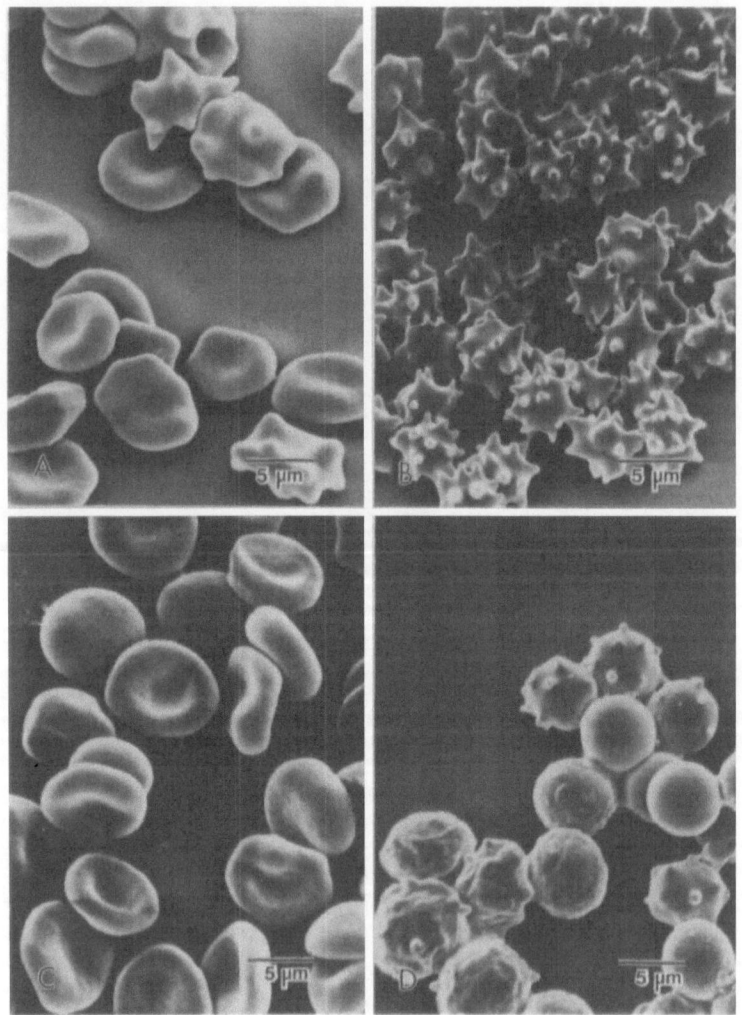

Fig. 2. Effect of T-2 toxin on morphology of guinea pig
erythrocytes. **A.** 0 hr, solvent control; **B.** 0 hr,
300 μg/ml T-2; **C.** 6 hr, solvent control;
D. 1 1/2 hr, 300 μg/ml T-2.

K. fragilis cells are examined, cells at all stages of cell cycle are observed, including mother cells, cells with buds and bud scars. Exposure to 10 μg of T-2 toxin per ml of culture causes changes in cell size and cell shape. As early as one hr after the treatment, collapsing of some portions of the cell membrane is observed (Fig. 7b-e). Exposures of progressively longer duration indicates further collapsing of portions of cell's membranes while other cells appear to be totally unaffected by T-2 toxin (Fig. 7f). Yeast cells contain phosphatidyl choline although the location of this in the inner or outer leaflets is not determined. It is therefore possible that either the swelling or shrinking of the cells and graded cytorhisis in the cell sizes occur as a physiological response related to cell membrane stretching and osmotic pressure. Such effects have been observed for several yeasts when placed in solutions of sucrose, glucose, fructose, sorbitol and glycerol or when treated with phenethyl alcohol for 2 hr.

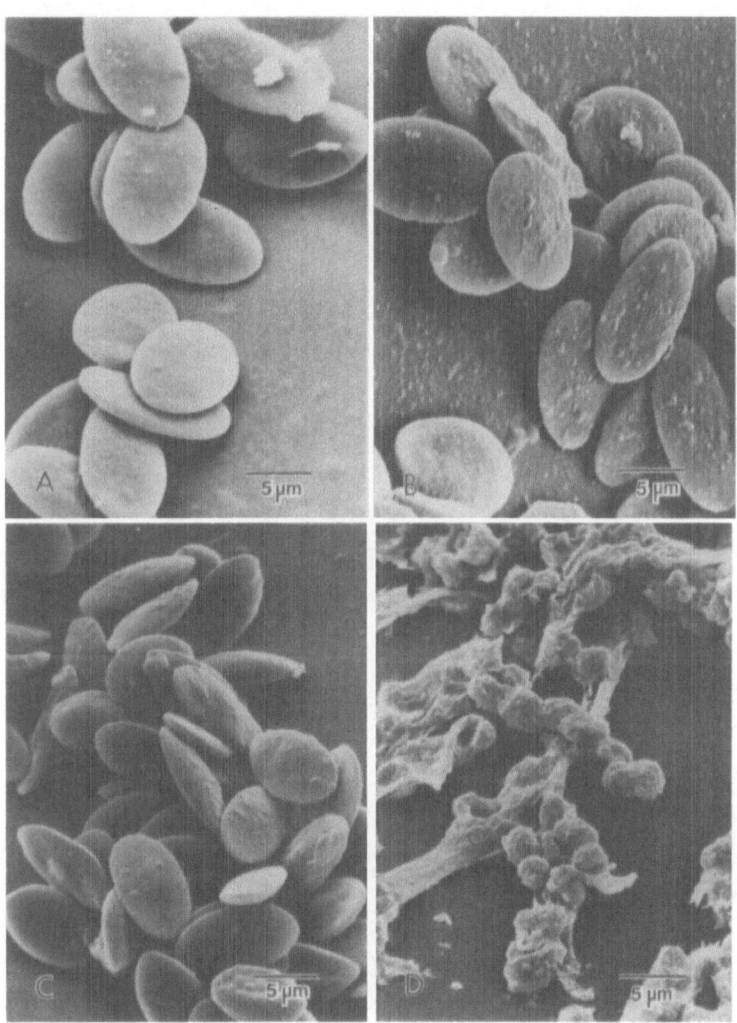

Fig. 3. Effect of T-2 toxin on morphology of chicken erythrocytes. **A.** 0 hr, solvent control; **B.** 0 hr, 300 μg/ml T-2; **C.** 6 hr, solvent control; **D.** 6 hr, 300 μg/ml T-2.

Effect of Phosphatidyl Choline on Sysceptibility of Erythrocytes to T-2 Toxin. All of the erythrocytes from ruminants contain little or no phosphatidyl choline[1,30,31] in their membrane. Hence, the cells compensate for phosphatidyl choline with increased sphingomyelin.[1] The ratio of phosphatidyl choline to sphingomyelin is high[3,7] for erythrocytes from rat, guinea pig, and mouse. A high ratio is thought to confer a high degree of fluidity to the membrane. However, sheep erythrocytes contain very little phosphatidyl choline, less than 0.03 mg per 10^{10} cells; rat and guinea pig erythrocytes contain approximately 1.0 mg per 10^{10} cells. The phosphatidyl choline to sphingomyelin ratio in sheep erythrocytes is 0.03.

Sheep erythrocytes are unique in that they contain a lecthinase enzyme capable of removing phosphatidyl choline from their membranes. Inhibition of the enzyme by EDTA, i.e., removal of CA^{+2}, can result in phosphatidyl choline accumulation in sheep erythrocyte membrane. Borochov et al.[1]

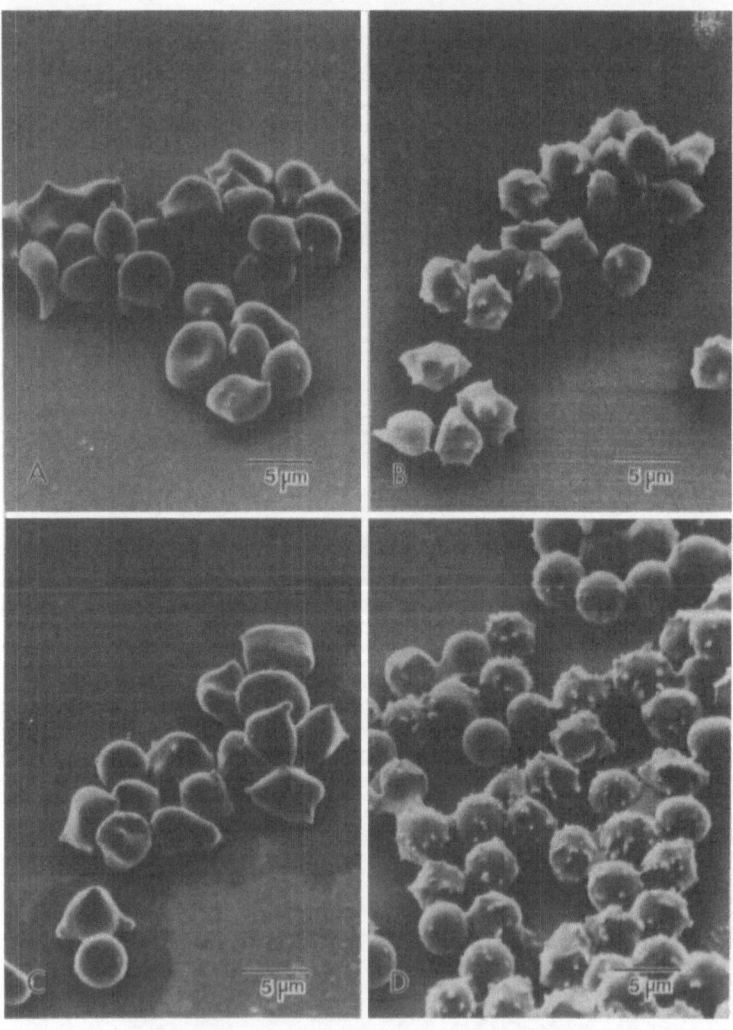

Fig. 4. Effect of T-2 toxin on morphology of sheep erythrocytes. **A.** 0 hr, solvent control; **B.** 0 hr, 300 μg/ml T-2; **C.** 6 hr, solvent control; **D.** 6 hr, 300 μg/ml T-2.

268

demonstrated the accumulation of 0.14 mg of phosphatidyl choline per 10^{10} cells by incubating washed sheep erythrocytes in the presence of 2 mM EDTA and human plasma, a rich source of phosphatidyl choline. Such phosphatidyl choline modified cells are slightly more osmotically fragile than normal cells but have greater membrane fluidity.

To test the hypothesis that T-2 dissolves in more fluid membranes containing phosphatidyl choline, sheep erythrocytes were prepared to contain phosphatidyl choline as described by Borochov et al.[1] Subsequently, control erythrocytes and phosphatidyl choline enriched erythrocytes were exposed to T-2 (300 μg/ml). Phosphatidyl choline enriched erythrocytes were susceptible to partial specific lysis by T-2 toxin (Table 4). Phospholipid analysis of phosphatidyl choline enriched cells confirmed the enrichment for phosphatidyl choline. Thus, a threshold amount of this phospholipid would appear to be a requirement for T-2 to lyse erythrocytes.

CONCLUSIONS

T-2 toxin behaves in a hydrophobic manner in solution.[19,24] Although,

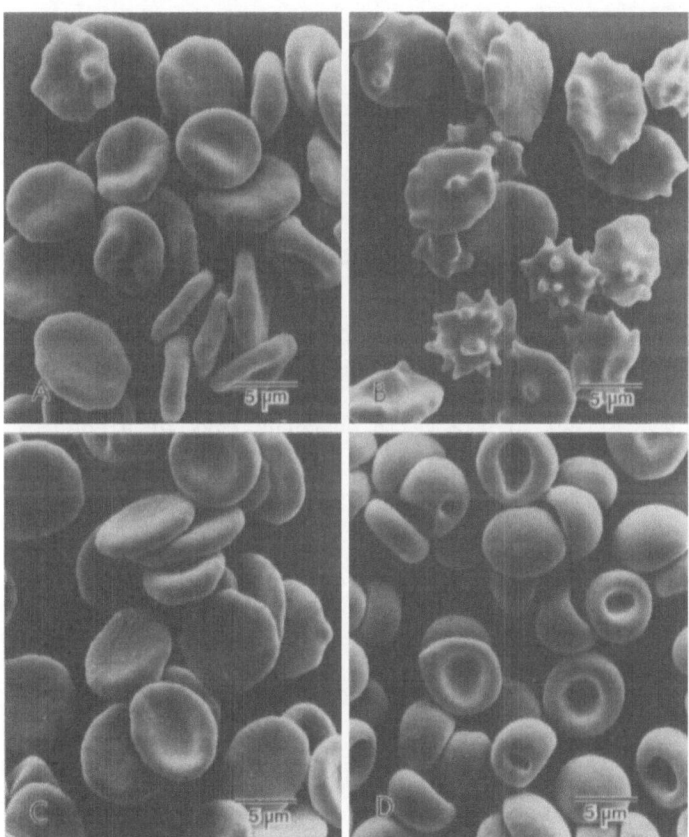

Fig. 5.   Effect of T-2 toxin on morphology of human
          erythrocytes. A. 0 hr, solvent control;
          B. 0 hr, 300 μg/ml T-2; C. 6 1/2 hr,
          solvent control; D. 6 1/2 hr 300 μg/ml T-2.

Table 4. Effect of phosphatidyl choline on the specific hemolysis
of sheep erythrocytes by T-2 toxin.

Erythrocytes	Phosphatidyl choline content mg/10^{10} cells	Specific hemolysis[a] (%)
Normal	0.03	0
Phosphatidyl choline enriched	0.14	18

[a]Hemolysis was for a 24 hr incubation at 37°C.

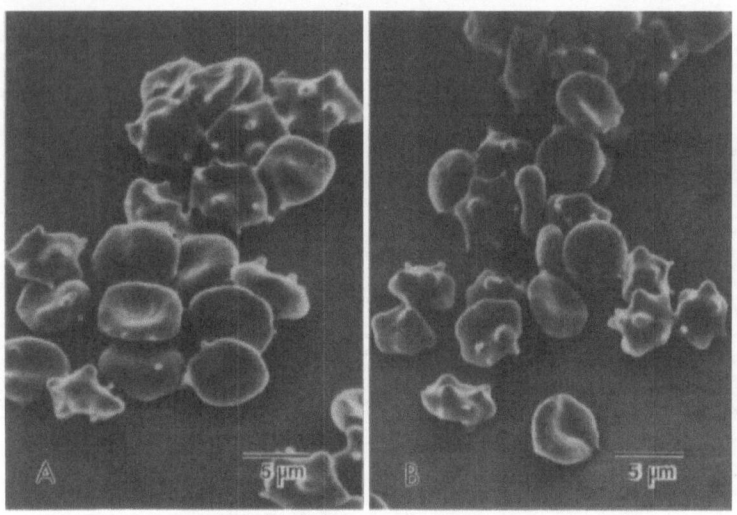

Fig. 6. Effect of T-2 toxin on morphology of bison
erythrocytes.
A. 6 hr, solvent control;
B. 6 hr, 300 μg/ml T-2.

its octanol:water partition coefficient is 25, its erythrocyte membrane:water
partition coefficient is 0.8.[24] T-2 toxin is an amphipath. The primary site
of action of T-2 toxin in whole cells[40] is at the level of protein synthesis, but much higher concentrations of T-2 are required to inhibit protein
synthesis in cell-free systems. Thus, T-2 toxin may act at sites other than
protein synthesis.[33]

The study of interaction of T-2 toxin with *Mycoplasma gallisepticum*[34]
has shown several interesting results in this regard. Growth inhibition of
this organism was shown to be most pronounced with hydrophobic derivatives of
T-2 acetate but very little with hydrophilic T-2 tetraol. In the study of the
inhibitory effects of T-2 toxin on the biosynthesis of either protein, DNA,
RNA, complex lipids, or intracellular pools there was very little or no
effect. The authors suggested that T-2 acetate, because of its hydrophobic
nature, accumulated within the lipid backbone affecting both cell membrane
functions and its permeability properties. Indeed these authors cite some
unpublished experiments of P. J. Davis which has suggested that electron

paramagnetic resonance studies demonstrated a lower freedom of motion of spin labelled phospholipid in membranes from T-2 acetate treated cells of *M. gallisepticum*.

In whole animals, dietary T-2 toxin causes some changes in hemostatis. In guinea pigs, a 10% decrease in hematocrit occurs over the first 3 days post-treatment.[6] A decrease in hematocrit is also seen with monkeys.[7] In mice, dietary T-2 toxin causes an increase in erythrocyte mean cellular volume.[5] With *in vitro* systems, T-2 toxin causes lysis of all mammalian erythcytes that contain phosphatidyl choline in their membrane.[14] T-2 toxin does not cause lysis of erythrocytes that contain little or no phosphatidyl choline. Phosphatidyl choline alters membrane fluidity and is probably required for T-2 to insert into the erythrocyte membrane.

Most of the erythrocyte phosphatidyl choline is located in the outer leaflet of the bilayer. The morphological changes that occur in erythrocytes

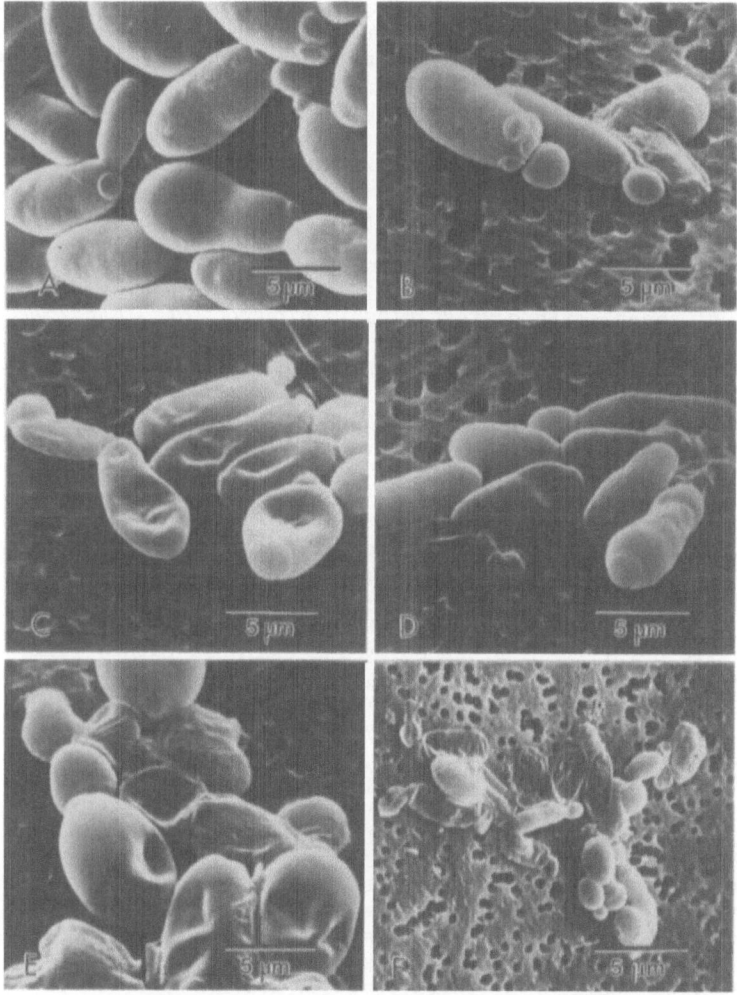

Fig. 7. Scanning electron micrographs of representative cells from a culture of *K. fragilis* exposed to T-2 toxin.
A. Untreated control cells;
B-E. T-2 toxin (10 μg/ml) for 1 hr; F. T-2 toxin for 10 hr.

exposed to T-2 toxin are consistent with the hypothesis that T-2 inserts into the outer leaflet of the bilayer.[21]

Further evidence that T-2 exerts effects on membranes and that phosphatidyl choline is a contributing factor was provided by the phosphatidyl choline enriched sheep erythrocytes. Sheep erythrocytes contain only about 3% of the phosphatidyl choline found in rat erythrocytes. When we incorporated phosphatidyl choline in sheep erythrocytes, they were rendered partially susceptible to T-2 toxin induced lysis. The time course of lysis implies that T-2 must first equilibrate in the bilayer before lesions in the membrane occur.

T-2 toxin's membrane interactive effects have recently been shown to occur in cell culture. Bunner and Morris[4] have shown that T-2 had multiple membrane effects on L-6 myoblasts. Others have shown T-2 toxin to protect Chinese hampster ovary cells in a chromium release assay.[26] Studies have also shown that flavonoids protect against T-2 cytotoxicity both *in vitro* and *in vivo* with murine thymocytes. These studies suggest that molecules such as the flavonoids interact with the membrane quenching T-2 effects by decreasing spontaneous lipid peroxidation.[29] Thus, there is increasing evidence that T-2 toxin membrane interaction is important to its overall toxicological effect on cells.[20,35]

REFERENCES

1. Borochiv H, Zahler P, Wilbrant W, Shinitzky M, 1977, The effect of phosphatidyl choline to sphingomyelin mole ratio on the dynamic properties of sheep erythrocytes, *Biochim Biophys Acta*, 838:252-256.
2. Buening GM, Mann DD, Hook B, Osweiler GB, 1982, The effect of T-2 toxin on the bovine immune system: cellular factors, *Vet Immunol Immunopathol.*, 3:411-417.
3. Bunner DL, Morris ER, Pace JG, Matson CF, 1986, Effect of T-2 mycotoxin on amino acid uptake in L-6 myoblasts, *Toxicon*, 25:136.
4. Bunner DL, Morris ER, 1988, Alteration of multiple cell membrane functions in L-6 myoblasts by T-2 toxin: an important mechanism of action, *Toxicol Appl Pharmacol.*, 92:113-121.
5. Chanin BE, Hamilton DL, Hancock DS, Schiefer HB, 1984, Biointeraction of dietary T-2 toxin and zinc in mice, *Canadian J Physiol Pharmacol.*, 10:1320-1326.
6. Cosgriff TM, Bunner DP, Waggemacher RW, Hodgson LA, Dinterman RE, 1984, The hemostatic derangement produced by T-2 toxin in guinea pigs, *Toxicol Appl Pharmacol.*, 76:454-463.
7. Cosgriff TM, Bunner DL, Wannemacher RW, Hodgson LA, Dinterman RE, 1986, The hemostatic derangement produced by T-2 toxin in cynomolgous monkeys, *Toxicol Appl Pharmacol.*, 82:532-539.
8. DeLoach JR, Harris RL, Ihler GM, 1980, An erythrocyte encapsulator dialyzer used in preparing large quantities of erythrocyte ghosts and encapsulation of a pesticide in erythrocyte ghosts, *Anal Biochem.*, 102:220-227.
9. DeLoach JR, 1985, Encapsulation potential of trypanocidal drug homidium bromide in bovine carrier erythrocytes, *Res Exp Med.*, 185:345-353.
10. DeLoach Jr, Droleskey RE, 1985, Interaction of anthracycline drugs with canine and bovine carrier erythrocytes, *J Appl Biochem.*, 7:178-187.
11. DeLoach JR, 1986, Carrier erythrocytes, *Med Res Rev.*, 6:487-504.
12. DeLoach JR, Andrews K, Naqi A, 1987, Erythrocytes as carriers of mycotoxins for targeting to the reticuloendothelial system, *Adv Biosci.*, 67:191-197.
13. DeLoach JR, Andrews K, Naqi A, 1987, Interaction of T-2 toxin with bovine carrier erythrocytes: effects on cells lysis permeability and entrapment, *Toxicol Appl Pharmacol.*, 88:123-131.

14. DeLoach JR, Mollenhauer HH, 1987, Interaction of T-2 mycotoxin with erythrocytes, *Fed Proc.*, 46:558.
15. DeLoach JR, Andrews K, Naqi A, 1988, Targeting of mycotoxins to reticuloendothelial system of mice with carrier erythrocytes, *Biotechnol Appl Biochem.*, 10:154-160.
16. DeLoach JR, Gyongyossy-Issa MIC, Khachatourians GG, 1989, Species specific hemolysis of erythrocytes by T-2 toxin, *Toxicol Appl Pharmacol.*, 97:107-112.
17. Friend SCE, Babruk LA, Schiefer HB, 1983, The effects of dietary T-2 toxin on the immunological function and herpes simplex reactivation in Swiss mice, *Toxicol Appl Pharmacol.*, 69:234-244.
18. Gerberick FG, Sorenson WG, Lewis DM, 1984, The effects of T-2 toxin on alveolar macrophage function *in vitro, Environ Res.*, 33:246-260.
19. Gyongyossy-Issa MIC, Christie EJ, Khachatourians GG, 1984, Charge shift electrophoretic behavior of T-2 toxin in agarose gels, *Appl Environ Micro.*, 47:1181-1184.
20. Gyongyossy-Issa MIC, Khachatourians GG, 1984, Interaction of T-2 toxin with murine lymphocytes, *Biochem Biophys Acta*, 803:197-202.
21. Gyongyossy-Issa MIC, Carol RT, Fergusson DJ, Khachataourians GG, 1986, Prehemolytic erythrocyte deformability changes by trichothecene T-2 toxin: an ektacytometer study, *Blood Cells*, 11:393-403.
22. Gyongyossy-Issa MIC, Khanna V, Khachatourians GG, 1985, Characterization of hemolysis induced by T-2 toxin, *Biochim Biophys Acta*, 838:252-256.
23. Gyongyossy-Issa MIC, Khanna V, Khachatourians GG, 1986, Changes induced by T-2 toxin in the erythrocyte membrane, *Fd Chem Toxicol.*, 24:311-317.
24. Gyongyossy-Issa MIC, Khachatourians GG, 1987, The octanol and membrane partitioning of trichothecene T-2 toxin, unpublished.
25. Hayes AW, 1981, "Mycotoxin Teratogenicity and Mutagenicity," CRC Press, Boca Raton, FL.
26. Holt PS, DeLoach JR, 1988, Cellular effects of T-2 mycotoxin on two different cell lines, *Biochim Biophys Acta*, in press.
27. Jagadeesan V, Rukmini C, Vijayaraghavan M, Tulpuli PG, 1982, Immune studies with T-2 toxin: effect of feeding and withdrawal in monkeys, *Fd Chem Toxic.*, 20:83-87.
28. Koshinsky H, Honour S, Khachatourians, 1988, T-2 toxin inhibits mitochondrial function in yeast, *Biochem Biophys Res Commun.*, 151:809-814.
29. Markhan RJF, Erhardt NP, Dininno VL, Penman D, Bhatti AR, 1987, Flavonoids protect against T-2 mycotoxins both *in vitro* and *in vivo, J Gen Microbiol.*, 133:1589-1592.
30. Nelson GL, 1967, Composition of neutral lipids from erythrocytes in common mammals, *J Lipids Res.*, 8:374-379.
31. Nelson GL, 1967, Lipid composition of erythrocytes in various mammalian species, *Biochim Biophys Acta*, 144:221-232.
32. Pace JG, 1983, Effect of T-2 mycotoxin on rat liver mitochondria electron transport system, *Toxicon*, 21:675-680.
33. Rosenstein Y, Lafarge-Frayssinet C, 1983, Inhibiting effect of fusarium T-2 toxin on lymphoid DNA and protein synthesis, *Toxicol Appl Pharmacol.*, 70:283-288.
34. Rotten S, Yagen B, Katznell A, 1984, Effect of trichothecenes on growth and intracellular pool size of *Mycoplasma gallisepticum, FEBS-Letters*, 175:189-192.
35. Schappert KT, Khachatourians GG, 1984, Influence of the membrane on T-2 toxin toxicity in Saccharomyces, *Appl Environ Microbiol.*, 47:681-684.
36. Schuster A, Hunder G, Fichtl B, Forth W, 1987, Role of lipid peroxidation in the toxicity of T-2 toxin, *Toxicon*, 25:1321-1328.
37. Segal R, Milo-Goldzweig I, Joffe A, Vagen B, 1983, Trichothecene-induced hemolysis. 1. The hemolytic activity of T-2 toxin., *Toxicol Appl Pharmacol.*, 70:343-349.
38. Thompson WL, Wannemacher RW Jr, 1986, Structure-function relationships of 12,13-epoxytrichothecene mycotoxins in cell culture: comparison to whole animal lethality, *Toxicon*, 24:985-994.

39. Ueno Y, Hosoya M, Morita V, Ueno I, Tutsuno T, 1968, Inhibition of the protein synthesis in rabbit reticulocytes by nivalenol, a toxic principle isolated from *Fusarium nivale* growing rice, *J Biochem.*, 64:478-485.

40. Ueno Y, Nakajima M, Sakai K, Sata IK, Shimado N, 1973, Comparative toxicology of trichothecene mycotoxins: inhibition of protein synthesis in animal cells, *J Biochem.*, 74:283-296.

41. Ueno Y, 1977, Mode of action of trichothecenes, *Pure Appl Chem.*, 49:1737-1745.

42. Yarom R, More R, Eldor A, Yagen B, 1984, The effect of T-2 toxin on human platelets, *Toxicol Appl Pharmacol.*, 73:210-217.

43. Yarom R, Sherman Y, More R, Ginsburg I, Borinsky R, Yagen B, 1984, T-2 toxin effect on bacterial infection and leukocyte function, *Toxicol Appl Pharmacol.*, 75:60-68.

BIOTRANSFORMATION AS A DETERMINANT OF SPECIES SUSCEPTIBILITY TO AFLATOXIN

B_1: *IN VITRO* STUDIES IN RATS, MOUSE, MONKEY, AND HUMAN LIVER*

David L. Eaton, Howard Ramsdell and David H. Monroe

Department of Environmental Health and
Institute for Environmental Studies
University of Washington
Seattle, WA 98195

INTRODUCTION

Susceptibility to the toxic and carcinogenic effects of aflatoxin B_1 (AFB_1) varies markedly between species.[3,4,47,48,49,62] The duck, rabbit, pig, and trout are highly susceptible to the acute toxic effects of AFB (LD_{50} 0.3-0.8 mg/kg) and the monkey is moderately susceptible (LD_{50} ~2.0 mg/kg). The rat, mouse, and hamster are generally considered to be relatively resistant to the acute effects of AFB, though some discrepancies in the published LD_{50}'s exist, possibly due to strain, sex, and/or age differences. The LD_{50} for rats has been reported to be between 1.0 and 17.9 mg/kg while published values for the mouse range from 9.0 to 60 mg/kg. The duck, trout, and rat are highly susceptible to the carcinogenic effects of AFB, while the mouse and hamster are relatively resistant. Lifetime dietary exposure to as little as 15 ppb has induced hepatocellular carcinoma in rats, and 100% incidence of liver cancer was obtained with 100 ppb of AFB for a lifetime.[64] Numerous other studies have demonstrated similarly potent carcinogenicity in rats.[60,63] In contrast to the rat, Swiss-Webster mice given a diet containing 150,000 ppb of an AFB+AFG mixture failed to develop any liver tumors.[64] However, liver tumors were obtained in C57BLxC3H mice injected with 1.25 mg/kg three times during the first 10 days after birth, demonstrating the importance of age-dependent factors in carcinogenic susceptibility.[55] Rainbow trout are perhaps the most sensitive of all species,[53] but the metabolic pathways for AFB in salmonids are substantially different than that in rodents and primates.[33,54] Susceptibility to the hepatocarcinogenic effects of AFB can also be altered by diet[56] and a variety of chemical pretreatment regimens. Rats maintained on a cabbage-supplemented diet showed reduced AFB-induced tumor incidence as compared to rats fed a control diet,[1] and cabbage and other cruciferous vegetables decrease binding of AFB to DNA.[52,59] Pretreatment of rats with ethoxyquin,[5,26,36] phenobarbital,[17,32,37,38] or BHA[42,50,61] has been shown to greatly reduce both covalent binding to DNA and hepatocarcinogenicity of AFB.

AFB is oxidized by the cytochrome P-450 mixed-function oxidase system to the highly reactive AFB-8,9-epoxide (formerly designated 2,3-oxide) which avidly binds to cellular macromolecules, including DNA[16] (Fig. 1).

*Supported by NIH grants ES-T07032, ES-03415, ES-03933, CA-47561 and American Cancer Society Grant IN-26W.

Covalent modification of cellular DNA is thought to be the first step in the initiation of chemical carcinogenesis.[34,39] The major DNA adduct formed in both rats and mice has been identified as AFB-N7-guanine[7,8,15,29]. AFB-8,9-epoxide may be inactivated directly by conjugation with glutathione via cytosolic glutathione S-transferase (GST).[11,14,30,44,46,51] Alternatively, the epoxide linkage may be hydrolyzed spontaneously or possibly via the action of epoxide hydrolase, yielding the dihydrodiol.[10,16] Approximately 50% of a dose of AFB, administered by i.p. injection to rats, is eliminated in the bile as polar metabolites, the major metabolite being the glutathione conjugate of AFB (AFB-GSH) which is formed via the action of cytosolic GST[11,23,42]. Trout apparently lack the ability to conjugate AFB-epoxide with GSH, which may in part account for their high susceptibility to AFB-induced hepatocarcinogenesis.[54]

These and many other studies support the hypothesis that differential susceptibility to AFB hepatotoxicity and carcinogenicity has a metabolic basis involving the relative rates of activation and detoxification of AFB.[11,45,48] These studies suggest that differences in biotransformation are reflected in changes in covalent binding of activated AFB metabolites to DNA, which in turn is a useful indicator of relative susceptibility to hepatocarcinogenesis.[6,7,35] Over the past several years, we have undertaken a series of investigations to more fully examine the metabolic basis for species and treatment-related differences in susceptibility to AFB hepatocarcinogenesis. The findings of our previously published data will be summarized, and some recent studies on both human and non-human primate liver will be reported.

Development of an assay to measure the rate of AFB-epoxide production and the rate of glutathione S-transferase conjugation of AFB-epoxide.

Our initial studies focused on the development of an HPLC-based assay that would allow for the determination of both the rate of formation of the putative AFB-8,9-epoxide intermediate, as well as the determination of GST activity directly toward the AFB-epoxide. An HPLC system which utilized a ternary gradient of ammonium phosphate buffer, methanol/THF (94:6), and water

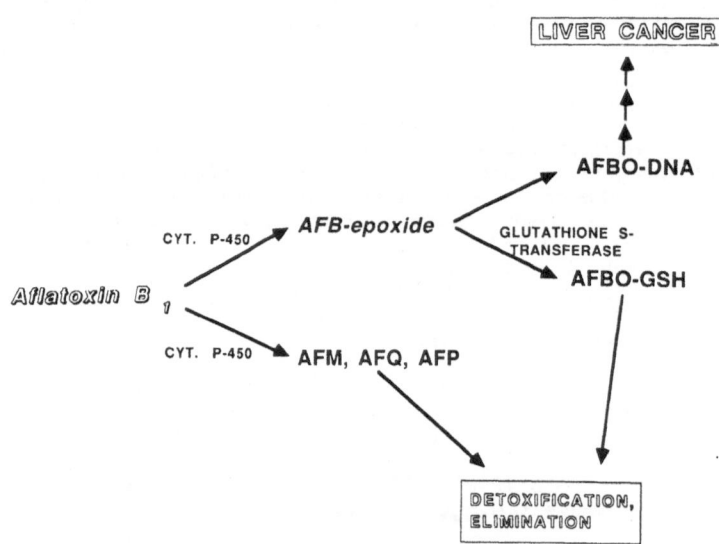

Fig. 1. Biotransformation of aflatoxin B₁ leading to the putative ultimate carcinogen aflatoxin B-8,9-epoxide, and various detoxification pathways.

was established which allowed for baseline resolution of all known aflatoxin metabolites, including the TRIS-dihydrodiol and GSH conjugates[13] (Fig. 2).

By utilizing thoroughly washed microsomes from BHA-treated mice, an *in vitro* AFB-epoxide generating system was developed that produced epoxide at a rate faster than its rate of conjugation. These microsomes convert over 90% of the total metabolized dose of AFB to the AFB-epoxide.[13] Determination of GST activity directly toward the AFB-epoxide can be achieved by addition of an appropriately diluted cytosolic fraction from the species of interest. If the ratio of diluted cytosol to mouse microsomes is such that the AFB-GSH peak is not greater than the TRIS-dihydrodiol peak, the production of AFB-GSH is linear with both incubation time and amount of added cytosolic protein.[40] We also found that cytosolic GST activity in Swiss-Webster mice was remarkably efficient at conjugating the AFB-epoxide, such that inclusion of mouse cytosol in an incubation of microsomes obtained from other species efficiently "trapped" almost 100% of generated epoxide. Using these methods, we were able to demonstrate that the production of AFB-epoxide was linear with both incubation time and amount of added microsomal protein.[40] Studies with radiolabeled AFB demonstrated that under these incubation conditions, less than 1% of the AFB-derived radioactivity was covalently bound to

REVERSED PHASE HPLC METHODS FOR AFB METABOLITES

Instrument: IBM 9533 Ternary Gradient HPLC
Column: Alltech 150 x 4.6 mm, 5 μ Econosphere C18
Detector: IBM variable wavlength, 360 nm
Integrator: Shimadzu CR3A
Temperature: 40° C

Flow Rate: 1.5 ml/min

Mobile Phases:

A - 0.05 M NH_4PO_3, pH 3.5
B - 94% methanol, 6% THF
C - water

Gradient Program:

Time	%A	%B	%C
0	90	10	0
0.5	90	10	0
2.0	0	24	76
13.0	0	38	62
20.0	0	80	20

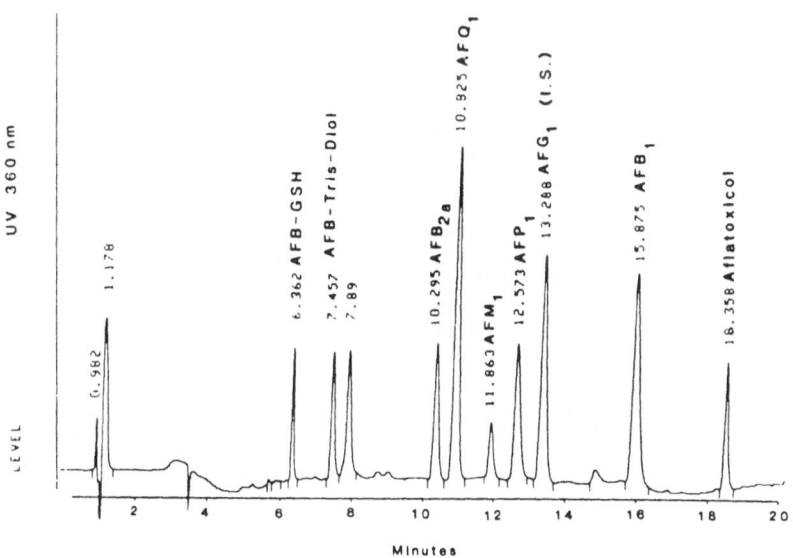

Fig. 2.   Analysis of aflatoxin B_1 metabolites by HPLC.

277

incubation protein, and no dihydrodiol was present in the incubation. Thus, it is possible to very closely approximate the actual rate of microsomal AFB-epoxide formation *in vitro* by simply measuring the extent of AFB-GSH conjugate formed in the presence of "excess" mouse liver cytosol.

Determination of the relative importance of the rate of AFB-epoxide formation and the rate of glutathione S-transferase mediated inactivation of AFB-epoxide in rats and mice.

To test the hypothesis that the difference in susceptibility to AFB carcinogenesis between rats and mice is determined by differences in the ratio of activation/detoxification,[10,12,16,20,24,30,48] we examined the extent of AFB-DNA adduct formation in adult Sprague-Dawley rats and Swiss-Webster mice two hr following a single i.p. dose of 0.25 mg/kg ^{3}H-AFB. AFB-DNA adduct formation in both species was also determined in animals given 0.75% BHA in the diet for 14 days,[40] a treatment shown previously to reduce carcinogenic activity of AFB in rats[61] and decrease AFB-DNA adduct formation.[31] To demonstrate the relative importance of glutathione conjugation, rats and mice were also depleted of hepatic glutathione by a combination of buthionine sulfoximine (BSO) and diethyl maleate (DEM) prior to administration of AFB.[22,41]

DNA adduct formation in control mice was only 1.2% of that in rats (Fig. 3), consistent with previous reports of the relatively low extent of AFB-DNA adduct formation in this species.[16] Associated with this remarkable species difference in AFB-DNA adduct formation was a 52-fold increase in GST activity toward AFB-epoxide, which was not reflected in activity toward CDNB (Table 1). Neal et al.[46] reported that one GST isoenzyme in the mouse has a much higher specific activity toward AFB-epoxide than toward CDNB. We have recently confirmed that the vast majority of mouse GST activity toward AFB-epoxide resides in a minor form of mouse GST, and found that this isoenzyme is not induced by BHA, although a different one with low but measurable activity toward AFB-epoxide is greatly induced (unpublished observation). It is evident from the data in Table 1 that the resistance of mice to AFB is not a result of the inability of mice to form AFB-epoxide, as mice formed the epoxide at a rate 3.2 times greater than rats. Furthermore, BHA treatment increased the rate of epoxide formation in mice by 3.3-fold, yet the extent of DNA binding was slightly lower than that seen in untreated mice (Fig. 3).

In contrast to the mouse, treatment of the rat with 0.75% BHA in the diet for 14 days produced a five-fold decrease in binding of AFB to DNA. Other investigators have also reported that dietary BHA inhibits the covalent binding of AFB to DNA in rats *in vivo*.[31] Ethoxyquin has been shown to inhibit the carcinogenicity of AFB *in vivo*,[5] and in a recent study 0.5% ethoxyquin in the diet for 14 days reduced covalent binding of AFB to DNA to only 3% of the untreated level.[36] In our studies, BHA treatment produced a 3.2-fold increase in hepatic GST activity toward AFB-epoxide, which was closely reflected in an increase in GST activity toward CDNB (Table 1). Mandel et al.[36] found that dietary treatment with ethoxyquin resulted in a 117-fold increase in the extent of AFB-GSH formed in 30 minutes with liver cytosol from Fisher 344 rats, using quail liver microsomes for *in situ* generation of AFB-epoxide. Interestingly, this dramatic increase in GST activity toward AFB-epoxide produced by ethoxyquin was not reflected in CDNB activity, as only a 4-fold increase in GST activity toward CDNB was observed. Whether the differences in effectiveness of GST-induction between BHA (measured in our study) and ethoxyquin (determined by Mandel et al.[36]) are chemical-related, species-related, or the result of differences in methodology remains to be determined.

To further demonstrate the critical importance of glutathione conjugation as a protective pathway for AFB-DNA adduct formation in the mouse, we attempted to reduce the effectiveness of GST conjugation of AFB-epoxide by

Table 1

	AFB-Epoxide formation Specific Activity (% of rat)[a]	Glutathione S- Transferase Specific Activity (% of rat)[a]		GSH (mM)
		AFB-epoxide	CDNB	
Control rat	100	100	100	4.1
BHA rat[b]	102	320	210	5.2
BSO-DEM rat[c]	100	-	98	0.8
Control mouse	340	5,200	230	5.2
BHA mouse[b]	1,120	7,900	720	7.5
BSO-DEM mouse[c]	325	5,000	-	0.2

[a]For comparative purposes, all values are expressed as a percent of the value for rats. Each value was determined from the mean of 4-6 animals. Details can be found in references 40 and 41.
[b]Rats and mice were maintained on a diet containing 0.75% BHA for 14 days prior to determination.
[c]Rats and mice were given BSO (1 gm/kg, rats; 0.6 gm/kg, mice) and DEM (0.75 ml/kg) 2 and 1.5 hr prior to removal of livers, respectively.

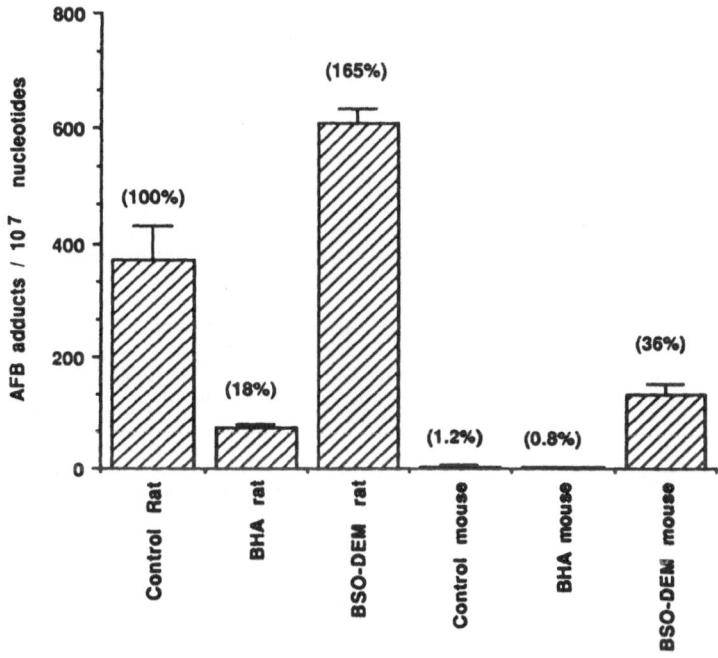

Fig. 3. Species and treatment-related differences in binding of AFB to hepatic DNA. Livers were removed 2 hr after a dose of 0.25 mg/kg, ^{3}H-AFB, and DNA was isolated. BHA was administered in the diet (0.75%) for 14 days prior to AFB dosing; BSO and DEM were administered 2 and 1.5 hr before AFB, respectively. Each bar represents the mean±SE of 4 animals.

depleting hepatic GSH in rats and mice.[41] A combination of 0.6 gm/kg of BSO (given 2 hr prior to AFB administration) and 0.75 ml/kg of DEM (given 1.5 hr prior to AFB administration) reduced hepatic GSH to 0.8 mM and 0.2 mM in rats and mice, respectively, but had no effect on the rate of AFB-epoxide production in $vitro$, compared to the respective untreated controls (Table 1). AFB-DNA adduct formation in the rat was increased 1.6-fold by GSH depletion, whereas adduct formation in the mouse was increased 30-fold, which was 36% of the binding observed in control rats. Because mice form the epoxide at a rate several times greater than rats, one might have expected the binding in mice to actually exceed that of rats if the GSH pathway were eliminated. However, even though hepatic GSH was greatly reduced by the BSO-DEM treatment, there was still a large molar excess of GSH in the liver (about 290 nmoles) relative to the amount of AFB administered (about 27 nmoles), and thus this treatment would not be expected to eliminate all conjugation of AFB-epoxide with GSH. Indeed, HPLC analysis of gallbladder bile from GSH-depleted mice identified the AFB-GSH conjugate as the major biliary metabolite, demonstrating that GSH conjugation of the AFB-epoxide was still functional in the GSH-depleted mice.[41] This study demonstrates the importance of GSH conjugation in protecting DNA from adduction with AFB, and strongly supports the role of GST as the principal determinant of species differences in susceptibility to AFB carcinogenesis between the rat and mouse.

Comparison of biotransformation of AFB in rodents to that of human and non-human primate liver

Although the previous data demonstrate that the rate of oxidative metabolism of AFB to the epoxide appears to be relatively less important than GST-mediated inactivation of the epoxide in mice, the rate of oxidative metabolism of AFB to other less toxic hydroxylated metabolites may serve as an important detoxification pathway in species with relatively low GST activity toward the AFB-epoxide. Gurtoo and co-workers[19,20,21,27] have proposed that oxidative metabolism of AFB to AFM may serve as an important detoxification pathway by diverting substrate away from epoxide formation. As evidence for this, they suggest that the anticarcinogenic effects of β-naphthoflavone toward AFB is a result of induction of AFB-4-hydroxylase activity (AFB-9a-hydroxylase activity using IUPAC numbering) and subsequent diversion of AFB from epoxide to AFM. AFM and conjugates of AFM have been identified as the principal AFB metabolites in human urine obtained from individuals exposed to AFB in the diet,[18,67] suggesting that this pathway may be an important route of detoxification. However, in $vitro$ studies have suggested that AFM formation is not a major route of metabolism of AFB in humans,[43,66] and this conclusion is supported by studies in non-human primates.[65] It appears that AFQ formation may, however, be a major oxidative pathway for AFB elimination.[2,43,66] As AFM and perhaps other monohydroxylated metabolites are probably at least 10-fold less potent carcinogens than AFB,[9] such oxidative pathways may contribute substantially to species resistance to AFB carcinogenesis, especially in species where GST activity toward AFB-epoxide is relatively low.

To further investigate the relative importance of oxidative metabolism of AFB to non-epoxide metabolites, we compared the biotransformation of AFB in hepatic microsomes obtained from Sprague-Dawley rats, Swiss-Webster mice, *Macacca nemestrina* monkeys, and humans. GST activity toward AFB-epoxide and CDNB was also determined in cytosolic fractions of these same samples. Fresh adult *Macacca* liver (samples from two males and one female) was obtained through the Regional Primate Research Center at the University of Washington. Microsomes and cytosol were prepared from fresh tissue and frozen at -80°C until use. Nine fresh human liver samples were obtained from organ donors. Seven male and two female livers were available. All had died from traumatic injuries to the head. With the assistance of the transplant surgeon, liver samples were obtained at the time of organ removal, immediately frozen in

liquid nitrogen and stored at -80°C. Microsomes and cytosol were prepared
from thawed tissue and immediately re-frozen in liquid nitrogen until use.
For time course and metabolite distribution studies, pooled samples of 3
Macacca and 9 human liver microsomal preparations were used for monkey and
human incubations. Incubations of liver microsomes were carried out, as previously described, in the presence of an excess of BHA-treated mouse liver
cytosolic GST to trap AFB-epoxide.[40] Samples of incubation media were
removed to an equal volume of methanol containing AFG$_1$ as an internal standard at 5, 10, 15, 20, 30 and 60 min. HPLC analysis of the protein free supernatant was performed to quantify individual AFB metabolites formed.

Figure 4 shows the time course of production of oxidative metabolites of

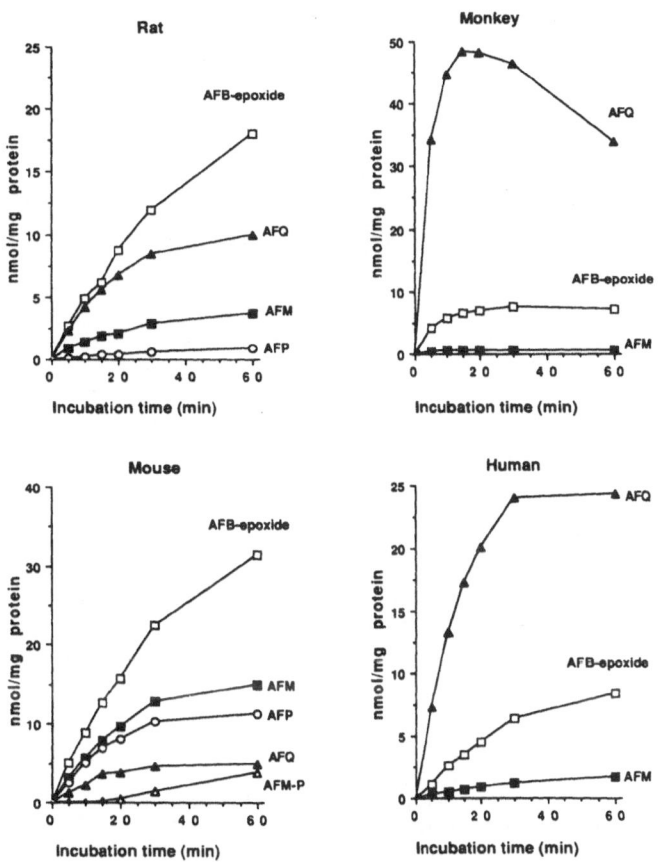

Fig. 4.   Species differences in microsomal metabolism of aflatoxin B$_1$. Microsomes (0.6 - 1.5 mg/ml) were incubated in the presence of an NADPH generating system,
GSH, and cytosol from BHA-treated mice to trap the
generated epoxide. Reactions were terminated with an
equal volume of methanol containing 10 μM AFG$_1$ as an
internal standard. AFB and metabolites were separated
and quantified by HPLC as shown in Figure 2. Each
point represents the average of duplicate analyses.

AFB in each species. Epoxide formation (determined as the GSH conjugate) was
the predominant oxidative pathway in both rats and mice at all time points.
In contrast, AFQ formation was the predominant metabolite formed at all time
points in both monkey and human liver, consistent with previous *in vitro*
studies with human liver microsomes.[2,43,66] Secondary metabolism of AFQ in
monkeys was evident, as the amount of AFQ present in the incubation declined
between 30 and 60 min, and two new peaks appeared, one eluting just prior to
AFB-GSH and one eluting prior to AFQ, but later than AFB-dihydrodiol.
Although the identity of these peaks has not been firmly established, their
chromatographic behavior is consistent with their identity as the GSH conjugate of AFQ and a secondary hydroxylation product of AFQ. We have demonstrated previously that AFP can undergo secondary oxidation to a dihydroxy
metabolite in BHA-induced mouse liver microsomes.[13]

Initial rates of formation of AFB-epoxide in mouse, monkey and human
liver were 189%, 150% and 40% of the rat, respectively (Fig. 5). Initial
rates of AFQ formation in the monkey and human were about 7-8-fold higher
than their respective rates of epoxide formation, whereas AFQ formation in

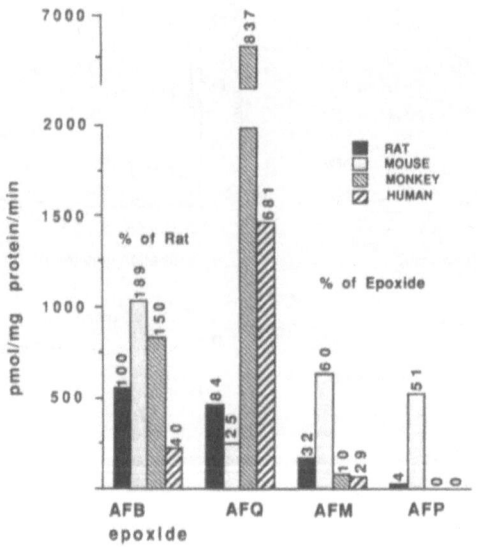

Fig. 5.   Initial rates of biotransformation.
Initial rates of oxidation of aflatoxin B_1 in hepatic microsomes
from different species. Values were
derived from the linear portion of
the time curves shown in Fig. 4.
Numbers above the AFB-epoxide bars
indicate the percent of the initial
rate for rats. Numbers above all
other metabolite bars represent, for
each metabolite, the percent of the
initial rate of epoxide generation for
the same species. Each bar represents
the mean of triplicate analyses incubations of pooled microsomal samples.

the mouse proceeded at a rate only 25% that of epoxide formation. The rate of AFM formation was substantially lower than the rate of epoxide formation for all species. Oxidative demethylation of AFB to AFP was a significant metabolic pathway in the mouse, proceeding at a rate 51% that of epoxide formation, but was not a significant pathway in any other species (Fig. 5). When considered by summation of initial rates for each metabolite, the total oxidative potential of mouse, monkey and human liver microsomes was 2.0, 6.4 and 1.5 times greater than that observed for the rat, respectively.

Because initial rates of biotransformation determined at saturating substrate conditions do not consider potential differences in Km values for each oxidative pathway, it is also important to examine the total distribution of metabolites after a dose of AFB is largely metabolized. Figure 6 represents the distribution of individual oxidative metabolites after 60 minutes of incubation with rat, mouse, monkey and human liver microsomes. In both the rat and mouse, approximately 50% of the dose is metabolized to AFB-epoxide, whereas in both monkeys and humans less than 25% of the dose is oxidized to the epoxide. It is evident from these data that AFQ serves as a major detoxification pathway in both monkeys and humans, whereas AFM formation accounts for less than 5% of AFB metabolites and thus may be a relatively unimportant detoxification pathway in humans. However, AFM formation may be an important detoxification pathway in rats and mice, as 11% and 22% of the dose was metabolized to AFM in rats and mice, respectively (Fig. 6).

Cytosolic GST activity toward AFB-epoxide and CDNB was also determined in all four species (Table 2). GST activity toward AFB-epoxide was readily measurable in monkey cytosol, and approached the activity seen in rats. Interestingly, the one female monkey had more than twice the GST activity (45.5 pmol/mg/min) than the two male monkeys (18.4 and 14.1 pmol/mg/min).

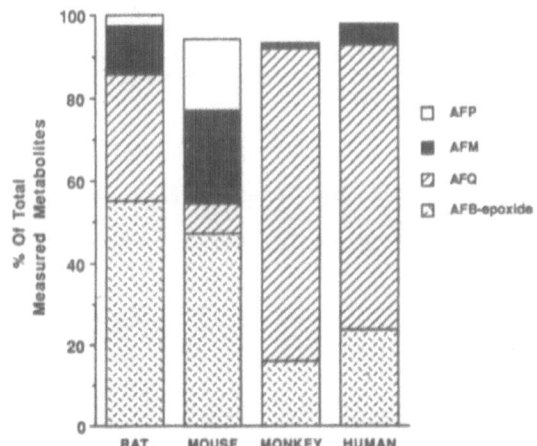

Fig. 6. *In vitro* metabolism of AFB at 60 min. Distribution of aflatoxin B$_1$ metabolites formed after 60 min incubations with microsomes from various species. Values were derived from the data in Fig. 4. The total amount of AFB metabolized after 60 min was: rat, 44%; mouse, 68%; monkey, 97%; human, 75%.

Table 2

| | AFB-Epoxide formation Specific Activity | Glutathione S-Transferase Specific Activity | | RATIO |
| | | AFB-epoxide | CDNB | AFBO CDNB |
	(pmol/mg/min)	(pmol/mg/min)	(μmol/mg/min)	
Rat[a]	419	30	1.9	16
Mouse[b]	559	2,400	5.0	480
Monkey[c]	674 ± 48 (n=3)	26 ± 10 (n=3)	1.20 ± 0.16 (n=3)	21
Human[d]	174 (pooled)[e]	8.7 ± 2.3[f]	0.77 ± 0.18 (n=9)	11 (n=9)

[a]Values represent average of numerous experiments in male
 Sprague-Dawley rats. SE are less than 10% of the mean.
[b]Values represent average of numerous experiments in male
 Swiss-Webster mice. SE are less than 10% of the mean.
[c]Represents the mean±SE of two males and one female adult
 Macacca nemestrina. Liver samples were obtained through the
 Regional Primate Research Center at University of Washington.
[d]Represents the mean±SE of 9 samples, except where noted. Human
 liver samples were obtained from 7 males and 2 females during
 organ transplantation. All organ donors died of traumatic injuries to the head.
[e]Liver microsomes from 9 individuals were pooled for oxidative
 metabolism assays. Individual determinations not completed.
[f]These values represent the sum of three glutathione-related peaks.
 The major AFB GSH peak identified in rat, mouse and monkey cytosol
 assays was below the limits of detection of about 2 pmol/mg/min in
 all human liver cytosol assays.

Whether this represents a characteristic sex difference will require additional studies. As noted previously, GST activity measured toward CDNB does
not fully reflect the large difference in activity toward AFB-epoxide between
mice and the other species. Indeed, the ratio of AFB-epoxide to CDNB activity is relatively similar between rats, monkeys and humans, and is about 20-40
times lower than that in mice. This further illustrates the apparently
unique characteristics of mouse cytosolic GST activity when measured toward
AFB-epoxide.

    In contrast to mice, rats and monkeys, human GST activity toward AFB-
epoxide was not readily detectable above the low, non-enzymatic rates of
formation (detection limit of 2 pmol/mg/min) when the principal 6.3 min GSH
peak was monitored. However, in all human cytosolic samples there were two
small but well resolved peaks at 5.9 and 6.1 min that were dependent upon
both GSH and cytosol. If these unidentified, GSH-dependent peaks are considered as GST-related activity toward AFB, then a measurable rate of 8.7
pmol/mg/min is obtained. Because the identity of these peaks is uncertain,
this estimated value for GST activity toward AFB-epoxide in humans represents
an upper limit of activity, and it is possible that it is actually less than
our 2 pmol/mg/min detection limit. This very low activity of human cytosolic
GST toward AFB-epoxide is consistent with the findings of Moss and Neal.[43]

These investigators were unable to detect any cytosolic GST activity toward
AFB-epoxide (generated *in situ* with quail liver microsomes) in two of three
human livers. One human liver sample was reported to have low but measurable
activity.

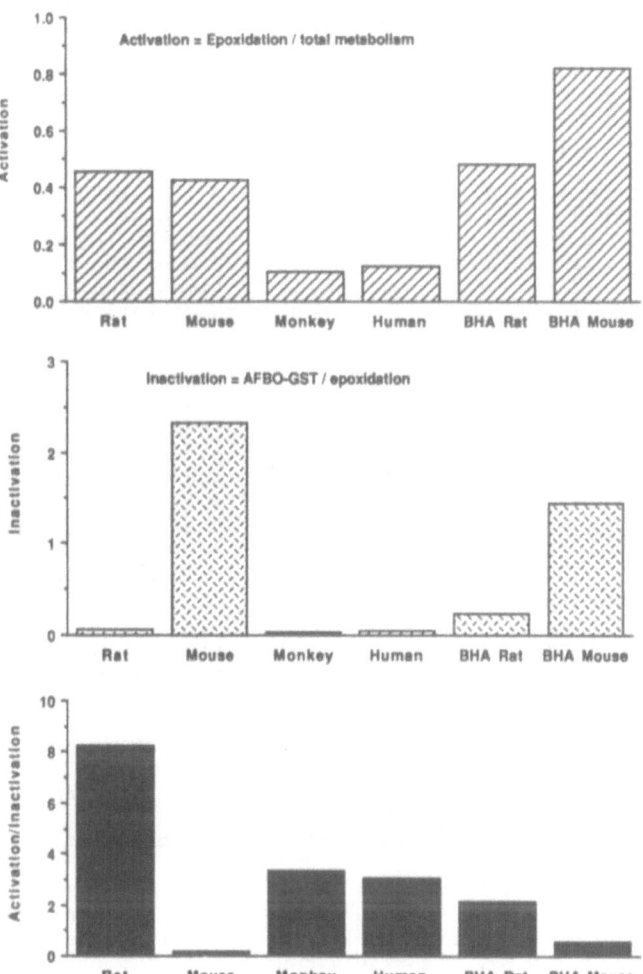

Fig. 7.   Comparison of the relative rates of activation and
          inactivation of aflatoxin B_1 in incubations of
          hepatic microsomes from different species. Initial
          rate values were determined from data in Fig. 5. The
          rate of activation was defined as the ratio of the
          rate of epoxide formation to the rate of total oxida-
          tive metabolism; the rate of inactivation was deter-
          mined as the rate of GST-mediated conjugation of AFB-
          epoxide to the rate of epoxide formation.[12] The
          ratio of activation to inactivation was determined
          as a relative index of species susceptibility to AFB
          hepatocarcinogenesis.

To compare relative sensitivities to AFB carcinogenesis, a useful, albeit simplistic, approach would be to compare the ratio of the rates of activation of AFB to the putative ultimate carcinogenic form, the AFB-epoxide, to the rate of inactivation of the epoxide intermediate. This approach was describ ed by Degen and Neumann,[12] who conducted *in vitro* studies on the metabolism of AFB in rats and mice. These authors defined metabolic activation as "the ratio of the amount of epoxide formed to total AFB metabolism", and defined the rate of inactivation as "the amount of AFB-GSH conjugate formed to the total amount of epoxide formed". In their study, rates of metabolism were not measured, but the distribution of metabolites after 30 min of incubation was used. We have applied this same approach to our data, using initial rates of biotransformation determined for each species (Fig. 7). Using the ratio of the rate of epoxide formation to the rate of total oxidative bio-transformation, the rat and mouse are seen to have nearly identical activation rates, with the monkey and human about 25% of the rat and mouse (Fig. 7, top). However, the rate of inactivation (ratio of the rate of GSH conjugation to the rate of epoxide formation) is 42 times greater in the mouse than the rat, whereas monkeys and humans have inactivation rates of 58 and 73% of the rat, respectively (Fig. 7, middle). When the ratio of the rate of activation/inactivation is determined, the value for mice is only 2.2% that of rats (Fig. 7, bottom). This is very close to the 1.2% relative difference in *in vivo* AFB-DNA adduct formation found between rats and mice (Fig. 3). BHA treatment produced a 3-fold decrease in the activation/inac-tivation ratio in rats, and a 5-fold decrease in DNA binding. BHA treatment of mice actually increased the ratio of activation/inactivation by 5.6-fold, yet DNA binding in BHA-treated mice was slightly less than that seen in untreated mice. This approach assumes that each detoxification pathway is equally important at protecting critical macromolecules from damage, and does not consider potentially important differences in kinetic parameters between the various biotransformation pathways.

Thus, using the ratio of activation to inactivation as a crude index of species susceptibility to AFB carcinogenesis, one would conclude that monkeys and humans are about 2.5 times less sensitive than the rat, but 17 times more sensitive than the mouse, to the hepatocarcinogenic actions of AFB. It is interesting to compare these data with the recent work of Cole et al.[6] who examined the binding of ^{3}H-AFB to DNA in cultured isolated hepatocytes from rats, mice and humans. These authors reported that binding of AFB to DNA in male CD-1 Swiss mice was 0.7% of that of male Sprague-Dawley rats, whereas 3 human liver samples were intermediate between rats and mice, with an average DNA binding level 20% that of rats, but 30 times greater than mice. Thus, their results suggest that humans are about 5 times less sensitive than the rat, but 30 times more sensitive than the mouse, to the carcinogenic effects of AFB. Given the large differences in experimental approach between these two studies, the results are remarkably consistent.

Of course, both of the above *in vitro* approaches to estimating differ-ences in species susceptibility to the carcinogenic effects of AFB are highly simplistic, and only consider differences in rates of biotransformation and DNA-binding from single doses of AFB. Many other steps are involved in the carcinogenic process, and there may well be important species differences in such process as DNA repair, immune surveillance, and tumor promotion/progres-sion factors that ultimately dictate whether a particular organism will con-tract cancer from exposure to AFB or other carcinogenic chemicals. Neverthe-less, it is now apparent that differences in metabolic pathways responsible for the activation and detoxification of AFB play a major role in determining species and individual susceptibility to the hepatocarcinogenic effects of AFB. These data strongly support the hypothesis that humans are of inter-mediate sensitivity compared to the highly susceptible rat and resistant mouse, and suggest that dietary factors could play an important role in reducing one's susceptibility to the carcinogenic effects of AFB.

REFERENCES

1. Boyd JN, Babish JG, and Stoewsand GS, 1982, Modification by beet and cabbage diets of aflatoxin B$_1$-induced rat plasma α-foetoprotein elevation, hepatic tumorigenesis, and mutagenicity of urine, *Fd Chem Toxic.*, 20:47-52.
2. Buchi GH, Muller PM, Roebuck BD and Wogan GN, 1974, Aflatoxin Q$_1$: A major metabolite of afltaoxin B$_1$ produced by human liver. *Res Commun Chem Pathol Pharmacol.*, 8:585-592.
3. Busby WF and Wogan GN, 1984, Aflatoxins, *In*: "Chemical Carcinogens", Chapter 16, American Chemical Society, Washington, DC.
4. Butler WH, 1969, Acute and chronic effects of aflatoxin on the liver of domestic and laboratory animals: a review, *Cancer Res.*, 29:236-250.
5. Cabral JRP and Neal GE, 1983, The inhibitory effects of ethoxyquin on the carcinogenic action of aflatoxin B$_1$ in rats, *Cancer Letters*, 19:126-132.
6. Cole KE, Jones TW, Lipsky MM, Trump BF and Hsu I-H, 1988, *In vitro* binding of aflatoxin B$_1$ and 2-acetylaminofluorene to rat, mouse and human hepatocyte DNA: the relationship of DNA binding to carcinogenicity. *Carcinogenesis*, 9:711-716.
7. Croy RG and Wogan, G N, 1981, Quantitative comparison of covalent aflatoxin-DNA adducts formed in rat and mouse livers and kidneys, *J Natl Can Inst.*, 66:761-768.
8. Croy RG, Essigmann JM, Reinhold VN and Wogan GN, 1978, Identification of the principal aflatoxin B$_1$-DNA adduct formed *in vivo* in rat liver, *Proc Natl Acad Sci USA*, 756:1745-1749.
9. Cullen JM, Ruebner BH, Hsieh LS, Hyde DM and Hsieh DPH, 1987, Carcinogenicity of dietary aflatoxin M$_1$ in male Fischer rats compared to aflatoxin B$_1$, *Cancer Res.*, 47:1913-1917.
10. Decad GM, Dougherty KK, Hsieh DPH and Byard JL, 1979, Metabolism of aflatoxin B$_1$ in cultured mouse hepatocytes: comparison with rat and effects of cyclohexene oxide and diethylmaleate, *Toxicol Appl Pharmacol.*, 50:429-436.
11. Degen GH and Neumann HG, 1978, The major metabolite of aflatoxin B$_1$ in the rat is a glutathione conjugate, *Chem-Biol Interact.*, 22:239-255.
12. Degen GH and Neumann HG, 1981, Differences in aflatoxin B$_1$-susceptibility of rat and mouse are correlated with the capability *in vitro* to inactivate aflatoxin B$_1$-epoxide, *Carcinogenesis*, 2:299-306.
13. Eaton DL, Monroe DH, Bellamy GM and Kalman DA, 1988, Identification of a novel dihydroxy metabolite of aflatoxin B$_1$ produced *in vitro* and *in vivo* in rats and mice, *Chem Res Toxicol.*, 1:108-114.
14. Emerole GO, Neskovic N and Dixon RL, 1979, The detoxification of aflatoxin B$_1$ with glutathione in the rat, *Xenobiotica*, 9:737-743.
15. Essigmann JM, Croy RG, Nadzan AM, Busby WF, Reinhold VN, Buchi G and Wogan GN, 1977, Structural identification of major DNA adducts formed by aflatoxin B$_1$ *in vitro*, *Proc Natl Acad Sci USA*, 74:1870-1874.
16. Essigmann JM, Croy RG, Bennett RA and Wogan GN, 1982, Metabolic activation of aflatoxin B$_1$: Patterns of DNA adduct formation, removal, and excretion in relation to carcinogenesis, *Drug Metab Rev.*, 13:581-602.
17. Garner RC, 1975, Reduction in binding of ^{14}C-aflatoxin B$_1$ to rat liver macromolecules by phenobarbitone pretreatment, *Biochem Pharamcol.*, 24:1553-1556.
18. Groopman JD, Donahue PR, Zhu J, Chen J and Wogan GN, 1985, Aflatoxin metabolism in humans: Detection of metabolites and nucleic acid adducts in urine by affinity chromatography, *Proc Natl Acad Sci USA*, 82:6492-6496.
19. Gurtoo HL, 1980, Genetic expression of aflatoxin B$_1$ metabolism:

Effects of 3-methylcholanthrene and 2,3,7,8-tetrachlorodibenzo-p-dioxin on the metabolism of aflatoxins B_1 and B_2 by various inbred strains of mice, *Mol Pharmacol.*, 18:296-303.

20. Gurtoo HL and Dahms RP, 1979, Effects of inducers and inhibitors on the metabolism of aflatoxin B_1 by rat and mouse, *Biochem Pharmacol.*, 28:3441-3449.

21. Gurtoo HL, Koser PL, Bansal SK, Fox HW, Sharma SD, Mulhern AI and Pavelic ZP, 1985, Inhibition of aflatoxin B_1-hepatocarcinogenesis in rats by β-naphthoflavone, *Carcinogenesis*, 6:675-678.

22. Holeski CJ and Eaton DL, 1986, Effects of glutathione depletion by buthionine sulfoximine (BSO) and diethyl maleate (DEM) on biliary excretion of aflatoxin B_1 (AFB) conjugates and covalent binding of AFB to hepatic DNA, *Toxicologist*, 6:469.

23. Holeski CJ, Eaton DL, Monroe DH and Bellamy GM, 1987, Effects of phenobarbital on the biliary excretion of aflatoxin P_1-glucuronide and aflatoxin B_1-S-glutathione in the rat, *Xenobiotica*, 17:139-153.

24. Hsieh DPH and Wong JJ, 1981, Metabolism and toxicity of aflatoxins, *Adv Exp Biol Med.*, 136:847-883.

25. Hsieh DPH, Wong ZA, Wong JJ, Michas C and Ruebner BH, 1977, Comparative metabolism of AFB, *In*: " Mycotoxins in Human and Animal Health", Rodricks JV, Hesseltine CW and Mehlmann MA, eds., Pathotox Publishing, Illinois.

26. Kensler TW, Egner PA, Davidson NE, Roebuck BD, Pikul A and Groopman JD, 1986, Modulation of aflatoxin metabolism, aflatoxin-N^7-guanine formation, and hepatic tumorigenesis in rats fed ethoxyquin: role of induction of glutathione S-transferases, *Cancer Res.*, 46:3924-3931.

27. Koser PL, Faletto MB, Maccubbin AE and Gurtoo HL, 1988, The genetics of aflatoxin B_1 metabolism, Association of the induction of aflatoxin B_1-4-hydroxylase with the transcriptional activation of cytochrome P_3-450 gene, *J Biol Chem.*, 263:12584-12595.

28. Lotlikar PD, Clearfield MS and Jhee EC, 1984b, Effect of butylated hydroxyanisole on *in vivo* and *in vitro* hepatic aflatoxin B_1-DNA binding in rats, *Cancer Letters*, 24:241-250.

29. Lin JK, Miller JA and Miller EC, 1977, 2,3-dihydro-2-(N^7-guanyl)-(guanyl-7)-3-hydroxy-aflatoxin B_1, a major acid hydrolysis product of aflatoxin B_1-DNA or ribosomal RNA adducts formed in hepatic microsome-mediated reactions and in rat aliver *in vivo*, *Cancer Res.*, 37:4430.

30. Lotlikar PD, Clearfield MS and JHee EC, 1984, Effect of butylated hydroxyanisole on *in vivo* and *in vitro* hepatic aflatoxin B_1-DNA binding in rats, *Cancer Letters*, 24:241-250.

31. Lotlikar PD, Jhee EC, Insetta SM and Clearfield MS, 1984, Modulation of microsome-mediated aflatoxin B_1 binding to exogenous and endogenous DNA by cytosolic glutathione S-transferases in rat and hamster livers, *Carcinogenesis*, 5:269-276.

32. Loury DJ, Hsieh DPH and Byard JL, 1984, The effect of phenobarbital pretreatment on the metabolism, covalent binding, and cytotoxicity of aflatoxin B_1 in primary cultures of rat hepatocytes, *J Toxicol Environ Health*, 13:145-159.

33. Loveland PM, Wilcox JS, Hendricks HD and Bailey GS, 1988, Comparative metabolism and DNA binding of aflatoxin B_1, aflatoxin M_1, aflatoxicol and aflatoxicol-M_1 in hepatocytes from rainbow trout (*Salmo gairdneri*), *Carcinogenesis*, 9:441-446.

34. Lutz WK, 1979, *In vivo* covalent binding of organic chemicals to DNA as a quantitative indicator in the process of chemical carcinogenesis, *Mutat Res.*, 65:289-356.

35. Lutz WK, Jaggi W, Luthy J, Sagelsdroff P and Schlatter C, 1980, *In vivo* covalent binding of aflatoxin B_1 and aflatoxin M_1 to liver DNA of rat, mouse and pig, *Chem-Biol Interac.*, 32:249-256.

36. Mandel HG, Mason MM, Judah DJ, Simpson JL, Green JA, Forrester LM, Wolf CR and Neal GE, 1987, Metabolic basis for the protective effect of the antioxidant ethoxyquin on aflatoxin B_1 hepatocarcinogenesis in the

rat, *Cancer Res.*, 47:5218-5223.

37. McLean AEM and Marshall A, 1971, Reduced carcinogenic effects of
    aflatoxin in rats given phenobarbitone, *Br J Exp Path.*, 52:322-329.

38. Mgbodile MUK, Holscher M and Neal RA, 1975, A possible protective role
    for reduced glutathione in aflatoxin B_1 toxicity: effect of pre-
    treatment of rats with phenobarbital and 3-methylcholanthrene on
    aflatoxin toxicity, *Toxicol Appl Pharmacol.*, 34:128-142.

39. Miller JA and Miller EC, 1981, Mechanisms of chemical carcinogenesis,
    Cancer, 47:1055-64.

40. Monroe DH and Eaton DL, 1987, Comparative effects of butylated
    hydroxyanisole on hepatic *In vivo* DNA binding and *In vitro* biotrans-
    formation of aflatoxin B_1 in the rat and mouse, *Toxicol Appl
    Pharmacol.*, 90:401-409.

41. Monroe DH and Eaton DL, 1988, Effects of modulation of hepatic glutathi-
    one on biotransformation and covalent binding of aflatoxin B_1
    to DNA in the mouse, *Toxicol Appl Pharmacol.*, 94:118-127.

42. Monroe DH, Holeski CJ and Eaton DL, 1986, Effects of single-dose and
    repeated-dose pretreatment with 2(3)-tert-butyl-4-hydroxyanisole (BHA)
    on the hepatobiliary disposition and covalent binding to DNA of
    aflatoxin B_1 in the rat, *Fd Chem Toxic.*, 24:1273-1281.

43. Moss EJ and Neal GE, 1985, The metabolism of aflatoxin B_1 by human
    liver, *Biochem Pharmacol.*, 34:3193-3197.

44. Neal GE and Green JA, 1983, The requirement for glutathione S-transferase
    in the conjugation of activated aflatoxin B_1 during aflatoxin hepato-
    carcinogenesis in the rat, *Chem-Biol Interact.*, 45:259-275.

45. Neal GE, Judah DJ, Stirpe F and Patterson DSP, 1981, The formation of
    2,3-dihydroxy-2,3-dihydro-aflatoxin B_1 by the metabolism of aflatoxin
    B_1 by liver microsomes isolated from certain avian and mammalian
    species and the possible role of this metabolite in the acute toxicity of
    aflatoxin B_1, *Toxicol Appl Pharmacol.*, 58:431-437.

46. Neal GE, Nielsch U, Judah DJ and Hulbert PB, 1987, Conjugation of model
    substrates or microsomally-activated aflatoxin B_1 with reduced glu-
    tathione, catalysed by cytosolic glutathione-S-transferase in livers
    of rats, mice and guinea pigs, *Biochem Pharmacol.*, 36:4269-4276.

47. Newberne PM and Butler WH, 1969, Acute and chronic effects of aflatoxin
    on the liver of domestic and laboratory animals: A review, *Cancer Res.*,
    29:236-250.

48. O'Brien K, Moss E, Judah D and Neal GE, 1983, Metabolic basis of the
    species difference to aflatoxin B_1 induced hepatotoxicity, *Biochem
    Biophys Res Comm.*, 114:813-821.

49. Patterson DSP, 1973, Metabolism as a factor in determining the toxic
    action of the aflatoxins in different animal species, *Fd Cosmet Toxicol.*,
    11:287-295.

50. Rahimtula AD and Martin M, 1984, Dietary administration of 2(3)tert-
    butyl-4-hydroxyanisole elevates mouse liver microsome-mediated DNA
    binding and mutagenicity of aflatoxin B_1, *Chem-Biol Interact.*,
    48:207-220.

51. Raj HG, Clearfield MS and Lotlikar PD, 1984, Comparative kinetic studies
    on aflatoxin B_1-DNA binding and aflatoxin B_1-glutathione conjugation
    with rat and hamster livers *in vitro*, *Carcinogenesis*, 5:879-884.

52. Ramsdell HS and Eaton DL, 1988, Modification of aflatoxin B_1 biotrans-
    formation *In vitro* and DNA binding *in vivo* by dietary broccoli in
    rats, *J Toxicol Env Health*, 25:265-280.

53. Roebuck BD and Wogan GN, 1977, Species comparison of *in vitro* metabolism
    of aflatoxin B_1, *Cancer Res.*, 37:1649-1656.

54. Sinnhuber RO, Hendricks JD, Wales JH and Putnam GB, 1977, Neoplasms in
    rainbow trout, a sensitive animal model for environmental carcinogenesis,
    Ann NY Acad Sci., 298:389-408.

55. Valsta LM, Hendricks JD and Bailey GS, 1988, The significance of gluta-
    thione conjugation for aflatoxin B_1 metabolism in rainbow trout
    and coho salmon, *Fd Chem Toxic.*, 26:129-135.

56. Vesselinovich SD, Mihailovich N, Wogan GN, Lombard LS and Rao KVN, 1972, Aflatoxin B$_1$, a heptocarcinogen in the infant mouse, *Cancer Res.*, 32:2289-2291.

57. Wattenberg LW, 1980, Inhibitors of chemical carcinogenesis, *J Environ Pathol Toxicol.*, 3:35.

58. Wei CI, Marshall MR and Hsieh DPH, 1985, Characterization of water-soluble glucuronide and sulphate conjugates of aflatoxin B$_1$. 1. Urinary excretion in monkey, rat and mouse, *Fd Chem Toxic.*, 23:809-819.

59. Whitham M, Nixon JE and Sinnhuber RO, 1982, Liver DNA bound *In vivo* with aflatoxin B$_1$ as a measure of hepatocarcinoma initiation in rainbow trout, *J Natl Cancer Inst.*, 68:623-630.

60. Whitty JP and Bjeldanes LF, 1987, The effects of dietary cabbage on xenobiotic-metabolizing enzymes and the binding of aflatoxin B$_1$ to hepatic DNA in rats, *Fd Chem Toxic.*, 25:581-587.

61. Williams GM, Reiss B and Weisburger JH, 1985, A comparison of the animal and human carcinogenicity of environmental, occupational and therapeutic chemicals, *In*: "Advances in Modern Environmental Toxicology", Flamm G and Lorentzen G, eds., Princeton Scientific Publications, Princeton, NJ.

62. Williams GM, Tanaka T and Maeura Y, 1986, Dose-related inhibition of aflatoxin B$_1$ induced hepatocarcinogenesis by the phenolic antioxidants, butylated hydroxyanisole and butylated hydroxytoluene, *Carcinogenesis*, 7:1043-1050.

63. Wogan GN, 1973, Aflatoxin carcinogenesis, *In*: "Methods in Cancer Research", Busch H, ed., Academic Press, New York.

64. Wogan GN, 1976, Aflatoxins and their relationship to hepatocellular carcinoma, *In*: "Hepatocellular Carcinoma", Okuda K and Peters RL, eds., Wiley, New York.

65. Wogan GN, Paglialunga S, Newbern PN, 1974, Carcinogenic effects of low dietary levels of aflatoxin B$_1$ in rats, *Fd Cosmet Toxicol.*, 12:681-685.

66. Wong ZA and Hsieh DPH, 1980, The comparative metabolism and toxico-kinetics of aflatoxin B$_1$ in the monkey, rat, and mouse, *Toxicol Appp Pharmacol.*, 55:115-125.

67. Yourtee DM, Bean TA and Kirk-Yourtee CL, 1987, Human aflatoxin B$_1$ metabolism: An investigation of the importance of aflatoxin Q$_1$ as a metabolite of hepatic post-mitochondrial fraction, *Toxicol Lett.*, 38:213-224.

68. Zhu J-Q, Zhnag L-S, Hu X, Chen J-S, Xu Y-C and Chu FS, 1987, Correlation of dietary aflatoxin B$_1$ levels with excretion of aflatoxin M$_1$ in human urine, *Cancer Res.*, 47:1848-1852.

THE GABA RECEPTOR AND THE ACTION OF TREMORGENIC MYCOTOXINS

Mohyee E. Eldefrawi, Daniel B. Gant and Amira T. Eldefrawi

Department of Pharmacology & Experimental Therapeutics
School of Medicine
University of Maryland
Baltimore, MD 21201

The gamma-aminobutyric acid (GABA) receptor is the major inhibitory receptor in vertebrate and invertebrate brains. Activation of this $GABA_A$ receptor, which traverses the membrane, results in its opening of an anionic channel that is part of the receptor protein and permeation of Cl^- along its concentration gradient. Thus, Cl^- influxes producing membrane hyperpolarization.[6,11] This $GABA_A$ receptor is inhibited competitively by bicuculline and allosterically by picrotoxinin and t-butylbicyclophosphorothionate (TBPS), which bind to another site that is closely associated with the Cl^- channel component of the receptor protein. The $GABA_A$ receptor is studied biochemically by the characteristics and specificity of its binding of $[^{35}S]$TBPS and also by monitoring its function in vitro, i.e. $^{36}Cl^-$ influx that is induced by binding of GABA to the receptor in membrane vesicles, is inhibited by competitive and noncompetitive inhibitors and is potentiated by certain modulators. $GABA_A$ receptors also have a site that binds benzodiazepines and another that binds depressant barbiturates.[14] Benzodiasepines are mostly potentiators of GABA receptor function (e.g., diazepam) increasing the receptor's affinity for GABA and facilitating the frequency of opening of its Cl^- channel, but a few benzodiazepines are inhibitors of diazepam action. There are also benzodiazepine modulators of receptor function, acting as convulsants or proconvulsants and thus are called inverse agonists

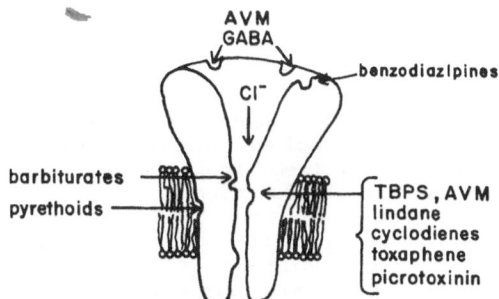

Fig. 1. GABA Receptor. Diagram of the $GABA_A$ receptor showing a variety of binding sites (from Eldefrawi and Eldefrawi[6]). AVM, avermectin B_{1a}.

Microbial Toxins in Foods and Feeds, Edited by A.E. Pohland et al.,
Plenum Press, New York, 1990

(e.g., ethyl β-carboline-3-carboxylate). There are as well convulsant barbitu rates, that are proposed to bind to the TBPS-binding site.[13] The binding of depressant barbiturates to the GABA$_A$ receptor also increases the receptor's affinity for GABA and benzodiazepines. Ethanol produces some of its effect via the GABA$_A$ receptor as shown by its potentiation of GABA-induced $^{36}Cl^-$ influx in primary cultured neurons of the spinal cord.[13] The GABA$_A$ receptor has also been discovered to be the molecular target for polychlorcycloalkane insecticides such as lindane and cyclodienes. Their actions are sterospe- cific and selective on the TBPS binding site.[3,9]

The GABA$_A$ receptor has been purified from bovine brain and the DNA sequence and predicted amino-acid sequence revealed.[12] Injection of RNA syn- thesized α or β subunits from their cloned DNAs into *Xenopus* oocytes produced functional receptors with the pharmacological properties characteristic of the GABA$_A$ receptor.[2] This suggested that the α or β subunits arose from gene duplication.

There is also a GABA$_B$ receptor, which is activated by GABA in presence of high Ca^{2+} concentration, but unlike the GABA$_A$ receptor is not activated by muscimol, is activated by baclofen, and not inhibited by bicuculline. The GABA$_B$ receptor does not have a Cl^- channel as part of the protein and its activation is suggested to increase K^+ and possibly Ca^{2+} conductances.[7] The GABA$_B$ receptor is proposed[12] to act by regulating cAMP production that is stimulated by other receptors acting via one of two mechanisms: One is asso- ciated with phospholipase A$_2$ (a Ca^{2+}-activated enzyme that catalyzes the release of arachidonic acid from membrane phospholipids) and the other with phospholipase C, which catalyzes the conversion of phosphatidyl-inositol 4,5- bisphosphate to the second messengers inositol triphosphate and diacylgly- cerol. Because inhibition of GABA$_A$ receptor may produce tremors and convulsions in animals, it was logical to consider it as a target for the action of tremorgenic mycotoxins. These have an active component derived from geranylgeraniol and tryptophan. They have been classified into groups based on their chemical structures: paspalitrems, territrems, fumitremorgen- verruculogens and tryptoquivalines.[4]

Fungi capable of producing tremorgenic mycotoxins can be found contami- nating a variety of agricultural commodities such as forages, corn, and silage. In cattle, these mycotoxins cause "staggers", a neurological syn- drome, whose clinical signs include muscle tremor, uncoordinated movements and general weakness in the hind legs with stiff, stilted movements of the forelegs. These symptoms become more evident when the animal is disturbed. Affected animals become debilitated and may die from dehydration, drowning or limb injury. However, mortalities are infrequent and the animals usually recover in 24-48 hours with no apparent pathologic change.[4]

The five mycotoxins studied were tested at 100 μM for their effect on binding of radioacative ligands to rat brain GABA$_A$ receptors. They had little or no effect on [^{3}H]muscimol binding to the receptor's GABA binding site, neither did they inhibit its binding of [^{3}H]flunitrazepam. However, the four tremorgenic mycotoxins (aflatrem, paspalinine, paxilline, and verru- culogen) inhibited binding of the convulsant ligand [^{35}]S-TBPS and GABA- induced $^{36}Cl^-$ influx into rat brain microsacs (resealed membranes), while the non-tremorgenic mycotoxin, verruculotoxin, had no effect up to 300 μM (Fig. 2). These data indicate that tremorgenic mycotoxins bind to the GABA$_A$ receptor in rat brain and inhibit its function.

The tremorgenic mycotoxins act like the convulsants TBPS, picrotoxinin and cyclodiene insecticides[1,3,9] in inhibiting [^{35}S]TBPS binding to, and GABA-induced $^{36}Cl^-$ influx into, rat brain microsacs (Fig. 2). However, although these insecticides inhibit [^{35}S]TBPS binding competitively, mycotoxins inhibit it in a complex manner as seen by the reduction by paspalinine of maximal binding from 0.93 to 0.73 pmol/mg protein, yet increase of K_d from 76 to 111 nM.[8] Also, there is a low correlation coefficient of 0.76 between potencies of tremorgenic mycotoxins in inhibiting

Table 1. Mycotoxin inhibition of GABA-induced $^{36}Cl^-$ influx and [^{35}S]TBPS binding to rat brain microsacs (from Gant et al.[8]).

Mycotoxin	IC$_{50}$ (μM)[a]	
	$^{36}Cl^-$	[^{35}S]TBPS binding
Aflatrem	38.5 ± 4.7	4.3 ± 0.1
Paxilline	39.2 ± 6.9	11.2 ± 1.3
Paspalinine	78.7 ± 5.4	3.5 ± 0.9
Verruculogen	129.0 ± 7.3	26.2 ± 6.0
Verruculotoxin	300	>100

[a]IC$_{50}$ values represent means ± standard errors of at least three separate experiments.

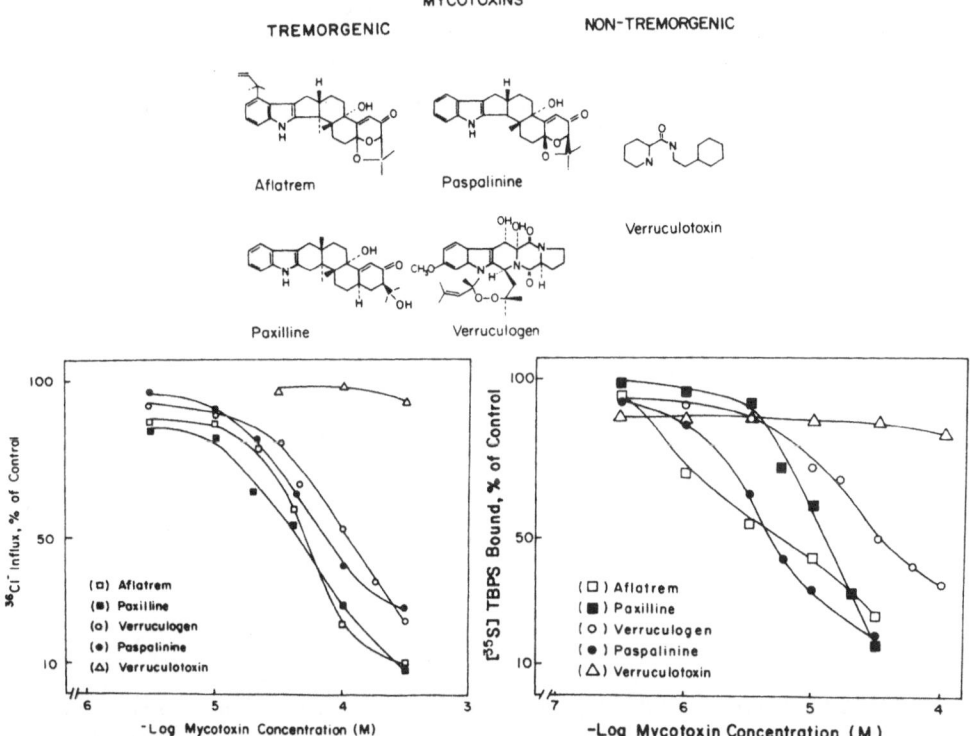

Fig. 2. Mycotoxins and their actions on GABA$_A$ receptor of rat brain.
Left-Inhibition of GABA (100 μM) induced $^{36}Cl^-$ influx into microsacs.
Right-Inhibition of binding of [^{35}S]TBPS (1 nM) to GABA receptors in these microsacs (from Gant et al.[8]).

293

[^{35}S]TBPS binding and GABA-induced ^{36}Cl$^-$ influx, suggesting that the site of action of these mycotoxins is not the [^{35}S]TBPS binding site. These tremorgenic mycotoxins also differ from GABA or barbiturate action on the GABA$_A$ receptor in not inducing ^{36}Cl$^-$ influx and not inhibiting [^{3}H]flunitrazepam binding. Thus, it is suggested that their binding site on the GABA$_A$ receptor is different from the GABA, benzodiazepine, barbiturate, or TBPS binding sites.

The higher potency of each mycotoxin in inhibiting [^{35}S]TBPS binding than GABA-induced ^{36}Cl$^-$ influx (Table 1) can be explained by the very high concentration (100 μM) of GABA that is used to activate all the GABA receptors and obtain maximal ^{36}Cl$^-$ influx compared to the low concentration of 1 nM [^{35}S]TBPS used which binds to only ~ 5% of the sites. Thus, a much higher concentration of the mycotoxin is required to inhibit the effect of the very high concentration of GABA. This concentration difference has been shown also for cyclodiene insecticides.[3]

The concentrations of aflatrem and verruculogen that inhibit 50% of [^{35}S]TBPS binding (4.3 and 26.2 μM, respectively) and GABA-induced ^{36}Cl$^-$ influx (38.5 and 129 μM, respectively) (Table 1) are well within the concentrations of these mycotoxins that produce tremors in mice. Aflatrem at doses ranging from 0.5-4 mg per kg i.p. in mice, elicited sharp tremors,[4] which in a 25 g mouse with 1.5 ml blood volume would translate to 17 and 133 μM of aflatrem. Similarly, the lowest concentration of verruculogen (0.78 mg per kg) that elicited tremors in four 20 g mice,[5] is calculated to be 30 μM.

Our findings contrast with the very recent report of the potentiation of aflatrem of GABA-induced Cl$^-$ current in *Xenopus* oocytes injected with chick brain mRNA.[15] However, these discrepencies may be related to differences in species (rat *vs* chicken) or factors related to expression of the mRNA in the oocyte. Since the aflatrem binding site on the GABA$_A$ receptor was proposed to be different from the GABA, benzodiazepine or TBPS binding sites,[15] similar to our proposal based on binding and ^{36}Cl$^-$ flux studies, we both agree that these mycotoxins bind to a different modulatory site on the GABA$_A$ receptor protein. Thus, it may be that a change in one or a few amino acids or different post-transcription modification in adult rat brain *vs* *Xenopus* oocyte could conceivably change their inhibitory effect to a potentiating one or vice versa. This is somewhat analogous to the inhibitory action of *d*-tubocuririne on skeletal muscle nicotinic receptor; but its partial agonist action on the receptor in embryonic muscle.[16]

ACKNOWLEDGEMENTS

The authors thank Dr. R. Cole (National Peanut Research Lab, Dawson, Georgia) for the supply of mycotoxins studied and Dr. J. Valdes (U. S. Army, CRDEC, Edgewood, Maryland) for his support and participation. The research conducted in the authors' laboratory was supported in part by NIH grant ES 02594.

REFERENCES

1. Abalis IM, Eldefrawi ME, Eldefrawi AT, 1985, High affinity stereospecific binding of cyclodiene insecticides and gamma-hexachlorocyclohexane to gamma-aminobutyric acid receptors of rat brain, *Pestic Biochem Physiol.*, 24:95-102.
2. Blair LAC, Levitan ES, Marshall J, Dionne VE, Barnard EA, 1988, Single subunits of the GABA$_A$ receptor form ion channels with properties of the native receptor, *Science*, 242:577-579.
3. Casida JE, Lawrence LJ, 1985, Structure activity correlations for

interactions of biocyclophosphorus esters and some polycholorocycloalkane
and pyrethroid insecticides with the brain specific *t*-butylbicyclophos-
phorothionate receptor, *Environ Health Perspect.*, 61:123-132.

4. Cole RJ, Dorner JW, 1986, Role of fungal tremorgens in animal disease,
In: "Mycotoxins and Phycotoxins," Steyn PS, Vleggaar R, eds.,
Elsevier, Amsterdam.

5. Cole RJ, Kirksey JW, Moore JH, Blankenship BR, Diener VL, Davis ND, 1972,
Tremorgenic toxin from *Penicillium verruculosum*, *Appl Microbiol.*,
24:248-250.

6. Eldefrawi AT, Eldefrawi ME, 1987, Receptors for gamma-aminobutryic acid
and voltage-dependent chlorida channels as targets for drugs and toxicants, *FASEB J.*, 1:262-271.

7. Enna SJ, Karbon EW, 1987, Receptor regulation: evidence for a relationship between phospholipid metabolism and neurotransmitter mediated cAMP
formation in brain, *Trends Pharmacol. Sci.*, 8:21-24.

8. Gant DB, Cole RJ, Valdes JJ, Eldefrawi ME, Eldefrawi AT, 1987, Action of
tremorgenic mycotoxins on GABA$_A$ receptor, *Life Sci.*, 41:2207-2214.

9. Gant DB, Eldefrawi ME, Eldefrawi AT, 1987, Cyclodiene insecticides
inhibit GABA$_A$ receptor-regulated chloride transport, *Toxicol Appl
Pharamcol.*, 88:313-321.

10. Maksay G, Ticky MK, 1985, Dissociation of [^{35}S*t*-butylbicyclophosphoro-
thionate binding differentiates convulsant and depressant drugs that
modulate GABAergic transmission, *J Neurochem.*, 44:480-486.

11. Olsen RW, 1982, Drug interactions at the GABA receptor-ionophore complex,
Ann Rev Pharmacol Toxicol., 22:245-277.

12. Schofield PR, Darlison MG, Fujita N, Burt DR, Stephenson FA, Rodriguez H,
Rhee LM, Ramachandran J, Reale V, Glencorse TA, Seeburg PH, Barnard EA,
1987, Sequence and functional expression of the GABA$_A$ receptor super-
family, *Nature*, 328:221-227.

13. Ticku MK, Lowrimore P, Lehoullier P, 1986, Ethanol enhances GABA-induced
^{36}Cl$^-$ influx in primary spinal cord cultured neurons, *Brain Res Bull.*,
17:123-126.

14. Trifiletti RR, Snowman AM, Snyder SH, 1985, Barbiurtrate recognition site
on the GABA/benzodiazepine receptor complex is distinct from the picro-
toxinin/TBPS recognition site, *Eur J Pharmacol.*, 106:441-447.

15. Yao Y, Peter AB, Baur R, Sigel E, The tremorigen aflatrem is a positive
allosteric modulator of the gamma-aminobutyric acid$_A$ receptor channel
expressed in *Xenopus* oocytes, *Mol Pharmacol.*, 35:319-324.

16. Ziskind L, Dennis MJ, 1978, Depolarising effect of curare on embryonic
rat muscles, *Nature*, 276:622-623.

APPLICATION OF MACROMOLECULAR SYNTHESIS MEASUREMENTS IN MYCOTOXIN TOXICITY STUDIES

Leonard Friedman, Edmund L. Peters, Dennis W. Gaines and
Robert C. Braunberg

Division of Toxicological Studies
Food and Drug Administration
Beltsville, MD 20708

INTRODUCTION

For a number of years an important aspect of work[*] in our laboratory
has been the utilization of in vitro and in vivo measurements of macromolecular synthesis and related phenomena as biochemical indices of the toxicity of food-related substances. The relevancy of studies involving mycotoxins
in this regard was particularly apparent because for many, if not most, of
these chemicals their biological potency is matched by their appreciable
effect at low concentrations on the metabolism of nucleic acid or proteins.
Because of their potent biochemical activity, e.g., T-2 toxin inhibits protein synthesis at a concentration of 10^{-8}M or a fraction of a nanogram per
milliter,[53] often at levels where cell damage is not discernible, and because
of the relative ease with which macromolecular synthesis measurements can be
carried out, we have a technique that can be applied readily to exploring
factors, such as diet, other chemicals, and physiological stress that may
regulate the acute and, most importantly, the chronic toxicity of these compounds. An example of this approach was presented by Moule et al. where the
authors reported on the differential effects of phenobarbital in modifying
the aflatoxin B_1 (AFB1) induced inhibition of transcription in mice and
rats.[52] This doesn't minimize the importance and usefulness of other types
of measurements, such as binding of toxins or their metabolites to macromolecules, but these types of measurements may not always apply (i.e., binding
has not always been found to constitute an obligatory correlate of toxicity),
and besides, binding studies require the availability of a highly pure and
usually expensive source of the radiolabeled toxin. Finally, modification of
macromolecular biosynthesis may reflect damage to the genome or at least to
related factors and thus may indicate functional modification of the genetic
machinery. Of course this is not always true, but in the case of the aflatoxins there is considerable evidence for this alteration of function: one
of the earliest in vivo effects of the toxin is inhibition of RNA synthesis.[31] This inhibition has been shown to be a consequence of both an
effect on RNA polymerase II and on the DNA or nucleolar chromatin template
function.[11,48,59,82] Also, neither inhibition of RNA synthesis[12,25]
nor protein synthesis[65] can be affected by AFB1 without

---

[*]This work has been presented in part at the 1974 annual meeting of the
Society of Toxicology and at the 1971, 1974, 1977 and 1980 annual meetings
of the Federation of American Societies for Experimental Biology.

Microbial Toxins in Foods and Feeds, Edited by A.E. Pohland *et al.,*
Plenum Press, New York, 1990

enzymatic conversion of the parent compound. Thus any treatment modifying the metabolic activation of AFB1 can be expected to produce changes in the biological potency of the toxin as reflected both by its inhibition of macromolecular synthesis and its carcinogenic activity.

Most of the work to be presented here deals with studies of aflatoxin's effects on macromolecular biosynthesis under conditions expected to produce perturbations in xenobiotic metabolism. In some of the studies of the macromolecular biosynthesis measurements were accompanied by *in vitro* measurements of the status of mixed function oxidase activity (MFO) of the liver using AFB1 and aminopyrene as the substrates. For comparison purposes, data will be presented from experiments with another mycotoxin, patulin, for which only negative evidence has been reported concerning its carcinogenic or other chronic effects.[19,56] There have been no reports of it binding to DNA, yet it inhibits macromolecular synthesis in a number of systems.[37,38,58,66]

MATERIALS AND METHODS

Animals

The following are sources and species of animals used for the studies:

> Rats - Osborne-Mendel (FDA colony), Sprague-Dawley
> Holtzman Co.), and hypophysectomized and sham-operated
> Holtzman rats (Hormone Assay Laboratories).
> Mice - ICRFRF$_2$ (Flander Research Farm)
> Rainbow Trout - (Maryland State Hatchery)
> Monkeys - Rhesus (Hazelton Laboratories)

Unless indicated otherwise the animals used were young adults (rats: 3-6 mo., mice: 4-5 mo., trout: 12-15 mo., and monkeys: 3-5 yrs.), were fed standard commercial chow diets or a semi-synthetic diet[#] *ad libitum*, and fasted overnight prior to dosing or sacrifice. The rodents were housed either singly or in small groups. Our preliminary results indicated no effects of this variable on the parameters measured. Monkeys were anesthetized with Sernylan and exsanguinated. The other animals were expired by stunning and exsanguination.

Test Chemicals

AFB1, purchased from Calbiochem, was shown to be free of fluorescent impurities by thin-layer chromatography (TLC) and 90-100% pure by ultraviolet absorption measurements.[30] Patulin was prepared and analyzed by the Division of Chemistry and Physics (FDA) and shown to be 98.7% pure by ultraviolet absorption, TLC, and infrared and mass spectroscopy. Actinomycin D was purchased from Sigma.

Other Chemicals

DDT (Chemical Compounding Corp.), SKF 525-A (Smith, Kline and French), Aroclor 1254 (Monsanto Co.), testosterone (Vitaline Co.), all other hormones (Sigma). *Sterculia foetida* oil was prepared and analyzed as previously described.[15] Other dietary oils and proteins were from commercial sources.

[#]The basic composition of the semi-synthetic diets used for these studies was as follows: vitamin-free casein, 20%; sucrose, 66%; salts, 4% (Jones-Foster Salt Mix, General Biochemicals); vitamin mix, 1% (Vitamin Fortification Mixture, General Biochemicals); cellulose, 4%; safflower oil, 5%.

In Vivo Measurements

Treatment of animals with aflatoxin and measurement of *in vivo* synthesis of RNA and protein was carried out as described previously.[28,29] Generally, AFB1 was administered ip as a solution in propylene glycol (1 ml/kg body weight) four hours prior to ip injections of radiolabeled precursors [6-^{14}C] orotic acid and/or L-[4,5-^{3}H (N)] leucine (New England Nuclear). One hour later animals were killed by decapitation, and tissues were removed and fractionated by a modification[28,29] of the method by Schneider.[68] Isolated fractions were counted by standard liquid scintillation techniques. For RNA synthesis results were either expressed as DPM(R)/DPM(L)/mg RNA x 10^3 (R = RNA fraction; L = Total liver homogenate) or DPM(R)/DPM(T)/mg RNA x 10^3 (T = cold TCA fraction). For protein synthesis, results were expressed as DPM/mg protein or DPM(P)/DPM(T)/mg protein (P = protein fraction). Polyribosomal analysis of liver postmitochrondrial supernatants was performed as described previously.[28]

In Vitro Measurements

Measurement of macromolecular synthesis using tissue slice techniques was carried out as described previously.[28] AFB1 was added to the *in vitro* systems as a solution in propylene glycol. Fractionation was carried out by a modification[29] of the Schmidt Thannhauser method.[67] Measurement of the *in vitro* metabolism of AFB1 (and aminopyrene) was also carried out as described previously.[29,30] Substrates were incubated with the 9,000Xg supernatent prepared from livers of individual animals or livers pooled from several animals in the case of neonates. Separation of AFB1 and its metabolites in defatted and dried chloroform extracts of the incubation mixtures was by TLC and quantitation by fluorescence densitometry. Reference standards used were aflatoxins B_1, M_1, B_{1a}, and P_1. Neither AFB1a nor AFP1 could be visualized as reaction products on the plates. Aflatoxicol was identified by its mobility relative to AFB1 and its metabolic characteristics as determined previously[30] and aflatoxin G_1 (AFG1) was tentatively identified from its chromatographic properties and green fluorescence (a reliable AFG1 standard was not available at the time of the study). However the metabolism of AFB1 was generally measured by the rate of production of the major chloroform soluble metabolite, AFM1, or disappearance of the parent compound itself. Binding of AFB1 to RNA, DNA, and protein of liver slices was determined by using the same system as described for measuring *in vitro* macromolecular synthesis except exhaustive washing with the solvents was carried out to remove residues of unbound metabolites. The [^{14}C] AFB1 used (Moravek Biochemicals, 180 mCi/mmol), had all carbons except the methoxy carbon labeled and a radiopurity of 98%.

Statistical Analysis

Differences between means of control and treated groups were analyzed for significance by Student's t test. Comparison among species for *in vitro* susceptibility to inhibition of hepatic RNA synthesis by AFB1 was by regression analysis and analysis of variance; the data (expressed as percent inhibition) were analyzed as if they were the result of a parallel line assay. A p value of 0.05 or less was accepted as an indication of statistical significance.

RESULTS AND DISCUSSION

As shown in Table 1, inhibition of hepatic RNA synthesis by AFB1 in rats fed a control and a protein free diet for seven days was approximately the same. This was so even though, as expected, mixed function oxygenase (MFO) activity and the *in vitro* metabolism of AFB1 were depressed in the protein-

Table 1. Effect of Protein Free Diets on Inhibition of RNA Synthesis in Livers of Male and Female Rats by AFB1.[a]

	RNA Synthesis		
Group	(DPM (R)/DPM (L)/ mg RNA x 10^3)		% Inhibition
Male	Control	Aflatoxin	
20% Casein	7.35 ± 0.70[b]	4.96 ± 0.44[c]	32.52
0% Casein	13.57 ± 1.38	9.70 ± 0.74[c]	28.52
Female			
20% Casein	8.83 ± 0.55	7.50 ± 0.53	15.06
0% Casein	12.75 ± 0.65	10.41 ± 0.63	18.35

[a]Six nonfasted young adult Osborne-Mendel rats per group used after being fed the diets for 8 days; animals sacrificed 5 hr after aflatoxin (0.25 mg/kg body weight) and one hr after [^{14}C]orotic acid (25 μCi/kg) both administered ip.
[b]Mean $\pm$ SEM.
[c]Significantly different from non-aflatoxin treated group ($p < 0.05$).

Table 2. Effect of Protein Free Diets on the *in vitro* Metabolism of AFB1 and Aminopyrene by Livers from Male and Female Rats.[a]

	Aflatoxin M1[b] (ng/g/30 min)	%	Aminopyrene demethylase (μg HCHO/g/30 min)	%
Male				
20% Casein	216.0 ± 45.8[c]		45.63 ± 2.33	
0% Casein	82.0 ± 10.7[d]	-62.04	15.29 ± 1.67	-66.49
Female				
20% Casein	144.0 ± 29.8		19.46 ± 0.96	
0% Casein	131.7 ± 15.1	-8.54	19.90 ± 1.55	+2.26

[a]Young adult Osborne/Mendel, non-fasted rats fed the diets for 8 days. The 9000Xg supernatant of liver homogenates was used for the assay (see text).
[b]Percent difference due to diet for each sex.
[c]Values are based on wet weight of tissue and are the mean $\pm$ SEM for 5-6 rats.
[d]Significantly different from 20% casein group ($p < 0.05$).

free diet (Table 2). The results of this early study indicated that at least the parent compound could not be the only active form of this toxin. In another study, however, where rats were fed diets containing varying levels of casein for 90 days it was apparent that the biochemical in vivo activity of AFB1 was diminished with the lower levels of dietary protein (Table 3). A similar finding was apparent when the *in vitro* effect of AFB1 on RNA synthesis was determined using liver slice preparations from rats also fed diets

containing increasing levels of casein (Figure 1). Apparently the optimum
dietary protein level for aflatoxin's activity was 20 percent. These results

correlate well with the effect of low protein diets on AFB1's carcinogenic[45]
and DNA-binding activity[61] in rats even though more recent studies[3,23] by
workers from this same laboratory have suggested greater influence of dietary
protein on the promotional than on the initiation phase of aflatoxin-induced
carcinogenesis. Another finding of the studies described above is that
simply restricting the amount of the high level protein diet to about one
third to one half that consumed by rats fed *ad libitum* reduced the *in vivo*
biochemical effect of AFB1 (Table 3). As with the protein free diet, feeding
the low protein diet (5% casein) also produced a diminution of hepatic MFO
activity and *in vitro* AFB1 metabolism relative to rats fed 20% casein (Table
4). AFB1 metabolism was higher in livers of rats fed the chow diet than in
those fed semi-synthetic diet containing 20% casein. This corresponds to a
consistent finding by us (illustrated in Table 3 and Figure 1) of a lower
AFB1 biochemical activity in rats fed the chow versus those fed the high
level protein semi-synthetic diet. The obvious explanation for protection by
the low protein diet would be a reduction of the metabolic transformation of
AFB1 to the presumed active expoxide metabolite.[17,44] Also, there have
been reports indicating an increase in activity of some detoxifying enzymes
in the livers of rats given diets low in protein,[7] although conflicting
information exists[1,8] We have found that not only protein level, but the
type of protein fed also influences the response to AFB1. The feeding diets
containing different proteins each at the 20% level for 10 weeks produces a
range of AFB1-induced inhibitions of RNA synthesis - from 63 to 84 percent in
one experiment and 28 to 56 percent in another (Table 5). There does not
seem to be a clear cut correlation between the degree of aflatoxin's effect
and the quality of the protein fed. Interestingly, there is an increased
effect with the addition of cysteine to the casein-containing diet. We have
seen this effect, but to a smaller degree,

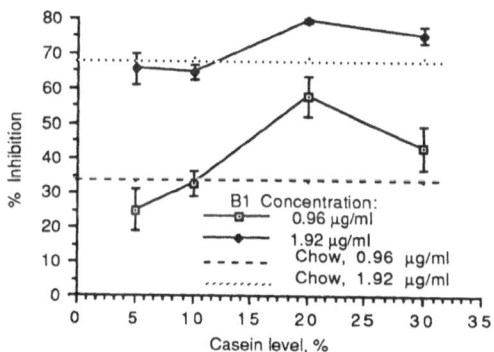

Fig. 1.   Effect of increasing levels of dietary casein and chow
          diet on inhibition of RNA synthesis in liver slices by
          AFB1 added *in vitro*. Sprague-Dawley young adult male rats
          were fed semi-synthetic diets containing indicated levels
          of casein for 13.5 mo. Incubation conditions and analyti-
          cal procedures for determination of RNA synthesis describ-
          ed in Methods. Results are expressed as % of control and
          each point represents the mean ± SE (vertical bars) for
          four separate experiments. SEs for chow diet = 4.22 and
          2.09, for AFB1 concentrations of 0.96 and 1.92 μg/ml,
          respectively. Differences in inhibition due to diet are
          significant (p < 0.05) for 5% casein vs. 20% casein and
          chow vs. 20% casein.

Table 3. Influence of Dietary Protein Level, Restricted Intake or Chow Diet Inhibition of Liver RNA and Protein Synthesis by AFB1.[a]

Treatment Casein %	Aflatoxin mg/kg	RNA Synthesis (DPM (R)/DPM (L)/ mg RNA x 10^3)	Protein Synthesis DPM/mg protein
5	0	12.70 ± 0.47^b	1545 ± 44
5	0.25	7.69 ± 0.59^c $(-39.5)^d$	1462 ± 76 (-5.4)
10	0	12.13 ± 0.33	1603 ± 79
10	0.25	6.36 ± 0.47^c (-47.6)	1259 ± 66^c (-21.5)
20	0	12.12 ± 0.46	1304 ± 69
20	0.25	3.08 ± 0.40^c (-74.6)	891 ± 55^c (-31.7)
20 R.I.[e]	0	11.42 ± 0.40	1097 ± 49
20 R.I.	0.25	6.00 ± 0.67^c (-47.5)	794 ± 69^c (-27.6)
Chow	0	9.38 ± 0.24	824 ± 24
Chow	0.25	7.13 ± 0.45^c (-24.0)	845 ± 40 $(+2.5)$

[a]7-8 young male adult Sprague-Dawley rats/group fed the diets for 90 days; animals sacrificed 5 hr after aflatoxin and one hr after [^{14}C]orotic acid (25.0 μCi/kg) and [^{3}H]leucine (50.0 μCi/kg) dosing, both administered ip.
[b]Mean ± SEM.
[c]Significantly different from non-aflatoxin treated control (p < 0.05).
[d]Numbers in parentheses = % inhibition.
[e]Rat given 10 g 20% casein diet per day; all other rats fed diets *ad libitum*.

Table 4. Effects of Diet on Hepatic Glutathione (GSH) Levels and *in vitro* Metabolism of Aminopyrene (AP) and AFB1 by Livers of Male Rats.

Diet	AP* (μg/HCHO/g/ 30 min)	Aflatoxin M1* (ng/g/30 min)	Microsomal* Protein (mg/g)	GSH** (μM/g)
5% Casein	47.0 *** $\pm 2.6^{ab}$	205 $\pm 96^b$	46.3 ± 4.4	2.03 $\pm 0.11^a$
20% Casein	68.9 $\pm 5.8^a$	541 $\pm 186^a$	39.9 ± 3.0	4.43 $\pm 0.14^a$
Chow	82.4 $\pm 4.6^b$	4449 $\pm 353^{ab}$	46.7 ± 1.8	$--$

* Sprague-Dawley rats, 5/group, started as 10 wk old rats; fed for 10 wk.
** Sprague-Dawley rats, 7/group started as 8 wk old rats; fed for 12 wk.
*** Mean ± SEM.
[a,b]Same letter = p < 0.05 between groups.

in a parallel experiment using female rats (31% vs 26.8% inhibition) and in another in which cysteinine HCl (500 mg/kg) was injected ip into rats prior to dosing with AFB1 (78.6% vs 75.5% inhibition of RNA synthesis and 24.9% vs 16.8% inhibition of protein synthesis). Wells et al.[78] have reported an exacerbation of the chronic effects of AFB1 in rats fed diets supplemented with cysteine. We note also in our studies a lesser effect of aflatoxin on RNA synthesis in chow fed rats in relation to those fed any of the semi-synthetic diets except the one containing lactoalbumin as the protein.

Looking at another macronutrient, fat, we first examined the effect of feeding rats semi-synthetic diets containing varying levels of safflower oil (an oil very high in polyunsaturated fatty acids) for 10 weeks on the in vivo inhibition of RNA synthesis by AFB1 (Figure 2). Somewhat consistent with another report,[77] aflatoxin's effect was low in rats fed a fat free diet (at least relative to the maximum effect seen at a dietary level of 5% oil). Increasing the content of fat in the diet above five percent produced decreasing levels of inhibition, with only a small degree of inhibition remaining at the 20% fat level. Several studies have shown high levels of dietary fat protect against the effects of AFB1. This was seen in rats, both in regard to mortality[53] and preneoplastic lesions (where high levels of selenium were also fed),[5] and in monkeys in regard to a lipid depressing[2] and anticoagulent[6] effect of the toxin. Marzuki and Norred have shown corn oil, in comparison to coconut oil, enhances the in vitro conversion of AFB1 to some of its metabolites.[47] Other investigators have also reported an increase of hepatic cytochrome P450 level and MFO activities with inclusion of polyunsaturated oils in the test animal diet.[46,53,63] Marzuki and Norred also reported a higher glutathione transferase activity in the livers of rats given diets containing corn oil relative to the activity seen with feeding coconut oil although this increase did not translate into a higher level of AFB1-conjugates in the bile.[47]

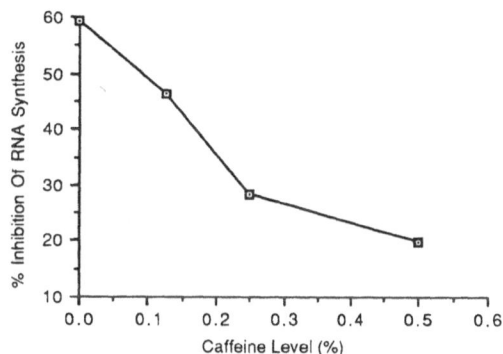

Fig. 2.   Effect of increasing levels of dietary safflower oil on inhibition of hepatic RNA synthesis in vivo by AFB1. Osborne-Mendel young adult male rats were fed semi-synthetic diets containing indicated levels of the oil for 10 wk and administered AFB1, 0.22 mg/kg body weight ip, 4 hr prior to sacrifice. Determination of RNA synthesis described in Methods. Results expressed as percent of control. Each point represents the mean for 8 animals and inhibition of RNA synthesis by AFB1 is statistically significant (p < 0.05) for all points. For 0% safflower oil the mean ± SE values for RNA synthesis [DPM(R)/DPM(L) /mg RNA x 10^3] are 11.964 ± 0.381 and 6.807 ± 0.767 for the control and aflatoxin groups respectively.

In another study we examined the effects of three different oils in the diets of rats. Where the effects of diets containing 20% lard and 20% safflower oil were compared, the ability of AFB1 to inhibit RNA synthesis *in vivo* was greater in the animals fed lard than in the animals fed safflower oil by a factor of two (Table 6). In agreement with this finding, Newberne reported a greater acute effect from AFB1 administered to rats consuming diets high in saturated fatty acids relative to the effect in rats fed a diet high in unsaturated fatty acids, but the effect on tumor incidence was the reverse.[53] One difference may be the source of the saturated fatty acid, which was beef fat (tallow) in Newberne's study, whereas in the present study the fat was lard. On the other hand, and somewhat consistent with our findings, Wells et al.[77] have reported greater renal tumorigenic effects from AFB1 in rats given diets high in saturated fatty acids derived from lard. Complicating the observations of the effects of dietary fat are the results of Marzuki and Norred[47] which showed no differences between the amount of nucleic acid-aflatoxin adducts in the livers of rats receiving dietary coconut oil and in the livers of those receiving dietary corn oil; both oils at the 20% level in the diet and both dosed with ^{14}C-AFB1. There appears to be general agreement on the relative effect of predominantly saturated fatty acid-containing diets and predominantly unsaturated fatty acid-containing diets on MFO activity in the livers of rats, with a higher activity found in the latter-type diet, but there are conflicting reports on the relative influence of these diets on AFB1's biological activity. In our studies the toxin was administered intraperitoneally. There is no information on the possible effect of dietary fat on absorption of AFB1 from the gastrointestinal tract. This may be relevant to interpretation of the feeding studies cited but possibly even more so because of the recent finding that metabolic alteration

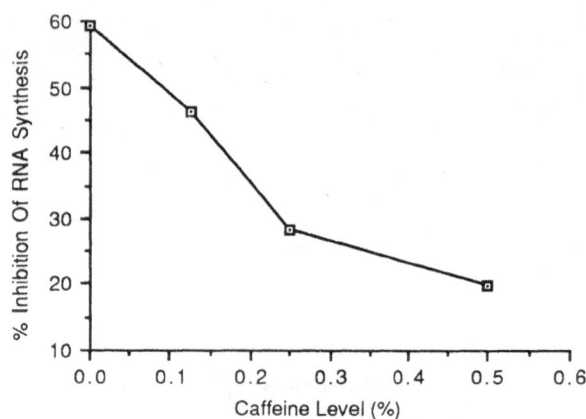

Fig. 3.   Effects of dietary PCB as Aroclor 1254 on inhibition of RNA synthesis in liver slices by AFB1 added *in vitro*. Sprague-Dawley young adult male rats fed chow diets containing indicated levels of PCB for 4 mo. Incubation conditions and analytical procedures for determination of RNA synthesis described in Methods. Results are expressed as percent of control and each point represents the mean ± SE (vertical bars) for four separate experiments. Inhibition was significantly different (p < 0.05) between control and PCB at the highest three levels of aflatoxin for 100 ppm PCB and for 25 ppm PCB only at the highest level of aflatoxin (p < 0.01 for two lower levels of aflatoxin).

Table 5.   Influence of Dietary Protein on Inhibition of Liver RNA Synthesis by AFBl in Male Rats.[a]

Protein	RNA Synthesis (DPM(R)/DPM(L)/ mg RNA 10^3)		% Inhibition
	-Aflatoxin	+Aflatoxin	
EXPERIMENT I			
Casein	11.12[b] ± 0.40	4.04 ±0.83	63.7
Casein + Cysteine (1%)	10.93 ± 0.38	2.74 ±0.38	74.9
Soy	12.39 ± 0.51	4.09 ±0.47	67.0
Egg White	12.76 ± 0.93	2.37 ±0.35	81.4
Wheat Gluten	13.40 ± 1.39	2.16 ±0.29	83.9
Chow	11.81 ± 0.15	4.88 ±0.73	58.7
EXPERIMENT II			
Casein	10.18 ± 0.24	4.42 ⊥0.21	56.6
Lactoalbumin	10.69 ± 0.51	7.68 ±0.20	28.2
Chow	10.80 ± 0.31	6.36 ±0.32	41.1

[a]6-8 three mo old Sprague-Dawley rats/group fed semi-synthetic diets containing 20% protein (by Kjeldahl analysis) for 10 wk; animals sacrificed 5 hr after AFBl (0.33 mg/kg body weight) and one hr after[^{14}C]orotic acid (11.1 μCi/kg for Exp. I and 21.4 μCi/kg for Exp. II), administered ip.
[b]Mean ± SEM.
All values for aflatoxin-treated groups are significantly different from those for corresponding non-aflatoxin treated (control) groups ($p < 0.05$).

of AFBl apparently also takes place in regions of the small intestine during the course of its absorption.[42]

Another result of our study shown in Table 6 is the effect of dietary *Sterculia foetida* seed oil, an oil rich in cyclopropenoid fatty acids, on inhibition of hepatic RNA synthesis by AFBl. Cyclopropenoid fatty acids enhance aflatoxin-induced carcinogenesis in trout,[69] but not in rats.[27,55] In agreement with this latter result was the failure of *Sterculia foetida* oil to enhance the biochemical activity of AFBl when added to either a semi-synthetic or to a chow diet.

Another interesting facet of aflatoxin's biological activity, at least in rats, is its dependence on the sex[81] and hormonal status of the

Table 6. Influence of Dietary Fat on Inhibition of Liver RNA Synthesis by
AFB1.[a]

Fat	RNA Synthesis (DPM(R)/DPM(L)/ mg RNA 10^3)		% Inhibition
	-Aflatoxin	+Aflatoxin	
Safflower, 20%	11.06 ± 0.34^b	8.45 ± 0.14	23.6
Lard, 20%	12.18 ± 0.68	6.54 ± 0.43	46.3
Safflower, 1%	12.12 ± 0.47	6.27 ± 0.73	48.3
Safflower, 1% + S.F. oil[c], 1%	10.60 ± 0.37	8.07 ± 0.37	23.9
Chow	10.80 ± 0.31	6.36 ± 0.32	41.1
Chow + S.F. oil, 1%	11.14 ± 0.40	7.74 ± 0.41	30.5

[a]Eight young adult Osborne-Mendel male rats/group fed semi-synthetic or
chow diets for 10 wk; animals sacrificed 5 hr after aflatoxin (0.22 mg/kg
body weight) and 0 hr after [^{14}C]orotic acid (11.1 μCi/kg) administered ip
[b]Mean $\pm$ SEM.
[c]*Sterculia foetida* seed oil.
All values for aflatoxin-treated groups significantly different ($p < 0.05$)
from those for non-aflatoxin-treated (control) groups.

animal.[9,33] As shown in Table 7, inhibition of RNA synthesis in the livers
of male rats by AFB1 is greater than that in the livers of female rats. Castrated male rats are as resistant in this regard as are female rats and
treatment of castrated rats with testosterone appears to partially reverse
the effect of castration. This is in agreement with similar experiments
using aflatoxin binding to liver DNA[34] and aflatoxin-induced mortality[62] as
the endpoints. Evidently seven day fasting of female rats increases their
response to AFB1 to approximately that of non-fasted male rats (Table 7).
Examining the *in vitro* metabolism of aminopyrene and AFB1 using livers from
male, castrated male and female rats, we observe (Table 8) that while the
metabolism of aminopyrene was approximately the same in female and castrated
males, but lower than in intact males, the production of AFM1 and AFB1 was
decreased by castration but was approximately the same for intact male rats
and female rats. Two other metabolites of AFB1, X1 and X2 (probably Q1),
were both less in female and castrated rats than in male rats. Aflatoxicol
production was not affected by sex or castration. On the other hand, fasting
did not affect the metabolism of aminopyrene, but did reduce the production
of AFM1 especially in those animals fasted and then "refed" with a sucrose
solution (Table 9) (this latter treatment is reported to further decrease the
MFO activity of liver[73]). Fasting produced little change in the rate of
production of other TLC-detectable metabolites of AFB1 relative to that seen
for AFM1 (data not shown). These results tend to confirm the distinctive
nature of the cytochrome involved with the enzymatic production of AFM1,[35]
but by themselves offer no explanation for the modification of AFB1's inhibition of RNA synthesis with these different treatments. Explanations for the
enchanced sensitivity to AFB1 due to fasting may reside in the severe depletion of hepatic glutathione levels with food deprivation noted by some
workers.[57] Conjugation of glutathione with electrophilic metabolites

Table 7. Effect of Castration and Fasting on Inhibition of RNA Synthesis in Livers of Adult Rats by AFB1.[a]

Group	RNA Synthesis $(DPM(R)/DPM(L)/mg$ RNA x $10^3)$		% Inhibition
	Control	Aflatoxin	
Normal Males	7.59 ± 0.40[b]	3.38 ± 0.98	55.46
Castrate Males[c]	7.81 ± 0.14	5.70 ± 0.83	27.02
Castrate Males + Testosterone[d]	6.75 ± 0.13	4.11 ± 0.53	39.11
Normal Females	6.77 ± 0.35	5.03 ± 0.29	25.70
Fasted Females (3 days)	10.54 ± 0.38	5.32 ± 0.46	49.53
Fasted-"Refed" Females[e]	12.98 ± 0.48	7.37 ± 0.42	43.22

[a]4-8 Osborne-Mendel rats per group sacrificed 5 hr after ip administration of aflatoxin (0.33 mg/kg body weight) and one hr after ip dosing with [^{14}C]orotic acid (25.0 mCi/kg).
[b]Mean $\pm$ SEM.
[c]Rats (10 mo old) castrated 18 days prior to use in experiment.
[d]Testosterone propionate administered im, 5 mg/kg for 7 days and 10 mg/kg for 2 days.
[e]Rats fasted for 3 days and given drinking water containing sucrose at a concentration of 35% for one day.
All values for aflatoxin-treated groups significantly different ($p < 0.05$) from those for non-aflatoxin-treated (control) groups.

Table 8. Effect of Castration and Sex on the *in vitro* Metabolism of AFB1 and Aminopyrene by Rat Liver.[a]

	Aminopyrene	Aflatoxin B_1[b]			
Substrate Metabolite	HCHO (μg/g/30 min)	M1	X1 (ng/g/30 min)	Ao	X2
Male	80.46 ± 6.37	2995 ± 255	1200 ± 192	363 ± 23	263 ± 55
Castrate	42.17[c] ± 4.37	1472[c] ± 197	372[c] ± 47	355 ± 43	< 80
Female	42.52[c] ± 3.27	3728 ± 289	552[c] ± 32	278 ± 24	< 80

[a]Young adult Osborne-Mendel rats fed chow diets; rats castrated ("Castrate" group) approximately 1 mo prior to sacrifice. 9000Xg supernatant of liver homogenate used for the assay (see text).
[b]Only metabolites M1 (aflatoxin M1) and Ao (aflatoxicol) were positively identified. X2 (green fluorescence and presumed to be aflatoxin Q1) and X1 (low mobility relative to M1) were unidentified and were quantified using AFB1 and AFM1 standards, respectively. All values based on wet weight of tissue and are the mean $\pm$ SEM of 5-6 rats.
[c]Significantly different from male group ($p < 0.05$).

Table 9. Effect of Fasting and "Refeeding" on the *in vitro* Metabolism of AFB1 and Aminopyrene by Livers from Female Rats.[a]

	Liver Weight (% body wt.)	Aflatoxin M1 (ng/g/30 min)	%[b]	Aminopyrene demethylase µg HCHO/g/30 min	%[b]
Fed	3.38 ±0.12	5517[c] ± 541	--	55.84 ± 2.63	--
Fasted (3 days)	2.66 ±0.33	4017 ± 552	-27.19	64.38[d] ± 1.92	+15.29
Fasted 3 days & "Refed" 1 day)	3.41 ±0.22	1536[d] ± 257	-70.16	61.72 ± 9.69	+10.53

[a]Young adult Osborne-Mendel rats were fed chow diet *ad libitum* ("Fed" group), had water but food was withheld for 3 days ("Fasted" group), or were fasted for 3 days with drinking water replaced by 35% sucrose solution ("Fasted-Refed" group) on day 4 (day prior to sacrifice). 9000Xg super-natant of liver homogenate was used for the assay (see text).
[b]Percent difference from fed group.
[c]Values are based on net weight of tissue and are the mean ± SEM of 5-6 rats.
[d]Significantly different from fed group ($p < 0.05$).

Table 10. Influence of Diethylstilbesterol (DES) on Relative Liver Weight and Inhibition of Liver RNA and Protein Synthesis by AFB1 in Male Rats.[a]

Treatment DES (mg/kg)	AFB1 (mg/kg)	Liver Wt. % body wt.	RNA Synthesis (DPM(R)/DPM(L)/ mg RNA x 10^3	Protein Synthesis (DPM/mg protein)
0	0.0	3.00 ± 0.12[b]	7.92 ± 0.29	2094 ± 182
0	0.5	----	2.84 ± 0.27[e] (-64.1)[f]	1556 ± 73[e] (-25.6)[f]
2.0	0.0	3.86 ± 0.13[c] (+28.7)[d]	12.38 ± 0.78	2588 ± 146
2.0	0.5	----	9.48 ± 0.35[e] (-23.4)[f]	2061 ± 155[e] (-20.4)[f]

[a]Six young adult Osborne-Mendel male rats/group administered either corn oil or DES in corn oil im for 8 days and either propylene glycol or AFB1 ip 5 hr prior to sacrificing the rats on the 9th day. All rats dosed ip with [^{14}C]orotic acid (25.0 µCi/kg) and [^{3}H]leucine (100 µCi/kg) one hr prior to sacrifice.
[b]Mean ± SEM.
[c]Significantly different from non-DES control group ($p < 0.05$).
[d]Values in parenthesis = percent increase of relative liver weight from non-DES control group.
[e]Significantly different from non-aflatoxin treated group ($p < 0.05$).
[f]Numbers in parenthesis = % inhibition.

Table 11. Effect of Hypophysectomy (Hypox) on Inhibition of Protein and RNA Synthesis by AFB1 in Livers of Male and Female Rats.[a]

Treatment	Protein Synthesis (DPM/mg protein)	RNA Synthesis (DPM(R)/DPM(L)/ mg RNA x 10^3
MALE		
Intact	1662 ± 110^b	9.40 ± 0.39
Intact + AFB1	1156 ± 12^c	5.53 ± 0.75^c
	$(30.4)^d$	(-41.2)
Hypox	846 ± 22	15.86 ± 1.16
Hypox + AFB1	857 ± 62	13.87 ± 0.91
	$(+1.3)$	(-12.5)
FEMALES		
Intact	1442 ± 54	12.71 ± 0.60
Intact + AFB1	1299 ± 66	10.61 ± 0.62^c
	(-9.9)	(-16.5)
Hypox.	1026 ± 184	16.29 ± 0.70
Hypox. + AFB1	1127 ± 88	13.55 ± 1.10^e
	$(+9.8)$	(-16.8)

[a]6-7 Sprague-Dawley rats/group treated as shown and sacrificed 5 hr after AFB1 (0.33 mg/kg body weight) and one hr after [^{3}H]leucine (50 μCi/kg body weight) and [^{14}C]orotic acid (18.75 mCi/kg body weight) ip dosing.
[b]Mean $\pm$ SEM.
[c]Significantly different from non-aflatoxin treated control ($p < 0.05$)
[d]Numbers in parenthesis = % inhibition (or stimulation (+)).
[e]Borderline significance ($0.05 < p > 0.01$).

electrophilic metabolites of aflatoxin is thought to represent an important detoxification mechanism for this toxin.[20,22]

Consistent with the finding of a lower AFB1 activity in the livers of female or castrated male rats is another finding by us and others[54] that the administration of diethylstilbesterol (DES) to male rats reduced the inhibition of liver RNA synthesis by AFB1. Inhibition of protein synthesis was also reduced but only moderately (Table 10).

We have previously reported on the relative resistance of hypophysecto-mized rats to inhibition of liver RNA synthesis by AFB1.[29] This resistance correlates well with the resistence of such treated rats to the carcinogenic activity of aflatoxin.[33] When male or female hypophysectomized rats or intact rats were treated with aflatoxin, and macromolecular synthesis mea-sured, aflatoxin-induced inhibition of both RNA and protein syntheis in male rats was reduced by the removal of the pituitary. Hypophysectomized female rats responded to aflatoxin treatment no differently than their intact coun-terparts (Table 11). A small but significant amount of inhibition (14.9%) of RNA synthesis (30.72 vs 26.15 DPM(R)/DPM(L)/mg RNA x 10^3) was seen in the kidneys of the intact male rats, but no inhibition was produced in the kid-neys of the hypophysectomized group of male rats (data not shown). The data tend to indicate a generalized effect of the gland ablation on the biochemi-cal response of the male rats to aflatoxin treatment. We have, in fact, reported evidence for an effect of hypophysectomy on the *in vitro* metabolism of AFB1 and negative evidence for this treatment on uptake of aflatoxin by the liver.[29] Since female rats did not respond as did male rats to the

Table 12. Effect of Hormone[a] Treatment on Response of Male Hypophysectomized (Hypox) Rats to Inhibition of Liver RNA and Protein Synthesis by AFB1.[b]

Treatment	RNA Synthesis (DPM(R)/DPM(L)/ mg RNA x 10^3	Protein Synthesis (DPM/mg protein)
Hypox	14.11 $\pm$ 0.29[c]	965.5 $\pm$ 61.8
Hypox + AFB1	13.31 $\pm$ 0.46 (-5.60)[d]	1091.7 $\pm$ 59.7 (+13.07)
Hypox + Hormones	16.21 $\pm$ 0.47	1306.7 $\pm$ 49.2
Hypox + Hormones + AFB1	13.58 $\pm$ 0.47[e] (-16.29)	1077.7 $\pm$ 187.1[e] (-17.52)

[a]Hormone treatment consisted of the following: thyroxin-0.15 mg, ACTH-10.5 units, somatotropin-1.5 mg, hydrocortisone succinate-7.5 mg and, and insulin 6.0 units, administered im (in saline) per rat daily for 5 days.
[b]6 Sprague-Dawley rats/group treated as shown and sacrificed on 5th day, 5 hr after AFB1 (0.25 mg/kg body weight) and 1 hr after [^{14}C]orotic acid (25.0 μCi/kg) ip dosing.
[c]Mean $\pm$ SEM.
[d]Numbers in parenthesis = % inhibition or stimulation (+).
[e]Significantly different from non-aflatoxin control (p < 0.05).

effect of hypophysectomy, hypophysectomized male rats were treated with physiological doses of testosterone (producing enlargement of epididymis) and tested, but this treatment failed to reverse the relative resistance of these rats to AFB1 (data not shown). Prompted by the finding of Moule et al.[51] showing enhancement of the effect of AFB1 in a resistant species with MFO inducer, phenobarbital, hypophysectomized rats were treated for ten weeks with 25 and 100 ppm of polychlorinated biphenyl (PCB). This treatment had no influence on the effect of AFB1 on RNA synthesis (data not shown). However, with a combination of hormones the sensitivity to the toxin was somewhat restored (Table 12). The equivocal nature of these results can probabaly be attribuated to the finding that hormonal treatment of intact rats itself may diminish the carcinogenicity of AFB1.[9] Finally, we can report that another potent liver carcinogen, dimethylnitrosamine (50 mg/kg), inhibited protein synthesis in the liver of male intact and hypophysectomized rats to about the same degree (43.4 vs 44.0%). This should not be suprising since hypophysectomized rats were reported to be sensitive to the hepatocarcinogenic effects of this compound.[32]

Similarly to hypophysectomized rats, neonatal rats were also relatively resistant to AFB1's inhibition of liver RNA synthesis (Table 13). This is in agreement with other reports that AFB1 binding to the DNA in liver of neonatal rats is also considerably less than seen for adult rats.[10,43] We also found that the in vitro metabolism of AFB1, as measured by production of AFM1 by liver homogenates, was considerably slower using the tissue preparation from neonatal rats (Table 14). All these data are contradictory to other findings that hepatocarcinoma can be produced in mice or in rats by their treatment with AFB1 at the neonatal stage.[76,80] Neoplastic nodules were also found in the livers of 15 month old rats treated at 7 days of age with AFB1 and phenobarbital[64]

Because of the well-documented evidence that the biological activity of

Table 13. Inhibition of RNA Synthesis in the Livers of Male Neonatal and
Young Adult Rats by AFB1 or Actinomycin D.[a]

| | RNA Synthesis (DPM(R)/DPM(L)/mg RNA x 10^3) | | Inhibition |
	Control	Toxin	%
	Experiment I (AFB1)		
Neonate	7.69 ± 0.48^b	7.41 ± 0.46	3.64
Adult	7.35 ± 0.70	4.96 ± 0.44^c	32.52
	Experiment II (AFB1)		
Neonate	9.96 ± 0.71	5.70 ± 0.64^c	42.77
Adult	8.13 ± 0.80	2.70 ± 0.41^c	66.79
	Experiment III (Actinomycin D)		
Neonate	7.48 ± 0.56	0.78 ± 0.07^c	89.57
Adult	8.78 ± 0.45	2.99 ± 0.09^c	65.95

[a] 5-6 pools of livers/treatment group from Osborne-Mendel 5-6 day old neonatal rats consisting of 2-3 livers/pool and livers from 5-6 adult rats/
group (except in Exp. III where 3 adult rats were used) were analyzed.
Rats (non-fasted) were sacrificed 5 hr after ip administration of AFB1
(Exp. I-0.25 mg/kg; Exp. II, 3.3 mg/kg, neonate and 0.33 mg/kg, adult) 6
hr after ip administration of actinomycin D (3 mg/kg-neonates; 1.5 mg/kg-
adult) and 1 hr after ip dosing with [^{14}C]orotic acid (0.1 ml/neonate and
25.0 μCi/kg adult).
[b] Mean $\pm$ SEM.
[c] Significantly different from non-toxin treated group ($p < 0.05$).

reactive aflatoxins, particularly that activity related to mutagenesis and
carcinogenesis, is dependent upon the metabolic transformation of the parent
compound to other species with electrophilic properties - most probably 2,3-
epoxides,[17,44] it is not surprising that chemicals known to affect MFO or
other drug metabolizing enzymes would also affect the biological and biochemical potency of this class of toxins. AFB1 administered to male rats fed
diets containing 25 or 100 ppm of the PCB, Aroclor 1254, for ten weeks caused
less inhibition of hepatic RNA and protein synthesis than were obtained in
rats fed the same chow diet containing no added PCB (Table 15). This finding
was confirmed by exposing liver slices from rats fed these same diets to AFB1
added in vitro and measuring RNA synthesis (Figure 3). The remarkable effect
of this chronic treatment with PCB on induction of the in vitro metabolism of
AFB1 by postmitochondrial liver preparations from these rats is shown in
Table 16. Aminopyrene metabolism was also increased but not to the same
extent as the rate of production of AFM1. These results alone probably could
not explain the protective effect of PCB on AFB1's biological[40] and biochemical activity. However, a report has since emerged demonstrating PCB
also induces the formation of aflatoxin-glutathione conjugates,[21] presumably a detoxification pathway for the toxin,[49,50] in the livers of rats
treated with these substances.

DDT, another polychlorinated hydrocarbon and inducer of MFO activity,[36]
did not produce an effect on aflatoxin-induced inhibition of RNA synthesis,
but it did enhance metabolism of AFB1 to AFM1. Another chlorinated pesticide,

dieldren, also failed to affect AFB1 toxicity in trout.[41] Likewise, SKF
525-A, an inhibitor of cytochrome P450 catalyzed activity,[16] had no *in vivo*
influence on AFB1's biochemical activity although it inhibited appreciably
the metabolism of the toxin (Table 17). In an *in vitro* system, however,
Gurto and Dahms reported an inhibition of metabolic activation of AFB1 by a
relatively high (1 mM) concentration of this same inhibitor.[35] When caffeine, a weak inducer of MFO activity,[26] was fed at varying levels to male
rats a dose-related diminution of the effect of AFB1 *in vivo* (Figure 4) and
in vitro (Figure 5) on RNA synthesis occurred. Other methylxanthines, theo-
phylline and theobromine, exerted a similar, though weaker, effect. The
reason for this effect is not known since MFO activity in livers from methyl-
xanthine treated rats did not appear to be affected. Also, the addition of
caffeine (1 mM) to liver slices failed to alter the inhibition of RNA synthesis induced by AFB1, making it unlikely that caffeine (itself) was competing
for enzyme or DNA sites.

To demonstrate that cytochrome P450 was essential for AFB1 to exert its
biochemical effects *in vivo*, rats were treated with cobalt chloride, a compound shown to interfere with the synthesis of this heme-protein.[74] The
treatment almost completely abolished the *in vivo* inhibition of RNA and protein synthesis by AFB1 (Table 18). As was found by others,[4,20,49] partial
depletion of liver glutathaione by diethylmaleate also increased AFB1's
effects (Table 19), but the effects were not proportional to the degree of
depletion. The maximum level of glutathione depletion attained (84%) would
not be expected to account for the magnitude of aflatoxin-induced effects
seen, based upon the very high (mg) levels of glutathione normally found in
the liver (Table 4) relative to the range of aflatoxin (ng) levels found in

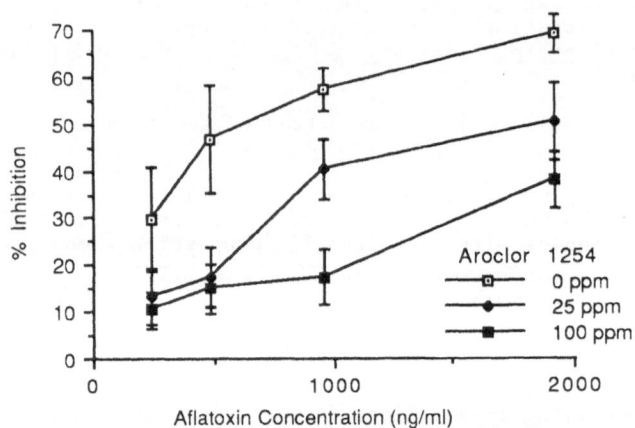

Fig. 4. Effect of increasing levels of dietary caffeine on inhibi-
        tion of hepatic RNA synthesis *in vivo* by AFB1. Weanling
        Osborne-Mendel rats fed chow diets containing indicated levels
        of caffeine for 6 mo and administered AFB1 (0.33 mg/kg body
        weight) 4 hr prior to injection with [^{14}C]orotic acid and 5
        hr prior to sacrifice. Determination of RNA synthesis des-
        dribed in Methods. Results are expressed as percent of con-
        trol. Each point represents the mean for 7 rats and inhibi-
        tion of RNA synthesis by AFB1 is statistically significant
        ($p < 0.05$) for all points. For 0% caffeine the mean $\pm$ SE
        values for RNA synthesis (DPM(R)/DPM(L)/mg RNA x 10^3) are
        14.044 $\pm$ 0.620 and 5.712 $\pm$ 0.497 for the control and aflatoxin
        groups, respectively.

Table 14. Metabolism of AFB1[b] and Aminopyrene *in vitro* by Livers from Adult and Neonatal Male Rats.[a]

Substrate Metabolite	Aminopyrene HCHO μg/g/30 min	Aflatoxin B_1[b] M1 ng/g/30 min	X1 ng/g/30 min	Ao
Neonate	7.51[c] ±0.26	186.0 ± 36.5	142.0 ± 8.6	478.0 ± 34.4
Adults	46.96 ±4.13	506.0 ± 8.1	442.0 ± 56.0	668.0 ± 24.2

[a]The 9000Xg supernatants of liver homogenates prepared from the livers of non-fasted 14 week old Osborne-Mendel rats (adults) and 4-5 day old rats (noenates) were used for the assay (see text).
[b]On the metabolites M1 (aflatoxin M1) and Ao (aflatoxicol) were identified. X1 is a major fluorescent metabolite migrating slower on TLC than AFM1 and quantitated using AFB1 standards.
[c]All values are based on wet weight and are the mean ± SEM of livers from 5 rats for the adults and 5 pooled samples of 5 livers per sample for the neonates.
All values for the neonates are significantly different ($p < 0.05$) from those of the adults.

Table 15. Effect of Dietary PCB[a] on Inhibition of RNA and Protein Synthesis in Livers of Male Rats by AFB1.[b]

Treatment			
Aflatoxin, mg/kg	PCB, ppm	RNA Synthesis (DPM(R)/ DPM(L)/mg RNA x 10^3	Protein Synthesis (DPM/mg protein)
0	0	11.57 ± 0.36[c]	1282 ± 57
0.5	0	4.02 ± 0.47[d] (-65.3[e])	931 ± 59[d] (-27.3)
0	25	9.96 ± 0.26	1311 ± 49
0.5	25	7.24 ± 0.34[d] (-27.3)	1330 ± 67 (+1.4)
0	100	9.94 ± 0.18	1798 ± 87
0.5	100	7.76 ± 0.33[d] (-21.9)	1697 ± 64 (-5.6)

[a]Young adult Osborne-Mendel rats fed chow diets containing indicated level of PCB as Aroclor 1254 for 10 weeks.
[b]Rats (6-7/group) sacrificed 5 hr after ip dose of AFB1 and 1 hr after ip dose of [^{14}C]orotic acid (12.5 μCi/kg) and [^{3}H]leucine (50 μCi/kg).
[c]Mean ± SEM.
[d]Significantly different from non-aflatoxin treated group ($p < 0.05$).
[e]Numbers in parenthesis = % change due to aflatoxin treatment.

Table 16. Effect of PCB on the *in vitro* Metabolism of AFB1 and Aminopyrene by Rat Liver.[a]

PCB[b]	Microsomal Protein	Aflatoxin		Aminopyrene
		B1 used	M1	Demethylase
ppm	mg/g/liver	(ng/g/30 min)		(μg HCHO/g/30 min)
0	51.0 ± 8.3	52,330 ± 10,171	940 ± 502	126.0 ± 15.5
25	56.0 ± 2.6	109,276 ± 1,019[c]	22,200 ± 6,027	250.8 ± 25.4[c]
100	73.1 ± 4.9[c]	119,224 ± 7,271[c]	23,680 ± 3,806[c]	294.5 ± 37.0[c]

[a]The 9000Xg supernatant of liver homogenates was used for the assays (see text). Values are based on wet weight of tissues and are the means ± SEM of 4-5 rats.
[b]Young mature male Osborne-Mendel rats were fed chow diets containing the indicated levels of PCB, as Aroclor 1254 for 2 mo.
[c]Significantly different from 0 ppm PCB group ($p < 0.05$).

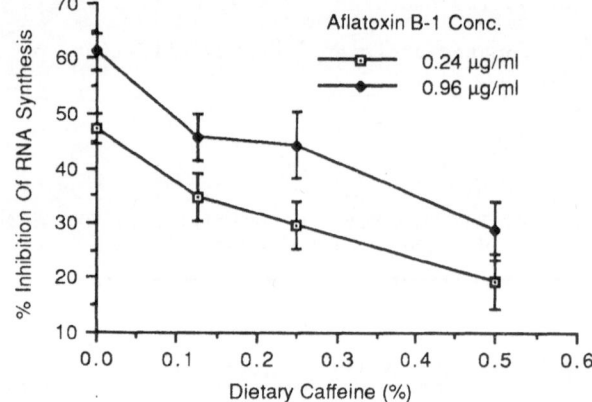

Fig. 5. Effect of dietary caffeine on inhibition of RNA synthesis in liver slices by AFB1 added *in vitro*. Young Sprague-Dawley male rats were fed chow diets containing 0, 0.125, 0.250 or 0.500% caffeine for 3 mo. Incubation conditions and analytical procedures for determination of RNA synthesis described in Methods. Results are expressed as percent of control and each point represents the mean ± SE (vertical bars) for 6 rats. Inhibition of RNA synthesis by both levels of AFB1, in the slices from control group, is significantly different ($p < 0.05$) from the inhibition obtained with slices from all groups of caffeine treated rats.

the livers of rats in such experiments.[29] Assumably, as was postulated by Appleton et al.,[4] diethylmaleate exerts its *in vivo* effect on the activity of AFB1 via another mechanism - possibly by modification of its metabolic activation. As a matter of fact, diethylmaleate failed to exert an effect on AFB1's binding to DNA or its toxicity when added to cultured hepatocytes.[39]

Liver slices for use in *in vitro* studies represent a better physiological

Table 17. Effect of Administering DDT and SKF 525-A (S) to Rats on *in vitro* Metabolism of AFB1 and its *in vivo* Inhibition of Hepatic RNA Synthesis.[a]

Group	RNA Synthesis[b] (DPM (R)/DPM (L)/ mg RNA x 10^3)	%I	Aflatoxin B1 Metabolism (ng M1/g/30 min)	% Effect
		Experiment I		
Control	5.92 ± 0.40^d		572 ± 98	
+ B1	2.07 ± 0.14^e	65.03		
DDT	4.26 ± 0.16		882 ± 51^f	+ 54.2
+ B1	2.13 ± 0.19^e	50.00		
		Experiment II		
Control	10.38 ± 0.13		2215 ± 413	
+ B1	6.69 ± 0.57^e	35.55		
DDT	10.12 ± 0.55		4475 ± 690^f	+102.0
+ B1	6.54 ± 0.43^e	35.38		
		Experiment III		
Control	7.78 ± 0.51		1290 ± 84	
+ B1	5.90 ± 0.69^e	24.16		
S-40 mg/kg	6.91 ± 0.34		776 ± 77^f	- 39.8
+ B1	5.29 ± 0.22	23.44		
S-80 mg/kg	7.64 ± 0.36		680 ± 52^f	- 47.3
+ B1	5.58 ± 0.63^e	26.96		

[a]4-6 young male Osborne-Mendel rats/group. Exp. I and II, rats either administered DDT 100 mg/kg body weight once/day for 3 days or diluent corn oil (4 ml/kg), ip. Exp. III, rats administered either SKF 525-A or the diluent saline (1 ml/kg one hr prior to sacrifice (metabolism) or to dosing with AFB1. Aflatoxin (0.33 mg/kg, Exp. I and II and 0.25 mg/kg for Exp. III), administered ip 5 hr prior to sacrifice.
[b][^{14}C]orotic acid (25 μCi/kg) administered ip one hr prior to sacrifice.
[c]9000Xg supernatant of liver homogenates was used for the assay (see text). Values are based on the wet weight of the tissue.
[d]Mean $\pm$ SEM.
[e]Significantly different from non-aflatoxin-treated group ($p < 0.05$).
[f]Significantaly different from non-treated control group ($p < 0.05$).

model than isolated hepatocytes.[70] Liver slices from male mice, monkeys, and humans (tissue from 51- and 66-year old males, organ-donor derived) were used and the results compared with slices from male rats. The effects of a range of AFB1 concentrations on RNA synthesis is shown in Figure 6. As expected livers from an alfatoxin resistant species, mice,[79] were affected to a much smaller degree than were the livers from rats, an aflatoxin sensitive species.[79] Livers from trout, also an aflatoxin sensitive species,[79] responded no differently than rat liver slices. Good agreement between the relative degree of binding of [^{14}C]AFB1 to macromolecular fractions and sensitivity to aflatoxin-induced inhibition of RNA synthesis was found among the three species tested (Figure 7). There was agreement between *in vitro* binding data and data obtained previously by measuring the *in vivo* binding of AFB1 to liver macromolecules of resistant and susceptible species.[75] Liver slices from monkeys and humans were highly resistant to the effects of AFB1 relative to the effects seen with the rat liver slices. While the source of the human livers was from individuals of late-middle age, a recent report indicates that in rats sensitivity to aflatoxin-DNA binding increases during senescene.[60] Application of other *in vitro* systems used for comparing species sensitivity to AFB1 produced results, as reviewed by Stoloff,[71] which generally concur with these findings. The parameters measured with these systems were AFB1 binding to DNA and mutagenesis, using liver homogenates as activating systems. A recent study, however, reporting on the results of measuring AFB1 binding to the DNA of hepatocytes of rats, mice, and humans (3 cases), indicated the highest degree of binding with rats, the least with mice; human cells occupyied an intermediate position.[14] Another very recent study using nucleolar segregation in cultured hepatocytes as the criteria of toxicity gave similar results, but the variability among samples from the five human cases was great.[13] Eaton and Monroe reported at the current conference that humans are potentially less susceptible to AFB1 carcinogenicity than rats, but more than mice based on *in vitro* biotransformation data, particularly the data showing relatively low amounts of glutathione transferase activity in human liver.[24] In general there is agreement on the probable susceptibility of human liver to AFB1 carcinogenicity relative to that of the

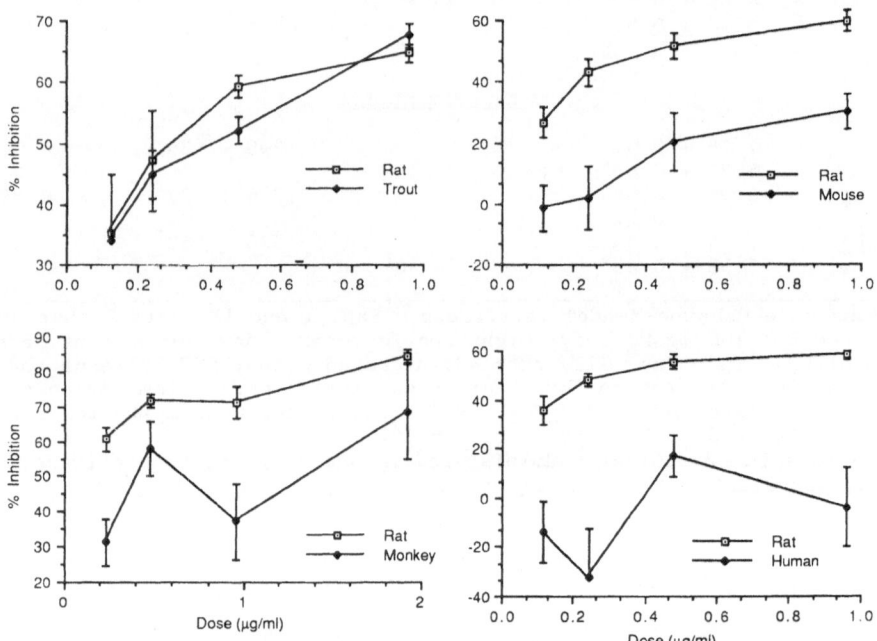

Fig. 6. Comparison of *in vitro* sensitivity to inhibition of RNA synthesis by AFB1 among liver slices from rats, mice, rainbow trout, rhesus monkeys, and humans. Incubation conditions and analytical procedures for determination of RNA synthesis described in Methods (incubation temperature for trout liver slices=18°C). Except for the experiment with human tissue where livers from two humans (organ donors) and two rats were used, a minimum of 5 animals of each species were used per experiment. Results are expressed as percent of control and each point represents the mean ± SE (vertical bars) for the number of total replicates as follows: Rat vs trout, 12-32 each; rat vs mouse, 12 each; rat vs monkey, 7-17 each; rat vs human, 5-6 each. Regression analysis and analysis of variance indicate statistical differences (p < 0.05) between each of the pairs.

rat (based on the referred to in vitro studies and epidemiologic evidence[72]), but the relationship of humans to mice in this regard is not quite clear.

Patulin

We[58] and others[37,66] have reported on the in vitro inhibition of macromolecular biosynthesis by patulin and Hatey and Moule have demonstrated in vivo inhibition of hepatic protein synthesis in rats by this mycotoxin.[38] However, in the latter case the patulin was administered intraperitoneally and thus the significance of the finding in regard to the potential toxicity of the toxin to humans would be questionable. We therefore administered patulin to male and female rats by oral gavage and measured RNA and protein synthesis in vivo two hours later. The results show (Table 20) that protein synthesis was inhibited at all levels (4.5 to 18.0 mg/kg) tested, both in the liver and kidneys of the animals. In another experiment, polysome profiles were determined in the livers of rats after oral administration of patulin. At a dose of 18 mg/kg, patulin produced a marked disaggregation of the hepatic polyribosomes (Figure 7). Since we failed to find an in vivo effect of patulin on RNA synthesis (data not shown), the effect on the polysomes and protein synthesis indicated a direct effect of patulin (or its metabolites) on translation and probably on the initiation process. These results are of interest for at least two reasons. First, because of the extreme reactivity of patulin with nucleophiles, especially glutathione, patulin probably could not reach the liver in a free, unconjugated state. In fact, Dailey reported that when oral doses of [14C]-patulin were used, none of the unmetabolized compound was detected in the blood, urine or tissues of the treated rats.[18] Secondly, none of the long term feeding studies with patulin have revealed any indications of specific overt effects on the tissues of treated animals in the form of necrotic, cancerous or other types of pathologic lesions.[19,55] This may point to threshold effects never being attained in the long-term

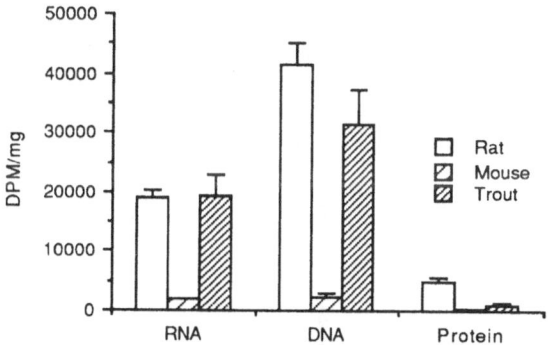

Fig. 7.   Comparison of in vitro binding of [14C]-AFB1 to RNA, DNA, and protein of liver slices from rats, mice and rainbow trout. Each replicate beaker contained 20 mg of liver protein and 0.65 μg AFB1 (856,125 DPM). Temperatures of incubation were 37°C for rats and mice, and 18°C for trout. Incubations conditions and analytical procedures described in Methods. Results are expressed as DPM/mg (adjusted for differeing macromolecular levels of liver tissue among the three species) and each bar represents the mean ± SE of three replicates (using pooled tissue from non-fasted animals).

Table 18. Influence of $CoCl_2$ Treatment on Inhibition of Macromolecular Synthesis by AFB1[a] in Livers of Young Adult Male Rats.

Treatment	RNA (DPM (R)/DPM (L)/ mg RNA x 10^3)	%I	Protein (DPM/mg protein)	%I
Control	11.42 ± 0.72[b]	53.2	1226 ± 41	33.2
Control + B1	5.35 ± 0.45		819 ± 53	
CoCl2	14.89 ± 0.65	19.0	1042 ± 22	9.9
CoCl2 + B1	12.06 ± 0.29		938 ± 30	

[a]14-16 Sprague-Dawley rats (375-436 g) administered either $CoCl_2 \cdot 2H_2O$, 6 mg/kg body weight or the diluent saline, ip, daily for 2 days. On third day half rats were administered AFB1, 0.33 mg/kg, 4 hr prior to ip dosing with [^{14}C]orotic acid (250 μCi/kg) and [^{3}H]leucine (100 μCi/kg) and 5 hr prior to sacrifice.
[b]Mean ± SEM.
All values for aflatoxin treated rats significantly different from values for corresponding control group ($p < 0.05$).

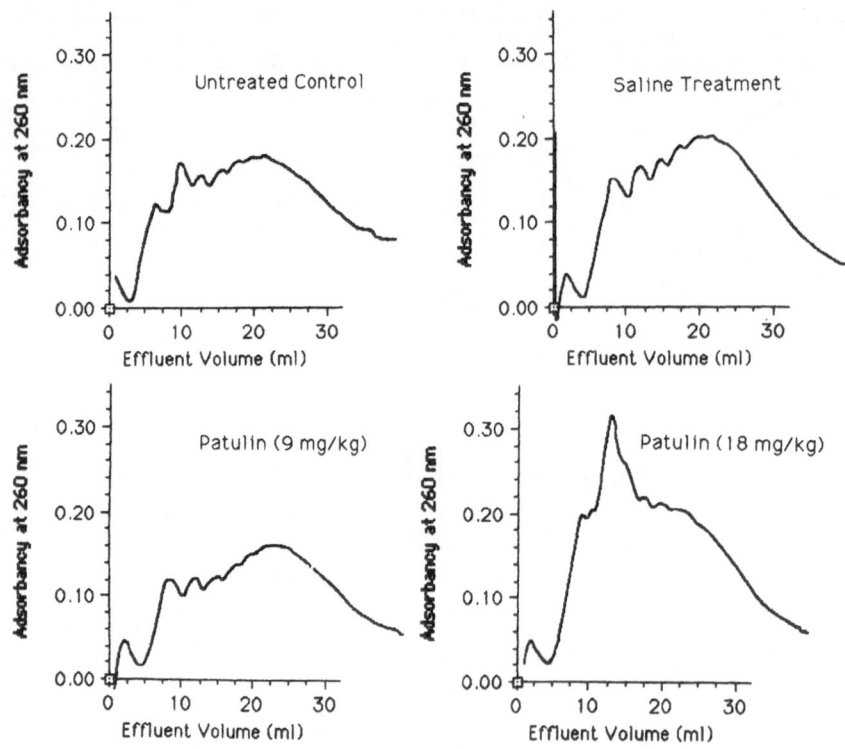

Fig. 8. Sucrose density pattern of polyribosomes prepared from livers of control, diluent (0.9% NaCl) and patulin-treated male (4.5 mo old, Sprague-Dawley) rats. Treated rats were given the saline (1 ml/kg body weight) or 9 or 18 mg of patulin/kg of body weight by gastric intubation 2 hr before sacrifice. Tracings are representative profiles from 3 experiments that were carried out. Heaviest particles or bottom of gradient are on right of each pattern.

Table 19. Influence of Glutathione Depletion by Diethylmaleate (DEM) on Inhibition of Liver RNA and Protein Synthesis by AFB1.[a]

Treatment		Inhibition of Synthesis		Glutathione
DEM (mg/kg)	AFB1 (mg/kg)	RNA (%)	Protein (%)	Decrease (%)
Experiment I - Female Rats				
0	0.50	35.3	----	
600	0.50	49.3 (39.7%)[b]	----	39.2[c]
Experiment II - Male Rats				
0	0.25	24.0	+ 2.5	
600	0.25	33.0 (37.5%)	34.8	16.0
Experiment III - Male Rats				
0	0.16	17.0	----	
1000	0.16	40.3 (137%)	----	44.7
Experiment IV - Male Rats				
0	0.25	29.2	----	
1000	0.25	41.1 (40.8%)	----	84.3

[a]7-8 young adult rats/group (Sprague-Dawley-Exp. I; Osborne-Mendel-Exp. II and III; 4-5 for Exp. IV) administered DEM one hr prior to aflatoxin, both ip. Rats were sacrificed 5 hr after aflatoxin and one hr after ip dosing with [^{14}C]-orotic acid (25 μCi/kg) and for Exp. II, [^{3}H]-leucine (100 μCi/kg).
[b]Numbers in parentheses indicate percent increase in aflatoxin's effect due to DEM treatment. All inhibition effects due to aflatoxin and depression of glutathione levels due to DEM are significant (p < 0.05).
[c]Range of control values for glutathione = 4.20-4.92 μmoles/g liver.

feeding studies or most probably compensating responses, such as induction of detoxifying enzymes, e.g., glutathione transferases, taking place. In any event, unlike with the aflatoxins, this is a case where major biochemical lesions themselves may not constitute appropriate predictors or indices of subsequent tissue lesions.

CONCLUSIONS AND FUTURE RESEARCH

Currently we are completing a series of studies designed to assess possible interaction among mycotoxins found together in foods and feeds using in vitro and in vivo techniques. We are also preparing to use more selective techniques to measure changes in specific proteins or m-RNAs.

However, based upon our many years of work using measurements of macromolecular biosynthesis to study the effects of toxins on animal tissue, we conclude the following:

Measurements of inhibition of macromolecular biosynthesis when used in

Table 20. Effects of Orally Administered Patulin on Protein Synthesis in
           Livers of Male Rats and Livers and Kidneys of Female Rats.[a]

Patulin (mg/kg)	Protein Synthesis (DPM (P)/DPM (T)/mg Protein x 10^3)		
		Female	
	Male Livers	Livers	Kidneys
0	8.476 ± 0.926[b]	20.023 ± 1.077	10.977 ± 0.501
4.5	5.570 ± 0.456 (34.29)[c]	15.319 ± 1.294 (23.49)	8.986 ± 0.751 (18.14)
9.0	4.765 ± 0.175 (42.34)	9.956 ± 0.539 (50.28)	5.809 ± 0.588 (47.08)
18.0	2.821 ± 0.173 (66.72)	8.523 ± 1.113 (57.43)	3.836 ± 0.229 (65.05)

[a]7-8 Osborne-Mendel rats/group sacrificed 3 hr after oral intubation with patulin (dissolved in 0.9% NaCl) and one hr after ip dose of [^{3}H]-leucine (50 μCi/kg body weight).
[b]Mean $\pm$ SEM.
[c]Numbers in parentheses = % inhibition due to patulin treatment.
All values for patulin-treated rats significantly different from values for corresponding control rats ($p < 0.05$).

conjunction with some other measurements of biochemical function such as mitochondrial respiratory activity, transport across and perturbation of membranes, or uptake, metabolism and binding of the toxin can be practically and rationally used to study interactions between toxins and other chemicals, diet, or different physiological states. Information leading to an under-standing of mechanism of toxicity will result. These studies can be useful in extrapolating toxicity data from animals to man under normal conditions and under conditions to which man is exposed, including stresses resulting from illness, lifestyle changes, or environmental conditions.

ACKNOWLEDGEMENTS

     The authors are grateful to the following for their expertise and assistance rendered in the conduct of the studies: Drs. Lillian W. Yinn, Robert Eppley, George C. Yang and Emilio A. Brouwer, James E. Keys, John Scalera, Charles L. Stone, Charles W. Thorpe, Matthew C. Smith, Kathy Davis, Lind D. Marra, Cynthia I. Stanislawski and Tommie L. Watkins, and to Katherine S. Hogye for statistical analysis of the portion of the data dealing with measurements of species differences. Special appreciation is given to Mrs. Dolores A. Price and Ms. Joan B. Shields for their patience and skill displayed in the typing of this manuscript, to Drs. Eugene Miller and Leonard Stoloff for their consultation and scientific review of this manuscript and to Mrs. Diana B. Friedman for valuable assistance in its preparation.

REFERENCES

1. Adekunle AA, Hayes JR, Campbell TC, 1977, Interrelationships of dietary protein level, aflatoxin B$_1$ metabolism, and hepatic microsomal epoxide

hydrase activity, *Life Sci.*, 21:1785-1792.

2. Alozie TC, Bassir O, 1979, Effect of diet on the lipid-depressing activity of aflatoxin B_1 in the Nigerian monkey, *Biochem Pharmacol.*, 28:2397-2398.

3. Appleton BS, Campbell TC, 1983, Effect of high and low dietary protein on the dosing and post dosing periods of aflatoxin B_1-induced hepatic preneoplastic lesion development in the rat, *Cancer Res.*, 43:2150-2154.

4. Appleton BS, Goetchins MP, Campbell TC, 1982, Linear dose-response curve for the hepatic macromolecular binding of aflatoxin B_1 in rats at very low exposure, *Cancer Res.*, 42:3659-3662.

5. Baldwin S, Parker RS, 1987, Influence of dietary fat and selenium in initiation and promotion of aflatoxin B_1-induced preneoplastic foci in rat liver, *Carcinogenesis*, 8:101-107.

6. Bassir O, Alozie TC, 1979, The effect of dietary fat on the anticoagulent activity of aflatoxin B_1, *Experientia*, 35:1087-1089.

7. Bitlack WR, Brown RC, Chandra C, 1986, Nutritional parameters that alter hepatic drug metabolism, conjugation, and toxicity, *Fed Proc.*, 45:142-148.

8. Butler LE, Dauterman WC, 1988, The effect of dietary protein levels on xenobiotic biotransformation in F344 male rats, *Toxicol Appl Pharmacol.*, 83:271-278.

9. Chedid A, Bundeally AE, Mendenhall CL, 1977, Inhibition of hepatocarcinogenesis by adrenocorticotropin in aflatoxin B_1-treated rats, *J Nat'l Cancer Inst.*, 58:339-349.

10. Chiao FF, Kado NY, Hsieh LS, Ruebner BH, Hsieh DPH, 1987, Effect of age on the metabolic activation and DNA binding of aflatoxin B_1 in Fischer rats, *The Toxicologist*, 7:34.

11. Ch'ih JJ, Wu JL, Feng H-W, 1980, Inhibition of poly(A)$^+$- and poly(a)$^-$-ribonucleic acid synthesis by a nonlethal dose of aflatoxin B_1 *in vivo*, *Biochem Pharmacol.*, 29:2247-2249.

12. Clifford JI, Rees KR, Stevens MEM, 1967, The effect of the aflatoxins B_1, G_1 and G_2 on protein and nucleic acid synthesis in rat liver, *Biochem J.*, 103:258-261.

13. Cole KE, Jones TW, Lipsky MM, Trump BF, Hsu I-C, 1989, Comparative effects of three carcinogens on human, rat and mouse hepatocytes, *Carcinogenesis*, 10:139-143.

14. Cole KE, Jones TW, Lipsky MM, Trump BF, Hsu I-C, 1988, *In vitro* binding of aflatoxin B_1 and 2-acetylaminofluorene to rat, mouse and human hepatocyte DNA: the relationship of DNA binding to carcinogenicity, *Carcinogenesis*, 9:711-716.

15. Coleman EC, Friedman L, 1971, Fatty acids in tissue lipids of rats fed *Sterculia foetida* oil, *J Agr Food Chem.*, 19:224-228.

16. Conney AN, 1971, Environmental factors influencing drug metabolism, *In*: "Fundamentals of Drug Metabolism and Drug Disposition," LaDu BN, Mandel HG, Way EL, eds., William and Wilkins Co., Baltimore.

17. Croy RG, Essignman JM, Reinold VN, Wogan GN, 1978, Identification of the principal aflatoxin B_1-DNA adduct formed *in vivo* in rat liver, *Proc Nat'l Acad Sci USA*, 75:1745-1749.

18. Dailey RE, Blashka AM, Brouwer EA, 1977, Absorption, distribution and excretion of [^{14}C] patulin by rats, *J Toxicol Environ Health*, 3:479-489.

19. Dailey RE, Brouwer E, Blashka AM, Reynaldo EF, Green S, Monlux WS, Ruggles DI, 1977, Intermediate-duration toxicity study of patulin in rats, *J Toxicol Environ Health*, 2:713-725.

20. Decad GM, Dougherty KK, Hsieh DPH, Byard JL, 1979, Metabolism of aflatoxin B_1 in cultured mouse hepatocytes: Comparison with rat and effects of cyclohexene oxide and diethyl maleate, *Toxicol Appl Pharmacol.*, 50:429-436.

21. Degen GH, Newmann H-G, 1981, Differences in aflatoxin B_1-susceptibility of rat and mouse are correlated with the capability *in vitro* to inactivate aflatoxin B_1-epoxide, *Carcinogenesis*, 2:299-306.

22. Degen GH, Newmann HG, 1978, The major metabolite of aflatoxin B_1 in the

rat is a glutathione conjugate, *Chem Biol Interact.*, 22:239-255.

23. Dunaif GE, Campbell TC, 1987, Relative contribution of dietary protein level and aflatoxin B_1 dose in generation of presumptive preneoplastic foci in rat liver, *J Nat'l Cancer Inst.*, 78:365-369.

24. Eaton D, Monroe D, 1988, Biotransformation as a determinant of species susceptibility to aflatoxin carcinogenesis, Symposium on Cellular and Molecular Mode of Action of Selected Microbial Toxins in Food and Feed, 23rd Annual Meeting of the UJNR Panel on Toxic Microorganisms, Oct. 31-Nov. 2, 1988, Chevy Chase, MD (Abstract).

25. Edwards GS, Wogan GN, 1970, Aflatoxin inhibition of template activity of rat liver chromatin, *Biochm et Biophys Acta*, 224:597-607.

26. Friedman L, 1980, Biochemical effect of methyxanthaines (a review), *FDA By-Lines*, 10:78-103.

27. Friedman L, Mohr H, 1968, Absence of interaction between cyclopropenoid fatty acids and aflatoxin in rats, *Fed Proc.*, 27:551.

28. Friedman L, Scalera J, Keys JE, Peters EL, Gaines DW, Stone CL, Kasza L, Balazs T, 1982, Some biochemical and histological effects of 2-chloro-ethanol in rats, 1982, *J Am Coll Tox.*, 1:37-55.

29. Friedman L, Yin L, 1973, Influence of hypophysectomy on the biochemical effects and metabolism of aflatoxin B_1 in rats, *J Nat'l Cancer Inst.*, 51:479-487.

30. Friedman L, Yin L, Verrett J, 1972, A new toxic fluorescent metabolite derived from the interaction of rat liver and aflatoxin B_1, *Toxicol Appl Pharmacol.*, 23:385-390.

31. Gelboin HV, Wortham JS, Wilson RG, Friedman M, Wogan GN, 1966, Rapid and marked inhibition of rat liver RNA polymerase by aflatoxin B_1, *Science*, 154:1205-1206.

32. Goodall CM, 1968, Endocrine factors as determinanats of the susceptibility of the liver to carcinogenic agents, *NZ Med J.*, 67(Suppl.):32-43.

33. Goodall CM, Butler WH, 1969, Aflatoxin carcinogenesis: Inhibition of liver cancer induction in hypophysecctomized rats, *Int J Cancer*, 4:422-429.

34. Gurtoo HL, Motycka L, 1976, Effect of sex difference on the *in vitro* and *in vivo* metabolism of aflatoxin B_1 by the rat, *Cancer Res.*, 36:4663-4771.

35. Gurtoo HL, Dahms RP, 1979, Effect of inducers and inhibitors on the metabolism of aflatoxin B_1 by rat and mouse, *Biochem Pharmacol.*, 28:3441-3449.

36. Hart LG, Fouts JR, 1965, Further studies on the stimulation of hepatic microsomal drug metabolizing enzymes by DDT and its analogs, *Arch Exp Pathol Pharmakol.*, 249:486-500.

37. Hatey F, Gaye P, 1978, Inhibition of translation in reticulocyte lysate by the mycotoxin patulin, *FEBS Letters*, 95:252-255.

38. Hatey F, Moule Y, 1979, Protein synthesis inhibition in rat liver by the mycotoxin patulin, *Toxicology*, 13:223-231.

39. Hayes MA, Murray CA, Rushmore TH, 1986, Influences of glutathione status on different cytocidal responses of monolayer rat hepatocytes exposed to aflatoxin B_1 or acetaminophen, *Toxicol Appl Pharmacol.*, 85:1-10.

40. Hendricks JD, Putnam TP, Bills DD, Sinnhuber RO, 1977, Inhibitory effect of a polychlorinated biphenyl (Aroclor 1254) on aflatoxin B_1 carcinogenesis in rainbow trout (*Salmo gairdneri*), *J Nat'l Cancer Inst.*, 59:1545-1551.

41. Hendricks JD, Putnam TP, Sinnhuber RO, 1979, Effect of dietary dieldrin on aflatoxin B_1 carcinogenesis in rainbow trout (*Salmo gairdneri*), *J Environ Path Toxicol.*, 2:719-728.

42. Kumagai S, 1989, Intestinal absorption and excretion of aflatoxin in rats, *Toxicol Appl Pharmacol.*, 97:88-97.

43. Kumagai S, Ito Y, Iwaki M, Hagahara A, Omata Y, 1988, Subcellular distribution and macromolecular binding of aflatoxin in fetal and neonatal rat liver, Symposium on Cellular and Molecular Mode of Action of Selected Microbial Toxins in Food and Feed, 23rd Annual Meeting of the UJNR Panel

on Toxic Microorganisms, Oct. 31-Nov. 2, 1988, Chevy Chase, MD (Abstract).

44. Lin JK, Miller JA, Miller EC, 1977, 2,3-Dihdro-2(Guan-7-yl)-3-hydroxy aflatoxin B_1, A major acid hydrolysis product of aflatoxin B-DNA or ribosomal RNA adducts formed in hepatic microsome mediated reactions and in rat liver *in vivo*, *Cancer Res.*, 37:4430-4438.

45. Madhavan TV, Gopalan C, 1968, The effect of dietary protein on carcinogenesis of aflatoxin, *Arch Path.*, 85:133-137.

46. Marshall WJ, McLean AEM, 1971, A requirement for dietary lipids for induction of cytochrome P-450 by phenobarbitone in rat liver microsome fraction, *Biochem J.*, 122:569-577.

47. Marzuki A, Norred WP, 1984, Effects of saturated and unsaturated dietary fat on aflatoxin B_1 metabolism, *Fd Chem Toxicol.*, 22:383-389.

48. Marshaly RI, Habib SL, Salem MH, Deeb E, Safwat MM, Sarhan F, 1988, *In vitro* effect of aflatoxin B_1 on the transcriptional activity of DNA template, chromatin and soluble DNA-dependent RNA polymerase in buffalo liver, *Tox Lett.*, 41:69-75.

49. McBodile MUK, Holscher M. Neal RA, 1975, A possible protective role for reduced glutathione in aflatoxin B_1 toxicity: Effect of pretreatment of rats with phenobarbital and 3-methylcholanthrene on aflatoxin toxicity, *Toxicol Appl Pharmacol.*, 34:128-142.

50. Monroe DH, Eaton DL, 1987, Comparative effects of butylated hydroxyanisole on hepatic *in vivo* DNA binding and *in vitro* biotransformation of aflatoxin B_1 in the rat and mouse, *Toxicol Appl Pharmacol.*, 90:401-409.

51. Moule Y, Lesage V, Darracq N, Rousseau N, 1975, Opposite effects of phenobarbital pretreatment on aflatoxin B_1-induced inhibition of transcription in rat and mouse liver, *Biochem Pharmacol.*, 24:1851-1854.

52. Murthy MRV, Radouco-Thomas S, Bharucha AD, Levesque G, Sithian P, Radouco-Thomas C, 1985, Effects of trichothecenes (T-2 toxin) on protein synthesis *in vitro* by brain polysomes and messenger RNA, *Prog Neuro-Psycopharmacol & Biol Psychiat.*, 9:251-258.

53. Newberne PM, Weigert J, Kula N, 1979, Effects of dietary fat on hepatic mixed-function oxidases and hepatocellular carcinoma induced by aflatoxin B_1 in rats, *Cancer Res.*, 39:3986-3991.

54. Newberne PM, Williams G, 1969, Inhibition of aflatoxin carcinogenesis by diethylstilbesterol in male rats, *Arch Environ Health*, 19:489-498.

55. Nixon JE, Sinnhuber RO, Lee DJ, Landers MK, Harr JR, 1974, Effect of cyclopropenoid compounds on the carcinogenic activity of diethylnitrosamine and aflatoxin B_1 in rats, *J Nat'l Cancer Inst.*, 53:453-458.

56. Osswald H, Frank HK, Komitowski D, Winter H, 1978, Long-term testing of patulin administered orally to Sprague-Dawley rats and Swiss mice, *Fd Cosmet Toxicol.*, 16:243-247.

57. Pessayre D, Dolder A, Aftigon J-Y, 1979, Effect of fasting on the metabolite-mediated hepatoxicity in the rat, *Gastroenterology*, 77:264-271.

58. Peters EL, Keys JE, Friedman L, 1977, Comparative *in vitro* biochemical effects of patulin and aflatoxin B_1, *Fed Proc.*, 36:397.

59. Prasanna HR, Gupta SR, Viswanathan L. Venkitasubramanian TA, 1978, Comparative study of the effects of aflatoxin B_1 metabolites and α-amanitin on rat liver RNA polymerase and chromatin template activities, *Toxicon*, 16:289-294.

60. Prasanna HR, Lotlikar PD, Hacobian N, Magee PN, 1986, Differential effects on the metabolism of dimethylnitrosamine and aflatoxin B_1 by hepatic microsomes from senescent rats, *Cancer Letters*, 33:259-267.

61. Preston RS, Hayes JR, Campbell TC, 1976, The effect of protein deficiency on the *in vivo* binding of aflatoxin B_1 to rat liver macromolecules, *Life Sci.*, 19:1191-1198.

62. Righter HF, Shalkop WT, Mercer HD, Leffel E, 1972, Influence of age and sexual status on the development of toxic effects in the male rat fed aflatoxins, *Toxicol Appl Pharmacol.*, 21:435-439.

63. Rowe L, Wills ED, 1976, The effect of dietary lipids and vitamin E on lipid peroxide formation, cytochrome P 450 and oxidative demethylation in the endoplasmic reticulum, *Biochem Pharmacol.*, 25:175-179.

64. Ruebner BH, Cullen J, Hsieh L, Hsieh DHP, 1986, Carcinogenicity of afla-toxin B$_1$ in infant rats: Effect of phenobarbital, *Fed Proc.*, 45:956.
65. Sarasin A, Moule Y, 1973, Inhibition of *in vitro* protein synthesis by aflatoxin B$_1$ derivatives, *FEBS Letters*, 32:347-350.
66. Schaeffer WI, Smith NE, Payne PA, Wilson DM, 1975, Physiological and bio-chemical effects of the mycotoxin patulin on Chang liver cell cultures, *In Vitro*, 11:69-77.
67. Schmidt G, Thannhauser SJ, 1945, A method for the detection of deoxyribo-nucleic acid, ribonucleic acid and phosphoproteins in animal tissues, *J Biol Chem.*, 161:83-89.
68. Schneider WC, 1957, Determination of nucleic acids in tissues by pentose analysis, *In:* "Methods in Enzymology," Colowick SP, Kaplan NO, eds., Academic Press, Inc., New York.
69. Sinnhuber RO, Lee DJ, Wales J, Ayres JL, 1968, Dietary factors and hepa-toma in rainbow trout (*Salmo gairdneri*), II. Cocarcinogenesis by cyclo-propenoid fatty acids and the effect of gossypol and altered lipids on aflatoxin-induced liver cancer, *J Nat'l Cancer Inst.*, 41:1293-1301.
70. Smith PF, Fisher R, Shubat PJ, Gandolfi AJ, Krumdieck CL, Brendel K, 1987, *In vitro* cytotoxicity of allyl alcohol and bromobenzene in a novel organ culture system, *Toxicol Appl Pharmacol.*, 87:509-522.
71. Stoloff L, 1986, A rationale for the control of aflatoxin in human foods, *In:* "Mycotoxins and Phycotoxins," Steyin PS, Vleggaar R, eds., Elsevier Science Publishers, R.V., Amsterdam.
72. Stoloff L, Friedman L, 1976, Information bearing on the evaluation of the hazard to man from aflatoxin ingestion, *PAG Bull.*, 6:21-32.
73. Strother A, Throckmorton JK, Herzer C, 1971, The influence of high sugar consumption by mice on the duration of action of barbiturates and *in vitro* metabolism of barbituates, aniline and p-nitroanisole, *J Pharmacol Exp Ther.*, 179:490-498.
74. Tephly TR, Hibbeln P, 1971, The effect of cobalt chloride administration on the synthesis of hepatic microsomal cytochrome P-450, *Biochem Biophys Res Commun.*, 42:589-595.
75. Ueno I, Friedman L, Stone CL, 1980, Species difference in the binding of aflatoxin B$_1$ to hepatic macromolecules, *Toxicol Appl Pharmacol.*, 52:177-180.
76. Vesselinovitch SD, Mihailovich N, Wogan GN, Lombard LS, Rao KVN, 1972, Aflatoxin B$_1$. A hepatocarcinogen in the infant mouse, *Cancer Res.*, 32:2289-2291.
77. Wells P, Aftergood L, Alfin-Slater RB, 1975, Effect of dietary fat upon aflatoxicosis in rats fed Torula yeast containing diet, *J Am Oil Chem Soc.*, 52:139-143.
78. Wells P, Aftergood L, Alfin-Slater RB, 1975, Effect of varying levels of dietary protein on tumor development and lipid metabolism in rats exposed to aflatoxin, *J Am Oil Chem Soc.*, 53:559-562.
79. Wogan G, 1973, Aflatoxin carcinogenesis, *In:* "Methods in Cancer Research," Busch H, ed., Academic Press, New York.
80. Wogan, GN, 1974, FDA Contract No. 223-74-2154, Massachusetts Institute of Technology.
81. Wogan GN, Newberne PM, 1967, Dose-response characteristics of aflatoxin B$_1$ carcinogenesis in the rat, *Cancer Res.*, 27:2370-2376.
82. Yu R-L, 1977, Mechanism of aflatoxin B$_1$ inhibition of rat hepatic nuclear RNA synthesis, *J Biol Chem.*, 252:3245-3251.

USE OF AFLATOXIN-DNA AND PROTEIN ADDUCTS FOR HUMAN DOSIMETRY

John D. Groopman

Department of Environmental Health Sciences
Johns Hopkins School of Hygiene and Public Health
615 North Wolfe Street
Baltimore, MD 21205

INTRODUCTION

The last twenty-five years has seen extensive efforts to investigate the association between aflatoxin exposure and human liver cancer. Studies using standard epidemiological methods have been hindered by the lack of adequate dosimetry data on aflatoxin intake, excretion and metabolism, as well as by the general poor quality of world-wide cancer morbidity and mortality statistics. Despite these difficulties, the aflatoxins are among the few strucrally identified environmental carcinogens for which quantitative risk assessments have been attempted. These efforts have spurred a number of investitors, in the last few years, to develop reliable, fast and accurate techniques to assess individual human exposure to this carcinogen.

This article will discuss many of the classical and molecular epidemiological studies used to associate dietary aflatoxin exposure with human liver cancer and cited by IARC in its reclassification of aflatoxin B_1 (AFB1) to a Category I carcinogen (IARC, Suppl. 7). The literature on the toxicology of aflatoxins has been extensively covered by Busby and Wogan[3] and reviews focused on biological monitoring and epidemiological considerations have been published by Groopman et al.[10] and Bosch and Munoz.[1] Well documented in this literature is that biological risk of exposure to aflatoxins is much lower in technologically developed countries than in developing ones. Parenthetically, prevention of dietary exposure to aflatoxins will improve the general health status of a developing nation's population. In fact, limiting intake of aflatoxins is a readily obtainable public health goal only requiring the allocation of appropriate economic resources to assure minimal mold contamination of foods and grains.

The development of a human tumor is modulated by many factors, both biological and chemical in nature. Since initiation, promotion and progression-like events are required prior to clinical diagnosis of a tumor, no one agent can be responsible or present at all critical stages during the growth of a tumor. However, by developing methods to permit the monitoring of an individual's aflatoxin induced genotoxic burden, the identification of people at high risk for developing disease long before clinical manifestation could be accomplished. Herein, particular attention will be paid to the current state of monitoring individual exposure to aflatoxins. Dosimetry data obtained using the techniques of monoclonal antibody affinity chromatography,

Microbial Toxins in Foods and Feeds, Edited by A.E. Pohland *et al.*,
Plenum Press, New York, 1990

polyclonal sera based enzyme linked immunoassay, synchronous fluorescence
spectroscopy and other immunological methods will be addressed. The technological advances of the last three years alone indicate that great strides
will be made in the near future to answer not whether aflatoxin is a cause of
human disease but how resources can be apportioned to prevent exposure to this
human toxin.

Aflatoxin and Human Liver Cancer Epidemiological Studies: Recent Studies

Primary liver cancer is one of the leading causes of cancer mortality in
Asia and Africa. In the People's Republic of China, this disease accounts for
120,000 deaths per year with an incidence rate in some areas of the country
approaching 100 cases per 100,000 per year. This malignancy is the third
leading cause of cancer mortality in males behind cancer of the esophagus and
stomach, as reported by the National Cancer Office of the Ministry of Public
Health, P.R.C.[13] In contrast, liver cancer incidence in the United States
is about 0.5 cases per 100,000 per year. Clearly, liver cancer varies world-
wide by at least 100 to 1,000 fold. Several epidemiological studies were conducted fifteen to twenty years ago to obtain information on the relationship
of estimated dietary intake of aflatoxin with the incidence of primary human
liver cancer in different parts of the world. These investigations showed
that increased aflatoxin ingestion from 3 to 222 ng per kg body weight per
day corresponded to increased liver cancer incidence values extending from a
minimum of 2.0 to a maximum of 35.0 cases per 100,000 population per year.
While these early studies could not account for confounding factors such as
hepatitis B (HBV) infection, this information provided a strong motivation to
further investigate the circumstantial relationship between aflatoxin
ingestion and liver cancer incidence.

Within this decade, epidemiologic studies have been published on the cor-
respondence of aflatoxin exposure and liver cancer. In one of these reports,
Bulatao-Jayme et al.[2] compared the dietary intakes of confirmed primary
liver cancer cases in the Phillipines against age-sex matched controls. By
using dietary recall, the frequency and amounts of food items consumed were
calcualted into units of aflatoxin load per day. These calculations revealed
that the mean aflatoxin load per day of the liver cancer cases was 4.5 times
higher than the controls. Alcohol intakes as a risk factor were also analyzed by subjectively allocating the subjects into heavy and light aflatoxin
exposure groups. These researchers combined aflatoxin load and alcohol
intake and determined that a synergistic and statistically significant effect
on relative risk with aflatoxin exposure and alcohol intake occurs. These
findings indicated a direct effect of alcohol, as a concurrent liver damaging
agent, upon aflatoxin consumption, especially among heavy drinkers, as a
probable synergistic factor in liver cancer development.

Van Rensburg[17] and his collaborators studied the occurrence and potential etiologies of hepatocellular carcinoma for the period 1968-74 in the
Province of Inhambane, Mozambique. These incidence rates were compared with
those observed in South Africa among mineworkers migrating from Inhambane.
Food samples were randomly collected and aflatoxin content determined in six
districts of Inhambane as well as from Manhica-Magude, a region of lower
liver cancer incidence. A third set of food samples were taken in Transkei
where an even lower incidence of liver cancer had been recorded. When all of
the calculations were completed, the mean aflatoxin dietary intake values
were significantly correlated to the varied liver cancer rates. These
studies provided evidence for a dose dependent increase of aflatoxin intake
corresponding to increased liver disease.

Peers et al.[14] extended the data base for Africa and published an epidemiological study conducted in Swaziland. The data collected were analyzed
for the relationship between aflatoxin exposure, hepatitis B infection, and

the incidence of liver cell carcinoma. Liver cancer is the most commonly
occurring malignancy among males in Swaziland. The levels of aflatoxin
intake were evaluated in dietary samples from households across the country,
and crop samples taken from farms. The prevalence of hepatitis B markers was
estimated from the serum of blood donors. Liver cancer incidence was recorded for the years 1979-83 through a national system of cancer registration.
Across four broad geographic regions, there was a more than five fold variation in the estimated daily intake of aflatoxin ranging from 3.1 to 17.5 μg.
The proportion of HBV-exposed males was very high, but varied relatively
little by geographic region. However, liver cancer incidence varied over a
five fold range, and was strongly associated with estimated levels of aflatoxin. In an analysis involving ten smaller subregions, aflatoxin exposure
emerged as a more important determinant of the variation in liver cancer
incidence than the prevalence of hepatitis infection.

The epidemiology of liver cancer in China was recently reviewed by Yeh
and Shen.[18] This article describes twenty years of research on the epidemiology of liver cancer in China. Their discussion covers the putative role
of HBV infection, dietary aflatoxin exposure and other potential etiologies
such as polluted drinking water, pesticide exposure, and nitrosamine contamination. One of the most extensive investigations described was done in
Guangxi Province. The staple food of people living in this region is corn
and much of the corn crop was determined to have high levels of aflatoxin.
In the heavily contaminated areas, aflatoxin content in corn ranged from 53.8
to 303 ppb, while the lightly contaminated regions showed aflatoxin levels in
grains of less than 5 ppb. It is important that, in the lightly contaminated
areas, rice was the predominant dietary grain. After five to eight years of
follow-up, liver cancer incidence was determined for these two aflatoxin contamination regions. Several thousand person years were observed for this
study. Those individuals who were HBV surface antigen positive and found to
have heavy aflatoxin exposure had a liver cancer incidence of 649.35 cases
per 100,000 compared with 65.92 cases per 100,000 in aflatoxin lightly contaminated areas. Those people who were HBV surface antigen negative and
eating heavily contaminated aflatoxin diets had a liver cancer rate of 98.57
per 100,000 compared with zero cases detected in the light contaminated
area. While these data indicate a strong interaction, in terms of relative
risk, for those people with HBV surface antigen exposure and aflatoxins in
the diet, these data also indicate that aflatoxin plays a significant risk
role in developing the disease in the absence of HBV infection.

Taken together, these epidemiological results provides the scientific
basis for the reclassification of aflatoxin as a known human carcinogen.
Biologically, the indication that hepatitis virus and aflatoxin can act in a
concerted fashion to increase cancer risk has been clearly shown. However,
aflatoxin exposure in the absence of chronic hepatitis B infection is also
etiologically associated with liver cancer. These findings provide the compelling basis to increase efforts both in hepatitis B virus immunization
programs and in the development of concerted programs to lower dietary aflatoxin exposure as means of lowering human cancer risk.

Aflatoxin Metabolism, DNA, and Protein Adduct Formation: The Molecular Basis for Molecular Epidemiological Studies

The aflatoxins are primarily metabolized by the microsomal mixed function
oxygenase system. These enzymes catalyze the oxidative metabolism of AFB1,
resulting in the formation of various hydroxylated derivatives, as well as an
unstable, highly-reactive epoxide metabolite. Detoxification of AFB1 is
accomplished by enzymatic conjugation of the hydroxylated metabolites with
sulfate or glucuronic acid to form water-soluble sulfate or glucuronide
esters that are excreted in urine or bile. An alternative route for removal
of AFB1 from the organism involves the enzyme-catalyzed reaction of the

327

epoxide metabolite with glutathione and its subsequent excretion in the bile. Some of the known detoxification pathways of AFB1 metabolism have been summarized in Figure 1.

Recently Shimada and Guengerich[16] reported on *in vitro* studies with human liver indicating that the major cytochrome P450 involved in the bioactivation of AFB1 to its genotoxic epoxide derivative is cytochrome P-450NF (P-450NF). This is a previously characacterized protein that also catalyzes the oxidation of nifedipine and other dihydropyridines, quinidine, macrolide antibiotics, various steroids, and other compounds. Evidence was obtained using activation of AFB1 as monitored by umuC gene expression response in *Salmonella typhimurium* TA1535/pSK1002, enzyme reconsitution, immunochemical inhibition, correlation of response with levels of P-450NF, and nifedipine oxidase activity in different liver samples. Liver samples with increasing levels of P-450NF also produced higher amounts of 2,3-dihydro-2-(N^7-guanyl)-3-hydroxyaflatoxin B_1 formed in DNA *in vitro*. Several drugs and conditions are known to influence the levels and activity of P-450NF in human liver, and the activity of the enzyme can be estimated by noninvasive assays. These findings provide another avenue for understanding the molecular basis of aflatoxin induced carcinogenesis.

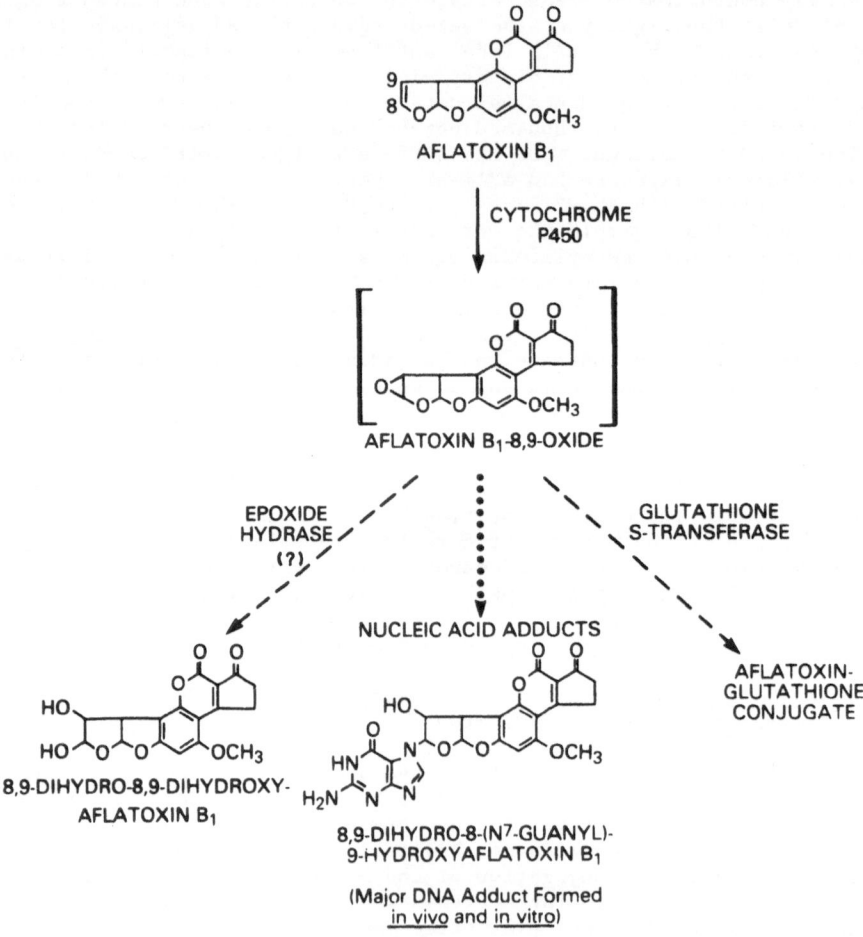

Fig. 1. Metabolic activation of aflatoxin B_1 to form DNA adducts.

The primary AFB1-DNA adduct was identified by Essigmann et al.[5] as 2,3-dihydro-2-(N^7-guanyl)-3-hydroxy-AFB1 (AFB1-N^7-Gua), the major product liberated from DNA modified *in vitro* with AFB1 and its presence was subsequently confirmed *in vivo*.[4] The binding of AFB1 residues to DNA *in vivo* was essentially a linear function of dose. A number of other DNA components including AFB-dihydrodiol, were isolated from nucleic acid hydrolysates activated *in vivo* and *in vitro* with AFB1.[11,12] These adducts were identified as 2,3-dihydro-2-(N^5-formyl-2,5,6-triamino-4-oxopyrimidin-N^5-yl)-3-hydroxy AFB1 (AFB1-FAPyr), a formamidopyrimidine derivative of AFB1-N^7-Gua which contained an opened imidazole ring. Apparently, between 95-98% of the aflatoxin residues bound to DNA have been accounted for by these chemical structures.

The investigation of the interactions and biological consequences of AFB1 with DNA has been an intensive area of study. Because DNA adduct formation is probably a requisite event in the initiation of carcinogenesis by AFB1, the efforts of many investigators to study AFB1-DNA adduct formation is well justified on mechanistic grounds. Despite the suggestions of many years ago, protein carcinogen interactions, sometimes called an epigenetic mechanism of initiation, has not been a research area favored by many laboratories. The interest to develop serum screening methods to assess human exposure to dietary aflatoxins has rejuvenated investigations into the mechanisms of aflatoxin binding to proteins in general, serum proteins in particular and specifically ablumin. Sabbioni et al.[15] have recently elucidated the structure of the major aflatoxin albumin adduct found *in vivo*. The adduct formed with serum albumin proved not to be an analog of the aflatoxin-DNA adduct, but rather an adduct formed by the binding of the epoxide with subsequent formation of the dihydrodiol and sequential oxidation to the dialdehyde and condensation with the epsilon amino group of lysine. This adduct is a Schiff base which undergoes Amadori rearrangement to an alpha-amino ketone (Figure 2). This protein adduct is a completely modified aflatoxin structure retaining only the coumarin and cyclopentenone rings of the parent compound. A human monitoring study will be described in the following section exploiting the structural knowledge of this adduct.

Monitoring Human Exposure to Aflatoxins

Immunological methods for the detection of low molecular weight substances has been extensively used over the last 30 years. Numerous antibodies and antisera have been generated because thousands of chemicals are antigenic. The development of the monoclonal antibody technology during the last decade has now permitted the refinement of antibody isolation to select and grow monoclonal antibodies having a unique specificity. Simply stated monoclonal antibodies are specific for a single epitope. Our research groups have applied these methods to combine chemical analytic and radiometric procedures with a monoclonal antibody-affinity chromatography column to purify the aflatoxin adducts and metabolites from urine.

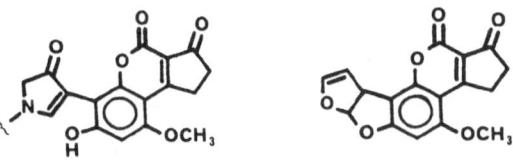

S A Adduct   Aflatoxin B₁

Fig. 2. Structure of the serum albumin lysine adduct of AFB1.

Our laboratory in collaboration with Dr. Chem Jun-shi (Chinese Academy of Preventive Medicine) and Dr. Gerald N. Wogan (MIT) have been studying people living in Guangxi Province, People's Republic of China for aflatoxin exposure. These studies were started five years ago and now are beginning to yield the requisite data to explore both dietary intake of the parent compound, AFB1, and the urinary output of aflatoxin metabolites in the same person. These pharmacokinetic data are essential to conduct the assessments that will address the question of the relationship between aflatoxin exposure and liver cancer.

People exposed to AFB1 from dietary sources were identified for pilot studies.[7,8,9] These urine samples were used to gain preliminary evidence of the applicability of the monoclonal antibody affinity column technique and HPLC analysis procedures for monitoring individuals for exposure to aflatoxins. For the initial study, 20 individuals were selected and two 25 ml aliquots of urine obtained from a morning voiding for each individual. The intake of AFB1 from the diet, primarily corn contaminated with AFB1 from 20 to 200 ppb (μg/kg), from the previous day (24 hrs) was calculated. The exposures ranged from 13.4 to 87.5 μg AFB1. Urine samples from four individuals who had been exposed to the highest level (87.5 μg) the previous day were prepared with the antibody affinity column and then measured by analytical HPLC (Figure 3). HPLC analysis demonstrated the presence of the major AFB1-DNA adduct, AFB1-N^7-Gua, at levels representing between 7-10 ng of the adduct. These data indicate that the monoclonal antibody columns, coupled with HPLC, can quantify aflatoxin-DNA adducts in human urine samples obtained from environmentally exposed people.

The experience from the first China samples and the experience gained from the Gambia work stimulated a more extensive study in Guangxi Province in 1985. To facilitate learning about the relationship between dose and excretion of AFB1 and its adducts in chronically exposed people, the following

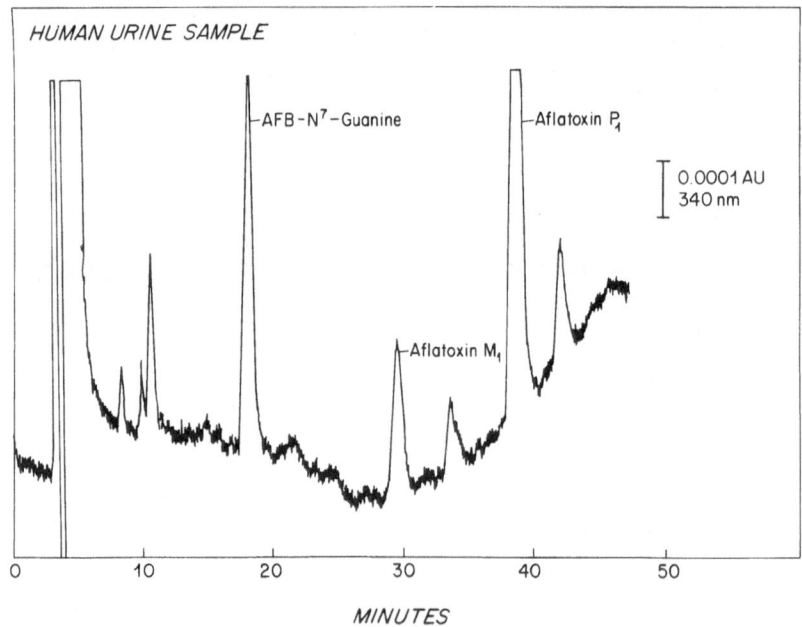

Fig. 3.   HPLC of human urine sample obtained from an individual environmentally exposed to 87 μg AFB1 the previous day.

protocol was developed. The diets of 30 males and 12 females, ages ranging from 25 to 64 years, were monitored for one week and total aflatoxin intake determined for each day. Urine was obtained in two 12-hour fractions for three consecutive days during the one week period. These urine samples were obtained only after dietary aflatoxin levels had been measured for at least three consecutive days. Therefore, the urine collections were initiated on the fourth day of the protocol. The average male intake of AFB1 was 48.4 μg per day for a total exposure over the 7 day period of 276.8 μg.

The average female daily intake was 92.4 μg per day. Immunoassays were performed on aliquots of the 12 hour urines following clean-up of the samples by C18 Sep-Pak and monoclonal antibody affinity chromatography.

Total AFB1 excretion for each 12 hour sample period was calculated by multiplying the urine volume by the concentration of AFB1 determined in the aliquot of urine. Figure 4 depicts a scatterplot comparison of aflatoxin intake with. aflatoxin metabolite excretion. All of the male and female data were combined for this analysis. The aflatoxin intake data represents the total integrated ingestion by an individual for the day prior to urine collection and during the three days of the urine collection. The excretion data are the composite of all aflatoxin metabolites excreted into the urine during the three days of urine sampling. These data reveal that despite a 20 fold range of AFB1 intake, the amount of aflatoxin excreted generally varies only over a three-fold range.

We also performed HPLC analysis of the urine samples for AFM1, AFP1, and the major AFB1-DNA adducts. The data shown in Figure 5 represents the afla- toxin DNA adduct excretion levels in urine plotted against the AFB1 intake amount in μg per day. Unlike the scatter diagram for total aflatoxin excretion, when an individual metabolite is measured, a dose dependent excre- tion is seen. Because the aflatoxin-DNA adduct has a relatively short half life in DNA, this dose dependent excretion pattern will reflect exposures

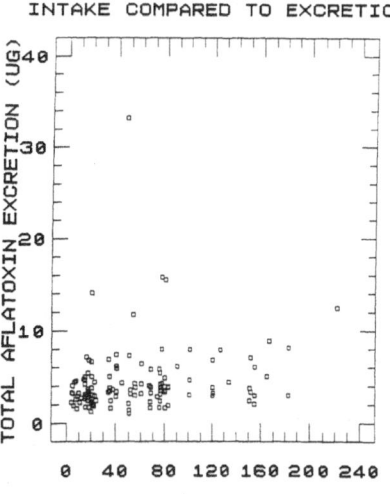

Fig. 4. Scatterplot of total aflatoxin intake and
excretion into urine of the studied population.

during the previous few days. The correlation coefficient for this association of adduct excretion and intake is 0.6 with a P value of less than
0.00001. Taken together, it appears that aflatoxin DNA adducts in urine is a
valid compartment to sample people for aflatoxin exposure, but more data must
be collected for developing a risk model for people.

In contrast with the DNA adduct work described, the significance of the
protein adduct work is that these adducts will represent the integrated dose
in people of aflatoxin exposure received over many previous weeks. The average half-life of albumin in people is about 20 days. Therefore, an accumulated dose of aflatoxin will be present in albumin long after dietary exposure has ceased. This hypothesis was tested by measuring aflatoxin-serum
albumin adducts in blood from the same people that we had collected urine.

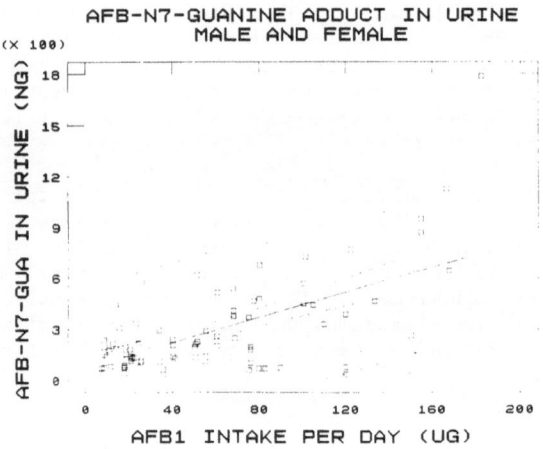

Fig. 5.   Linear regression of excreted aflatoxin N^7-guanine
adduct in urine and AFB1 dose in the diet.

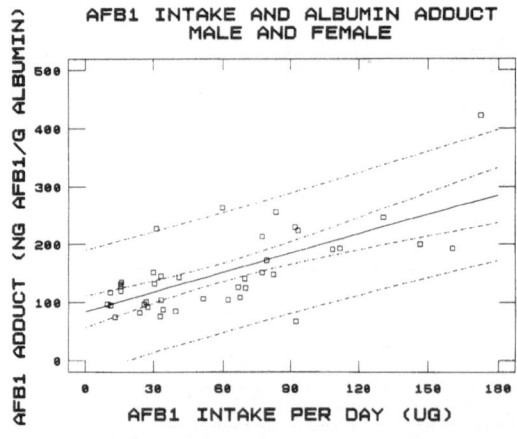

Fig. 6.   Linear regression of serum albumin adduct formation
by AFB1 in people dietarily exposed to the carcinogen.

Blood specimens were obtained during the same period that urine was collected and diet was sampled. Serum albumin was isolated from blood by affinity chromatography on Reactive Blue 2-Sepharose and subjected to enzymatic proteolysis using pronase. Immunoreactive products were purified by immunoaffinity chromatography and quantified by competitive radioimmunoassay. A highly significant correlation of adduct level with intake ($r = 0.69$, P less than 0.000001) was also observed (Figure 6). From the slope of the regression line for adduct level as a function of intake, it was calculated that 1.4-2.3% of ingested AFB1 becomes covalently bound to serum albumin, a value very similar to that observed when rats are administered AFB1. These studies indicate serum albumin adduct formation and DNA adduct excretion can be used as a screen to identify populations at risk and that measurements of these adducts become useful dosimeters in exposed people.

SUMMARY

Primary hepatocellular carcinoma is one of the most lethal and common cancers in the world, and is particularly prevalent on the continents of Africa and Asia. A number of epidemiological studies have associated the exposure status of people to AFB1 as being important in the etiology of liver cancer. However, to date these studies have relied upon the criteria of presumptive intake data, rather than relying upon quantitative analyses of aflatoxin DNA adduct and metabolite content obtained by monitoring biological fluids from exposed people. Information obtained by monitoring exposed individuals for specific DNA adducts and metabolites will define the pharmacokinetics of AFB1 in people, thereby facilitating risk assessments. However, findings do support the concept that measurement of the major, rapidly excised AFB-N^7-Gua adduct in tissues and fluids and the more persistent aflatoxin albumin adduct are appropriate dosimeters for estimating exposure status and risk in individuals consuming this mycotoxin. Finally, while there is a large body of scientific data about the presumptive role of aflatoxins in the etiology of human liver cancer, one must not lose sight of the importance of providing the resources necessary to protect the food supply from aflatoxin contamination as a readily obtainable and important goal for protecting the public's health in high exposure regions of the world.

ACKNOWLEDGEMENTS

This work was supported by grants from the USPHS U01 CA48409, P01 ES00597 and R01 CA39416.

REFERENCES

1. Bosch FX, Munoz N, 1988, Prospects for epidemiological studies on hepatocellular cancer as a model for assessing viral and chemical interactions. *IARC Sci Publ.*, 89:427-438.
2. Bulatao-Jayme J, Almero EM, Castaro CA, Jardeleza TR, Salamat L, 1982, A case-control dietary study of primary liver cancer risk from aflatoxin exposure, *Intnl J Epidemiol.*, 11(2):112-119.
3. Busby WF, Wogan GN, 1985, Aflatoxins, *In*: "Chemical Carcinogens," Searle CE, ed., American Chemical Society, Washington, DC.
4. Croy RG, Essigmann JM, Reinhold VN, Wogan GN, 1978, Identification of the principle aflatoxin B1-DNA adduct formed *in vivo* in rat liver, *Proc Natl Acad Sci USA*, 75:1745-1749.
5. Essigman JM, Croy RG, Nadzan AM, Busby WF, Reinhold VN, Buchi G, Wogan GN, 1977, Structural identification of the major DNA adduct formed by aflatoxin B_1 *in vitro*, *Proc Natl Acad Sci USA*, 74:1870-1874.
6. Gan LS, Skipper PL, Peng XC, Groopman JD, Chen JS, Wogan GN, Tannenbaum

SR, 1988, Serum albumin adducts in the molecular epidemiology of aflatoxin carcinogenesis: correlation with aflatoxin B_1 intake and
urinary excretion of aflatoxin M_1, *Carcinogenesis*, 9:1323-1325.

7. Groopman JD, Donahue PR, Zhu J, Chen J, Wogan GN, 1985, Aflatoxin metabolism in humans: Detection of metabolites and nucleic acid adducts in
urine by affinity chromatography, *Proc Natl Acad Sci USA*, 82:6492-6497.

8. Groopman JD, Busby WF, Donahue PR, Wogan GN, 1986, Aflatoxins as risk
factors for liver cancer: an application of monoclonal antibodies to
monitor human exposure, *In*: "Biochemical and Molecular Epidemiology
of Cancer," Harris, CC, ed., Alan R. Liss, Inc., New York.

9. Groopman JD, Donahue PR, Zhu J, Chen J, Wogan GN, 1987, Temporal patterns
of aflatoxin metabolites in urine of people living in Guangxi Province,
P.R.C., *Proc Amer Assoc Cancer Res.*, 28:36.

10. Groopman JD, Cain LG, Kensler TW, 1988, Aflatoxin exposure in human populations: Measurements and relationships to cancer, *CRC Critical Reviews
in Toxicology*, 19:113-145.

11. Hertzog PJ, Lindsay-Smith JR, Garner RC, 1982, Production of monoclonal
antibodies to guanine imidazole ring-opened aflatoxin B_1-DNA, the
persistent DNA adduct *in vivo*, *Carcinogenesis*, 3:723-725.

12. Lin JK, Miller JA, Miller EC, 1977, 2,3-Dihydro-2-(guan-7-yl)-3-hydroxy-
aflatoxin B_1, a major acid hydrolysis product of aflatoxin B_1-DNA
or -ribosomal RNA adducts formed in hepatic microsome mediated reactions
in rat liver *in vivo*, *Cancer Res.*, 37:4430-4438.

13. National Cancer Office of the Ministry of Public Health, P.R.C., 1980,
In: "Studies on Mortality Rates of Cancer in China," People's
Publishing House, Beijing.

14. Peers F, Bosch X, Kaldor J, Linsell A, Pluijmen M, 1987, Aflatoxin exposure, hepatitis B virus infection and liver cancer in Swaziland, *Intnl J
Cancer*, 39:545-553.

15. Sabbioni G, Skipper P, Buchi G, Tannenbaum SR, 1987, Isolation and characterization of the major serum albumin adduct formed by aflatoxin B_1
in vivo in rats, *Carcinogenesis*, 8:819-824.

16. Shimada T, Guengerich FP, 1989, Evidence for cytochrome P-450NF, the
nifedipine oxidase, being the principal enzyme involved in the bioactivation of aflatoxins in human liver, *Proc Natl Acad Sci USA*, 86:462-
465.

17. Van Rensburg SF, Cook-Mozaffari P, Van Schalkwyk DJ, Van der Watt JJ,
Vincent TJ, Purchase IF, 1985, Hepatocellular carcinoma and dietary aflatoxin in Mozambique and Transkei, *Br J Cancer*, 51:713-726.

18. Yeh FS, Shen KN, 1986, Epidemiology and early diagnosis of primary liver
cancer in China, *Adv in Cancer Res.*, 47:297-329.

ANALYSIS OF AFLATOXIN B$_1$-DNA ADDUCTS

Dennis P. H. Hsieh

Department of Environmental Toxicology
University of California
Davis, CA 95616

INTRODUCTION

DNA modification is now known as an initial step in the induction of mutagenesis and carcinogenesis by chemicals.[21] The formation of DNA adducts with some model chemical carcinogens and their characterization have been extensively studied for elucidation of the mechanism of chemical carcinogenesis.[12] Carcinogen-DNA adducts have also been used as internal biomarkers for monitoring human exposure to environmental mutagens and carcinogens.[11]

The papers in this section regard some details about these two aspects of the aflatoxin B$_1$-DNA adducts (ADA). In the present communication, a brief review on the formation, metabolic fate, and analysis of ADA is presented to facilitate the reading of subsequent papers.

Metabolic Activation of Aflatoxin B$_1$

Aflatoxin B$_1$ (AFB$_1$, structure shown in Fig. 1) is a potent genotoxic carcinogen to certain species of laboratory animals such as the rat and rainbow trout.[28] It forms covalent adducts effectively with DNA in the liver, which is the major target organ in sensitive species. Although AFB$_1$ is a direct-acting mutagen in the Ames Salmonella/microsomal test (unpublished data from DPH Hsieh's laboratory), its covalent binding to the DNA requires metabolic activation.[17,26] The metabolism pathways of AFB$_1$ by the various hepatic drug-metabolizing enzymes have been well characterized. The profile of the phase I metabolism of AFB$_1$ is summarized in Fig. 1. AFB$_1$ undergoes oxidative hydroxylation, O-demethylation, and epoxidation to form aflatoxins M$_1$, Q$_1$, P$_1$, and the 8,9-oxide of AFB$_1$, respectively.[30] AFB$_1$ can also be reduced to aflatoxicol. Except for the 8,9-oxide of AFB$_1$, all the other phase I metabolites have been isolated and characterized as the detoxification products.[29] The putative 8,9-oxide, or the AFB$_1$-epoxide, is generally accepted as the active electrophilic form of AFB$_1$ that exclusively reacts with the nucleophilic site at the N-7 position of the guanine residues of the DNA.[5,17] The activation of AFB$_1$ is mediated by a specific microsomal and nuclear cytochrome P-450 associated monoxygenase.[27] The primary metabolites of AFB$_1$, such as AFM$_1$, AFQ$_1$, and AFP$_1$, also modify the DNA through epoxide formation at the same vinyl ether double bond of the molecules.[4] When a susceptible animal such as rat is given AFB$_1$, the hepato-cellular DNA would be modified to a large extent by the epoxide of AFB$_1$ and to a small extent by the

Microbial Toxins in Foods and Feeds, Edited by A.E. Pohland *et al.*,
Plenum Press, New York, 1990

AFB$_1$ metabolites.[6] The principal adduct, 8,9-dihydro-9-hydroxy-(N7-guanyl) AFB$_1$, accounts for more than 90% of the total ADA.

The requirement of metabolic activation for AFB$_1$ to modify DNA has been well demonstrated by the *in vitro* enzymatic and chemical syntheses of ADA using the hepatic cytochrome P-450 associated monoxygenase and 3-chloroper-benzoic acid as activation agents, respectively.[17,20] The *in vitro* formation ADA and its chemical reactions were elegantly elucidated by Lin et al.[17]

Stereochemistry of AFB$_1$-epoxide

Molecular modeling analysis[18] has revealed that the AFB$_1$-epoxide is not a planar molecule. There is an angle between the terminal epoxidized furan ring and the remaining planar, aromatic, coumarin ring system. The conformation of the terminal, saturated furan ring on the aflatoxin moiety when bound to DNA is determined to be most likely chair-like rather than boat-like. The chair-like conformation does not allow the AFB$_1$-epoxide molecule to intercalate between base-pair stacks of the DNA molecule. This conformation and the steric restriction imposed by the DNA backbone therefore dictates that the epoxides of aflatoxins exclusively bind the DNA at the

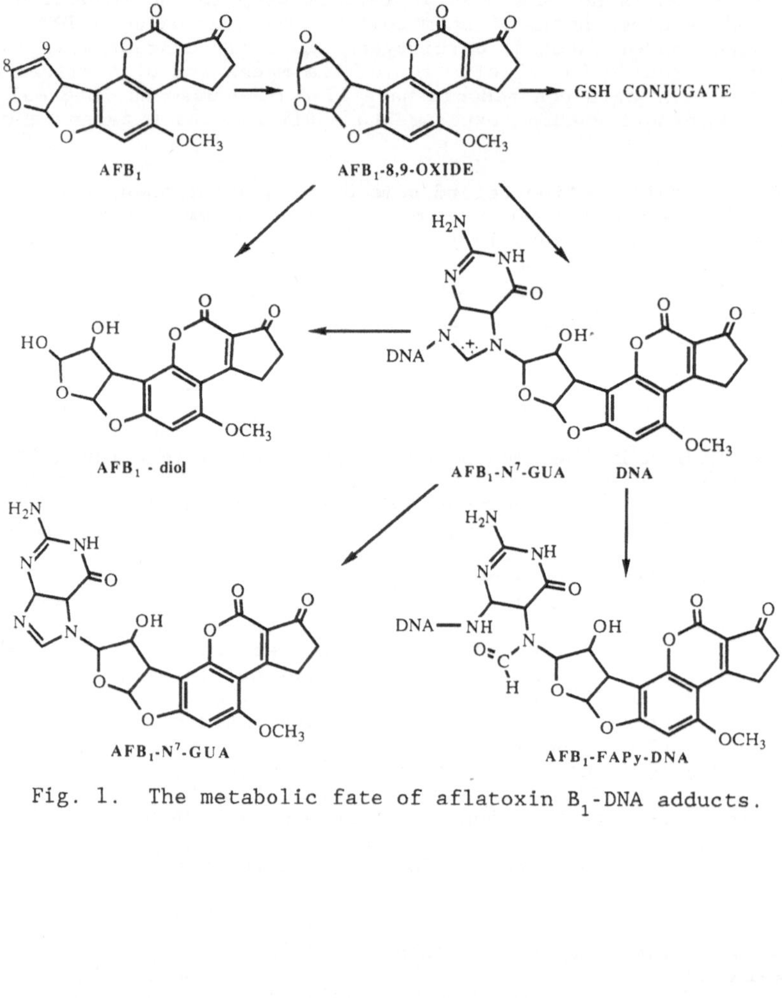

Fig. 1. The metabolic fate of aflatoxin B$_1$-DNA adducts.

N-7 site of guanine in the major grooves of DNA, from outside of the double helix. The binding from outside of the DNA molecule, rather than intercalation, is shown by the lack of an upfield shift of nuclear magnetic resonance signals (unpublished data) for adenine-thymine and guanine-cytosine base pairs characteristically produced by intercalating agents.[7]

Metabolic Fate of AFB_1-DNA Adducts

The ADA formed in the liver of the exposed animals is unstable and subject to transformation through three fate processes: (a) the normal DNA excision repair, (b) the removal of the AFB_1-modified guanyl moiety of DNA, and (c) the opening of the imidazole ring of the guanine moiety to stabilize the adduct. In the rat liver, a large proportion of ADA is readily repaired. The AFB_1-modified guanyl moiety of part of the remaining ADA can be removed spontaneously or enzymatically to give rise to apurinic sites in the DNA. The removed AFB_1-guanine adducts are excreted into the urine to serve as a biomarker for exposure to AFB_1.[1,8] The last part of the unrepaired ADA is transformed into a persistent formamidopyrimidine derivative (AFB_1-FAPy-DNA), which upon acid hydrolysis yields 8,9-dihydro-8-(N5-formyl-2',5',6'-triamino-4,-oxo-N5-pyrimdyl)-9-hydroxy AFB_1 (AFB_1-FAPy).[17] Both the apurinic sites and the repair-resistant AFB_1-FAPy-DNA are premutational lesions. Thus, in the liver of a rat that has received a single dose of AFB_1, the primary ADA in the liver is formed rapidly after dosing and then disappears exponentially with time, with part of it being transformed into AFB_1-FAPy-DNA.[4] The latter therefore would increase with time after dosing and then persist in the liver. If multiple doses of AFB_1 are given to the rat in a period of time, a steady state level of modified DNA, mostly in the form of AFB_1-APy-DNA, will be reached. The steady state levels of AFB_1-modification of DNA in the liver of rats given various daily doses of AFB_1, as determined by Lutz and coworkers,[2,19] is shown in Table 1.

Table 1. Steady state levels of AFB_1-FAPy-DNA (ADA) in the liver of rats exposed to AFB_1.

Dose (ng/Kg/day)	Duration (weeks)	Steady state ADA level (#/10 nuc)
2.2	8	9.1
73	8	320
2110	6	8500

Data taken from Ref. 2

Selected Binding of AFB_1 to DNA

AFB_1, like other chemical carcinogens, binds to DNA in a selective, non-random manner. For example, it binds preferentially to rat liver mitochondrial DNA,[23] ribosomal RNA gene sequences of rat liver DNA,[15] and transcriptionally active regions of rat liver nucleolar chromatin.[32] This specificity of binding is consistent with the accessibility of these regions of DNA to the carcinogen due to less protection by histones. The mitochondrial DNA is known to lack association with histones, while the rDNA regions maintain a diffuse conformation due to their high transcriptional activity. The detailed interactions of AFB_1 with specific transcribed genes is presented in this section by Irvin. Other factors that may contribute to this preferential binding include specific neighboring base composition and nucleotide sequence.[16] The most preferred guanine to be attacked by the AFB_1-oxide is the one followed by another guanine or by cytosine in the nucleotide sequence.[22,31] The specific binding of AFB_1 on a human oncogene of low efficiency is described in this section by Yang and coworkers.

Analysis of ADA in Liver Tissues

The steady state levels of ADA found in the animals chronically exposed to AFB_1 suggests that ADA could be used as a biomarker to assess human exposure to AFB_1. Monitoring the ADA levels in human hepatoma and other liver tissues would allow estimation of doses of AFB_1 received by individuals to strengthen the validity of epidemiological evidence and to confirm the involvement of AFB_1 in the etiology and to confirm the involvement of AFB_1 in the etiology of human liver cancer. ADA in liver tissues can be detected and measured by three analytical techniques: (1) 32P-postlabeling, (2) enzyme-linked-immuno-sorbent assay (ELISA), and (3) high-performance-liquid-chromatography. The usefulness of these techniques in ADA analysis will be compared as follows:

32P-Postlabeling Technique

This technique, developed by Randerath and coworkers,[24] offers a sensitive and nonspecific method to detect the presence of adducts formed between the DNA and any chemical carcinogens. Briefly, the method uses a mixture of micrococcal endonuclease and spleen exonuclease to digest completely the carcinogen-modified DNA to mononucleotides. The carcinogen-modified mononucleotides are isolated from normal mononucleotides by chromatographic techniques and are labeled with 32P at the 5' position of the nucleotides using 32P-labeled ATP as the labeling reagent. The 32P-labeled, carcinogen-modified mononucleotides are then developed on a two-dimensional thin-layer chromatographic plate for autoradiography. The characteristic positions of the radioactive spots on the autoradiograph are used as "fingerprints" for the identification of the DNA adducts of the carcinogen in question.

Owing to the strong radioactivity of 32P and the availability of 32P-ATP with high specific activity as the donor of the label, this method is capable of detecting very low levels of carcinogen adducts on DNA, not requiring the knowledge of what carcinogens are in question and the availability of the radioisotope-labeled carcinogens. Typically, a modification level of one carcinogen adduct per billion nucleotides is detectable from about 10 μg of DNA, making the method especially suitable for the monitoring of ADA in human liver tissues.

Detection of ADA in liver tissues of rats that were dosed with AFB_1 has been described by Randerath et al.[24] The authors point out that the bulky AFB_1 residues in ADA hinder the action of the micrococcal endonuclease and spleen exonuclease and render the digestion of the modified DNA incomplete. Therefore, AFB_1-modified DNA can only be digested to AFB_1-modified oligonucleotides rather than mononucleotides to constitute the "fingerprints". A similar observation was reported of DNA modified by sterigmatocystin, which is a mycotoxin structurally and biosynthetically related with AFB_1.[25]

ELISA Method

The ELISA method takes advantage of the availability of specific antibodies developed against DNA- or protein-conjugates of AFB_1-epoxide or AFB_{2a}, which have high affinity for ADA. The extent to which a sample of DNA inhibits the binding between the antibodies and their respective antigens is measured as an index of the level of AFB_1 adducts in the DNA, using established ELISA procedures.[3]

Using a competitive-inhibition ELISA method involving a mouse monoclonal antibody, Garner et al.[9] detected and estimated the ADA levels in the hepatoma tissues of eight patients in Czechoslovakia. They found levels of ADA in the range of 0.2 to 1.1 AFB_1 adducts per million nucleotides in seven of the eight liver samples analyzed. The 50% inhibition value of AFB_1 in this ELISA method is 0.3 ng per well or 3 ng AFB_1 per ml. Since the antibody used in this investigation has high affinity to ADA and other AFB_1 analogs, the exact identify of the aflatoxin moiety in DNA, whether AFB_1 or its metabolites, is not known.

Similarly, Hsieh et al.[13] report the use of an ELISA method involving a monoclonal antibody developed against AFB_1-FAPy-DNA in the detection of ADA levels in the hepatoma and vicinal normal liver tissues of nine Taiwan patients. ADA was found in all nine hepatoma tissues at levels ranging from 1.2 to 3.5 AFB_1 adducts per million nucleotides. In contrast, only two normal liver tissues were found to contain ADA at the level of 1.2 - 1.7 AFB_1 adducts per million nucleotides. The 50% inhibition value of AFB_1-FAPy-DNA is only about 0.05 ng per well, indicating a very high affinity between the antibody and ADA. Since the monoclonal antibody used in this investigation was developed against AFB_1-FAPy-DNA, the ADA detected is most likely the antigen itself. The finding that ADA is present in the hepatoma tissues at levels considerably higher than in the normal tissues is particularly intriguing. It suggests that the metabolic fate of AFB_1 in hepatoma tissues is quite different from that in the neighboring normal liver tissues. Whether AFB_1 modification of DNA is playing a role in the progression of hepatoma remains to be investigated.

An attempt to develop a similar immunological as well as immunohisto-chemical method for detection of ADA in the livers of rats and rainbow trout is described in this section by Ishikawa and coworkers.

HPLC Method

The operational conditions for separation of the various hydrolysis products of ADA have been well established. The guanine and FAPy adducts of AFB_1 and its metabolites can be readily separated and identified by reverse-phase HPLC.[10] These adducts in urine samples have been analyzed as an estimation of exposure to AFB_1. The HPLC method for ADA analysis is greatly facilitated by the use of an immuno-affinity column as a cleanup step in the analytical procedure. The immuno-affinity column serves to effectively isolate the molecular species containing AFB_1 and its analogs from the complex impurities in the urine samples to greatly increase the sensitivity of the HPLC determination.

This combined procedure is currently being used in large scale epidemiological studies in China for the assessment of human exposure to AFB_1 [as described by Groopman in this Section]. The HPLC method, however, is not as useful in monitoring ADA levels in human liver tissues as in the urine, because there usually are not large enough human liver samples available to provide sufficient amounts of ADA for hydrolysis and HPLC analysis.

CONCLUSION

ADA and its hydrolysis products have proven to be useful biomarkers for the assessmet of human dosimetry of AFB_1. 32P-postlabeling analysis and ELISA are, in combination, a sensitive and efficient method to monitor ADA levels in human liver tissues. Existing evidence indicates that most of the hepatoma patients in Czechoslovakia and Taiwan have been exposed to AFB_1 and that the level of ADA in human hepatoma tissues is considerably higher than in normal liver tissues. The difference in the metabolism of AFB_1 in these two types of tissues and the possible role of ADA in the progression of hepatoma are interesting areas of further investigation.

ACKNOWLEDGEMENT

The author thanks the financial support provided by the Foundation of Promotion for Cancer Research (FPCR), Tokyo, Japan, for a research project related to the subject of this presentation. He was FPCR visiting fellow during July-November 1988.

REFERENCES

1. Autrup H, Bradley KA, Shamsuddin AKM, Wakhisi J, Wasunna A, 1983, Detection of putative adduct with fluorescence characteristics identical to 2,3-dihdro-2-(7'-guanyl)-3-hydroxyaflatoxin B_1 in human urine collected in Murang's district, Kenya, *Carcinogenesis*, 4:1193-1195.
2. Buss P, Lutz WK, 1988, Steady-state DNA adduct level in rat liver after chronic exposure to low doses of aflatoxin B_1 and 2-acetylaminofluorene, *Proc Amer Assoc Cancer Res.*, 29:#380.
3. Chu FS, 1986, Recent studies on immunochemical analysis of mycotoxins, *In*: "Mycotoxins and Phycotoxins", Steyn PS, ed., Elsevier, Amsterdam.
4. Croy RG, Wogan GN, 1981, Temporal patterns of covalent DNA adducts in rat liver after single and multiple doses of aflatoxin B_1, *Cancer Res.*, 41:197-203.
5. Essigmann JM, Croy RG, Nadzan AM, Busby WR, Reinhold VN, Buchi G, Wogan GN, 1977, Structural identification of the major DNA adduct formed by aflatoxin B_1 *in vitro*, *Proc Natl Acad Sci USA*, 74:1870-1874.
6. Essigmann JM, Croy RG, Bennett RA, Wogan GN, 1982, Metabolic activation of aflatoxin B_1: patterns of DNA adduct formation, removal and excretion in relation to carcinogenesis, *Drug Metabol Rev.*, 13:581-602.
7. Feign J, Denny WA, Liupin W, Kearns DR, 1984, Interactions of antitumor drugs with natural DNA: 1H NMR study of binding mode and kinetics, *J Med Chem.*, 27:450-465.
8. Garner RC, Ryder R, Montesano R, 1985, Monitoring of aflatoxins in human body fluids and application to field studies, *Cancer Res.*, 45:922-927.
9. Garner RC, Dvorackova I, Tursi F, 1988, Immunoassay procedures to detect exposure to aflatoxin B_1 and benzo(a)pyrene in animals and man at the DNA level, *Int Arch Occup Environ Health*, 60:145-150.
10. Groopman JR, Croy RG, Wogan GN, 1981, *In vitro* reactions of aflatoxin B_1-adducted DNA, *Proc Natl Acad Sci USA*, 78:5445-5449.
11. Harris CC, 1985, Future directions in the use of DNA adducts as internal dosimeters for monitoring human exposure to environmental mutagens and carcinogens, *Env Health Persp.*, 62:185-191.
12. Hemminki K, 1983, Nucleic acid adducts of chemical carcinogens and mutagens, *Arch Toxicol.*, 52:249-285.
13. Hsieh LL, Hsu SW, Chen DS, Santella RM, 1988, Immunological detection of aflatoxin B_1-DNA adducts formed *in vivo*, *Cancer Res.*, 48:6328-6332.
14. International Agency for Research on Cancer, 1987, "IARC Monographs of the Evaluation of Carcinogenic Risks to Humans - Overall Evaluation of Carcinogenicity: An Updating of IARC Monographs Vol. 1-42", Supplement 7, pp 82-87.
15. Irvin TR, Wogan GN, 1984, Quantitation of aflatoxin B_1 adduction within the ribosomal RNA gene sequences of rat liver DNA, *Proc Natl Acad Sci USA*, 81:664-668.
16. Kadlubar FF, Beland FA, Beranek DT, Dooley KL, Heflich RH, Evans FE, 1982, *In*: "Environmental Mutagens and Carcinogens", Sugimura T, Kondo S

and Takebe H, eds., Alan Liss, New York.

17. Lin J-K, Miller JA, Miller EC, 1977, 2,3-Dihydro-2-(guan-7-yl)-3-hydroxy-aflatoxin B_1, a major acid hydrolysis product of aflatoxin B_1-DNA or -ribosomal RNA adducts formed in hepatic microsome-mediated reactions and in rat liver *in vivo*, *Cancer Res.*, 37:4430-4438.

18. Loechler EL, Teeter MM, Whitlow MD, 1988, Mapping the binding site of aflatoxin B_1 in DNA: molecular modeling of the binding sites for the N(7)-guanine adduct of aflatoxin B_1 in different DNA sequences, *J Biomol Struct Dyn.*, 5:1237-1257.

19. Lutz WK, 1987, Quantitative evaluatioan of DNA-binding data *in vivo* for low-dose extrapolations, *Arch Toxicol Suppl.*, 11:66-74.

20. Martin CN, Garner RC, 1977, Aflatoxin B-oxide generated by chemical or enzymic oxidation of aflatoxin B_1 causes guanine substitution in nucleic acids, *Nature*, 267:863-865.

21. Miller EC, Miller JA, 1981, Mechanism of chemical carcinogenesis, *Cancer*, 27:2327-2345.

22. Modali R, Yang SS, 1986, Specificity of aflatoxin B_1 binding on human proto-oncogene nucleotide sequence, In: "Monitoring Occupational Genotoxicants", Sorsa M and Norppa H, eds., Liss, New York.

23. Niranjan BG, Bhat NK, Avadhani NG, 1982, Preferential attack of mitochondrial DNA by aflatoxin B_1 during hepatocarcinogenesis, *Science*, 215:73-75.

24. Randerath K, Randerath E, Agrawal HP, Reddy MV, 1984, Biochemical (postlabeling) methods for analysis of carcinogen-DNA adducts, *IARC Sci Publ.*, 59:217-231.

25. Reddy MG, Irvin TR, Randerath K, 1985, Formation and persistence of sterigmatocystin-DNA adducts in rat liver determined via 32P-post-labeling analysis, *Mutation Res.*, 152:85-96.

26. Swenson DH, Miller JA, Miller EC, 1975, The reactivity and carcinogenicity of aflatoxin B_1-2,3-dichloride, a model for the putative 2,3-oxide metabolite of aflatoxin B_1, *Cancer Res.*, 35:3811-3823.

27. Ueno Y, Ishii K, Omata Y, Kamataki T, Kato R, 1983, Specificity of hepatic cytochrome P-450 isoenzymes from PCB-treated rats and participation of cytochrome b5 in the activation of aflatoxin B_1, *Carcinogenesis*, 4:1071-1077.

28. Wogan GN, 1973, Aflatoxin carcinogenesis, *Meth Camcer Res.*, 7:309-344.

29. Wong JJ, Hsieh DPH, 1976, Mutagenicity of aflatoxins related to their metabolism and carcinogenic potential, *Proc Nat Acad Sci USA*, 73:2241-2244.

30. Wong ZA, Hsieh DPH, 1980, The comparative metabolism and toxicokinetics of aflatoxin B_1 in the monkey, rat, and mouse, *Toxicol Appl Pharmacol.*, 55:115-125.

31. Yang SS, Taub JV, Modali R, Vieira W, Yasei P, Yang GC, 1985, Dose dependency of aflatoxin B_1 binding on human high molecular weight DNA in the action of proto-oncogene, *Env Health Persp.*, 62:231-238.

32. Yu FL, 1983, Preferential binding of aflatoxin B_1 to the transcriptionally active regions of rat liver nucleolar chromatin *in vivo* and *in vitro*, *Carcinogenesis*, 4:889-893.

PRODUCTION AND CHARACTERIZATION OF A MONOCLONAL ANTIBODY TO

TETRAACETYLNIVALENOL

Hideharu Ikebuchi[a], Reiko Tesima[a], Masakatu Ichinoe[b]
and Tadao Terao[a]

Division of Radiochemistry[a]
Division of Microbiology[b]
National Institute of Hygienic Sciences
1-18-1, Kamiyoga, Setagaya-Ku, Tokyo 158
Japan

INTRODUCTION

Nivalenol (3,4,7,15-tetrahydroxy-12,13-epoxytrichothec-9-en-8-one: NIV)
(Fig. 1) is a toxic metabolite produced by *Fusarium graminearum* and was often
detected along with deoxynivalenol (DON) in cereals and health food in Japan[1]
and Korea.[2]

Furthermore, it is known that the acute toxicity of NIV is greater than
that of DON.[3] In view of heavy contamination of NIV in 1968 and 1972 wheat
grown in the western part of Japan[4,5] and of the potential health hazard to
both humans and animals, intense research on NIV has been carried out in
Japan in the last several years.

Although progress has been made, research on NIV is still hindered by the
lack of a good sensitive, rapid detection method. We tried to develop immuno
assays for NIV in our laboratory. We already established the radioimmunoassay system using rabbit polyclonal antibody to tetraacetylnivalenol (Ac-NIV)
(unpublished data). However, we did not have reliable results with an
enzyme-linked immunoassay system using polyclonal antibody. This prompted us
to produce monoclonal antibody to Ac-NIV and develop an ELISA system.

In this paper, we describe the preparation and characterization of a
monoclonal mouse anti-Ac-NIV antibody.

MATERIALS AND METHODS

Materials

Nivalenol (NIV), deoxynivalenol (DON), T-2 toxin, and diacetyl-scirpenol
(DAS) were purchased from Wako Chemical Co. (Tokyo, Japan). Nivalenol-
tetra-acetate (Tetra-Ac-NIV), T-2-tetra-acetate (Tetra-Ac-T-2), deoxyni-
valenol-triacetate (Tri-Ac-DON) and Triacetylscirpenol (TAS) were prepared as
described in the previous paper. Bovine serum albumin (BSA: VI grade) and
ovalubumin (OVA: V grade) were purchased from Sigma Chemical Co. (St. Louis,
MO). Tri-n-butylamine and isobutyl chlorocarbonate and glutaric anhydride

Microbial Toxins in Foods and Feeds, Edited by A.E. Pohland *et al.,*
Plenum Press, New York, 1990

and succinic anhydride were purchased from Wako Chemical Co.   Complete
Freund's adjuvant was obtained from Difco Laboratories (Detroit, MI).   All
chemicals and organic solvents were reagent grade or better.

Preparation of hemiglutarate of Tetra-Ac-NIV

The hemiglutarates of Tetra-Ac-NIV were prepared as follows: 6 mg of
8-H-Ac-NIV, obtained by the reduction of Tetra-Ac-NIV with sodium borohydride, was dissolved in 3 ml of dry pyridine with 48 mg glutaric anhydride
and 2 mg N,N,dimethylaminopyridine and allowed to react for 4 hr at 110 C.
After reaction, the mixture was diluted with distilled water.   The aqueous
solution was acidified with 1 N HCl and extracted with 10 ml chloroform.   The
organic phase was washed, with 5 ml of 3% NaHCO$_3$ aqueous solution, three
times, dried over anhydrous sodium sulfate and concentrated.

Further purification was achieved by a preparative TLC method using a
silica gel plate.   The silica gel band containing Tetra-Ac-NIV-HG was removed
from the TLC plate and extracted with 10% methanol in chloroform and yielded
3.5 mg Tetra-Ac-NIV-HG.   GC-MASS spectral analysis (CI) of the methylated
derivative showed a molecular ion of 610 (M+) which was consistent with the
molecular weight of methyl-Tetra-Ac-NIV-HG.

Preparation of conjugate

Hapten-carrier conjugate was prepared by the mixed anhydride method.
Briefly, 3 mg of Tetra-Ac-NIV-HG in 100 μl of dioxane and 3.5 μl of
tri-n-butylamine and 1 μl of isobutylchlorocarbonate were added to a
solution containing 5 mg of BSA.   The reaction proceeded overnight at 4 C
with stirring.   After reaction, the solution was loaded on a Sephadex G-25

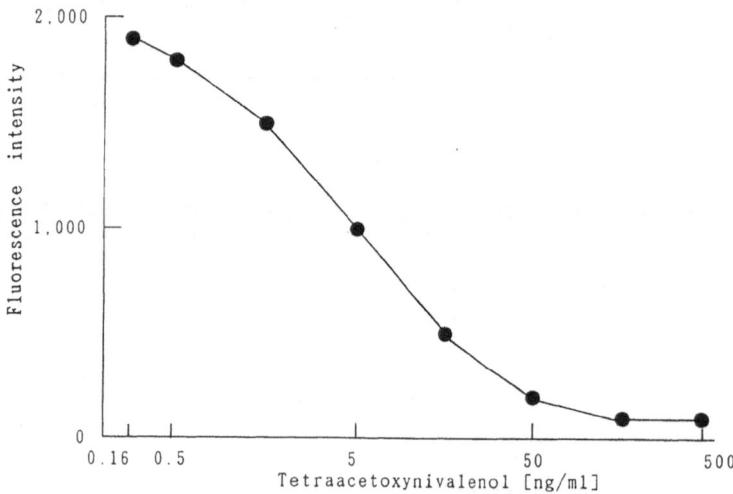

Fig. 1.   CI-ELISA inhibition curve for D18.102.59
          of Ac-NIV as inhibitor.   Nonsaturating
          antibody concentration chosen for use in
          the CI-ELISA was a dilution of the ascites
          fluid of 1:10,000.   Solid phase antigen was
          Ac-NIV-HS-OVA.

column (1 x 15 cm) equilibrated in PBS, then eluted with PBS and the void volume was collected.

ELISA

Fifty µl of Tetra-Ac-NIV-hemisuccinate, (HS)-OVA prepared as shown in the previous paper by the water soluble 1-ethyl-3-(3-dimethyl-aminopropyl) carbodiimide (EDPC) method (1 µl/ml in 0.15 M carbonate buffer, pH 9.6), was added to each well of a 96-well polyvinyl microtiter plate (Costar, 96 well EIA plate, flat bottom, 3590) and incubated overnight at 4 C. Wells were washed four times with 0.2 ml of 0.01 M phosphate buffer in 0.15 M saline including 0.5% Tween 20 (NaCl/Pi/Tween). For blocking of unbound solid-phasae sites and minimizing nonspecific binding, 100 µl of casein in NaCl/Pi was added and incubated for 1 hr at 25 C. Wells were washed four times with 1.2 ml NaCl/Pi/Tween and then incubated for 1 hr at 25 C with 50 µl of culture supernatant. Wells were washed four times and allowed to react for 1 hr at 25 C with 50 µl of sheep-anti-mouse-Ig-b-galactosidase conjugate (10^{-3} conc., Amersham International plc, Amersham UK) in casein in NaCl/Pi. Wells were washed four times with NaCl/PI/Tween. The wells were incubated for 1 hr at 25 C with 100 µl of a solution of 0.1 mM 4-methylum-belliferyl-b-galactoside (Sigma Chemical Co.). Finally fluorescence was monitored by titertek Fluoroscan.

Competitive-inhibition ELISA

The competitive-inhibition ELISA (CI-ELISA) was identical to the ELISA procedure except that a 50 µl sample of trichothecenes dissolved in 5% (v/v) methanol in NaCl/Pi was simultaneously incubated overnight at 4 C with 50 µl of a nonsaturating solution of monoclonal antibody in casein in NaCl/Pi. The amount of the bound antibody was determined as described above.

Immunization

Six 9-week old female BALB/c mice were immunized 2-3 weeks by injecting 0.2 ml of immunogen at multiple sites on the back. The immunogen was prepared by emulsifying 100 µg of the antigen in 0.1 ml sterilized phosphate-saline buffer with 0.1 ml complete adjuvant. Three mice with high titers to Tetra-Ac-NIV received another intraperitoneal booster injection without adjuvant and three days later spleen cells were prepared.

Hybridization and cloning

Immune spleen cells and myeloma cells (NS-1) were combined in the ratio of 3:1 and treated with 50% polyethylene glycol 4000. Fused cells were seeded in 24-well culture plates and HAT (hypoxanthine/aminopterin/thymidine) medium containing RPMI-1640 plus 10% fetal calf serum (Boehringer Manneheim, W. Germany), 100 µM hypoxanthaine, 16 µM thymidine, 0.1 µM aminopterine, 2 mM glutamine, 0.1 mg/ml kanamycine and 1 x 10^{-5}M 2ME. Hybridomas were visible microscopically, tested again, and then cloned by limiting dilution in the presence of thymocytes as the feeder layer.

Antibody specificity and immunochemical characterization

Antibody specificity was determined by competitive inhibition binding in the CI-ELISA as described above with various NIV-related trichothecenes as inhibitors and Tetra-Ac-NIV-HS carrier conjugates as solid phase antigen. The Ig class, subclass and light-chain composition of the anti-Tetra-Ac-NIV in hybridoma culture fluids were determined by the use of rabbit antibodies specific for mouse IgM, IgG1, IgG2a, IgG2b, or IgG3 and antibodies against antibodies against mouse K or λ light chains (10^{-5} dil., Miles Laboratory, Elkhart, IN, USA) as the second antibody in the ELISA.

RESULTS AND DISCUSSION

The response of BALB/c mice immunized with tetraacetylnivalenol hemi-glutarate (Ac-NIV-HG) BSA conjugate was demonstrated as early as 4 weeks after immunization. High antibody titers were observed 10 weeks after immunization. Subsequently, the spleens of three of these immunized mice were used for fusion 3 days after the last immunization. Of 288 cultures seeded with fused cells, 158 (55%) yielded viable hybridoma clones. The supernatants were screened in ELISA with a solid phase antigen (Ac-NIV-HS-OVA) prepared by the EDPC method. The positive hybridoma cells (2% of viable hybridoma clones) were expanded and cloned. Only three clones were stable and were designated clone D18.102.59, J22.3.34, and L21.191.52. The low fusion efficiency might be explained by the release of the immunotoxic haptens from the hapten-carrier conjugate during immunizations as described recently for the preparation of monoclonal anti-T-2 toxin antibodies by Hunter et al.[6]

Specificity of the monoclonal anti-Ac-NIV antibody

The association constants (K_a) of these monoclonal antibodies were determined from the effect on inhibition by unlabelled Ac-NIV in radioimmu-noassay. These antibodies had a K^a of the order of 10^8. Their heavy- and light-chain isotypes were determined by ELISA. The heavy chains of all of the antibodies consisted of $\gamma-_1$. Either *kk-* or λ-chains were used as light chains in these mAb.

Table 1. Specificity of the monoclonal anti-Ac-NIV antibodies. IC_{50}: Concentration of 50% inhibition of the antibody binding in the CI-ELISA.

Trichothecene	IC_{50} of Ac-NIV/IC_{50} of trichothecene		
	D18.102.59	J22.3.34	L21.191.52
Tetraacetylnivalenol (Ac-NIV-HG)	1.6	2.5	2.2
Ac-NIV-HS	1.2	1.4	1.2
Ac-NIV	1.0	1.0	1.0
Nivalenol (NIV)	<0.001	<0.001	<0.001
T-2 Toxin	<0.001	<0.001	<0.001
T-2 Toxin triacetate	0.02	0.16	0.2
Deoxynivalenol (DON)	<0.001	<0.001	<0.001
Triacetoxy-DON	<0.001	<0.001	<0.001
Diacetoxyscirpenol (DAS)	<0.001	<0.001	<0.001
Triacetoxyscirpenol (TAS)	0.03	0.23	0.04
Fusarenon X	<0.001	<0.001	<0.001

The cross-reactivities of the mAb obtained were examined by competitive ELISA. The concentrations required to give 50% inhibition (IC_{50}) relative to that of Ac-NIV is shown in Table 1. The reactivity of three mAb with Ac-NIV, T-2 toxin triacetate (Ac-T-2) and TAS was in the same range. No inhibi-tion was observed when 1,000 ng/ml of NIV, DON, DAS, and T-2 toxin was tested. The reactivity of the mAbs with Ac-DON decreased about 500 times as compared with Ac-NIV. These indicated that the mAb recognized acetyl groups at C-3, C-4, and C-6 positions. Of the three antibodies, D18.102.59 was the least cross-reactive with other closely related trichothecenes, and showed the highest affinity with Ac-NIV.

The concentration of Ac-NIV causing 10% inhibition of binding of mAb was around 0.5 to 1.0 ng/ml (Fig. 1). Since the standard deviation in the assay was normally within this range, it is reasonable to assume that the lower limits for the detection of Ac-NIV in the CI-ELISA also fall within this range.

The mAbs to Ac-NIV together with an ELISA solid phase can be used effectively as assay reagents for detection of NIV in grains after the extracted samples are acetylated. We also showed that a mAb recognized Ac-T-2 toxin and TAS as well as Ac-NIV. However, Tanaka et al.[1] reported that in, Japan, nivalenol, deoxynivalenol and zearalenone were detected but there was no detection of other trichothecene mycotoxins in foodstuffs. The object of this study was to measure NIV separately from DON in domestic foodstuffs. These data show that the D18.102.59 is a valuable analytical reagent for the detection of Ac-NIV.

REFERENCES

1. Tanaka T, Hasegawa A, Matsuki Y, Ueno Y, 1985, A survey of the occurrence of nivalenol, deoxynivalenol and zearalenone in food stuffs and health foods in Japan, *Food Add Contam.*, 2:259-265.
2. Lee US, Jang HS, Tanaka T, Toyasaki N, Sugiura Y, OH YJ, Cho M, Ueno Y, 1986, Mycological survey of Korean cereals and production of mycotoxins by *Fusarium* isolates, *Appl Environ Microbiol.*, 52:1258-1260.
3. Ueno T, Sato N, Ishii K, Sakai K, Tsunoda H, Enomoto M, 1973, Biological and chemical detection of trichothecene mycotoxins of *Fusarium* species, *Appl Environ Microbiol.*, 25:699-704.
4. Tatsuno T, Fujimoto Y, Morita Y, 1969, Toxicological research on substances from *Fusarium nivale*. III. The structure of nivalenol and its monoacetate, *Tetrahedron Lett.*, 33:2823-2826.
5. Morooka N, Uratsuji N, Yoshizawa T, Yamamoto H, 1972, Toxic substances in barley infected with *Fusarium* spp., *J Food Hyg Soc Japan*, 13:368-375.
6. Hunter KW, Brimfield AA, Miller M, Funkelman FD, Chu FS, 1985, Preparation and characterization of monoclonal antibodies to the trichothecene mycotoxin T-2, *Applied Environ Microbiol.*, 49:168-172.

INTERACTION OF AFLATOXIN B$_1$ WITHIN SPECIFIC GENE SEQUENCES

T. Rick Irvin

Laboratory of Toxicology
Veterinary Anatomy Department
Texas A & M University
College Station, Texas 77843

INTRODUCTION

The irreversibility of tumor initiation, as well as the heritability of the tumor phenotype, have drawn attention to DNA as the critical macromolecular target in chemical carcinogenesis.[9] Changes in DNA sequences, brought about by carcinogen-DNA interactions, could constitute molecular bases for observed alterations in gene expression that accompany neoplastic transformation by chemicals. Evidence supporting a putative role of carcinogen-induced DNA sequence change in initiating tumor development has been provided by both qualitative and quantitative correlations between the carcinogenic and mutagenic potency exhibited by many carcinogens and the potency of chemicals as carcinogens compared to the total levels of covalent DNA modification observed in treated animals.[20,36]

Identification of adduct moieties resulting from carcinogen-DNA interactions has permitted not only kinetic studies of DNA adduct appearance and removal within cells, but also mechanistic studies seeking to relate specific DNA adduct patterns with tumor development in target tissues. Relationships have been established between target organ specificity and concomitant sensitivity of target organ DNA in carcinogen modification and damage, and also between persistence of specific DNA adduct lesions in target organs and the ultimate appearance of tumors in these organ sites.[5,6,11]

Many factors are known to affect the kinds and amounts of DNA adducts formed when cells are exposed to carcinogens. For example, enzymatic competence for carcinogen activation is an important determinant of adduct localization within specific tissues and cell types under *in vivo* conditions; further, adduct persistence can be strongly affected by DNA repair capacities of specific cells types.[27] Under *in vitro* conditions, neighboring base comsition can influence the reactivity of specific nucleophilic centers with respect to distribution of carcinogen adducts within defined nucleotide sequences.[25,26]

Elucidation of the molecular bases for mutation and tumor initiation by carcinogens will, however, require detailed knowledge of DNA modification within specific base sequences that can be related to alterations in gene function. Earlier experiments with this objective have included attempts to relate mutation frequency with DNA adduct formation by mutagens/carcinogens such as benzo[a]pyrene and aflatoxin B_1. While this approach has provided quantitative data on apparent mutagenic efficiency of various DNA adducts, structural modifications could not be localized within the marker genes in which mutation was scored. Furthermore, this quantitative approach assumes that DNA-carcinogen adducts are randomly distributed. Recent studies have indicated the susceptibility to carcinogen modification is not random but is affected by factors such as transcriptional activity, specific neighboring base composition, and nucleotide sequence.

The organization and structure of chromatin also modulates the suscep- tibility of DNA to carcinogen modification and damage. Morphological studies suggest that transcribed gene sequences are maintained in a discrete, open conformation distinguishable from nontranscribed DNA.[22,23] Biochemical investigations, in which transcribed regions undergo selective digestion by DNase I or II, further support the ultrastructural observations.[12,16]

To probe the structural and functional characteristics of chromatin important in the genomic location of carciogen-DNA interactions, our group has developed molecular and toxicological tools to isolate specific liver gene sequences and monitor the formation and removal of adducts formed by the hepatocarcinogenic mycotoxin aflatoxin B_1 (AFB_1). AFB_1 has proven a superior model carcinogen in these studies for several reasons. First, the products of AFB_1 reactions with DNA have been thoroughly characterized, and the only known site of reaction is at the N^7 position of guanine as shown in Figure 1.[6] Second, extensive information about AFB_1-DNA adduct sta- bility and removal has facilitated development of hybridization technology permitting isolation of AFB_1-adducted gene sequences and quantitation of aflatoxin residues within them.

In this report, the formation and removal of AFB_1-DNA residues is described within two expressed liver gene sequences, the ribosomal RNA (rRNA) and transfer RNA (tRNA) genes. The results demonstrate that AFB_1-DNA resi- dues are preferentially formed and removed from these DNA regions as compared to unfractionated nuclear DNA.

MATERIALS AND METHODS

Chemicals

[^{3}H]AFB_1 (14 Ci/mmol) was obtained from Moravek Biochemicals (Brea, CA). Ultrapure sucrose (DNase and RNase free) and cesium chloride were obtained from BRL, Inc. (Gaithersburg, MD). Tris buffer, AFB_1, cesium sulfate (Grade 1), and ribonuclease A (R-4875) were purchased from Sigma Chemical Co. (St. Louis, MO). All glassware used in hybridization steps was washed with 0.01% diethyl pyrocarbonate and autoclaved for 4 h to minimize endogenous nuclease activity.

Animals

Male Fischer rats were obtained as weanlings from Charles River Breeding Laboratories (Wilmington, MA) and Harlan-Sprague Dawley (Houston, TX). They were housed in pairs in suspended, wire-bottom cages and fed Ziegler rodent chow and water *ad libitum*.

Isolation of Ribosomal DNA

The experimental procedures used for rDNA isolation are summarized in Figure 2 and can be summarized as follows. Male Fischer rats (125 to 150 g) were given injections of 1.0 mg [^{3}H]AFB_1/kg in 0.05 ml of dimethyl sul- foxide and were killed 2 to 12 h later. After perfusion *in situ* with buffer [25 mM Tris-HCl (pH 6.9): 0.25 M sucrose:3.3 mM CaCl$_2$], livers were minced and homogenized in the above buffer plus 1% Triton X-100 (v/v). A crude DNA preparation, isolated from liver nuclear fractions via modification of the phenol extraction method of Marmur, was sheared by homogenization; this suspension was subjected to cesium salt density centrifugation yielding a final nuclear DNA isolate relatively free of RNA and protein contamination.[21] Each gradient was fractionated while monitoring 254 nm absorbance, and DNA fractions of maximum density were retained. These fractions, which accounted on average for approximately 10% of the total fractionated DNA, were briefly dialyzed, and the DNA, after precipitation in cold ethanol, was solubilized in 0.11 M glycine:NaOH (pH 10.5). After incubation for 30 min at 37°C, the alkali-treated DNA was again recovered by ethanol precipitation. Samples of DNA collected before and after glycine treatment were analyzed to quantify the loss of aflatoxin-bound radioactivity due primarily to tritium exchange during base treatment. AFB$_1$-tritium bound to DNA was determined by liquid scintillation. DNA concentrations were quantified by the diphenylamine centrimetric reaction as modified by Giles and Myers.[10]

Ribosomal DNA sequences were selectively isolated from these nuclear DNA

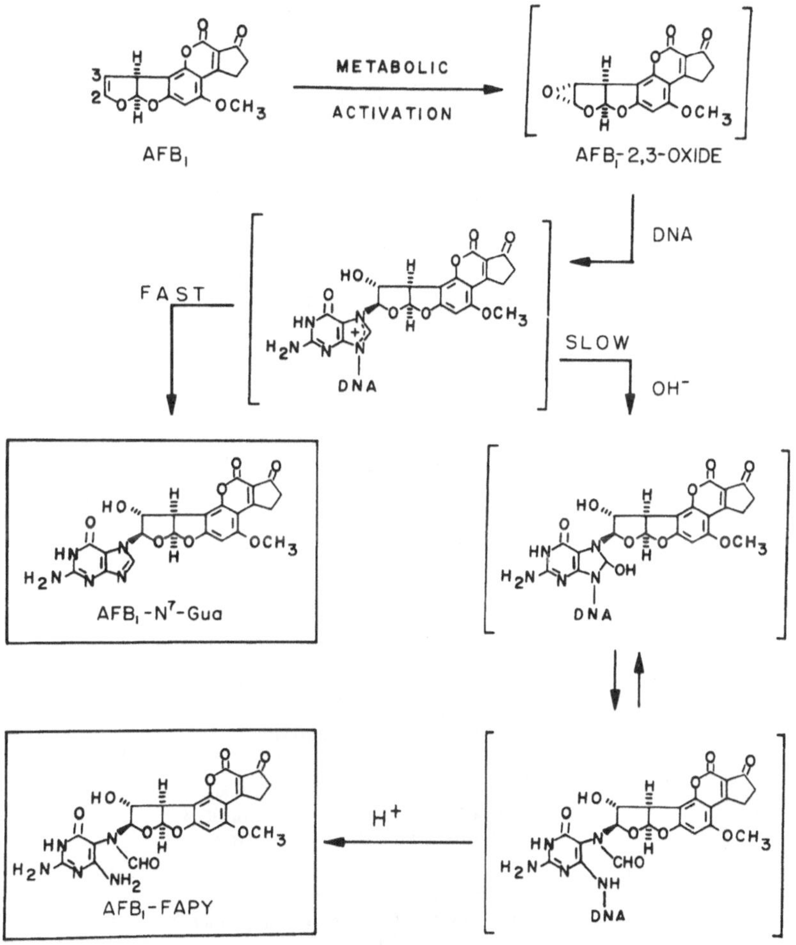

Fig. 1. Formation of AFB$_1$-DNA adducts subsequent to metabolic activation of AFB$_1$ *in vitro* and *in vivo*.

reactions by hybridization with rRNA using methods adapted from those of Wellauer and Dawid.[33,34] DNA isolates were incubated in 70% formamide buffer at a concentration of 0.015 mg/ml; 18S and 28S rRNA were added to a final concentration of 0.015 mg/ml for each RNA species, and the nucleic acid mixture was incubated at 50°C for 3 h. The hybridization mixture was chilled, dialyzed and subjected to 2 successive CsCl gradient centrifugations. The rDNA fractions, identified by filter hybridization of cesium gradient aliquots with [^{32}P]rRNA, were pooled and digested with ribonuclease A to remove DNA-bound rRNA. Samples of the rRNA isolate were taken for DNA quantitation (via UV absorbance at 260 nm and by colorimetry) and DNA-bound radioactivity determination (as described above).

Analysis of rDNA Adduct Populations

Ribosomal DNA isolates were made 0.1 N in HCl by the addition of 1 N HCl, heated at 90°C for 15 min, and cooled on ice. Potassium acetate and $ZnCl_2$ were subsequently added to final concentrations of 50 mM, respec-

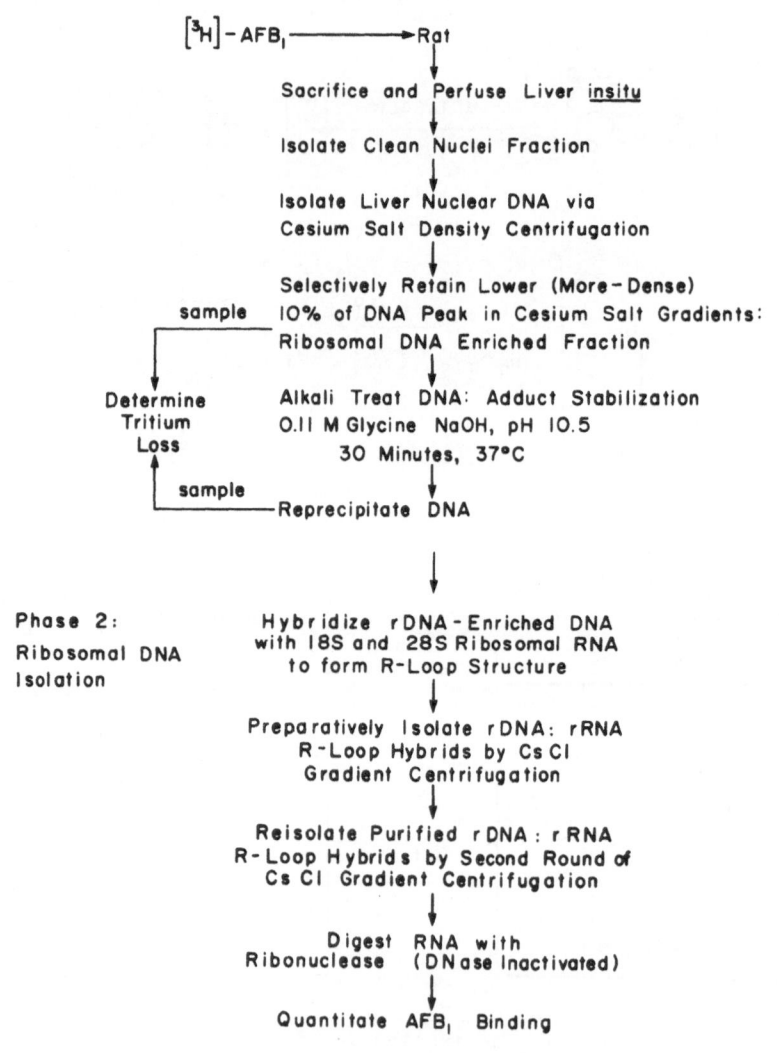

Fig. 2. Protocol for isolation of ribosomal RNA gene sequences from liver nuclear DNA of AFB$_1$-treated animals.

tively, and the depurinated, denatured DNA was digested for 3 h at 37°C
with nuclease P_1 (Sigma Chemical Co.: 1.0 mg enzyme protein/50 mg DNA). The
DNA hydrolysate was made 5% in ethanol and injected onto a C_{18}-Bondapak
column (Waters Associates) eluted at ambient temperature within 45-min linear
gradient (flow rate, 1.0 ml/min) of 13 to 18% ethanol adjusted to a pH of 5.0
by the addition of formic acid.

A second set of nuclear DNA preparations, enriched in rDNA sequences by
one round of cesium salt density centrifugation, was incubated in 20 mM
sodium phosphate: 1 mM Na_2EDTA (pH 6.8) at 85-88°C for 20 min at a DNA concentration of 0.6 to 0.7 mg/ml. As originally described by Lawley and Brooks,
heat treatment of N^7-substituted guanines causes depurination under
near neutral conditions: after heat treatment of AFB_1-adducted DNA, only
AFB_1-FAPyr adduct derivatives (formed by scission of the N^7, C^8, N^9 ring
of AFB_1-substituted guanines) should remain.[26] The neutral buffer-treated
DNA preparations were subsequently hybridized to rRNA; the resulting rDNA
isolates were depurinated via mild acid treatment, hydrolyzed with nuclease
P_1, and analyzed by HPLC, as described above.

Isolation of Transfer DNA

Chromatin was prepared from purified liver nuclei, digested with DNase II
(100 units/ml), and DNase II-sensitive chromatin was recovered as described
by Rodriques and Becker.[14] Purified DNA was isolated from DNase-sensitive
chromatin by cesium salt density centrifugation yielding 6-9% of initial
nuclear DNA. DNase II-sensitive chromatin DNA was incubated for 30 min in
0.11 M glycine-NaOH buffer, pH 10.5 at 37°C, and subsequently recovered by
ethanol precipitation. Transfer RNA gene sequences (tDNA) were selectively
isolated from chromatin DNA by hybridization with tRNA probes to form R-loops
utilizing methods adapted from those of Wellauer and Dawid described above
employing transfer RNA as the RNA probe.

In vitro Adduction of Nuclear DNA

Rat liver DNA was adducted in vitro with [^{3}H]AFB_1 via incubation in a
buffered NADPH-generating system with phenobarbital-induced rat liver microsomes as described by Essigmann and coworkers.[7] AFB_1-modified DNA was
reisolated from the in vitro reaction mixture via cesium salt density
centrifugation.

RESULTS

AFB^1 Modification of Ribosomal versus Nuclear DNA

Ribosomal DNA, the DNA sequences coding for the 45S precursor to 18S and
28S rRNA, was initially investigated since these genes are present in
approximately 300 copies per cell in the rat, and the presence of multiple
copies made it feasible in the early experiments of this study to isolate
sufficient rDNA for quantitative adduct analysis. In addition, rRNA synthesis is preferentially inhibited (as compared to the synthesis of mRNA and
tRNA) in AFB_1-treated animals, due exclusively to the impairment of rDNA
template function.[37] These biochemical findings suggested that rRNA genes
were preferential targets for AFB_1 modification and thus appropriate model
gene sequences with which to study localized carcinogen modification.

Figure 3 compares the levels of adduction within ribosomal DNA and
unfractionated nuclear DNA over a 12 h period after treatment with AFB_1.
rDNA contained more AFB_1 residues per mg DNA than did total nuclear DNA
over the entire period; 2 h after toxin administration, 4.8-fold more AFB_1
residues were bound per mg DNA to rDNA than to nuclear DNA, and this ratio
was similar (4.1) at 12 h postdosing.

As is evident from the slopes of the 2 curves in Figure 3, the rates of AFB$_1$ adduct removal differ between ribosomal and nuclear DNA fractions. Over the 12-h period of observation, the half-life of bound AFB$_1$ residues within nuclear DNA was calculated to be 11 h, while that of adducts within rDNA was 6 h.   The AFB$_1$-N^7-guanine moiety is quite stable *in vitro* with a half-life of approximately 100 h;[13] this observed decrease in the *in vivo* as compared to the *in vitro* half-life of AFB$_1$ adducts (11 as compared to 100 h, respectively) suggests an involvement of enzymatic activity in the removal of these adduct species from DNA.

Further examination of the data in Figure 3, in terms of the rate of adduct removal, reveals marked preferential removal of AFB$_1$:DNA moieties from within rRNA gene sequences. Table 1 summarizes the rates of removal of bound AFB$_1$ moieties from ribosomal and unfractionated liver nuclear DNA during the time intervals between points of observation. From 2 to 12 h after AFB$_1$ residues were initially stabilized by base-catalyzed conversion

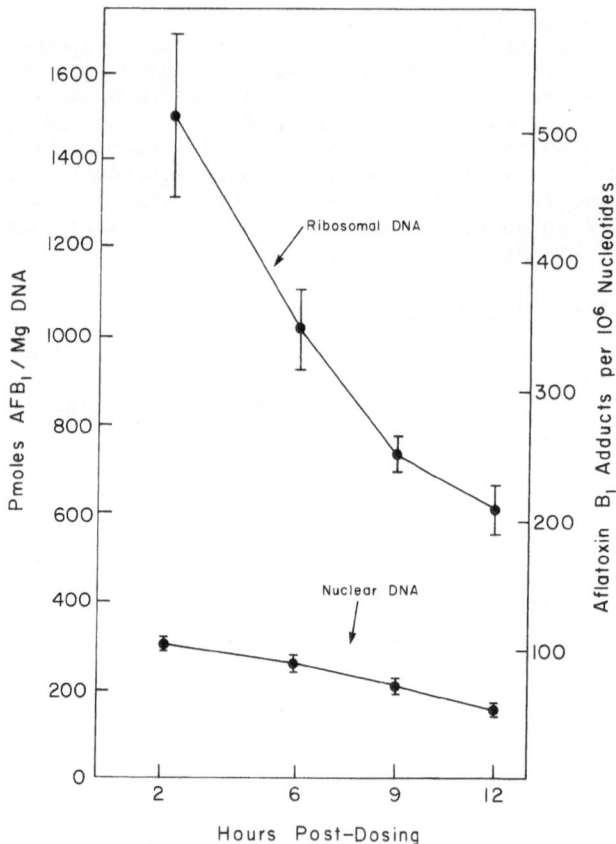

Fig. 3.   AFB$_1$ adduction of ribosomal DNA and total nuclear DNA isolated from animals administered 1 mg [^{3}H]AFB$_1$/kg body weight and sacrificed 2-12 h after dosing. Points, average of 3 experiments bars, standard equivalent of the mean.

to formamidopyrimidine derivatives (Fig. 1). This treatment, however, precluded detection of the parent N^7-guanine adduct. The procedure used to
estimate the amounts of the 3 major AFB_1 adduct species present in ribosomal DNA fractions was, in principle, to quantify in each sample the total
AFB_1-bound radioactivity that could be liberated by heat-induced depurination under neutral pH conditions as an index of the amount of the $AFB_1:N^7$-
guanine adduct present. The residual, stable radioactivity would thus represent AFB_1:FAPyr and AFB_1:Peak F adducts.

This heat-induced depurination protocol was used to estimate levels of
the 3 principal AFB_1:DNA adducts in ribosomal and nuclear DNA samples isolated at intervals during a 12-h period following aflatoxin dosing as described earlier. The results are presented in Table 2. After 2 h, $AFB_1:N^7$-gua-
nine was the major adduct in nuclear DNA, comprising 87% of the bound AFB_1
radioactivity. The relative proportion of this adduct decreased with time,
and after 12 h represented only 56% of the total. Over the same period, the
stable AFB_1:FAPyr derivatives increased proportionally, as the total
bound adduct levels decreased from 288 to 158 pmol/mg DNA. Although the
total adduct levels in rDNA were elevated by a factor of 3 or 4 over those
in nuclear DNA, there was no difference in the distribution of individual
adducts or in the rate at which they were removed.

AFB_1 Modification of Transfer versus Nuclear DNA

The ribosomal DNA studies summarized above employ a gene model which
could be isolated to bulk quantity due to enrichment in either G-C or A-T
bases. Methods similarly are needed for the isolation of specific, expressed
genes from carcinogen-treated cells and tissues which would permit the qualitative as well as the quantitative description of DNA adduct formation and
removal. To extend these studies to other, non-GC rich genomic sequences, we
developed a generic approach in which nuclear chromatin, enriched in transcribed genes by DNase II treatment, is hybridized with RNA probes to form
DNA:RNA R-loop hybrids; recovery of double-stranded DNA from purified hybrids
permits the comparative analysis of carcinogen distribution within various
nuclear genes.

Figure 4 compares AFB_1 adduction of tDNA versus total nuclear DNA
within liver tissue of AFB_1-treated animals. The values given summarize
the levels of AFB_1 modification within these two DNA populations from
animals injected with 0.25-2.0 mg of [^{3}H]AFB_1 per kg of body weight and
sacrificed after 2 h. Binding levels extended from 110 to 590 pmoles of
AFB_1/mg DNA (37-190 AFB_1 adducts per 10^6 nucleotides) in nuclear DNA and
from 480 to 3390 (170-1180) in tDNA. AFB_1 residues were found to occur preferentially within tDNA at each dose level studied; on a pmol/mg basis, 5.3-
6.8-fold more AFB_1 residues were bound to tDNA than to nuclear DNA over this
dose range.

Two factors could account for the preferential formation of AFB_1-DNA
adducts in tDNA as compared to total nuclear DNA: chromatin morphology as
well as nucleotide sequence. Humayun and coworkers have reported preferential modification of specific guanine residues by AFB_1 is determined in
part by the nature of flanking nucleotides; the nucleotide sequence and composition within transfer RNA regions could therefore influence the formation
of AFB_1 adducts preferentially within these genes.[14] To differentiate
between these two explanations, the susceptibility of tDNA sequences to
AFB_1-DNA adduct formation was determined within rat liver DNA adducted with
AFB_1 in vitro; in the absence of native in vivo chromatin conformation, the
effect of nucleotide content and sequence on tDNA AFB_1 adduct formation
could be monitored.

Figure 5 compares the levels of AFB_1 adduction within in vitro modified

Table 1.   Rate of removal of AFB$_1$:DNA adducts from ribosomal
           and nuclear DNA at various time intervals after dosing.

AFB$_1$:DNA residues removed from nuclear and ribosomal DNA
isolated from rat liver 2-12 h after administration of a
1.0 mg/kg AFB$_1$ dose.

Time interval (h)	Rate of removal (pmol AFB$_1$ mg/DNA/h)	
	rDNA	Nuclear DNA
2-6	118	14
6-9	98	12
9-12	43	20
Mean (2-12 h)	90	15

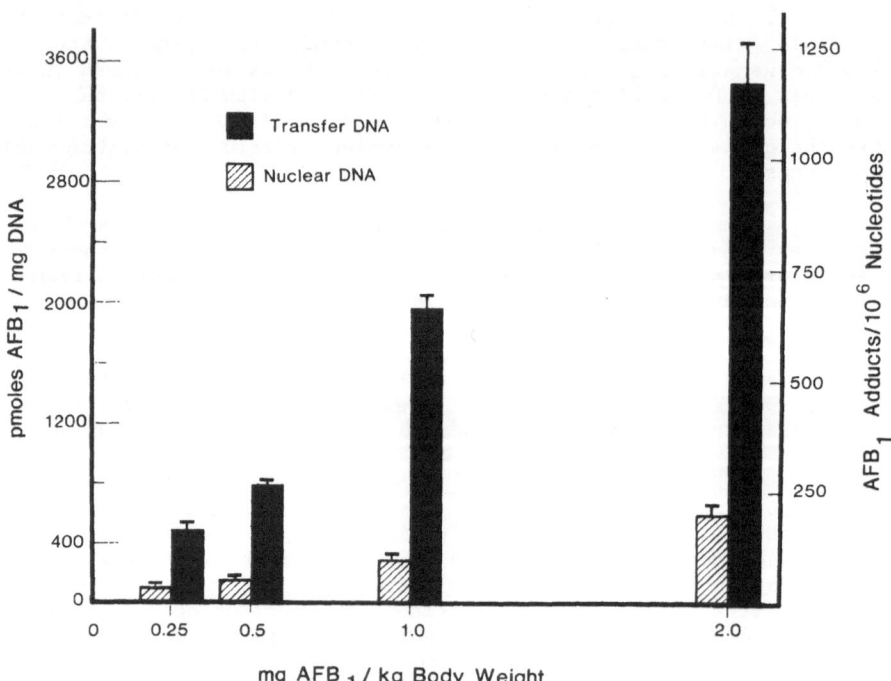

Fig. 4.   AFB$_1$ adduction of tDNA and total nuclear DNA isolated
          from animals administered 0.25-2.0 mg of [^{3}H]AFB$_1$ per
          kg of body weight and sacrificed 2 h after dosing.   Each
          bar represents the Mean $\pm$ SEM of three experiments.

Table 2. Quantitation of acid hydrolysis products of AFB$_1$-modified ribosomal and nuclear DNA at various times after dosing.

Concentrations of aflatoxin derivatives hydrolyzed from AFB$_1$-modified DNA fractions isolated from rat liver 2 to 12 h after administration of AFB$_1$ (1.0 mg/kg). In parallel studies using tritium and carbon-14 labeled AFB$_1$-adducted DNA, it was found that 12% of the AFB$_1$-bound tritium was liberated from AFB$_1$:FAPyr and AFB$_1$:Peak F under these conditions. The levels of bound AFB$_1$ radioactivity in rDNA prepared from nuclear DNA were thus adjusted to reflect this heat-induced tritium loss.

Time (h)	Concentration (pmol/mg DNA)							
	AFB$_1$:N^7-guanine		AFB$_1$:FAPyr		AFB$_1$:Peak F		Total	
	nDNA[a]	rDNA	nDNA	rDNA	nDNA	rDNA	nDNA	rDNA
2	251(87)[b]	1030(85)	28(10)	110(9)	9(3)	72(6)	288	1212
6	169(74)	550(69)	44(19)	190(23)	14(6)	60(8)	227	800
12	88(56)	330(55)	55(35)	200(34)	15(9)	60(11)	158	590

[a]nDNA, nuclear DNA.

[b]Numbers in parentheses, percentages of the hydrolysis products represented by each adduct peak.

357

rat liver DNA versus transfer DNA isolated from these same DNA preparations. Over the 6-fold range of AFB_1 modification, transfer DNA contained 10-20% more bound AFB_1 moieties than the unfractionated nuclear DNA sample from which it was isolated. Over this same range of AFB_1 modification (as shown in Fig. 1), liver tDNA isolated from AFB_1-treated animals contained 5-7 times more adduct moieties than unfractionated nuclear DNA. Given the 10-20% preferential modification of tDNA *in vitro* due to both guanine content and nucleotide sequence, these values suggest a principal role for chromatin ultrastructure in the preferential adduction of these DNA regions *in vivo*.

DISCUSSION

Differential rates of carcinogen-DNA adduct formation and removal have been described within specific, expressed regions of the nuclear genome. The data presented demonstrate preferential formation and removal of AFB_1:DNA residues with rRNA gene sequences, in the absence of qualitative differences in AFB_1:rDNA adduct populations, as compared to nuclear DNA. These findings also were extended to show similar preferential formation of AFB_1-DNA adducts within transfer RNA genes.

The results summarized above emphasize the importance of chromatin structure and conformation as a determinant in the localization of putative precarcinogenic and premutagenic lesions within carcinogen-treated cell populations. Recent investigations probing structural bases for gene activation and expression have established clear relationships between the transcriptional activity within expressed gene sequences and both the presence of specific chromatin proteins and DNase sensitivity of the respective DNA regions.[16,17,19] Extension of these observations in describing strutural associations with localized chemical-DNA damage further suggests important roles for chromatin ultrastructure and morphology in determining carcinogen-DNA adduct formation and repair.

The eucaryotic genome is organized into basic subunits consisting of DNA periodically folded around a histone core, creating series of DNA-protein particles (nucleosomes) connected by lengths of DNase-sensitive DNA (linker DNA). The relative accessibility of nucleosome core versus linker DNA to DNase digestion has been shown to correlate well with the susceptibility of these DNA regions to carcinogen modifications. Metzger et al. monitoring the release of micrococcal nuclease-soluble radioactivity from the liver nuclei of [^{3}H]acetylaminoflueorene-treated animals, found 2.2-fold greater levels of adduct residues within linker than in core DNA; Ramanathan and coworkers using a similar approach studying the release of radioactivity from the liver nuclei of [^{3}H]-dimethylnitrosamine-treated animals by DNase I, found 2.6-fold greater levels of methylation in linker as compared to core DNA.[24,30] Similar studies monitoring [^{3}H]AFB_1 distribution within core and linker DNA isolated from hepatic tissue have shown preferential linker DNA modification in the rainbow trout.[1]

Investigations examining the importance of chromatin structure in the localization of carcinogen adduct removal have additionally provided direct evidence for the role of DNA-protein interactions in the distribution of DNA repair. Lieberman and coworkers[28,32] have evaluated the relationship between the formation of carcinogen-DNA moieties and the subsequent distribution of DNA repair-incorporated nucleotides within DNase-sensitive (linker) and -resistant (core) regions of chromatin. Pulse chase and continuous label experiments indicated that, in cells treated with acetylaminofluorene of 7-bromomethylbenz(a)anthracene, a rapid and extensive redistribution of repair-incorporated nucleotides occurs creating with time a more uniform nucleotide distribution with respect to DNase sensitivity. This time-dependent redistribution of DNA repair products within chromatin was found to be associated with the rearrangement of histones along DNA.

These observations concerning chromatin ultrastructure suggest that following the onset of carcinogen-induced DNA repair, nontranscribed regions of chromatin would become increasingly susceptible with time to DNA repair processes. Examination of the rates of AFB_1:DNA adduct removal from rDNA versus nuclear DNA given in Table 1 support this hypothesis. During the first 6 h subsequent to AFB_1 administration, AFB_1:DNA residues were found to be removed at an 8-fold greater rate from rDNA as compared to unfractionated nuclear DNA; however, 9-12 h postdosing, the ratio of AFB_1:DNA adduct removal from rDNA versus nuclear DNA decreased to 2. Thus AFB_1-induced DNA repair synthesis, initially localized within transcribed DNA regions, might extend to non-transcribed regions as repair-induced chromatin conformational changes increase the accessibility of AFB_1:DNA adducts within these regions to enzymatic repair activity.

Left unresolved in these and other studies of AFB_1:DNA interactions *in vivo* is the quantitation of specific adduct removal and repair processes

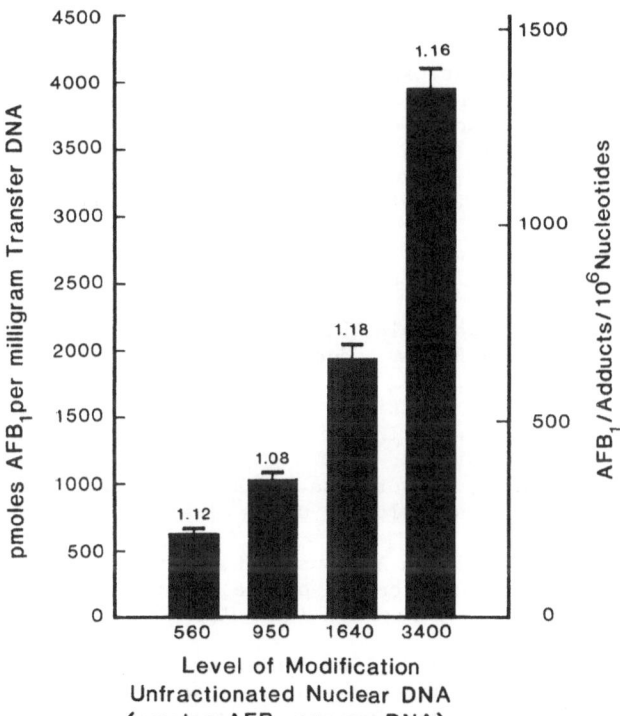

Fig. 5. Levels of AFB_1-DNA adducts within transfer DNA isolated from rat liver DNA modified *in vitro* via incubation with rat liver microsomes. Level of AFB_1-DNA adducts within the parent liver DNA sample is given on the X axis and adduct levels within tDNA are given on the Y axis. The ratio of DNA-bound AFB_1 moieties within tDNA/nuclear DNA is given above each bar figure. Each bar figure represents the average ± SEM of three experiments.

operative in AFB_1-treated cell populations. Spontaneous removal of $AFB_1:N^7$-guanine in the urine of AFB_1-treated animals further supports this mechanism of $AFB_1:DNA$ adduct loss.[2] Separation of AFB_1-modified DNA through sucrose density gradient centrifugation under alkaline conditions did not reveal the presence of persistent alkali-labile apurinic sites or single-strand breaks.[3] Thus the extensive loss of $AFB_1:DNA$ lesions through depurination appears to be accompanied by cellular processes which efficiently repair AFB_1-induced apurinic lesions.

Also left unresolved in these and other mechanistic studies of AFB_1-induced mutation is the identification of the specific premutagenic lesions(s) formed in AFB_1-modified DNA. Spontaneous or enzymatic removal of the AFB_1-N^7-guanine adduct should give rise to apurinic sites at a high frequency. The coexistence of apurinic sites together with stable $AFB_1-FAPyr$ adducts represent two types of potentially premutagenic lesions in the DNA of AFB_1-treated cell populations. The possibility thus exists that multiple mechanisms may account for the mutagenic potency of N^7-substituted aklylating agents such as AFB_1.

REFERENCES

1. Bailey GS, Nixon JE, Hendricks JD, Sinnhuber RO, Vanholde KE, 1980, Carcinogen aflatoxin B_1 is located preferentially in internucleosomal deoxyribonucleic acid following exposure *in vivo* in rainbow trout, *Biochemistry*, 119:5836-5842.
2. Bennett RA, Essigmann JM, Wogan GN, 1981, Excretion of an aflatoxin-guanine adduct in the urine of aflatoxin B_1-treated rats, *Cancer Res.*, 41:650-654.
3. Croy RG, 1979, Interaction of aflatoxin B_1 with DNA *in vivo* in the rat and mouse, *PhD Dissertation*, Massachusetts Institute of Technology, Cambridge.
4. Croy RG, Essigmann JM, Reinhold VN, Wogan GN, 1978, Identification of the principal aflatoxin B_1-DNA adduct formed *in vivo* in rat liver, *Proc Natl Acad Sci USA*, 75:1745-1749.
5. Croy RG, Wogan GN, 1981, Quantitative comparison of covalent aflatoxin-DNA adducts formed in rat and mouse livers and kidneys, *J Natl Cancer Inst.*, 66:761-768.
6. Croy RG, Wogan GN, 1981, Temporal patterns of covalent DNA adducts in rat liver after single and multiple doses of aflatoxin B_1, *Cancer Res.*, 41:197-203.
7. Essigman JM, Croy RG, Nazadan AM, Busby WF, Reinhod VN, Buchi G, Wogan GN, 1977, Structural identification of the major DNA adduct formed by aflatoxin B_1 *in vitro*, *Proc Natl Acad Sci USA*, 74:1870-1874.
8. Fahl WE, Scarpelli DG, Gill K, 1981, Relationship between B(a)-induced DNA base modification and frequency of reverse mutations in mutant strains of *Salmonella typhimurium*, *Cancer Res.*, 41:3400-3406.
9. Farber E, Cameron R, 1980, Sequential development of cancer, In: "Advances in Cancer Research Vol. 31," Klein G, Weinhouse S, eds., Academic Press, New York.
10. Giles KW, Meyers A, 1965, An improved diphenylamine method for the estimation of deoxyribonucleic acid, *Nature London*, 206:93.
11. Goth R, Rajewsky MF, 1974, Molecular and cellular mechanisms associated with pulse carcinogenesis in the nervous system by ethylnitrosourea: ethylation of nucleic acids and elimination rates of ethylated bases from the DNA of different tissues, *Z Krebforsch*, 82:37-42.
12. Gottesfeld JM, Partington GA, 1977, Distribution of messenger RNA-coding sequences in fractionated chromatin, *Cell*, 12:;953-962.
13. Groopman JD, Crou RG, Wogan GN, 1981, *In vitro* reaction of aflatoxin B_1-adducted DNA, *Proc Natl Acad Sci USA*, 78:5445-5449.

14. Irvin TR, Wogan GN, 1984, Quantitation of AFB_1 adduction within the ribosomal RNA gene sequences of rat liver DNA, *Proc Natl Acad Sci USA*, 81:664-668.

15. Kadlubar FF, Beland FA, Berenek DT, Dooley KL, Heflich RH, Evans FE, 1982, Formation of DNA damage by aromatic amines, *In*: "Environmental Mutagens and Carcinogens," Sugimura T, Kondo S, Takebe H, eds., Liss, New York.

16. Levy WB, Dixon G, 1977, Renaturation kinetics of cDNA complementary to cytoplasmic polyadenylated RNA from rainbow trout testes. Accessibility of transcribed genes to pancreatic DNase, *Nucleic Acid Res.*, 4:883-898.

17. Levinger L and Varshavsky A, 1982, Selective arrangement of ubiquit-inated and D_1 protein-containing nucleosomes within the Drosophila genome, *Cell*, 28:375-385.

18. Levy-Wilson B, Connor W, Dixon GH, 1979, A subset of trout testes nuclesomes enriched in transcribed DNA sequences contains high mobility group proteins as major structural components, *J Biol Chem.*, 254:609-620.

19. Levy-Wilson B, Gjerset RA, McCarthy BJ, 1977, Acetylation and phosphorylation of Drosophila histones: distribution of acetate and phosphate groups in fractionated chromatin, *Biochem Biophys Acta.*, 475-168-175.

20. Lutz WK, 1982, Dose dependency of DNA adduct formation by benzo(a)-pyrene, *Adv Exp Med Biol.*, 136:1349-1365.

21. Marmur J, 1981, A procedure for the isolation of DNA from microor-ganisms, *J Mol Biol*, 3:208-218.

22. McKnight SL, 1979, Visualization of transcribed nucleolar chromatin, *In*: "The Cell Nucleus," Busch H, Academic Press, New York.

23. McKnight SL, Miller OJ, 1976, Ultrastructural patterns of RNA synthesis during early embryogenesis of Drosphilia melanogaster, *Cell*, 8:305-319.

24. Metzger G, Wilhelm FX, Wilheim ML, 1977, Nonrandom binding of a chemical carcinogen to DNA in chromatin, *Biochem Biophys Res Commun.*, 75:703-710.

25. Misra RP, Muench KF, Humayun MZ, 1983, Covalent and noncovalent interactions of aflatoxin with defined deoxyribonucleic acid sequences, *Biochemistry*, 22:3351-3359.

26. Muench KF, Misra RP, Humayun MZ, 1983, Sequence specificity in aflatoxin B_1-DNA interactions, *Proc Natl Acad Sci USA*, 80:6-10.

27. Nebert DW, 1981, Genetic differences in susceptibility to chemically induced myelotoxicity and leukemia, *Environ Health Perspect*, 39:11-22.

28. Oleson FB, Mitchell BL, Dipple A, Lieberman MS, 1979, Distribution of DNA damage in chromatin and its relation to repair in human cells treated with 7-bromomethylbenz(a)anthracene, *Nucleic Acids Res.*, 7:1343-1361.

29. Pegg AE, Perry W, 1983, Alkylation of nucleic acids and metabolism of small doses of dimethylnitrosamine in the rat, *Cancer Res.*, 41:3128-3132.

30. Ramanathan R, Sharma DSR, Farber E, 1976, Nonrandom nature of *in vivo* methylation by dimethylnitrosamine and subsequent removal of methylated products from rat liver chromatin DNA, *Cancer Res.*, 36:2073-2078.

31. Schaaper RM, Loeb LA, 1982, Heat-induced mutagenesis of QX174 DNA *in vitro*, *Mutat Res.*, 104:75-78.

32. Tisty TD, Lieberman MW, 1978, The distribution of DNA repair synthesis in chromatin and its rearrangement following damage with N-acetoxy-2-acetylaminofluorene, *Nucleic Acids Res.*, 5:3261-3273.

33. Wellauer PK, Dawid IB, 1977, The structural organization of ribosomal DNA in Drosophilia melanogaster, *Cell*, 10:193-212.

34. Wellauer PK, Dawid IB, 1979, Isolation and sequence organization of

human ribosomal DNA, *J Mol Biol.*, 128:289-303.

35. Wogan GN, Croy RG, Essigmann JM, Groopman JD, Thilly WG, Skopek TR, Liber HL, 1979, Formation and removal of aflatoxin B_1 - DNA adducts, *In*: "Environmental Carcinogenesis: Occurrence, Risk, Evaluation and Mechanisms," Emmelot PE, Kreik E, eds., Elsevier/North-Holland Biomedical, New York.

36. Wong JJ, Hsieh DPH, 1976, Mutagenicity of aflatoxins related to their metabolism and carcinogenic potency, *Proc Natl Acad Sci USA*, 73:2241-2244.

37. Yu FL, 1977, Inihibition of RNA polymerases by aflatoxin B_1, *J Biol Chem.*, 252:3245-3251.

IMMUNOLOGIC *IN VIVO* DETECTION OF AFLATOXIN B$_1$-DNA ADDUCTS IN RATS AND RAINBOW TROUT

Takatoshi Ishikawa, Yoko Nakatsuru, Xiusheng Qin and
Prince Masahito

Department of Experimental Pathology
Cancer Institute
Tokyo, 170, Japan

INTRODUCTION

A relationship between mycotoxins and hepatocarcinogenesis has been suspected since aflatoxins were first isolated from peanut meal following an outbreak of liver disease among ducklings, turkeys, and pigs during 1960 in England.[2] A world wide epizootic occurrence of rainbow trout liver tumors also came to the attention of fish pathologists in 1960,[16] and nearly all cases could be attributed to hatchery diets containing aflatoxin-contaminated cottonseed meal.[16] Experimental studies have subsequently demonstrated that related mycotoxins are potent hepatocarcinogens in the rat, monkey, duck, and rainbow trout,[1,2,15,16] and convincing epidemiological evidence suggesting an association between primary liver cancer in man and aflatoxin contamination in some areas of Asia and Africa has been well documented.

Aflatoxins themselves are not active, requiring enzymatic activation before DNA binding can occur, and it is now considered that the principal ultimate product is aflatoxin B$_1$-8,9-oxide, which enters the nucleus and covalently binds to the DNA.[3] DNA adduct formation appears to be an important initial event for the process of initiation in carcinogenesis and it therefore follows that detection of carcinogen-DNA adducts in organs or tissues is a valid approach to monitoring exposure and for assessment of the importance of the interplay between DNA binding and DNA repair capacity.

Considerable advances have been made during recent years in the identification of carcinogen adducts in DNA by the introduction of various new methods.[5,7] For example, specific antibodies have been raised against DNA adducts produced with several kinds of carcinogens.[8,9,13,14] Since highly sensitive enzyme immuunoassays have been developed to measure adducts they can be used as probes in detection assays for DNA modification caused by low level exposure.[5,7] In this paper we will focus on immunological detection of aflatoxin B$_1$ adducts as a model. To establish assay conditions, immunologic detection levels were quantitated by concurrent radioactive analysis using tritium labeled aflatoxin B$_1$. Both the rat and rainbow trout were used, as convenient animal models in the present experiments.

Anti-aflatoxin B$_1$-DNA adduct antibodies

Both polyclonal and monoclonal antibodies against aflatoxin-DNA adducts

Microbial Toxins in Foods and Feeds, Edited by A.E. Pohland *et al.,*
Plenum Press, New York, 1990

were produced in this study. For technical reasons, calf thymus DNA was chemically modified *in vitro* by aflatoxin B_1 according to the method described by Martin and Garner.[11] Under the condition used, about one molecule of aflatoxin B_1 is bound per 100 base molecules.

Monoclonal antibodies. Mice were immunized with the aflatoxin B_1 modified DNA and monoclonal antibodies produced by the standard procedure. Titration and specificity of antibodies were checked by the ELISA method. A competitive inhibition assay using chemically modified DNA (also used for antibody production) and enzymatically modified DNA was carried out. Both DNAs demonstrated similar effective inhibition, whereas calf thymus DNA or benz(a)pyrene diol epoxide(BPDE)-I modified DNA were ineffective. However, when the antibody was applied to detection of liver DNA isolated from animals treated with aflatoxin B_1 *in vivo*, it proved insufficiently sensitive for detailed analysis. It was concluded that the monoclonal antibody recognizes a special type of DNA modification, which may not be the main adduct produced *in vivo*.

Polyclonal antibodies. Polyclonal antibodies were therefore raised. Here the same chemically modified DNA that had been used for monoclonal antibody production was applied. Seven rabbits were immunized several times with aflatoxin B_1 modified DNA plus adjuvant. To select suitable sera for use, competitive inhibition assays were carried out and only one serum was found to be specific. For competitive inhibition assays of this rabbit serum various kinds of DNA modified by aflatoxin B_1 were applied. Among them chemically modified DNA showed the highest inhibition. Interestingly, the imidazole ring opened (iro)-aflatoxin B_1-DNA[11] was less inhibitory, showing a similar inhibition curve to that obtained by DNA modified by aflatoxin B_1 enzymatically using rat microsomes. One antibody found to be specific was used in the further experiments.

Comparative DNA binding studies between rat and rainbow trout

To standardize the immunological detection of aflatoxin B_1-DNA adducts, radioactive analysis was simultaneously conducted. For this, male F344 rats were injected with tritium labelled aflatoxin B_1 at doses of 1 to 2 mg per kilogram and their livers removed 2 to 4 hours later. Similarly, rainbow trout (*Salmo gairdneri*) of both sexes received an intraperitoneal injection of tritium labeled aflatoxin B_1 at doses of 0.3 to 0.5 mg per kilogram and sacrificed after 9 to 24 hours. The fish were maintained in laboratory aquaria at 15-20°C. DNA was extracted from the liver and radioactivities compared. The adduct levels measured by radioactivity were approximately the same in rat and rainbow trout livers even though smaller aflatoxin B_1 doses were administered in the rainbow trout than in the rat.

Immunological detection of aflatoxin B_1-DNA adducts in animal tissues

Similar competitive inhibition curves were obtained using aflatoxin B_1-modified DNA from both rat and rainbow trout livers and it was thus apparent that the polyclonal antibody could recognize liver DNA modified with aflatoxin B_1 *in vivo*. This suggests that equivalent DNA modifications might be common to rat and rainbow trout.[3,4] The curves observed with DNA chemically modified by aflatoxin B_1 were also similar. Assessment of the sensitivity of immunodetection by comparing immunologic results with those obtained from radioactive analysis revealed a lowest limit of detection of one adduct per 10^5 nucleotides, about 10^5 adducts per cell.

Standardization of the aflatoxin B_1-DNA adduct levels by radioactive analysis and ELISA method

For future assessment purposes it is necessary to estimate absolute values for DNA samples from animals for which the level of exposure to aflatoxin B_1 is unknown. In order to validate the assay conditions, several aflatoxin B_1-modified DNA samples obtained from animals treated under different conditions were therefore tested. For rats, several DNA samples with different adduct levels calculated from radioactive analysis were used. For immunological detection, the value corresponding to 50% inhibition was taken as the adduct level. The values for each DNA sample calculated from radioactive analysis and ELISA methods were in good accord. Similarly, when studies on DNA samples from rainbow trout liver treated with aflatoxin B_1 were performed, adducts levels obtained by radioanalysis and ELISA were within the same order of magnitude. However, the ratio between ELISA method and radioactive analysis values tended to diminish with decreasing adduct levels.

Comparative studies on aflatoxin B_1-DNA adduct formation between rainbow trout and coho salmon

Interestingly different species of salmonid fish show wide variation in their response to aflatoxins.[12] One possible explanation is that the high susceptibilities of some species might be related to a high capacity for specific carcinogen metabolism and a correspondingly high rate of DNA modification. Comparative biochemical or adduct measurement in different species can give us a clue to whether this is indeed the case or whether the second possibility that species differences could be due to some genetic resistance or proneness to cancer unrelated to carcinogen activation is more likely. In rainbow trout, which are known to be especially sensitive to aflatoxins, liver cancers were induced even after dietary exposure to 20 ppb for only one day. The epizootic occurrence of liver cancers in hatchery reared rainbow trout in 1960 has been mentioned above.[16] Halver and Ashley[1] first reported that aflatoxins were effective in experimental hepatic cancer induction in rainbow trout. Their comparative studies showed that coho salmon (Oncorhynchus kisutch) are 10-30 times more resistant to aflatoxins than rainbow trout and do not develop liver tumors under identical conditions to these giving good yields in the trout.[6] Coho salmon are anadromous migratory fish in the United States and Canada, spending their first year after hatching in fresh water and then going to the sea to mature. This fish was imported into Japan and has been cultured commercially in some hatcheries. The coho salmon examined in this study had been kept throughout their lives in fresh water. For comparative studies rainbow trout and coho salmon were injected intraperitoneally with two doses of tritium labeled aflatoxin B_1 (0.1 and 0.5 mg per kilogram). Fish were maintained at 15°C in our laboratory aquaria. Their livers were removed 24 hours after carcinogen treatment, DNA was extracted and radioactivities compared. The results revealed adduct levels in coho salmon to be significantly lower (about 1/5) than in the rainbow trout case. Competitive inhibition experiments using DNA from coho salmon treated with aflatoxin B_1 did not reveal detetable levels of inhibition as was also found with control DNA. Although it must be borne in mind that it has not yet been established that our antibody specifically recognizes DNA modifications produced in coho salmon liver, the present results do suggest that variation in metabolism and binding of active species might be directly involved in the observed differences in species susceptibility.

Immunohistochemical detection of aflatoxin B_1-DNA adducts

Preliminary studies on immunohistochemical detection of aflatoxin B_1-modified DNA in individual rat and rainbow trout liver cells have already been commenced in our laboratory. In previous studies we reported that the antibodies against hydroxyaminoquinoline 1 oxide (4HAQO) and benzo(a)pyrene diol epoxide(BPDE) I DNA adducts could be effectively used for detection of adducts in paraffin-embedded tissue sections after whole body carciogen application.[13,14] Immunohistochemical studies have the advantage of allowing specific morphological detection of carcinogen-DNA adducts at the cellular level, although ELISA and radioimmunoassays are probably more sensitive and therefore suitable for determining absolute quantities of carcinogen-DNA adducts.

SUMMARY AND CONCLUSIONS

1. Immunological assessment of aflatoxin B_1-DNA adduct formation was performed using both monoclonal and polyclonal antibodies in two convenient animal models, utilizing rats and rainbow trout. The polyclonal antibody proved useful for screening aflatoxin B_1-modified DNA *in vivo*.

2. All adduct measurements were performed relative to standard carcinogen-modified DNA preparations with known levels of adduct (calculated from radioactive labeling). The lowest limit of detection with this assay was about one adduct per 10^5 nucleotides, about 10^5 adducts per cell.

3. Our polyclonal antibody was also useful for detection of aflatoxin B_1-modified DNA from rainbow trout liver. Thus the antibody probably recognizes DNA modifications common to both rat and rainbow trout.

4. Aflatoxin B_1-DNA adduct levels between rainbow trout and coho salmon were compared. The adduct levels in coho salmon were much lower than in rainbow trout suggesting that species differences in susceptibility to aflatoxin B_1 may be partly due to variation in ability to metabolize the carcinogen to DNA reactive species.

ACKNOWLEDGEMENTS

1. Supported by Grants-in-aid for Cancer Research from the Ministry of Education, Science, and Culture, the Ministry of Health and Welfare of Japan, and the Smoking Research Foundation.

2. We are grateful to Dr. Junya Yonezawa (Tokyo Metropolitan Fisheries Experiment Station Okutama Branch, Tokyo) and Dr. Ryozo Sato (National Research Institute of Aquaculture, Nikko Branch, Nikko city) for providing fish used in this study. We thank Mr. Hironori Murayama for technical assistance and Miss Tomoko Toyama for help in preparation of the manuscript.

REFERENCES

1. Ashley LH, Halver JE, 1961, Hepatomagenesis in rainbow trout, *Fed Proc.*, 20:290.
2. Butler WH, Barnes JM, 1963, Toxic effect of groundnut meal containing aflatoxin to rat and guinea pigs, *Brit J Cancer*, 17:699-710.
3. Croy RG, Essigmann JM, Reinhold VN, Wogan GN, 1978, Identification of the principal aflatoxin-DNA adduct formed *in vivo* in rat liver, *Proc Natl Acad Sci USA*, 75:1745-1749.
4. Croy RG, Nixon JE, Sinnhuber RO, Wogan GN, 1980, Investigation of covalent aflatoxin B_1-DNA adducts formed *in vivo* in rainbow trout (*Salmo gairdneri*) embryos and liver, *Carcinogenesis*, 1:903-909.
5. de Serres FJ, 1988, Banbury Center DNA adduct workshop, *Mutation Res.*, 203:55-68.
6. Halver JE, Ashley LM, Smith RR, 1969, Aflatoxicosis in coho salmon, *Natl*

Cancer Inst Monogr., 31:141-155.

7. Harris CC, 1985, Future directions in the use of DNA adducts as internal dosemeters for monitoring human exposure to environmental mutagens and carcinogens, *Environ Health Perspect.*, 62:185-191.

8. Haugen A, Groopman JD, Hsu IC, Goodrich GR, Wogan GN, Harris CC, 1981, Monoclonal antibody to aflatoxin B_1-modified DNA detected by enzyme immunoassay, *Proc Natl Acad Sci USA*, 78:4124-4127.

9. Hertzog PJ, Smith JR, Garner RC, 1982, Production of monoclonal antibodies to guanine imidazole ring opened aflatoxin B_1 DNA, the persistent DNA adduct *in vivo*, *Carcinogenesis*, 3:825-828.

10. Loveland PM, Wilcox JS, Pawlowski NE, Bailey GS, 1987, Metabolism and DNA binding of aflatoxicol and aflatoxin B_1 *in vivo* and in isolated hepatocytes from rainbow trout (*Salmo gairdneri*), *Carcinogenesis*, 8:1065-1070.

11. Martin CN, Garner RC, 1977, Aflatoxin B_1-oxide generated by chemical or enzymic oxidation of aflatoxin B_1 causes guanine substitution in nucleic acids, *Nature*, 267:863-865.

12. Mashito P, Ishikawa T, Sugano H, 1988, Fish tumors and their importance in cancer research, *Jpn J Cancer Res (Gann)*, 79:545-555.

13. Nakagawa K, Tada M, Morita T, Utsunomiya T, Ishikawa T, 1988, Immunohistochemical detection of 4-hydroxyaminoquinoline 1-oxide-DNA adducts in mouse tissues *in vivo*, *J Natl Cancer Inst.*, 80:419-425.

14. Nemoto N, Nakatsuru Y, Nakagawa K, Tozawa A, Ishikawa T, 1988, Immunohistochemical detection of anti-(+)-trans-7,8-dihydroxy-9,10-epoxy-7,8,9,10-tetrahydrobenzo(a)pyrene bound adduct in nuclei of cultured Hela cells and mouse lung tissue, *J Cancer Res Clin Oncol.*, 114:225-230.

15. Sieber SH, Correa P, Dalgard DW, Adamson RH, 1979, Induction of osteogenic sarcomas and tumors of the hepatobiliary system in nonhuman primates with aflatoxin B_1, *Cancer Res.*, 39:4545-4554.

16. Wales JH, 1970, Hepatoma in rainbow trout, *In:* "A Symposium on Diseases of Fish and Shellfishes", Snieszko SF, ed., Amez Fish Soc., Washington, DC.

THE MODE OF ACTION OF ANTHRAQUINONE MYCOTOXINS ON ATP SYNTHESIS IN

MITOCHONDRIA

Kiyoshi Kawai,[a] Jiro Kitamura,[b] Takashi Hamasaki[c] and
Yoshinori Nozawa[d]

Chukyo Women's University, Ohbu 474, Japan[a]
Gifu Pharmaceutical University, Gifu 502, Japan[b]
Tottori University, Tottori 680, Japan[c]
Gifu University School of Medicine, Gifu 500, Japan[d]

INTRODUCTION

A variety of quinone and quinoid compounds have been isolated from fungi[1] and knowledge of their biological activities have been accumulated little by little. For anthraquinones, however, there is no such clear-cut biological activity as for the contribution of ubiquinones in the electron transport system of oxidative phosphorylatiaon. The genotoxicity, including mutagenicity, carcinogenicity, and anti-carcinogenicity of various anthraquinone compounds have been examined[2-8] (and references sited therein). The toxicity to cellular function and to mitochondrial function has also been studied in several anthraquinones from fungi.[9-20] The molecular basis of these *in vivo and in vitro* toxic activities is, however, not fully understood. In the present article, the results of our recent research on the interaction of anthraquinones with ATP synthesis in mitochondria are reviewed.

The Uncoupling Effect of Anthraquinones on Oxidative Phosphorylation in Mitochondria

Several hydroxyanthraquinones have been found to impede mitochondrial function by uncoupling oxidative phosphorylation and by depressing mitochondrial respiration.[12-20] Among 1-hydroxyanthraquinone and several dihydroxyanthraquiones tested (Fig. 1), only alizarin (1,2-dihydroxyanthraquinone) exerted the uncoupling effect and depression on mitochondrial respiration.[16] Emodin and its dimer quinone, skyrin (Fig. 1) from a yellow-rice mold, *Penicillium islandicum*, uncoupled the oxidative phosphorylation in mitochondria and their O-methyl derivatives at β-position were devoid of the uncoupling action.[12] Chrysophanol (Fig. 2), which has no hydroxyl group at the β-position, was negative in the uncoupling effect. Islandicin and its dimer quinone iridoskyrin (Fig. 1) from *P. islandicum* did not show the uncoupling effect.[20] Erythroglaucin (Fig. 1), which has been recently reported to uncouple mitochondrial respiration[19] against our expectation, did not exhibit reproducible results in our experiment using a synthesized sample.[20] These results indicate that only anthraquinones which have a hydroxyl group at β-position are capable of uncoupling mitochondrial respiration to inhibit ATP biosynthesis.

Averufin, nidurufin, norsolorinic acid and versicolorin A (Fig. 3), which

Microbial Toxins in Foods and Feeds, Edited by A.E. Pohland *et al.,*
Plenum Press, New York, 1990

are biosynthetic intermediates of aflatoxin B_1, were found to be potent uncouplers of oxidative phosphorylation[14,15] (and unpublished results for nidurufin and norsolorinic acid). The uncoupling effect of averufin and ver sicolorin A was strongest among those of natural products possessing the uncoupling activity (Fig. 4). Again, O-methyl and dehydroxy-derivatives of both anthraquinones were devoid of uncoupling activity.[14,15] Thus, we con- cluded that the presence of the hydroxyl group at the β-position is essen- tial for the uncoupling activity of anthraquinones.

It is now well documented that the chemicals capable of uncoupling

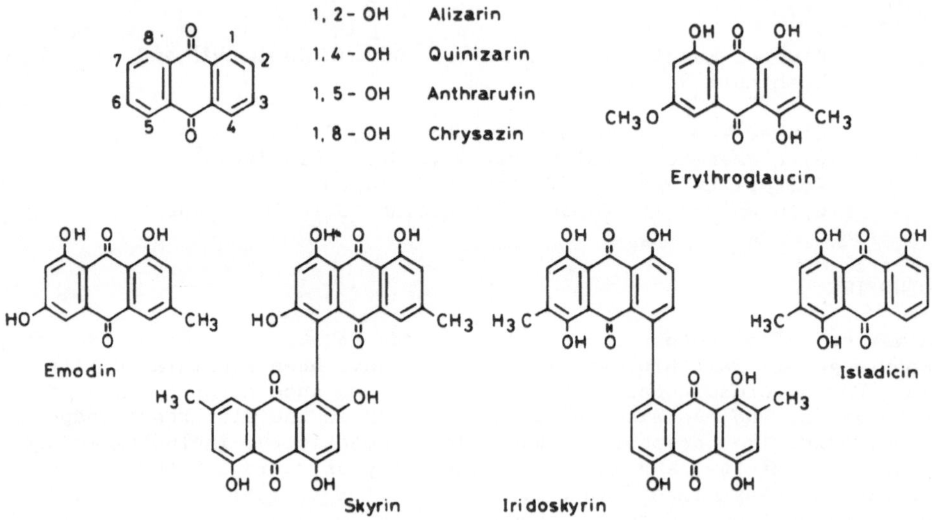

Fig. 1. Structures of dihydroxyanthraquinones, emodin, skyrin, erythroglaucin, islandicin, and iridoskyrin.

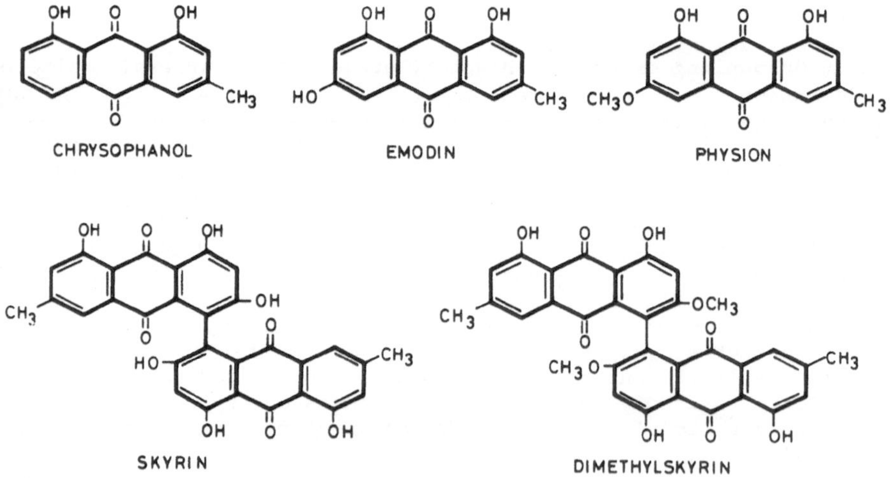

Fig. 2. Derivatives of emodin and skyrin.

oxidative phosphorylation in mitochondria are characterized by strongly lipophilic and weakly acidic properties by which they can conduct as protonophores carrying proton(s) across the mitochondrial inner membrane to cancel
the electrochemical potential induced by the electron transport system.
Spectrophotometric titration of hydroxylanthraquinones revealed that the pK-
values were localized around the physiological pH range in anthraquinones
that had of uncoupling activity, strongly suggesting their protonophore
activity in the mitochondrion.

The Interaction of Anthraquinones with the Respiratory Chain in Mitochondria

All of anthraquinones, which exhibited the uncoupling effect, caused a
marked depression of mitochondrial respiration (the oxidation of succinate
and NAD-linked substrates) at a little higher concentrations than those for

OH O   OH O

HO   OH
      O

NORSOLORINIC ACID

OH O   OH

HO   CH₃
      O

R = H   AVERUFIN
  = OH   NIDURUFIN

OH O   OH

HO   O   O
      O

VERSICOLORIN A

Fig. 3.   Structures of averufin, nidurufin,
          norsolorinic acid, and versicolorin A.

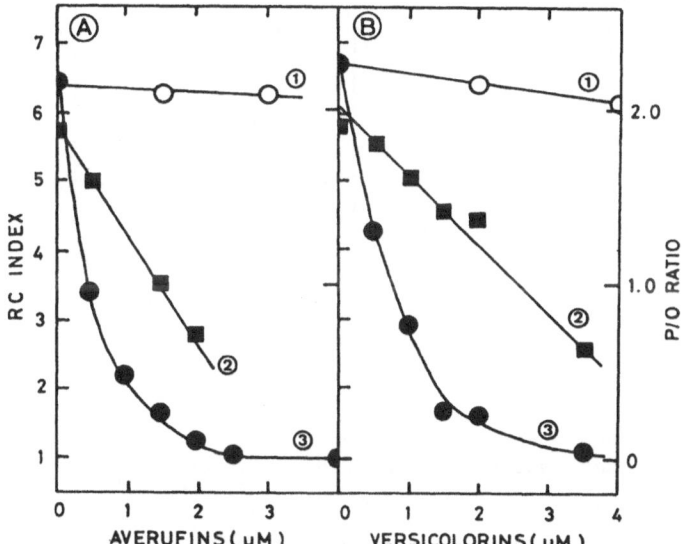

Fig. 4.   Uncoupling effect of averufin (A) and
          versicolorin A (B). Curves (1) show the
          effects of O-methyl derivatives of averufin
          and versicolorin A on RC ratio. Curves (2)
          and (3) show the effects of averufin and
          versicolorin A on P/O ratio and RC ratio,
          respectively. P/O and RC ratios were
          calculated from oxygraph.

the uncoupling effect. Except for averufin and versicolorin A, they did not exert any depressive effect on respiration of electron transport particles (ETP), indicating no direct interaction with the electron transport system (respiratory chain).

Averufin and versicolorin A inhibited both respiration in mitochondria and ETP (Fig. 5), implying their direct inhibition of the respiratory chain. The mode of inhibition of averufin, which showed a stronger effect than versicolorin A, was studied by means of ETP prepared from beef heart mitochondria. The reduced-minus oxidized difference spectrum of ETP with averufin gave a typical spectrum of the reduced form of cytochrome b (Fig. 6) as observed for the inhibition by antimycin A which was a selective inhibitor of the electron transport system at cytochrome bc_1 region. The activities of NADH-CoQ reductase (complex I) using duroquinone as an artificial electron acceptor in the presence of antimycin A, succinate-CoQ reductase (complex II) using phenazine methosulfate (PMS) as an artificial electron acceptor in the presence of antimycin A, and cytochrome c oxidase (complex IV) in the presence of TMPD were not affected by averufin. The activities of succinate-cytochrome c reductase (complex II and III) and durohydroquinone oxidase (Complex III and IV) were strongly inhibited by averufin. These results clearly substantiate that averufin selectively inhibited the enzyme activity of complex III at the cytochrome bc_1 region like antimycin A (Fig. 7).

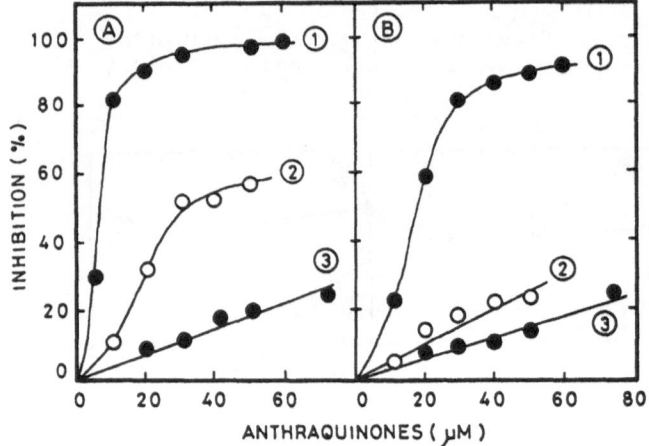

Fig. 5.   The effects of averufin, versicolorin A, and emodin on NADH (A) and succinate (B) oxidases in ETP. Curves (1), (2), and (3) show the effects of averufin, versicolorin A, and emodin, respectively. The inhibition rate was calculated from oxygraph.

Many complicated anthraquinoid compounds have been isolated such as luteoskyrin, rugulosin, rubroskyrin, flavoskyrin (Fig. 8) and anthracyclines. Luteoskyrin and flavoskyrin from *P. islandicum* have been demonstrated to impair mitochondrial function.[13,17] Rugulosin also shows a similar effect, but is conspicuously less toxic than luteoskyrin to mitochondrial respiration. Recently we have found that rubroskyrin interacts with the respiratory chain producing an electron transport shunt from complex I to oxygen (unpublished results). Luteoskyrin and rugulosin did not exhibit such redox-response in the respiratory system. Anthracyclines such as adriamycin and daunomycin are known to interfere with the mitochondrial respiratory system mitochondrial respiratory system by producing oxygen radicals or hydroxyl radical(s)[8] (and references sited therein).

These toxic effects of anthraquinone compounds to mitochondrial function may lead to the depression of cellular and organ functions when they are absorbed from the digestive tract and are conveyed to intracellular fluid without undergoing detoxication by enzymatic and/or chemical processes.

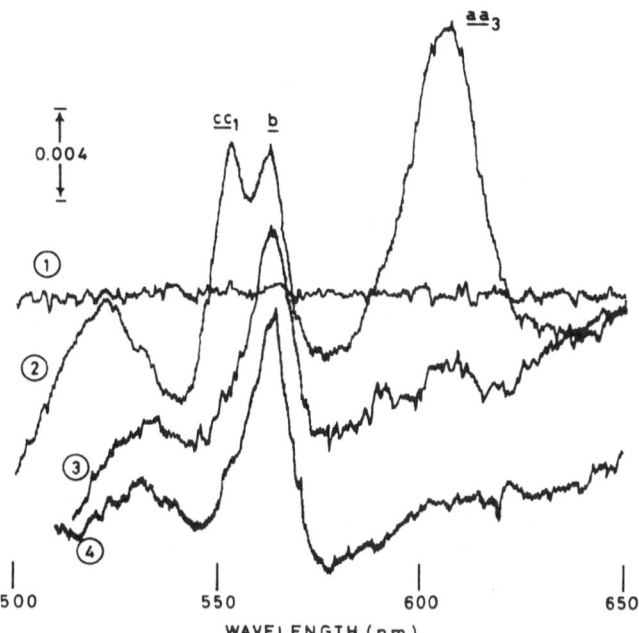

Fig. 6. The reduced-minus oxidized difference spectra of ETP. Curves (1), (2), (3) and (4) show control without succinate, reduction by succinate in the presence of cyanide, averufin, and antimycin A, respectively.

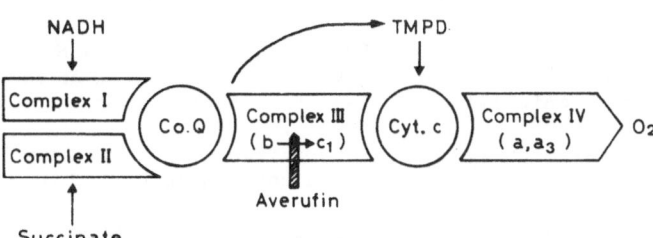

Fig. 7. The enzyme complexes of mitochondrial respiratory chain and proposed site of inhibition by averufin.

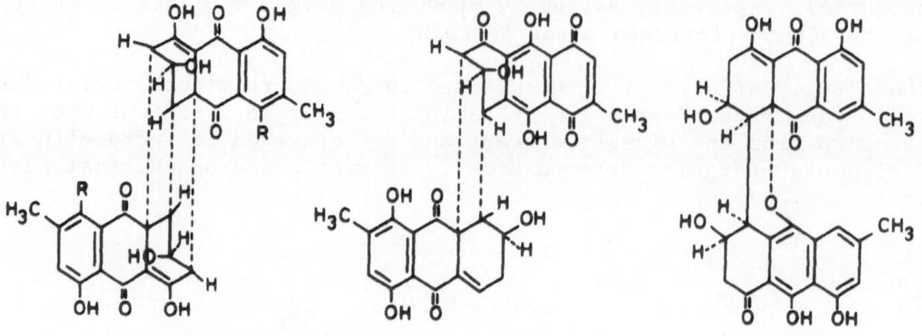

R = OH (-)Luteoskyrin
R = H   (-)Rugulosin

(-)Rubroskyrin

(-)Flavoskyrin

Fig. 8. Structures of luteoskyrin, rugulosin,
rubroskyrin, and flavoskyrin.

ACKNOWLEDGEMENTS

Authors are grateful to Dr. T. Ozawa of Nagoya University and Dr. Y. Ebizuka of Tokyo University for their kind supply of beef heart ETP and erythroglaucin, respectively.

REFERENCES

1. Thomson RH, 1971, "Naturally Occurring Quinones," Academic Press, New York.
2. Mori H, Kawai K, Ohbayashi F, Kuniyasu T, Yamazaki M, Hamasaki T, Williams GM, 1984, Genotoxicity of a variety of mycotoxins in the hepatocyte primary culture/DNA repair test using rat and mouse hepatocytes, *Cancer Res.*, 44:2918.
3. Mori H, Kitamura J, Sugie S, Kawai K and Hamasaki T, 1985, Genotoxicity of fungal metabolites related to aflatoxin B_2 biosynthesis, *Mutation Res.*, 143:121-125.
4. Brown JP and Brown RJ, 1976, Mutagenesis by 9,10-Anthraquinone derivatives and related compounds in *Salmonella typhimurium, Mutation Res.*, 40:203-224.
5. Brown JP and Dietrich PS, 1979, Mutagenicity of anthraquinone and benzanthrone derivatives in the *Salmonella*/microsome test: Activation of anthraquinone glycosides by enzymic extracts of rat cecal bacteria, *Mutation Res.*, 66:9-24.
6. Enomoto M and Ueno I, 1974, *Penicillium islandicum* (toxic yellowed rice) - luteoskyrin - islanditoxin - cyclochlorotine, *In*: "Mycotoxins," Purchase IFH, ed., Elsevier Scientific Publ. Co., New York.
7. Ueno Y, Sato N, Ito T, Ueno I, Enomoto T, and Tsunoda H, 1980, Chronic toxicity and hepatocarcinogenicity of (+) rugulosin, an anthraquinoid mycotoxin from penicillium species: Preliminary surveys in mice, *J Toxicol Sci.*, 5:295.
8. Porumb H and Petrescu I, 1986, Interaction with mitochondria of the anthracycline cytostatics adriamycin and daunomycin, *Prof Biophys Molec Biol.*, 48:103-125.
9. Ueno Y, Ueno I, Sato N, Iitoi Y, Saito M, Enomoto M and Tsunoda H, 1971, Toxicological approach to (-)rugulosin, an anthraquinoid mycotoxin of *Penicillium rugulosum, Japan J Exp Med.*, 41:177-188.
10. Umeda M, Saito M and Shibata S, 1974, Comparison of cytotoxic effects of

various anthraquinonoids on cultured mammalian cells, *Japan J Exp Med.*, 44:249-255.

11. Ueno Y, Sato N, Ito T, Ueno I, Enomoto M and Tsunoda H, 1980, Chronic toxicity and hepatocarcinogenicity of (+)rugulosin, an anthraquinoid mycotoxin from penicillium species: Preliminary surveys in mice, *J Toxicol Sci.*, 5:295.

12. Kawai K, Kato T, Mori H, Kitamura J and Nozawa Y, 1984, A comparative study on cytotoxicities and biochemical properties of anthraquinone mycotoxins emodin and skyrin from *Penicillium islandicum* Sopp., *Toxicol Lett.*, 20:155.

13. Ueno I, 1966, *Seikagaku* (in Japanese), 38:741.

14. Kawai K, Nozawa Y, Maebayashi Y, Yamazaki M and Hamasaki T, 1984, Averufin, an anthraquinone mycotoxin possessing a potent uncoupling effect on mitochondrial respiration, *Proc Jap Assoc Mycotoxicol.*, 18:35.

15. Kawai K, Nozawa Y, Maebayashi Y, Yamazaki M and Hamasaki T, 1984, Averufin, an anthraquinone mycotoxin possessing a potent uncoupling effect on mitochondrial respiration, *Appl Environ Microbiol.*, 47:481.

16. Kawai K, Mori H, Sugie S, Yoshimi N, Inoue T, Nakamaru T, Nozawa Y and Matsushima T, 1986, Genotoxicity in the hepatocyte/DNA repair test and toxicity to liver mitochondria of 1-hydroxyanthraquinone and several dihydroxy-anthraquinones, *Cell Biol Toxicol.*, 2:457.

17. Kawai K, Nozawa Y, Mori H and Ogihara Y, 1986, Inhibition of mitochondrial respiration by falvoskyrin, a toxic metabolite of *Penicillium islandicum* Sopp., *Toxicol Lett.*, 30:105-111.

18. Ubbink-Kok T, Anderson JA and Konings WN, 1986, Inhibition of electron transfer and uncoupling effects by emodin and emodinanthrone in *Escherichia coli*, *Antimicrob Agents Chemotheraphy*, 30:147-151.

19. Betina V and Kuzela S, 1987, Uncoupling effect of fungal hydroxyanthraquinones on mitochondrial oxidative phosphorylation, *Chem-Biol Interactions*, 62:179-191.

20. Kawai K, Ishizaki K, Hisadaka K, Nakamaru T and Nozawa Y, 1987, *Mycotoxin* (in Japanese), 26:43.

SUBCELLULAR DISTRIBUTION AND MACROMOLECULAR BINDING OF AFLATOXIN IN

FETAL AND NEONATAL RAT LIVER

S. Kumagai, Y. Ito, M. Iwaki, A. Nagahara, and Y. Omata

National Institute of Health
Kamiosaki 2-10-35, Shinagawa-ku
Tokyo 141, Japan

INTRODUCTION

The aflatoxins are mycotoxins produced by several species of *Aspergillus*, aflatoxin B_1 (B_1) being the most potent hepatotoxic and hepatocarcino-genic compound among these metabolites. There is a wide variation in suscep-tibility to B_1 among animal species and sexes.[23] Considerable research has been undertaken on biochemical mechanisms of the action of aflatoxin in rela-tion to the species and sex differences. Formation of B_1-8,9-epoxide by the cytochrome P-450 enzyme system and subsequent binding to DNA are regarded as critical steps of the carcinogenic effect of B_1.[1,21,24,29,31] The importance of formation of the B_1-glutathione conjugate by cytosolic glutathione S-trans-ferase in inactivation of B_1-epoxide has been well documented in relation to species and sex differences of the B_1 effects.[3,4,18,22,27]

Effects on offspring of pregnant or lactating rats exposed to aflatoxin have also been studied. Administration of B_1 to pregnant animals caused malformation, death and growth retardation in fetuses of hamsters, mice and rats.[5,6,8,28] When B_1 was given to female mice and rats during gestation or lactation, their offspring showed higher incidence of liver tumors than con-trol animals.[10,15,32] However, biochemical processes involved in such effects of aflatoxin remain uncertain. To gain a better understanding of the effects of aflatoxin in fetal and suckling rats, we studied the metabolic activity of the liver during development and the transfer of $[^3H]$-B_1 from preg-nant and lactating rats to the liver of their offspring.

MATERIALS AND METHODS

Animals and Chemicals

Wistar rats were maintained under 14 hr lighting and given food and water *ad lib*. The day of mating (=proestrus) was designated as day 0 of pregnancy and the day of parturition was day 0 of lactation. Parturition took place on day 22-23 of pregnancy. $[^3H]$-B_1 (30 Ci/mole) was purchased from Moravek Biochemicals Inc. (California, USA), and non-radiolabelled aflatoxins were from Makor Chemical Ltd. (Jerusalem, Israel).

Microbial Toxins in Foods and Feeds, Edited by A.E. Pohland *et al.,*
Plenum Press, New York, 1990

days after [3H]-B1 administration, and then decreased. Serum concentrations
of [3H] in suckling rats tended to be lower when their mothers were given
[3H]-B1 on day 9 oflactation than when given on day 0 of lactation (Fig.

Treatment of Rats and Sampling of Tissues

After ip injection of [3H]-B1 dissolved in DMSO (43 μCi/0.5 mg/kg, 1
mg/ml) or B1 dissolved in corn oil (1 mg/kg, 0.25 mg/ml) to rats, tissue
samples were taken from them under light ether anesthesia. Fetal blood was
taken from the umbilical vein through a polyethylene tube inserted into it,
and amniotic fluids from the amniotic cavity by a syringe. For studying subcellular distribution of [3H] and metabolic activity of the liver, livers of
adult rats were removed after perfusion with cold saline, and those of fetuses
and suckling rats were cut into pieces and washed briefly after being removed. Milk was collected from the stomach of suckling rats.

Determination of Metabolic Activity of Liver

Livers were homogenized with ice-cold 1.15% KCl, the homogenates were
centrifuged at 9000 g for 15 min, and supernatant was then centrifuged at
420,000 g for 15 min. The supernatant was used as cytosol. After being
resuspended in KCl and recentrifuged at 420,000 g for 15 min, the precipitate
was used as microsomes. Glutathione S-transferase (GST) activity in the cytosol was determined using 2,4-dinitrochlorobenzene as a substrate at 30°C[13].
Microsomal activity to bind B1 to DNA was determined by incubating calf
thymus DNA and [3H]-B1 with the microsomes.[36]

Preparation of Subcellular Fractions of Liver

Liver homogenates were separated into mitochondrial, cytosolic, microso-
somal and nuclear fractions, and then TCA insoluble components were obtained
from them according to the method of Mainigi and Campbell.[20]

Isolation of Hepatic Macromolecules

DNA and rRNA were isolated by the method of Lijinsky and Ross,[17] and protein was by the method of Swenson et al.[29] Nucleic acids were quantitated by
their absorbance at 260 nm. Quantity of protein was estimated from its dry
weight.

Determination of Radioactivity

Before being mixed with Atomlight (New England Nuclear, Mass., USA),
tissue homogenates and isolated nucleic acids were digested with solubilizer
(New England Nuclear) and TCA precipitates were with 1N NaCl. Protein recovered from phenol layer of the tissue homogenates, and milk from the rat
stomach, were combusted in an Oxidizer (Model 306, Packard Inst. Co., Ill.,
USA) to dissolve [3H] into scintillation medium (MONOPHASE S, Packard Inst.
Co., Ill., USA). The other samples were mixed with Atomlight without pretreatment. Radioactivity was determined in a scintillation counter
(LSC-700R, Aloka Co., Tokyo, Japan).

RESULTS

Changes in Metabolic Activity of the Liver with Age

Data from both sexes of infant rats were combined and are shown in Fig. 1
and 2, because no sex differences were noted in either DNA binding or GST
activity. Microsomal activity to bind B1 to DNA increased from fetal age
to 9 days of age (Fig. 1). The activity on 9 and 12 days of age was similar
to that in adult male rats, when expressed on the basis of amount of microsomal protein, but higher than the adult level when expressed on the basis of
liver weight (Fig. 1). GSA activity of liver cytosol remained nearly constant until 12 days of age. The levels were 1/2-1/3 of those in adult

male rats, when expressed on the basis of either liver weight or amount of cytosolic protein (Fig. 1).

B_1 administration caused significant reduction in B_1-DNA binding activity of microsomes at 9 days of age, but no noticeable changes in the activity on day 0. Cytosolic GST activity was not significantly affected by B_1 treatment (Fig. 2).

Concentrations of [³H] in Serum and Amniotic Fluid and Contents of [³H] in Liver and Milk

[³H]-B_1 was injected ip into pregnant rats of day 19, lactating rats on day 0 or day 9, and young adult male rats. Tissue samples were taken from the pregnant rats and their fetuses 3, 24, or 72 hrs after [³H]-B_1 adminis-tration to the pregnant rats, and from young adult male rats with the same time schedule after [³H]-B_1 administration. Tissue and milk samples were taken from suckling rats 2-7 days after their mothers were given [³H]-B_1.

The serum concentration of [³H] decreased rapidly in the pregnant and male rats during the 72 hrs after [³H]-B_1 administration, whereas the [³H] concentration in fetal serum was almost unchanged during the first 24 hrs and then decreased slowly compared to the maternal serum concentration (Fig. 3). Concentrations of [³H] in amniotic fluid, which could be taken at 3 and 24 hrs, were similar to, or slightly higher than those in fetal serum. When mother rats were given [³H]-B_1 on day 0 of lactation, serum concentra-tions of [³H] in suckling rats increased gradually to the fetal level by 4

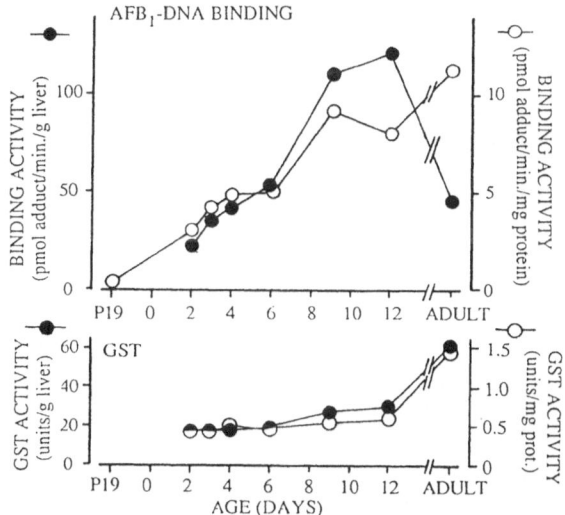

Fig. 1. Activity of hepatic microsomes to bind [³H]-B_1 to calf thymus DNA and gluta-thione S-transferase activity of hepatic cytosol. A point of day 19 of pregnancy (P-19) was the value determined using pooled fetal livers. Each point of the other ages indicates mean value for 6-8 livers.

Detectable amounts of [3H] was contained in milk taken from the stomach of the suckling rats (29-266 pmole equivalent B_1/g).

Liver contents of [3H] in the pregnant rats and male rats decreased markedly after [3H]-B_1 administration (Fig. 4). Consistent with the changes in serum [3H] concentrations, [3H] contents in the liver of the fetuses and suckling rats remained relatively constant during the sampling periods, being 1/2.3 - 1/38.5 of that in the pregnant or male rats. The [3H] content was similar in both fetuses and suckling rats at 3.6 days of age. Suckling rats at 11-17 days of age tended to show smaller contents of [3H] than those of 3-6 days of age.

Subcellular Distribution of [2H]

Subcellular distribution pattern of [3H] in liver was similar in pregnant rats and young adult male rats. In these rats, the proportion of [3H] was highest in nuclear fraction than in other fractions and a large part of [3H] in each fraction was associated with TCA insoluble components (Fig. 5). In contrast, the proportion of [3H] in the fetuses of mothers given [3H]-B_1

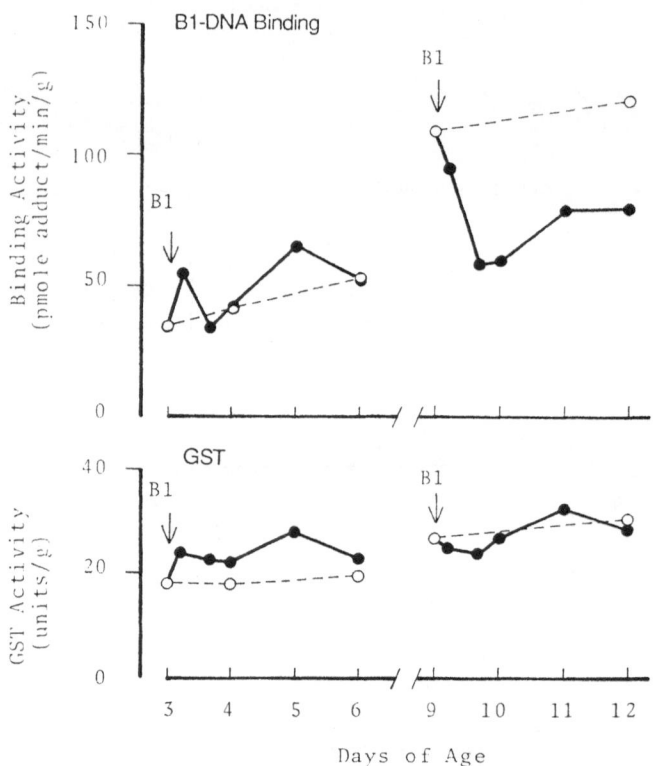

Fig. 2.   Changes in B_1-DNA binding activity of hepatic microsomes and glutathione S-transferase activity of hepatic cytosol in infant rats given B_1 (1 μg/ml). Each point indicates mean value for 3-8 livers.

was highest in the cytosolic fraction (Fig. 5). TCA insoluble fractions of fetal liver contained significant amounts of [³H], although its proportion was far lower than that in pregnant or young adult male rats (Fig. 5). male rats (Fig. 5). Distribution patterns in suckling rats or mothers given [³H]-B₁ resembled that in the fetuses (Fig. 6). Proportion of [³H] in TCA insoluble fractions was slightly higher in rats of 11-16 days than in those of 3-6 days of age.

For comparison, subcellular distribution of [³H] was studied in suckling rats which received an ip injection of [³H]-B₁ on 0 day or 9 days of age and were sacrificed 24 hrs later. Marked differences in distribution were noted between the rats of different ages (Fig. 7). Distribution pattern in rats of 10 days of age was nearly the same as that seen in the adult rats, whereas the pattern in rats of 1 day of age resembled that noted in the fetuses and suckling rats exposed to [³H] via their mother (Fig. 5, 6, 7).

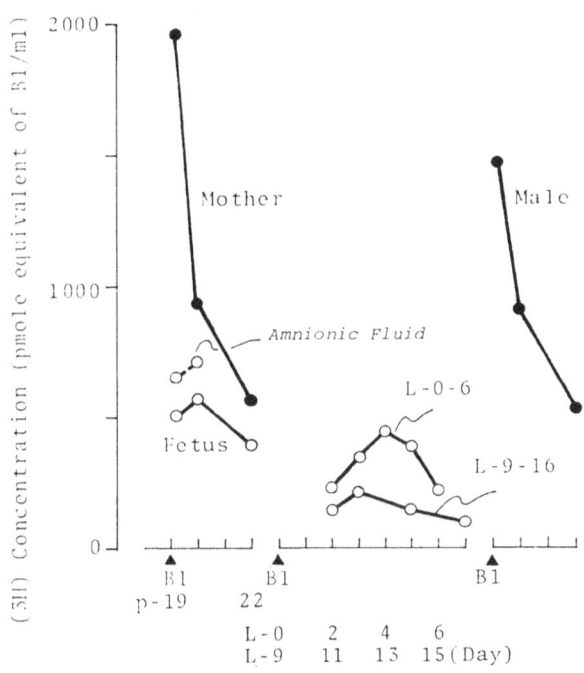

Fig. 3. Concentrations of [³H] in serum and amniotic fluid in fetal, suckling, pregnant and young adult rats. Samples were taken from pregnant rats and their fetuses, 3, 24 or 72 hrs after ip injection of [³H]-B₁ to the pregnant rats, and from young adult male rats with the same time schedules after ip injection of [³H]-B₁. Samples from suckling rats were taken 2-7 days after their mothers were injected ip with [³H]-B₁ on day 0 (L-0) or day 9 (L-9) of lactation. L-0 to L-6 and L-9 to L-16 indicate data of suckling rats of the mothers given [³H]-B₁ on day 0 and those of the mothers given on day 9, respectively. Pooled sera and amniotic fluids from 3-4 fetuses were used for a fetal sample. Each point indicates mean value for 2-4 samples.

Levels of [³H]-Aflatoxin Binding to Hepatic Macromolecules

Tritium was detected in hepatic macromolecules from offspring of mothers which were given [³H]-B$_1$ on day 19 of pregnancy or during lactation (Table 1). The maximal level of DNA binding found in suckling rats was close to 1/5 of that in young adult male rats.

The levels of [³H]-B$_1$ binding to hepatic macromolecules were far lower in the fetal and suckling rats whose mothers were given [³H]-B$_1$ during gestation or lactation than in the young adult male rats. When mothers were given [³H]-B$_1$ on day 19 of pregnancy, their offspring tended to show higher binding levels in the postnatal period than in the prenatal period (Table 1).

DISCUSSION

Sera of fetal and suckling rats contained [³H] derived from [³H]-B$_1$ given to their mothers. Detectable amounts of [³H] were associated with macromolecules and acid-insoluble components of the liver of these fetal and suckling rats. Also milk taken from stomachs of suckling rats contained significant amounts of [³H]. These results demonstrate clearly that aflatoxin binds to hepatic macromolecules of fetal and infant rats after transferring through the placenta and milk from their mothers.

Examination of metabolic activity of liver showed that hepatic microsomes' ability to bind B$_1$ to calf thymus DNA increased remarkably from fetal age to 9 days of age, while cytosolic GST activity remained nearly

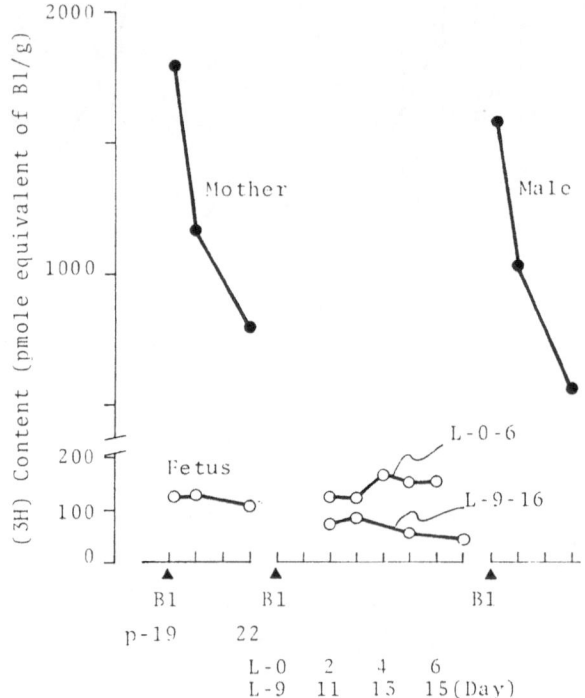

Fig. 4. Liver [³H] content in fetal, pregnant, suckling and young adult rats. Livers (n=2-4) were taken from the rats in Fig. 3. Each point indicates mean value for 2-4 livers.

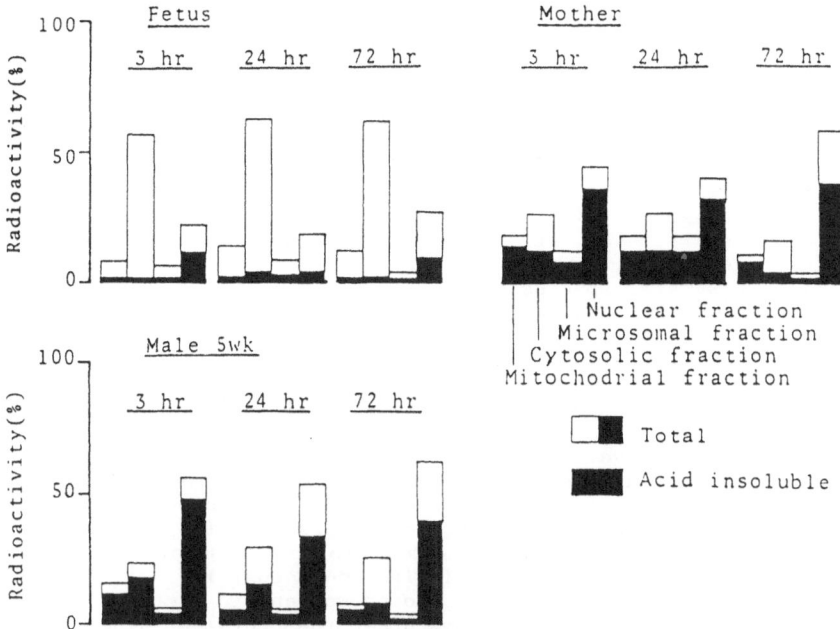

Fig. 5. Subcellular distribution of [³H] in livers of fetal,
pregnant and young adult rats. Livers are the same as
those in Fig. 4. Each point indicates mean value for
2-4 rats.

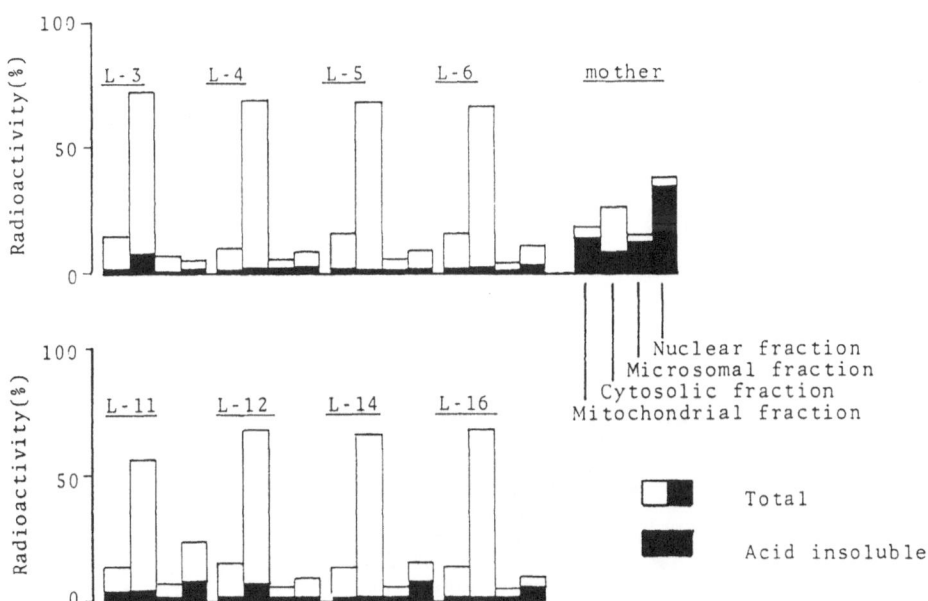

Fig. 6. Subcellular distribution of [³H] in livers of suckling and lac-
tating rats. Livers of suckling rats were the same as those in
Fig. 4. Livers of lactating rats were taken as described in
"Materials and Methods". Mean values for 2-4 livers are shown.

constant during these stages of rat growth. Consistent with the increase in the B_1-DNA binding activity of microsomes with rat growth, infant rats of different ages showed markedly different patterns of [^{3}H] distribution in the liver after they were injected with [^{3}H]-B_1. A large part of the liver [^{3}H] was contained in TCA insoluble fractions in rats of 10 days of age, whereas only a minute amount of [^{3}H] was found in these fractions in neonatal period.

In contrast, when suckling rats were exposed to [^{3}H] aflatoxin through milk from the rats given [^{3}H]-B_1, the proportion of [^{3}H] in TCA insoluble fractions of these rats was very low compared to that in adult rats, regardless of their age. M_1 and other metabolites have been observed to be secreted into milk in several species of animals given B_1.[1,16,30] Less efficient binding of M_1 than of B_1 to hepatic macromolecules has been observed in rats and primary cultures of rat hepatocytes exposed to these metabolites.[9,19] Therefore, the low proportion of [^{3}H] in TCA insoluble fractions despite high activity of B_1-DNA binding activity of microsomes at 11-16 days of age may reflect less efficient binding of the metabolites secreted in milk.

As was described above, the result of direct injection of [^{3}H]-B_1 into neonatal rats showed that the proportion of [^{3}H] in TCA insoluble fractions of these rats' livers was far smaller compared to that of adults. Neonatal rats are known to be more sensitive to acute toxicity of B_1 than adult rats.[14,23] Aflatoxin binding to hepatic macromolecules or protein has been postulated to play an important role in its acute toxicity.[1,21,24,29,31] However, the low levels of aflatoxin binding to hepatic macromolecules in neonatal rats indicates that mechanisms other than covalent binding of aflatoxin to hepatic macromolecules may be involved in high toxicity of aflatoxin to these rats. The mechanisms could be specific in this stage of rat growth, because other mycotoxins such as rubratoxin B and ochratoxin A are also more highly toxic to neonatal rats than to adult rats.[14]

Mice are regarded as a resistant species with respect to aflatoxin carcinogenicity,[23] but they are susceptible to the carcinogenic effect of aflatoxin if exposed during the neonatal period.[32] Also, tumorigenic effects of aflatoxin have been observed in offspring when their mothers were exposed to

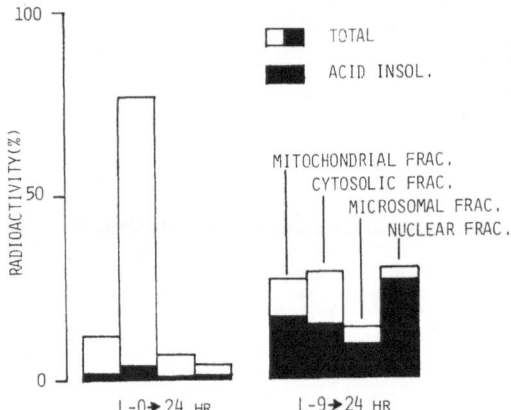

Fig. 7. Subcellular distribution of
[^{3}H] in infant rat livers
taken 24 hrs after ip injection of [^{3}H]-B_1 on 0 day
or 9 days of age. Mean values
for 3 livers are shown.

Table 1. Hepatic Macromolecule-bound Aflatoxin in Fetal and Suckling
Rats of Mothers Given $[^3H]$-B_1 (0.5 mg/kg)

Time of $[^3H]$-B_1	Time of sacrifice		pmoles aflatoxin bound/mg macromolecule $(\times 10^{-2})$		
			DNA	RNA	Protein
p 19[a]	p 20-22	Fetus	16-30	15-160	6-11
		Mother	1056-5120	4384-6080	160-374
p 19	L 1-4	Suckling	31-99	80-1792	4-9
L 0[b]	L 1-2	Suckling	80-1405	27-435	2-8
L 9	L 10-11	Suckling	566-886	141-378	13-16
D 0[c]	D 1-3	Male 5wks	3206-6730	11221-26930	866-1603

[a], p 19 indicates day 19 of pregnancy (day 0=the day of mating).
[b], L 0 indicates 0 day of age of offspring.
[c], D 0 indicates the day of $[^3H]$-B_1 administration.

B_1 during gestation or lactation.[10,15] The results of our study showed that
$[^3H]$ aflatoxin binding to hepatic DNA was less than 1/5 the level of that
noted in young adult male rats. Other studies have shown similar modification of DNA in adult rats when they were given M_1 or B_2.[19,29] In those
studies, levels of binding of M_1 and B_2 to hepatic DNA were 1/5 and 6/1000-
11/1000, respectively, of those of B_1. If it is considered that both M_1 and
B_2 are hepatocarcinogenic to rats, although their potency is less than
B_1,[2,33,34] modification of DNA to such small extent may not be less important
in neonatal rats than in adult rats, in aflatoxin carcinogenicity.

REFERENCES

1. Campbell TC, Hayes JR, 1976, The role of aflatoxin metabolism in its
   toxic lesion. *Tox Appl Pharmacol*. 35:199-222.
2. Cullen JM, Ruebner BH, Hsieh LS, Hyde DM, Hsieh DP, 1987, Carcinogenicity
   of dietary aflatoxin M_1 in male Fischer rats compared to aflatoxin B_1.
   Cancer Res. 47:1913-1917.
3. Degen GH, Neumann H-G, 1978, The major metabolite of aflatoxin B_1 in
   the rat is a glutathione conjugate. *Chem-Biol Interact*., 22:239-255.
4. Degen GH, Neumann H-G, 1981, Differences in aflatoxin B_1-susceptibility
   of rat and mouse are correlated with the capacity *in vitro* to inactivate
   aflatoxin B_1-epoxide. *Carcinogenesis*, 2:299-306.
5. DiPaolo JA, Elis J, Erwin HM 1967, Teratogenic response by hamsters,
   rats, and mice to aflatoxin B_1. *Nature*, 215:638-639.
6. Elis J, DiPaolo JA, 1967, Aflatoxin B_1 induction of malformations.
   Arch Path., 83:53-57.
7. Essigman JM, Croy RG, Bennett RA, Wogan GN, 1982, Metabolic activation of
   aflatoxin B_1: Patterns of DNA adduct formation, removal, and execution
   in relation to carcinogenesis. *Drug Metab Rev*., 13:581-602.
8. Geissler F, Faustman EM, 1988, Developmental toxicity of aflatoxin B_1 in
   the rodent embryo *in vitro*: Contribution of exogenous biotransformation
   systems to toxicity. *Teratology*, 37:101-111.
9. Green CE, Rice DW, Hsieh DPH, Byard JL, 1982, The comparative metabolism
   and toxic potency of aflatoxin B_1 and aflatoxin M_1 in primary cultures

of adult-rat hepatocytes. *Fd Chem Toxic.*, 20:53-60.

10. Grice HC, Moodie CA, Smith DC, 1973, The carcinogenic potential of aflatoxin or its metabolites in rats from dams fed aflatoxin pre- and post-
partum. *Cancer Res.*, 33:2262-268.

11. Gurtoo HL, Dave CV, 1975, *In vitro* metabolic conversion of aflatoxins and
benzo(a)pyrene to nucleic acid-binding metabolites. *Cancer Res.*,
35:382-389.

12. Gurtoo HL, Motycka L, 1976, Effect of sex difference on the *in vitro* and
in vivo metabolism of aflatoxin B_1 by the rat. *Cancer Res.*, 36:4663-
4671.

13. Hales BF, Neims AH, 1976, Developmental aspects of glutathione S-transferase (Ligandin) in rat liver. *Biochem J.*, 160:231-236.

14. Hayes AW, Cain JA, Moore BG, 1977, Effect of aflatoxin B_1, ochratoxin
A, and rubratoxin B on infant rats. *Fd Cosmet Toxic.*, 15:23-27.

15. Homma S, Oka M, Goto N, Tomizawa M, Miura K, Imamura K, 1984, Effects of
phenobarbital on the transplacental tumorigenesis by aflatoxin B_1 in
mice DDK strain. *In:* "Toxigenic Fungi-their toxins and health hazard,"
Ueno Y, ed., Kodansha, Tokyo.

16. Li YK, Chu FS, 1982, Kinetics of transformation of aflatoxin B_1 into
aflatoxin M_1 in lactating mouse: An ELISA analysis. *Experientia*,
38:842-843.

17. Lijinsky W, Ross AE, 1969, Alkylation of rat liver nucleic acids not
related to carcinogenesis by N-nitrosamines. *J Nat Cancer Inst.*, 42:1095-
1100.

18. Lotlikar PD, Jhee EC, Insetta SM, Clearfield MS, 1984, Modulation of
microsome-mediated aflatoxin B_1 binding to exogenous and endogenous DNA
by cytosolic glutathione S-transferases in rat and hamster livers.
Carcinogenesis, 5:269-276.

19. Lutz WK, Jaggi W, Luthy J, Sagelsdorff P, Schlatter C, 1980, *In vivo* covalent binding of aflatoxin B_1 and aflatoxin M_1 to liver DNA of rat,
mouse and pig. *Chem-Biol Interact.*, 32:249-256.

20. Mainigi KD, Campbell TC, 1981, Effect of sex differences on subcellular
distribution of aflatoxin B_1 in F-344 rats treated with various risk-
modifying factors. *Tox Appl Pharmacol.*, 58:236-243.

21. Neal GE, Judah DJ, Stirpe F, Patterson DSP, 1981, The formation of
2,3-dihydroxy-2,3-dihydro-aflatoxin B_1 by the metabolism of aflatoxin
B_1 by liver microsomes isolated from certain avian and mammalian
species and the possible role of this metabolite in the acute toxicity
of aflatoxin B_1. *Tox Appl Pharmacol.*, 58:431-437.

22. Neal GE, Green JA, 1983, The requirement for glutathione S-transferase in
the conjugation of activated aflatoxin B_1 during aflatoxin hepatocarcinogenesis in the rat. *Chem-Biol Interact.*, 45:259-275.

23. Newberne PM, Butler WH, 1969, Acute and chronic effects of aflatoxin on
the liver of domestic and laboratory animals: a review. *Cancer Res.*,
29:236-250.

24. O'Brien K, Moss E, Judah D, Neal G, 1983, Metabolic basis of the species
difference to aflatoxin B_1 induced hepatotoxicity. *Biochem Biophys Res
Commun.*, 114:813-821.

25. Pong RS, Wogan GN, 1971, Toxicity and biochemical and fine structural
effects of synthetic aflatoxins M_1 and B_1 in rat liver. *J Nat Cancer
Inst.*, 47:585-592.

26. Portman RS, Plowman KM, Campbell TC, 1970, On mechanisms affecting
species susceptibility to aflatoxin. *Biochim Biophys Acta*, 208:487-495.

27. Raj HG, Clearfield MS, Lotlikar PD, 1984, Comparative kinetic studies on
aflatoxin B_1-DNA binding and aflatoxin B_1-glutathione conjugation
with rat and hamster livers *in vitro*. *Carcinogenesis*, 5:879-884.

28. Schmidt RE, Panciera RJ, 1980, Effects of aflatoxin on pregnant hamsters
and hamster foetuses. *J Comp Path.*, 90:339-347.

29. Swenson DH, Lin J-K, Miller ED, Miller JA, 1977, Aflatoxin B_1-2,3-oxide
as a probable intermediate in the covalent binding of aflatoxins B_1 and

B$_2$ to rat liver DNA and ribosomal RNA *in vivo*. *Cancer Res.*, 37:172-181.

30. Trucksess MW, Richard JL, Stoloff L, McDonald JS, Brumley WC, 1983, Absorption and distribution patterns of aflatoxicol and aflatoxins B$_1$ and M$_1$ in blood and milk of cows given aflatoxin B$_1$. *Am J Vet Res.*, 44:1732-1756.

31. Ueno I, Friedman L, Stone CL, 1980, Species difference in the binding of aflatoxin B$_1$ to hepatic macromolecules. *Tox Appl Pharmacol.*, 52:177-180.

32. Vesselinovitch SD, Mihailovich N, Wogan GN, Lombard LS, Rao KVN, 1972, Aflatoxin B$_1$, a hepatocarcinogen in the infant mouse. *Cancer Res.*, 32:2289-2291.

33. Wogan GN, Edwards GS, Newberne PM, 1971, Structure-activity relationship in toxicity and carcinogenicity of aflatoxins and analogs. *Cancer Res.*, 31:1936-1942.

34. Wogan GN, Paglialunga S, 1974, Carcinogenicity of synthetic aflatoxin M$_1$ in rats. *Fd Cosmet Toxic.*, 12:381-384.

35. Wong ZA, Hsieh DPH, 1980, The comparative metabolism and toxicokinetics of aflatoxin B$_1$ in the monkey, rat, and mouse. *Tox Appl Pharmacol.*, 55:115-125.

36. Yoshizawa H, Uchimaru R, Kamataki T, Kato R, Ueno Y, 1982, Metabolism and activation of aflatoxin B$_1$ by reconstituted cytochrome P-450 system of rat liver. *Cancer Res.*, 42:1120-1124.

THE IMMUNOSUPPRESSIVE ACTION OF GLIOTOXIN AND RELATED

EPIPOLYTHIODIOXOPIPDERAZINES

Arno Mullbacher, Ronald D. Eichner and Paul Waring

Division of Virology and Cellular Pathology
John Curtin School of Medical Research
Australian National University
Canberra, Australia

INTRODUCTION

The epipolythiodioxopiperazines (ETP) including gliotoxin (GT) are a class of secondary fungal metabolites with a bridged polysulfide ring structure as their common distinctive feature (Figure 1). Early studies of these unusual compounds concentrated on their anti-microbial properties and possible anti-tumor activities. These early studies have been extensively reviewed.[14,16]

Although their biological properties, especially their antiviral activity in vitro[15] looked promising, their therapeutic application has been curtailed due to unacceptably high levels of toxicity. Research into these compounds was largely confined to the elucidation of action and mechanisms(s) of toxicity of sporidesmin, an ETP compound involved in the aetiology of facial eczema.[6] The serendipitous discovery by one of us (A.M.) of the antiphagocytic and immunosuppressive activity of a fungal product found in Aspergillus fumigatus-contaminated lymphocyte cultures and its subsequent identification as GT[10] led to renewed interest in these compounds.

Three areas of interest in fungal immunosuppressants can be easily identified. Firstly, they may themselves play a part in the aetiology of fungal infections[2] such as aspergillosis and cryptococcosis both of which are on the increase due primarily to man's modern medical practices, i.e. the use of immunosuppressants required in transplantation of organs, the use of chemotherapeutic agents to combat neoplastic diseases and the HIV epidemic. Secondly, fungal immunosuppressants are themselves employed as therapeutic

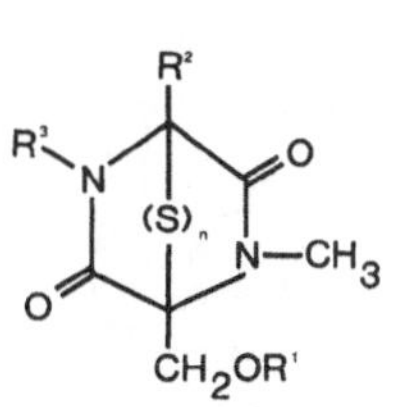

Fig. 1. Generalized chemical structure
of epipolythiodioxopiperazines.

Microbial Toxins in Foods and Feeds, Edited by A.E. Pohland *et al.,*
Plenum Press, New York, 1990

agents to prevent or delay graft rejection after transplantation of tissues or organs and to alleviate autoimmune disorders. Cyclosporin A, a recently discovered fungal metabolite, is the most prominent compound of this type in use today.[14] Finally, immunosuppressant compounds may prove useful as tools probing fundamental parameters of immune induction, similar to the role antibiotics played in the elucidation of macromolecular biosynthesis.

The effect of GT and other ETP compounds on macrophage function

The original observation, namely that GT inhibited macrophages to adhere to plastic surfaces has been exploited to become a rapid colorimetric bioassay for estimating concentrations of GT and related compounds. The phagocytic property of macrophages is directly related to their ability to adhere to glass and plastic surfaces.[5] The effect of GT and other ETP compounds on phagocytosis has been investigated directly using electron and light microscopy of treated and untreated cells. GT was observed to inhibit the phagocytosis of carbon, carbonyl iron and fluorescent beads with ED_{50} values in the range of 30 - 300 nM.[3] A number of morphological changes to the macrophages were also observed following GT treatment, including loss of microvilli and condensation of chromatin. In these studies, GT of quantities up to 3 μM had little effect on cell viability, as assessed by trypan blue exclusion, i.e. the effects were occurring at concentrations well below those producing the general toxic effects of GT. GT also inhibited RNA synthesis and protein synthesis. However, GT induced a 75% enhancement of ^{14}C-glucose conversion to CO_2 in macrophages and a decrease (95%) into radiolabelled glycogen. Additionally, the basal rate of hydrogen peroxide production by human polymorphonuclear cells was inhibited 65% by GT at 300 nM. Therefore, to find that the bactericidal activity of macrophages was abrogated at similar concentration[3] is not surprising.

The effect of GT on immune induction

Besides the macrophage's own property of defense against microbial invasion of the host, they play a central role in the regulation of the immune response by their capacity to function as antigen-presenting cells (APC) and by their ability to release soluble mediators with immunoregulatory properties.[8]

Pathways of immune induction may be explained by the simplified two-signal model shown in Figure 2. APCs, e.g., macrophages, process and present nominal antigen (Ag) as a result of microbial infection, or - as in a foreign graft situation - their self antigens on the cell surface may constitute the antigen per se. Antigen on APC is recognized by the clonally distributed antigen receptor on T cells which provides "Signal 1" to the

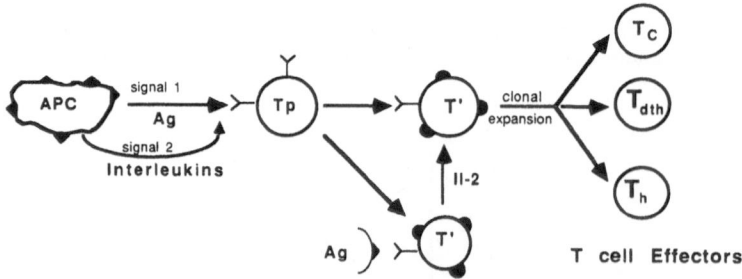

Fig. 2. T Cell Activation. The two-signal concept for immune induction.

Treatment of T and B cells with ETP compounds abrogates their ability to proliferate in response to mitogens, e.g., concanavalin A and lipopolysaccharide, respectively.[11] Furthermore, T' cells do not proliferate in response to lymphokines, nor are they able to secrete lymphokines after treatment with low concentrations of ETP compounds. This effect is most likely the result of DNA fragmentation due to the induction of apoptosis by ETP compounds referred to in a separate article in this volume.[16] A summary of all the effects of GT on leukocyte function is given in Table 1.

Applications, implications and conclusions

As detailed above, ETP compounds act on all stages of immune induction and most importantly in an irreversible fashion.[1] Furthermore GT and other ETP compounds affect cells in a tissue specific manner with mature cells of haemopoietic origin being one of the most sensitive. Thus, ETP compounds may prove of therapeutic value in the prevention of graft rejection after allogeneic organ or tissue transplantation by pre-treating organs of tissues with GT prior to transplantation thereby inactivating APC.

We have used a murine thyroid allograft model to test the proposition that gliotoxin treatment of the tissue prior to transplantation may selectively inactivate stimulator cells present in the tissue and thereby render them non-immunogenic. The techniques involved the removal of thyroid tissue of one strain of mouse, culturing the tissue for 16 hr in media containing GT and then grafting it under the kidney capsule of a genetically different strain of mouse. When allogeneoc thyroids were treated with 1 μM GT for 16 hr, 60-70% of such were accepted long term (more than 3 months) and were functional as determined by ^{125}I uptake, there was little or only transient infiltration of lymphocytes in the graft; the latter being one criterium for graft rejection. It should be stressed here that the animals that received the transplants were totally without the need of immunosuppressive therapy and therefore not at increased risk of infections.

Besides this tissue specificity, GT has been shown to act selectively on mature cells of haemopoietic origin.[12] Mitogen-induced proliferation of mature B and T cells was affected and totally inhibited at 100 nM GT at a cell concentration of 10^6 cells per ml. However, pluripotent stem cells still responded to growth factors at 10-fold higher concentrations.[12] This differential anti-proliferative activity of GT can be correlated with DNA fragmentation on polyacrylamide gels. Furthermore, using a spleen colony forming assay, bone marrow (BM) treated with GT (up to 300 nM) showed no change in its ability to form spleen surface colonies as compared to untreated controls. GT concentrations as high as 1 μM reduced this by only 0.5 log. BM treated with 300 nM GT has been transferred into fully allogeneic, lethally irradiated recipient animals. Such animals become fully reconstituted with donor-derived haemopoietic cells, i.e., they were fully chimeric without signs of graft versus host disease, which in turn was prevalent in control animals that received untreated BM.

Finally, GT and other ETP compounds might be involved in the aetiology of opportunistic fungal infections and mycotoxicoses. The role of sporidesmin in facial ezema in ruminants has been extensively documented.[6] We have recently established a murine model for disseminated aspergillosis using *Aspergillus fumigatus* as the pathogen. The fungus was detected by light micro scopy in the peritoneal cavity, kidney and spleen of these animals. More specifically, GT and its characteristic dimethylthioether derivative have been isolated and identified from peritoneal lavages from these animals.[4] Thus, the normal host defense mechanisms involved in protection against invading microbes (macrophage and T cell mediated mechanisms) could be abrogated by the known immunosuppressive effects of GT.

In conclusion, GT and related ETP compounds exert potent immunosuppressive effects on mature cells of the immune system in an irreversible fashion. This may potentially allow them to be used in tissue pretreatment prior to organ transplantation to prevent graft rejection. The demonstration that GT is produced early *in vivo* during disseminated aspergillosis may potentially allow much needed early detection of certain opportunistic fungal infections and suggest novel treatments.

REFERENCES

1. Braithwaite AW, Eichner RD, Waring P, Mullbacher A, 1987, The immunomodulating agent gliotoxin causes genomic DNA fragmentation, *Mol Immunol.*, 24:47-55.
2. Eichner RD, Mullbacher A, 1984, Fungal toxin involvement in aspergillosis and AIDS, *Aust J Exp Biol Med Sci.*, 62:479-484.
3. Eichner RD, Al Salami M, Wood PR, Mullbacher A, 1986, The effects of gliotoxin upon macrophage function, *J Immunopharmarc.*, 8:789-798.
4. Eichner RD, Tiwari-Palni U, Waring P, Mullbacher A, 1988, Detection of the immunomodulating agent gliotoxin in experimental aspergillosis, *In*: "Proceedings of the Xth International Congress of Human and Animal Mycology," Barcelona.
5. Grinnel R, 1978, Cellular adhesiveness and extracellular substrata, *Int Rev Cytol.*, 53:65-144.
6. Jordan TW, Cordiner J, 1987, Fungal epipolythiodioxopiperazine toxins have therapeutic potential and roles in disease, *Trends in Pharmacol Sci.*, 8:144-148.
7. Lafferty KJ, Prowse SJ, Simeonovic CJ, Waren HS, 1983, Immunobiology of tissue transplantation: A return to the passengere leukocyte concept, *Ann Rev Immunol.*, 1:143-174.
8. Mizuno D, Cohn ZA, Takeya K, Ishida N, 1982, "Self defense mechanism role of macrophages," University of Tokyo Press, Tokyo.
9. Mullbacher A, Eichner RD, 1984, Immunosuppression *In vitro* by a metabolite of a human pathogenic fungus, *Proc Natl Acad Sci, USA*, 81:3835-3837.
10. Mullbacher A, Waring P, Eichner RD, 1985, Identification of an immunosuppressive agent in cultures of *Aspergillus fumigatur*, *J Gen Microbiol.*, 131:1251-1258.
11. Mullbacher A, Waring P, Tiwari-Palni U, Eichner RD, 1986, Structural relationship of epipolythiodioxopiperazines and their immunomodulating activity, *Mol Immunol.*, 23:231-235.
12. Mullbacher A, Hume D, Braithwaite AW, Waring P, Eichner RD, 1987, Selective resistance of bone marrow derived hemopoietic progenitor cells to gliotoxin, *Prod Natl Acad Sci, USA*, 84:3822-3825.
13. Mullbacher A, Moreland AF, Waring P, Sjaarda A, Eichner RD, 1988, Prevention of graft versus host disease by treatment of bone marrow with gliotoxin in fully allogeneic chimeras and their cytotoxic T cell repertoire, *Transplantation*, 46:120-126.
14. Nagarajan P, 1984, Gliotoxin and epipolythiodioxopiperazines, *In*: "Mycotoxins, Production, Isolation, Separation and Purification," Betina V, Ed., Elsevier, Amsterdam.
15. Rightsel WA, Schneider HGF, Sloan BS, Graf PR, Miller FA, Bartz QR, Ehrlich J, Dixon GJ, 1964, Antiviral activity of gliotoxin and gliotoxin acetate, *Nature*, 204:1333-1335.
16. Taylor A, 1971, The toxicology of sporidesmins and other epipolythiodioxopiperazines, *In*: "Microbial Toxins VII," Ajl, SJ, Ed., Academic Press, New York.
17. Thomson AW, 1983, Immunobiology of cyclosporin A -- A review, *Aust J Exp Biol Med Sci.*, 61:147-172,
18. Waring P, Eichner RD, Mullbacher A, 1988, The molecular mechanism of toxicity of gliotoxin and related epipolythiodioxopiperazines,

In: "Cellular and Molecular Mode of Action of Selected Microbial Toxins in Foods and Feeds," Pohland AE, Dowell VR, Richard JL, Eds., Plenum Publishing Inc., New York.

CELLULAR EFFECTS OF CYTOCHALASINS[a]

Shinsaku Natori[b] and Ichiro Yahara[c]

Meiji College of Pharmacy
Tanashi-shi, Tokyo 188, Japan[b]

The Tokyo Metropolitan Institute of Medical Science
Bunkyo-ku, Tokyo 113, Japan[c]

INTRODUCTION

Recent progress in cell biology and biochemistry has clarified the proteins involved in cell structure, function, and movement; microtubules composed of tubulin and microfilaments composed of actin play the most important roles in these cellular phenomena. Some of the cytotoxic agents known so far are now found to affect the cytoskeletal proteins. Examples of those interacting with tubulin are colchicine, podophyllotoxins, *Catharantus* (*Vinca*) alkaloids, maytanshinoids and griseofulvin. Drugs known to interact with actin are phallotoxins from *Amanita* spp. and some mycotoxins. The group of mycotoxins known as cytochalasins[a] fall into this category. The name "cytochalasin" refers to the cytological effects of the mold metabolites and originates from the Greek words, Cytos (cell) and chalasis (relaxation). The discovery of the cytochalasins attracted the attention of cell biologists because of their distinctive effects, such as inhibition of cytoplasmic cleavage, resulting in the formation of polynucleate cells, and inhibition of cell movement.[4,6,12] Forty-eight compounds of this class have so far been isolated from diverse fungal sources. The common structural feature of the group is expressed by a perhydroisoindol-1-one bearing a benzyl (cytochalasin), *p*-methoxybenzyl (pyrichalasin), indol-3-ylmethyl (chaeteoglobosin), or 2-methylpropyl (aspochalasin) group at the C_3 position and an 11-, 13- or 14-membered carboxylic (or oxygen-containing) ring between the C_8 and C_9 position (Fig. 1). The variation in the structure is due to the size of the macrocyclic ring which has several functional groups beyond C_{17}, different substituents at the C_3 position, and the oxygen function and double bond at C_5-C_7.[5]

Chemistry of Novel Cytochalasins

In the course of our screening tests for mycotoxin production by food-borne fungi, based on cytotoxicity to HeLa cells, extracts of the mycelia and filtrates of *Chaetomium globosum* and *C. cochliodes* were found to cause polynucleation and multipolar division of the cells. Eight metabolites, called chaetoglobosins A-G and J, were isolated as the causative agents of the cell phenomena and the structures (1-8) were elucidated as a novel type of

---

[a]In the chemical literature, cytochalasins are now accepted as trivial names of 10-phenyl compounds and the group of compounds as a whole should be called cytochalasans. However, the commonly used term "cytochalasins" will be adopted to cover all the compounds discussed in this paper.

Microbial Toxins in Foods and Feeds, Edited by A.E. Pohland *et al.,*
Plenum Press, New York, 1990

cytochalasin in which the phenyl group in about ten cytochalasins known at that time was replaced by the indol-3-yl group (Fig. 2).[6]

Recently we reexamined molds exhibiting cytotoxicity to HeLa cells. From culture on wheat of *Phomopsis* sp. (68-GO-164), four known cytochalasins (9-12), epoxycytochalasins H and J and cytochalasins H and J, and six new compounds (13-18), called cytochalasins N, O, P, Q, R and S, were isolated and the structures of these compounds were elucidated by the physical data and reactions as shown in Fig. 3. Cytochalasins P, Q, R and S (15-18) have novel diol-type structures in the cyclohexane part of the molecules.[13]

Effect of Cytochalasins on Cellular Structure and Actin in vitro

Cytochalasins inhibit a variety of cellular movements, including cell division, motility, secretion and phagocytosis, and produce a change in cell shape. Some cytochalasins inhibit sugar transport by competing with sugars for binding to high-affinity sites in the plasma membrane, but this effect has not proved to be related to the effect on cellular structures containing actin and other contractile proteins.[12]

The interaction between rabbit muscle actin and cytochalasins *in vitro* was studied and the enhancing effects on the polymerization of rabbit muscle G-actin by different doses of cytochalasins B, D, E and G and chaetoglobosins A, B, C, E, F and J were recognized. The polymers thus formed differ widely in their viscosity and equal levels of viscosity are attained by interaction of F-actin with the same drugs.[2]

In human erythrocyte membranes, there are three classes of cytochalasin B binding sites (sites I, II and III), each of which shows a distinct affinity and specificity. Site I is competitively inhibited by glucuose carrier

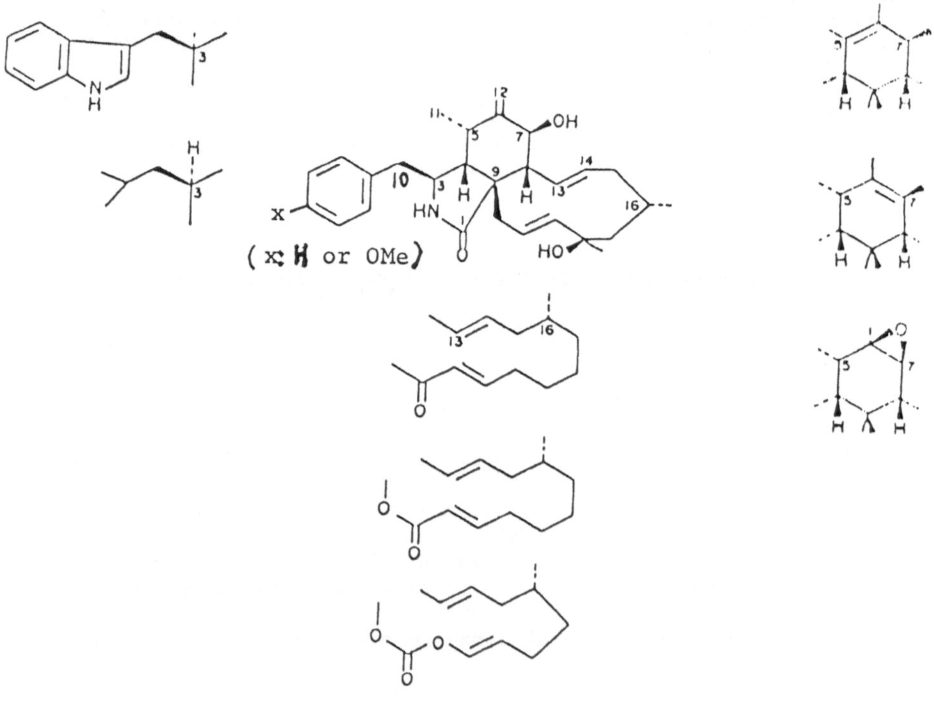

Fig. 1. The skeletal structure of the cytochalasins

substrates and inhibitors, while site II is considered to be related to cell
motility and cell morphology. When displacement of radioactive cytochalasin
B from erythrocyte membranes by twenty cytochalasin derivatives was tested,
only nine of them were found to bind at site I and to inhibit glucose trans
port. All but three of the cytochalasins bind at site II.[7] The reason for
the exceptional behavior of the negative compounds is not clear.

Since cytochalasin A inhibits self-assembly of bovine brain tubulin,
nineteen cytochalasins were investigated by viscometry to examine their
effects on *in vitro* polymerization and depolymerization of microtubules. All
except cytochalasin A showed very weak or insignificant effects and the wide
range of cellular effects of cytochalasins was concluded not to be related to
interaction with microtubules.[8]

Due to these observations, the confusion which has existed about the
effects of the cytochalasins has been removed.

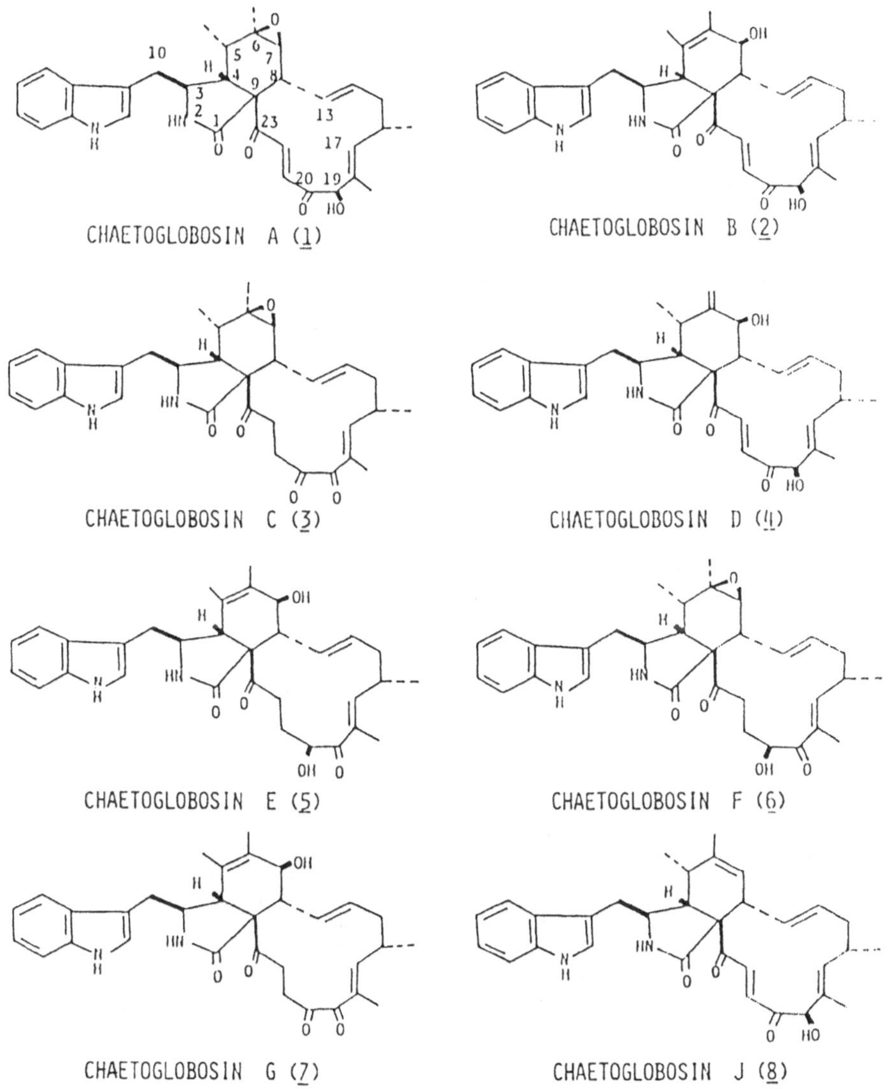

CHAETOGLOBOSIN A (1)

CHAETOGLOBOSIN B (2)

CHAETOGLOBOSIN C (3)

CHAETOGLOBOSIN D (4)

CHAETOGLOBOSIN E (5)

CHAETOGLOBOSIN F (6)

CHAETOGLOBOSIN G (7)

CHAETOGLOBOSIN J (8)

Fig. 2.   The structure of chaetoglobosins

To compare the effects of cytochalasins at the cellular level with those at the molecular level, thirty natural cytochalasins and their derivatives were examined using C3H-2K cells (a mouse fibroblastic cell line), mouse splenic lymphocytes, and depolymerized and polymerized rabbit skeletal muscle actin. Four high-dose (2-20 μM) effects at the cellular level (rounding up of fibroblastic cells, contraction of actin cables (Fig. 4), formation of hairy filaments containing actin, and inhibition of lymphocyte capping), a low-dose (0.2-2 μM) effect (inhibition of membrane ruffling), and two *in vitro* effects [inhibition of actin filament elongation (the high-affinity [low-dose] effect *in vitro*) and alteration of the viscosity of actin filaments (the low-affinity [high-dose] effect *in vitro*)] were tested. The

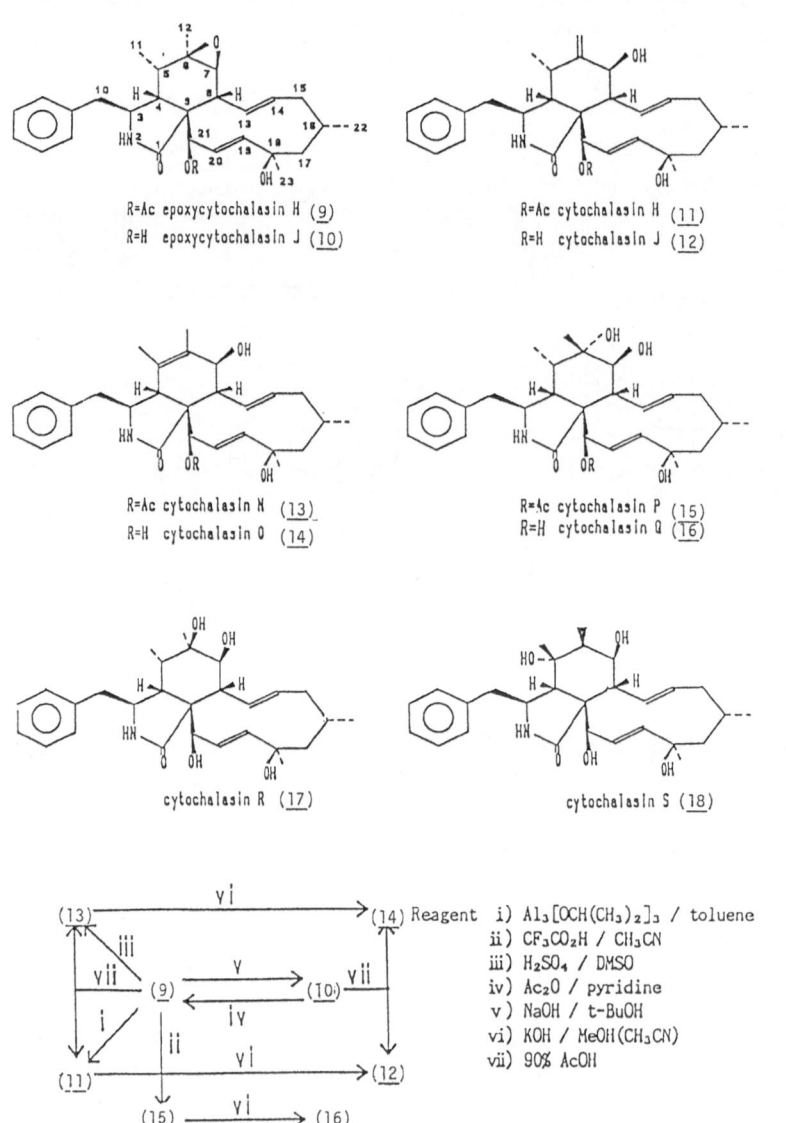

Fig. 3.   Structures and Correlation Reactions of the Cytochalasans from *Phomopsis* sp. (68-GO-164)

results shown in Table I indicate that there is an almost hierarchic order of relative effectiveness of different cytochalasins from low-dose to high-dose effects and from cellular to molecular effects (Fig. 5). The strong positive correlation suggests that most of the effects caused by a cytochalasin, irrespective of dose or affected phenomena, might be attributed to the interaction between the drug and the common target protein, actin.[14]

Binding between cytochalasins and actin is not covalent because of the rapid reversibility of the cell phenomena after removal of the cytochalasins. It is known that polymerization of actin monomers occurs more rapidly at the barbed end of the filament than at the pointed end and that cytochalasins selectively block the barbed end. Effects of chaetoglobosin J (8) on the G-F transformation of actin was precisely examined.[3]

Structure-Activity Relationships of Cytochalasins

Our work with NMR[9,10,13] and work in the United States with X-ray crystallography[1] have clarified the conformations of the cytochalasins.

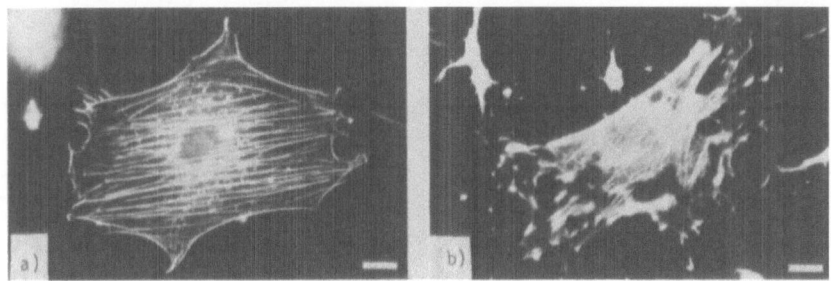

Fig. 4. Immunofluorescence staining with mouse C3H cells untreated or treated with cytochalasin at 37°C. a) untreated cells, b) treated with 20 μM chaetoglobosin A.

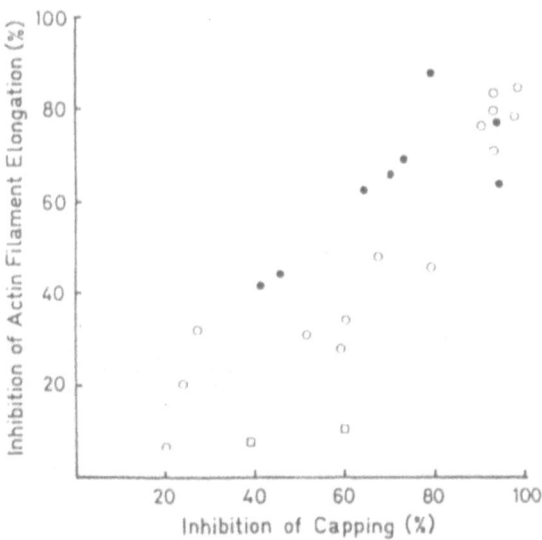

Fig. 5. Correlation between inhibition of capping and inhibition of actin filament elongation by cytochalasins.
● 10-phenylcytochalasins; o 10-indolylcytochalsins;
▢ 10-isopropylcytochalasins.

Table 1. Effect of cytochalasins on cellular structure and actin *In Vitro*

Drugs	Cell rounding up	Inhibition of membrane ruffling	Actin-cables contraction	Formation of hairy structures	Inhibition of capping (%)	Inhibition of actin filament elongation (%)	Decrease in viscosity of actin filaments (%)
10-Phenylcytochalasans							
Cytochalasin A	++	++	+	±	95	63	22
Cytochalasin B	+	+	+	+	47	44	9
Cytochalasin C	++	+	+	+	64	63	27
Cytochalasin D	++	++	+	+	79	88	62
Cytochalasin E	++	++	+	-	94	77	52
Cytochalasin H	++	+	+	+	73	69	25
Cytochalasin J	+	+	+	+	41	42	23
Deoxaphomin	+	+	+	+	70	66	12
Epoxycytochalasin H				+	79		
Epoxycytochalasin J				+	55		
Cytochalasin H				+	57		
Cytochalasin J				+	58		
Cytochalasin N				+	72		
Cytochalasin O				+	50		
Cytochalasin P				-	7		
Cytochalasin Q				-	0		
Cytochalasin R				-	38		
Cytochalasin S				+	60		
10-Indolylcytochalasans							
Cytochalasin G	+	+	+	+	27	32	5
Chaetoglobosin A	++	++	+	-	98	85	43
Chaetoglobosin B	++	++	+	±	93	80	36
Chaetoglobosin C	+	+	+	+	59	28	24
Chaetoglobosin D	++	++	+	-	93	84	49
Chaetoglobosin E	+	+	+	+	49	33	5
Chaetoglobosin F	+	+	+	+	67	48	6
Chaetoglobosin J	++	++	+	-	90	77	38
Chaetoglobosin K	-	++	-	-	93	71	31
ChA monoacetate	++	++	+	+	97	79	40
10-Isopropylcytochalasans							
Aspochalasin B	+	+	+	±	39	8	0
Aspochalasin D	+	+	+	+	60	13	0

Recently we prepared some derivatives of chaetoglobosins and studied the structure-activity relationships of cytochalasins using 30 natural compounds, seven prepared derivatives, and seven synthetic synthons. The results can be summarized as follows (cf. Fig. 1): Compounds with aromatic substituents at C_{10} (phenyl or indolyl; cytochalasins and chaetoglobosins) show the same magnitude of effect, while the isolpropyl compounds (aspochalasins) show little effect; the perhydroisoindol-3-one nucleus is a most important factor in activity, but the position of the double bond and the presence or absence of the oxygen function (6-ene, 6,7-epoxide, 6(12)-en-7-ol, or 5-en-7-ol) do not influence the effects; the macrocyclic ring starting from C_8 and running to C_9 is essential since synthetic synthons lacking this ring did not show these effects; however, no definite functional group or carbon number of the ring appeared to be necessary for the effects.[11] The absence of effects by the novel diol-type compounds (15, 16) again suggests the importance of this part of the molecules (unpublished results of our laboratories).

As a whole, the core composed of the perhydroisoindolone moiety and the macrocyclic ring and the hydrophobic region from C_{13} to C_{16} are assumed to be essential for the effects, while the aromatic ring, the units at C_{16} on the macrocyclic ring including the conjugated enedione system, and the physical properties of the compounds such as lipophilicity presumably influence their relative effectiveness.[11]

CONCLUSIONS

Although little epidemiological evidence has been presented for the toxic effects of cytochalasins on livestock, the rather severe acute toxicities exhibited by some cytochalasins and the wide occurrence of the fungi that produce cytochalasins should be considered in future studies on mycotoxins.

Furthermore, the role of cytochalasins as cell protein modifiers is obvious and their application in the fields of biology and medicine can be expected.

ACKNOWLEDGEMENTS

This work has been conducted with the collaboration of many scientists shown in the references.

REFERENCES

1. Griffin JF, Rampal AL, Jung CY, 1982, Inhibition of glucose transport in human erythrocytes by cytochalasins, *Proc Natl Acad Sci.*, 79:3759-3763.
2. Low I, Jahn W, Wieland TH, Sekita S, Yoshihira K, Natori S, 1979, Interaction between rabbit muscle actin and several chaetoglobosins or cytochalasins, *Anal Biochem.*, 95:14-18.
3. Maruyama K, Oosawa M, Tashiro A, Suzuki T, Tanikawa M, Kikuchi M, Sekita S, Natori S, 1986, Effects of chaetoglobosin J on the G-F transformation of actin, *Biochim Biophys Acta*, 874:137-143.
4. Natori S, 1977, Toxic cytochalasins, In: "Mycotoxins in Human and Animal Health," Rodricks JV, Hesseltine CW, Mehlman MA, eds, Pathotox Publisher, Park Forest South, IL, pp 558-581.
5. Natori S, Iida K, Yahara I, 1983, Agents affecting cytoskeletal proteins, *Tanpakushitsu Kakusan Koso*, 28:789-800; Natori S, 1983, Chemical Surveys on mycotoxins using cytotoxicity testing with special reference to cytochalasans, *Yakugaku Zasshi* 103:1009-1028.
6. Pendse GS, ed, 1986, "Recent Advances in Cytochalasans," Indian Drug Research Associates, Pune, India.

7. Rampal AL, Pinkofsky HB, Jung CY, 1980, Structures of cytochalasins and cytochalasin B binding sites in human erythrocyte membranes, *Biochemistry*, 19:679-683.
8. Sato Y, Saito Y, Tezuka Y, Sekita S, Yoshihira K, Natori S, 1982, Viscometric analysis of effects of cytochalasans on *in vitro* polymerization and depolymerization of microtubules, *J Pharmacobio Dyn.*, 5:418-422.
9. Sekita S, Yoshihira K, Natori S, Kuwano H, 1982, Chaetoglobosins, cytotoxic 10-(indol-3-yl)-[13]cytochalasans from *Chaetomium* spp. III. Structures of chaetoglobosins C, E, F, G and J, *Chem Pharm Bull.*, 30:1629-1638.
10. Sekita S, Yoshihira K, Natori S, 1983, Chaetoglobosins, cytotoxic 10-sindol-3-yl)-[13]cytochalasans from *Chaetomium* spp. IV. ^{13}C-nuclear magnetic resonance spectra and their application to a biosynthetic study, *Chem Pharm Bull.*, 31490-498.
11. Sekita S, Yoshihira K, Natori S, Harada F, Iida K, Yahara I, 1985, Structure-activity relationship of thirty-nine cytochalasans observed in the effects on cellular structures and cellular events and actin polymerization *in vitro*, *J Pharmacobio Dyn.*, 8:906-916.
12. Tanenbaum SW, ed, 1978, "Cytochalasins, biochemical and cell biological aspects," North-Holland, Amsterdam.
13. Tomioka T, Izawa Y, Koyama K, Natori S, 1987, Three new 10-phenyl-[11] cytochalasans, cytochalasins N, O and P from *Phomopsis* sp., *Chem Pharm Bull.*, 35:92-95; Izawa Y, Hirose T, Shimizu (nee Tomioka) T, Koyama K, Natori S, 1988, Six new 10-phenyl-[11]cytochalasans, cytochalasins N-S from *Phomopsis* sp. *Tetrahedron*, in press.
14. Yahara I, Harada F, Sekita S, Yoshihira K, Natori S, 1982, Correlation between effects of 24 different cytochalasins on cellular structures and cellular events and those on actin *in vitro*, *J Cell Biol.*, 92:69-78.

SELECTIVE ANTITUMOR ACTIVITY OF T-2 TOXIN-ANTIBODY CONJUGATES

Katsumi Ohtani and Yoshio Ueno

Faculty of Pharmaceutical Sciences
Science University of Tokyo
Ichigaya, Shinjuku-ku, Tokyo 162, Japan

INTRODUCTION

T-2 toxin (T-2), a trichothecene mycotoxin produced by *Fusarium* spp., is a potent inhibitor of protein and DNA syntheses in mammalian cells.[12] It is not mutagenic.[5,11] Diacetoxyscirpenol (Anguidin), another trichothecene, was investigated clinically as an antitumor agent.[1,6,9,14] However, this toxin has not been put to practical use due to side effects such as nausea. This suggests that attention must be paid to the side effects resulting from the clinical use of trichothecene mycotoxins. To specify the anti-tumor activity and reduce side effects, the authors tried to prepare T-2-tumor specific monoclonal antibody (mAb) conjugates as a suitable immunotoxin.

In this proceeding, we report that T-2-conjugated anti-EL-4 mAb (T-2-mAb conjugates) inhibited the growth of EL-4 thymoma cells both *in vitro* and *in vivo*. Additionally, several metabolic inhibitors were employed to investigate the mechanism of cytotoxcity of T-2-mAb conjugates.

MATERIALS AND METHODS

Passage medium was made from 100 ml of Dulbecco's modified Eagle medium, 15 ml of fetal bovine serum, 1 ml of MEM nonessential amino acid, 1 ml of AB mixture (kanamycin 5 mg/ml and gentamicin 1 mg/ml), and 1 ml of 2.5% of L-glutamine. Murine thymoma EL-4 cells and myeloma SP2/0-Ag 14 cells were maintained *in vitro* in a passage medium. All the cells were cultured at 37°C in a humidified atmosphere of 7% CO_2 in air.

Monoclonal antibody (mAb) and normal gamma globulin (nGG)

Anti-EL-4 mAb (5C10) was obtained from the globulin fraction of ascitic fluids by repeated precipitation with ammonium sulfate solution. Normal gamma globulin (nGG) was obtained from C57BL mice sera in a similar way. The specificity of 5C10 mAb was determined by the antibody-dependent complement-mediated cell lysis method.

Preparation of T-2-antibody conjugates

T-2[13] and T-2 hemiglutarate (T-2 HG)[7] were prepared in our laboratory

Microbial Toxins in Foods and Feeds, Edited by A.E. Pohland *et al.,*
Plenum Press, New York, 1990

T-2 HG was activated by N-hydroxysuccinimide in the presence of dicyclohexyl-carbodiimide (DCC) according to the methods reported.[2,3] One, two, three, and four mg of activated T-2 HG (T-2-G-OSu) in 0.05 ml of dimethylformamide were separately mixed with 10 mg of anti-EL-4 mAb (5C10) dissolved in 5 ml of 0.05 M phosphate buffered saline, pH 7.4 (PBS), and the mixtures were allowed to stand at 4°C overnight. After centrifugation, the supernatant was dialyzed against PBS. T-2-mAb conjugates were obtained from the non-dialyzable portion. T-2-conjugated nGG (T-2-nGG conjugates) were prepared using nGG instead of anti-EL-4 mAb. The estimation of the molecular ratio of T-2 to immunoglobulin was performed as described previously.[7]

Determination of binding activity of T-2-mAb conjugates to EL-4 cells by the indirect enzyme-linked immunosorbent assay (ELISA)

PBS containing 0.01 mg/ml poly-L-lysine (0.1 ml/well) was added to each well of a 96-well microtiter plate (Corning, Corning, NY USA), and the plate was incubated at 37°C for 1 h. Wells were washed 3X with PBS. One hundred microliter of EL-4 cell suspension (10^5 cells/ml) in PBS was added to each well and the plate was centrifuged at 500 rpm for 5 min. After removing the PBS, 0.05 ml of 0.1% glutaralaldehyde in PBS was added to each well, and the plate was incubated at room temperature for 10 min. Wells were washed 4X with PBS. To block aldehyde residues and unbound solid phase sites, 0.1 ml of PBS containing both bovine serum albumin (0.1 mg/ml) and glycine (100 mM) was added to each well, and the plate was incubated at room temperature for 1 h. The wells were washed 4X with PBS, and then reacted at 37°C for 1 h with T-2-mAb conjugates or the mAb not conjugated with T-2, which were step-wise diluted with PBS. Wells were washed 4X with PBS, and then reacted at 37°C for 1 h with 0.05 ml of alkaline phosphatase labeled sheep anti-mouse IgG (diluted 75-fold in PBS). Wells were washed 4X with PBS, and then incubated with 0.1 ml of p-nitrophenylphosphate disodium chloride (1 mg/ml) dissolved in 1 M diethanolamine-HCl buffer (pH 9.8) at 37°C for 30 min. The absorbance in each well was estimated at 405 nm by a microplate reader (Microplate Photometer MTP-22, Corona Electrics Co., Ltd., Ibaraki).

The cytotoxicity of T-2-antibody conjugates

One hundred microliter of EL-4 and SP2/0 cells in passage medium (10^6 cells/ml) was added to 96-well culture plates. T-2-mAb and T-2-nGG conjugates were previously diluted with sterile PBS. Fifty microliter of each of the conjugates in PBS, as well as PBS alone, was added to each well and the cells were cultured at 37°C for 20 h. Each test was done in quadruplicate. After culturing, 0.05 ml of 0.3% trypan blue in PBS was added to each well, and the viable cells, based on trypan dye exclusion, were counted.

Effects of metabolic inhibitors on cytotoxicity of T-2-mAb conjugates

EL-4 cells in passage medium were added to each well of 96-well, flat-bottom, microtiter plates (10^5 cells/well). Ammonium chloride in PBS (40 mM, 0.05 ml) and monensin in PBS containing 10% ethanol (10% EtOH-PBS) (40 mM, 0.05 ml) were added separately to each well, and the plates were incubated at room temperature for 15 min. T-2-mAb conjugates (0.05 ml) in PBS was added to each one of the cultures and incubated at 37°C for 20 h. PBS alone and 10%-EtOH-PBS were used as control vehicles.

Five-hundred microliters of EL-4 cells in passage medium (10^6 cells/ml) were added to each well of a 24-well tissue culture plate. Sodium azide (2 mM) and 2-deoxyglucose (2-dG, 10 mM) together in PBS (0.25 ml) were added to each culture as endocytosis inhibitors[8], followed by incubation at room temperature for 15 min. After the addition of 0.25 ml of T-2 mAb conjugates in PBS or PBS alone (as control), the cells were incubated for 1 h, and washed to remove the inhibitors. The cells were further incubated at 37°C for 20 h, and the viable cells were counted using the dye exclusion method and 50% inhibitory concentrations (IC_{50}) were calculated.

The statistical differences between viable cell numbers resulting from cells that had been treated with metabolic inhibitors, and those of the control vehicles were assessed by the Student's t test.

Antitumor activity in in vivo system

C57BL/6 female mice aged 4 weeks were i.p. inoculated on day 0 with 1 x 10^7 EL-4 cells in 0.5 ml of PBS. All of the mice were then given i.p. injections with either T-2-mAb conjugates, anti-EL-4 mAb, free T-2, or vehicle alone, in 0.5 ml of PBS or in 0.5 ml of 10% EtOH-PBS, as follows. In Experiment 1, the mice were given injections with T-2-mAb conjugates, 0.6 mg/mouse, 1.5 mg/mouse in PBS, and PBS alone (as control), on days 2, 4, and 6. In Experiment 2, the mice were given injections with T-2-mAb conjugates, 0.9 mg/mouse, non-conjugated mAb 0.9 mg/mouse in PBS, free T-2 0.0071 μg/mouse (0.5 mg/kg) in 10% EtOH-PBS, PBS only, and 10% EtOH-PBS only on days 2, 4, and 6. Mice were monitored for 90 days after tumor inoculation, and the mean survival time (MST), the prolongation rate of MST, and the final survival rate were calculated.

RESULTS

Preparation of T-2-antibody conjugates

T-2 HG was converted to the active ester T-2-G-OSu by N-hydroxysuccinimide/DDC. The conjugation efficiency of T-2-G-OSu with anti-EL-4 mAb is summarized in Table 1. When the weight ratio of T-2-G-OSu/mAb was 0.1/1, T-2-G-OSu was linked to the mAb with best recovery of protein, T-2-mAb conjugates had the highest binding activity to EL-4 cells. The above mentioned T-2-mAb conjugates were used in in vitro studies.

Specificity of 5C10 mAb

Using anti-EL-4 mAb (5C10) and rabbit complement, the antibody-dependent, complement-mediated, cell lysis was examined. More than 90% of EL-4 cells were dead. Such killing was not observed with spleen cell stimulated with Con A and SP2/0 cells (data not presented).

Table 1. Conjugation efficiency of T-2 with anti-EL-4-mAb.

Weight rate of T-2-G-OSu/mAb*	Molar ratio of T-2/mAb	Recovery (%)	
		Protein	Binding activity to EL-4 cells
0.1	14	94	100
0.2	50	85	33
0.3	75	81	0
0.4	100	75	0

* T-2-G-OSu and mAb were mixed at the weight ratio indicated, and incubated at 4°C for 12 h.

Cytotoxicity of T-2-mAb conjugates to EL-4 cells

The cytotoxicities of T-2-mAb and T-2-nGG conjugates to EL-4 and SP2/0 cells were compared (Table 2). T-2-mAb conjugates had a high cytotoxicity to only EL-4 cells (IC_{50} was 20 μg/ml), T-2-nGG conjugates had a weak cyto- toxicity to both EL-4 and SP2/0 cells.

Cytotoxicities of T-2-mAb conjugates in the presence of metabolic inhibitors

All the data are summarized in Table 3. When exposed to T-2-mAb conju- gates in the presence of 10 mM NH_4Cl or 10 nM monensin, the cytotoxic effect of the conjugates was enhanced in both cases (IC_{50}, 7.8 and 6.8 μg/ml, respectively). However, the treatment of the cells with 0.5 mM NaN_3 and 2.5 mM 2-dG together reduced the cytotoxicity of T-2-mAb conjugates (IC_{50} was 80.8 μg/ml). When comparing EL-4 cells pretreated with several inhibitors and those not treated, the differences between the numbers of viable cells after the addition of the same dosages of T-2-mAb conjugates were statis- tically significant ($P < 0.05$).

In vivo antitumor activities of T-2-mAb conjugates, non-conjugated mAb and free T-2

In Experiment 1, all 8 mice given i.p. inoculations of 1 X 10^7 EL-4 cells and then treated with PBS alone (as control), died by day 21. Of 8 tumor cell-implanted mice given i.p. injections with 0.6 mg/mouse of T-2-mAb conjugates on days 2, 4, and 6, only one survived. Of 8 mice likewise given injections with 1.5 mg/mouse of T-2-mAb conjugates, five survived. Their

Table 2. IC_{50}* (μg/ml) of T-2-mAb and T-2-nGG conjugates.

Conjugates	SP2/0 cells	EL-4 cells
T-2-nGG	> 1,000	> 1,000
T-2-mAb	> 1,000	20

* IC_{50} - 50% growth inhibition concentration of conjugates *in vitro*.

Table 3. Effect of inhibitors on IC_{50}* of T-2-mAb conjugates to EL-4 cells.

Inhibitors (mM)	IC_{50} (μg/ml)
None	20.0
Ammonium chloride (10)	7.8
Monensin (0.01)	6.7
2-deoxyglucose (2.5) and NaN_3 (0.5)	80.8

* IC_{50} - 50% growth inhibition concentration of conjugates *in vitro*.

MST, prolongation rate of the MST, and the final survival rate are summarized in Table 4.

In Experiment 2, all 5 mice given i.p. injections of 7.1 µg/mouse X3 of T-2, died by day 37. The 7 mice given i.p. injections with 0.9 mg/mouse X3 of anti-EL-4 mAb not conjugated with T-2, died by day 58. However, 2 out of 7 mice treated with 0.9 mg/mouse X3 of T-2-mAb conjugates survived. Their MST, prolongation rate of the MST, and the final survival rates are summarized in Table 5.

Table 4. The antitumor activity of T-2-mAb conjugates.

T-2-mAb (mg/A X3)	Number of mice	MST* (days)	Prolongation rate** (%)	Survival rate (%)
--	8	20	0	0
0.6	8	38	94	13
1.5	8	70	257	63

*   Median survival time.
** Prolongation rate of MST.

Table 5. The antitumor activity of T-2-mAb conjugates, non-conjugated mAb, and T-2 toxin.

Antitumor agent (mg/A X3)	Number of mice	MST* (days)	Prolongation rate** (%)	Survival rate (%)
-	8	20	0	0
T-2 toxin (0.0071)	5	31	33	0
Non-conjugated mAb (0.9)	7	28	39	0
T-2-mAb conjugates (0.9***)	7	44	120	29

 * Median survival time.
 ** Prolongation rate of MST.
*** 0.0071 mg/A equivalent to T-2 toxin.

DISCUSSION

We herein report the antitumor activity of T-2-mAb conjugates *in vitro* and *in vivo*. First, T-2-mAb conjugates were prepared efficiently. This method included an introduction of a carboxylic acid residue to the T-2 molecule by the anhydride method, and the transformation of the conjugation of T-2-G-OSu with immunoglobulin via lysyl residues (Fig. 1). The T-2-mAb conjugates prepared in this way exhibited both perfect protein preservation and binding activity to EL-4 cells (Table 1). In various laboratories, antitumor agents such as mitomycin C[3] and daunomycin[10] were conjugated to antitumor antibodies. However, these compounds are mutagenic. Even though these immunotoxins possess high antitumor activities, the tumorigenesis can be increased with their extended use. Because T-2 exhibited a potent cytotoxicity without mutagenic activity,[12] it seemed to be an effective anticancer reagent when employed as an immunotoxin. The T-2-mAb conjugates selectively inhibited the growth of EL-4 cells *in vitro* (Table 2). The antitumor

activity presumably results from an antigen-antibody reaction. Regarding th
uptake of the conjugates, it is likely that the conjugate enters into EL-4
cells by receptor-mediated endocytosis. Endocytosis is reported to be an
energy-dependent process.[8] Therefore, this process is inhibited by 2-dG and
sodium azide. The evidence that sodium azide and 2-dG reduced the cytotoxicity of T-2-mAb conjugates are internalized by endocytosis. Furthermore,
ammonium chloride and monensin promoted the cytotoxic effect of T-2-mAb conjugates (Table 3). These compounds inhibit lysosomal enzyme activities by
changing the pH level within the lysosomes.[14] This suggests that T-2-mAb
conjugates integrated within the EL-4 cells are hydrolyzed by lysosomal
enzymes to give rise to free T-2, which may interfere with the protein synthesis in the target EL-4 cells.

In the previous papers,[1,6,9,14] the antitumor activity of diacetoxyscirpenol, a trichothecene chemically close to T-2, was evaluated. However,
the antitumor activity was poor, and side effects such as nausea, vomiting,
fever, and skin erythema were notable in the tumor patients. This suggests
that the free trichothecene alone was not effective as an antitumor agent.
In mice transplanted with EL-4 cells, T-2-mAb conjugates increased both the
prolongation rate and the final survival rate, dose-dependently (Table 4).
The T-2-mAb conjugates were more effective than anti-EL-4 mAb and free T-2 i
prolongation rate of MST and the final survival rate of the tumor-bearing
(Table 5). The data strongly suggest that the *in vivo* antitumor effect of
anti-EL-4 mAb against EL-4 cells was promoted upon conjugation with T-2, a
potent cytotoxic trichothecene mycotoxin. Thus, T-2 may be applicable for
cancer chemotherapy as an immunotoxin.

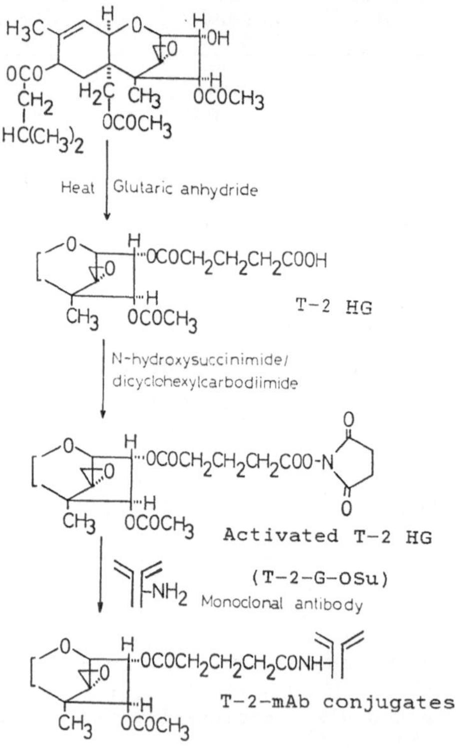

Fig. 1.   Preparation of T-2-mAb conjugates.

REFERENCES

1. Goodwin W, Hass CD, Fabian C, 1987, Phase I evaluation of anguidine, *Cancer*, 42:23-26.
2. Hosoda H, Yokohama H, Ishii K, Ito Y, Nambara T, 1983, Preparation of haptens for use in immunoassay of tetrahydro-11-deoxycortisol and its glucuronides, *Chem Pharm Bull.*, 31:4001-4007.
3. Kato Y, Tsukada Y, Hara T, Hirai H, 1983, Enhanced antitumor activity of mitomycin C conjugated with anti-α-fetoprotein antibody by novel method of conjugation, *J Appl Biochem.*, 5:313-319.
4. Kronke M, Schlik E, Waldmann TA, Vitetta ES, Green WC, 1986, Selective killing of human T-lymphotropic virusss-I infected leukemic T-cells by monoclonal anti-interleukin 2 receptor antibody-ricin A chain conjugates: Potentiation by ammonium chloride and monensin, *Cancer Res.*, 46:3295-3298.
5. Kuczuk MH, Benson PM, Health H, Hayes AW, 1978, Evaluation of the mutagenic potential of mycotoxins using *Salmonella typhimurium* and *Saccharomyces cereviseae*, *Mut Res.*, 53:11-20.
6. Murphy WK, Burgess MA, Valdivieso M, 1978, Phase I clinical evaluation of anguidine, *Cancer Treat Rep.*, 62:1497-1502.
7. Ohtani K, Kawamura O, Ueno Y, 1988, Improved preparation of T-2 toxin-protein conjugates, *Toxicon*, 26: - (in press).
8. Smyth MJ, Pietersz GA, McKenzie IFC, 1986, Potentiation of the *in vitro* cytotoxicity of chlorambucil by monoclonal antibodies, *J Immunol.*, 137:3361-3366.
9. Thigpen JT, Vaughn C, Stuckey WJ, 1981, Phase II trial of anguidine in patients with sarcaomas unresponsive to prior chemotherapy, *Cancer Treat Rep.*, 65:881-882.
10. Tsukada Y, Kato Y, Umemoto N, Hara T, Hirai H, 1984, An anti-α-feto-protein antibody-daunorubicin conjugate with a novel poly-L-glutamic acid derivation as intermediate drug carrier, *J Natl Cancer Inst.*, 73:721-729.
11. Ueno Y, Kobota K, 1976, DNA-attacking ability of carcinogenic mycotoxins in recombination-deficient mutant cells of *Bacillus subtilis*, *Cancer Res.*, 36:445-451.
12. Ueno Y, 1987, Trichothecenes in food, *In*: "Mycotoxins in Food," Krogh P, ed., Academic Press, London.
13. Ueno Y, Sawano M, Ishii K, 1975, Production of trichothecene mycotoxins by *Fusarium* species in shake culture, *Appl Microbiol.*, 30:4-9.
14. Yap H-Y, Murphy WK, Distefano A, Blumenschein GR, Bodey GP, 1979, Phase II study of anguidine in advanced brest cancer, *Cancer Treat Rep.*, 63:789-791.

EFFECTS OF CYCLOPIAZONIC ACID: GUINEA PIG SKELETAL MUSCLE

W. Michael Peden

National Animal Disease Center
Agricultural Research Service
U. S. Department of Agriculture
P. O. Box 70, Dayton Road
Ames, Iowa 50010

LITERATURE REVIEW

Cyclopiazonic acid (CPA) is a mycotoxin originally isolated from *Penicillium cyclopium* Westling by Holzapfel in 1968. CPA has been characterized as a toxic indole tetramic acid, similar to the mycotoxins tenuazonic acid and erythroskyrine, and acts as a lipophilic monobasic acid.[25,26,27,28,61]

Cyclopiazonic acid has been isolated as a naturally occurring mycotoxin from corn, peanuts, and cheese.[18,30,32] Several fungi in the *Penicillium* and *Aspergillus* genera have been shown to produce CPA, among them *Penicillium cyclopium, P. patulum, P. viridicatum, P. puberulum, P. crustosum, P. camembertii, Aspergillus versicolor, A. oryzae, A. tamarii*, and *A. flavus.*[8,9,31-33,35,46,47,62] CPA may be a more common metabolite of *A. flavus* than aflatoxin, as one study has shown that 19 of 31 isolates of *A. flavus* from dried food products produced CPA, while only 6 of the 31 isolates produced aflatoxin.[63]

Experimental work with cyclopiazonic acid in animals began in 1971, when Purchase determined that the intraperitoneal (ip) LD_{50} for male rats was 2.3 mg/kg while the LD_{50} per os (po) was 36 mg/kg for males and 63 mg/kg for females. Oral administration of CPA resulted in dose-related degenerative changes in the liver, kidney, spleen, pancreas, and myocardium.

Other studies with rats have shown similar results, with toxicity and lesions dependent on dose and frequency.[23,38,42,52,65] Histopathologic changes in liver, kidney, pancreas, adrenal gland, salivary gland, testes, and GI tract have been reported. These lesions range from vacuolated and granular hepatocytes and dilated renal tubules with pyknotic nuclei and a few scattered casts, to small foci of coagulative necrosis and hepatocyte degeneration in the liver, splenic necrosis and hemorrhage, mucosal erosions in the gastrointestinal tract, and occasional foci of ulceration and necrosis in the GI tract. Salivary gland changes consisting of swollen serous and ductular epithelial cells, with clear nucleoplasm and prominent nucleoli are also reported. Also diarrhea, crusted eyelids, and abnormal posture and movement have been noted.[38,42]

Microbial Toxins in Foods and Feeds, Edited by A.E. Pohland *et al.*,
Plenum Press, New York, 1990

Cytoplasmic vesiculation due to dilatation of the endoplasmic reticulum (ER) was the major ultrastructural change in hepatocytes of rats dosed with 4 mg/kg po of CPA. The increase in width of ER was dose related. Mitochondrial swelling was also present, but mitochondrial lysis was not observed.[24]

No teratogenic effects have been found in offspring of pregnant Fischer 344 rats or mice given CPA.[29,37] In mice, only mild fetal growth retardation was noted when dams were given CPA.[29]

Neurological effects of CPA in mice include catalepsy, decreased spontaneous motor activity and lowered body temperature. Chemical analysis of brain tissue revealed that CPA altered concentratioans of neurotransmitters and metabolites, suggesting CPA may affect dopaminergic and serotonergic systems.[40,41] The time of death in mice was related to the amount of CPA given, with earliest deaths (24-259 min) associated with a dose of 12.5 mg/kg or greater. Delayed death appeared to be due to cachexia, as food and water intake was severely curtailed. Although tremors associated with forced exercise were noticed in mice dosed with CPA, the tremors were not as severe as those seen with tremorine or harmaline.[40]

In dogs, CPA given at 1.0 mg/kg po twice a day (BID) resulted in moribundity and death within 48 hr.[45] Clinical signs included anorexia, vomiting, diarrhea, dehydration, weight loss, pyrexia, and depression. Gross lesions were most frequently seen in the gastrointestinal (GI) tract and kidneys. Hyperemia and ulceration of soft palate, esophagus, and GI tract were seen with ulcers most numerous near Peyer's patches.

Kidneys of most dogs given 0.5 or 1.0 mg/kg BID had raised circular infarcts. This lesion was not seen in animals that survived. Lesions were also seen in other organs, including uterus, epididymis, adrenal glands, skin, and urinary bladder.

Microscopic lesions were categorized as either vascular, ulceration and necrosis, or necrosis. Most ulcerative and necrotic lesions were associated with vascular damage. Nuclear enlargement was also described. Lymphoid tissues were the exception, as extensive necrosis of lymphocytes and lymphoid depletion occurred in the spleen, tonsils, mesenteric lymph nodes, Peyer's patches, and other lymphoid follicles in the GI tract. Areas rich in B-lymphocytes (germinal centers, white pulp) were more severely affected than those rich in T-lymphocytes.

Erosion and ulceration in the GI tract were accompanied by a hyperemic and edematous lamina propria, and frequently by a heavy infiltrate of neutrophils.

Vascular lesions were random and segmental and ranged from swollen, vacuolated endothelial cells with infiltrates of a granular acidophilic material between the endothelium and internal elastic membrane, to loss of endothelial cells, fibrin in the tunica media, vacuolated myocytes of the tunica media, and infiltrates of neutrophils and eosinophils. The most severely affected vessels had foci of necrosis, hemorrhage and neutrophils in the intima and media. The lesions in the vascular system were suggested to be due to a direct toxic effect of CPA resulting in endothelial damage, leakage of fibrin, erythrocytes, and other blood components into the intima and media of vessels, progressing to vasculitis or necrosis. The ulceration and necrosis in the various organs were attributed to ischemia from vascular damage and thrombosis.

Males and spayed females succumbed to the toxin sooner than intact females, suggesting a protective effect for estrogen. Similar observations were made by Purchase,[52] who noted male rats had a lower LD_{50} than female rats.

Diarrhea, weakness, inactivity, rough hair coats, anorexia, and weight loss were observed in pigs given 10 mg/kg/day po.[34] Animals given 1.0 mg/kg/day had rough hair coats and decreased activity.

Lesions occurred only in pigs given 10 and 1.0 mg/kg/day and, as in dogs, were most prominent in the GI tract and kidney. Gross lesions of gastric ulceration, mucosal hyperemia, and hemorrhages were seen in the small and large intestines in pigs given 10 mg/kg/day CPA.

Microscopically, the lesions consisted of gastric mucosal necrosis and inflammation, villous blunting, and mucosal necrosis and inflammation in the small intestine and large intestine. The liver lesions varied from mild hepatocellular vacuolization to severe, diffuse hepatic necrosis. Renal lesions consisted of necrosis of tubular epithelial cells and focal purulent tubulointerstitial nephritis. Lesions in animals given 1.0 mg/kg/day of CPA consisted of mild focal gastric mucosal necrosis.

Chickens fed CPA-contaminated feed at 100 ppm had high mortality, decreased weight gain and poor feed conversion.[9] Lesions were restricted to birds fed 100 ppm or 50 ppm CPA, and were primarily noted in the GI tract. Proventriculi from birds in the 100 ppm group had mucosal erosions and hyperemia. Birds given 50 ppm had thick proventricular mucosa and dilated proventricular lumens. Mucosal epithelial necrosis and inflammation of the crop, proventriculus, and gizzard, hepatocellular vacuolization, necrosis and inflammation were seen also in birds fed 100 ppm CPA. Similar but less severe changes were seen in birds given 50 ppm in the feed. Other lesions noted were splenic necrosis and myocardial inflammation in birds fed 50 or 100 ppm CPA.

Cullen et al.[5] reported necrosis, hemorrhage, and hyperplasia of pro-ventriculus mucosa, hepatocellular vacuolation, and skeletal muscle degeneraon in chickens given CPA by gavage.

Norred et al.[44] demonstrated that CPA was found in muscle of chickens given a single dose of CPA, and that excretion of CPA was dependent on the amount of CPA given. Weight gains were also decreased for up to 96 hr following the single high dose.

[14]C radiolabeled CPA is rapidly distributed and excreted in Sprague-Dawley rats, with a half-life of 33±12 h for ip administration and 43±25 h for intragastric administration. Blood contained the highest levels of CPA, followed by skeletal muscle and liver. CPA was excreted in both feces and urine, with fecal excretion the major route. This suggested that biliary excretion plays a major role in CPA excretion.[43]

Guinea pigs given multiple doses of CPA per os had clinical signs of anorexia, rough hair coats, diarrhea, and slight incoordination.[54] Lesions and mortality were confined to high dose groups (1.6 and 1.95 mg/day). Liver enzymes (AST, ALT) increased slightly in high dose groups. Lesions were similar to those reported previously for rats, with hepatocellular vacuolation and superficial necrosis of the gastric mucosa the only significant lesions.

Immunological responses, as measured by cutaneous hypersensitivity and complement activity, were not affected in guinea pigs given multiple doses of CPA per os.[54] CPA also had minimal short term effects on immune response in rats, resulting primarily in an initial lag in antibody production to sheep red blood cells. Results of cell-mediated tests were inconclusive.[23]

413

A recent report on "kodua poisoning" implicated CPA as a possible factor.[53] Kodua poisoning is associated with ingestion of moldy millet (*Paspalum scrobiculatum*).[2] Clinical signs in livestock include depression, nervousness, incoordination, and spasms.[39] In man, ingestion of infected grain results in clinical signs of tremors and sleepiness. An extract from the affected grain produced depression and loss of mobility when injected into mice, and CPA was identified in the extract. Two fungi *Aspergillus flavus* and *A. tamarii* were consistently isolated from the seed and both produced CPA.[53]

In summary, the primary organs involved in CPA intoxication in most species affected are the liver, kidney, and GI tract. The dog has unusual lesions associated with vascular damage. Lymphoid organs also appear to be a target of CPA, although immune response (antibody titer, cell-mediated response) does not appear to be significantly affected. Other tissues, including pancreas, salivary gland, and myocardium can also be variably involved. The response to CPA was both species and dose related. Skeletal muscle degeneration has been noticed in experiments investigating the effect of CPA in guinea pigs and CPA effects on striated muscle in other species have also been reported.[5,43,49,52]

GUINEA PIG STUDIES: AN INTRODUCTION

Because CPA has been found in high concentration in muscle, a potential for mycotoxicity and contamination of meat and meat products exists. Guinea pig vastus lateralis, gastrocnemius and soleus muscles have been well characterized using histochemical and biochemical techniques. Certain muscle and muscle groups in guinea pigs consist primarily of fast twitch-glycolytic (FG) (v. lateralis white), fast twitch oxidative-glycolytic (FOG) (v. lateralis red), or slow twitch-oxidative (SO) (soleus) fibers, as well as more mixed muscles (gastrocnemius).[1,50] Additionally, morphometric methods have been applied to guinea pig muscle to quantitate the various components of these muscles and myofibers.[11,12,13,14,15] Therefore, experiments were conducted to assess the ability of CPA to cause muscle degeneration, to characterize the clinical signs, to determine the fiber type(s) involved using histochemical and ultrastructural methods, and to detect the site of injury using electron miroscopy.[48]

MATERIALS AND METHODS

CPA was dissolved in chloroform, dispensed into gelatin capsules, and the solvent was allowed to evaporate overnight. Guinea pigs were given 4 mg/kg/day of CPA po for 1, 2, 3, or 4 days. Animals were anesthetized with CO_2, and blood was drawn for serum chemistry. Guinea pigs were killed by pentobarbital overdose, and muscle was collected for histologic, histochemical, and ultrastructural anlaysis.

RESULTS

Increased serum enzyme concentrations of creatine phosphokinase (CPK) and aspartate aminotransferase (AST) indicated acute necrosis and enzyme leakage from muscle associated with CPA intoxication, most severe in animals given CPA for 4 days (Table 1, Figs. 1, 2). Serum phosphorus levels increased in animals given CPA, as did the amount of blood urea nitrogen (BUN) (Figs. 3, 4). Serum protein and albumin levels were increased in toxin-treated animals on day 4 (Fig. 5). Body weight decreased in guinea pigs given CPA (Fig. 6).

Excellent differentiation of normal myofibers was achieved with Gomori

Table 1.   Groups of guinea pigs given cyclopiazonic acid.

Group	CPA/day	Days	Total Dose Received
1	4 mg/kg	1	4 mg/kg
2	4 mg/kg	2	8 mg/kg
3	4 mg/kg	3	12 mg/kg
4	4 mg/kg	4	16 mg/kg
5	Control		

trichrome, NADH-TR, and ATPase stains, but most necrotic and degenerate fibers had lost both normal orientation and some staining specificity (Figs. 7, 8, 9, 10). Myofiber degeneration was most severe in the gastrocnemius; there was less involvement of other muscles. Frozen sections of muscle stained with NADH-TR and myosin ATPase techniques[3,60] indicated that affected myofibers were type IIa (fast oxidative-glycolytic or FOG) fibers (Figs. 8, 9). Affected type II fibers were classified as FOG fibers because of intense staining with ATPase at pH 10.2, indicative of Type II versus Type I, and because of large accumulations of formazan granules in peripheral areas, and a more intense overall staining using NADH-TR (indicative of Type IIa versus IIb). Lesions were not seen in the soleus muscle (100% type I-slow oxidative or SO muscle fibers in the guinea pig).[15,50]

Ultrastructural observations indicated there was an abrupt transition from normal to a severely degenerate and necrotic appearance in affected fibers. Early changes were restricted to swelling of mitochondria and

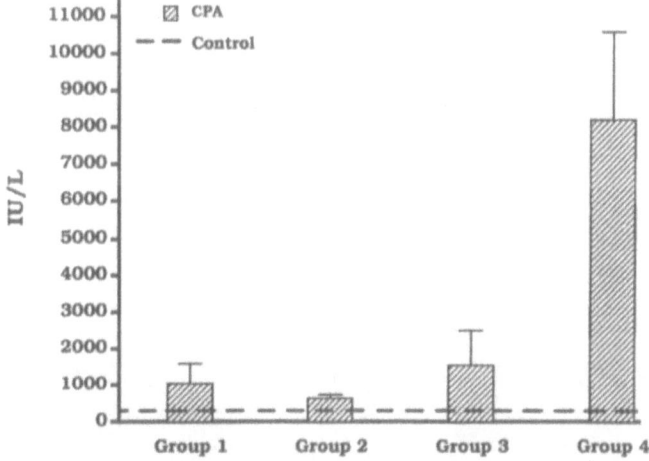

Fig. 1.   Average serum creatine phosphokinase activity in guinea pigs given CPA. Control value is average of 6 control animals over 3 days, two animals each day sacrificed with CPA-treated animals in Groups 2, 3, and 4.

415

sarcoplasmic reticulum (SR). Lesions were observed in gastrocnemius and vastus lateralis; no similar degenerate fibers were noted in soleus muscles of CPA-intoxicated guinea pigs. In gastrocnemius and vastus lateralis, mitochondria were swollen and contained fragmented cristae and multiple granules. Many affected fibers had focal loss of plasmalemma. Sarcoplasmic reticulum in degenerate fibers varied from normal morphology to dilated tubules (Fig. 11). Condensed myofilament debris was interspersed with foci of short filaments, granular debris, swollen mitochondria and SR (Fig. 12). Lysis of myofibrils and edema was seen within affected fibers and within intact myofibers adjacent to more severely affected myofibers (Fig. 13). Fiber type was not easily determined using electron microscopy due to the very limited number and size of samples with early changes without necrosis and hypercontraction, both of which made fiber identification difficult. Occasionally blood vessels scattered among degenerate fibers had swollen endothelial cells.

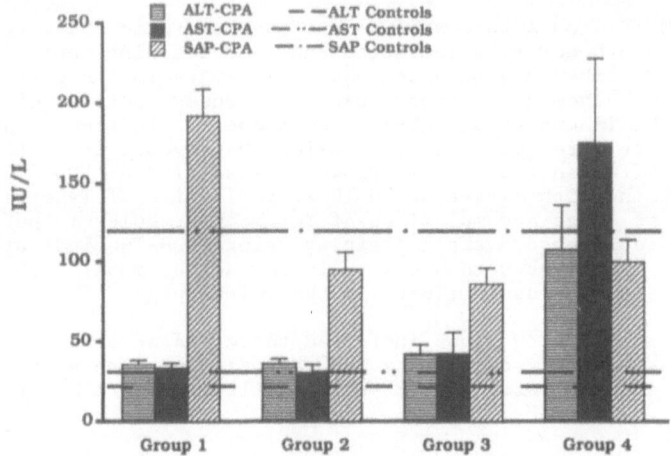

Fig. 2.   Liver Enzymes.   Average of serum enzymes in guinea pigs given CPA.   Control values as in Fig. 1.

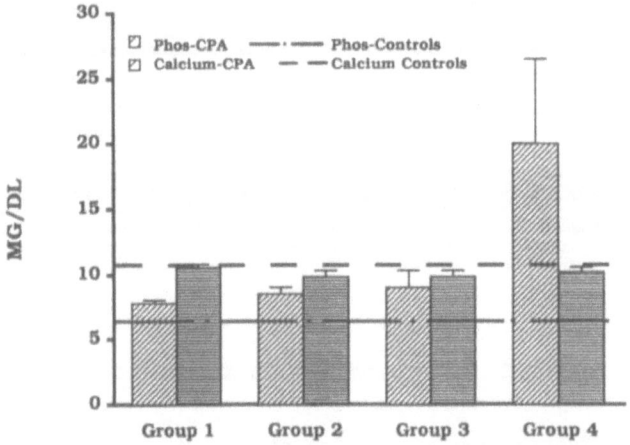

Fig. 3.   Calcium and Phosphorus.   Average serum calcium and phosphorus in guinea pigs given CPA. Control values as in Fig. 1.

DISCUSSION

Accumulations of mitochondria and prominent lipid vacuoles in degenerate and necrotic fibers suggests that either SO or FOG fibers were involved. FOG fibers in guinea pigs contain a higher percentage of mitochondria than FG and SO fibers, and also more lipid than FG fibers.[12,13,15]

Morphometric studies indicated the total SR volume was similar in SO, FG, and FOG fibers.[11,13,15] Calcium transport by SR of FG and FOG fibers was similar.[17] These findings suggest that something other than SR volume or activity was responsible for the preferential effect of CPA on Type IIa fibers found in the present study. FOG fibers have the highest oxidative capacity in guinea pigs and other rodents, and have a high glycolytic enzyme activity as well.[50,59] In addition SO and FOG fibers have a marginally higher capillary to fiber ratio than FG fibers.[36] There is little information on the comparable surface volume of mitochondrial membranes in SO, FOG, or FG fibers, although muscles comprised predominantly of FOG and SO fibers have a greater volume percentage of mitochondria than muscle with predominantly FG fibers.[11] Perhaps the metabolic makeup of FOG fibers (high oxidative capacity and increased mitochondrial content compared to FG fibers) or the increased delivery of toxin to the fiber (due to greater blood vessel to fiber ratio) may make these fibers more susceptible to this particular mycotoxin. A significant percentage of ingested or parenterally administered CPA was delivered to muscle, although it did not accumulate there.[43]

Calcium uptake into SR may be slowed and intracellular calcium levels raised within susceptible myofibers following muscle contraction in CPA-intoxicated animals. CPA inhibited Ca-ATPase and decreased calcium uptake in CPA-treated SR vesicles.[19] In the present study SR did not have the only or most marked morphologic changes in spite of the *in vitro* evidence that SR Ca-ATPase was decreased by CPA.[19] The sarcolemma, mitochondria and myofilaments were also involved in affected fibers.

Increased intracellular calcium is a common event in cell death and necrosis.[16,68] In muscle, if SR is unable to take up the increased load of calcium, mitochondria must remove the cation from the sarcoplasm. Calcium transport and accumulation by mitochondria is at the expense of energy

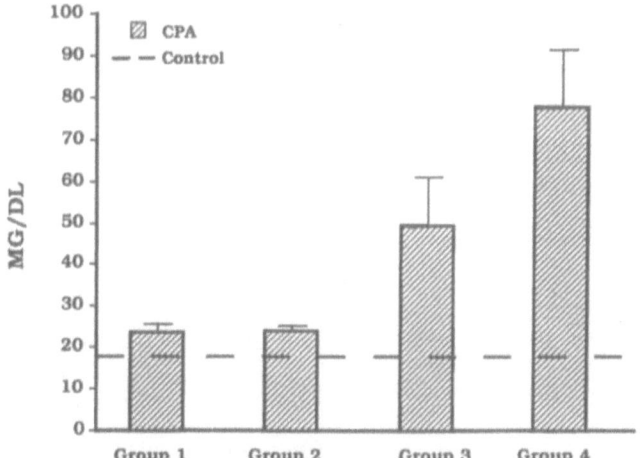

Fig. 4. Average serum BUN in guinea pigs given CPA. Control values as in Fig. 1.

production (oxidative phosphorylation), leading to decreased levels of ATP within the myocyte.[68]

An increase in intracellular calcium may activate calcium activated proteinase (calpain, calcium activated factor or CAF),[6,7,20,21,22] or lysosomal or nonlysosomal proteases.[58,69] Calcium activated proteinase has been implicated in normal myofibrillar protein turnover as well as in pathogic conditions.[51] CAF removes Z-lines *in vitro*.[6,7] In the present study, myofibers adjacent to necrotic fibers as well as degenerate fibers had evidence of lysis of myofibrils and Z-lines. Calpain or CAF activity is greater for some proteins associated with contractile proteins and structural proteins (titin, desmin, C-protein), than the primary contractile proteins in muscle (myosin, actin).[21,22] Degradation of these proteins along with

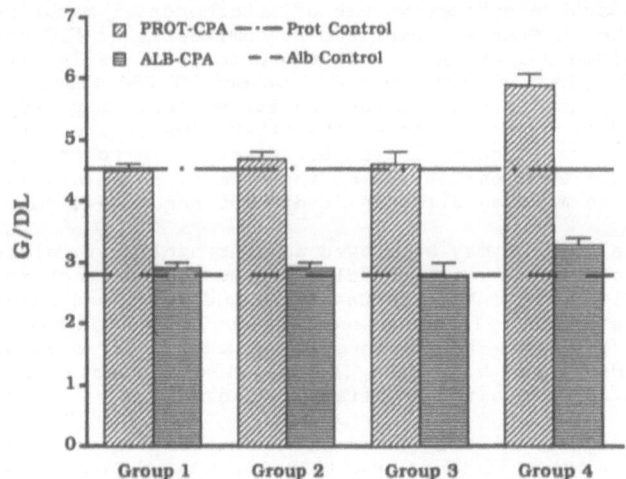

Fig. 5. Average serum protein and albumin. Control values as in Fig. 1.

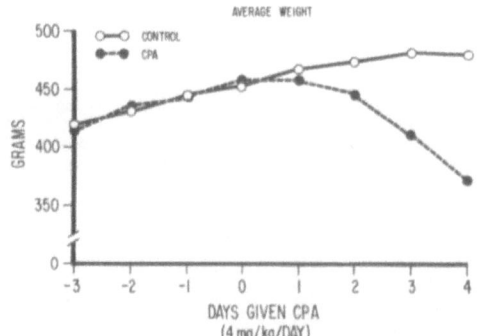

Fig. 6. Average weight for control and guinea pigs given 4 mg/kg/day of CPA. CPA was first given on Day 0.

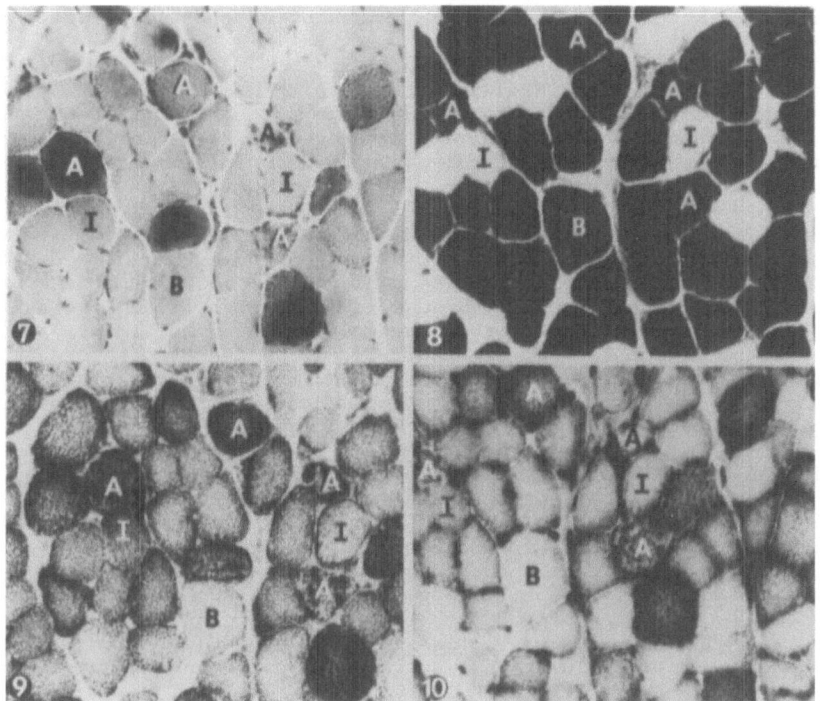

Figs. 7, 8, 9, and 10.   Muscle.   Guinea pigs given 4 mg/kg/day CPA for
                         4 days.   Serial frozen sections.

Fig. 7.   Necrotic myofibers and myofibers with beginning degenera-
           tion.   Myofibers labeled A are Type IIa or FOG fibers, B
           are Type IIb or FG fibers, and I are Type I or SO fibers.
           Several shrunken, slightly hyaline, necrotic fibers are
           visible, as well as dark, variably stained fibers which are
           undergoing beginning degeneration.   Gomori trichrome.

Fig. 8.   Lightly stained fibers are Type I fibers.   Darkly stained
           fibers are Type II fibers.   No degenerate or necrotic fibers
           stain similar to Type I fibers.   ATPase, pH 10.2 preincu-
           bation.

Fig. 9.   Three patterns of staining, 1) intense with large accumula-
           tions of formazan granules within and at the periphery of
           myofibers (Type IIa), 2) light staining, with few formazan
           granules at the periphery or in the middle of the myofiber
           (Type IIb), and 3) intermediate staining, with small, uni-
           formly dispersed granules in the center, as well as at the
           periphery of the myofiber (Type I).   Necrotic and degenerate
           fibers have intense aggregates of formazan granules and are
           representative of Type IIa fibers.   NADH-TR.

Fig. 10.   Fibers stained more intensely are Type IIa fibers.   Type I
           and IIb fibers stain less intensely.   Some Type IIa fibers
           appear degenerate as in Figs 7,8,9.   ATPase, pH 4.7 prein-
           cubation.

hypercontraction might explain the high number of whorled fibers seen in affected muscle.

Tetraphenylphosphonium (TPP+) accumulated within cultured kidney cells by a non-saturable mechanism dependent on membrane potential difference. CPA increased the accumulation and apparent binding of TPP+ to intracellular membrane components, particularly the mitochondrial and plasma membrane compartments of epithelial cell homogenates. The mechanism involved is unknown.[56,57] Extraction of soluble proteins and lipids from L6 muscle cell cultures prevented the binding of TPP+, suggesting that the phenomenon required biomembranes.[55] Riley et al.[57] suggested that the binding of CPA within membranes may affect the function of such membranes, causing electrical alterations in the membranes themselves leading to increased TPP+ binding. They proposed that the site of action between CPA and TPP+ was on the cytoplasmic side of the plasma membrane, and possibly the mitochondrial membrane. The effects on membranous organelles (sarcolemma, mitochondria, SR) in the present study also suggested a membrane-localized site of action for CPA.

Binding of CPA to myofiber membranes may activate endogenous phospholipases, increasing membrane permeability and intracellular calcium levels. Phospholipase-A has been shown to be activated by increased intracellular calcium levels via calmodulin in platelets.[67] Increased muscle membrane permeability may be due to subsequent calcium-mediated activation of phospholipase-A and generation of free fatty acids and lysophospholipids causing membrane damage. Oxidized products from fatty acids produced from phospholipase activity on membranes may act as calcium ionophores,[64,66] which could lead to increased intracellular calcium levels and further phospholipid turnover.

Although endothelial cells were swollen and degenerate in small blood vessels adjacent to affected myofibers, it could not be determined whether this was a primary event in the myofiber degeneration because vascular changes were occasionally seen 1 or 2 fibers removed from a necrotic myofiber.

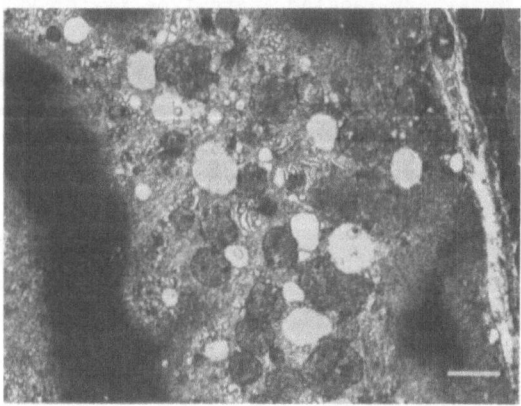

Fig. 11. Muscle. Guinea pig given 4 mg/kg/day CPA for 4 days. Aggregates of swollen mitochondria and SR are interspersed between condensed accumulations of myofilament debris. Many mitochondria contain granules. High numbers of lipid vacuoles are present indicating a Type I or IIa myofiber. Part of an intact myofiber is seen in one corner of the micrograph. TEM Bar = 1 micron.

The increase in serum phosphate and BUN were most likely caused by a decrease in glomerular filtration rate associated with dehydration.[4,10] Liver involvement was also shown by the slight increase in ALT and dose-related vacuolation of hepatocytes. All other clinical pathology results were compatible with a toxin induced necrosis of skeletal muscle and leakage of muscle specific enzymes.

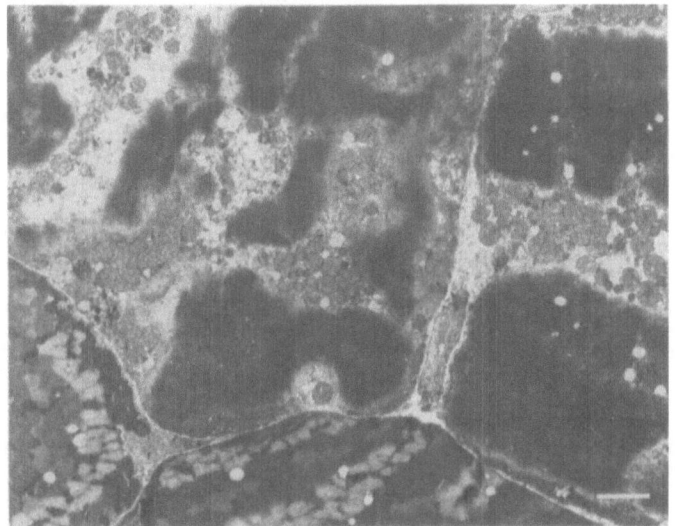

Fig. 12.   Muscle.   Guinea Pig given 4 mg/kg/day CPA for 4 days. Areas of four myofibers are present, two of which are necrotic.   The typical aggregations of organelles and hypercontracted, condensed myofilament debris are easily seen, as is the amount of lipid in these cells.

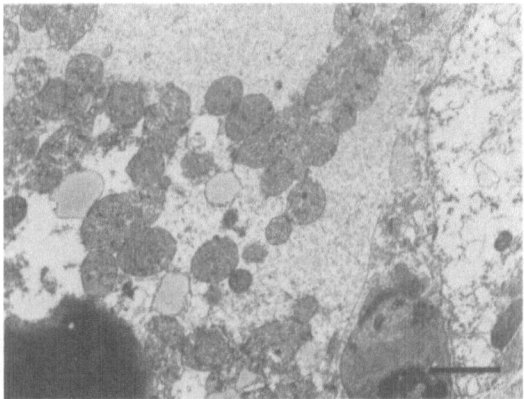

Fig. 13.   Muscle.   Guinea pig given 4 mg/kg/day for 4 days.   Two adjacent myofibers, one severely necrotic with swollen mitochondria with multiple granules, edema, and condensed myofilament debris, and the other with beginning myofilament lysis and edema. TEM Bar = 2 microns.

In summary, muscle degeneration and necrosis in CPA intoxication are likely caused by either a direct toxic effect on membranous organelles, such as mitochondria, SR, or sarcolemma, or is a functional disturbance such as intereference with sarcoplasmic reticulum Ca-ATPase. Either lesion would result in disturbances in calcium metabolism, raising intracellular calcium levels, and leading to calcium-activated degeneration of myofibers.

REFERENCES

1. Barnard RJ, Edgerton VR, Furukawa T, Peter JB, 1971, Histochemical, biochemical, and contractile properties of red, white, and intermediate fibers, *Am J Physiol.*, 220:410-414.
2. Bazlur M, 1960, Probable mona grass (*P. commersoni*) poisoning, *Ind Vet J.*, 37:43-47.
3. Brooke MH, Kaiser KK, 1969, Some comments on the histochemical characterization of muscle adenosine triphosphatase, *J Histochem Cytochem.*, 17:431-432.
4. Coles EH, 1986, "Veterinary Clincal Pathology" W. B. Saunders Co., Philadelphia, PA.
5. Cullen JM, Wilson M, Hagler WM, Ort JF, Cole RJ, 1988, Histologic lesions in broiler chicks given cyclopiazonic acid orally, *Am J Vet Res.*, 49:728-731.
6. Dayton WR, Goll DE, Stromer MH, Reville WJ, Zeece MG, Robson RM, 1975, Proteases and biological control, *In*: Cold Spring Harbor Conference on Cell Proliferation," Reich E, Rifkin DB, Shaw E, eds., Cole Spring Laboratory, Cold Spring, New York.
7. Dayton WR, Reville WJ, Goll DE, Stromer MH, 1976, A Ca++ activated protease possibly involved in myofibrillar protein turnover. Partial characterization of the purified enzyme, *Biochem.*, 15:2159-2167.
8. Dorner JW, 1983a, Production of cyclopiazonic acid by *Aspergillus tamarii* Kita, *Appl Environ Microbiol.*, 46:1435-1437.
9. Dorner JW, Cole RJ, Lomax LG, Gosser S, Diener UL, 1983b, Cyclopiazonic acid production by *Aspergillus flavus* and its effects on broiler chickens, *Appl Environ Microbiol.*, 46:698-703.
10. Duncan JR, Prasse KW, 1986, "Veterinary Laboratory Medicine: Clinical Pathology," Iowa State University Press, Ames, IA.
11. Eisenberg BR, 1983, Quantitative ultrastraucture of mammalian skeletal muscle, *In*: "Handbook of Physiology Section 10. Skeletal Muscle," Peachey LD, Adrian RH, Geiger SR, eds., American Physiological Society, Bethesda, MD.
12. Eisenberg, BR, Kuda AM, 1975, Sterological analysis of mammalian skeletal muscle. II. White vastus muscle of the adult guinea pig, *J Ultrastruct Res.*, 51:176-187.
13. Eisenberg BR, Kuda AM, 1976, Discrimination between fiber populations in mammalian skeletal muscle by using ultrastructural parameters, *J Ultrastruct Res.*, 54:;76-88.
14. Eisenberg BR, Kuda AM, 1977, Retrieval of cryostat sections for comparison of histochemistry and quantitative electron microscopy in a muscle fiber, *J Histochem Cytochem.*, 25:1169-1177.
15. Eisenberg BR, Kuda AM, Peter JB, 1974, Sterological analysis of mammalian skeletal muscle. I. Soleus muscle of the adult guinea pig, *J Cell Biol.*, 60:732-754.
16. Farber JL, 1982, Membrane injury and calcium homeostasis in the pathogenesis of coagulative necrosis, *Lab Invest.*, 47:114-123.
17. Fiehn W, Peter JB, 1971, Properties of the fragmented sarcoplasmic reticulum from fast twitch and slow twitch muscles, *J Clin Invest.*, 50:570-573.
18. Gallagher RT, Richard JL, Stahr HM, Cole RJ, 1978, Cyclopiazonic acid production by aflatoxigenic and non-aflatoxigenic strains of *Aspergillus flavus*, *Mycopathologia*, 66:31-36.

19. Goeger DE, Riley RT, Dorner JW, Cole RJ, 1988, Cyclopiazonic acid inhibition of the Ca^{2+}-transport ATPase in rat skeletal muscle sarcoplasmic reticulum vesicles, *Biochem Pharmacol.*, 37:978-981.

20. Goll DE, Edmunds T, Kleese WC, Sathe SK, Shannon JD, 1985, Some properties of the Ca^{2+}-dependent proteinase, *Prog Clin Biol Res.*, 180:151-164.

21. Goll DE, Otsuka Y, Nagainis PA, Shannon JD, Sathe SK, Muguruma M, 1983, Role of muscle proteinases in maintenance of muscle integrity and mass, *J Food Biochem.*, 7:137-177.

22. Goll DE, Shannon JD, Edmunds T, Sathe S, Kleese WC, Nagainis PA, 1983, Properties and regulation of the calcium-dependent proteinase, *In:* "Calcium-Binding Proteins," de Bernard B, Sottocase GL, Sandria G, Carafoli E, Taylor AN, Vanaman TC, Williams RJP, eds., Elsevier Science Publishers, Amsterdam.

23. Hill JE, Lomax LG, Cole RJ, Dorner JW, 1986, Toxicologic and immunologic effects of sublethal doses of cyclopiazonic acid in rats, *Am J Vet Res.*, 47:1174-1177.

24. Hinton DM, Morrissey RE, Norred WP, Cole RJ, Dorner JW, 1985, Effects of cyclopiazonic acid on the ultrastructure of rat liver, *Toxicol Letters*, 35:211-218.

25. Holzapfel CW, 1968, The isolation and structure of cyclopiazonic acid, a toxic metabolite of *Penicillium cyclopium* Westling, *Tetrahedron*, 24:2101-2119.

26. Holzapfel CW, 1980, The biosynthesis of cyclopiazonic acid and related tetramic acids, *In:* "The Biosynthesis of Mycotoxins," Steyn PS, ed., Academic Press, Inc., New York.

27. Holzapfel CW, Wilkins DC, 1971, On the biosynthesis of cyclopiazonic acid, *Phytochemistry*, 10:351-358.

28. Howard BH, Raustrick H, 1954, Studies in the biochemistry of micro-organisms 92. The colouring matters of *Penicillium islandicum* Sopp., *Biochem J.*, 57:212-222.

29. Khera KS, Cole RJ, Whalen C, Dorner JW, 1985, Embryotoxicity study on cyclopiazonic acid in mice, *Bull Environ Contam Toxicol.*, 34:423-426.

30. Lansden JA, Davidson JI, 1983, Occurrence of cyclopiazonic acid in peanuts, *Appl Environ Microbiol.*, 45:766-769.

31. Le Bars J, 1979a, Cyclopiazonic acid bioproduction by *Penicillium camembertii* Thom. Effect of temperature on individual strains, *Ann Rech Vet.*, 10:601-602.

32. Le Bars J, 1979b, Cyclopiazonic acid production by *Penicillium camembertii* Thom and natural occurrence of this mycotoxin in cheese, *Appl Environ Microbiol.*, 38:1052-1055.

33. Leistner L, Pitt JI, 1977, Miscellaneous *Penicillium* toxins, *In:* "Mycotoxin in Human and Animal Health," Rodricks JV, Hesseltine CW, Mehlman MA, eds., Pathotox Publishers, Inc., Park Forest South, IL.

34. Lomax LG, Cole RJ, Dorner JW, 1984, The toxicity of cyclopiazonic acid in weaned pigs, *Vet Pathol.*, 21:418-424.

35. Luk KC, Kobbe B, Townsend JM, 1977, Production of cyclopiazonic acid by *Aspergillus flavus* Link, *Appl Environ Microbiol.*, 33:211-212.

36. Mai JV, Edgerton VR, Barnard RJ, 1970, Capillarity of red, white, and intermediate muscle fibers in trained and untrained guinea pig, *Experientia*, 26:1222-1223.

37. Morrissey RE, Cole RJ, Dorner JW, 1984, The effects of cyclopiazonic acid on pregnancy and fetal development of Fischer rats, *J Toxicol Environ Health*, 14:585-594.

38. Morrissey RE, Norred WP, 1985, Toxicity of the mycotoxin, cyclopiazonic acid, to Sprague-Dawley rats, *Toxicol Appl Pharmacol.*, 77:94-107.

39. Nayak NC, Misra DB, 1962, Cattle poisoning by *Paspalum scrobiculatum* (kodua poisoning), *Ind Vet J.*, 39:501-504.

40. Nishie K, 1985, Toxicity and neuropharmacology of cyclopiazonic acid, *Food Chem Toxic.*, 23:831-839.

41. Nishie K, Porter JK, Cole RJ, Dorner JW, 1985, Neurochemical and

pharmacological effects of cyclopiazonic acid, chlorpromazine and
reserpine, *Res Comm Psychol Psych Behav.*, 10:291-302.

42. Norred WP, Morrissey RE, Cole RJ, Dorner JW, 1984, Hepatotoxic effects of
cyclopiazonic acid in rats, *Fed Proc.*, 43:478.

43. Norred WP, Morrissey RE, Riley RT, 1985, Distribution, excretion and
skeletal muscle effects of the mycotoxin [^{14}C]cyclopiazonic acid in
rats, *Food Chem Toxic.*, 23:1069-1076.

44. Norred WP, Porter JK, Dorner JW, Cole RJ, 1988, Occurrence of the
mycotoxin cyclopiazonic acid in meat after oral administration to
chickens, *J Agric Food Chem.*, 36:113-116.

45. Nuehring LP, Rowland GN, Harrison LR, Cole RJ, Dorner JW, 1985,
Cyclopiazonic acid mycotoxicosis in the dog, *Am J Vet Res.*,
46:1670-1676.

46. Ohmomo S, Sugita M, Abe M, 1973, Isolation of cyclopiazonic acid,
cyclopiazonic acid imine and bisseco-dehydrocyclopiazonic acid from
cultures of *Aspergillus versicolor* (Vuill.) Tiraboschi, *J Agric
Chem Soc Japan*, 47:57-63.

47. Orth R, 1977, Mycotoxins of *Aspergillus oryzae* strains for use in the
food industry as starters and enzyme producing molds, *Ann Nutr
Aliment*, 31:617-624.

48. Peden WM, Cheville NF, 1989, Ultrastructure, histology, and histochemistry of an acute myopathy in guinea pigs given cyclopiazonic acid,
Submitted.

49. Peden WM, Richard JL, Thurston JR, 1986, Comparative histopathologic
changes in cyclopiazonic acid toxicosis and rubratoxicosis, *In*:
"Diagnosis of Mycotoxicoses," Richard JL, Thurston JR, eds., Martinus
Nijhoff Publishers, Dordrecht.

50. Peter JB, Barnard RJ, Edgerton VR, Gillespie CA, Stempel KE, 1972,
Metabolic profiles of three fiber types of skeletal muscle in guinea pigs
and rabbits, *Biochem.*, 11:2627-2633.

51. Pontremoli S, Melloni E, 1986, Extralysosomal protein degradation, *Ann
Rev Biochem.*, 55:455-481.

52. Purchase IFH, 1971, The acute toxicity of the mycotoxin cyclopiazonic
acid to rats, *Toxicol Appl Pharmacol.*, 18:114-123.

53. Rao BL, Husain A, 1985, Presence of cyclopiazonic acid in kodo millet
(*Paspalum scrobiculatum*) causing "Kodua poisoning" in man and its
production by associated fungi, *Mycopathologia*, 89:177-180.

54. Richard JL, Peden WM, Fichtner RE, Cole RJ, 1986, Effect of cyclopiazonic
acid on delayed hypersensitivity to Mycobacterium tuberculosis,
complement activity, serum enzymes, and bilirubin in guinea pigs,
Mycopathologia, 96:73-77.

55. Riley RT, Goeger DE, Norred WP, Cole RJ, Dorner JW, 1987, Age and
growth-related changes in cyclopiazonic acid-potentiated lipophilic
cation accumulation by cultured cells and binding to freeze-thaw lysed
cells, *J Biochem Toxicol.*, 251-264.

56. Riley RT, Norred WP, Dorner JW, Cole RJ, 1985, Increased accumulation of
the lipophilic cation tetraphenylphosphonium+ by cyclopiazonic
acid-treated renal epithelial cells, *J Toxicol Environ Health*,
15:779-788.

57. Riley RT, Showker JL, Cole RJ, Dorner JW, 1986, The mechanism by which
cyclopiazonic acid potentiates accumulation of tetraphenylphosphonium in
cultured renal epithelial cells, *J Biochem Toxicol.*, 1:13-29.

58. Rodemann HP, Waxman L, Goldbery AL, 1982, The stimulation of protein
degradation in muscle by Ca^{2+} is mediated by prostaglandin E_2
and does not require the calcium-activated protease, *J Biol Chem.*,
257:8716-8723.

59. Saltin B, Gollnick PD, 1983, Skeletal muscle adaptability: Significance
for metabolism and performance, *In*: "Handbook of Physiology. Section
10. Skeletal Muscle," Peachey LD, Adrian RH, eds., American Physiological
Society, Bethesda, MD.

60. Sarnatt HB, 1983, "Muscle Pathology and Histochemistry," American Society

of Clinical Pathologists Press, Chicago.

61. Stickings CE, 1959, Studies in the biochemistry of micro-organisms. 106. Metabolites of *Alternaria tenuis* Auct.: The structure of tenuzaonic acid, *Biochem J.*, 72:332-340.

62. Still P, Eckardt C, Leistner L, 1978, Bildung von Cyclopiazonsaure durch *Penicillium camembertii*-isolate von Kase, *Die Fleischwirtschaft*, 58:876-878.

63. Trucksess MW, Mislivec PB, Young K, Bruce VR, Page SW, 1987, Cyclopiazonic acid production by cultures of *Aspergillus* and *Penicillium* species isolated from dried beans, corn meal, macaroni, and pecans, *J Assoc Off Anal Chem.*, 70:123-126.

64. Trump BF, Berezesky IK, Phelps PC, 1981, Sodium and calcium regulation and the role of the cytoskeleton in the pathogenesis of disease: a review and hypothesis, *Scanning Electron Microscopy*, 11:435-454.

65. Van Rensburg SJ, 1984, Subacute toxicity of the mycotoxin cyclopiazonic acid, *Food Chem Toxic.*, 22:993-998.

66. Weismann G, 1980, "Prostaglandins in acute inflammation – Current Concepts," The Upjohn Co., Kalamazoo, MI.

67. Wong, PY-K, Cheung WY, 1979, Calmodulin stimulates human platelet phospholipase A_2, *Biochem Biophys Res Comm.*, 90:473-480.

68. Wrogemann K, Pena SDJ, 1976, Mitochondrial calcium overload: a general mechanism for cell-necrosis in muscle diseases, *Lancet*, 27:672-674.

69. Zeman RJ, Kameyama T, Matsumoto K, Bernstein P, Etlinger JD, 1985, Regulation of protein degradation in muscle by calcium, *J Biol Chem.*, 260:13619-13624.

IMMUNOGLOBULIN A NEPHROPATHY AS A MANIFESTATION OF VOMITOXIN

(DEOXYNIVALENOL) IMMUNOTOXICITY

James J. Pestka, M. A. Moorman, R. L. Warner
M. F. Witt, J. H. Forsell, and J-H. Tai

Department of Food Science and Human Nutrition
Michigan State University
East Lansing, MI 48824

INTRODUCTION

The trichothecenes, mycotoxins produced by members of the genus *Fusarium*, are of immense concern because of their frequent presence in agricultural staples such as wheat, corn, barley and oats.[19,30,42] These compounds are sesquiterpenoids that are characterized by a trichothecane nucleus and that include some of the most potent inhibitors of protein synthesis known.[27,45] Acute exposure to trichothecenes results in severe damage to actively dividing cells in tissues such as bone marrow, lymph nodes, spleen, thymus and intestinal mucosa.   Trichothecenes have been implicated as causative agents in numerous episodes of fatal human and animal toxicoses.[5,20,44]

While the acute primary effects of the trichothecenes are well characterized, the secondary effects are less clear.   Immunosuppression is a frequently observed effect in field cases of low-level mycotoxin exposure in livestock.[35] Experimentally, repeated injection of animals with model trichothecenes such as T-2 toxin and diacetoxyscirpenol results in markedly increased susceptibility to candidiasis[38] and cryptococcosis.[18] Trichothecene impairment of host resistance is dramatically illustrated by the observation that concurrent exposure to T-2 toxin lowers the oral LD_{50} of *Salmonella typhimurium* from 5×10^6 to 5×10^0 organisms per mouse.[40] Decreased humoral reponse to T-dependent antigens and increased response to T-independent antigen,[36] increased skin-graft rejection times,[37] and diminished resistance to lipopolysaccharide[41] have also been attributed to T-2 toxin exposure.   Similarly, lymphocytes from T-2 toxin-treated animals have decreased B- and T-cell mitogen responses.[17,25] T-2 toxin induces DNA single-strand breakage in lymphoid tissue *in vitro* and *in vivo* but not in hepatic tissue.[24]

Although over 50 trichothecenes have been identified, vomitoxin (deoxynivalenol) appears most commonly in cereal grains produced in North America.[30,46] For example, 80% of midwestern feed samples submitted for diagnostic testing during 1981 in Illinois were shown to be contaminated by vomitoxin.[5] This mycotoxin is resistant to inactivation during milling and processing, and this facilitates its entry into foods and feeds.   Thus, outbreaks of *Fusarium* infection in North American cereals in recent years have resulted in significant low-level contamination of the food chain by vomitoxin.

Microbial Toxins in Foods and Feeds, Edited by A.E. Pohland *et al.,*
Plenum Press, New York, 1990

Although it is well-established that injections of high doses of model trichothecenes affect immune function, much less information exists on the immune effects of dietary administration of naturally occurring levels of the most common trichothecene, vomitoxin. We have sought to systemically evaluate the immunotoxic potential of vomitoxin. The purpose of this review is to highlight features of this previously reported research that have emphasized *in vitro* structure:function analyses in human lymphocytes, as well as pathotoxicological and immune function assessment in the genetically defined B6C3F1 mouse model. These investigations have led to the realization that vomitoxin and possibly other trichothecenes may have particular significance in their ability to alter mucosal immune function, specifically the immunoglobulin a (IgA) repsonse. Vomitoxin-induced dysregulation of the IgA response contributes to manifestations in the mouse that are analogous to human IgA nephropathy, the most common form of glomerulonephritis worldwide.

In vitro structure: function studies

To assess the relative potency of vomitoxin in an immunologically relevant *in vitro* system, we evaluated various trichothecene analogues *in vitro* in the mitogen-induced human lymphocyte blastogenesis assay.[13,14,32] Using the dose required for 50% inhibition of (^{3}H)-thymidine incorporation (ID_{50}) as a basis for comparison, trichothecene concentrations required for inhibition of human lymphocytes blastogenesis were approximately 10-fold less than those required to inhibit cell-free protein synthesis (Table 1). The rank order of potency was macrocyclic group > type A group > type B group and was predictable based on the known toxic and biochemical properties of these compounds. Lymphotoxicity of both the type A trichothecenes and the type B trichothecenes (8-ketotrichothecenes) is dependent on the degree of acylation in substituent groups. Inhibition studies indicate that fusarenon X and nivalenol are more toxic than vomitoxin and 15-acetyldeoxynivalenol. Replacement of the position C-4 acetyl of fusarenon X with the hydroxyl in nivalenol results in an approximately 4-fold decrease in toxicity, whereas replacement of the position C-4 hydroxyl of nivalenol with the hydrogen in vomitoxin results in a further two-fold decrease in toxicity. In an analogous manner, hydrolysis of T-2 toxin at the position C-4 acetyl to HT-2 toxin and hydrolysis of 3'OH T-2 toxin at the position C-4 acetyl to 3'OH HT-2 decreases lymphotoxicity 2.3-fold and 12.5-fold, respectively. Substitution of the position C-15 hydroxyl of vomitoxin with acetyl in 15-acetyldeoxynivalenol results in a 1.7-fold decrease in *in vitro* toxicity. In contrast, replacement of the position C-15 hydroxyl of T-2 triol with the acetyl in HT-2 toxin results in a 43-fold increase in toxicity. This suggests that dependence of lymphotoxic activity on the C-15 substituent is very different for the type A and B trichothecenes.

Although the relative *in vivo* and *in vitro* toxicities within the three trichothecene groups are qualitatively similar, the magnitude of difference among the ID_{50}s for the macrocyclic group and the type A and type B groups was unexpected in view of previously reported structure:function studies as summarized by Ueno.[45] Mitogen-induced lymphoblastogenesis involves cell proliferation and synthesis of interleukins by various lymphocyte subsets and accessory cells that in turn stimulate additional lymphocyte transformation and interleukin synthesis. The net result is further amplification of lymphocyte transformation. Because trichothecenes are likely to inhibit not only protein and DNA synthesis, but the entire amplification process as well, blastogenesis is an extremely sensitive assay that represents a more complex system than HeLa cell cytotoxicity, rabbit reticulocyte protein synthesis, or splenic lymphocyte protein synthesis and is one that may amplify differences in toxicity among trichothecenes. This contention is further supported by the observation that ID_{50} concentrations for individual trichothecenes in the HeLa cell and protein synthesis assays are one or more orders of magnitude higher than those found for the same compounds in the blastogenesis

Table 1. Comparative toxicities of trichothecenes in the mitogen-induced human lymphocyte blastogenesis assay.

Type	Trichothecene	ID_{50}[a] (pg/ml)
A[b]	T-2 toxin	1.5×10^3
	HT-2 toxin	3.5×10^3
	3' OH T-2 toxin	4.0×10^3
	3' OH HT-2 toxin	5.0×10^4
	T-2 triol	1.5×10^5
	T-2 tetraol	1.5×10^5
B[c]	Fusarenon X	1.8×10^4
	Nivalenol	7.2×10^4
	Deoxynivalenol	1.4×10^5
	15-Acetyldeoxynivalenol	2.4×10^5
Macrocyclic[d]	Roridin A	2×10^1
	Verrucarin A	9×10^0

[a] Inhibitory dose that caused 50% reduction of [^{3}H]TdR incorporation. Represents average of mitogen-stimulated lymphocyte subsets.
[b] From Forsell et al.[14]
[c] From Forsell and Pestka[13]
[d] From Pestka and Forsell[32]

assay. Lymphocyte blastogenesis might be particularly applicable to predicting direct trichothecene-lymphocyte interactions such as might occur between dietary vomitoxin and the gut-associated lymphoid tissue.

Acute and chronic toxicity

Toxicity determinations as well as immune function experiments were conducted with the B6C3F1 mouse (C57BL/6 female X CeH/HeN male) model. The strain is used as the basic murine model for carcinogen testing by the National Cancer Institute and has been previously used in studies of the immunotoxic effects of diethylstilbestrol, benzo(a)pyrene, and phorbol esters.[7] This hybrid strain was chosen over inbred strains because of the increased hardiness and longevity characteristics of heterosis and because of the wider genetic diversity as characteristic of human populations.[3] The hardiness is especially critical for conduct of extended feeding studies and the use of genetically identical mice minimizes variability encountered in immunological experiments.

A novel purification procedure was devised for the efficient production of vomitoxin in gram quantities.[47] Vomitoxin is elaborated in rice culture of F. graminearum, extracted, and placed on a column of water-saturated silica gel. The column is washed with methylene chloride and vomitoxin eluted in a small volume of water. Eluted vomitoxin is easily crystallized. Using the abbreviated procedure of Lorke,[26] LD_{50} values for purified vomitoxin were estimated to be 78 mg/kg (oral) and 49 mg/kg (ip) in the female B6C3F1 mouse.[16] Acute doses resulted in extensive necrosis of the gastrointestinal tract, bone marrow and lymphoid tissues, and focal lesions in kidney and cardiac tissue. The minimum doses required for these histopathological effects were consistent with LD_{50} estimates.

To evaluate the chronic effects of vomitoxin, weanling female B6C3F1 mice were fed semi-purified diets containing 0, 0.5, 2, 5, 10 or 25 ppm vomitoxin for 8 wk and assessed for effects on feed intake, body-weight gain, terminal organ weights, histopathology, and hematology.[15] To determine whether vomitoxin effects were potentiated by the estrogen zearalenone, a mycotoxin frequently found to occur with vomitoxin in cereals, two additional groups of mice were fed diets containing either 10 ppm zearalenone or 10 ppm zearalenone plus 5 ppm vomitoxin. The rate of body-weight gain was significantly reduced for all mice consuming feed containing 2.0 ppm or more of vomitoxin (Fig. 1b), whereas only the mice ingesting the diet containing 25 ppm vomitoxin showed a significantly decreased rate of feed consumption (Fig. 1b). Gross and histopathological evaluation of thymus, spleen, liver, kidney, uterus, small intestine, colon, heart, brain, lungs, and bone marrow from control and all mycotoxin-exposed mice revealed that these tissues were normal. Vomitoxin amended diets did cause dose-dependent decreases in terminal organ weights (thymus, spleen, liver, kidney, and brain). In the vomitoxin-treated groups, statistically significant dose-dependent decreases in the counts of total circulating white blood cells were associated with an increase in polymorphonuclear neutrophils and a decrease in lymphocytes and monocytes. In none of the above instances was 10 ppm zearalenone shown to act synergistically or antagonistically with 5 ppm vomitoxin.

Host resistance

Infection with *L. monocytogenes* has been used extensively as a model for assessing the effects of chemicals on host resistance. *Listeria* challenge elicits an early immune response, allowing rapid quantitative assessment of the course of the infection. Tryphonas et al.[43] reported that dietary vomitoxin caused a reduced, dose-related, time-to-death interval following challenge with *L. monocytogenes*. The potential of vomitoxin and zearalenone to alter *Listeria* resistance in the mouse was assessed by monitoring effects on bacterial counts in the spleen on days 1 and 4 after infection,[34] the time points when macrophage and cell-mediated immunity, respectively, are likely to be expressed.[8] Experiments using restricted and pair-feeding design indicated that 2-wk exposure to vomitoxin alone, zearalenone, and vomitoxin with zearalenone, significantly increased splenic *Listeria* counts on day 1 or day 4 (Table 2). Thus macrophage and cell-mediated immune function might both be altered by the mycotoxin treatments. Notably, although the splenic counts of vomitoxin-treated animals were consistently higher than those of the restricted controls, these differences were not significant. Diminished *Listeria* resistance in vomitoxin-treated mice may, in part, be a result of a nutritional effect associated with vomitoxin-induced feed refusal rather than a reflection of direct immunotoxicity of this mycotoxin. Furthermore, even though we observed significantly decreased resistance to *Listeria* after 2 wk dietary exposure to vomitoxin and zearalenone, similar immunosuppression by those toxins was not evident after the 8-wk exposure (Table 2). This observation may be the result of age-related differences in immune response or ability to metabolize the toxins.

Cell mediated immunity

Cell-mediated immunity is essential for resistance to various infections, tumour immunity and allograft rejection. Delayed hypersensitivity is a slowly evolving inflammatory response that can serve as a model for cell-mediated immune function. Significant effects of 25 ppm vomitoxin on delayed hypersensitivity were observable after 3 wk when compared to values for either *ad lib* or restricted controls (Table 2).[34] Thus, in contrast to the *Listeria* results, depressed cell-mediated immunity in this case was likely to be the direct result of immunotoxicity of vomitoxin rather than an indirect effect of reduced food intake. Notably, co-administration of 10 ppm zearalenone significantly decreases the inhibitory effects of 25 ppm vomitoxin. After 8 weeks, food restriction apparently diminishes the

Table 2. Significant immunological alterations in the B6C3F1 mouse following dietary exposure to vomitoxin and/or zearalenone or after dietary restriction (modified from ref 34).[a]

Dose Treatment	Duration of feeding (ppm) (wks)		Listerial resistance		Delayed hypersensitivity response		PFC response to SRBC[b]	
	(ppm)	(wks)	2 wk	8 wk	3 wk	8 wk	2 wk	8 wk
Vomitoxin	5	8	no effect	no effect	no effect	no effect	no effect	no effect
Vomitoxin	25	8	decreased[c]	no effect	decreased[d]	no effect	decreased[c]	no effect[d]
Zearalenone	10	8	decreased	no effect	no effect	no effect	no effect	no effect
Vomitoxin + Zearalenone	5 10	8	decreased	no effect	no effect	no effect	no effect	no effect
Vomitoxin + Zearalenone	25 10	8	decreased	no effect	no effect	no effect	--	--
5 ppm reduced feed intake equivalent		8	no effect	no effect	no effect	no effect	no effect	no effect
25 ppm reduced feed intake equivalent		8	no effect	no effect	no effect	no effect	no effect	increased

[a]Summary based on comparison with the *ad lib* control.
[b]PFC = Plaque forming cell. SRBC = Sheep red blood cells.
[c]No effect when based on comparison with the restricted control.
[d]Decreased when compared to the restricted control.

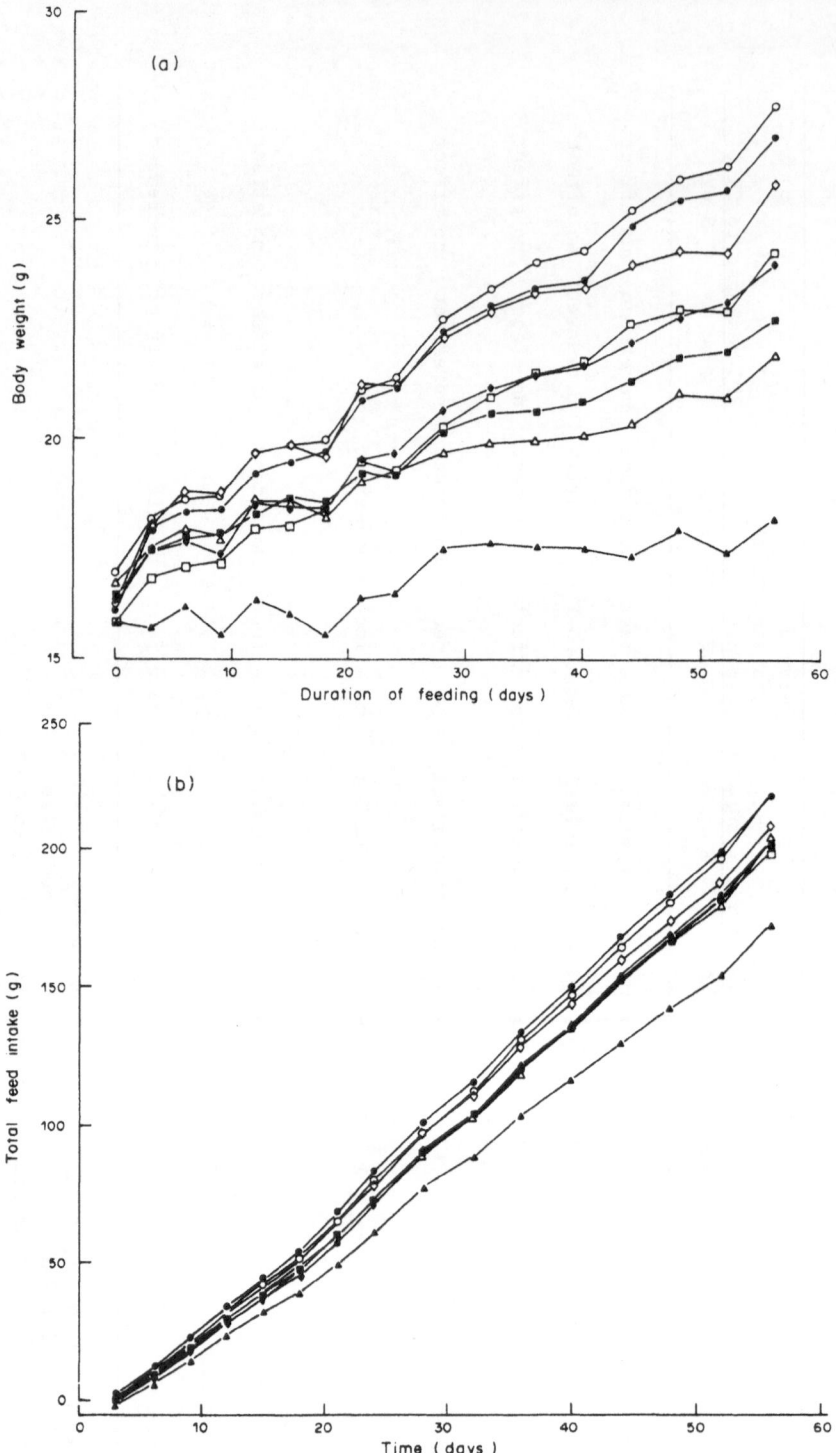

Fig. 1. Cumulative weight gains (a) and cumulative feed intake (b) in B6C3Fl
mice fed 0 (O), 0.5 (●), 2 (ɹ), 5 (■), 10 (△) or 25 (▲) ppm vomitoxin, 10 ppm
zearalenone or 5 ppm vomitoxin plus 10 ppm zearalenone (◆). Points are means
for groups of 26 controls and 8 treated mice. Five lower curves (a) and
bottom curve (b) differ significantly from the control: P < 0.01 by
Dunnett's t test. [15]

delayed hypersensitivity response, but not significantly. As was found with the 8-wk *Listeria* study, exposure to vomitoxin does not decrease the delayed hypersensitivity response. Again, the age-related factors mentioned above for *Listeria* resistance may have contributed to this observation.

Humoral immune response

The effect of vomitoxin on humoral immune function was assessed by measuring the splenic plaque-forming response to the T-dependent antigen SRBC. This response is lower in mice ingesting 25 ppm vomitoxin for 2 wk than in mice in the *ad lib* and restricted control groups.[34] Our inability to detect an effect by vomitoxin at the 5 ppm level is consistent with a similar observation made by Tryphonas et al.[43] The response of either restricted control group after 2 wk is not significantly different from that of the *ad lib* control group. After exposure for 8-wk, both restricted control groups had a significanat increase in PFC/10^6 cells compared to *ad lib* controls and the percentage increase is proportional to the degree of feed restriction (Table 2). Values for PFC/spleen are not significantly different, because of decreased total spleen counts in the restricted groups. Moderate dietary restriction previously has been reported to enhance some immune functions, including the PFC response to SRBC.[22] Both groups of mice ingesting vomitoxin alone for 8 wk had a decrease in response compared to their respective restricted control group (the effect with 25 ppm vomitoxin was statistically significant), but not compared to the *ad lib* control group. Thus depression by vomitoxin actually restores humoral responses to the level of the *ad lib* controls, illustrating the importance of dietary controls in trichothecene immunotoxicity studies.

Elevation of serum IgA

The most interesting effect of vomitoxin in the B6C3F1 mouse, that we have observed, is the dramatic elevation of serum IgA.[15] Elevation of serum IgA is maximal in B6C3F1 mice fed 25 ppm as compared to 2, 10, and 50 ppm of the compound.[33] Significant increases in total serum IgA are detectable in 4 wks in animals fed 25 ppm vomitoxin, with IgA levels increasing more than 17-fold after 24 wks exposure (Fig. 2a). No mortalities occurred. Because significant decreases in total serum IgG and IgM occur at the 25 ppm level, the effect apparently is specific for IgA (Fig. 2b). When intake of control diet is reduced by restricting available feed to a level equivalent to that caused by vomitoxin feed refusal, elevation of serum IgA was not observed (Fig. 2b). Thus, although vomitoxin causes feed refusal at the 25 ppm level,[15] apparently alteration of serum IgA does not result from decreased feed intake. Reduced feed intake did result in significantly decreased serum IgG suggesting that this may be a nutritional effect.

Murine IgA exists in a polymeric (pIgA) form (primarily dimeric) within mucosal secretions while both pIgA and monomeric IgA (mIgA) are found in serum.[9] Western blots revealed that serum from control animals exhibit a small band corresponding to pIgA and a relatively larger band corresponding to mIgA.[33] In contrast, pIgA is the predominant IgA form in sera from animals exposed to 25 ppm vomitoxin for only 8 weeks. Thus elevated serum IgA was primarily polymeric in nature.

Regulation of the IgA response in the mucosal and systemic compartments is extremely complex and thus the mechanism by which vomitoxin increases serum IgA is unclear. Elevated serum IgA might result from impaired catabolism and removal. Although diminished liver function[21] or blockage of the hepatobiliary tract[9,12] inhibit removal of polymeric IgA and results in an increased ratio of pIgA to mIgA and total serum IgA, histopathological assessment of B6C3F1 mice exposed to 25 ppm vomitoxin did not show evidence of impaired hepatobiliary function. Possibly, IgA elevation resulted from impaired secretion of dimeric IgA at the mucosal level.[28] Here vomitoxin might interfere with secretory component production by inhibiting protein synthesis at the mucosal epithelium and this might result in diversion of IgA from mucosal sites such as the lamina propria into the systemic compartment. In several attempts, we failed to show a significant decrease in intestinal or salivary IgA following vomitoxin exposure that should be evident if such an impairment occurred.

Dysregulation of IgA production

IgA production is significantly increased in both spontaneous and lipopolysaccharide-stimulated splenocyte cultures of animals exposed to vomitoxin

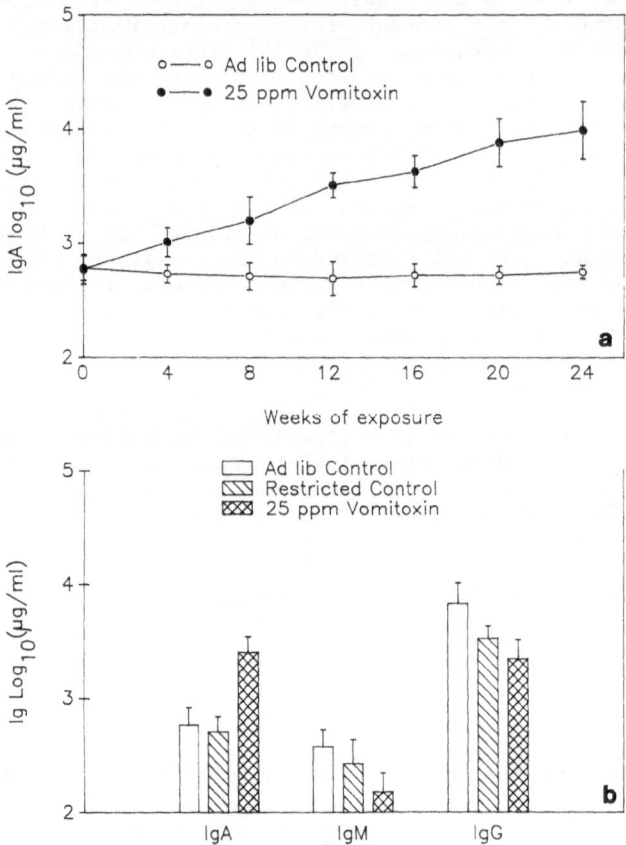

Fig. 2.   Elevation of serum IgA in B6C3F1 mouse during dietary vomitoxin exposure. Data are geometric means ± S.D. (a) Kinetics of vomitoxin-induced elevation of serum IgA. Treatment and control groups were significantly different from wk 4 to wk 24 (P<0.01) (13 mice/group). (b) Effect of dietary vomitoxin and food restriction in serum Ig isotype distribution. Mice were fed treatment and control diets for 12 wks. Serum IgA in vomitoxin-treated groups were significantly different from controls (P<0.01) (6-8 mice/group).[33]

but not in *ad lib* or restricted controls (Fig. 3).[33] Although addition of concanavalin A with or without lipopolysaccharide-reduced total IgA yield, levels of IgA produced by lymphocytes from vomitoxin-treated animals are significantly higher than controls. Depression of IgA production in splenocyte cultures by concanavalin A has been observed previously and attributed to stimulation of T suppressor cells.[11] Our results suggest that this concanavalin A sensitive lymphocyte population is still functional in vomitoxin-treated mice. Consistent with our *in vivo* observations, concurrent stimulation of IgM or IgG production following vomitoxin feeding is not observable in spontaneous or mitogen-stimulated cultures.

In the gut, isotype switching, activation, and differentiation of IgA progenitors upon antigen exposure are regulated by accessory cells, T helper, and T suppressor cells[11,23,28] that are localized in the Peyer's patches. Recently, a third level of regulation for IgA synthesis has been described that involves putative contrasuppressor cells.[39] It is now widely accepted that B cells originating at Peyer's patches migrate via the systemic compartment (including lymph nodes and spleen) to distant mucosal sites such as the lamina propria. The *in vitro* splenocyte data suggest that at least part of the elevated IgA arises from the systemic compartment although it is not known whether this is a result of increased numbers of IgA-producing B cells or due to an increased rate of IgA synthesis by splenic B cells.

Increased IgA production observed in splenocytes isolated from vomitoxin-treated mice relative to controls might suggest that vomitoxin interferes with normal regulation of the IgA response. Here, it is possible that dietary vomitoxin modulates regulation at the Peyer's patch level and that this could lead to increased activation, proliferation, and finally migration of IgA-producing B cells into the systemic compartment. As shown in earlier investigations, *in vivo* and *in vitro* exposure to trichothecenes stimulate or inhibit lymphocyte proliferation in a concentration dependent-manner[13,25,29]. Trichothecenes and other protein synthesis inhibitors, specifically stimulate both IL-1[29,31] production by macrophage and IL-1 formation by human lymphocytes.[10,29,31] Therefore, interaction of dietary vomitoxin with the Peyer's patch could result either in superinduction of interleukins by accessory cells and T helper cells or in inhibition of T suppressor cells. Both effects theoretically could cause the dysregulation of IgA production observed here. The possibility of dysregulation is strongly supported by the marked preponderance of serum pIgA and large degree of spontaneous IgA production by splenocytes observed following vomitoxin treatment.

Accumulation of mesangial IgA

The observation that serum IgA becomes elevated in mice during dietary exposure to vomitoxin is potentially important because dysregulation of the normal IgA response is believed to contribute to IgA nephropathy (Berger's disease),[1] the most common form of glomerulonephritis worldwide.[6] The nephropathy is characterized by accumulation of IgA in the mesangial region of the kidney glomerulus and between 20-50% of patients exhibiting mesangial IgA run a progressive course that includes renal failure.[4] Because elevation of serum IgA, pIgA to mIgA ratio, and *in vitro* spontaneous and polyclonal IgA production are associated with IgA nephropathy,[4,6] we sought to assess the effect of dietary vomitoxin on glomerular IgA accumulation. When cryostat sections of kidneys from treated and control mice were evaluated by immunofluorescence, the vomitoxin-exposed animals exhibited marked mesangial IgA accumulation compared to controls (Fig. 4). Immunofluorescence staining did not reveal accumulation of IgG or complement component C3. Electron microscopy indicated electron dense mesangial deposits in vomitoxin-exposed animals that were consistent with IgA nephropathy.

Conclusions

The above investigations confirm and extend previous work implicating trichothecene mycotoxins in immune modulation. This is particularly significant because vomitoxin is the most frequently occurring trichothecene, and because the dietary levels of vomitoxin used realistically represent the concentrations encountered in cereals after natural infection by *F. graminearum*. Furthermore, exposure of the B6C3F1 mouse at these levels does not result in detectable histopathological damage[15] but can cause subtle immunosuppressive effects. As demonstrated in these investigations, several factors must be included in the overall assessment of vomitoxin modulated immune function. First, some immune effects associated with vomitoxin are undoubtedly the result of decreased nutrient intake caused by food refusal. Second, other mycotoxins such as zearalenone that occur concurrently with vomitoxin can contribute to immune modulation. Third, although most investigations specifically describe effects on systemic immune parameters, our work suggests that trichothecenes have their greatest effect on gut-associated lymphoid tissue, before they are absorbed and subsequently metabolized by detoxification enzymes. Acute doses of vomitoxin cause severe lesions in the gastrointestinal tract of the B6C3F1 mouse,[16] but chronic effects may be more subtle. The potential for effects of dietary vomitoxin on mucosal immunity have been verified by our observation that dietary exposure to vomitoxin increases serum IgA levels and causes deposition of IgA in kidney.

We first reported elevated IgA production caused by a naturally-occurring dietary toxicant. D'Amico[6] estimated that up to 40% of all cases of glomerulonephritis and 10% of maintenance dialysis patients[1,4] exhibit IgA nephropathy. Although the etiology of this disease or group of related diseases is unknown, factors such as diet, prior mucosal infection, and genetic predisposition all have been suggested to be contributory.[4,6] Elevated serum IgA, increased ratio of pIgA to mIgA in serum, increased spontaneous and mitogen-stimulated IgA production by isolated lymphocytes, and mesangial IgA accumulation as described here are identical to observations found in human IgA nephropathy. It may be feasible to identify factors that affect IgA

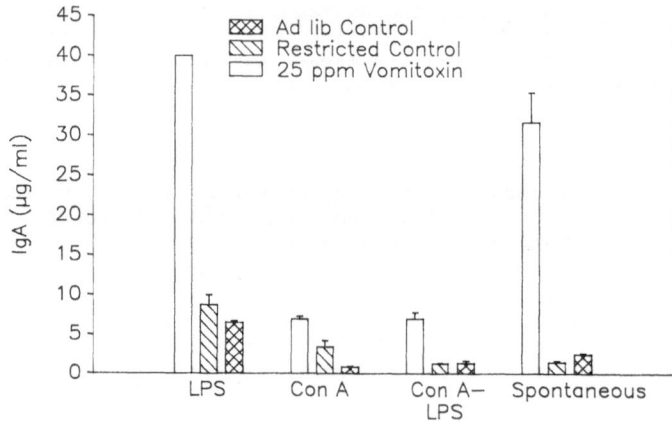

Fig. 3.   IgA production by spenocytes following vomitoxin exposure. B6C3F1 mice were fed control and vomitoxin diets for 21 wks. Results expressed as mean ± S.E. and are representative of 5 separate experiments. Vomitoxin-treated mice were significantly different from controls (P < 0.01) (8 determinations/group).[33]

immune deposition in the glomerulus, by varying rate, dose, and challenge mode of both vomitoxin and selected mucosal antigens.

On a very speculative level, contamination of food by trichothecenes might alter normal regulation of IgA production in humans and actually be an etiological factor in IgA nephropathy. Vomitoxin occurs worldwide in cereal grains and is recalcitrant to processing.[42] A U.S. survey of 228 samples of grain and grain products revealed detectable vomitoxin in greater than 80% of samples with a maximal level of 9 ppm.[2] Although dysregulation was characterized following exposure to 25 ppm vomitoxin (a level where serum IgA elevation was maximal), we have observed significant IgA elevation at dietary vomitoxin levels as low as 2 ppm.[15] The possibility that extended low level exposure to vomitoxin, or as demonstrated in structure:function studies, other much more potent trichothecenes could cause pertubations in IgA production and mesangial IgA accumulation merits further investigation.

Clearly, dietary *Fusarium* toxins can significantly affect immune function. Future investigations should be directed towards evaluation of the long term cumulative effects of dietary trichothecenes, a thorough understanding of the interactions between dietary trichothecenes and specific cell sub-populations of the mucosal and systemic immune systems, and the assessment of the potential for trichothecenes to modify host defense against infectious agents acting at the gastrointestinal level.

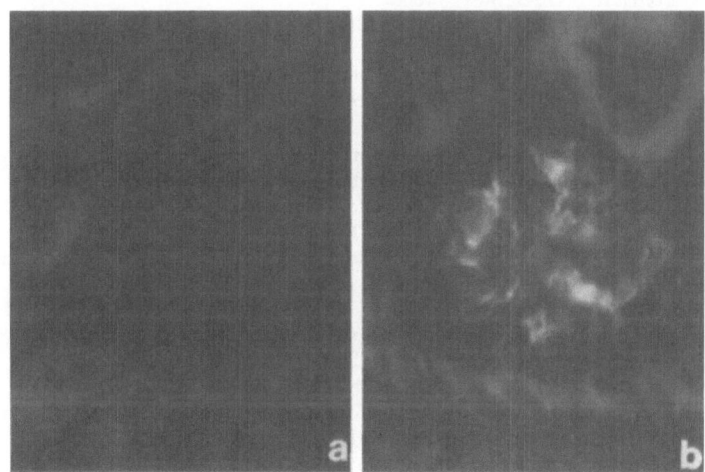

Fig. 4. Immunofluorescence detection of mesangial IgA deposits following dietary vomitoxin exposure. (a) Glomerular section from mouse fed the control diet for 22 wks and stained with fluorescein conjugated anti-mouse IgA. (b) Glomerular section from mouse fed diet containing 25 ppm vomitoxin for 22 wks and stained as in (a).[33]

REFERENCES

1. Berger J, 1969, IgA glomerular deposits in renal disease, *Proc.*, 1:939-941.
2. Brumley WC, Truckcess MW, Adler SH, Cohen CK, White KD, Sphon JA, 1985, Negative ion chemical ionization mass spectrometry of deoxynivalenol (DON): Application to identification of DON in grains and snack foods

after quantitation/isolation by thin-layer chromatography, *J Agric Food Chem.*, 33:326-333.

3. Cameron RP, Hickman RL, Lornreich MR, Tarone RE, 1985, History, survival and growth patterns of B6C3F1 mice and F344 ratios in the National Cancer Institute carcinogenesis program, *Fund Appl Toxicol.*, 5:526-538.

4. Clarkson AR, Woodroffe AJ, Aarons I, Hiki Y, Hale G, 1987, IgA nephropathy, *Ann. Rev. Med.*, 38:157-168.

5. Cote LM, Reynolds JD, Vesonder RF, Buck WB, Swanson SP, Coffey RT, Brown DC, 1984, Survey of vomitoxin-contaminated feed grains in mid-western United States, and associated health problems in swine, *J Am Vet Med Ass.*, 184:189-192.

6. D'Amico GD, 1987, The commonest glomerulonephritis in the world: IgA nephropathy, *Quart J Med.* New Series 64, 247:709-727.

7. Dean JH, Luster MI, Boorman GA, Lauer LD, 1982, Procedures available to examine the immunotoxicity of chemical and drugs, *Pharm Reviews*, 34:137-148.

8. Dean JF, Luster MI, Boorman GA, Luebke RW, Lauer LD, 1980, The effect of adult exposure to diethylstilbestrol in the mouse, alterations in tumour susceptibility and host resistance parameters, *J Reticuloendothel Soc.*, 28:571.

9. Delacroix DL, Malburny GN, Vaerman JP, 1985, Hepatobiliary transport of plasma IgA in the mouse: contribution to clearance of intravascular IgA, *Eur J Immunol.*, 15:893-899.

10. Efrat S, Zelig S, Yagen B. Kaempfer R, 1984, Superinduction of human interleukin-2 messenger RNA by inhibitor of translation, *Biochem Biophys Res Commun.*, 123:842-848.

11. Elson CO, Herk JA, Strober W, 1979, T-cell regulation of murine IgA synthesis, *J Immunol.*, 149:632-643.

12. Emancipator SN, Gallo, GR, Razaboni R, Lamm ME, 1983, Experimental cholestasis promotes the deposition of glomerular IgA immune complexes, *Am J Pathol.*, 113:19-26.

13. Forsell JH, Pestka JJ, 1985, Relation of 8-ketotrichothecene and zearalenone analog structure to inhibition of mitogen-induced human lymphocyte blastogenesis, *Appl Environ Microbiol.*, 50:1304-1309.

14. Forsell JH, Kately JR, Yoshizawa T, Pestka JJ, 1985, Inhibition of mitogen-induced blastogenesis in human lymphocytes by T-2 toxin and its metabolites, *Appl Environ Microbiol.*, 49:1523-1526.

15. Forsell JH, Witt MF, Tai J-H, Jensen R, Pestka JJ, 1986, Effects of 8-week exposure of the B6C3F1 mouse to dietary deoxynivalenol (vomitoxin) and zearalenone, *Fd Chem Toxic.*, 24:213-219.

16. Forsell JH, Jensen R, Rai J-H, Witt M, Lin WS, Pestka JJ, 1987, Comparison of acute toxicities of deoxynivalenol (vomitoxin) and 15-acetyldeoxynivalenol in the B6C3F1 mouse, *Fd Chem Toxic.*, 25:155-162.

17. Friend SCE, Babiuk LA, Schiefer HB, 1983, The effects of dietary T-2 toxin on the immunological function and herpes simplex reactivation in Swiss mice, *Toxicol Appl Pharmacol.*, 69:2324-244.

18. Fromentin H, Salazar-Mejicanos S, Mariat F, 1981, Experimental crypto-coccossis in mice treated with diacetoxyscirpenol, a mycotoxin of *Fusarium, Sabouraudia*, 19:311-313.

19. Ichinoe M, Kurata H, Sugiura Y, Ueno Y, 1983, Chemotaxonomy of *Gibberella zeae* with special reference to production of tricho-thecenes and zearalenone, *Appl Environ Microbiol.*, 46:1364-1369.

20. Joffe A, 1978, *Fusarium poae* and *F. sporotrichoides* as principal causal agents of alimentary toxic aleukia, *In*: "Mycotoxic Fungi, Mycotoxins, Mycotoxicoses," Wyllie T, Morehouse L, eds., Marcel Dekker, New York.

21. Kaartinen M, 1978, Liver damage in mice and rats causes ten-fold increase of blood and immunoglobulin A, *Scand J Immunol.*, 7:519-522.

22. Keusch GT, Wilson CS, Waksal SD, 1983, Nutrition, host defenses, and the

lymphoid system, *In*: "Advances in Host Defense Mechanisms," Gallin and Fauvi, eds., Raven Press, New York.

23. Kiyono H, McGhee JR, Wannemuehler MJ, Frangakis MV, Spalding DM, Michalek SM, Koopman WJ, 1982, *In vitro* immune responses to a T cell-dependent antigen by cultures of disassociated Peyer's patch, *Proc Natl Acad Sci USA*, 79:596-600.

24. Lafarge-Frayssinet C, Decloitre F, Mousset S, Martin M, Frayssinet C, 1981, Induction of DNA single-strand breaks by T-2 toxin, a trichothecene metabolite of *Fusarium*: effect on lymphoid organs and liver, *Mutation Res.*, 88:115-123.

25. Lafarge-Frayssinet C, Lespinats G, Lafont P, Loisillier F, Mousset S, Rosenstein Y, Frayssinet C, 1979, Immunosuppressive effects of *Fusarium* extracts and trichothecenes: blastogenic response of murine splenic and thymic cells to mitogens, *Proc Soc Exp Biol Med.*, 160:302-311.

26. Lorke O, 1983, A new approach to practical acute toxicity testing, *Arch Toxicol.*, 54:275-287.

27. McLaughlin CS, Vaughan MH, Campbell IM, Wei CM, Stafford Me, Hansen BS, 1977, Inhibition of protein synthesis is by trichothecenes, *In*: "Mycotoxins in Human and Animal Health," Rodricks J, Hesseltine C, Mehlman M, eds., Pathotox, Park Forest South, IL.

28. Mestecky J, McGhee JR, 1987, Immunoglobulin AS (IgA): Molecular and cellular interactions involved in IgA biosynthesis and immune response, *Adv Immunol.*, 40:153-245.

29. Miller K, Atkinson HAC, 1986, The *in vitro* effects of trichothecenes on the immune system, *Fd Chem Toxic.*, 24:545-549.

30. Mirocha CJ, Pathre SV, Schauerhamer B, Christensen CM, 1976, Natural occurrence of *Fusarium* toxins in feedstuff, *Appl Environ Microbiol.*, 32:553-556.

31. Mizel SB, Mizel DJ, 1981, Purification to apparent homogeneity of murine interleukin 1, *J Immunol.*, 126:834-837.

32. Pestka JJ, Forsell JH, 1988, Inhibition of human lymphocyte transformation by the macrocyclic trichothecenes roridin A and verrucarin A, *Toxicol Lett.*, 41:215-222.

33. Pestka JJ, Moorman MA, Warner R, 1989, Dysregulation of IgA production and IgA nephropathy induced by the trichothecene vomitoxin, *Fd Chem Toxicol.*, (submitted).

34. Pestka JJ, Tai J-H, Witt MF, Dixon DE, Forsell JH, 1987, Suppression of immune response in the B6C3F1 mouse after dietary exposure to the *Fusarium* toxins deoxynivalenol (vomitoxin) and zearalenone, *Fd Chem Toxicol.*, 25:297-304.

35. Pier AC, Richard JL, Cysewski SJ, 1980, Implications of mycotoxins in animal disease, *J An Vet Med Ass.*, 176:719-724.

36. Rosenstein Y, Kretschmer RR, Lafarge-Frayssinet C, 1981, Effect of *Fusarium* toxins, T-2 toxin and diacetoxyscirpenol on murine T-independent immune responses, *Immunology*, 44:555-560.

37. Rosenstein Y, Lafarge-Frayssinet C, Lespinats G, Loisillier F, Lafont D, Frayssinet C, 1979, Immunosuppressive activity of *Fusarium* toxins. Effects on antibody synthesis and skin grafts of crude extracts, T-2 toxin, and diacetoxyscirpenol, *Immunology*, 36:111-117.

38. Salazar S, Fromentin H, Mariat F, 1980, Effects of diacetoxyscirpenol on experimental candidasis of mice, *Cr Adad Sci, Paris*, 290:877-878.

39. Suzuki I, Kiyono H, Kitamura K. Green DR, McGhee JR, 1986, Abrogation of oral tolerance by contrasuppressor T cells suggests the presence of regulatory T-cell networks in the mucosal immune system, *Nature*, 320:451-454.

40. Tai J-H, Pestka JJ, 1988a, Impaired murine resistance to *Salmonella typhimurium* following oral exposure to the trichothecene T-2 toxin, *Fd Chem Toxic.*, (in press).

41. Tai J-H, Pestka JJ, 1988b, Synergistic interaction between the trichothecene T-2 toxin and *Salmonella typhimurium* lipopolysaccharide

in C3H/HeN and C3H/HeJ mice, *Toxicol Lett.*, 44:191-200.

42. Tanaka T, Hawegawa A, Yamamoto S, Lee U-S, Suguira Y, Ueno Y, 1988,
    Worldwide contamination of cereals by the *Fusarium* mycotoxins
    nivalenol, deoxynivalenol, and zearalenone, *J Agr Food Chem.*,
    36:979-983.

43. Tryphonas H, Iverson F, So Y, Nera EA, McGuire PF, O'Grady L, Clayson DB,
    Scott PM, 1986, Effects of deoxynivalenol (vomitoxin) on the humoral and
    cellular immunity of mice, *Toxicol Lett.*, 30:137-150.

44. Ueno Y, 1983a, Historical background of trichothecene problems, *In*:
    "Trichothecenes: Chemical, Biological and Toxicological Aspects," Ueno
    Y, ed., Elsevier, New York.

45. Ueno Y, 1983b, General toxicology, *In*: "Trichothecenes: Chemical,
    Biological and Toxicological Aspects," Ueno Y, ed., Elsevier, New York.

46. Vesonder RF, Ciegler A, Rohwedder WK, Eppley R, 1979, Reexamination of
    1972 midwest corn for vomitoxin, *Toxicology*, 17:658-662.

47. Witt MF, Hart LP, Pestka JJ, 1985, Purification of deoxynivalenol
    (vomitoxin) by water-saturated silica gel chromatography, *J Agr Food
    Chem.*, 33:745-748.

COMBINED CYCLOPIAZONIC ACID AND AFLATOXIN B$_1$ EFFECTS ON SERUM
BACTERIOSTASIS, COMPLEMENT ACTIVITY, GLYCOCHOLIC ACID, AND ENZYMES AND
HISTOPATHOLOGIC CHANGES IN GUINEA PIGS

J. L. Richard, W. M. Peden and J. R. Thurston

National Animal Disease Center
U. S. Department of Agriculture, Agricultural Research Service
P. O. Box 70
Ames, IA 50010

INTRODUCTION

Cyclopiazonic acid is an indole-tetramic acid and produced by several
species of fungi such as *Penicillum cyclopium, P. patulum, P. viridicatum,
P. puberulum, P. crustpsum, P. camembertii, Aspergillus flavus, A. versicolor, A. oryzae* and *A. tamarii.*[3,9] The toxin was first reported to occur
naturally in corn[4] and subsequently has been reported in cheese,[6] peanuts,[5]
and kodo millet.[11]

Gallagher et al.[4] found that 14 of 54 isolates of *A. flavus* produced
cyclopiazonic acid as well as aflatoxins and that 14 produced CPA alone and
4 produced aflatoxins only.

Toxicological studies have been conducted in a number of animal species
using CPA.[2,10] Generally, the effects of CPA on animals involved necrosis
of liver or gastrointestinal tissue with lesions also present in skeletal
muscle and kidney, depending upon the animal species selected for toxicological studies.[9]

The effects of aflatoxins on animals have been studied extensively
resulting in voluminous literature; the major effects, including the primary
changes induced by these hepatotoxic compounds, have been reviewed recently
by Norred.[8] One study using high concentrations of CPA and AFB$_1$ for a short
period of time in rats, reported that neither toxin potentiates the other
in any of the parameters measured.[7] Similarly, Yates et al.[16] found
that CPA and AFB$_1$ appeared to have independent modes of action in mutagenesis
and cytotoxic activities in combination studies utilizing bioluminescence
of *Photobacterium phosphoreum.*

Although both CPA and AFB$_1$ are capable of being produced under identical environmental conditions by *A. flavus* isolates, little emphasis has
been given to studies concerned with the toxicological aspects (as tested
in animals) of the interaction of these two mycotoxins at relative quantities of the two toxins produced by *A. flavus*. Therefore, we investigated
the relative concentration of CPA and AFB$_1$ produced by several isolates
of *A. flavus* and used these concentrations in studies of the interaction
of these toxins on toxicity to guinea pigs when administered orally.

Microbial Toxins in Foods and Feeds, Edited by A.E. Pohland *et al.,*
Plenum Press, New York, 1990

MATERIALS AND METHODS

Aspergillus flavus isolates

Eight isolates were selected for use in this study because they produced both CPA and AFB₁ in a previous study.[4] The isolate identifications were: MC-48, MC-92, MC-38, MC-16, MC-77, MC-22, NRRL 1290 and 63-IV-78. The MC isolates were obtained from stored corn,[13] NRRL 1290 was from a human throat swab (obtained from J. J. Ellis, Northern Regional Research Center, Peoria, IL), and 63-IV-78 was cultured by one of us (JLR) from a sheep adrenal gland. All isolates were stored as frozen cultures (-40°C) on Czapek's agar.

Toxin production by A. flavus isolates

Rice medium was prepared by placing 75 g polished rice and 30 ml distilled water in 1 L cotton stoppered Erlenmeyer flasks. The flasks were placed at room temperature for 2 hr to allow for imbibition and then autoclaved for 15 min. After the flasks were cool, a 1 ml conidial suspension (5×10^8 conidia) of an isolate in phosphate buffered (pH 7.4) saline solution was seeded on the medium in the respective flasks. All cultures were incubated at 27°C for 1 week and each culture was shaken by hand once each day.

Extraction of toxins

Two hundred ml of chloroform was placed in each culture flask and allowed to extract overnight without shaking; the chloroform extract was collected after filtration through Whatman No. 1 filter paper. Each culture was extracted again for 8 hr with an additional 300 ml of chloroform, the extract collected as before and combined with the first extract. Each culture extract was evaporated to dryness and redissolved in 5 ml chloroform.

Thin-layer chromatography

For analysis of CPA, 25 µl of each extract was spotted on silica gel 60 analytical thin-layer plates (E. Merck, Darmstadt, Germany) that were pretreated by spraying until wet with 2% aqueous oxalic acid and dried 1 hr at 100°C. Internal and external standards of CPA were included on each plate and the plates were developed in chloroform:methylisobutylketone (4:1 v/v). After development to 20 cm, the plates were air dried, sprayed with 1% p-dimethylaminobenzaldehyde in 95% ethanol, dried with warm air and placed in a tank containing HCL vapors for 10 min to allow for the lavender to purple color of CPA to appear. Quantitation of the cyclopiazonic acid was based on visual comparison with known amounts of standard applied to each plate.

For aflatoxin analysis, 25 µl of each extract or appropriate dilutions thereof were spotted on silica gel 60 analytical thin-layer plates. Appropriate internal and external standards were spotted on each plate, the plate was developed 20 cm in chloroform:methanol:formic acid (97:2:1 v/v/v), and dried at room temperature. The aflatoxins were observed under long wavelength U.V. light. Quantitation was based on U.V. densitometry (Shimadzu CS-920, Shimadzu Corp., Kyoto, Japan) of the aflatoxin on each plate.

Animals

Seventy female guinea pigs weighing between 350-450 g were obtained (Bio-Lab Corporation, P. O. Box 10589, St. Paul, MN 55110) and randomly divided into 14 groups of 5 guinea pigs each. The guinea pigs were given feed and water ad libitum throughout the experimental period.

Toxins and Dosages

Aflatoxin B_1 was obtained in crystalline form (Sigma Chemical Company, P. O. Box 14508, St. Louis, MO 63178) and the toxin was dissolved in chloroform at a concentration to yield the desired amount of AFB_1 in 0.1 ml, which was placed in a gelatin capsule (No. 5 gelatin capsule, Eli Lilly and Co., Indianapolis, IN 46285). Cyclopiazonic acid was obtained as a gift (R. Cole, USDA/ARS, National Peanut Research Laboratory, Dawson, GA). The CPA was >97% pure and was dispensed in capsules as described for AFB_1. All solvent used for filling capsules was allowed to evaporate in the dark overnight before the top was replaced on the capsules. The combination dosages as described in Table 1 were each included in a single capsule so that only one capsule/day was given to each guinea pig. The combinations of toxins selected were based on the ratios of AFB_1 to CPA found in the isolates examined for production of these two toxins (Table 1). Only two concentrations of aflatoxin B_1 were used; 0.03 mg/day and 0.01 mg/day for the 3 week experimental period. These dosages were selected because 0.03 mg/day in this size guinea pig in previous studies[12] was the minimum effective dosage and 0.01 μg/day was the maximum ineffective dosage. The amounts of CPA included in the study was then based on the ratios as described above with these two dosages of aflatoxin (Table 2).

Experimental Design

All guinea pigs were given their respective daily (in the AM) dosages, orally by capsule for 3 weeks (Table 2). Each guinea pig was weighed each week and all guinea pigs were bled and then euthanized at the end of the third week. Serum was harvested and stored at -60°C, and analyzed for aspartate amino transferase (AST), alanine amino transferase (ALT), and sorbitol dehydrogenase (SDH) using an automated analyzer (Rotochem Centrifical Chemistry Analyzer, Travenol Laboratories, Inc., Deerfield, IL) and reagent kits (Worthington Diagnostic Systems, Freehold, NJ). Serum glycocholic acid was determined by radioimmunoassay according to the method supplied with the commercially available kit (Nuclear Diagnostics, Inc., 575 Robbins Dr., Troy, MI 48083). Complement activity was assessed using a 50% hemolytic endpoint according to established procedures.[15] The bacteriostatic activity of serum against a strain of *Escherichia coli* was conducted according to a previously described method.[14] Data were analyzed using the analysis of variance.

Histopathology

Tisues, including heart, skeletal muscle, liver, kidney, adrenal gland, pancreas, spleen, stomach, jejunum, cecum, and colon were collected at necropsy and fixed in 10% neutral formalin. Tissues were processed routinely, embedded in paraffin, sectioned and stained with haematoxylin and eosin.

RESULTS

The analysis of AFB_1 and CPA from the eight selected isolates of *A. flavus* (Table 1) yielded a ratio of aflatoxin to CPA among the isolates of 4:1, 2:1, 1:1, 1:2, and 1:4. The greatest yield of aflatoxin among the isolates was 169.5 μg/g rice and the lowest yield of this toxin was 22.3 μg/g; for CPA, the greatest and lowest yields were 126 μg/g and 8.2 μg/g, respectively.

No increases were found in any of the serum enzymes tested in guinea pigs given CPA except in SDH in guinea pigs given 0.12 mg/day of CPA only (Table 3). There was a two-thirds times increase in AST in serum of

Table 1. Concentrations of AFB$_1$ and CPA produced by isolates of *Aspergillus flavus* grown on rice.

Isolates	AFB$_1$ (μg/g rice)	CPA (μg/g rice)	Ratio (approximate) AFB$_1$:CPA
MC-48	169.5	88.2	2:1
MC-92	73.6	20.5	4:1
MC-38	64.7	126.0	1:2
MC-16	69.9	94.8	1:2
MC-77	67.1	53.5	1:1
MC-22	34.3	18.2	2:1
63-IV-78	22.3	84.4	1:4
NRRL 1290	44.4	20.0	2:1

Table 2. Experimental Design - Combinations of AFB$_1$ and CPA in Guinea pigs.

Group*	Toxins (mg/day)	
	AFB1	CPA
1	0.01	0
2	0.01	0.0025
3	0.01	0.005
4	0.01	0.01
5	0.01	0.02
6	0.01	0.04
7	0.03	0
8	0.03	0.0075
9	0.03	0.015
10	0.03	0.03
11	0.03	0.06
12	0.03	0.12
13	0	0.12
14	0	0

*Five guinea pigs each group, female, 350-450 g. Three week experimental period; daily dosing with encapsulated toxins.

guinea pigs given 0.03 mg AFB$_1$/day compared to those given 0.01 mg AFB$_1$/day. Also, a significant increase (P $\leq$ 0.05) in AST was noted with increasing amounts of CPA given in combination with 0.03 mg AFB$_1$/day when examined by regression analysis.

Bacteriostatic activity of serum was less in guinea pigs given 0.03 mg AFB$_1$/day than in those given 0.01 mg AFB$_1$/day (Figs. 1 and 2) when compared with the bacteriostatic activity of normal guinea pig serum. Also, the addition of CPA at the ratios selected to the dosage of 0.03 mg AFB$_1$/day caused a decrease in the amount of bacteriostatic activity particularly when 0.12 mg of CPA/day was given. All of the treatments caused a decrease in the amount of serum bacteriostatic activity compared to that of normal serum and only the combination of 0.03 mg AFB$_1$ and 0.12 mg CPA significantly (P $\leq$ 0.05) reduced the bacteriostatic activity to near that of the medium control. Cyclopiazonic acid at the highest dosage of 0.12 mg/day caused a decrease in bacteriostatic activity of serum when compared with normal serum.

Table 3. Mean concentrations of aspartate amino transferase, alanine amino transferase and sorbitol dehydrogenase in serum of guinea pigs (five per dose group) given combinations of aflatoxin B_1 and cyclopiazonic acid for three weeks.

Enzyme	AFB_1 (mg/day)	0	0.0025	0.0050	0.0075	0.01	0.015	0.02	0.03	0.04	0.06	0.12
AST	0	45	--*	--	--	--	--	--	--	--	--	39
	0.01	44	53	34	--	49	--	--	--	40	--	--
	0.03	68	--	--	46	--	54	--	95	--	80	97
ALT	0	44	--	--	--	--	--	--	--	--	--	28
	0.01	35	30	26	--	26	--	29	--	29	--	--
	0.03	28	--	--	26	--	35	--	35	--	27	34
SDH	0	61	--	--	--	--	--	--	--	--	--	93
	0.01	64	115	60	--	78	--	92	--	64	--	--
	0.03	67	--	--	61	--	57	--	106	--	75	85

* = Not determined (due to experimental design).

445

The glycocholic acid concentrations in guinea pigs given 0.03 mg afla-toxin often were more than twice that of guinea pigs given 0.01 mg aflatoxin (Table 4). No differences in glycocholic acid concentrations in serum were noted in guinea pigs given CPA plus these dosages of AFB_1 to those of guinea pigs given AFB_1 alone. No increased concentrations of glycocholic acid were found in serum of guinea pigs given the highest dosage of CPA.

There was no change in the complement activity of serum associated with increasing dosages of CPA in guinea pigs. The only difference noted was that all guinea pigs given 0.03 mg AFB_1/day had a serum complement titer <1:100 (Table 4).

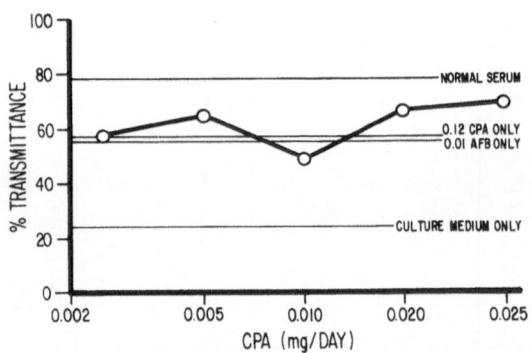

Fig. 1.   Reduction in growth of *Escherichia coli*, as measured by light transmission (540 mμ), by serum from guinea pigs (n=5) given various doses of cyclopiazonic acid with 0.01 mg/day of aflatoxin BH_1.

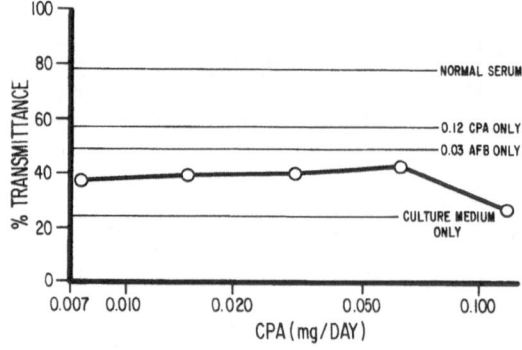

Fig. 2.   Reduction in growth of *Escherichia coli*, as measured by light transmission (540 mμ), by serum from guinea pigs (n=5) given various doses of cyclopiazonic acid with 0.03 mg/day of aflatoxin B_1.

Table 4. Mean serum glycocholic acid concentrations and complement
activity (C') in serum of guinea pigs (five per dose
group) given AFB_1 and CPA alone and in combination for
three weeks.

AFB_1 mg/day	CPA mg/day	Glycocholic Acid $\mu Mol/L$	C' titer*
.01	0	7.20	1:155.4
.01	0.0025	7.90	1:134.0
.01	0.005	4.40	1:143.6
.01	0.01	10.58	1:141.6
.01	0.02	4.75	1:160.6
.01	0.04	6.04	1:159.0
0.03	0	10.77	<1:100
0.03	0.0075	21.88	<1:100
0.03	0.015	15.95	<1:100
0.03	0.03	10.75	<1:100
0.03	0.06	15.80	<1:100
0.03	0.12	19.57	<1:100
0	0.12	2.12	1:165
0	0	1.77	1:219

*Complement activity based on 50% hemolysis assay of sensitized sheep
red blood cells.

Microscopic changes in guinea pigs given either CPA or AFB_1 or combinations thereof were noted primarily in the liver and lungs.

Liver lesions were related to the dose of aflatoxin given. Guinea pigs
given 0.03 mg/day of AFB_1 had mild to moderate bile ductule proliferation,
mild to moderate vacuolation of centrilobular and periportal hepatocytes,
mild periportal infiltrates of mononuclear cells, swollen and occasionally
degenerate hepatocytes, disrupted and disorganized hepatic plates, and rare
hepatocellular necrosis. These changes varied in severity among animals
within a given AFB_1-CPA-treated group, with some animals having mild to
moderate lesions and others no histological lesions. The effect of CPA on
liver lesions in guinea pigs given 0.03 mg AFB_1/kg was not discernible.

Animals given 0.01 mg/day of AFB_1 had little or no evidence of bile
ductule proliferation. Some guinea pigs had mild vacuolation of centrolobular hepatocytes or a slight infiltrate of mononuclear cells around portal
areas, but these changes were minimal and inconsistent. CPA did not alter
this effect in guinea pigs given this dose of aflatoxin.

Guinea pigs given CPA alone had minimal hepatic changes consisting prin-
cipally of very mild centrolobular vacuolation, but was observed only in
guinea pigs given 0.12 mg/day of CPA. Lesions in the lungs were found in all
guinea pigs and consisted of perivascular and peribronchiolar lymphoid hyperplasia, usually associated with a mild histiocytic interstitial pneumonia.
The interstitial pneumonia was usually mild and located around areas of lymphoid hyperplasia. Associated bronchioles occasionally contained an exudate
of sloughed epithelial cells mixed with a few heterophils.

DISCUSSION

Information concerning the relative amounts of AFB_1 and CPA produced by
Aspergillus flavus could not be found in a search of the literature. Thus,

the findings of the amounts of these two toxins produced by the 8 isolates of *A. flavus* used in this study are believed to be the first such report. Of interest, is that the ratios of the two toxins produced on rice varies from approximately 1:4 to 4:1.

The increase in AST in serum of guinea pigs was expected because Baetz and McLoughlin[1] found similar increases in this serum enzyme using the same doses of aflatoxin as equivalents of AFB_1; in the present study, we used pure AFB_1. The only increase in any of the parameters measured by the combination of AFB_1 and CPA was in AST when CPA was given with 0.03 mg AFB_1/day to guinea pigs for 3 weeks.

There appeared to be no interaction of CPA with AFB_1 in reducing bacteriostasis of serum of guinea pigs except where 0.12 mg of CPA/day was given with 0.03 mg of AFB_1/day. Some toxic interaction was evident in liver lesions with this combination of toxins and may subsequently affect basteriostasis as well by interfering with a substance involved in bacteriostasis that is produced by the liver. The effect of aflatoxin on the reduction of serum bacteriostasis was evident when the guinea pigs were given 0.03 mg AFB_1/day and was slightly reduced when 0.01 mg AFB_1/day was given. These dosages were used to determine if CPA could increase the effect observed when given in combination with AFB_1. Thurston et al.[14] determined that because bacteriostasis decreased without a concomitant decrease in complement activity in steers given small daily doses of aflatoxin, bacteriostasis may be dependent in cattle upon a serum substance other than complement. Because of insufficient amounts of serum, we were unable to determine the complement titers below 1:100. However, we determined that complement activity was decreased concomitantly with reduced serum bacteriostasis in guinea pigs given 0.03 mg AFB_1/day.

The increase in serum glycocholic acid concentrations were dependent upon the concentrations of aflatoxins given and were similar to the results obtained in previous studies in guinea pigs given the same dosages in AFB_1 equivalents.[1] Additionally, rats given combinations of AFB_1 at 2.0 mg/kg/day and CPA at 4 mg/kg/day orally had no significant increase in serum glycocholic acid over that of rats given this dosage of AFB_1 alone.[7] However, Peden et al.[10] did not detect significant increases in glycocholic acid of serum of guinea pigs given approximately 0.02 mg AFB_1 equivalents/day for 21 days.

Interaction of CPA and AFB_1 was not noted in this study by morphologic changes in selected tissues. There did not appear to be any interaction of either AFB_1 or CPA in modulating immunologic function in these animals, although animals given high doses had a slightly greater incidence of lung lesions characteristic of *Mycoplasma* infection.

REFERENCES

1. Baetz AL, McLoughlin ME, 1984, Serum concentration of bile acids in guinea pigs as an indicator of liver damage caused by aflatoxins, *Am J Vet Res.*, 44:1971-1972.
2. Cole RJ, 1986, Occurrence and clincal manifestations of rubratoxins A and B and cyclopiazonic acid, *In*: "Diagnosis of Mycotoxicoses," Richard JL, Thurston JR, eds., Martinus Nijhoff Publishers, Dordrecht.
3. Dorner JW, 1983, Production of cyclopiazonic acid by *Aspergillus tamarii* Kita, *Appl Environ Microbiol.*, 46:1435-1437.
4. Gallagher RT, Richard JL, Stahr HM, Cole RJ, 1978, Cyclopiazonic acid production by aflatoxigenic and non-aflatoxigenic strains of *Aspergillus flavus*, *Mycopathol.*, 66:31-36.

5. Lansden JA, Davisdon JL, 1983, Occurrence of cyclopiazonic acid in peanuts, *Appl Environ Microbiol.*, 45:766-769.

6. LeBars J, 1979, Cyclopiazonic acid production by *Penicillium camembertii* Thom and natural occurrence of this mycotoxin in cheese, *Appl Environ Microbiol.*, 38:1052-1055.

7. Morrissey RE, Norred WP, Hinton DM, Cole RJ, Dorner JW, 1987, Combined effects of the mycotoxins aflatoxin B_1 and cyclopiazonic acid on Sparque-Dawley rats, *Fd Chem Toxic.*, 25:;837-842.

8. Norred WP, 1986, Occurrence and clinical manifestations of aflatoxicosis, *In*: "Diagnosis of Mycotoxicoses," Richard JL, Thurston JR, eds., Martinus Nijhoff Publishers, Dordrecht.

9. Peden WM, Richard JL, Thurston JR, 1986, Comparative histopathologic changes in cyclopiazonic acid toxicosis and rubratoxicosis, *In*: Diagnosis of Mycotoxicoses," Richard JL, Thurston JR, eds., Martinus Nijhoff Publishers, Dordrecht.

10. Peden WM, Richard JL, Thurston JR, Sacks JM, 1987, Effects of pre-treatment with aflatoxin on a second aflatoxin treatment in guinea pigs, *Mycopathol.*, 99:107-114.

11. Rao BL, Husain A, 1985, Presence of cyclopiazonic acid in kodo millet (*Paspalum scrobiculatum*) causing "Kodua poisoning" in man and its production by associated fungi, *Mycopathol.*, 89:177-180.

12. Richard JL, Thurston JR, Pier AC, 1978, Effects of mycotoxins on immunity, *In*: "Toxins: Animal, Plant and Microbiol," Rosenberg P, ed., Pergamon Press, New York.

13. Richard JL, Tiffany LH, Pier AC, 1969, Toxigenic fungi associated with stored corn, *Mycopathol.*, 38:313-326.

14. Thurston JR, Cook W, Driftmier K, Richard JL, 1986, Decreased complement and bacteriostatic activities in the sera of cattle given single or multiple doses of aflatoxin, *Am J Vet Res.*, 47:836-849.

15. Thurston JR, Richard JL, Cysewski SF, Pier AC, Graham CK, 1972, Effect of aflatoxin on complement activity in guinea pigs, *Proc Soc Exp Biol Med.*, 139:300-303.

16. Yates IE, Cole RJ, Giles JL, Dorner JW, 1987, Interaction of aflatoxin B_1 and cyclopiazonic acid toxicities, *Molecular Toxicol.*, 1:95-106.

MYCOTOXIN-INDUCED ALTERATIONS IN ION TRANSPORT ACROSS CELL MEMBRANES

R.T. Riley,[a] D.E. Goeger[b] and D.M. Hinton[a]

Toxicology and Mycotoxin Research Unit
R. B. Russell Agricultural Research Center
U.S. Department of Agriculture, Agricultural Research Service
Athens, GA 30613[a]

Dept. of Preventive Medicine & Community Health
University of Texas Medical Branch
Galveston, TX 77550[b]

INTRODUCTION

Patulin and cyclopiazonic acid are fungal metabolites produced by certain species of *Penicillium* and *Aspergillus*. Patulin is also produced by certain *Byssochlamys* species. Both compounds are known to contaminate numerous agricultural products at low levels and are commonly consumed by both animals and man. Although food and feed safety is a prime concern, we are also interested in these compounds because understanding their interaction with biological systems can reveal useful information that could be the basis for future technological advances in controlling agricultural productivity. For the past several years we have studied these two compounds with regards to their effects on membrane function and in particular to their abilities to alter ion transport.[6,7,9,30,31,32,33] Basic information about the mode of action of mycotoxins at the level of cell membranes could lead to a better understanding of the role these chemicals may play in 1) the growth and development of the fungus and competing species, 2) the fungal-plant relationship, and 3) the fungal-plant-animal relationship. Additionally, these studies can identify new applications for fungal products as biochemical probes of membrane function and potential therapeutic applications in the areas of plant and animal health and productivity. The U.S. Department of Agriculture/Agricultural Research Service is well aware of the potential of membrane research in developing new agricultural frontiers[11] and thus we have chosen as a point of departure the study of the mechanisms by which fungal products alter membrane activity. The cell membrane plays a pivotal role in transduction of extracellular signals into intracellular action. Fungal compounds that perturb normal cellular activity often do so through alterations in ion movement across cell membranes. The purpose of this paper is to summarize our recent research on how patulin and cyclopiazonic acid alter membrane function. Much of what we will present has been published or is in the process of being published. Some new data will be presented to illustrate certain points that are not covered in published literature.

Microbial Toxins in Foods and Feeds, Edited by A.E. Pohland *et al.*,
Plenum Press, New York, 1990

451

METHODS

The following description of materials and methods is abbreviated and the interested reader is encouraged to seek the original publications for a complete description of the experimental details. All protein measurements were by the method of Lowry et al.[15] and radioactivity was measured by standard liquid scintillation counting techniques.

Materials

Cyclopiazonic acid, estimated to be greater than 95% pure based on its UV extinction coefficient at 284 nm, was produced and purified from cultures of *Penicillium griseofulvum* (or obtained from Sigma Chemical Co., St. Louis, MO). Patulin was obtained from Sigma Chemical Co. and its purity was calculated to be 95%. All other reagents and chemicals were obtained from either Sigma Chemical Co. or Aldrich Chemical Co. (Milwaukee, WI). [^{3}H]Tetraphenyl-phosphonium bromide [phenyl-3-H] was from Nuclear Research Centre-Negev (Beersheba, Israel); all other radioisotopes were from NEN Research Products (Boston, MA), or Amersham Corp. (Arlington Heights, IL).

Assays Using Cultured Cells

Renal epithelial cells (LLC-PK$_1$) and rat skeletal muscle myoblasts (L6) were obtained from the ATCC (Rockville, MD). Cells were grown and maintained as previously described.[30] Our methods for use of the membrane potential probe, [^{3}H]tetraphenylphosphonium bromide, are based on those originally developed by Lichtshtein et al.[14] [^{3}H]Tetraphenylphosphonium, [^{3}H]oua-bain, ^{22}Na$^+$, ^{86}Rb$^+$, and ^{45}Ca^{2+} uptake and efflux by LLC-PK$_1$ and L6 cells was determined using subcultures grown in 24-well plates or 10 cm^2 dishes. Assays were conducted directly in plates or dishes as previously described for LLC-PK$_1$ cells and in some cases suspended cells and freeze-thaw lysed cells were used in assays.[30,31,33] Cells used for x-ray microanalysis were grown on Thermanox^R (Miles Scientific, Naperville, IL) plastic tissue culture coverslips. Relative intracellular Na$^+$ and K$^+$ concentrations were determined as described in Hinton et al.[9] using a scanning electron microscope (SEM) equipped with an energy dispersive spectroscopy x-ray microanalysis system. The effects of patulin and cyclopiazonic acid on cell viability were evaluated by lactate dehydrogenase release.[31] A combination of phase microscopy and SEM (Philips 505T SEM) was used to determine the effects of patulin or cyclopiazonic acid on cell morphology. The site of tetraphenyl phosphonium accumulation in suspensions of LLC-PK$_1$ cells was determined by fractionating cells into crude-nuclei, mitochondrial, plasma membrane, microsomal and cytosolic fractions using a method originally developed for rat liver.[5]

Preparation of, and Assays with, Rat Sarcoplasmic Reticulum Vesicles

A crude intermediate sarcoplasmic reticulum (SR) fraction (10,000-45,000g fraction in 10% sucrose) was prepared from skeletal muscle of the hind leg of mature male Sprague-Dawley rats.[7] Ca^{2+} uptake was measured by using either a rapid filtration method similar to that of Martonosi and Feretos[17] or using the metallochromic Ca^{2+} indicator arsenazo III according to a method similar to that of Beeler and Gable.[2] ATPase activity was terminated by filtration and inorganic phosphate was assayed by the method of Baginski et al.[3] as modified by Ottolenghi.[22] The binding of Ca^{2+} to SR vesicles and the effect of cyclopiazonic acid on binding was determined by equilibrium dialysis.[6]

RESULTS AND DISCUSSION

Cyclopiazonic Acid Induced Alterations in Membrane Function in Cultured Cells

Potentiated tetraphenylphosphonium accumulation. Perhaps the most drastic and reproducible effect of cyclopiazonic acid on cultured renal cells

LLC-PK$_1$) and muscle myoblasts (L6) is the potentiated accumulation of the quaternary cationic lipophilic membrane potential probe, tetraphenylphosphonium bromide (Fig. 1a). Interestingly, primary cultures of rat liver hepatocytes did not show cyclopiazonic acid potentiated tetraphenylphosphonium accumulation.[30] Because potentiated tetraphenylphosphonium accumulation was prevented by co-incubation with protonophores (carbonylcyanide-m-chlorophenylhydrazone, dinitrophenol) and metabolic inhibitors (n-ethylmaleimide) we initially suspected, that the potentiated accumulation of tetraphenylphosphonium was dependent on the transmembrane potential difference resulting from stimulation of the electrogenic plasma membrane Na$^+$/K$^+$ pump.[32] However, we soon discovered that in LLC-PK$_1$ cells treated with cyclopiazonic acid the accumulated tetraphenylphosphonium was not in passive equilibrium with the transmembrane potential difference (Fig. 1b). In cyclopiazonic acid treated LLC-PK$_1$ cells, tetraphenylphosphonium was recovered tightly bound in the plasma membrane and mitochondrial fractions (Table 1[33]). Further investigations[30] revealed that L6 cells (Fig. 2) as well as hepatocytes and LLC-PK$_1$ cells (data not shown) permeabilized by freeze-thaw lysis, all exhibited cyclopiazonic acid potentiated tetraphenylphosphonium partitioning, even in the presence of carbonylcyanide-m-chlorophenylhydrazone. This result indicated that both tetraphenylphosphonium and cyclopiazonic acid must have access to the intracellular space for potentiated tetraphenylphosphonium partitioning to be observed. We concluded that the site of interaction between cyclopiazonic acid and tetraphenylphosphonium was intracellular and possibly due to electrostatic alterations on the cytoplasmic side of the plasma membrane and mitochondria.

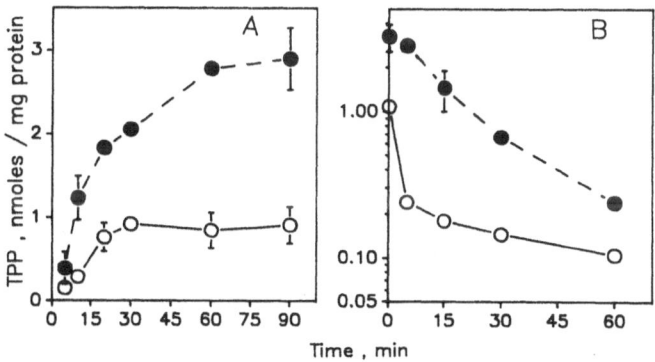

Fig. 1.   a) Uptake and b) efflux of tetraphenylphosphonium (2 μM by 4-day-old L6 cells with (--●--) and without (--O--) 50 μM cyclopiazonic acid. Tetraphenylphosphonium uptake was corrected for potential insensitive tetraphenylphosphonium accumulation by subtracting accumulation by cells depolarized with 5 μM carbonylcyanide-m-chlorophenylhydrazone. Lines are plotted through the means ± S.D., n=3 at each time. In the efflux experiment (b) the cells were allowed to accumulate tetraphenylphosphonium for 60 minutes, rinsed with phosphate buffered saline followed by incubation in the same solution plus 5 μM carbonylcyanide-m-chlorophenylhydrazone to initiate efflux of tetraphenylphosphonium that had accumulated in passive equilibrium with the transmembrane potential difference. Values in (b) are the amount of tetraphenylphosphonium remaining in cells after the period of time given.[30]

Table 1. Distribution of tetraphenylphosphonium (TPP) in subcellular
fractions of LLC-PK$_1$ cells dosed with the protonophore
uncoupler carbonylcyanide-m-chlorophenylhydrazone (CCCP) or
cyclopiazonic acid (CPA).[33]

	Picomoles TPP/mg protein (60 min)		
Additions	Homogenate	25K g pellet	105K g super
None	210	10	199
CCCP	15	2	13
CPA	685	528	67

Percent of TPP in subcellular fractions of CPA-treated
cells compared to control cells (additions="None")

	Additions	
	CPA	None
Cytosolic	14%	88%
Microsomal	1%	2%
Nuclear	8%	7%
Plasma membrane	56%	1%
Mitochondrial	21%	1%

[a]Concentrations of additions, "None"= 2 μM tetraphenylphosphonium;
"CPA"= 2 μM tetraphenylphosphonium plus 50 μM cyclopiazonic acid;
"CCCP"= 2 μM tetraphenylphosphonium plus 5 μM carbonylcyanide-m-
chlorophenylhydrazone.

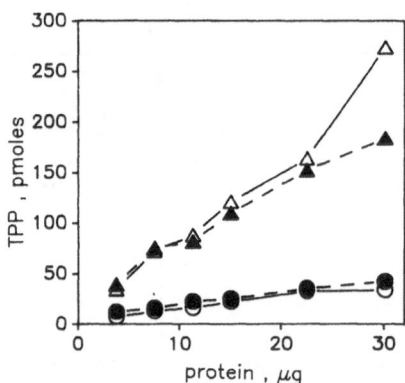

Fig. 2. Relationship between cell protein and cyclopiazonic acid-potentiated
tetraphenylphosphonium accumulation (tetraphenylphosphonium=10 μM,
cyclopiazonic acid=50 μM) by 3-day old L6 cells subjected to freeze-
thaw ($\triangle$,$\blacktriangle$) in hypotonic buffer (10 mM HEPES, pH 7.2) or held for 20
min at 22°C (0,$\bullet$) in hypotonic buffer and with ($\bullet$,$\blacktriangle$) or without (0,$\triangle$)
5 μM carbonylcyanide-m-chlorophenylhydrazone. Aliquots of cells were
suspended in cellulose propionate tubes with test solutions and lysed
by freezing at -80°C for 10 min followed by thawing at room temperature for 10 min. Binding of tetraphenylphosphonium to lysed cells
and cell fragments was determined by centrifuging the suspensions at
165,000 g in a Beckman Airfuge for 10 min at room temperature and
measuring radioactivity in the pellets. In the absence of cyclopiazonic acid there was no potentiated tetraphenylphosphonium accumulation (data not shown).[30]

Alterations in sodium, potassium, and calcium. Cyclopiazonic acid at
50 μM, a concentration that potentiated tetraphenylphosphonium accumulation
in renal cells and muscle myoblasts by 200% to 400%, had no effect on intracellular Na^+ and K^+ content of LLC-PK$_1$ cells (Fig. 3a, unpublished data),
slightly decreased $^{86}Rb^+$ flux, and had no effect on $^{22}Na^+$ flux (Table 2).[33]
These results suggested that in renal cells, cyclopiazonic acid had little
effect on the electrogenic plasma membrane Na^+/K^+ pump. However, in L6 cells,
cyclopiazonic acid at 50 μM had dramatic effects on $^{45}Ca^{2+}$ flux (Fig. 4,
unpublished data). The mechanism by which cyclopiazonic acid induced these
alterations in $^{45}Ca^{2+}$ flux is unknown. It is possible that cyclopiazonic
acid-induced electrical alterations in the plasma membrane may be causally
related to the effects of cyclopiazonic acid on Ca^{2+} flux in L6 cells.
There are electrical alterations in hydrophobic regions of sarcoplasmic reticulum vesicles, and the Ca^{2+} ATPase, that alter the binding and release of
calcium.[13] Thus, it is possible that cyclopiazonic acid induced electrical
alterations in the vicinity of the plasma membrane Ca^{2+} extrusion pump

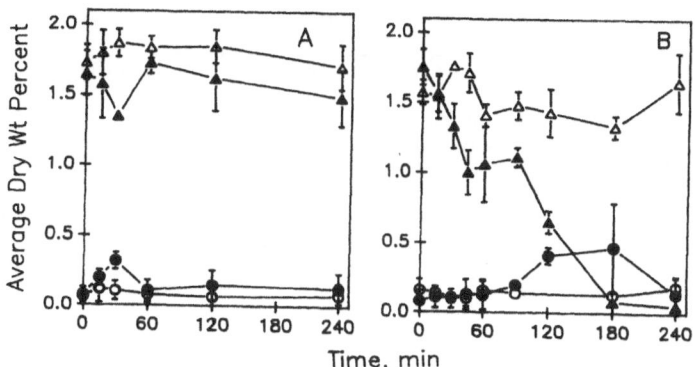

Fig. 3. Cyclopiazonic acid (a) and patulin (b) induced alterations in sodium
(0, ●) and potassium (△, ▲) content of LLC-PK$_1$ cells. Controls (0, △)
and treated cells (●, ▲) were grown on Thermanox coverslips. Values
are means of 4 replicates at each time expressed relative to phosphorus. After aspirating the growth medium, cells on coverslips
were equilibrated at 37°C for 1 hr in Dulbecco's phosphate buffered
saline plus 10 mM glucose. This buffer was then replaced with test
agents in solution in the same buffer (37°C). Each experiment was
terminated by rapidly rinsing the coverslip 3 times in 3 beakers (9
sec total) of deionized water (4°C) or 140 mM ammonium acetate.
Excess water was removed using filter paper, taking care to avoid
contact with the cells. This method was chosen because it has been
shown to effectively remove extracellular ions without osmotic damage to cells. Coverslips were air dried in a dust free environment,
mounted on aluminum stubs using carbon base adhesive, and carbon
coated in a vacuum vaporator. X-ray spectra were obtained from
standard and unknown samples on a Philips 505T SEM equipped with a
Tracor Northern 30 mm^2 Microtrace Detector and Tracor Northern 5500
MCA System. All spectra were obtained at 10 Kv accelerating voltage. A beam current of 0.20 nm at the specimen was measured using
a Faraday cup and a Keithley 485 autoranging picoammeter. Spectra
were acquired for 200 sec per cell. At least four cells were analyzed from each coverslip. Lithium borate glass x-ray standards
were obtained from C.E. Fiori (Biomedical Engineering and Instrumentation Branch, Division of Research Services, National Institute of
Health, Bethesda, MD 20205) and used to standardize the Tracor
Northern x-ray system. Results are expressed as average dry weight
percent which is the proportion of the mass of the analyzed volume
represented by each ion.

Table 2. Effect of various test agents on $^{22}Na^+$ and $^{86}Rb^+$ accumulation at 60 min by LLC-PK$_1$ cells.[a,33]

Additions[c]	% of Control[b]	
	$^{22}Na^+$	$^{86}Rb^+$
None	100 ± 12 (3/8)	100 ± 5 (6/8)
Cyclopiazonic Acid	102 ± 20 (3/8)	89 ± 7 (6/8)*
None	100 ± 16 (1/12)	100 ± 4 (1/12)
Cyclopiazonic Acid	101 ± 10 (1/12)	83 ± 2 (1/12)*
None	100 ± 24 (3/8)	ND
Ouabain	224 ± 36 (3/8)*	ND

[a]Accumulation corrected for binding of $^{22}Na^+$ or $^{86}Rb^+$ to culture plates.
[b]Mean values ± SD, with the number of experiments (first number) and replicates per experiment (second number) given in parentheses.
[c]Concentrations in test solutions; cyclopiazonic acid = 50 μM, ouabain = 5 mM, tetraphenylphosphonium = 2 μM.
Symbols, * = significantly different from control, P < 0.05 (Student's "t" test); ND = not done.

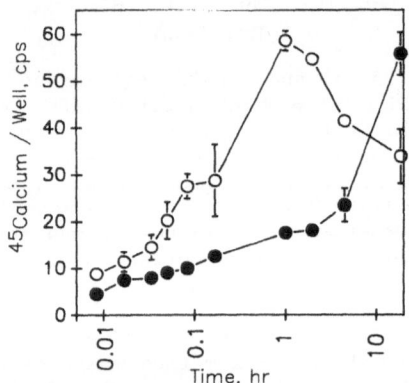

Fig. 4. $^{45}Ca^{2+}$ uptake by 7 day old fused cultures of L6 cells grown and assayed in 24 well plates. Growth medium was 1:1 Ham's F12:- Dulbeco's modified Eagles medium, 5% fetal calf serum. Growth medium was aspirated and replaced with HEPES buffered saline (25 mM HEPES, 132 mM NaCl, 6 mM KCl, 10 mM glucose, 100 μM CaCl$_2$, 100 μM MgCl$_2$) andthe cells pre-incubated at room temperature, 21°C, for 1 hr. The preincubation buffer was then replaced with HEPES buffered saline with (●) and without (O) 50 μM cyclopiazonic acid plus a trace of $^{45}CaCl_2$. At the times indicated, the $^{45}Ca^{2+}$ containing solutions were aspirated and the cells washed twice with cold HEPES buffered saline which was 2 mM CaCl$_2$ and 2 mM MgCl$_2$ and then the cells were digested in 0.2 N NaOH and radioactivity determined.

pump could alter the transport of Ca^{2+}. However, this cannot explain the kinetics of $^{45}Ca^{2+}$ accumulation documented by Figure 4. The regulation of Ca^{2+} by cells is extremely complex[12] and thus intact cells did not provide the best model for studying the possible interactions of cyclopiazonic acid with the Ca^{2+}-transport process. Sarcoplasmic reticulum vesicles are a much better preparation for studying effects on the Ca^{2+}-transport system.

Cyclopiazonic Acid Induced Alterations in Membrane Function of Rat Skeletal Muscle Sarcoplasmic Reticulum Vesicles

Inhibition of calcium transport and calcium dependent ATPase activity. Cyclopiazonic acid inhibited both Ca^{+2} transport and Ca^{+2}-dependent ATPase activity of SR vesicles isolated from rat skeletal muscle (Table 3).[7] The concentration which reduced oxalate-assisted Ca^{2+} uptake by 50% was between 20 and 50 nmoles cyclopiazonic acid/mg SR protein (Fig. 5)[7] or approximately 0.6 to 1 μM cyclopiazonic acid. This makes cyclopiazonic acid one of the most potent inhibitors of SR Ca^{2+} transport known. The effects of cyclopiazonic acid on Ca^{2+} transport and Ca^{2+}-dependent ATPase activity was not reversed by either dialysis or gel filtration of cyclopiazonic acid treated SR vesicles nor was it prevented by pretreatment of vesicles with excess dithiothreitol or glutathione.[6]

Table 3. Cyclopiazonic acid inhibition of oxalate-assisted Ca^{2+} uptake and concurrently measured Ca^{2+}-dependent ATPases activity of rat skeletal muscle sarcoplasmic reticulum vesicles (0.1 mg protein/ml) during the first 90 sec of incubation.[a,7]

Cyclopiazonic Acid nmol/mg protein	% Inhibition	
	Ca^{2+} Uptake	Ca^{2+}-Dependent ATPase Activity
25	50	42
50	66	60
100	85	73

[a]Values are means of 4 observations. Control Ca^{2+} uptake and Ca^{2+}-dependent ATPase activities were 0.72 ± 0.07 ($\pm SE$) μmol Ca^{2+}/mg protein/min and 0.82 ± 0.44 μmol P_i (inorganic phosphate) liberated/mg protein/min, respectively.

Inhibition of calcium binding at the high affinity calcium binding site. Cyclopiazonic acid (25 μM) inhibited the binding of Ca^{2+} to the SR when the free Ca^{2+} concentration was 100 μM or less.[6] Both Ca^{2+} uptake and Ca^{2+}-dependent ATPase activity were also completely inhibited by cyclopiazonic acid at this Ca^{2+} concentration. A Scatchard plot of measurable Ca^{2+} binding (Fig. 6) shows that cyclopiazonic acid inhibited the binding of Ca^{2+} to a high affinity binding site on crude SR vesicles. Cyclopiazonic acid at 1 μM, the lowest concentration tested, inhibited Ca^{2+}-dependent phosphorylation by approximately 15%, whereas 5 μM cyclopiazonic acid caused 95% inhibition (Fig 7). When Ca^{2+}-independent phosphorylation (approximately 0.1 nmole phosphate/mg protein) is taken into account, cyclopiazonic acid at 25 μM completely inhibited phosphorylation. It is generally accepted that Ca^2 binding to the high affinity site is obligatory to the phosphorylation of the ATPase.[18]

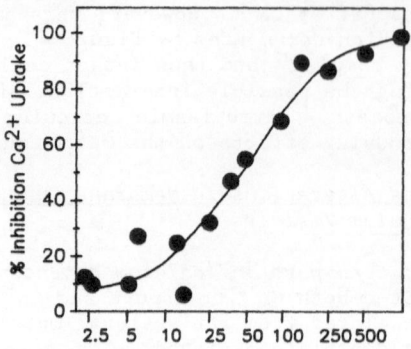

Fig. 5. Cyclopiazonic acid inhibition of oxalate-assisted Ca^{2+} uptake using
0.02 mg SR protein/ml. Each value represents the percent inhibition
of the initial rate of Ca^{2+} uptake vs cyclopiazonic acid concentration (20-90 sec, 4 time points, 2-5 observations per time point).
Ca^{2+} uptake was measured by a rapid filtration method. The reaction buffer was 5 mM MOPS, 0.1 KCL, 4 mM $MgCl_2$, 5 mM oxalate and
0.1 mM $^{45}CaCl_2$, pH 7.0. SR vesicles were preincubated 5 min at
25°C and reaction was started by addition of 10 mM ATP (pH 7.0) to
a final concentration in the reaction buffer of 4 mM ATP. Radioactivity retained by 0.45 μm Millipore filters and in aliquots of the
filtrate was determined by standard liquid scintillation methods.[7]

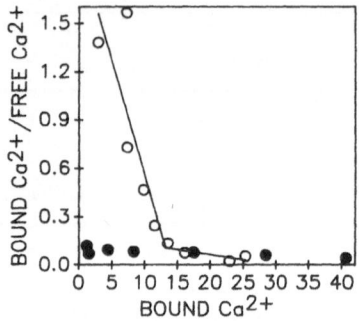

Fig. 6. Scatchard plot of data for Ca^{2+} binding. Bound and free Ca^{2+} are in
nmole and μmole units, respectively. Binding of Ca^{2+} to rat skeletal muscle SR vesicles with and without 25 μM cyclopiazonic acid was
determined by dialyzing 1 ml of SR vesicles (0.5 mg protein/ml) for
4 hr against 25 ml of 0.1 M KCl, 4 mM $MgCl_2$, 10 mM MOPS, pH 7.0,
and various levels of $^{45}Ca^{2+}$ with (●) and without (O) 25 μM
cyclopiazonic acid. Aliquots of both the SR vesicles and buffer
were then analyzed for $^{45}Ca^{2+}$ with the difference between
the two being the Ca^{2+} bound to the SR vesicles. Triplicate
samples were taken at each Ca^{2+} concentration.[6]

Patulin Induced Alterations in Membrane Function of LLC-PK₁ cells

Alterations in intracellular sodium and potassium. Unlike cyclopiazonic
acid, patulin induces significant alterations in Na^+ and K^+ content of
LLC-PK₁ cells (Fig. 3b). We compared the effects of patulin with ouabain, a
known specific inhibitor of the Na^+/K^+ pump.[10] The concentrations of
ouabain (60 min) which resulted in a 50% change in both Na^+ influx and K^+
efflux was between 10 and 50 μM (Fig. 8a and 8b).[9] Whereas for patulin

the concentrations resulting in a 50% change in intracellular Na^+ and K^+ was 250 μM and 100 μM, respectively (Fig. 8). However, even low concentrations (10 μM) of patulin brought about total efflux of K^+ if cells were exposed for sufficient periods of time.[31]

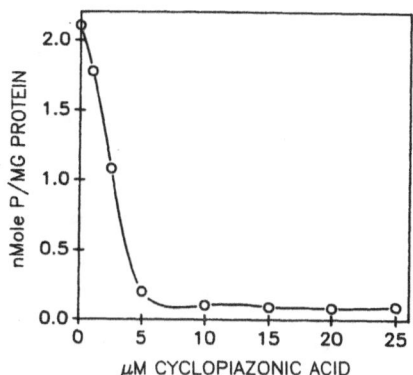

Fig. 7.   Effect of cyclopiazonic acid on phosphorylation of rat skeletal muscle SR vesicles.   SR vesicles (0.5 mg protein/ml) were phosphory lated with 0.2 mM $[\gamma-^{32}P]$-ATP (12,000 cpm/nmole) in 0.1 M KCl, 4 mM $MgCl_2$, 100 μM Ca^{2+}, 10 mM MOPS, pH 7.0, with and without various levels of cyclopiazonic acid.   An ATP regenerating system of 5 mM phospho(enol)pyruvate and 23 units/ml pyruvate kinase was also included, final volume was 2.0 ml.   Reactions were initiated by ATP addition and terminated after 30 sec by addition of 7 ml of 5% tri- chloroacetic acid containing 2 mM inorganic phosphate and 0.1 mM ATP.   The 10,000 g pellet was washed 3 times, digested in 1 N NaOH by heating for 1 hr in a boiling water bath and radioactivity and protein determined.[6]

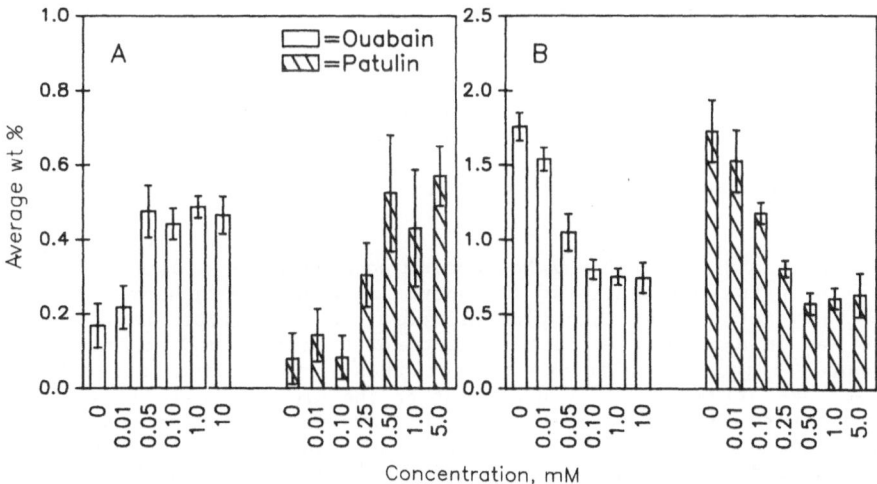

Fig. 8.   Dose response (60 min) for sodium influx (A) and potassium efflux (B) in cells treated with patulin and ouabain.   Results normalized to phosphorus.   Experimental details are the same as described in Fig. 3.[9]

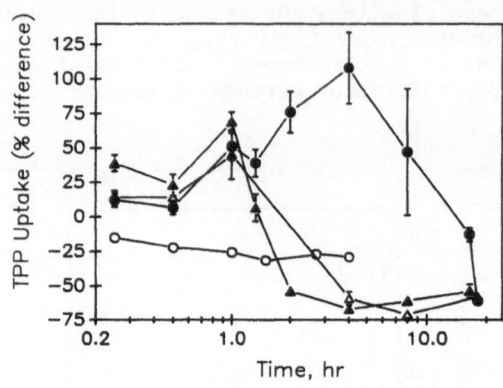

Fig. 9. Effect of patulin at 10 μM (●), 50 μM (△) and 250 μM (▲) and
ouabain at 50 μM (O) on uptake of [^{3}H]tetraphenylphosphonium by
pig renal epithelial cells (LLC-PK$_1$) grown in 10 cm^2 culture
dishes. Uptake was measured directly in the culture dishes.
Values are means of 3 to 6 replicates at each time and expressed
as the percent difference between treated cultures and concurrent controls at each time point. This value is calculated as
$([cpm]_t/[cpm]_c)-1)100)$ where $[cpm]_t$ and $[cpm]_c$ are the [^{3}H]radio-
activity accumulated by patulin-treated and control cells, respectively. Experimental details were as described in Fig. 1
except that values were not corrected for potential insensitive
accumulation in the same manner as described. Instead, because
we were only interested in passively accumulated tetraphenylphosphonium, we allowed cells to accumulate tetraphenylphosphonium,
rinsed the cells to remove free tetraphenylphosphonium, and then
added Dulbecco's phosphate buffered saline plus 10 mM glucose
plus 5 μM carbonylcyanide-m-chlorophenylhydrazone to depola-
rize all membrane potential differences and measured the amount
of radioactivity in the release buffer after 60 min. Preliminary studies indicated that the time required for 50% of the
accumulated tetraphenylphosphonium to be released was less than
5 min for both control and patulin-treated cultures. Carbonylcyanide-m-chlorophenylhydrazone (5 μM) had no effect on cell
viability at 60 min (trypan blue exclusion). The passively
accumulated tetraphenylphosphonium is proportional to the sum of
the transmembrane potentials for all serially connected compartments within the cell.[34] Liquid scintillation counting of
digests of attached cells after treatment with 5 μM carbonylcyanide-m-chlorophenylhydrazone indicated that all accumulated
tetraphenylphosphonium in the presence of patulin was in passive
equilibrium with the transmembrane potential difference.[31]

Alterations in tetraphenylphosphonium accumulation. Like cyclopiazonic acid, patulin potentiated the accumulation of tetraphenylphosphonium
Fig. 9), however, unlike cyclopiazonic acid the accumulated tetraphenylphosphonium was in a passive equilibrium and not tightly bound to intracellular
components.[31] Thus, patulin appears to induce an increased intracellular electronegativity that is dependent on the electrochemical potential
gradient across the plasma membrane. The patulin induced potentiated accumulation of tetraphenylphosphonium was transient and was followed by a loss of
all accumulated tetraphenylphosphonium to a level which was the same as in
cells that had been treated with the depolarizing agent, carconylcyanide-m-
chlorophenylhydrazone. The duration of the transient increase in tetraphenylphosphonium accumulation was concentration dependent (Fig. 9).

Blockage of the Na^+/K^+ pump with ouabain resulted in only a partial decrease in tetraphenylphosphonium accumulation even at concentrations as high as 1.0 mM ouabain for 4 hr.[31] Patulin at concentrations as high as 0.5 mM had little or no effect on [3]ouabain (0.1 μM) binding.[31] Thus it appears that patulin does not interact directly with the ouabain binding site, not does patulin stimulate turnover of the pump itself. We have yet to investigate patulin's effects on Ca^{2+} homeostasis in cells, however it is well established that Na^+ influx via either inhibition of the Na^+/K^+ pump or membrane damage is usually accompanied by Ca^{2+} influx.[12]

Cytoskeletal alterations and cellular toxicity. Patulin was much more toxic to LLC-PK$_1$ cells than cyclopiazonic acid based on lactate dehydrogenase release after exposure to the toxins for 20 hr (Fig. 10a and 10b, unpublished data). Patulin, but not cyclopiazonic acid (unpublished data), induced significant cytoskeletal alterations (blebbing) in LLC-PK$_1$ cells (Fig. 11b). Patulin-treated cells that were co-incubated with dithiothreitol (a sulfhydryl protecting agent) did not exhibit blebbing (unpublished data) and appeared similar to untreated cells (Fig. 11a). Ouabain, a specific inhibitor of the Na^+/K^+ pump alone is insufficient to explain the cytoskeletal alterations observed with LLC-PK$_1$ cells. Cyclopiazonic acid at $\leq$100 μM for 20 hr had no apparent effect on cell morphology (Fig. 11c), but induced significant swelling of cells at doses of 250 μM and 500 μM after 20 hr (Fig. 11d).

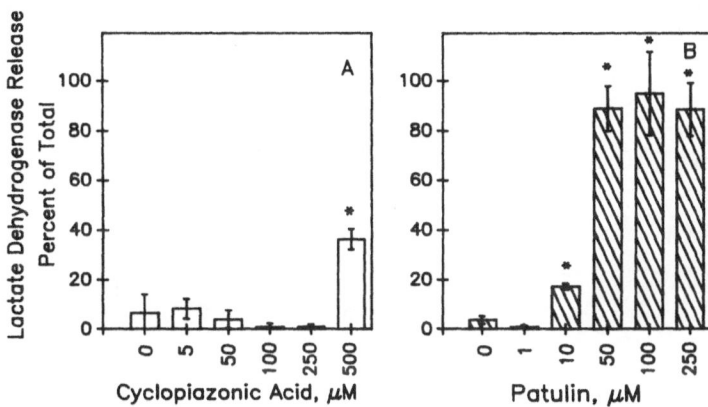

Fig. 10. Cell viability using lactate dehydrogenase release in pig renal epithelial cells treated with different concentrations of cyclopiazonic acid (a) or patulin (b). LLC-PK$_1$ cells were grown in 10 cm^2 dishes. The results are presented as the amount of lactate dehydrogenase in the media at 20 hr divided by the total lactate dehydrogenase after lysis with Triton X-100. Values are means of 3-4 replicates at each concentration. At 1 hr prior to an assay the growth medium was replaced with Dulbecco's phosphate buffered saline plus 10 mM glucose and without penicillin or streptomycin. The cultures were transferred to a 37°C, >70% RH incubator. After 1 hr fresh buffer (2 ml) plus various concentrations of patulin or cyclopiazonic acid was added. Samples of the buffer (100 μl) were removed after 20 hr and stored at 4°C. Intracellular lactate dehydrogenase was released by lysis of cells with 0.5 ml of 3% Triton X-100. Lactate dehydrogenase in all samples was determined on a Multistat III Microcentrifugal Analyzer.

CONCLUSIONS

Cyclopiazonic acid and patulin are both fungal metabolites that are common contaminants of food products. They are both produced by certain species of *Aspergillus* and *Penicillium* and both have been shown to alter ion movement and affect transport ATPases. However, their mechanisms of action at the level of cell membranes are quite different. Patulin is sulfhydryl reactive while cyclopiazonic acid is not, patulin has dramatic effects on Na^+ and K^+ flux in cells while cyclopiazonic acid has little effect. Patulin inhibits the Na^+/K^+ ATPase *in vitro*[25,26,27] whereas cyclopiazonic acid is a potent inhibitor of the sarcoplasmic reticulum Ca^{2+}-dependent ATPase. Patulin is much more toxic to cells than cyclopiazonic acid and patulin induces cytoskeletal alterations in LLC-PK$_1$ cells while cyclopiazonic acid does not.

Many of the pathological and clinical signs of cyclopiazonic acid toxicity involve skeletal muscle and the symptoms include muscular inco-ordination and altered motor activity.[19,21] Cyclopiazonic acid toxicity may be due to a direct effect on muscle.[22,24] Possibly the inhibitory effect of cyclopiazonic acid on Ca^{2+} transport and Ca^{2+}-dependent ATPase activity of the SR may be responsible. Alternatively, many of the effects of cyclo-piazonic acid parallel those of drugs which alter neurotransmitter levels in tissues.[20,21] It has been hypothesized that alterations in cellular Ca^{2+} homeostasis may be a target for neurotoxic agents[12] and it is well estab-lished that Ca^{2+} plays a pivotal role in both muscle contraction[16] and in hormone and neurochemical signalling processes.[28] Although we have not directly examined the effect of cyclopiazonic acid on the plasma membrane Na^+/K^+ ATPase, the fact that cyclopiazonic acid has little effect on Na^+

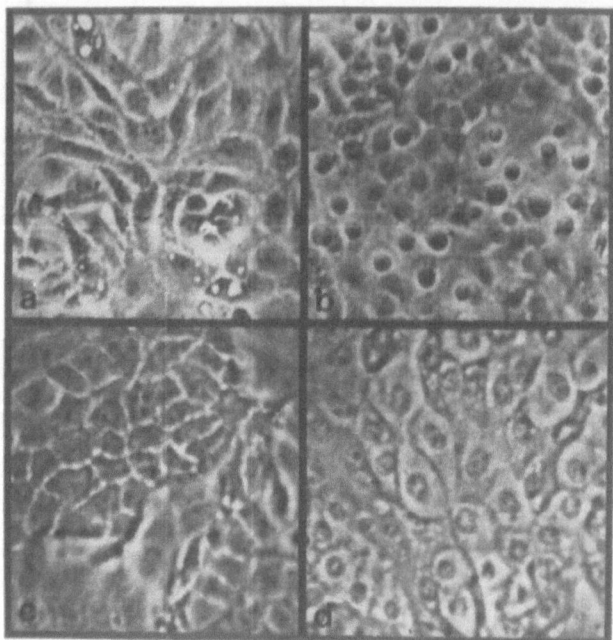

Fig. 11.   Phase contrast photomicrographs (x 3680) of LLC-PK$_1$ cells
          grown in 10 cm^2 culture plates and treated for 3 hr with
          250 μM patulin plus 1 mM dithiothreitol (a) or with only
          250 μM patulin (b). Also, control LLC-PK$_1$ cells after 20 hr
          in phosphate buffered saline plus 10 mM glucose (c) and 250 μM
          cyclopiazonic acid treated cells after 20 hr (d).

or K^+ flux but significant effects on Ca^{2+} flux suggests that it may inhibit Ca^{2+}-dependent processes specifically. We are currently testing this hypothesis. With regards to patulin, we hypothesize that in addition to inhibition of the Na^+/K^+ ATPase[1,25,26,27] and inhibition of carrier mediated transport systems[36] the biological activity of patulin at the level of cell membranes involves the development of cation leaks. The increased intracellular electronegativity (hyperpolarization) in LLC-PK$_1$ cells occurs as a result of the increased membrane permeability to K^+ resulting in a net efflux of positive charge. At low patulin doses the membrane remains relatively intact and thus the apparent hyperpolarization is sustained. We have not yet examined the effect of patulin on Ca^{2+}-dependent ATPases. However, because sulfhydryl reactive agents, such as mersalyl acid inhibit the SR Ca^{2+}-ATPase[8] and the Na^+/K^+ATPase[35], it is likely that patulin, also a sulfhydryl reactant, may inhibit Ca^{2+}-ATPases. Many of the manifestations of patulin toxicity in LLC-PK$_1$ cells parallel those of sulfhydryl reactive chemicals and conditions that favor radical-mediated peroxidative damage.[4]

DISCLAIMER

The mention of a trade name, vendor or proprietary name does not imply its preference by the U.S. Department of Agriculture to the exclusion of others that may also be suitable.

REFERENCES

1. Andraud J, Andraud G, 1971, Effect of some lactones on the activity of (Na^+-K^+)-adenosine triphosphatases of human erythrocytes, *CR Seances Soc Biol Filiales*, 165:301-305.
2. Beeler T, Gable K, 1985, Effects of halothane on Ca^{2+} release from sarcoplasmic vesicles isolated from rat skeletal muscles, *Biochin Biophys Acta*, 821:142-152.
3. Baginski ES, Foa, PP, Zak B, 1967, Determination of phosphate: Study of labile organic phosphate interference, *Clin Chim Acta*, 15:155-158.
4. Deuticke B, 1986, The role of membrane sulfhydryls in passive, mediated transport processes and for the barrier function of the erythrocyte membrane, *Membr Biochem.*, 6:309-326.
5. Fleischer S, Kervina M, 1974, Subcellular fractionation of rat liver, *In*: "Methods in Enzymology," Vol. XXXI, Biomembranes, Part A, Fleischer S, Packer L, eds., Academic Press, New York.
6. Goeger DE, Riley RT, 1989, Interaction of cyclopiazonic acid with rat skeletal muscle sarcoplasmic reticulum vesicles: Effects on Ca^{2+} binding and Ca^{2+} permeability, *Biochem Pharmac.*, 38:3995-4003.
7. Goeger D, Riley R, Dorner J, Cole J, 1988, Cyclopiazonic acid inhibition of Ca^{2+}-transport ATPase in rat skeletal muscle sarcoplasmic reticulum vesicles, *Biochem Pharmac.*, 37:978-981.
8. Hasselbach W, Seraydarian K, 1966, The role of sulfhydryl groups in calcium transport through sarcoplasmic membranes of skeletal muscle, *Biochem Z.* 345,159-172.
9. Hinton DM, Riley RT, Showker JL, Rigsby W, 1989, Patulin induced ion flux in renal cells and reversal by dithiothreitol and glutathione: A scanning electron microscopy (SEM) x-ray microanalysis study, *J Biochem Toxicol.*, 4:47-54.
10. Hootman SR, Ernst SA, 1988, Estimation of Na,K-pump numbers and turn over in intact cells with [^{3}H]ouabain, *In*: "Methods in Enzymology," Vol. 156, Biomembranes, Part P, Fleischer S, Fleischer B, eds., Academic Press, New York.
11. Kinney TB, 1985, Introduction: The importance of basic research to agriculture, *In*: "Frontiers of Membrane Research in Agriculture," St. John JB, Berlin E, Jackson PC, eds., Rowman & Allanheld, Totowa, NJ.
12. Komulainen H, Bondy SC, 1988, Increased free intracellular Ca^{2+} by

toxic agents: An index of potential neurotoxicity?, *TIPS*, 9:154-156.

13. Levitsky DO, Loginov VA, Lebedev AV, 1986, Charge changes in sarco-plasmic reticulum and Ca^{2+}-ATPase induced by calcium binding and release: A study using lipophilic ions, *Membr Biochem.*, 6:291-307.

14. Lichtshtein D, Kaback HR, Blume AJ, 1979, Use of a lipophilic cation for determination of membrane potential in neuroblastoma-glioma hybrid NG108-15, *Proc Natl Acad Sci USA*, 76:650-654.

15. Lowry OH, Rosebrough N, Farr AL, Randall RJ, 1951, Protein measurement with the Folin phenol reagent, *J Biol Chem.*, 193:265-275.

16. Martonosi AN, 1984, Mechanisms of Ca^{2+} release from sarcoplasmic reticulum of skeletal muscle, *Physiol Rev.*, 64:1240-1320.

17. Martonosi A, Feretos R, 1964, Sarcoplasmic reticulum I. The uptake of Ca^{++} by sarcoplasmic reticulum fragments, *J Biol Chem.*, 239:648-658.

18. de Meis L, Vianna AL, 1979, Energy interconversion by the Ca^{2+}-depen-dent ATPase of the sarcoplasmic reticulum, *Ann Rev Biochem.*, 48:275-292.

19. Nishie K, Cole RJ, Dorner JW, 1985, Toxicity and neuropharmacology of cyclopiazonic acid, *Fd Chem Toxic.*, 23:831-839.

20. Nishie K, Cole RJ, Dorner JW, 1986, Effects of cyclopiazonic acid on the contractility of organs with smooth muscle, and frog ventricles, *Res Comm Chem Pathol Pharmacol.*, 53:23-37.

21. Nishie K, Porter JK, Cole RJ, Dorner JW, 1985, Neurochemical and pharma-cological effects of cyclopiazonic acid, chlorpromazine and reserpine, *Res Comm Psychol, Psychiat and Behav.*, 10:291-302.

22. Norred WP, Morrissey RE, Riley RT, Cole RJ, Dorner JW, 1985, Distribu-tion, excretion and skeletal muscle effects of the mycotoxin [^{14}C]cyclo-piazonic acid in rats, *Fd Chem Toxic.*, 23:1069-1076.

23. Ottolenghi P, 1975, The reversible delipidation of a solubilized sodium-plus-potassium ion-dependent adenosine triphosphatase from the salt gland of the spiny dogfish, *Biochem J.*, 151:61-66.

24. Peden WM, Richard JL, Thurston JR, 1986, Comparative histopathological changes in cyclopiazonic acid toxicosis and rubratoxicosis, *In*: "Diagnosis of Mycotoxicoses" Richard JL, Thurston JR, eds., Martinus Nijhoff, Dordrecht.

25. Phillips TD, Hayes W, 1977, Effects of patulin on adenosine triphos-phatase activities in the mouse, *Toxicol Appl Pharmacol.*, 42:175-187.

26. Phillips TD, Hayes W, 1978, Effects of patulin on the kinetics of sub-strate and cationic ligand activation of adenosine triphosphatase in mouse brain, *J Pharmacol Exp Ther.*, 205:606-616.

27. Phillips TD, Hayes W, 1979, Inhibition of electrogenic sodium transport across toad bladder by the mycotoxin patulin, *Toxicology*, 13:17-24.

28. Rasmussen H, Zawalich W, Kojima I, 1985, Ca^{2+} and cAMP in the regula-tion of cell function, *In*: "Calcium and Cell Physiology", Marme D, ed., Springer-Verlag, Berlin.

29. Riley RT, Goeger DE, Norred WP, 1989, Lipophilic cations as membrane potential probes in cultured renal cells (LLC-PK$_1$) and muscle cells (L6), *In*: "Nephrotoxicity: Extrapolation from *in vitro* to *in vivo* and from animals to man," Bach PH, Lock EA, eds., Plenum Press, New York.

30. Riley RT, Goeger DE, Showker JL, Cole RJ, Dorner J, 1987, Age and growth related changes in cyclopiazonic acid-potentiated lipophilic cation accu-mulation by cultured cells and binding to freeze-thaw lysed cells, *J Biochem Toxicol.*, 2:251-264.

31. Riley RT, Hinton DM, Showker JL, Rigsby W, Norred WP, 1990, Chronology of patulin induced alterations in membrane function of cultured renal cells, LLC-PK$_1$, *Toxicol Appl Pharmacol.*, 102:128-141.

32. Riley RT, Norred WP, Dorner JW, Cole RJ, 1985, Increased accumulation of the lipophilic cation tetraphenylphosphonium by cyclopiazonic acid-treated renal epithelial cells, *J Toxicol Envir Hlth.*, 15:779-788.

33. Riley RT, Showker JL, Cole RJ, Dorner JW, 1986, The mechanism by which cyclopiazonic acid potentiates accumulation of tetraphenylphosphonium in cultured renal epithelial cells, *J Biochem Toxicol.*, 1:13-30.

34. Ritchie RJ, 1984, A critical assessment of the use of lipophilic cations

as membrane potential probes, *Prog Biophys Molec Biol.*, 43:1-32.

35. Schwartz A, Lindenmayer GE, Allen JC, 1975, The sodium-potassium adenosine triphosphatase: Pharmacological, physiological and biochemical aspects, *Pharmacol Rev.*, 27:3-134.

36. Ueno Y, Matsumoto H, Ishii K, Kukita K-I, 1976, Inhibitory effects of mycotoxins on Na^+-dependent transport of glycine in rabbit reticulocytes, *Biochem Pharmacol.*, 25:2091-2095.

BIOTRANSFORMATION OF TRICHOTHECENES: THE ROLE OF INTESTINAL MICROFLORA IN

THE METABOLISM AND TOXICITY OF TRICHOTHECENE MYCOTOXINS

S. P. Swanson[a], C. Helaszek and H. D. Rood, Jr.[b]

Department of Veterinary Biosciences
University of Illinois
Urbana, IL

INTRODUCTION

The trichothecene mycotoxins are a group of tetracyclic sesquiterpenoid compounds characterized by a six-membered oxygen ring, an epoxide in the 12,13 position and an olefinic bond in the 9,10 position. They are produced by a variety of fungi including *Stachybotrys*, *Myrothecium*, *Trichothecium*, and particularly species of the genus *Fusarium*.[21]

Three of the more agriculturally important members of the trichothecenes are T-2 toxin, diacetoxyscirpenol (DAS) and deoxynivalenol (DON). All three compounds have been detected as naturally occurring contaminants of feeds and food products.[11,21] Mycotoxicoses caused by exposure of animals to trichothecenes have been reported worldwide. Adverse health effects resulting from trichothecene consumption include: reduced weight gain and feed consumption, feed refusal, diarrhea, emesis, immune suppression, gastrointestinal irritation, oral lesions and death.[11,21,22,29,34]

Trichothecenes are rapidly eliminated from animals experimentally administered these mycotoxins. The mean plasma disappearance half-lives of T-2 toxin in cattle, swine and dogs are 17.4, 13.8 and 5.3 minutes, respectively.[1,35] Diacetoxyscirpenol is even more rapidly cleared with halflives of 11.6 and 6.4 minutes in swine and cattle, respectively.[7] Very little parent compound is excreted intact, and elimination is achieved primarily through rapid and extensive biotransformation. Four different biotransformation pathways have been described for the trichothecenes including: ester hydrolysis,[9,16,23,42] conjugation with glucuronic acid,[8,10,13,25] aliphatic hydroxylation at the C-3' position on the isovaleryl side chain,[8,16,46] and by reduction of the 12,13 epoxide group (deepoxidation) to yield a carbon-carbon double bond.[3,9,10,19,32,45-47]

Previously, we demonstrated that anaerobic bovine rumen microflora reduced the 12,13 epoxy group of the three trichothecenes, T-2 toxin, DAS and DON.[38] Purified deepoxy metabolites of T-2 and DAS were shown to be non-toxic to brine shrimp.[39] In the present study, we investigated the

[a]Present address: Eli Lilly and Co., Lilly Corporate Center, Department of
    Drug Metabolism and Disposition, MC 909 28/2, Indianapolis, IN 46285.
[b]Present address: J & W Scientific, 91 Blue Ravine, Folsom, CA 95630.

role intestinal and fecal microflora play in the biotransformation and toxicity of trichothecene mycotoxins.

EXPERIMENTAL

Chemicals. T-2 toxin and diacetoxyscirpenol (DAS) were extracted from cultures of *Fusarium sporotrichiodes* grown and purified in our laboratory. Metabolites were prepared as described previously.[38,39] See Figure 1 for chemical structures.

Animals. Male 200-250 g Sprague-Dawley rats and male 20-25 g Balb/c mice were obtained from Harlan Sprague-Dawley (Indianapolis, IN). Swine, chickens, horses, dogs and cattle used in these experiments were healthy control animals housed at the College of Veterinary Medicine, University of Illinois.

Preparation of inoculum. Fresh feces were collected, purged with nitrogen and used within 15 minutes. Inoculum was prepared by diluting 1 part of fresh feces with 5 parts of yeast extract-peptone (YEP) media. For

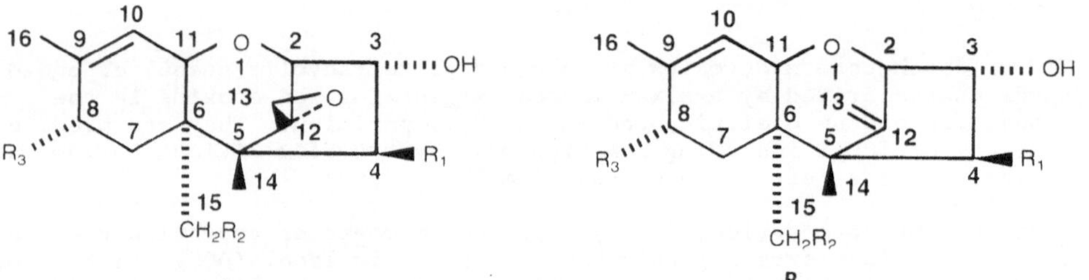

Fig. 1. Structure of T-2 toxin, diacetoxyscirpenol and their metabolites formed *in vitro* under anaerobic incubation conditions.

Compound	Skeleton	R1	R2	R3
DAS	A	OAC	OAC	H
15-MAS	A	OH	OAC	H
SCP	A	OH	OH	OH
DE MAS	B	OH	OAC	H
DE SCP	B	OH	OH	H
T-2	A	OAC	OAC	ISV
HT-2	A	OH	OAC	ISV
TRIOL	A	OH	OH	ISV
TOL	A	OH	OH	OH
DE T-2	B	OAC	OAC	ISV
DE HT-2	B	OH	OAC	ISV
DE TRIOL	B	OH	OH	ISV
DE TOL	B	OH	OH	OH

ISV = $OCOCH_2CH(CH_3)_2$, OAC = $OCOCH_3$, DAS = diacetoxyscirpenol, MAS = 15-monoacetoxyscirpenol, SCP = scirpentriol, DE DAS = deepoxy monoacetoxyscirpenol, DE SCP = deepoxy scirpentriol, TRIOL = T-2 triol, TOL = T-2 tetraol, DE T-2 = deepoxy T-2, DE HT-2 = deepoxy HT-2, DE TRIOL = deepoxy T-2 triol, DE TOL = deepoxy T-2 tetraol.

comparison of deepoxidation activity in different sections of the rat gastro-intestinal tract, the GI tracts were divided into 4 parts; the upper and lower halves of the small intestine, cecum and colon. Sections were removed through a small ventral incision, the contents removed and rapidly trans-ferred to YEP media as described for feces.

Incubation conditions. All inoculum and transfers were prepared and con-ducted under an oxygen-free nitrogen atmosphere. Incubations were conducted by adding 1 mg of toxin to test tubes containing 8 ml of YEP media. The tubes were then inoculated with 2 ml of intestinal content or fecal suspen-sions and incubated for 96 hours at 37°C.

Chemical analysis of metabolites. Metabolites were extracted from the incubation mixtures with C18 cartridges (Analytichem International, Harbor City, CA; 1 g) preconditioned with 2 column volumes of methanol followed by 2 column volumes of water. The cartridges were rinsed with water (1 column volume) and the toxins eluted with methanol (2 ml). The methanol eluate was concentrated to near dryness, the residue redissolved in ethyl acetate and transferred to a minicolumn packed with 0.65 g of Florisil. Six ml of ethyl acetate were then added to the column. The total eluate was collected, con-centrated over a gentle nitrogen stream and the residue redissolved in ethanol.

Aliquots were removed and concentrated to dryness for gas chromatographic analysis. Derivatization to the corresponding trimethylsilyl ether deriva-tives (TMS) and subsequent capillary gas chromatographic analysis was accom-plished as previously described.[40] See Figure 2 for a chromatogram dis-playing separation of trichothecenes and metabolites.

Synthesis of deepoxy T-2. Deepoxy T-2 was synthesized from T-2 toxin using a modification of the method described by Colvin and Cameron[6] for the deoxygenation of triacetoxyscirpenol as described by Swanson et al.[40] Syn-thetic and biosynthetic DE TOL were prepared by alkaline hydrolysis of syn-thesized DE T-2 and microbially derived DE HT-2, respectively.

Nuclear magnetic resonance and mass spectrometry. Mass spectrometry was performed on an Extranuclear Simulscan 300 series mass spectrometer using electron impact and methane chemical ionization. Proton NMR spectra were obtained with a General Electric QE-300 spectrometer operating at 300 MHz.

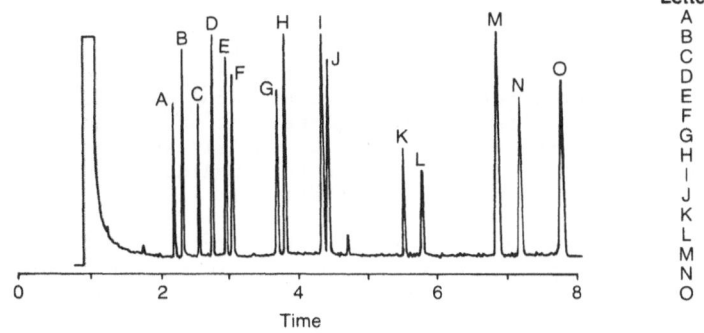

Column: 25m × 0.25mm DB1701
250° C − 275° C at 5° C/min

Letter	Compound
A	DE STRIOL
B	DOM-1
C	DE TETRAOL
D	DE MAS
E	DON
F	STRIOL
G	TETRAOL
H	MAS
I	DAS
J	DE TRIOL
K	DE HT-2
L	TRIOL
M	C_{30} (ISTD)
N	HT-2
O	T-2

Fig. 2. Separation of TMS ether derivatives of tricho-thecenes by capillary gas chromatography.

DE T-2 and DE TOL were dissolved in CDCL$_3$ and deuterated DMSO, respectively.

Rat dermal toxicity. A section of hair (5 x 3 cm) on the back of each animal was clipped and the area divided into 6 separate sections. To each skin section was applied 3 μl of the test compound dissolved in ethyl acetate. Dermal reaction to the test compounds was observed 48 hours following administration. Scores ranged from 0 (no reaction) to 4 (severe reaction) as described by Hayes and Shiefer.[14] Each of 5 rats received all six treatments. Treatments were as follows: 1) 120 ng T-2 toxin, 2) 480 ng T-2 toxin, 3) 480 ng DE T-2, 4) 4,800 ng DE T-2, 5) 48,000 ng DE T-2 and 6) ethyl acetate vehicle control.

Forty-eight hours after dosing, the rats were killed, and the treated skin was removed and fixed in 10% neutral buffered formalin. The fixed tissue was embedded in paraffin, sectioned and stained with hematoxylin and eosin for microscopic examination.

Intraperitoneal toxicity of DE T-2. Mice were administered either T-2 toxin or deepoxy T-2 (dissolved in 0.5 ml of corn oil) by intraperitoneal injection. The animals were randomly assigned to four treatment groups with seven animals per treatment group, and administered the test compounds as follows: 1) DE T-2 (12 mg/kg), 2) DE T-2 (60 mg/kg), 3) T-2 toxin (12 mg/kg) and 4) corn oil vehicle control. Animals were observed daily for 5 days after administration of the toxins.

LD$_{50}$ determination. Eighty-four male NIH Swiss mice weighing 24-30 g were purchased from Harlan Sprague-Dawley Inc. (Indianapolis, IN). Animals were housed in groups of six and were acclimatized in cages for 1 week prior to the experiments. One week prior to dosing with T-2 toxin the animals were divided into two treatment groups. Group 1 received water only and group 2 a solution containing a mixture of antibiotics (4 mg/ml bacitracin, 4 mg/ml neomycin sulfate and 0.3 mg/ml penicillin G). The antibiotic solutions were prepared fresh every 48 hours and treatment was continued for 1 week after dosing. Seven days after initiation of antibiotic therapy, both treatment groups were randomly administered T-2 toxin (dissolved in 10% ethanol) by oral gavage at dosages of 0, 6, 8, 10, 12, 14 and 16 mg/kg body weight. Each dosage was administered to six mice. Deaths were recorded daily for 7 days.

LD$_{50}$ calculations and 95% confidence intervals were calculated by a computerized program using the moving average method.[44] In addition, LD$_{50}$ values were also calculated by the trim-logit method as described by Sanathana et al.[33] A Z test was used for statistical comparison of the two treatment groups with an alpha level = 0.05.[37]

RESULTS

Metabolism of DAS by fecal microorganisms. Metabolism of DAS by fecal microorganisms from horses, cattle, dogs, swine, rats and chickens were compared under anaerobic conditions. Fecal microflora from dogs, horses and chickens did not produce detectable concentrations of deepoxy metabolites. By contrast, microflora from rats, cattle and swine produced significant amounts of deepoxy products. See Figure 3 for the percent of DAS converted to total deepoxy metabolites by the six species. Of the six species examined, rats were the most efficient at reducing the epoxide group, converting 100% of the parent compound to deepoxy metabolites.

The individual metabolites produced by fecal microflora from six species are shown in Table 1. In contrast to deepoxidation reactions, which were detected only with rats, cattle and swine, esterase activity was observed

with microflora from all species. The deacylated deepoxy compounds, DE MAS
and DE SCP, were the major products found in incubations with rat, swine and
cattle fecal microorganisms. In contrast, the major products produced upon
incubation with horse, dog and chicken fecal microflora were the simple C-4
deacylated product, MAS, in addition to unaltered parent DAS.

Deepoxidation activity in intestinal segments. No deepoxy metabolites of
DAS were detected in any incubations using rat small intestine contents as
inoculum. Microflora obtained from both the cecal and colon contents of rats
efficiently converted DAS to deacylated deepoxy metabolites (data not shown).

Biotransformation of T-2 toxin by cecal microflora. T-2 toxin was completely biotransformed by rat cecal microflora, to C-4 and C-15 deacylated,
deepoxy products (Table 2). The major product, DE HT-2, was derived from
both epoxide reduction and C-4 deacylation, although minor quantities of
DE TRIOL were also detected. No compounds requiring C-8 ester hydrolysis
(TOL, 4-deacetylneosolaniol or their deepoxy analogs) were detected in any

Table 1. Metabolism of diacetoxyscirpenol by anerobic fecal
microorganisms obtained from rats, chickens, dogs,
cattle, horses and swine.

Species	Recovered (Mean ± SEM)[a]				
	DAS	MAS	SCP	DE MAS	DE SCP
Rat	0	0	0	66.5± 6.2	33.5± 6.2
Swine	0	12.6± 7.5	2.2±1.1	61.7± 5.9	23.5± 3.1
Cow	7.9± 2.4	15.8±11.3	3.7±2.4	32.4±13.0	40.1±16.1
Dog	11.2± 1.4	88.7± 1.2	0.2±0.2	0	0
Horse	36.1± 6.8	60.1± 5.9	3.8±3.9	0	0
Chick	43.5±13.7	48.6± 8.8	7.9±5.2	0	0

[a]Molar percent of total metabolites recovered ± standard error
of the mean. Number of animals = rat (4), swine (4), cattle
(6), dog (3), horse (3), chicken (4).

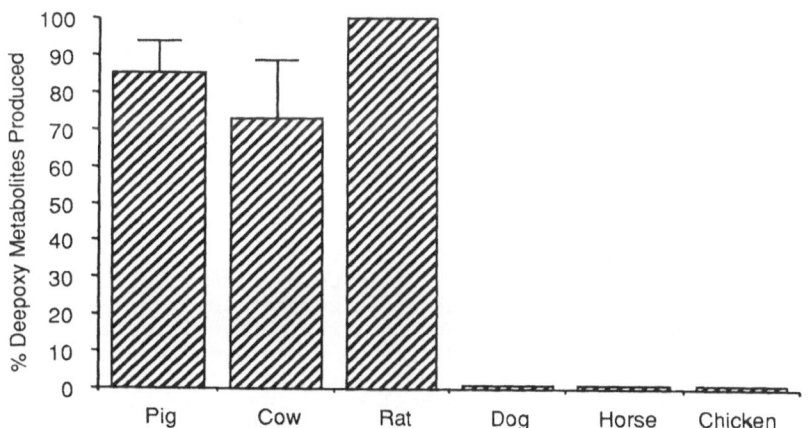

Fig. 3. Conversion of diacetoxyscirpenol (DAS) to total deepoxy
metabolites by fecal microflora from six species.

of the T-2 toxin incubation mixtures. Diacetoxyscirpenol was biotransformed in a manner analogus to T-2 toxin, yielding DE MAS and DE SCP. Traces (less than 0.5%) of the fully hydrolyzed product, scirpentriol, were also detected.

The hydrophilic trichothecenes, T-2 tetraol (TOL) and scirpentriol (SCP), were incubated with rat cecal microorganisms in order to determine whether esterification and overall lipophilicity were important factors in microbial deepoxidation. The majority of both TOL and SCP were reduced by the anaerobic microflora to yield their corresponding deepoxy analogs, DE TOL and DE SCP (Table 2).

Synthesis of DE T-2. Purity of the synthesized deepoxy T-2 was greater than 99% as determined by capillary gas chromatography of the corresponding TMS ether derivative. More importantly, no traces of T-2 were detected in the purified product.

Table 2. Metabolism of T-2 toxin, diacetoxyscirpenol, T-2 tetraol and scirpentriol *in vitro*, by anaerobic rat cecal microorganisms.

Toxin added	Products	% Recovered
T-2	T-2	0
	HT-2	0
	TRIOL	0
	DE HT-2	97.6±1.2
	DE TRIOL	2.4±1.2
DAS	DAS	0
	MAS	0
	SCP	0.3±0.2
	DE MAS	81.9±1.6
	DE SCP	17.8±1.7
TOL	TOL	12.7±2.4
	DE TOL	87.3±2.3
SCP	SCP	4.3±2.2
	DE SCP	95.7±2.3

[a]Mean percent of total metabolites recovered ± SEM (three replications).

Electron impact mass spectra of DE T-2, both by direct inlet and GC/MS of the TMS Ether derivative, showed extensive fragmentation, and no molecularons were observed. The direct inlet methane positive chemical ionization mass spectrum of DE T-2 (Fig. 4A, 4B, 4C) displayed a M+H pseudomolecularon at m/z 391, 349, 289 and 229. The positive chemical ionization mass spectrum of DE T-2 TMS ether derivative showed a pseudomolecular ion at m/z 523, with major fragments at m/z 507, 463, 421, 361 and 301. These fragments were 16 mass units less than the epoxy congener, T-2 toxin. By contrast, the negative chemical ionization mass spectrum of DE T-2 TMS derivative showed little fragmentation, with m/z 521 (M-H) the major fragment ion observed. These data confirmed that the molecular weight of DE T-2 was 450.

The proton NMR spectrum of DE T-2 was similar to that of T-2. However, instead of doublet resonances at 2.81 and 3.05 ppm due to the C-13 methylene protons of the epoxide group in T-2 toxin, singlet resonances were observed

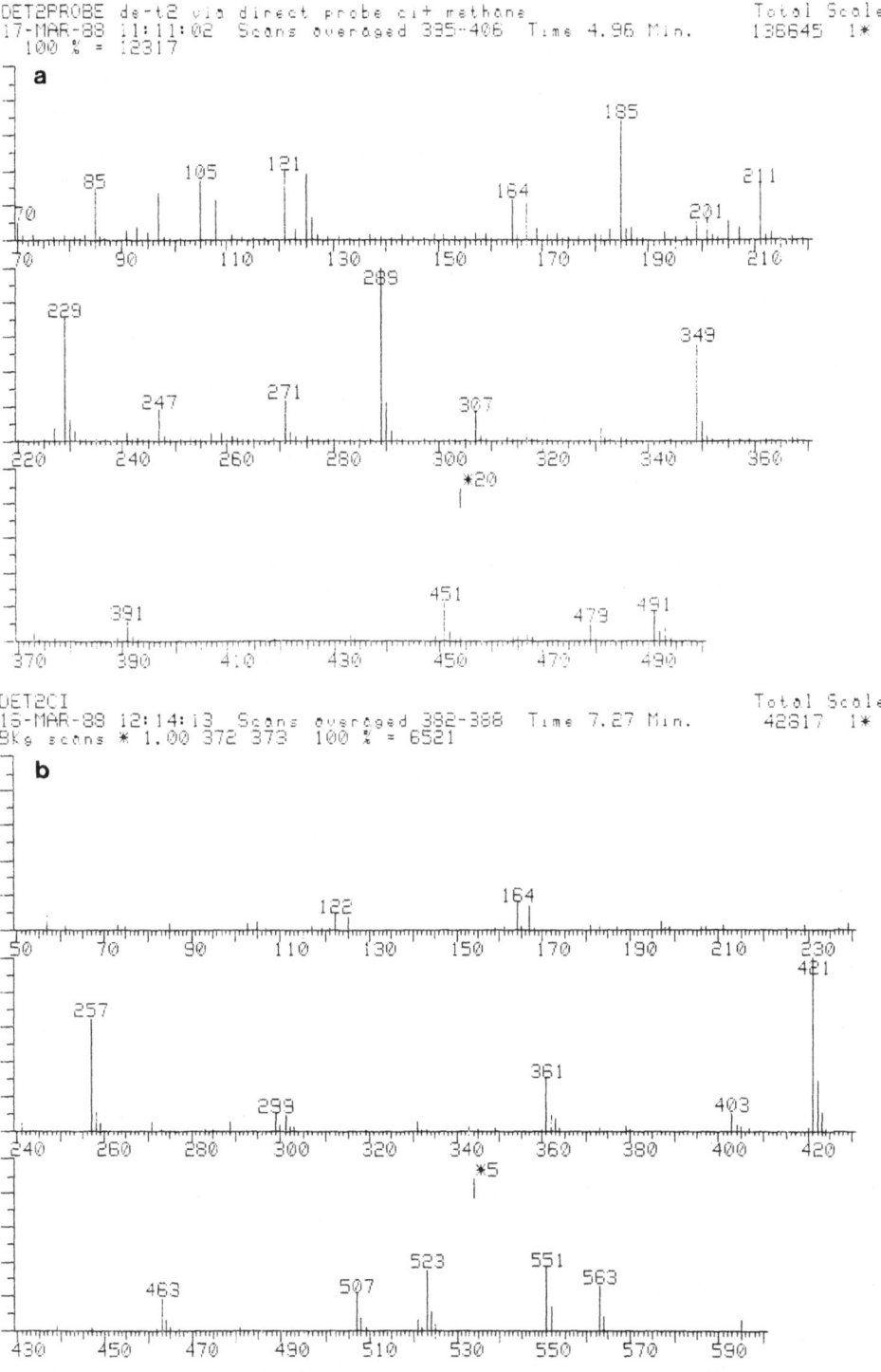

Fig. 4A and B.   Mass spectra of deepoxy T-2.
A) Direct inlet positive chemical ionization of
DE T-2.
B) Methane positive chemical ionization of
DE T-2 TMS ether derivative.

at 4.77 and 5.11 ppm in DE T-2. These singlets were assigned to the terminal
vinyl protons at the C-13 position.

Hydrolysis of DE T-2 with sodium hydroxide yielded synthetic DE TOL.
Proton NMR of synthetic DE TOL and biosynthetic DE TOL (derived from hydrolysis of DE HT-2 obtained from incubations of T-2 with rat cecal microflora)
were identical (data not shown). These data confirmed that both synthetic
and microbial deoxygenation of trichothecenes yield deepoxy products which
retain the same basic skeleton and stereochemical configuration.

Dermal irritation of deepoxy T-2. The dermal irritant toxicity of DE T-2
and T-2 toxin following topical exposure were compared in the rat skin bioassay. Results are shown in Table 3. Erythema was the only major response
observed for both test compounds, however, edema was observed at the highest
T-2 toxin dose (480 ng). DE T-2 induced a grossly detectable response only
at the highest dosage, 48,000 ng. At this dose, the reaction produced by DE
T-2 was equivalent to that caused by 120 ng of T-2 toxin (P < 0.05).

T-2 toxin caused moderate fibrinosuppurative exudation with local epidermal ulceration at a dosage of 480 ng. The intact epidermis was moderately
thickened and hyperplastic. Moderate edema and infiltration by neutrophils
were present in both the epidermis and dermis. Similar, but less severe
lesions were observed at a dosage of 120 ng, however, no epidermal ulceration
was seen. DE T-2 produced microscopically detectable lesions only at the
highest dosage, 48,000 ng, and at this dose the lesions were less severe than
those induced by 120 ng of T-2 toxin.

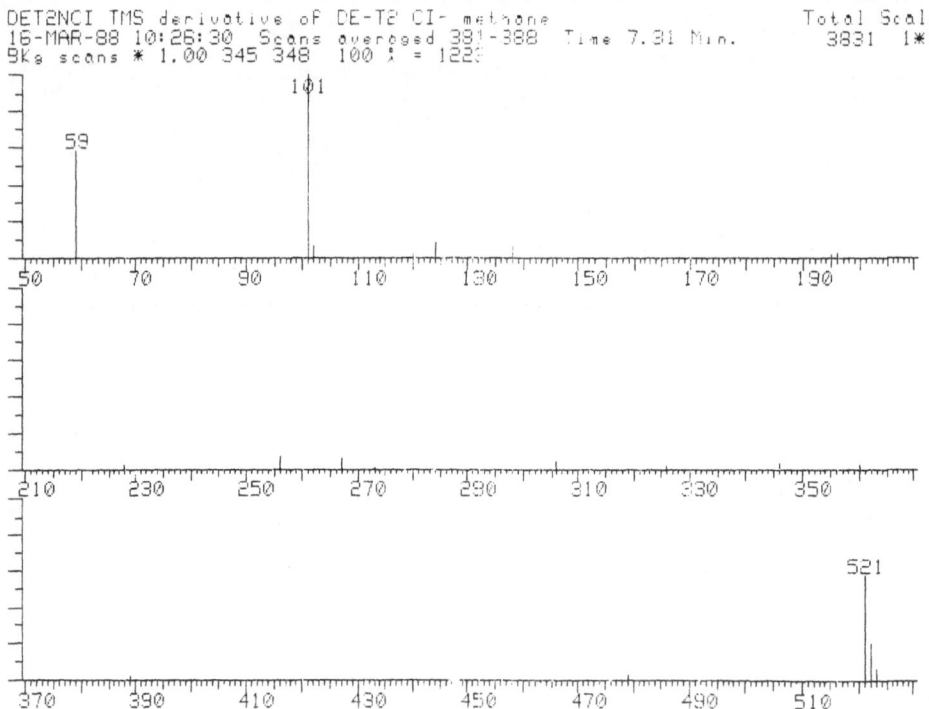

Fig. 4C.   Mass spectra of deepoxy DE T-2.   Methane negative
           chemical ionization of DE T-2 TMS ether derivative.

Intraperitoneal toxicity of deepoxy T-2. Neither lethality nor clinical signs of toxicity were observed in mice administered DE T-2 at either dosage. Seventy-one percent of the mice administered 12 mg/kg T-2 toxin died by 96 hours after toxin administration, and all mice displayed clinical signs consistent with T-2 intoxication (rough hair coats, diarrhea, decreased startle response and lethargy) (Table 4).

Effect of antibiotic therapy on T-2 induced toxicity. No deaths were observed in any of the vehicle control groups. The calculated LD_{50}'s (95% confidence interval) for the T-2 toxin plus antibiotics and the T-2 toxin only groups calculated by the moving average method were 12.53 (12.16-13.14) and 9.42 (9.15-9.72) mg/kg body weight, respectively. The LD_{50}'s (95% confidence interval) calculated by the trim-logit method were 13.09 (11.58-14.79) and 9.79 (8.17-11.74) for the T-2 toxin plus antibiotic and T-2 toxin alone groups, respectively. The differences in the two treatments were statistically significant at $P < 0.05$ using both the moving average and trim-logit methods for LD_{50} calculation. See Table 4 for details.

Table 3. Rat skin irritation caused by topical administration of T-2 toxin and its deepoxy analog.

Compound	Dose (ng)[b]	Score[b]
Control	0	0 ± 0
T-2	480	2.10 ± 0.22
T-2	120	0.90 ± 0.55
DE T-2	480	0 ± 0
DE T-2	4,800	0 ± 0
DE T-2	48,000	0.70 ± 0.67

[a]Three μl of each solution were applied to the shaved skin of the rats.
[b]Dermal irritation score for erythema; mean ± SD (N=5). 0 = no reaction, 4 = severe erythema.

Table 4. Percentage of mice surviving after administration of a single oral dose of T-2 toxin following combined antibiotic therapy.

Dose[a]	T-2	T-2 + AB[b]
0	100	100
6	83	100
8	83	100
10	67	100
12	0	67
14	0	33
16	16	0
Total	50	86
	(21/42)	(36/42)

[a]Dosage of T-2 toxin in mg/kg.
[b]Animals treated with combined antibiotic therapy. See Experimental section for details.

DISCUSSION

DE DAS was not detected as a metabolite after incubation of DAS with fecal microflora from any of the six species. In addition, neither DE DAS nor DE T-2 were detected as products following incubation of T-2 toxin or DAS with rat cecal microflora. The failure to detect direct deepoxidation products upon incubation of these two esterified trichothecenes with intestinal microflora is consistent with previous studies utilizing bovine rumen microflora as inoculum.[38] Presumably, the rapid hydroxysis of the C-4 esters present in DAS and T-2, coupled with the relatively slow rate of deepoxidation, precludes the production of DE T-2 or DAS as metabolites. However, steric hinderance of this reduction reaction by the C-4 acetyl groups cannot be completely ruled out. Future experiments, in which esterase inhibitors are added to the microbial incubation mixtures, may shed light on this question.

The rat cecal microflora were not able to hydrolyze the C-8 ester position in T-2 toxin to yield TOL or its deepoxy analog, DE TOL. T-2 tetraol and DE TOL are major metabolites detected in the urine and feces of cattle, guinea pigs and rats orally administered T-2 toxin.[2,24,32] The lack of C-8 ester hydrolysis by rat cecal microflora in this study confirms that products of ester cleavage at the C-8 position observed in while animal studies are not the result of metabolism by microflora inhabiting the gut, but rather are exclusively the products of metabolism by animal carboxyesterases.

The polar trichothecenes TOL and SCP were extensively biotransformed to their deepoxy analogs, DE TOL and DE SCP. Both TOL and SCP lack ester side groups and are water-soluble, polar metabolites. The fact that the epoxide group in these very polar trichothecenes was reduced to a carbon-carbon double bond, suggests ester side groups and lipid solubility are not important factors in deepoxidation reactions by anaerobic intestinal microflora.

The species variations in conversion of trichothecenes to deepoxy products as observed in this *in vitro* study correlate with previously published whole animal studies. Rats, cattle and swine administered trichothecene mycotoxins have previously been shown to excrete a significant portion of the dose as deepoxy metabolites. Rats orally administered DAS excrete DE MAS and DE SCP as the major products,[32] and rats given T-2 toxin orally[13,24] excrete a variety of deepoxy metabolites. Rats also are able to reduce the epoxide group of DON to its deepoxy analog, DOM-1, *in vivo*.[19,47]

Cattle orally administered either T-2 toxin or DON excrete deepoxy metabolites as major excretion products in both the feces and urine.[2,3,10,45] Previous work demonstrated that bovine rumen microflora have the capacity to reduce the epoxide group of several trichothecenes, including DAS, DON and T-2 toxin.[17,38] The present study demonstrates that epoxide reduction can occur not only through the action of anaerobic bovine rumen microflora, but also through the action of anaerobic intestinal microflora present in the large intestine of cattle.

Fecal microorganisms from chickens, dogs and horses displayed no capacity for epoxide reduction in the present study. Although data is lacking for horses, several *in vivo* studies with T-2 in dogs[35] and both T-2 toxin and DON in poultry[18,42] have been conducted. Deepoxy metabolites of trichothecenes were not detected in any of the whole-animal dog or poultry studies reported to date, supporting the *in vitro* data presented in the present study that both chickens and dogs lack the prerequisite intestinal microflora necessary for epoxide reduction.

Although swine fecal microflora yielded predominantly deepoxy metabolites of T-2 toxin in the present study, only minor amounts of deepoxy metabolites were detected as excretion products following intravascular administration of tritium-labeled T-2 toxin to swine.[9] However, swine orally administered DON did not excrete detectable quantities of deepoxy DON.[27] The discrepancy in the ability of swine to perform trichothecene deepoxidation reactions in the *in vitro* studies presented here and previously reported *in vivo* data has yet to be resolved and further work is needed.

Both T-2 toxin and diacetoxyscirpenol are potent dermal irritants and dermal irritation has been used as a bioassay for the detection of trichothecenes.[14] Histopathologic examination of the skin following topical application of the toxins was more sensitive compared to gross evaluation. Histologically, 48,000 ng DE T-2 produced a less severe lesion than 120 ng T-2 toxin, demonstrating that deepoxy T-2 is at least 400-times less toxic than T-2 toxin in terms of dermal irritation.

No lethality was observed after IP administration of DE T-2 to mice at dosages up to 60 mg/kg, approximately six times the LD_{50} of T-2 toxin. In addition, no adverse clinical signs were noted in any of the mice administered DE T-2. Information on the toxicity of deepoxy trichothecenes is limited. We recently demonstrated that the metabolites DE HT-2 and DE MAS were at least 50-fold less toxic to brine shrimp than their corresponding epoxy analogs, HT-2 and MAS.[39] These data confirm that the epoxide group in trichothecene mycotoxins is essential for toxicity and demonstrate that reduction of the epoxide group by anaerobic intestinal microflora to yield a carbon-carbon double bond is an effective, single-step detoxification reaction.

The dramatic decrease in toxicity following reduction of the epoxide group in T-2 toxin suggests that species with intestinal microflora containing high deepoxidation activity (rats, swine and cattle) should display reduced sensitivity to the trichothecenes, compared to species lacking microflora with deepoxidation activity (horses, chickens and dogs). This relationship, however, is not supported by existing acute toxicity data. The oral LD_{50} of T-2 toxin in rats (4-5.3 ng/kg),[20,34] a species displaying high deepoxidation activity, is not appreciably different than the LD_{50} values of 3.6 to 4.9 mg/kg for chickens, a species displaying no deepoxidation activity.[4,34] The LD_{50} of T-2 toxin in swine is the lowest of all species examined, 1.2 mg/kg,[39] yet in the present study swine fecal microflora displayed relatively high deepoxidation activity with DAS. If deepoxidation were the sole detoxification pathway involved, one would expect species with microflora capable of deepoxidation to be significantly less sensitive to the trichothecene toxins and display correspondingly higher LD_{50} values.

The metabolism of trichothecenes in whole animals is very complex, involving up to 4 different competing pathways. T-2 toxin undergoes the most complex biotransformation of the major trichothecenes. All four pathways simultaneously compete in several animal species, rendering a simple explanation of the importance of any single pathway impossible. To date 29 different metabolites have been identified by a variety of researchers (Table 5).

Deepoxidation is a detoxification reaction accomplished via the action of anaerobic intestinal microflora in a number of animal species. Antibiotic therapy should result in decreased populations of intestinal microflora and an overall reduction in the level of this detoxification reaction. As a result one would expect to see an increase in toxicity if deepoxidation (a detoxification reaction) were the predominant biotransformation pathway involved in toxicity of T-2 toxin. This is in contrast to what was observed in the present study.

Alterations in biotransformation as a reslt of antibiotic therapy may be at least partially responsible for the observed effect. One of the most

Table 5.   Identified metabolites of T-2 toxin.

HT-2	Deepoxy HT-2
T-2 triol	Deepoxy T-2 triol
Neosolaniol	Deepoxy T-2 tetraol
4-Deacetylneosolaniol	Deepoxy 4-deacetylneo-solaniol
T-2 tetraol	Deepoxy 3'OH HT-2
3'OH T-2	Deepoxy 3'OH T-2 triol
3'OH HT-2	T-2 glucuronide
3'OH T-2 triol	HT-2 glucuronide
4'OH T-2	Neosolaniol glucuronide
4'OH HT-2	4-Deacetylneosolaniol glucuronide
3-acetoxy-3'OH HT-2	3'OH T-2 glucuronide
8-acetoxy tetraol	3'OH HT-2 glucuronide
3'OH-7-OH HT-2	T-2 tetraol glucuronide
4'acetoxy T-2 tetraol	
Acetyl T-2	
Acetyl HT-2	

pronounced effects of antibiotic therapy, in general, is the reduction of overall microflora present in the gastrointestinal tract and alterations in the species of bacteria populating the intestines. Alterations in intestinal bacteria would most likely affect two major biotransformation pathways, deepoxidation and glucuronide hydrolysis. Glucuronide conjugation has been demonstrated to be a major pathway of metabolism for trichothecenes in several animal species, either with the parent compound directly or a Phase I metabolite. Glucuronides of T-2 toxin and/or metabolites have been detected as biotransformation products in swine,[8,9] guinea pigs[23] and rats.[13,31] Glucuronides of MAS and/or scirpentriol have been reported as major metabolites of DAS in rats[13,30] and mice[30] and DON glucuronides have been found along with its deepoxy metabolite, DOM-1, as elimination products in cattle,[10,45] sheep[26] and rats.[19] Most animal species investigated to date conjugate trichothecenes or their phase I metabolites, at least to a limited extent.

It is well documented that intestinal microflora produce beta-glucuronidase within the gut and that this enzyme is responsible for the hydrolysis of many different xenobiotic conjugates secreted into the intestines via the bile.[36] Specific examples of compounds shown to undergo enterohepatic recirculation include morphine and the estrogens (estriol, estradiol and estrone).[12,28] Hydrolysis of xenobiotic glucuronide conjugates by bacterial β-glucuronidase in the intestinal tract causes liberation of the aglycone. If the aglycone is sufficiently lipophilic, reabsorption may take place resulting in enterohepatic recirculation. In instances where the xenobiotic in question is a toxic compound, disruption of the enterohepatic cycle can reduce toxicity by increasing the rate of elimination.

T-2 toxin is excreted in the bile to a significant extent as the glucuronide conjugate and as glucuronide conjugates of related toxic metabolites such as HT-2 and 3'OH T-2. T-2 toxin and metabolites undergo enterohepatic recirculation in the rat.[5] Antibiotic therapy decreased toxicity in the present study. Therefore, deepoxidation most likely plays only a minor role in the overall toxicity of T-2 toxin. Reduction in the enterohepatic recirculation of toxic metabolites and consequently, enhanced rates of elimination may be the main mechanism by which antibiotic therapy reduces toxicity of orally administered T-2 toxin. Additional studies comparing metabolite excretion profiles in animals exposed to T-2 toxin with and without antibiotic therapy are needed to determine which reaction (deepoxidation or conjugation) is more important in the overall metabolism and toxicity of trichothecenes.

REFERENCES

1. Beasley VR, Swanson SP, Corley RA, Buck WB, Koritz GD, Burmeister HR, 1986, Pharmacokinetics of the trichothecene mycotoxin, T-2 toxin, in swine and cattle, *Toxicon*, 24:13-23.
2. Chatterjee K, Pawolsky R, Treeful L, Mirocha CJ, 1986, Kinetic study of T-2 toxin metabolites in a cow, *J Food Safety*, 8:25-34.
3. Chatterjee K, Visconti A, Mirocha CJ, 1986, Deepoxy T-2 tetraol: a metabolite of T-2 toxin in cow urine, *J Agric Fd Chem.*, 34:695-697.
4. Chi MS, Robison TS, Mirocha CJ, Reddy KR, 1978, Acute toxicity of 12,13-epoxytrichothecenes in one-day-old broiler chicks, *Appl Environ Microbiol.*, 35:636-640.
5. Coddington K, Swanson SP, Hassan AS, Buck WB, 1989, Enterohepatic circulation of T-2 toxin metabolites in the rat, Submitted for publication.
6. Colvin EW, Cameroon S, 1986, Chemical deoxygenation of the trichothecenes, diacetoxyscirpenol and deoxynivalenol, *J Chem Soc Chem Comm.*, 467:1084-1085.
7. Coppock RW, Swanson SP, Gelberg HB, Koritz GD, Buck WB, Hoffman WB, 1987, Pharmacokinetics of diacetoxyscirpenol in cattle and swine: effects of halothane, *Am J Vet Res.*, 48:691-695.
8. Corley RA, Swanson SP, Buck WB, 1985, Glucuronide conjugates of T-2 toxin and metabolites in swine bile and urine, *J Agric Fd Chem.*, 33:1085-1089.
9. Corley RA, Swanson SP, Gullo G, Johnson L, Bealsey VR, Buck WB, 1986, Disposition of T-2 toxin, a trichothecene mycotoxin, in intravascularly dosed swine, J Agric Fd Chem., 34:868-875.
10. Cote LM, Dahlem AM, Yoshizawa T, Swanson SP, Buck WB, 1986, Excretion of deoxynivalenol and its metabolite, DOM-1, in milk, urine and feces of lactating dairy cattle, *J Dairy Sci.*, 69:2416-2423.
11. Cote LM, Reynolds JD, Vesonder RF, Buck WB, Swanson SP, Coffey RT, Brown DC, 1984, Survey of vomitoxin-contaminated feed grains in mid-western United States, and associated health problems in swine, *J Am Vet Med Assoc.*, 2:189-192.
12. Dutton GJ, 1980, "Glucuronidation of Drugs and Other Compounds," CRC Press Inc., Boca Raton, FL.
13. Gareiss M, Hashem A, Bauer J, Gedek B, 1986, Identification of glucuronide metabolites of T-2 toxin and diacetoxyscirpenol in the bile of isolated perfused rat liver, *Toxicol Appl Pharmacol.*, 84:168-172.
14. Hayes MA, Shiefer HB, 1979, Quantitative and morphological aspects of cutaneous irritation by trichothecene mycotoxins, *Fd Cosmet Toxicol.*, 17:611-621.
15. Hoerr FJ, Carlton WW, Yagen B, 1981, Mycotoxicosis caused by a single dose of T-2 toxin or DAS in broiler chicks, *Vet Path.*, 18:652-664.
16. Knupp CA, Swanson SP, Buck WB, 1987, Comparative *in vitro* metabolism of T-2 toxin by hepatic microsomes prepared from phenobarbital-induced or control rats, mice, rabbits and chickens. *Fd Chem Toxicol.*, 25:859-865.
17. King RR, McQueen RE, Levesque D, Greenhalgh R, 1984, Transformation of deoxynivalenol (vomitoxin) by rumen microorganisms, *J Agric Fd Chem.*, 32:1181-1183.
18. Kubena LF, Swanson SP, Harvey RB, Fletcher OJ, Rowe LD, Phillips TD, 1985, Effects of feeding deoxynivalenol (vomitoxin)-contaminated wheat to growing chicks, *J Poultry Sci.*, 64:1649-1655.
19. Lake BG, Phillips JC, Walters DG, Bayley DL, Cook MW, Thomas LV, 1987, Studies on the metabolism of deoxynivalenol in the rat, *Fd Chem Toxicol.*, 25:589-592.

20. Marasas WFO, Bamburg JR, Smalley EB, Strong FM, Ragland WL, Degurse PE, 1969, Toxic effects on trout, rats, and mice of T-2 toxin produced by the fungus *Fusarium tricinctum* (CD.), Snyd. et Hans., *Toxicol Appl Pharmacol.*, 15:471-482.

21. Mirocha CJ, Pathre SV, Christiansen CM, 1977, Chemistry of Fusarium and Stachbotrys mycotoxins, *In*: "Mycotoxic Fungi, Mycotoxins and Myco-toxicoses," Wyllie TD, Morehouse LG, eds., Marcel Dekker, Inc., New York.

22. Obara T, Matsuda E, Takemoto T, Tatsuno T, 1984, Immunosuppressive effect of a trichothecene mycotoxin, Fusarenone-X, *In*: "Developments in Food Science. Toxigenic Fungi-Their Toxins and Health Hazard," Kurata H, Ueno Y, eds., Elsevier/North-Holland Publishing Co., New York.

23. Pace JG, Watts MR, Burrows EP, Dinterman RE, Matson E, Hauer EC, Wannemacher RW, 1985, Fate and distribution of [^{3}H]-labeled T-2 mycotoxin in guinea pigs, *Toxicol Appl Pharmacol.*, 80:377-385.

24. Pfeiffer R, Swanson SP, Buck WB, 1988, Metabolism of T-2 toxin in rats: effects of dose, route and time, *J Agric Fd Chem.*, 36:1227-1232.

25. Prelusky DB, Trenholm HL, Lawrence GA, Scott PM, 1984, Nontransmission of deoxynivalenol (vomitoxin) to milk following oral administration to dairy cows, *J Environ Sci Health*, B19:593-609.

26. Prelusky D, Hamilton RMG, Trenholm HL, Miller JD, 1986, Excretion pro-es of the mycotoxin deoxynivalenol, following oral and intravenous admin-ration to sheep, *Fund Appl Toxicol.*, 6:645-655.

27. Prelusky D, Hartin KE, Trenholm HL, Miller JD, 1988, Pharmacokinetic fate of [^{14}C]-labeled deoxynivalneol in swine, *Fund Appl Toxicol.*, 10:276-286.

28. Renwick AG, 1986, Gut bacteria and the enterohepatic circulation of foreig compounds, *In*: "Microbial Metabolism in the Digestive Tract," Hill MJ, ed. CRC Press, Boca Raton, FL.

29. Rosenstein Y, Lafarge-Frayssinet C, Lespinats G, Loisillier F, Lafont P, Frayssinet C, 1979, Immunosuppressive activity of Fusarium toxins: Effect on antibody synthesis and skin grafts of crude extracts, T-2 toxin and diacetoxyscirpenol, *Immunology*, 36:111-117.

30. Rousch WR, Marletta MA, Rodriguez SR, Recchia J, 1985a, Trichothecene metabolism studies: Isolation and structure determination of 15-acetyl-3(1'-glucopyranosidurohyl)-scirpen-3,4,5-triol, *J Am Chem Soc.*, 107:3354-3355.

31. Rousch WR, Marletta MA, Rodriguez SR, Recchia J, 1985b, Trichothecene metabolism studies: structure of 3α-(1''β-D-glucopyranosiduronyl)-8α-isovaleryloxy-scirpen-3,4β,15-triol 15-acetate produced from T-2 toxin *in vitro*, *Tetrahedron Let.*, 26:5231-5234.

32. Sakamoto T, Swanson SP, Yoshizawa T, Buck WB, 1986, Structures of new metabolites of diacetoxyscirpenol in the excreta of orally administered rats, *J Agric Fd Chem.*, 34:698-701.

33. Sanathana LP, Gaade ET, Shipkourtz NL, 1987, Trimmed logit method for estimating the ED_{50} in quantal bioassay, *Biometrics* 43:225-232.

34. Sato N, Ueno Y, 1977, Comparative toxicities of trichothecenes, *In*: "Mycotoxins in Human and Animal Health," Rodricks JV, Hesseltine CW, Mehlman MA, eds., Pathotox Publishers, Park Forest South, IL.

35. Sintov A, Bialer M, Yagen B, 1986, Pharmacokinetics of T-2 toxin and its metabolite HT-2 toxin, after intravenous administration in dogs, *Drug Metab Disp.*, 14:250-254.

36. Sipes, IG and Gandlofi AJ, 1986, Biotransformation of toxicants, *In*: "Caserett and Doull's Toxicology: The Basic Science of Poisons," Klaassen DC, Amdur MO and Doull J, eds., Macmillan Pupl. Co., New York.

37. Sprague JB, Fogels A, 1974, Proceedings of the 3rd Aquatic Toxicity Workshop, "EPA Technical Report No. EPS-5-AR-77-1," Halifax, Canada.

38. Swanson SP, Nicoletti J, Rood HD, Buck WB, Cote LM, Yoshizawa T, 1987a, Metabolism of three trichothecenes T-2 toxin, diacetoxscirpenol and deoxynivalenol, by bovine rumen microorganisms, *J Chromatogr Biomed Appl.*, 414:335-342.

39. Swanson SP, Rood HD, Behrens JC, Sanders PE, 1987b, Preparation and

characterization of the deepoxy trichothecenes: deepoxy HT-2, deepoxy T-2 triol, deepoxy T-2 tetraol, deepoxy monoacetoxyscirpeneol and deepoxy scirpentriol, *Appl Environ Microbiol.*, 53:2821-2826.

40. Swanson SP, Helaszek C, Buck WB, Rood HD, Haschek WM, 1988, The role of intestinal microflora in the metabolism of trichothecene mycotoxins, *Fd Chem Toxicol.*, 26:823-829.

41. Visconti A, Treeful L, Mirocha CJ, 1985a, Identification of ISO-TC-1 as a new T-2 toxin metabolite in cow urine, *Biomed Mass Spectr.*, 12:689-694.

42. Visconti A, Mirocha CJ, 1985b, Identification of various T-2 toxin metabolites in chicken excreta, *Appl Environ Microbiol.*, 49:1246-1250.

43. Weaver GA, Kurtz HJ, Bates FY, Chi MS, Mirocha CJ, Behrens JC, Robison TS, 1978, Acute and chronic toxicity of T-2 mycotoxin in swine, *Vet Rec.*, 103:531-535.

44. Weil CS, 1952, Tables for convenient calculation of median effective dose (LD_{50} or ED_{50}) and instructions for their use, *Biometrics*, 8:249-263.

45. Yoshizawa T, Cote LM, Swanson SP, Buck WB, 1986, Confirmation of DOM-1, a de-expoydation metabolite of deoxynivalenol in biological fluids of lactating cows, *J Agric Biol Chem.*, 50:227-229.

46. Yoshizawa T, Sakamoto T, Kumamura K, 1985, Structures of deepoxy trichothecene metabolites of 3'hydroxy HT-2 and T-2 tetraol in rats, *Appl Environ Microbiol.*, 50:676-679.

47. Yoshizawa T, Takeda H, Ohi T, 1983, Structure of a novel metabolite of deoxynivalenol, a trichothecene mycotoxin, in animals, *Agric Biol Chem.*, 47:2133-2135.

DEREGULATION OF C-MYC GENE AND MODULATION OF GLUCOCORTICOID RECEPTOR BY

AFLATOXIN B_1

Fumio Tashiro,[a] Shigeru Morimura,[b] Nobuo Horikoshi [b]
Kazuko Kato[b] and Yoshio Ueno[b]

Dept. of Biological Science & Technology
Faculty of Industrial Science & Technology
Science University of Tokyo, 2641, Yamazaki, Noda-Shi
Chiba 278, Japan[a]

Dept. of Toxicology & Microbial Chemistry
Faculty of Pharmaceutical Sciences
Science University of Tokyo, Ichigaya, Shinjuku-ku
Tokyo 162, Japan[b]

INTRODUCTION

Aflatoxin B_1 (AFB$_1$), a metabolite of *Aspergillus flavus*, is well known for its potency in hepatocarcinogenesis in experimental animals.[19] AFB$_1$ as well as hepatitis B virus has been implicated in the etiology of human liver cancer, on the basis of significant statistical correlation between the increased occurrence of liver cancer in areas of Asia and Africa and the consumption of grains contaminated with AFB$_1$.[20]

We have demonstrated the over-expression of c-myc and c-Ha-ras mRNAs in AFB$_1$-induced rat hepatocellular carcinomas and cultured cell lines.[17,18] Subsequently, the activation of c-ras oncogene family by point mutation in AFB$_1$-induced rat liver has been reported by McMahon et al.[7,8] and Shinha et al.[16] Most recently we have also demonstrated the modulation of hormonal induction of tyrosine aminotransferase (TAT) and glucocorticoid receptors (GR) by AFB$_1$ and sterigmatocystin (STC), a dehydrobisfuranoid mycotoxin similar to AFB$_1$, in a rat hepatoma cell line.[4] Here we summarize the recent results from our experiments and outline future directions of research needed in this area.

Activation of Oncogenes in AFB$_1$-induced Rat Hepatomas

Rat liver tumors were induced according to Wogan et al.[19] Male 5-week-old Fisher rats were used in this study. AFB$_1$, dissolved in dimethyl sulfoxide, was given to rats by stomach tube at a total dose of 1.5 mg/rat in 40 equal doses over an 8 week period. After 14 to 15 months, fully developed liver tumors were observed in 8 of 9 rats. Histological examination indicated that the tumors were well-differentiated hepatomas.[17]

Expression of cellular-oncogenes in seven hepatocellular carcinomas

Microbial Toxins in Foods and Feeds, Edited by A.E. Pohland *et al.,*
Plenum Press, New York, 1990

from AFB$_1$-treated rats was examined. Both c-myc and c-Ha-ras transcripts were highly elevated in all hepatomas. In one tumor (TS-1), the c-myc gene was amplified without significant rearrangement. This amplified c-myc gene was specifically localized in a tumor portion of the liver, but not in a normal section of the liver, kidney, lung, and spleen. Amplification and rearrangement of c-ras gene were not detected in these hepatomas.[18] N-ras mRNA were detected as faint bands about 3 and 5 Kb in the control liver and all hepatomas. However, no increase in the expression of N-ras was observed in any tumors examined.

The expression of c-fos and albumin genes in two tumors, T1-2 and T2-1, was also examined. The transcript of c-fos was not detected in either tumor or control liver. Albumin gene, which was actively transcribed in normal adult liver, was expressed in both tumors, but at a reduced level.

Our data are in agreement with the conclusions reached by other groups.[1,6,22] They suggested that the elevated expression of c-myc and c-Ha-ras genes, which belong to different complimentation groups, are associated with rodent hepatocarcinogenesis. McMahon et al.[7,8] and Sinha et al.[16] have previously reported the point mutation of ras gene family in AFB$_1$-induced rat liver tumors. Taking into account these reports and our results, it is conceivable that the c-myc and c-ras oncogene families play an important role in AFB$_1$ mediated rodent hepatocarcinogenesis.

Characterization of Hepatoma Cell Lines Established from AFB$_1$-treated Rats

To elucidate the detailed biological characteristics of AFB$_1$-induced liver tumors, we established two cell lines, Kagura-1 and Kagura-2, from T2-1 and T1-2 tumors, respectively.[10] Both cell lines formed colonies in 0.3% agar and tumors in nude mice. The expression of c-myc gene in these cells was more active than that of the primary tumors, except for the C-Ha ras gene. On the otherhand, the expression of the albumin gene was not detected in these cell lines (Table 1). Interestingly the active expression of α-fetoprotein gene, which was not detected in normal liver and AFB$_1$-induced tumors, was observed in Kagura-1 cells. Activity of one of the liver specific enzymes, TAT, was considerably lower in these cells when compared to that of H4-II-E cells, which are well known as one of the minimum deviation cell lines (Table 1). Moreover, enzyme activity was not induced by the addition of 1×10^{-6} M dexamethasone.

As mentioned above, Kagura-1 and Kagura-2 cells actively expressed the c-myc gene but lost hepatocyte-specific characters such as the expression of albumin and TAT genes, and the response to glucocorticoid hormone. The reciprocal relationship between c-myc and albumin gene expression may provide important information for the determination of the malignant potential of liver cells.

Effects of AFB$_1$ and STC on the Hormonal Induction of TAT and GR in H4-II-E cells

It is well known that gene expression, cellular differentiation, and proliferation of tumors are regulated by many cellular factors including oncogene products and hormones. We therefore examined the effects of AFB$_1$ and STC on the cellular regulatory system in H4-II-E cells, which possess many characteristics of mature liver cells.[13]

Our basic approach to the induction of liver specific enzymes revealed that TAT activity was induced by various hormones such as hydrocortisone (HC), insulin, and dibutyryl cyclic AMP (Bt$_2$cAMP). This induction was dependent on the concentrations of hormones added and the exposure times. AFB$_1$ markedly inhibited HC-induced TAT activity with an IC$_{50}$ estimated

Table 1. Liver specific functions of rat hepatoma cell line

Cells	Albumin		TAT	
			Enzyme Activity[b] (nmol products/min/mg protein)	
	mRNA[a]	mRNA[a]	Basal	Induced[c]
Normal Liver	++[d]	+	N.T.	N.T.
H4-II-E[e]	+	+	13.9 ± 3.5	45.5 ± 9.6
Kagura-1	-	-	1.5 ± 0.4	2.0 ± 0.4
Kagura-2	-	-	1.9 ± 0.5	1.5 ± 0.4

[a]Total RNA was extracted by guanidium thiocyanate/CsCl method and
 analyzed by Northern blot. Inserted fragments of prAlbI[5] and
 pcTAT3[15] were used as probes for albumin and TAT, respectively.
[b]TAT activity was assayed according to method of Granner and
 Tomkins[2]. Mean$\pm$S.D. from 4 experiments. N.T.=not tested.
[c]Cells were treated for 6 h with 1×10^{-6} M dexamethasone.
[d]Relative ratio of mRNA: + = <x 10; ++ = >x 10.
[e]H4-II-E cells derived from the rat H35 Reuber hepatoma were
 obtained from the American Type Culture Collection (Rockville, MD).

at 0.2 μg/ml. Insulin- and Bt$_2$AMP-dependent inductions were less sensitive to the effects of 5 μg/ml of AFB$_1$ with about 20 and 40% inhibition
observed, respectively. The inhibitory effect of AFB$_1$ on HC-induced TAT
activity, which accompanied the net synthesis of mRNA, was different from
that of Bt$_2$AMP-dependent induction of TAT, which is also known to accompany
the synthesis of mRNA.[3] Our preliminary experiments indicated that AFB$_1$
decreased the HC- and BT$_2$cAMP-dependent induction of TAT mRNA levels in
H4-II-E cells (Fig. 1). Therefore, it is conceivable that the effects of
AFB$_1$ on the hormonal induction of enzymes were multiple, and that the
inhibition observed as not simply limited to the activity of template DNA[21]
and RNA polymerase.[23]

   As for STC, inhibition of HC-dependent induction of TAT activity was dose
dependent with the IC$_{50}$ of 3.5 μg/ml, or about 18 times higher than that
of AFB$_1$. No significant depressions were observed in the insulin- and
Bt$_2$cAMP-inducible TAT activities. The data clearly showed that the mode of
action of STC was markedly different from that of AFB$_1$ in the hormonal induction of TAT activity in H4-II-E cells.

   Based on the observation that AFB$_1$ interfered with HC-dependent induction of TAT, we supposed that there were very sensitive molecules for AFB$_1$
in the cascade of TAT induction by HC. The detailed mechanism of AFB$_1$-
induced disturbance of HC-inducible TAT activity was then investigated.

   Glucocorticoid receptor-hormone complelxes (GRCs) of H4-II-E cells
treated with 5 μg/ml AFB$_1$ for 6 h were analyzed by sucrose density gradient centrifugation. The S values of nonactivated and activated forms of the
cytosolic GRC were estimated to be 6.5 and 4.6, respectively. AFB$_1$ induced
no significant changes in these S values, but the number of GRCs was reduced
to 65% of the control. Further experiments with cycloheximide revealed that
the AFB$_1$-induced reduction of cytosolic GRCs was not affected by 2.5 μg/ml
cycloheximide, a level which inhibited protein synthesis by more than 98% in
the control. The decrease therefore was caused by the reduction of the preexisting glucocorticoid receptor (GR) and not by an inhibition of new

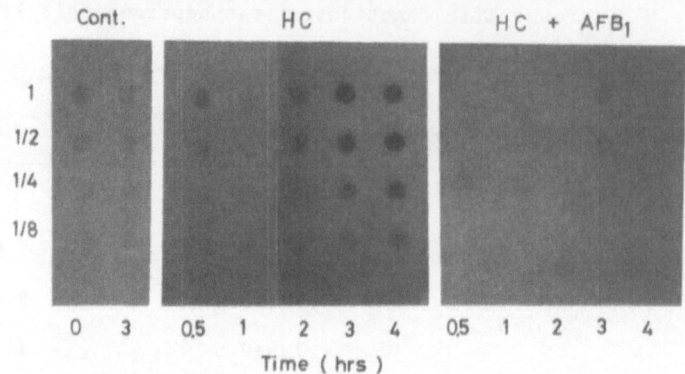

Fig. 1. Inhibition of HC-induced TAT mRNA by AFB$_1$.
The cells, prior treated with or without 5 µg/ml
of AFB$_1$ for 6 h, were cultured in the absence or
presence of 1 x 10^{-7} M HC for desired times
indicated, and the expression of TAT gene was
analyzed by Quick Blot method.

synthesis of GR. Miesfeld et al.[9] suggested that GR number in the cytosol
was important in glucocorticoid induction, and that the induced level of TAT
was increased in cytosolic GR in a number dependent manner.

Although STC possesses comparable cytotoxicity and mutagenicity to
AFB$_1$,[12] it is far less active in hepatocarcinogenicity[14] because no pronounced effect on the cytosolic GR was observed in H4-II-E cells treated with
5 µg/ml of STC for 6 h. The nuclear acceptor sites and the Kd values were
reduced about 67 and 55% of the control, respectively. Therefore, the mode
of action of STC is presumed to be different from that of AFB$_1$ with regard
to TAT induction and the effect on GR.

DISCUSSION

The experiments described here support the view that the deregulation of
c-myc gene expression plays an essential role in AFB$_1$-induced rat hepatocarcinogenesis in addition to the role of point mutation of c-ras gene family
reported by McMahon et al.[7,8] and Sinha et al.[16]

We have also demonstrated a marked difference between the potent hepatocarcinogen, AFB$_1$, and the far less potent carcinogen, STC, in regard to
modulation of GRs in cultured hepatoma cells. It is well known that AFB$_1$ and
other carcinogens exert a modifying influence on the physiological actions of
steroids in vivo, and their carcinogenic potentials are also modified after
hypophysectomy and endogenous factor.[11]

We hope that future experiments will clarify the precise hormonal action
on AFB$_1$-induced hepatocarcinogenesis.

ACKNOWLEDGEMENTS

We wish to thank Drs. S. M. Tilghman and G. Schutz for kindly providing prAlbI and pcTAT3, respectively. This work was supported in part by a Grant-in-Aid for Cancer Research from the Ministry of Education, Science and Culture of Japan.

REFERENCES

1. Corcos D, Defer N, Raymandjean M, Paris B, Corral M, Tichonnicky L, Kruth J, Glaise D, Saulnier A, Guguen-Guillouze C, 1984, Correlated increase of the expression of the c-ras genes in chemically induced hepatocarcinomas, *Biochem Biophys Res Commun.*, 122:259-264.
2. Granner DK, Tomkins GM, 1970, Tyrosine aminotransferase (rat liver), *In*: "Methods in Enzymology, Vol. 17A", Tabor H and Tabor CW, eds., Academic Press, London.
3. Hashimoto S, Schmid W, Schutz G, 1984, Transcriptional activation of rat liver tyrosine aminotransferase gene by cAMP, *Proc Natl Acad Sci USA*, 81:6637-6641.
4. Horikoshi N, Tashiro F, Tanaka N, Ueno Y, 1988, Modulation of hormonal induction of tyrosine aminotransferase and glucocorticoid receptors by aflatoxin B_1 and sterigmatocystin in Reuber hepatoma cells, *Cancer Res.*, 48:5188-5192.
5. Kioussis D, Hamilton R, Hanson RW, Tilghman SM, Taylor JM, 1979, Construction and cloning of rat albumin structural gene sequences, *Proc Natl Acad Sci USA*, 76:4370-4374.
6. Makino R, Hayashi K, Sato S, Sugimura T, 1984, Expressions of the c-Ha-ras and c-myc genes in rat aliver tumors, *Biochem Biophys Res Commun.*, 119:1096-1102.
7. McMahon G, Hanson L, Lee JJ, Wogan GN, 1986, Identification of an activated c-Ki-ras oncogene in rat liver tumors induced by aflatoxin B_1, *Proc Natl Acad Sci USA*, 83:9418-9422.
8. McMahon G, Davis E, Wogan GN, 1987, Characterization of C-Ki-ras oncogenes alles by direct sequencing of enzymatically amplified DNA from carcinogen-induced tumors, *Proc Natl Acad Sci USA*, 84:4974-4978.
9. Miesfeld R, Rusconi S, Godwski PJ, Maler BA, Okret S, Wikstrom AC, Gustafessen J-A, Yamamoto KR, 1987, Genetic complementation of a glucocorticoid receptor deficiency by expression of cloned receptor cDNA, *Cell*, 46:389-399.
10. Morimura S, Tashiro F, Ueno Y, 1988, Establishment and characterization of rat hepatoma cell lines (Kagura-1 and Kagura-2) induced by aflatoxin B_1 (Submitted).
11. Neal GR, Judah DJ, 1978, Effect of hypophysectomy and aflatoxin B_1 on rat liver, *Cancer Res.*, 38:3460-3467.
12. Noda K, Umeda M, Ueno Y, 1981, Cytotoxic and mutagenic effects of sterigmatocystin on cultured Chinese hamster cells, *Carcinogenesis (London)*, 2:945-949.
13. Pitot HC, Peraino C, Morse PA, Potter VA, 1964, Hepatoma in tissue culture compared with adapting liver *in vivo*, *Natl Cancer Inst Monogr.*, 13:229-245.
14. Purchase IFH, van der Watt JJ, 1969, Acute toxicity of sterigmatocytin to rats, *Fd Cosmet Toxicol.*, 7:135-139.
15. Scherer G, Schmid W, Strange CM, Rowekamp W, Schutz G, 1982, Isolation of cDNA clones coding for rat tyrosine aminotransferase, *Proc Natl Acad Sci USA*, 79:7205-7208.
16. Sinha S, Webber C, Marshall CJ, Knowles MA, Proctor A, Barrass NC, Neal GE, 1988, Activation of ras oncogene in aflatoxin-induced rat liver carcinogenesis, *Proc Natl Acad Sci USA*, 85:3673-3677.
17. Tashiro F, Morimura S, Hayashi K, Makino R, Kawamura H, Horikoshi N, Nemoto K, Ohtsubo K, Sugimura T, Ueno Y, 1986, Expression of protoconcogenes in aflatoxin B_1-induced rat hepatocellular carcinomas and cultured cell line, Kagura-1, *Proc Jpn Assoc Mycotoxicol.*, 23:29-33.

18. Tashiro F, Morimura S, Hayashi K, Makino R, Kawamura H, Horikoshi N, Nemoto K, Ohtsubo K, Sugimura T, Ueno Y, 1986, Expression of the c-dHa-ras and c-myc genes in aflatoxin B_1-induced hepatocellular carcinomas, *Biochem Biophys Res Commun.*, 138:858-864.

19. Wogan GN, Edward BS, Newberne PM, 1971, Structure-activity relationships in toxicity and carcinogenicity of aflatoxins and analogs, *Cancer Res.*, 31:1936-1942.

20. Wogan GN, 1973, Aflatoxin carcinogenesis, *In*: "Methods in Cancer Research, Vol. III", Busch H, ed., Academic Press, New York.

21. Wogan GN, Croy RC, Essigmann JM, Croopmann JD, Tilly WN, Skopek TR, Liberm HL, 1979, Mechanism of action of aflatoxin B_1 and sterigmatocystin: relationship of macro-molecular binding to carcinogenicity and mutation, *In*: "Environmental Carcinogenesis," Emmelot P and Kriek E, Eds., Elsevier, Amsterdam.

22. Yaswen P, Goyette M, Shank PR, Fausto N, 1985, Expression of c-Ki-ras, C-Ha-ras, and c-myc in specific cell types during hepatocarcinogenesis, *Mol Cell Biol.*, 5:780-786.

23. Yu F-L, Cases M, Rokusek L, 1982, Tissue, sex, and animal species specificity of aflatoxin B_1 inhibition of nuclear RNA polymerase II activity, *Carcinogenesis (London)*, 3:1005-1009.

SELECTIVE EFFECTS OF T-2 TOXIN AND OCHRATOXIN A ON IMMUNE FUNCTIONS

Michael J. Taylor,[a] Raghubir P. Sharma[b] and
Michael I. Luster[a]

NIH, NIEHS, National Toxicology Program
P. O. Box 12233
Research Triangle Park, NC   27709[a]

Toxicology Program, Utah State University
Logan, UT[b]

INTRODUCTION

Mycotoxins have been reported to alter immune responses in experimental animals.[12,19] The selective effects of various trichothecene mycotoxins and the ochratoxins on the immune system have gained attention recently.[13,14,23] Herein, we have reviewed some of our findings on the trichothecene mycotoxins, particularly T-2 toxin (T-2). Data on the Ochratoxins A and B will also be discussed.

Trichothecenes

The current data suggests that the trichothecene mycotoxins exert their immunomodulatory actions through disruption of various T-cell functions. Modulation of the delayed-type-hypersensitivity response, immunologic tolerance, host resistance, allograft rejection, and antibody response have all been observed following exposure to T-2.[24] Exposure to T-2 can result in thymic involution, characterized by cortical depletion.[24] As the thymus is responsible for T-cell maturation, the thymic effects of T-2 may be linked to its observed disruption of various, T-cell functions.

The immunomodulatory activities of several trichothecene mycotoxins on host resistance, the acute-phase response, and antibody production are summarized in Table 1. The immunomodulatory activities of the trichothecene mycotoxins may, in part, be dependent upon a temporal relationship between toxin exposure and antigen challenge to the immune system, and is readily evident following challenge with infectious agents. For example, when T-2 is administered subsequent to challenge with either *Mycobacterium bovis* or *Listeria monocytogenes*, mortality is increased. Increased mortality has also been reported in mice exposed to diacetoxyscirpenol (DAS) following infection with *Candidia albicans*. However, in mice pretreated with T-2, mortality following *Listeria* inoculation is decreased. A similar phenomenon has been observed for serum-amyloid-P component, an acute-phase, hepatic protein whose level is increased when T-2 treatment precedes *Listeria* infection or decreased if toxin treatment follows infectious challenge.[28] Modulation of antibody response by T-2 appears to be due to altered T-cell regulation. The

Microbial Toxins in Foods and Feeds, Edited by A.E. Pohland *et al.,*
Plenum Press, New York, 1990

Table 1. Summary of trichothecene mycotoxin effects on antibody production, acute-phase response, and host resistance.

	Toxin	Dose Duration	Infecting Organism	Mortality	Ref.
Host Resistance					
	DAS	1.12 mg/kg, ip days 3-9 after infection	*Candidia albicans*	Increased	8
	T-2	0.1 mg po, days 8,10,...,18 after infection	*M. bovis*	Increased	10
	T-2	4 mg/kg, po same day as infection	*L. monocytogenes*	Increased	7
	T-2	4 mg/kg, po days 4 or 2 before infection	as above	Decreased	7
Acute-Phase Response					
	T-2	4 mg/kg, ip 4 days before infection	as above	Increased serum-amyloid-P component	28
	T-2	4 mg/kg, po same day as infection	as above	Decreased	2/

	Toxin	Dose Duration	Antigen[a]	Effect	Ref.
Antibody Response					
	T-2	0.5-2 mg/kg, ip days 1-3 before and days 4-7 after antigen	TI[b,c]	Increased, dose-related	17
	DAS	as above	TI[b,c]	Increased, dose-related	17
	T-2	2.5 mg/kg, po days 1,4,6,8 before and days 1 and 3 after antigen	TI[b]	Increased	26
	T-2	as above	TD[d]	Decreased	26
	T-2	0.5-2 mg/kg ip, days 1-3 before and days 4-7 after antigen	TD[d]	Decreased	16

[a]TI, thymic independent; TD, thymic dependent; [b]DNP-Ficoll; [c]polyvinyl-pyrrolidone; [d]SRBC.

490

T-cell-dependent response to sheep-red-blood cells is decreased following exposure to either T-2 or DAS, while T-independent responses are increased. The experimental designs employed thus far for the evaluation of antibody response in trichothecene-mycotoxin-treated animals have not allowed for an interpretation of a temporal relationship.

Trichothecene mycotoxins are skin irritants. The ability of trichothecene mycotoxins to induce erythema after topical application has been used as a sensitive bioassay for their presence. We hypothesized that T-2's immunomodulatory effects begin with tissue injury, followed by perturbations in homeostasis, leading to alterations in immune function. Our hypothesis emphasized the stress paradigm activation of the hypothalamic-pituitary-adrenal (HPA) axis and, in initial experiments, both the percentage of animals with detectable serum levels of endotoxin and serum corticosterone, the major glucocorticoid hormone in rodents, increased following acute exposure to T-2 (Fig. 1).[26] These initial observations were extended as the presence of endotoxin and elevated corticosterone persisted through 1 and 4 weeks, respectively. Endotoxin was detected on day 7, in the sera of animals treated po with 2.5 mg/kg on days 1, 3, and 5 of the first week. The alternate-day-dosing schedule continued for 4 weeks. Endotoxin was not detected in either 2 or 4 weeks. However, increased serum-corticosterone levels persisted in animals treated with 0.5 or 2.5 mg/kg T-2 for 2 weeks and in animals treated with 0.1, 0.5, or 2.5 mg/kg T-2 for 4 weeks. The level of hypothalamic norepinephrine was also elevated following 2-week exposure (on alternate days) to 2.5 mg/kg T-2. Adrenergic stimulation of the hypothalamus has been purported as an important signal for the release of corticotropin-releasing hormone from the hypothalamus,[3,22] which induces the release of adrenocorticotropic hormone, the adrenal signal for glucocorticoid hormone production, from the pituitary. Taken together, these data suggest that T-2 had exerted a stimulatory influence on the HPA axis.

As the glucocorticoid hormones have potent immunological effects, and are suspected to play a regulatory role in the developing immune response,[2,5] we proceeded to evaluate the role of the adrenal gland in T-2-induced immunomodulation. Mice were treated po with 2.5 mg/kg T-2 on days 1, 3, 5, 8, 10, and 12 and challenged with either dinitrophenyl-Ficoll (DNP-Ficoll) or SRBC antigens on day 9 or 10, respectively, of the treatment period. The T-dependent antibody response to SRBC decreased with the type-I, T-independent response to DNP-Ficoll increased following exposure to T-2.[27] After adrenalectomy, the effects of T-2 on both of these antibody responses decreased, as neither the increased T-independent nor decreased T-dependent effects (observed in the non-operated, T-2-treated groups) were observed in adrenalectomized, T-2-treated mice. Adrenalectomy itself did not alter the antibody responses. These experiments indicated that a significant part of T-2's immunomodulatory effect on antibody production is associated with disturbance of adrenal gland function, as its removal markedly decreased T-2's immunomodulatory effect on antibody production. The dichotomy of T-2's effects on antibody production (i.e., increased T-independent and decreased T-dependent responses) may be due, in part, to differences in the relative sensitivities of T-independent and T-dependent antibody responses to adrenal hormones.

The observation that endotoxemia and elevated corticosterone levels occurred concomitantly following T-2 exposure led us to examine the role of endotoxin in T-2-induced immunomodulation. As endotoxin can activate the HPA axis,[4] its increase may be a significant factor in our observations of HPA activation. Additionally, endotoxin has a spectrum of biological activities,[9] many of which are shared with the trichothecene mycotoxins, including: poly clonal-B-cell activity[6] and enhanced resistance to infection, leucoytosis, leycopenia, and bone marrow necrosis.[24] Combined treatment of mice with endotoxin and T-2 results in an apparent synergy of their toxicities, as evidenced by increased mortality.[23] Our results also indicate

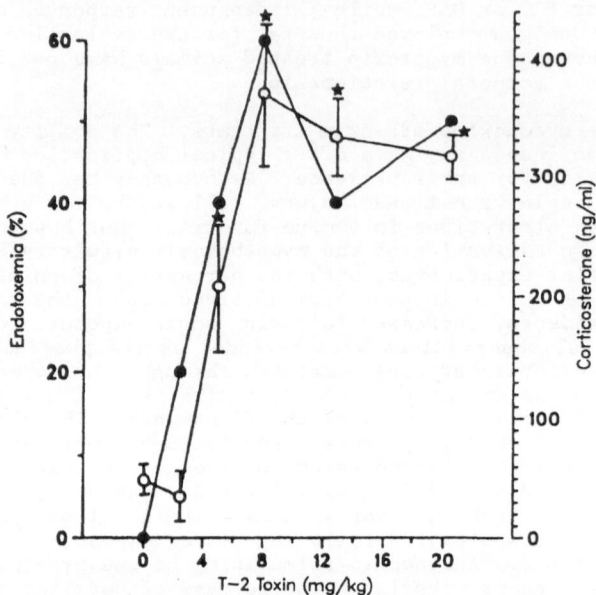

Fig. 1. Relationship between the percentage of mice and endotoxemia (●) and serum corticosterone levels (O) 24 h after a single, oral dose of T-2. Presence of endotoxin was assayed for with *Limulus* amebocyte lysate.[25]

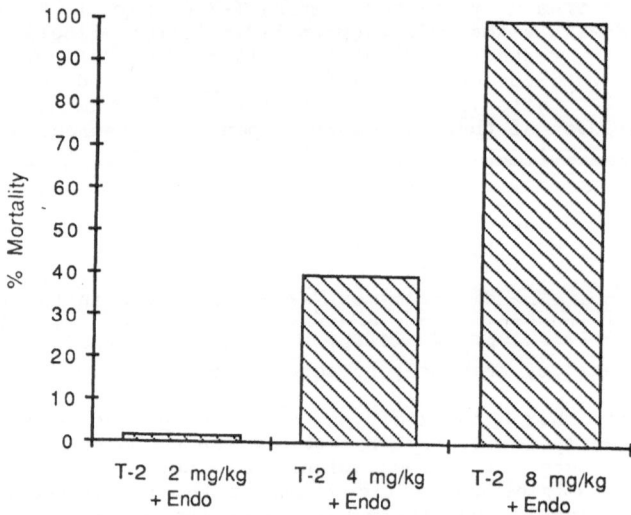

Fig. 2. Percent mortality after 24 h in mice exposed simultaneously to T-2 (po) and endotoxin (ip, 3 μg/mouse. Neither T-2 nor endotoxin alone resulted in mortality.

that T-2 enhances endotoxin sensitivity in a dose-related manner (Fig. 2), and that mortality is significantly reduced if the mice are pretreated with endotoxin, thus initiating a state of endotoxin tolerance. Endotoxin sensitivity can be increased by treatment with either D-galactosamine or α-amanitin.[11,18] The mechanism for this has been linked to inhibition of hepatocellular RNA synthesis. Though T-2 has been demonstrated to decrease both hepatic DNA and protein synthesis in vivo, RNA synthesis was not decreased,[15,20] moreover, it was reported[21] that the level of both total and microsomal hepatic RNA increased following daily exposure to 0.75 mg/kg T-2 for 7 days. T-2 has a compromising affect on endotoxin resistance. As endotoxin has many and potent effects on the immune system, decreased resistance to endotoxin may play a significant role in the interpretation of T-2's immunotoxicity. The relationship between T-2 (and other trichothecenes) and endotoxin sensitivity requires further investigation.

Ochratoxins

Ochratoxin A (OCA) is a carcinogen and reported to increase the incidences of both hepatocellular neoplasms and renal carcinomas in mice fed a diet containing 40 ppm for a two-year period.[1] As the immune system is one mechanism whereby neoplastic changes are monitored and controlled, it was of interest to evaluate immune function subsequent to exposure to this carcinogen. We have previously reported on the immunotoxic effects of OCA and OCB in female (C57BL/6N x C3H)F$_1$ mice.[13] The animals were dosed every other day for a total of six doses at dose levels ranging from 3.4-13.4 mg/kg body weight. All studies were conducted within two days of the final dose. The immunological findings are summarized in Table 2.

OCA, but not OCB, decreased natural killer (NK) cell activity following in vivo exposure. The effects were similar following either po or sc treatment. Considering interferon plays an important, inductive role in the generation of NK cells, stimulation of basal interferon levels with poly I:C was assessed as a potential therapy for the depression of NK activity in

Table 2.   Selective immunosuppressive effects of Ochratoxin A[a]

Parameter	Dose	Response
Natural killer cell activity	≥6.7 mg/kg	Decreased
Natural killer cell activity following Ochratoxin A and Poly I:C treatment	13.4 mg/kg	Natural killer cell activity restored to control values
Poly I:C induced interferon	≥3.4 mg/kg	Decreased
Tumor growth	3.4-13.4 mg/kg	Increased
Macrophage-mediated anti-tumor activity	3.4-13.4 mg/kg	No significant effect
Lymphoproliferative responses to various mitogens	3.4-13.4 mg/kg	No significant effect
Cytotoxic-T-lymphocyte response	3.4-13.4 mg/kg	No significant effect

[a]Luster et al.[12]

OCA-treated mice. Indeed, poly I:C treatment diminished the suppressive effect of OCA on NK activity, suggesting that basal interferon levels were decreased by OCA and that this phenomenon may be the mechanism whereby exposure to OCA led to decreased NK cell activity. Though poly I:C-induced interferon levels were lower in OCA-treated animals than in controls, we suspect that the level of interferon in the OCA-exposed, poly I:C-treated mice was sufficient to account for the reversal of OCA's suppressive effect on NK activity. As NK cell activity can be an important element in tumor cell surveillance and the early stages of their destruction, the affect of the ochratoxins on tumor progression was assessed. The percentage of tumor-bearing animals, following injection of animals with PYB6-tumor cells, increased with OCA treatment. These data, viewed in conjunction with the NK cell data, suggest that reduced NK cell activity may play a role in the increased susceptibility to tumor growth in OCA-treated mice. Additionally, T-cell and macrophage-mediated, antitumor activities, and lymphoproliferative responses were less sensitive indicators of OCA's immunosuppressive effect. These findings indicated that within the dose range investigated, OCA, but not OCB, is an immunosuppressive compound and that its effects are associated with decreased NK cell activity.

REFERENCES

1. Bendele AM, Carlton WW, Krogh P, Lillehoj EB, 1985, Ochratoxin carcino-genesis in the (C57BL/6J)F$_1$ mouse, *J Natl Cancer Inst.*, 75:733-742.
2. Besedovsky HO, Del Rey A, Sorkin E, 1983, Neuroendocrine immuno-regulation, *In*: "Immunoregulation," Fabaris W, Garaci E, Hadden J, Michison NA, eds, Plenum Press, New York.
3. Calogero AE, Gallucci WT, Chrousos GP, Gold PW, 1988, Catecholamine effects upon rat hypothalamic corticotropin-releasing hormone secretion *in vivo*, *J Clin Invest.*, 82:839-846.
4. Carroll BJ, Person MJ, Martin FIR, 1986, Evaluation of three acute tests of hypothalamic-pituitary-adrenal function, *Metabolism*, 18:476-483.
5. Claman HN, 1972, Corticosteroids and lymphoid cells, *New Engl J Med.*, 287:388-397.
6. Cooray R, Lindahl-Kiessling K, 1987, Effect of T-2 toxin on the spontaneous antibody-secreting cells and other non-lymphoid cells in the murine spleen, *Fd Chem Toxic.*, 25:25-29.
7. Corrier DE, Ziprin RL, Mollenhauer HH, 1987, Modulation of cell-mediated resistance to Listeriosis in mice given T-2 toxin, *Toxicol Appl Pharmacol.*, 89:323-331.
8. Frometin H, Salazar-Mejicanos S, Mariat F, 1980, Pouvoir pathogene de *Canidia albicans* pour la souris normale ou deprimee par une mycotoxine: le diacetoxyscirpenol, *Ann Microbiol (Inst Past.)*, 131B:39-46.
9. Kabir S, Rosenstreich DL, Mergenhagen SE, 1978, Bacterial endotoxins and cell membranes, *In*: "Bacterial Toxins and Cell Membranes," Academic Press, New York.
10. Kanai K, Kondo E, 1984, Decreased resistance to mycobacterial infection in mice fed a trichothecene compound (T-2 toxin), *Jap J Med Sci Biol.*, 37:97-104.
11. Keppler D, Pausch J, Decker K, 1974, Selective uridine triphosphate deficiency induced by D-galactosamine in liver and reversed by pyrimidine nucleotide precursors: Effect on ribonucleic acid synthesis, *J Biol Chem.*, 249:211-216.
12. Luster MI, Blank JA, 1987, Molecular and cellular basis of chemically induced immunotoxicity, *Ann Rev Pharmacol.*, 27:23-49.
13. Luster MI, Germloc Dr, Burleson GR, Jameson CW, Ackerman MF, Lamm KJ, Hayes HT, 1987, Selective immunosuppression in mice of natural killer cell activity by ochratoxin A, *Cancer Res.*, 47:2259-2263.

14. Otokawa M, 1983, Immunological disorders, *In*: "Trichothecenes, Chemical, Biological and Toxicological Aspects, Developments in Food Science, Vol. 3," Ueno Y, ed., Elsevier Press, New York.

15. Rosenstein Y, Lafarge-Frayssinet C, 1983, Inhibitory effect of *Fusarium* T2-toxin on lymphoid DNA and protein synthesis, *Toxicol Appl Pharmacol.*, 70:283-288.

16. Rosenstein Y, Lafarge-Frayssinet C, Lespinats G, Loisillier F, Lafont P, Frayssinet C, 1979, Immunosuppressive activity of *Fusarium* toxins. Effects on antibody synthesis sand skin grafts of crude extracts, T_2-toxin and diacetoxyscirpenol, *Immunology*, 36:111-117.

17. Rosenstein Y, Kretschmer RR, Lafarge-Frayssinet C, 1981, Effect of Fusarium toxins, T-2 toxin and diacetoxyscirpenol on murine T-Independent immune responses, *Immunology*, 44:555-560.

18. Seyberth HW, Schmidt-Gayk H, Hackenthal E, Toxicity, clearance and distribution of endotoxin in mice as influenced by actinomycin D, cycloheximide, α-amanitin and lead acetate, *Toxicon*, 10:491-500.

19. Sharma RP, Reddy RV, 1987, Toxic effects of chemicals on the immune system, *In*: "Handbook of Toxicology," Haley JJ, Bernt WO, eds., Hemisphere, Washington, DC.

20. Suneja SK, Ram GC, Wagle DS, 1983, Effects of feeding T-2 toxin on RNA, DNA and protein contents of liver and intestinal mucosa of rats, *Toxicol Lett.*, 18:73-76.

21. Suneja SK, Wagle DS, Ran GC, 1987, Effects of T-2 toxin gavage on the synthesis and contents of rat liver macromolecules, *Fd Chem Toxic.*, 25:387-392.

22. Szafarczyk A, Guillaume V, Conte-Devoix B, Alonso G, Malaval F, Pares-Herbute N, Oliver C, Assenmacher I, 1988, Central Catecholaminergic system stimulates secretion of CRH at different sites, *Am J Physiol.*, 255:E463-E468.

23. Tai JH, Pestka JJ, 1988, Synergistic interaction between the trichothecene T-2 toxin and *Salmonella typhimurium* lipopolysaccharide in C3H/HeN and C3H/HeJ mice. *Toxicol Lett.*, 44:191-200.

24. Taylor MJ, Pang VF, Beasley VR, 1989, The immunotoxicity of trichothecene mycotoxins, *In*: "Trichothecene Mycotoxicosis: Pathophysiologic Effects," Beasley VR, Ed., CRC Press Inc., Boca Raton, FL.

25. Taylor MJ, Reddy RV, Sharma RP, 1985, Immunotoxicity of repeated low level exposure to T-2 toxin, a trichothecene mycotoxin, in CD-1 mice, *Myco Res.*, 1:57-64.

26. Taylor MJ, Smart RA, Sharma RP, 1989, Relationship of the hypothalamic-pituitary-adrenal axis with chemically induced immunomodulation. I. Stress-like response after exposure to T-2 toxin, *Toxicology*, 56:179-197.

27. Taylor MJ, Warren RP, Sharma RP, 1989, Relationship of the hypothalamic-pituitary-adrenal axis with chemically induced immunomodulation. II. Effects of T-2 toxin on T-dependent and T-independent antibody responses in adrenalectomized mice. *Toxicology*, (in press).

28. Ziprin RL, Holt PS, Mortensen RF, 1987, T-2 toxin effects of the serum amyloid P-component (SAP) response of *Listeria monocytogenes*- and *Salmonella typhimurium*-infected mice, *Toxicol Lett.*, 39:1;77-184.

EMESTRIN, A NEW MYCOTOXIN, INTRODUCED INJURIES IN VARIOUS ORGANS OF MICE

Kiyoshi Terao and Emiko Ito

Research Institute for Pathologic Fungi
  and Microbial Toxicoses
Chiba University
1-8-1 Inohana, Chiba 280 Japan

INTRODUCTION

Emestrin (EMS) is an epidithiodioxopiperazine mycotoxin first described as a metabolite of *Emericella striata*[6] (Fig. 1). EQ-1 named by Maebayashi et al.[3] for a toxic metabolite of *E. quadrilineata* and *E. paravathecia* was identified as EMS by Seya et al.[6] Species of the genus *Emericella* are disibuted widely throughout tropical and subtropical countries. EMS-producing species have been isolated from various speices imported into Japan from several counries in Asia.[7] In contrast to chemical studies and surveillance of food-born EMS-producing fungi, only limited information is available regarding the biological activities of EMS. In an *in vitro* system, K. Ishizaki and K. Kawai[2] observed potent suppressive effects of EMS on the respiratory system in mitochondria. In an *in vivo* system, however, no precise toxicological studies have been reported regarding the target organs of experimental animals after administration of EMS. Our present study was undertaken to investigate the morphological changes in the target organs of mice after the administration of EMS.

MATERIALS AND METHODS

EMS was prepared according to the method of Seya et al.[6] The lethality of the toxin was assayed in male ICR mice (20-25 g) by i.p. injection. The LD_{50} values were calculated according to the method of Lichfield and Wilcoxon at 24 or 48 hr after EMS injection. Male ICR mice were also used for the determination of LD_{50} values and for morphological alteration caused by EMS.

RESULTS AND DISCUSSION

LD_{50}

The LD_{50} values of EMS were 17.7 mg/kg of body weight at 24 hr after the injection, and 13.0 mg/kg at 48 hr. Most of the dead animals showed a marked retention of ascites and effusion of fluid into the thorax and peritoneal cavity. Severe congestion of whole organs was also noted. For sequential studies of EMS intoxication, EMS was dissolved in a 50% aqueous

Microbial Toxins in Foods and Feeds, Edited by A.E. Pohland *et al.,*
Plenum Press, New York, 1990

N,N-dimethylformamide solution. Animals receiving 30 mg/kg or more died
within 2 hr. The most severely affected organ was the heart. Figures 2a and
2b show, respectively, a light and an electron micrograph of single cell
necrosis in the myocardium of the right ventricle. Swollen mitochondria with
an irregular arrangement of the cristae were abundant in the necrotizing
cell. Marked edema within the cytoplasm was usually observed in the less
affected muscle cells neighboring the necrotic lesion.

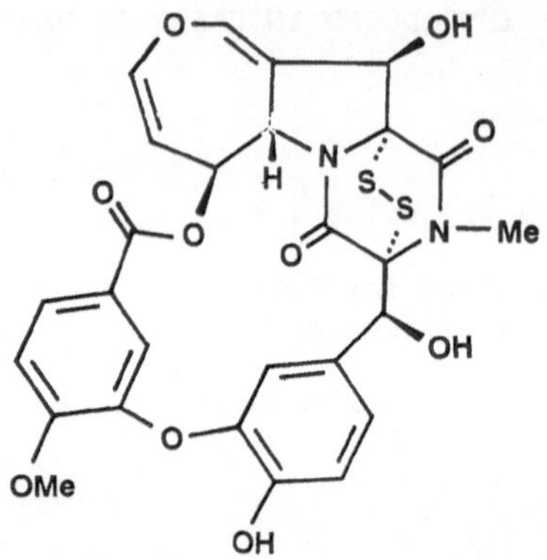

Fig. 1. Chemical structure of emestrin.

One of the characteristic features of EMS intoxication was severe cen-
ilobular necrosis of the liver (Figs. 3, 4). In the centrilobular region of
the liver there was pronounced necrosis of the hepatocytes after i.p. injec-
on of 20 mg/kg of EMS (Fig. 3a). An electron microscopic observation showed
considerable mitochondrial swelling of the hepatocytes of centrilobular
regions (Fig. 3b). In the cytoplasm of hepatic cells near the necrotic
lesion there was notable proliferation of rough endoplasmic reticulum and
ribosome-like granules (Figs. 3c, 3d). The ribosome-like granules associated
with or without membrane structure occupied almost one third of the cytoplasm
of most hepatocytes (Fig. 4a). In cytoplasm of the hepatocytes there were
occasional inclusion bodies containing clusters of free ribosomes and rough
endoplasmic reticulum (Fig. 4b). Membrane whorls associated with ribosomes
were also found in several hepatocytes (Fig. 4c). In contrast to the pro-
feration of rough endoplasmic reticulum, careful observation of the lesions
showed no proliferation of smooth endoplasmic reticulum in the hepatocytes
either centrilobular or in peripheral regions. Therefore, it is suggested
that ERS effects are primarily on the RER system of the hepatocytes.

Degeneration of lymphoid tissue was another prominent result of EMS
intoxication. Twenty-four hr after the injection of the toxin, there was
massive necrosis of lymphocytes in the cortical layer of the thymus (Fig.
5a). In contrast to the lymphocytes, however, epithelial reticular cells in
the cortical layer always survived (Fig. 5b). Forty-eight hr after a single
injection of EMS, almost all lymphocytes had disappeared from the cortical
layer of the thymus (Fig. 5c). Simultaneously the weight of the spleen was
reduced to about 50% of the average weight of the control animals. Histologically, a reduction in the number of lymphocytes in the periarterial lymphoid sheath as well as in the red pulp was noted (Fig. 5d).

To clarify the effects of EMS on lymphoid tissue, six bilateral adrenalectomized mice were given injections of 20 mg/kg of EMS, and the effect of
the toxin on the lymphoid tissue was compared with that of control mice.
Figure 4a is a photomicrograph of the thymus of a control mouse treated with

20 mg/kg of EMS for 24 hr. In contrast to control mice, mice treated with bilateral adrenalectomy 2 weeks prior to the i.p. injection showed no discernible pathologic changes in the thymus after i.p. injection of the same dose of EMS (Fig. 6). In EMS injection of adrenalectomized mice, histological appearance of the spleen was similar to that of the control mice. Table 1 summarizes the effects of EMS on the thymus and spleen of adrenalectomized and control mice. Thymic lesions were not induced in adrenalectomized animals. Therefore, this suggests that the thymic lesion was not caused from a direct effect of EMS, but by some indirect effect via an adrenocortical route. Several investigators have reported the acute toxicity of mycotoxins belonging to the epipolythiodioxopiperazine such as gliotoxin,[4] sporidesmin,[1] and chetomin.[5] Nagarajan[4] has confirmed that the bridged disulfite structure of the gliotoxin molecule may play an important role in the biological activities of the mycotoxin. The bridged disulfite structure in EMS molecule may also play an important role in the mode of action of hepatic lesions induced by EMS. However, the mechanism involved in thymic lesion is different from that of the hepatic injuries, although the precise mechanism remains unclear.

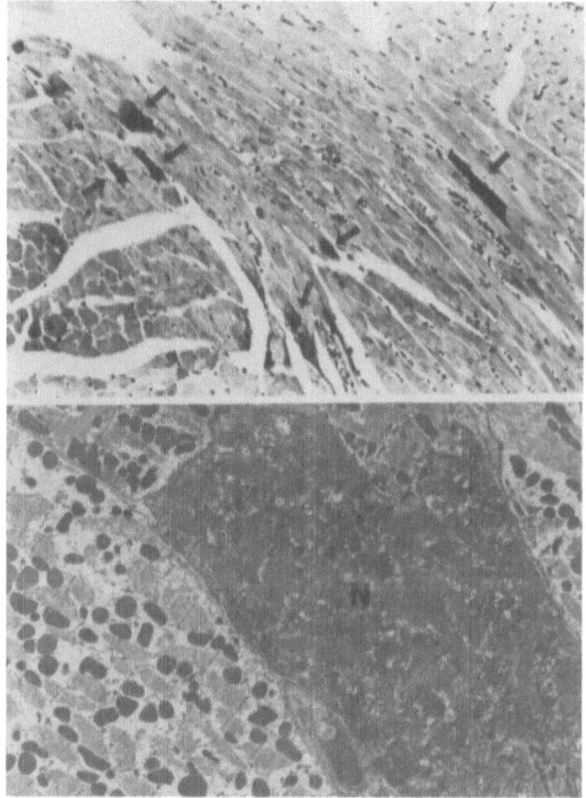

Fig. 2 (a) A light micrograph of mycocardium from the same mouse of Fig. 1a. All mitochondria in a necrotizing cell (N) are degenerated and cells around the necrotizing area are edematous, X 4,000; (b) An electron micrograph of mycocardium from the same mouse in Fig. 1a. All mitochondria in a necrotizing cell (N) are degenerated and cells around the necrotizing area are edematous, X 4,000.

Results depicted in Figs. 3 to 5 are from mice that received 20 mg/kg of EMS 24 hr prior to being killed.

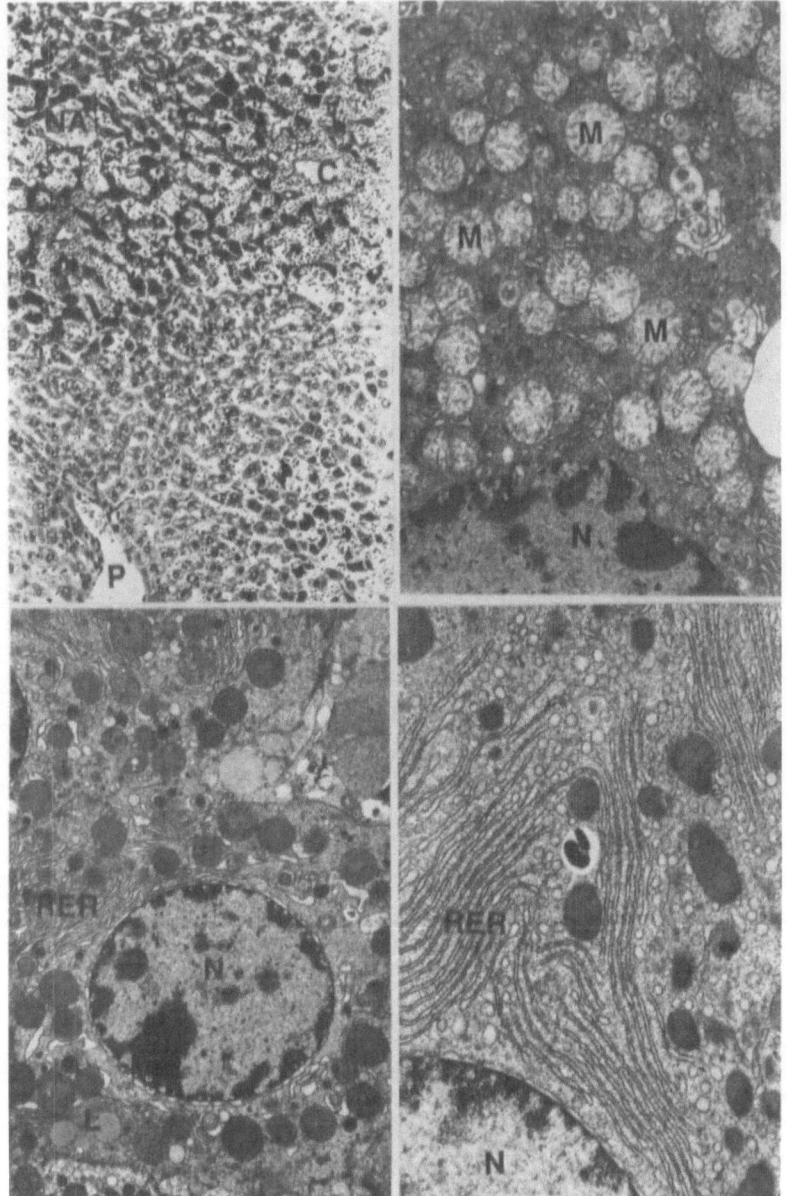

Fig. 3.   (a) A light micrograph of the liver. Around the cen-
           tral vein (C) there is necrotic area (NA). P: portal
           vein, HE X 500; (b) An electron micrograph of a hepato-
           cyte near the necrotic region. All mitochondria (M) are
           swollen. N: nucleus of a hepatocyte, X 5,000; (c) An
           electron micrograph of a hepatocyte around the lesion.
           Proliferation of rough endoplasmic reticulum (RER) is
           prominent. L: fat droplet, N: nucleus, X 5,000;
           (d) An electron micrograph of a hepatocyte near the
           necrosis. Abundant RER and free ribosomes are present
           in the cytoplasm, X 12,000;

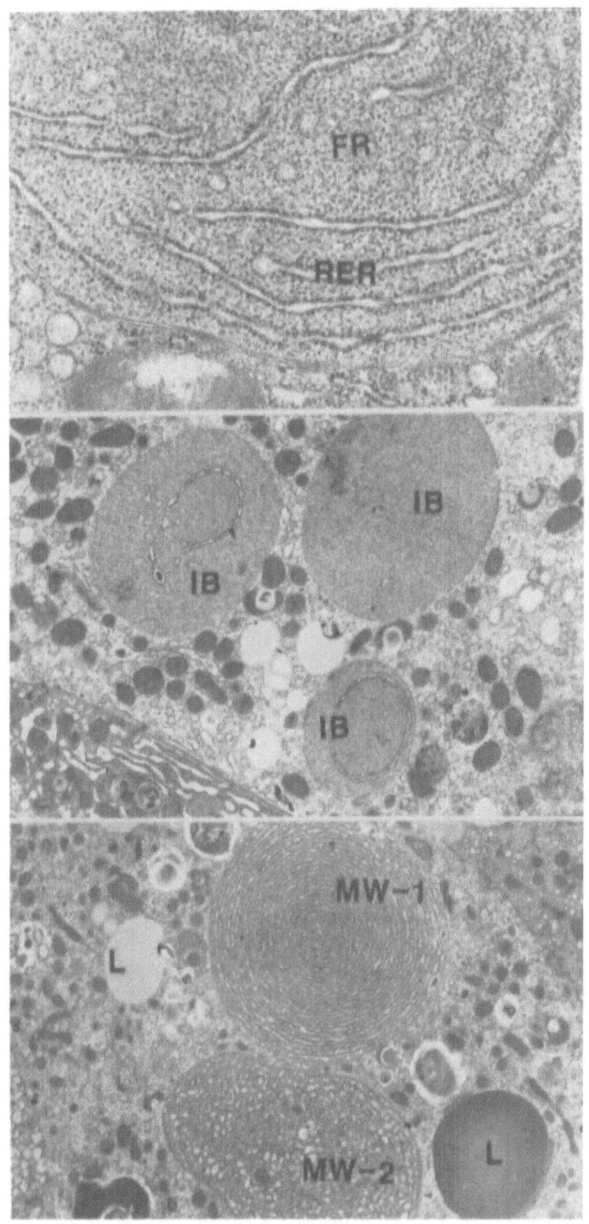

Fig. 4.   (a) An electron mirograph of a hepato-
          cyte. A part of Fig. 3c, X 42,000;
          (b) An electron micrograph of a hepato-
          cyte in centrilobular region of the liver.
          Three inclusion bodies (IB) are seen.
          Two of them contain RER, X 8,000;
          (c) An electron micrograph of a hepato-
          cyte in centrilobular region. There are
          two membrane whorls associated with ribo-
          somes (MW-1 & 2) in the cytoplasm. L: fat
          droplet, X 7,000.

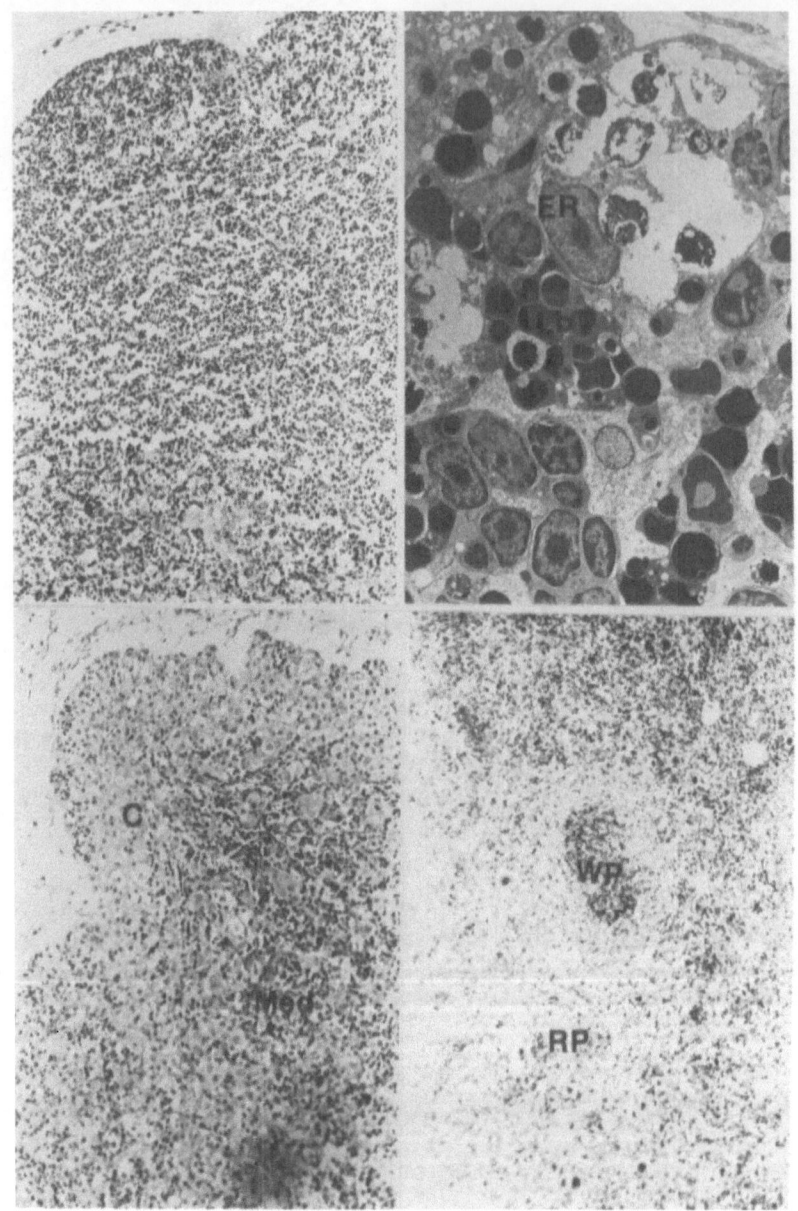

Fig. 5. (a) A light micrograph of the thymus. Numerous lymphocytic debris can be seen, HE X 250;
(b) An electron micrograph of the cortical layer of the
thymus. Epithelial reticular cells (ER) are resistant to
EMS. Lb: lymphocytic debris, X 2,300;
(c) A light micrograph of the thymus from a mouse given
20 mg/kg of EMS 48 hr prior to sacarifice. There are no
lymphocytes in the cortical layer (C). Med: medulla, HE
X 500.
(d) A light micrograph of the spleen from the same mouse
in Fig. 5a. Marked reduction of lymphocytes from white
(WP) as well as red pulp (RP), HE X 250.

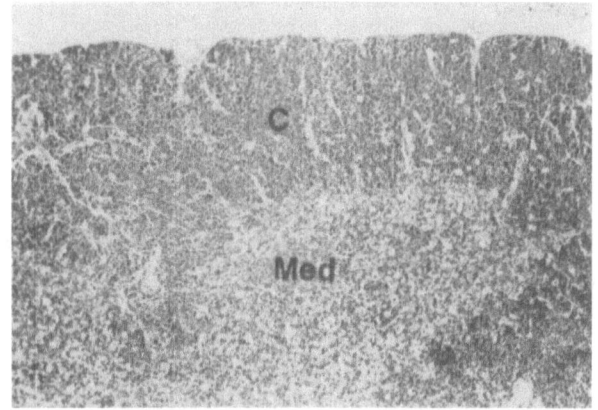

Fig. 6. A light micrograph of the thymus from
an adrenalectomized mouse injected
20 mg/kg of EMS 24 hr prior to sacrifice.
No discernible pathological changes are
seen. C: cortical layer, Med: medulla,
HE X 250.

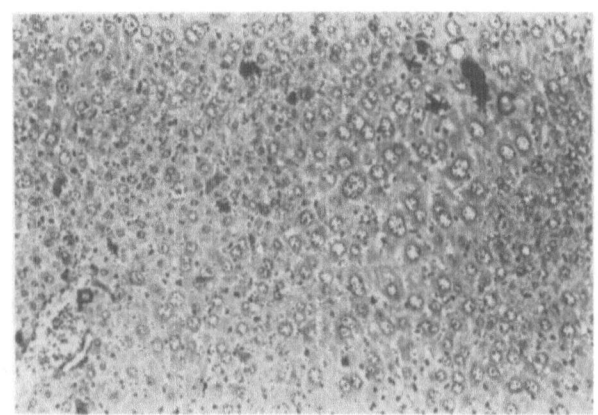

Fig. 7. A light micrograph of the liver from a
mouse receiving 1 mg/kg of EMS for 50
times. Marked nuclear atypia is seen.
C: central vein, P: portal area,
HE X 750.

Repeated injections into the peritoneal cavities of mice resulted in an
increase in the volume of hepatocyte nuclei in centrilobular regions of the
hepatic lobules (Fig. 7). There are also many giant and bizarre nuclei in
the centrilobular region. Table 2 shows the area of hepatocyte nuclei in the
liver after 50 i.p. injections of 1 ppm of EMS. The area of the hepatocyte
nuclei located around the central vein is significantly larger than that of
the periportal region. Oral administratioan of 10 ppm of diet-containing EMS
for 44 weeks resulted in one myeloic leukemia out of 10 mice. The survivors
showed no discernible pathological changes and are now still under
observation.

Table 1. The effect of emestrin on the thymus and spleen
          of control and adrenalectomized mice.

	Injuries in Lymphoid Organs	
Pretreatment	Thymus	Spleen
Non-treated	6*/6	6**/6
Adrenalectomized	0/6	0/6

*Massive necrosis of lymphocytes in the cortical layer.
**Marked reduction of lymphocytes in both white and red
 pulp.

Table 2. Distribution pattern of nuclei of hepatocytes
          in the hepatic lobules (μmm^2)

	Centrilobular	Periportal
Control	28.6 ± 5.9	28.6 ± 5.0
EMS*	156.8 ± 12.0**	91.9 ± 18.3

*1 mg/kg of emestrin was injected i.p. for 50 times.
**Effect of treatment is significant ($p < 0.05$).

REFERENCES

1. Carlton WW, Scech GM, 1977, Mycotoxicosis in laboratory animals, *In*:
   "Mycotoxic Fungi, Mycotoxins, Mycotoxicosis, Vol. 3," Wyllie TD,
   Morehouse LG, eds., Marcel Dekker, New York.
2. Ishizaki K, Kawai K, Kawai K, 1988, The effect of emestrin from
   Emericella striata on mitochondrial reaction, *In*: Symposium -
   "7th International symposium on mycotoxins and phycotoxins," August
   16-19, 1988, Tokyo (abstract).
3. Maebayashi Y, Horie Y, Yamazaki M, 1984, Productivity of some mycotoxins
   in fungi isolated from Iraqi soil, *Prod Jpn Soc Mycotoxicol.*,
   20:28-30.
4. Nagarajan R, 1984, Gliotoxin and epipolythiodioxopiperazines, *In*:
   "Mycotoxins, Production, Isolation, Separation and Purification," Betin
   V., ed., Elsevier, New York.
5. Saito T, Suzuki Y, Koyama K, Natori S, Iitaka Y, Kinoshita K, 1988,
   Chetracin A and Chaetocins B and C, three new epipolythiodioxopiperazines
   from *Chaetomium* spp., *Chem Pharm Bull.*, 36:1942-1956.
6. Seya H, Nakajima S, Kawai K, Udagawa S, 1985, Structure and absolute
   configuration of emestrin, a new macrocyclic epidithiodioxopiperazine
   from *Emericella striata*, *J Chem Soc Chem Commun.*,   :657-658.
7. Udagawa S, 1986, Fungal contamination of imported spices, with special
   reference to ascomycetous fungi, *Jpn J Food Microbiol.*, 3:46-52.

THE MOLECULAR MECHANISM OF TOXICITY OF GLIOTOXIN AND RELATED

EPIPOLYTHIODIOXOPIPERAZINES

Paul Waring, Ronald D. Eichner and Arno Mullbacher

Division of Virology and Cellular Pathology
John Curtin School of Medical Research
Australian National University
P. O. Box 334, Canberra 2601, Australia

INTRODUCTION

Gliotoxin (Fig. 1a) belongs to the epipolythiodioxopiperazine (ETP) class of fungal metabolites having a bridged polysulphide piperazine ring in common.[10,11,14] Gliotoxin was the first member of this class of metabolites to be described being isolated from a strain of *Trichoderma lignosum* in 1932.[17] The tri[12] and tetra[13] sulphide analogues of gliotoxin have also been described. Gliotoxin and a number of related ETP compounds have been shown to be responsible for a number of mycotoxicoses following ingestion of contaminated feed by ruminants.[3]

Gliotoxin attracted early attention because of its antimicrobial properties. As early as 1942 the activity of gliotoxin against a wide range of gram positive bacteria was described and structure activity studies using *Bacillus subtilis* was carried out by Brewer et al.[2] in 1966. Gliotoxin was also found to inhibit the multiplication of viruses[5], and this led to a number of studies of the effects of ETP compounds on virus replication. Unfortunately the systemic toxicity of gliotoxin (LD_{50}: 25 mg/kg in mice) has precluded any clinical use.

Renewed interest in the biological activity of ETP compounds followed the observation that gliotoxin displayed antiphagocytic and other immunomodula ting properties *in virto*[6,7,8] including inhibition of macrophage adherence to plastic, a process related to phagocytosis, inhibition of phagocytosis of particulate matter by macrophages, and inhibition of proliferation of mitogen stimulated spleen lymphocytes. All of these effects occur at concentrations having little or no effect on cell viability over the time course of assay as determined by trypan blue exclusion.

Mechanism of Action

Oxidative. Treatment of sensitive cells with gliotoxin causes fragmentation of cellular DNA and this may account for the antiproliferative properties of ETP compounds. Cells treated with varying concentrations of gliotoxin for 1 h were washed and cultured overnight. DNA was extracted and examined using agarose gel electrophoresis in the presence of ethidium bromide.[1] Double stranded scission was detected as a smear of fluorescence

Microbial Toxins in Foods and Feeds, Edited by A.E. Pohland *et al.,*
Plenum Press, New York, 1990

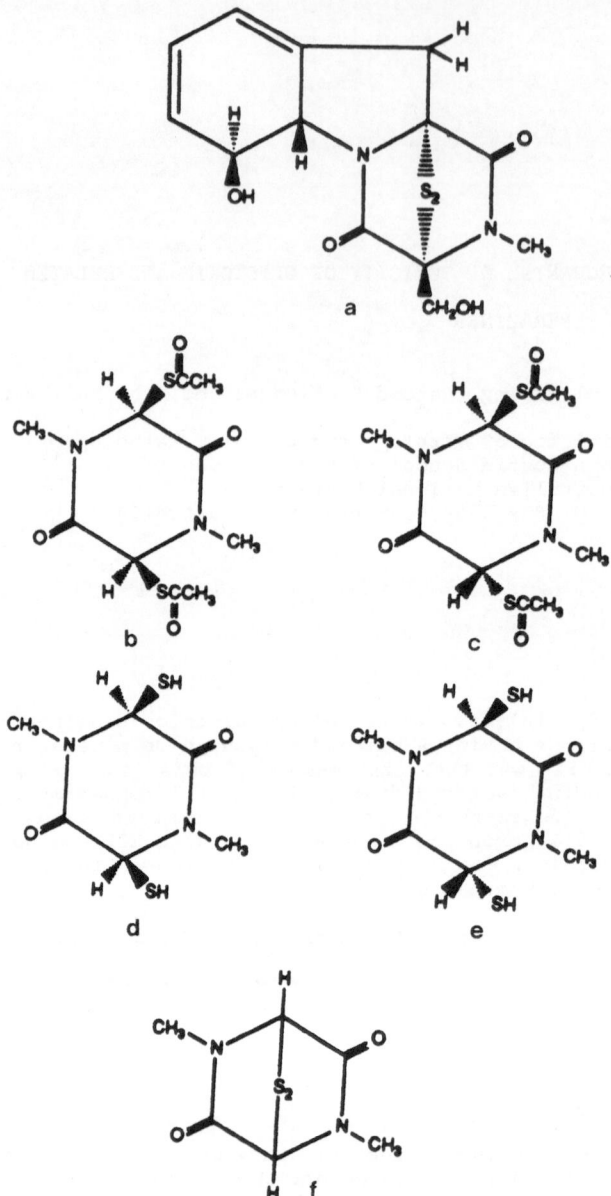

Fig. 1. Compounds tested for their antiphagocytic activity and ability to fragment DNA.

(a) gliotoxin
(b) *cis* DEP(SAc)$_2$
(c) *trans* DEP(SAc)$_2$
(d) *cis* DEP(SH)$_2$
(e) *trans* DEP(SH)$_2$
(f) DEPS$_2$

of fluorescence (Fig. 2). The DNA fragmentation appears to correlate with the antiproliferative effects of gliotoxin on a range of cell types (Table 1).

To investigate the possible molecular mechanism(s) of action of glio-toxin, we examined the effect of gliotoxin and its analogues and reduced forms of the molecule on plasmid DNA in a cell free system.[4] The con-version of closed circular (CC) plasmid DNA to the open circular (OC) and finally the linear (L) form requiring single and multiple strand scission, respectively, forms the basis of a sensitive assay for the production of reactive oxygen species. Incubation of the CC form of plasmid DNA with the simple ETP compound (Fig. 1f) in the presence of Fe^{+3} produced no effect. However, under the same conditions, the reduced form of this compound (Fig. 1d) induced the rapid conversion of the CC form to OC and L forms (Fig. 3). This process required the presence of iron suggesting the possible inter-mediacy of the hydroxyl radical or similar species. Gliotoxin alone had no effect on plasmid DNA but the presence of a reducing agent such as glutathi-one or DTT promoted the iron dependent scission of the DNA. The reducing agents alone had no or minimal effect on DNA. This suggests that reduction of the disulphide to the dithiol form is required for plasmid scission and is consistent with the generation of reactive oxygen species following autooxi-dation of the thiol. This hypothesis is consistent with the work of Munday[9] who demonstrated oxidative stress in erythrocytes exposed to sporidesmin.

The nature of the reactive oxygen species in the single and double stranded scission of plasmid DNA was evaluated by examining the effects of radical scavangers and metal chelating agents (Table 2). In particular, cata lase at 20 μg/ml, inhibited plasmid DNA degradation. Superoxide dismutase at 10 μg/ml had no effect suggesting the intermediacy of hydrogen peroxide which was confirmed by the direct observation of the production of hydrogen peroxide during autooxidation of both cis and trans $DEP(SH)_2$ at neu-tral pH (Fig. 4).

Apoptosis. The direct effect of intracellular free radicals formed by redox cycling of reduced ETP compounds with subsequent damage to DNA could account for the general toxicity and the antiproliferative effects of these compounds. A number of observations suggested that the direct effects of free radicals may not be sufficient to explain all the immunosuppressive pro-perties of ETP compounds, in particular, the effect on phagocytosis. Clas-sical radical scavangers such as BHA and trolox have no effect on the anti-phagocytic action of gliotoxin. In fact DTT protected macrophages from the effects of gliotoxin. In a later study of the effects of ETP compounds on macarophage DNA,[15] short incubations (4-6 h) resulted in discrete fragments of DNA (Fig. 5). These fragments had molecular weights that were multiples of 170 bp. Such fragmentation is characteristic of so-called programmed cell death or apoptosis. This was confirmed in a study of the morphological changes in cells following treatment with ETP compounds. The observed pres-ence of chromatin condensation giving rise to dense crescent shaped bodies and the observation of apoptotic bodies are classical features of apoptosis. The fragmentation of the DNA during apoptosis is thought be be mediated by the induction/activation of an endonuclease responsible for internucleosomal cutting of the DNA molecule, hence, giving rise to fragments of multiple molecular weights. This process is inhibited by Zn^{+2} salts presumably by inhibition of the Ca^{+2}/Mg^{+2} requiring endonuclease. Consistent with this, in the presence of Zn^{+2} salts apoptosis induced by gliotoxin was prevented.[16] Interestingly zinc salts have been used successfully in the treatment of facial eczema caused by ingestion of sporidesmin, although the rationale for its success is not yet fully understood.

To establish the role of apoptosis in the antiphagocytic action of gliotoxin, we examined a number of analogues of gliotoxin for their ability

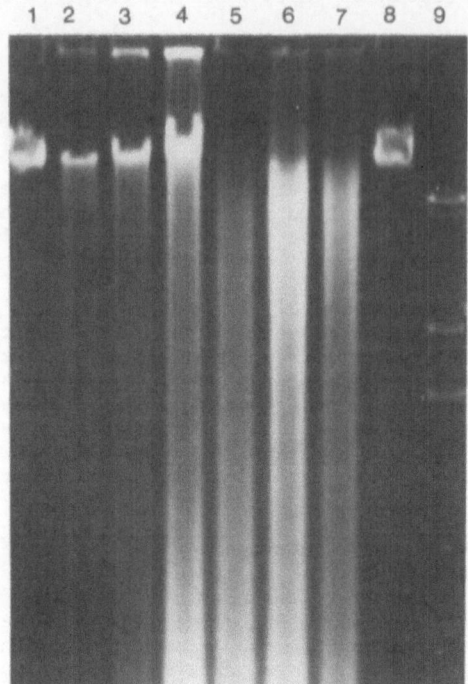

Fig. 2. Fragmentation of spleen cell DNA following treatment with gliotoxin. Lanes 1
and 8: no treatment; lanes 2-7: DNA after
treatment with 3, 10, 30, 100, 300 and
1000 ng/ml gliotoxin; lane 9: MW markers.

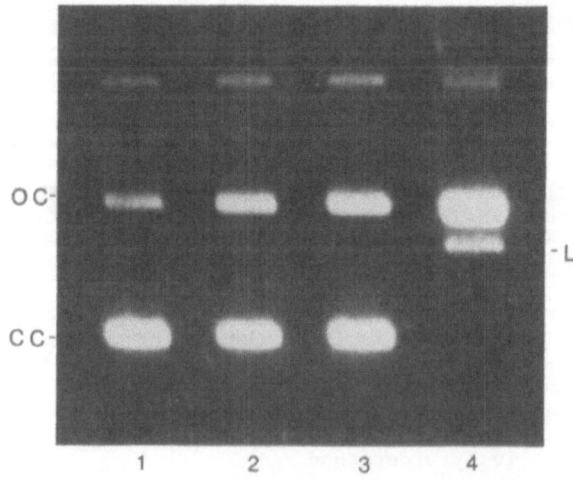

Fig. 3. Scission of plasmid DNA following treatment with *cis* DEP(SH)$_2$; 50 μM *cis*
DEP(SH)$_2$ + Fe+^{3} (30 μM); 50 μM
cis DEP(SH)$_2$ + Fe+^{3} (300 μM).

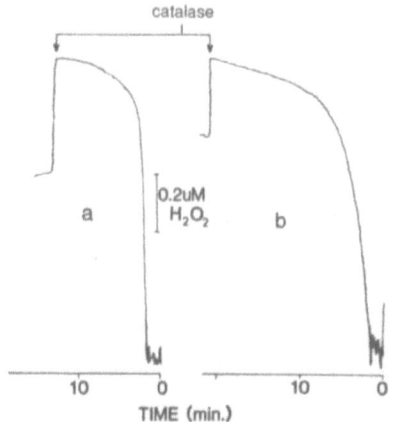

Fig. 4. The production of hydrogen
peroxide following autooxi-
dation of *cis* (a) and *trans*
(b) DEP(SH)$_2$ in phosphate
buffer pH 7.5.

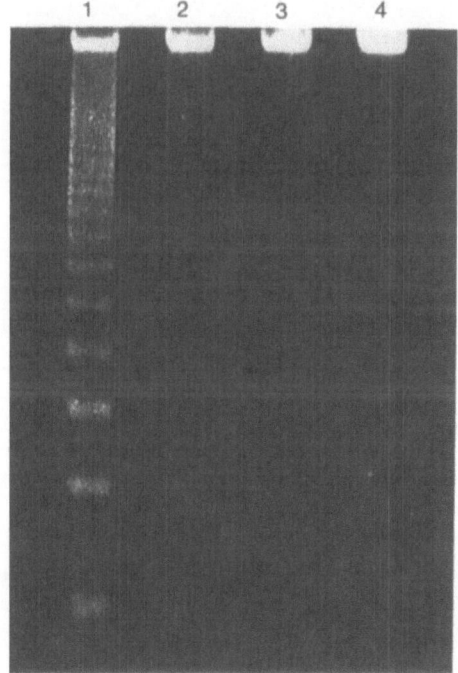

Fig. 5. Apoptosis induced in elicited murine macrophages following treatment with gliotoxin.
Lanes 1, 2, 3: 3 μM, 1 μM
and 0.3 μM gliotoxin, respectively; lane 4: no treatment.

Table 1. Comparison of the ability of gliotoxin to fragment DNA and inhibit proliferation

Cell type	Origin	ds DNA break	Inhibition of Proliferation
Chlamydomonas	Plant	--	--
Embryo	Mosquito	--	--
Vero	Monkey kidney	--	--
MAT	Rat carcinoma	--	--
SQC	Sheep carcinoma	++	++
L929	C3H mouse	++	++
P815	DBA/2 mouse	--	--
Spleen(T,B)	Mouse	++	++
Macrophage	Mouse	++	++

++ = produces defined effect; -- = does not produce defined effect.

Table 2. The effect of various agents on the disappearance of the CC form of plasmid DNA in the presence of cis DEP(SH)$_2$

Conditions*	Agent	% CC Form
Control	None	85
DES(SH) + Fe^{+3}	None	16
DES(SH) + Fe^{+3}	EtOH, 1 mM	48
Control	None	95
DEP(SH) + Fe^{+3}	None	3
DEP(SH) + Fe^{+3}	Desferal, 200 μM	57
Control	None	91
DEP(SH) + Fe^{+3}	None	11
DEP(SH) + Fe^{+3}	Catalase, 20 μg/ml	86
	10 μg/ml	45
	10 μg/ml (boiled)	14
DEP(SH) + Fe^{+3}	SOD, 10 μg/ml	1
	2 μg/ml	5
	0.1 μg/ml	16

*DEP(SH)$_2$ at 2.5 mM; Fe^{+3} 0.2 mM.

Table 3. DNA fragmentation and antiphagocytic activity of compounds in Fig. 1

Compound	ED$_{50}$[a] DNA Fragmentation	ED$_{50}$[b] Macrophage Adherence
Gliotoxin	300 nM	120 nM
DEPS$_2$	300 nM	120 nM
cis DEP(SH)$_2$	300 nM	250 nM
$trans$ DEP(SH)$_2$	500 nM	2400 nM

[a]Concentration giving 50% DNA fragmentation
[b]Concentration giving 50% cell loss.

510

Table 4.   Scission of plasmid DNA by *cis* DEP(SAc)$_2$ in the presence of
macrophage lysate.

Concentration DEP(SAc) μM	Macrophate Lysate μL	% CC Form
600	0	100
0	5	100
50	5	80
200	5	40
600	5	<20

to cause DNA breakage and for their effects on macrophage phagocytosis.   The
compounds shown in Fig. 1 were tested for their ability to fragment DNA and
effect macrophage phagaocytosis.   DNA fragmentation was quantified and expressed as a fraction of the "intact" DNA.   All compounds tested (Fig. 1) caused
the same degree of double stranded DNA scission in macrophages.   However,
only those compounds possessing the intact disulpide bridge or those capable
of extracellular conversion to this form had demonstrable antiphagocytic
activity (Table 3).   Thus, the *trans* isomer and both *cis* and *trans* S acetylated compounds were ineffective in preventing macrophage adherence to
plastic.   This indicated the requirement for the extracellular presence of
the disulphide as the active form of the molecule affecting macrophage phagocytosis[15] and was consistent with a second mechanism involving formation of
mixed disulphides between gliotoxin and an essential cell surface receptor.

   The initiation for the induction of apoptosis by ETP compounds may be the
intracellular production of free radicals by redox cycling as described
above.   The *trans* isomer, as a thiol, can still autooxidize giving rise to
superoxide and hydrogen peroxide (Fig. 4).   The S-acetylated compounds are
masked forms of the *cis* and *trans* thiols that are unmasked following cellular
uptake and intracellular hydrolysis by endogenous hydrolase(s) of the acetyl
group.   This hypothesis was tested using a crude macrophage lysate, the S
acetylated compound, and plasmid to assay for any thiol produced following
incubation of the acetylated compound with partially purified macrophage
cytosolic proteins.   In the presence of both the *cis* S-acetylated compound
and the macrophage lysate there was scission of plasmid DNA in an iron
requiring reaction (Table 4).   This showed that the active thiol was produced
following hydrolysis by hydrolases present in the crude macrophage preparation.

   Mixed disulphide formation. The possibility that mixed disulphides are
formed between gliotoxin and thiol groups of essential cell surface receptors
on macrophages and T cells is currently being explored using ^{35}S labeled
gliotoxin.   Preliminary results have shown that gliotoxin will covalently
bind to a number of proteins but further work will be necessary to establish
the functional relationships between these effects.[14]

CONCLUSIONS

   Gliotoxin has been shown to induce double stranded DNA scission in
cellular DNA and to be capable of similar damage to plasmid DNA in a cell
free system by redox cycling to produce deleterious reactive oxygen species.
The effects on cellular DNA does not appear to be due to direct free radical
attack but rather to the induction of programmed cell death or apoptosis by
an as yet unidentified mechanism, but one which seems likely to involve free
radicals.   The effect of gliotoxin on macrophage function such as

phagocytosis may be mediated by covalent interaction with cell surface thiols via mixed disulphide formation.

REFERENCES

1. Braithwaite AW, Eichner RD, Waring P, Mullbacher A, 1987, The immuno-modulating agent gliotoxin causes genomic DNA fragmentation, *Molec Immunol.*, 27:47-55.
2. Brewer D, Hannah DE, Taylor A, 1966, The biological properties of 3,6-Epidithiadiketopiperazines: Inhibition of growth of *Bacillus subtilis* by gliotoxin, sporidesmins, and chetomin, *Can J Microbiol.*, 12:1187-1195.
3. Done J, Mortimer PH and Taylor A, 1960, Some observations on field cases of facial eczema, *Res Vet Sci.*, 1:76-83.
4. Eichner RD, Waring P, Geue A, Braithwaite AW, Mullbacher A, 1988, Gliotoxin causes oxidative damage to plasmic and cellular DNA, *J Biol Chem.*, 263:3772-3777.
5. Miller PA, Milstrey KP and Trown PW, 1968, Specific inhibition of viral ribonucleic acid replication by gliotoxin, 1968, *Science*, 159:431-432.
6. Mullbacher A, Eichner RD, 1984, Immunosuppression *in vitro* by a metabolite of a human pathogenic fungus, *Proc Natl Acad Sci USA*, 81:3835-3837.
7. Mullbacher A, Waring P, Eichner RD, 1985, Identificatiaon of an agent in cultures of *Aspergillus fumigatus* displaying antiphagocytic and immuno-modulating activity, *J Gen Microbiol.*, 131:1251-1258.
8. Mullbacher A, Waring P, Tiwari-Palmi U, Eichner RD, 1986, Structure relationships of epipolythiodioxopiperazines and their immunomodulating activity, *Molec Immunol.* 23, 231-235.
9. Munday R, 1982, Studies on the mechanism of toxicity of the mycotoxin sporidesmin, *Chem Biol Interact.*, 41:361-374.
10. Nagarajan R, 1984, Gliotoxin and epipolythiodioxopiperazines, *In*: "Mycotoxins - Production, Isolation and Purification," Vol. 8 - "Developments In Food Science," Betina V, ed., Elsevier Science Publishing, New York.
11. Taylor A, 1971, The toxicology of sporidesmins and other epipolythadioxo-piperazines, *In*: "Microbial Toxins, Vol. 7", Kadis S, Ciegler A, Ajl SJ, eds., Academic Press, New York.
12. Waring P, Eichner RD, Tiwari-Palni U, Mullbacher A, 1986, The isolation and identification of a new metabolite from *Aspergillus fumigatus* related to gliotoxin, *Tetrahedron Letters*, 27:735-738.
13. Waring P, Eichner RD, Tiwari-Palni U, Mullbacher A, 1987, Gliotoxin E: a new biologically active epipolythiodioxopiperazine isolated from *Penicillium terlikowskii*, *Aust J Chem.*, 40:991-997.
14. Waring P, Eichner RD and Mullbacher A, 1988a, The chemistry and biology of the immunomodulating agent gliotoxin and related epipolythiodioxo-piperazines, *Medicinal Research Reviews*, 8:499-524.
15. Waring P, Eichner RD, Mullbacher A, Sjaarda A, 1988b, Gliotoxin induces apoptosis in macrophages unrelated to its antiphagocytic properties, *J Biol Chem.*, 263:18493-18499.
16. Waring P, Sjaarda A, Braithwaite AW, Geue A, Eichner RD, Mullbacher A, personal communication, Zinc protects against DNA fragmentation induced by sporidesmin and relation epipolythiodioxoxpiperazines, (in prep.).
17. Weindling R, 1932, Trichoderma lignorum as a parasite of other soil fungi, *Phytopathology*, 22:837-845.

ACTIVATION AND SPECIFICITY OF AFLATOXIN B_1 BINDING ON hhc^M, A HUMAN

ONCOGENE OF LOW EFFICIENCY, AND CELL TRANSFORMATION

Stringner S. Yang,[a] Ke Zhang[a], George C. Yang[b] and Janet Taub[a]

Laboratory of Cellular Oncology
National Cancer Institute
Bethesda, MD 20892[a]

Division of Contaminant Chemistry
Food and Drug Administration
Washington, DC 20204[b]

INTRODUCTION

Aflatoxin B_1 (AFB_1), a metabolite of *Aspergillus flavus*, chemically classified as a furocoumarin, is known to be the most potent hepatocarcinogen experimentally tested in various species including trout, rat, hamster, dog and rhesus monkey.[13,21] The carcinogenic effect resides in the mutagenic potential of AFB_1 upon metabolic conversion into the epoxide form by liver mixed function oxidases. By mode of nucleophilic attack, AFB_1-epoxide binds to DNA forming covalent bonds with the N^7 of deoxyguanine (dG), thus inducing an erroneous base-pairing with deoxyadenine (dA) instead of deoxycytidine (dC) (Fig. 1). In eukaryotes, such as hepatocytes, a strong DNA repair enzyme system usually safeguards against such mistakes by excising the mis-paired base, and an organism depends on continual repair of DNA (Fig. 2). If this erroneous base-pairing eluded the cell's DNA repair enzyme, at subsequent DNA replication a conversion of dG.dC --> T.dA ensures, thus resulting in a point-mutation.[7] Such a mutation generates an irreversible DNA lesion (Fig. 2) and if it is of survival value, it usually produces an altered gene product. On the other hand, a disruption of the open reading frame results if the DNA repairing process excised the mis-paired base without replacement with the correct base, i.e. generated a deletion mutation, or if the T.dA substitution of dG.dC generated a nonsense, i.e. stop codon. This usually results in a lethal mutation. Thus within the first 24-48 hours of AFB_1 consumption, 95% of the AFB_1-dG adduct becomes excised and excreted in the urine. The surviving 5% takes a longer course for total elimination. Meanwhile the evolution of an irreversible transformation as a result of this point mutation ensures the initiation of a number of phenotypic changes culminating into uncontrollable cell replication and eventually autonomous tumor growth (Fig. 2).

The mechanism of AFB_1 carcinogenesis was once speculated as the consequence of a cascade of events initiated by the synthesis of an altered protein bearing one single amino acid change due to a point-mutation. This interpretation was based on observations made in an analogous case, e.g. the activated c-Ha-ras-1, which suffered a val (GTC) substitution of the normal gly (GGC) in the 12th codon of the p21 ras protein.[5,18,19]

Microbial Toxins in Foods and Feeds, Edited by A.E. Pohland *et al.*,
Plenum Press, New York, 1990

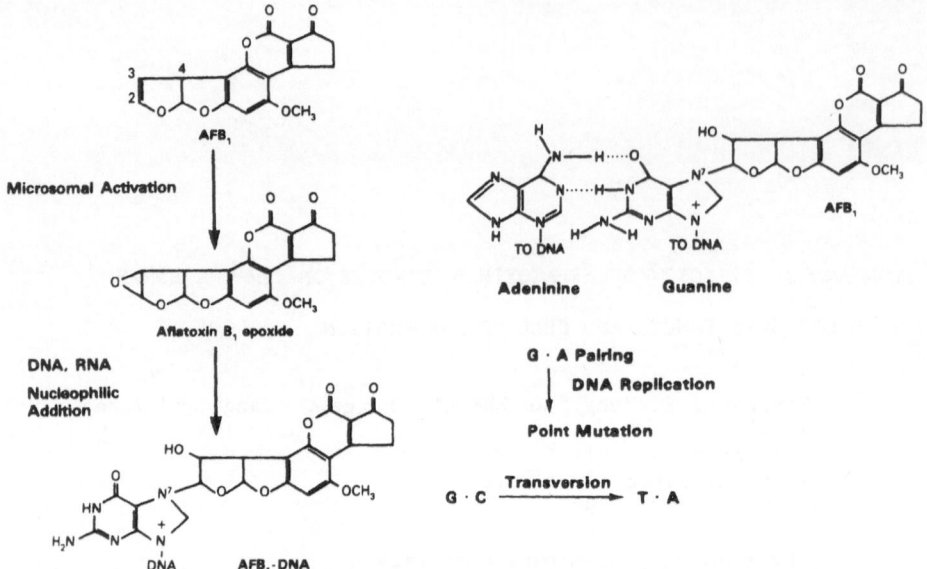

Fig. 1. Interaction of AFB$_1$ with DNA and mechanism of AFB$_1$ induced mutation.

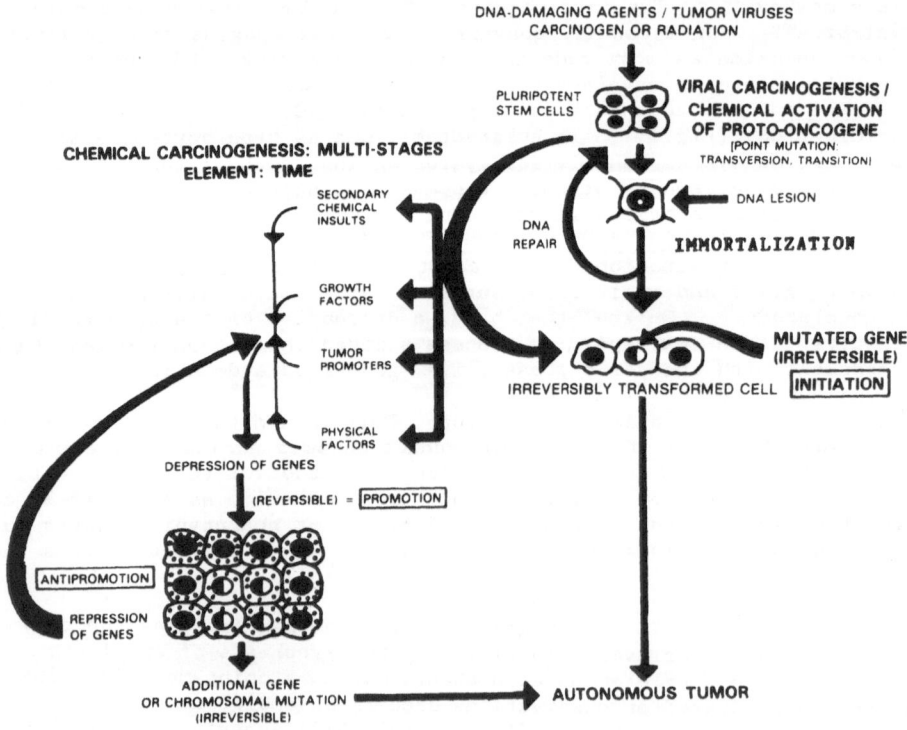

Fig. 2. Mechanism of oncogenesis and tumor promotion.

However, later reports suggested that chemical carcinogen treatment might permit the cell to become immortalized and predisposed to further evolution; but progression and cell transformation required subsequent alteration such as activation of c-Ha-ras-1.[16] Further insight on the role of multistage chemical carcinogenesis was gained by the work of Balmain and Pragnell,[1] in which they showed that c-Ha-ras-l was activated in fully developed mouse skin tumors induced by carcinogen dimethyl benzanthracene and tumor promoter 12-O-tetradecanoyl-phorbol-13-acetate. However, c-Ha-ras-l was also activated in premalignant papillomas, which normally regressed and would not progress to frank malignant tumor without treatment with the promoter. These results argued that the activation of a ras gene by chemical carcinogen alone may not be sufficient to induce both immortalization and malignant transformation. Elevated expression of the cellular ras proto-oncogene also led to cell transformation in tissue culture cells predisposed to immortality.[4] Altogether, even though it is not yet clear if c-Ha-ras-1 gene activation involved the initiation event, its full expression is certainly necessary in the course of tumor development.

In our study with the human hepatoma DNA activation we found that mutational changes induced by AFB_1-epoxide binding on dG also involve sequences other than structural genes, such as regulatory consensus sequences that might lead to other physiological implications. It was noted earlier that AFB_1-epoxide bound preferentially to certain subgenomic restriction DNA fragments of human Mahlavu hepatocellular carcinoma (HHC). In addition, within a finite dosimetry, AFB_1-epoxide activation of one particular Mahlavu HHC 3.1 kb DNA sequence, potentiated its cell-transforming capability when assayed by DNA-mediated gene transfer (transfection) technique on NIH/3T3 cells.[15,23] We have subsequently molecularly cloned the 3.1 kb Mahlavu HHC DNA sequence, and referred to it as hhc^M after extensive screening by [^{3}H]AFB_1 binding,[22] transfection assay on NIH/3T3 cells, and by DNA-DNA hybridization analysis with [^{32}P]Mahlavu HindIII digested DNA fragments and/or human [^{32}P]Alu sequence.[10] Details of the isolation of the hhc^M are described in the Methods. Using the hhc^M, the following questions pertaining to the mechanism by which AFB_1-epoxide activation led to oncogenesis are asked:

(1) What is the nature of the dG targets seen by AFB_1 within a double stranded DNA such as hhc^M? In other words, do the vicinal nucleotides affect the recognition of the dG target by AFB_1 epoxide?
(2) Does the intrinsic secondary structure of the polynucleotide strand affect the accessibility of the AFB_1 epoxide?
(3) Within what particular kind of sequence resides the specific dG target responsible for the potentiation of the hhc^M cell-transforming capability? In short, does this particular sequence encode a structural protein or a regulatory sequence?
(4) What is the relevance of AFB_1-epoxide activation of dG within hhc^M to oncogenesis?

METHODS AND MATERIALS

Molecular cloning of hhc^M

Genomic DNAs purified from human normal liver and Mahlavu (African) hepatocellular carcinoma (HHC), as described below, were subjected to complete digestion by HindIII restriction endonuclease.* The DNA samples, both

---

*Other restriction endonucleases, including BamHI, EcoRI and PstI, were also used for isolating genomic DNA fragments from HHC and liver DNAs in our attempts to clone HHC DNA sequences; the clones isolated from these efforts were not successful with respect to transfection studies.

[³H]AFB₁-epoxide bound (as described below) and unbound, were separated into 180 fractions by polyacrylamide gel electrophoresis. Specificity of [³H]AFB₁-epoxide per μg of DNA was determined. Fractions with significant [³H]AFB₁-epoxide specific activity were used in DNA transfection assay on NIH/3T3 cells as described below. Fractions showing positive focus formation indicating cell transformation, were identified and the parallel *unbound* DNA fractions were molecularly cloned by ligation onto the HindIII site of pBR 322, pBR 325 and/or Puc 8 plasmid DNAs for transformation of *E. coli* HB101 cells as described elsewhere.[23] Primary selection of the resultant clones was thus based on (1) the sensitivity to tetracycline, and/or color change associated with the disruption of the *lac z* operon containing the B-galactosidase coding sequence of the plasmid; and (2) the capability of cell-transformation in transfection assays on NIH/3T3 cells with or without AFB₁ binding, and (3) the presence of human DNA sequence in colony-hybridization and DNA-DNA hybridization against [³²P]probes prepared from human Alu sequence[10] and also [³²P] labelled HindIII disgested MAH HHC DNA fragments. After screening over 30,000 clones by these technical approaches including [³H]AFB₁ binding, transfection assay on NIH/3T3 cells and DNA-DNA hybridization against the [³²P]Alu and [³²P]HindIII MAH HHC DNA probes, three clones were isolated. Two had identical 3.1 kb restriction endonuclease map and one has a slightly larger human DNA insert.

Preparation of plasmid DNA and AFB₁ binding

The clone used in this study, has been referred to as PM-1. Plasmid DNA was prepared by the Holmes' method, i.e. the rapid heating method, followed by CsCl₂-ethidium bromide isopycnic centrifugation at 180,000xg for 20 hrs.[11] The banded PM-1 DNA was then purified free of ethidium bromide by isopropanol extraction and exhaustive dialysis against Tris-EDTA-NaCl (TEN) buffer. A yield of 25 to 50 μg of total plasmid DNA per 5 ml of culture was generally obtained. The 3.1 kb hhc^M DNA was then separated from PUC 8 DNA and other contaminants by digesting the PM-1 DNA with HindIII endonuclease and then subjecting to agarose gel electrophoresis and electroelution of the separated 3.1 kb band. The resultant 3.1 kb hhc^M DNA was homogeneously purified and used in AFB₁ activation experiments.

The hhc^M 3.1 kb DNA was also cloned into a pSVneo vector that carried a murine retroviral (Moloney) LTR, SV40 promoter and part of the T antigen besides the neomycin resistance gene. This clone, rpMpN-1, is expressed at significantly higher level when transfected into cells and offers special advantage for transfection assay.

[³H]AFB₁ at 15 Ci/mmole specificity was acquired from Morales Laboratory, CA. It was further purified by HPLC to homogeneity and the resultant single peak of [³H]AFB₁ had the specific activity of 9,250 cpm/pmole. It was then used in activation reaction with either mixed function oxidases freshly prepared from liver microsomal preparation or by the chemical peroxidation reaction using perchlorobenzoic acid and methylene chloride as described earlier.[2,9] Binding of [³H]AFB₁ epoxide with either high molecular weight HHC or plasmid DNA was monitored by kinetic analysis.[15,23] Samples withdrawn at each time point were washed free of unbound [³H]AFB₁ epoxide with chloroform and ethanol precipitated prior to redissolving the [³H]-AFB₁-DNA in TEN buffer for transfection assay on sequence analysis.

Cells, tissue culture and transfection assay

NIH/3T3 cells, passage 6 to 11, and Buffalo rat liver cells (BRL-1) for transfection assays, were maintained in Dulbecco's modified Eagle's media supplemented with 10% heat-inactivated fetal calf serum, penicillin (50 units ml⁻¹) and streptomycin (25 μg ml⁻¹) (DMEM) in a 5% CO_2 atmosphere, 37°C. All tissue culture materials were purchased from GIBCO, NY.

DNA transfection was carried out as described earlier.[23] Optimal condition was achieved by carefully titrating the pH curve for the DNA-cal-
cium phosphate complex mixture and in our experiments, it was usually found
at pH 6.75 to ensure a fine complex precipitation during transfection.

Preparations of DNA and RNA from tissue culture cells and tumor tissues

Total high molecular weight (HMW) DNA was extracted and purified from
tissue culture cells and tumor tissues as described elsewhere.[23] The HMW
DNA thus purified, has been subjected to proteinase K digestion, first
sequential chemical purification with phenol-cresol, chloroform-isoamyl
alcohol, ether and ethanol-NaCl precipitation, followed by RNase digestion
and a second sequential chemical purification. The purified DNAs were then
dialyzed against TEN buffer for use in experiments. Total RNA was extracted
from tissue culture cells and prepared as described previously.[11] Poly A
rich RNA was obtained by affinity separation with oligo dT cellulose (Collaborative Research, MA) column elution.

Tumorigenesis

Transformed cells, cloned out from the transfected cell culture by either
cloning cylinder method or terminal dilution method, were expanded and inoculated at 10^4 to 10^6 cells into athymic Swiss nu/nu mice subcutaneously.
Tumorigenesis in the challenged mice was monitored closely.

Nucleotide sequence analysis and site-targeted mutagenesis

Nucleotide sequencing of the hhc^M 3.1 kb and variants produced by site-
targeted mutagenesis were carried out by the standard Maxam-Gilbert[12] and
the Sanger (M13) dideoxy sequencing methods.[11]

Specified oligonucleotide sequence of 20 mers carrying the targeted
dG-->T mutation were synthesized by the Applied Biosystem oligonucleotide
synthesizer. They were used as templates in generating the mutated clon·
Mutant DNA clones were produced in accordance with the protocol provided
by and using the oligonucleotide-directed in vitro system of Amersham
(Arlington Heights, IL). DNAs of the mutated clones were verified by nucleotide sequencing. Effects of these site-targeted mutagenized DNA were analyzed by potentiation of cell-transformation in transfection assay on NIH/3T3
cells and RNA expressions in transfected cells using the BRL dot-blot technique (Bethesda Research Laboratory, Rockville, MD).

RESULT AND DISCUSSION

Dosimetry of AFB$_1$ binding and potentiation of hhc^M cell-transformation capability on NIH/3T3 cells

AFB$_1$ epoxide binds high molecular weight DNAs prepared from human hepatoma, human liver and mouse NIH/3T3 cells efficiently (Fig. 3). The initial
rates in each binding kinetic were extremely rapid. The rates of AFB$_1$
epoxide binding to human normal liver or hepatoma DNA and to murine NIH/3T3
cell DNA became significantly different after one minute of binding reaction.
The MAH HHC DNA showed a greater rate of binding than normal liver DNA and
all the dG targets became saturated earlier, whereas AFB$_1$ epoxide bound the
normal liver DNA at a slower rate but eventually saturated all the dG targets
at a slightly lower level. The human DNAs showed a higher level of AFB$_1$
binding than the murine NIH/3T3 cell DNA, the significance of this has yet to
be understood. The overall AFB$_1$ specific activity, i.e. AFB$_1$-dG adduct, was
found to be about one dG bound per 10^5 nucleotides among these high molecular weight double stranded DNAs. This overall specificity also took

into consideration the existence of secondary or tertiary structure of the high molecular weight DNAs. AFB_1 epoxide binding on linearized 3.1 kb double stranded hhc^M DNA was consistently found to be 4 to 8 dG bound per 10^4 nucleotides. This higher binding capability reflects the relatively easy accessibility of dG within the linearized double stranded PM-1 DNA by AFB_1 epoxide and should not be compared with the efficiency of AFB_1-dG adduct formation with high molecular weight native double-stranded DNA.

Within a finite dosimetry the binding of AFB_1 epoxide with dG potentiates the cell-transformation capability of hhc^M by 10 to 20 fold as seen in the experiment illustrated in Table 1. Whereas the efficiency of unbound PM-1 DNA in transforming NIH/3T3 cells was usually observed at about 15 FFU/μg, the efficiency of AFB_1 epoxide activated PM-1 DNA was optimized at 38 FFU/100 ng DNA, representing an increase of 20 fold. The possibility of non-specific mutagenization accounting for this potentiation was considered. That this potentiation effect was due to free AFB_1 which diffused into the cell and/or due to the recycling of AFB_1 adducts has been ruled out earlier with the appropriate control experiments which showed that activation of normal liver or *E. coli* DNA *at the same dosimetry* failed to activate any cell-transforming capability.[23] Moreover in this experiment with AFB_1 activated DNA from c-ras^k-1 or c-hhc, a normal human liver homolog to hhc^M as the appropriate controls, no cell-transformation of NIH/3T3 cells was obtained suggesting that AFB_1 epoxide activated PM-1 DNA was not a random phenomenon. Moreover the AFB_1 dose-dependency of PM-1 DNA in cell-transformation efficiency (Table 1) further substantiated the

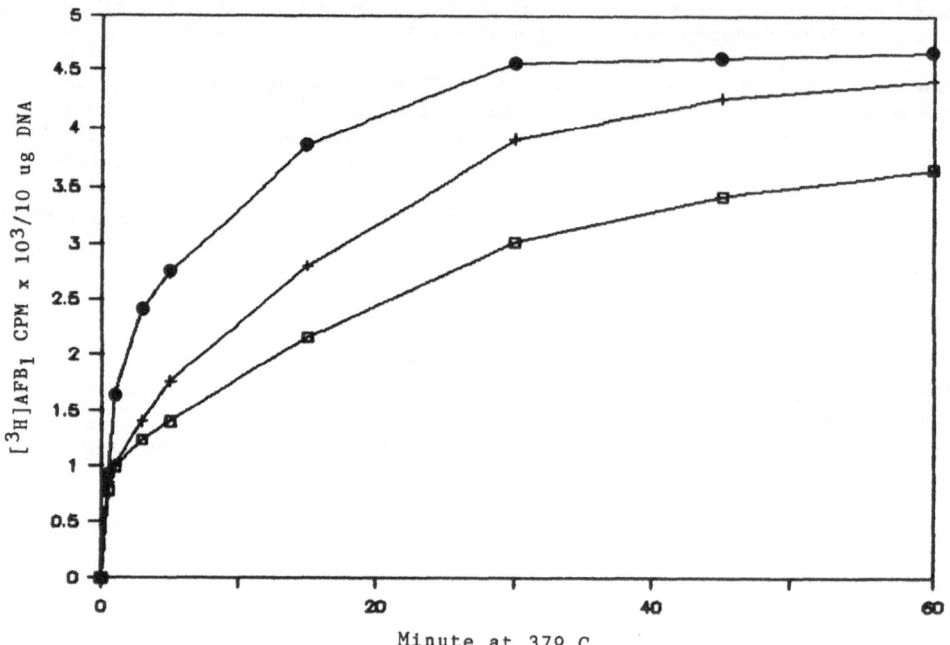

Minute at 37º C

Fig. 3. Aflatoxin B_1 epoxide binding on high molecular weight DNAs prepared from human hepatocellular carcinoma (Mahlavu), ●--●, human normal liver, +--+, and from murine NIH/3T3 fibroblasts, ❑--❑. All experimental details pertaining to the purification of the DNAs and AFB_1-epoxide binding were as described in the Methods.

Table 1. AFB$_1$ Dose-dependent Activation of PM-1 DNA in the transformation of NIH/3T3 Cells.

DNA Source	AFB$_1$ femtomole per 100 ng DNA	Number of Foci per 100 ng DNA
hhc^M (PM-1)	0	15 x 10^{-1}
c-Ha-ras-1	0	465
c-K-ras-1	0	0
c-hhc (human liver homolog)	0	0
E. coli	0	0
hhc^M (PM-1)	0	15 x 10^{-1}
hhc^M (PM-1)	5	18
	14	26
	24	36
	35	3
c-hhc	0	0
	8	0
	15	0
	30	0
	40	0

AFB$_1$ binding and transfection assayswere as described in Methods.
Data were calculated on the basis of per 100 ng. In the assay with
unbound hhc^M DNA the transfection assays were carried out with
500 ng to 1.5 μg of DNA in order to obtain reasonable foci formation on NIH/3T3 cells. Transfection with AFB$_1$-epoxide bound
DNA was carried out at a range of 50 to 500 ng DNA. Data were normalized to show potentiation of hhc^M cell-transformation capability by AFB$_1$-epoxide activation.

specificity of AFB$_1$ epoxide binding in conferring the potentiation of
cell-transformation. Whereas optimal dosimetry was seen at 24 femtomole of
AFB$_1$/100 ng of PM-1 DNA, at dosimetry beyond 45 femtomole per 100 ng of
PM-1 DNA, an overkill effect was observed. No transformed foci were obtained
in NIH/3T3 cells transfected with AFB$_1$ epoxide bound PM-1 DNA although
human DNA was incorporated into the NIH/3T3 cells in a degraded form.[15,23]
This observation suggested that over activation of PM-1 DNA not only generated scissions in the molecule but possibly degradations leading to a loss
of biological activity. It was also evident from these results that no more
than one, or at most a few, AFB$_1$-dG adduct per PM-1 DNA molecule could be
tolerated by the hhc^M DNA before the biological activity of the hhc^M DNA
became compromised and at the risk of survival. Moreover the potentiation of
hhc^M DNA in cell transformation probably necessitates no more than one or
at most a finite number of AFB$_1$ bindings.

In order to understand the mechanism of AFB$_1$ epoxide potentiation of
hhc^M cell transformation capability it became imperative to understand the
nature of the DNA sequence accessible by and targeted by AFB$_1$ epoxide.

Specificity of the AFB$_1$ epoxide binding on dG's of PM-1 DNA

Deoxyguanine nucleotide of native DNA, when bound by AFB$_1$ epoxide, became alkali and therefore could be identified by piperidine cleavage, whereas unbound deoxyguanine nucleotide within the same native DNA would not cleave without dimethyl sulfide (alkali) treatment. Figure 4 shows the dG targets within the PM-1 DNA bound at a saturation condition. When the targeted sequences are evaluated in sets of tetranucleotides, an empirical formulation can be derived on the basis of the binding pattern of AFB_1 epoxide with the dG's in PM-1 DNA. Table 2 summarizes the nucleotide sequences in sets of tetranucleotides that were seen and targeted by AFB_1 epoxide. As shown in Figure 4, the dG within a sequence of any one of the following tetranucleotides of AGAG, AGAT, TGTT, TGAT, or AGAA escaped attack by AFB_1 epoxide (Category III in Table 2) and hence showed no cleavage in the sequence without prior DMS treatment. This is contrasted by the distinct cleavage of dG as a result of AFB_1 epoxide attack on N^7 of dG in a sequence of GGGC, CGGC, AGGC, TGGC, or CGGG. Upon evaluating the various seuqence in which a dG target could be accessed by AFB_1 epoxide, it can thus be concluded that within a double stranded DNA, the most preferred dG target would be that flanked by dG and/or dC, i.e. category I. And that tetranucleotide sequences in which dG is either preceded by dA or T and followed by dG or dC would be the moderately preferred targets of AFB_1 epoxide, i.e. category II (not listed in Table 2). This, of course, did not take into consideration the secondary or the tertiary structure of the DNA in its natural state since these analyses were done on linearized double-stranded DNA It should also be mentioned that whereas the dG binding affinity of AFB_1 epoxide was greatly affected by the vicinal nucleotides in the double-stranded PM-1 DNA, no specificity was observed with respect to AFB_1 epoxide binding to dG in single stranded DNA. Our observations[23] were basically in agreement with others working on AFB_1 binding on ØX174 and pBR 322 DNAs.[14]

Within the past two years the nucleotide sequence of hhc[M] has been resolved by a combination of Maxam-Gilbert nucleotide sequencing technique and the M13 dideoxy method using the BRL kilobase sequencing system. Applying these empirical rules in a computer analysis of the hhc[M] nucleotide sequence, we predicted the most and moderately preferred dG targets within the various loci of hhc[M] (Table 3). Although a maximum number of 60 dG targets was predicted on the basis of AFB_1 epoxide binding studies with linearized 3.1 kb hhc[M] DNA, it was evident upon examining the possible secondary and tertiary structure of hhc[M] sequence, that a much lower number of dG targets would be accessible by AFB_1 epoxide. Moreover, only a few such induced mutations would produce any effect of survival value.

In order to analyze the possible effect of any such AFB_1 induced dG-->T mutation, site-targeted mutagenesis study of the hhc[M] DNA was initiated using polynucleotides of 20 mers that carried a predicted dG-->dT point-mutation, presumably the result of an AFB_1-epoxide mutagenesis. Thus far, only a few of the predicted dG-->dT mutagenesis sites have been analyzed and these are summarized in Table 4. The recombinant construct carrying hhc[M] sequence in the SV40 T antigen vector plus a neomycin resistance marker, rpN[r]pM-1 was used in this study since it offered the advantage of selecting the transfected cells by its resistance to Gentamicin sulfate (G418), an analog of neomycin. Using expression of hhc[M] specific mRNA as a criterion, we analyzed by Northern dot-blot in a semi-quantitative assay of the mRNA, i.e. poly A enriched RNA, expressed in the G418 resistant NIH/3T3 cells after transfection with the mutagenized hhc[M] sequence. Focal transformation in these cells was monitored for 4 to 6 weeks.

Results from seven mutagenized clones, for which nucleotide sequence confirmation was available, suggested that, thus far, mutation leading to a structural protein alteration did not seem to potentiate the cell-transformation of hhc[M] (Table 4). Alternatively the introduced dG-->T mutations which led to amino acid substitution, so far, have not altered cell-transformation or expression of mRNA levels. These included mutation at 577 which

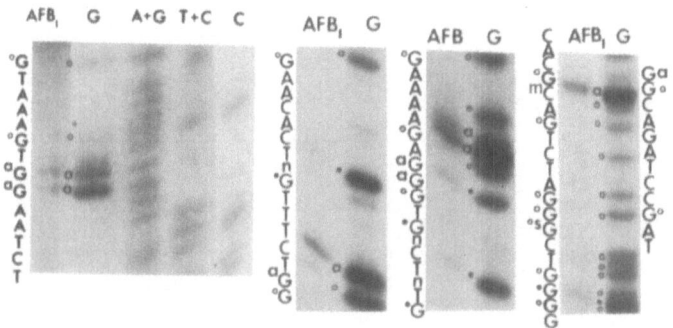

Figure 4. Identification of the dG bound by AFB₁ epoxide within
the hhc^M (PM-1) DNA by a modified Maxam–Gilbert
sequencing method. Nucleotide sequences are specified
on the side. The left panels illustrates ladder for
all four deoxynucleotides and AFB₁-dG; only native dG
and AFB₁-dG were given in all other three panels on
the right. aG = AFB₁ bound dG at all time; ^OG = dG
that was not reacted with AFB₁; whereas •G =moderately
preferred dG. s^O= In this case, methylation of dG prevented
AFB₁ binding on dG and hence no cleavage was obtained.

Table 2. Vicinal Nucleotide Sequence Dictates the dG Targets
of AFB_1 epoxide Binding[@]

Preferred Targets Category I	Least Favored Targets Category III
*	*
GGGG	AGAG
GGGC	AGTG
GGGA	AGAA
GGGT	AGAC
	AGAT
CGGG	TGAG
AGGG	TGAC
TGGG	TGAA
	TGTA
CGGC	TGTG
AGGC	TGAT
TGGC	TGTC
	TGTT
CGGA	
AGGA	
TGGA	
CGGT	
AGGT	
TGGT	

*

[@]This table represents the dG targets of AFB_1 epoxide binding
observed in studies with linearized double stranded PM-1 DNA.
Moderately preferred dG targets, i.e. Category II, are omitted
here but are described elsewhere[15].

Table 3. Predicted dG Targets within the Nucleotide Sequence of hhc^M
Preferrentially Attached by AFB$_1$-epoxide.

*	*	*	*	*	*	*	*	*	*	*
GGGG	CGGC	GGCC	GGGC	GGGA	AGGA	TGCC	TGCG	TGGA	TGGG	GGAG
	* CGGC			* GGGA	* AGGA			* TGGA	* TGGG	
					73					
					74					
						84				
								97		
								98		
				125						
				126						
									140	
								221		
					223					
					224					
					307					
					308					
								371		
								391		
			472							
									481	
									492	
				494						
				495						
					539					
										550
				560						
				561						
626			577							
				692						
									860	
					901					
					1125					
				1320						
				1321						
								1330		
									1354	
									1404	
	1405									
				1431						
	1543									
	1588									
									1637	
						1652				
				1765						
					1815					
	1853									
					1862					
										1868
							1878			
					1986					
				2064						
					2094					
	2205									
								2315		
								2331		
				2352						
								2460		
2482										
								2718		
							2797			
								2884		
							2926			

Table 4. The Effect of dG -->dT Mutation Induced by Site-targetted Mutagenesis within the hhc^M DNA Sequence

# on hhc^M	Sequence	mRNA Synthesis$^&$	Cell Transformation$^&$
73	AGGA --> ATGA	+	_@1
74	AGGA --> AGTG	+	_@1
492	TGGG --> TGTG	+	_@2
550	GGAG --> GTAG	+	_@2
577	GGGC --> GTGC	+	+
626	GGGG --> GTGG	++	++@3
1005	TGCA --> TTCA	+	+

@1 Disruption of ribosomal RNA (16S) binding site: AGGA.

@2 Creation of a stop codon: UGA.

@3 Creation of an enhancer sequence: GGTGTGGTAAAG[8,17,20] and hence increases expression.

$^&$Cell transformation was determined by transfection analysis as described in Methods and mRNA synthesis in transfected cells was determined by Northern dot-blot analysis with [^{32}P]3.1 kb hhc^M DNA.

caused an amino acid substitution of Gly-->Val, and mutation at 1005 which resulted in no amino acid substitution because of the wobbling code.

Within the hhc^M nucleotide sequence there exists an apparent open reading frame, ORF, coding for a polypeptide of about 467 amino acids. This was in good agreement with a 53-55 kD protein and some smaller polypeptides including one 53 kD protein observed in cell-free protein synthesis using hhc^M-specific mRNA in a rabbit reticulocyte lysate system. dG-->T mutations at nucleotide 73 and 74 in the 5' terminus, which bears the consensus sequence for ribosomal RNA binding site just 5' ahead of the first methionine codon, blocked cell transformation although hhc^M specific mRNA level showed no difference. This could be interpreted as a result of blocking protein synthesis. Likewise, mutations at 492 and 550 also blocked cell-transformation since a stop codon (UGA) was introduced in each case to stop protein synthesis prematurely.

It was of interest to note that dG-->T mutation at 626 generated a sequence resembling the enhancer sequence for RNA polymerase II, which was reported to function even within the coding sequence (footnote of Table 4). The level of mRNA was increased by 1.5 fold and cell-transformation seemed to be enhanced by a slight increase in the number of foci per μg of DNA. This observation suggested that one possible action by which AFB$_1$ induced mutation in hhc^M, which itself is a moderately transforming DNA sequence, led to an increase in its transformation potential is probably mediated through augmentation of hhc^M expression. This is analogous to other observations which also indicated that an elevated expression of the cellular ras protooncogene driven by a murine LTR sequence, containing both promoter and enhancer sequence, also led to cell transformation in tissue culture cells predisposed to immortality.[4,6]

CONCLUSION

We have earlier molecularly cloned a moderately transforming DNA sequence, hhc^M, from an African hepatocellular carcinoma cell line. It was observed that related DNA sequences homologous to hhc^M were readily detected in both Korean and Chinese hepatomas. hhc^M was identified and molecularly cloned from hepatoma DNA by virtue of its relatively high accessibility by AFB$_1$ epoxide binding to its dG residues. Populations in Central Africa and in countries like Korea, China and Japan, were exposed to AFB$_1$ regularly by their food intake, and that epidemiological data implicated AFB$_1$ as the etiological agent for the high cancer mortality observed among males in these areas. This investigation on the DNAs prepared from hepatomas of patients originated from these countries intended to glean information on the possible molecular mechanims of AFB$_1$ mutagenesis.

Although AFB$_1$ mutagenesis is no doubt an important factor in the hepatocarinogenesis process among these populations, the role of other infectious agents, such as human hepatitis B virus (HBV), has also been implicated on the basis of epidemiological studies. In view of the multi-stage nature of hepatocarcinogenesis it is not surprising that both chemical carcinogen, like AFB$_1$, and biological carcinogen, like HBV, may play a significant role, whether independently or synergistically, in hepatocarcinogenesis. HBV infection early in the life of human, has been etiologically proposed as a crucial factor in the initiation of oncogenesis, which then remained dormant until re-activation later in life. Evidence against this thesis came from the fact that female infants were exposed to the same extent of HBV infection as male infants. Yet, hepatocarcinogenesis is a neoplasm striking predominantly male.

AFB$_1$ exerts its carcinogenic effect probably by repeated chemical insults.[1] Our study shows that not all the dG's within the human liver DNA are accessible by AFB$_1$ epoxide. Results obtained in this analysis of the nature of the AFB$_1$ dG targets indicate that some degree of specificity dictated by the vicinal nucleotide sequence, seems to prevail. AFB$_1$ epoxide binding of dG within a critical codon ultimately leads to an alteration of biological function, whether it be a mutation within a sequence encoding a structural protein(e.g. proto-oncogene) or a regulatory sequence in transcription. In the latter, it leads to the heightened expression of a similar proto-oncogene or hhc^M, driven by a promoter-enhancer like sequence, causing imbalance synthesis, and eventually cell-transformation.

REFERENCES

1. Balmain A, Pragnell IB, 1983, Mouse skin carcinomas induced *In vivo* by chemical carcinogens have a transforming Harvey-ras oncogene, *Nature*, 303:72-74.
2. Bennett RA, Essingmann JM, Wogan GN, 1981, Excretion of an aflatoxin-guanine adduct in the urine of aflatoxin B$_1$ rats, *Cancer Res.*, 41:650-654.
3. Beasley RP, Hwang LY, 1984, Epidemiology of hepatocellular carcinoma, *In*: "Viral Hepatitis and Liver Disease," Vyas GN, Dienstag JL, eds., Grune and Stratton, Inc., New York.
4. Blair DG, Oskarsson M, Wood TG, McClements WC, Fischinger PJ, van de Woude GG, 1981, Activation of the transforming potential of a normal cell sequence: a molecular model of oncogenesis, *Science*, 212:941-943.
5. Capon DJ, Chen EY, Levinson AD, Seeburg, PH, Goeddel DV, 1983, Complete nucleotide sequences of the T24 human bladder carcinoma oncogene and its normal homologue, *Nature*, 302:33-37.
6. Chang EH, Furth ME, Scolnick EM, Lowy DR, 1982, Tumorigenic transformation of mammalian cells induced by a normal human gene homologous to the oncogene of Harvey murine sarcoma virus, *Nature*, 297:479-483.
7. Croy RG, Essingmann JM, Reinhold VN, Wogan GN, 1978, Identification of the principal aflatoxin B$_1$-DNA adduct formed *in vivo* in rat liver, *Proc Natl Acad Sci USA*, 75:1745-1749.
8. Dynan WS, Tjian R, 1985, Control of eukaryotic messenger RNA synthesis by sequence-specific synthesis DNA binding proteins, *Nature*, 316:774-777.
9. Garner RC, Martin CN, Smith JRN, Coles BF, Tolson MR, 1979, Comparison of aflatoxin B$_1$ and aflatoxin G$_1$ binding to cellular macromolecules *in vitro*, *in vivo* and after peracid oxidation; characterisation of the major nucleic acid adducts, *Chem-Biol Interact.*, 26:57-73.
10. Lawn RM, Fritsch EF, Parker RC, Blake G, Maniatis R, 1978, The isolation and characterization of linked d- and b-globin genes from a cloned library of human DNA, *Cell*, 15:1157-1174.
11. Maniatis T, Fritsch EF, Sambrook J, 1982, "Molecular Cloning", Cold Spring Harbor Laboratory, Cold Spring Harbor, NY.
12. Maxam AM, Gilbert W, 1980, *Methods Enzymology*, 65:499-506.
13. Miller EC, Miller JA, 1981, Mechanisms of chemical carcinogenesis, *Cancer*, 27:2327-2345.
14. Misra RP, Muech KF, Hamayun MZ, 1983, Covalent and noncovalent interactions of aflatoxin with defined deoxyribonucleic acid sequences, *Biochemistry*, 22:3351-3359.
15. Modali R, Yang SS, 1986, Specificity of aflatoxin B$_1$ binding on human proto-oncogene nucleotide sequence, *In*: "Monitoring of Occupational Genotoxicants", Alan R. Liss, Inc., New York.
16. Newbold RF, Overell RW, 1983, Fibroblast immortality is a prerequisite for transformation by EJ c-Ha-ras oncogene, *Nature* 304:648-651.
17. Schaffner WE, Serfling E, Jasin M, 1985, Enhancers and eukaryotic gene transcription, *In*: "Trends in Genetics I", Elsevier Science Publisher, Amsterdam.

18. Tabin CJ, Bradley SM, Bargmann CI, Weinberg RA, Papageorge AG, Scolnick EM, Dhar R, Lowy DR, Chang EH, 1982, Mechanism of activation of a human oncogene, *Nature*, 300:143-149.
19. Taparowsky E, Saurd Y, Fasano O, Shimizu K, Goldfarb MP, Wigler M, 1982, Activation of the T24 bladder carcinoma transforming gene is linked to a single amino acid change, *Nature*, 300:762-765.
20. Watson JD, Hopkins NH, Roberts JW, Steitz JA, Weiner AM, 1987, "Molecular Biology of the Gene," The Benjamin/Cummings Publishing Co., Menlo Park, CA.
21. Wogan GN, 1973, Aflatoxin Carcinogenesis, *In:* "Methods in Cancer Research, Vol. III," Busch H, ed., Academic Press, New York.
22. Yang SS, Modali R, Parks J-B, Taub J., 1988, Transforming DNA sequences of human hepatocellular carcinomas, their distribution and relationship with hepatitis B virus sequence in human hepatomas, *Leukemia*, 2:102s-113s.
23. Yang SS, Taub J, Modali R, Vieira W, Yasei P, Yang GC, 1985, Dose dependency of aflatoxin B_1 binding on human high molecular weight DNA in the activation of proto-oncogene, *Environmental Health Perspective*, 62:231-238.

ALGAL TOXINS

ALGAL TOXINS: GENERAL OVERVIEW

A. E. Pohland

Division of Contaminants Chemistry
Center for Food Safety and Applied Nutrition
Food and Drug Administration
200 C Street, S.W.
Washington, DC 20204

INTRODUCTION

One of the most serious, widespread, and difficult to control health problems associated with the consumption of foods is that presented by toxins of
algal origin in seafoods. This problem is especially difficult to manage
because of: (a) the extreme toxicity of the toxins; (b) the difficulty of
obtaining the toxins in sufficient quantity for study, and for the development and application of analytical methods; (c) the difficulty inherent in
the sampling of seafood; and (d) the fact that the resulting toxicoses are
frequently misdiagnosed. Nevertheless, in recent years considerable progress
has been made in the study of the seafood toxins and their control.

Generally, fish and shellfish sometimes become toxic by feeding on microscopic, planktonic organisms called dinoflagellates. These dinoflagellates
(algae) can elaborate toxins (phycotoxins: Gr. *phykos* = seaweed = L. *algae*),
in much the same way that molds generate mycotoxins. Over 2500 dinoflagellate species are known; fortunately, only a few of these species (ca. 2
dozen) seem to elaborate toxins.[2] Under certain environmental conditions,
at present poorly defined, these organism undergo a period of rapid growth (a
bloom), causing a phenomenon known descriptively as a "red tide", because the
large number ($1x10^6$/l) of organisms in the water often color the water various shades of red. Note, however, that as few as 200 "dinos"/l may result in
shellfish too toxic for human consumption.[9] Neither the presence nor the
absence of a red tide guarantees that the shellfish growing in such an area
will be toxic or non-toxic. We now know that shellfish, and especially
clams, mussels, oysters and scallops, as well as certain fish (carnivores,
herbivores, detritus feeders) that feed on the dinoflagellates tend to absorb
and concentrate the toxins; in some cases the toxins are metabolized to toxin
derivatives.

Toxicology

In terms of human suffering, the most common phycotoxicosis is ciguatera,
a toxicosis resulting from ingestion of certain species of tropical fish.
This toxicosis is characterized by both gastrointestinal (nausea, vomition,
diarrhea, abdominal pain) and neurological disorders; the latter include, in
order of increasing severity, cold sensitivity, itching, asthenia, muscular

Microbial Toxins in Foods and Feeds, Edited by A.E. Pohland *et al.*,
Plenum Press, New York, 1990

ache, tingling/numbness in the extremities, metallic taste, dizziness, cyano-
sis, chills/profuse sweating, and reversal of the sensations of hot and
cold.[17] In severe cases slowing of the heart and lowering of blood pressure
leads to circulatory collapse, coma and death.[24] As is frequently the case
with lipid soluble toxins, the effects are long lasting, again dependent on
dose, sometimes incapacitating an individual for years. The average duration
of illness is highly variable from patient to patient depending on dose, with
gastrointestinal symptoms of shorter duration (diarrhea - 6.8 days ave.) than
neurological symptoms (cold sensitivity, oral - 53 days ave.).[17] Fortunately, indications are that intravenous administration of mannitol may be
useful in the treatment of ciguatera.[46]

Paralytaic shellfish poisoning (PSP) is the second most frequently encountered (and documented) human phycotoxicosis. It is clearly associated
with the ingestion of toxic shellfish (clams, mussels, oysters, scallops).
The effects of ingestion of such shellfish are predominantly neurological;
symptoms include a tingling, burning, numbness sensation, drowsiness, inco-
herent speech, and respiratory paralysis.[68] The toxins are relatively fast
acting compared with ciguatera, death occurring within 24 hours when no respiratory support is provided. On the other hand, because of the largely
lipophobic nature of the toxins involved, the patient recovers completely
within 24 hours, when respiratory support is provided, the toxins are completely eliminated from the body with no long-lasting, negative effects.

Diarrheic shellfish poisoning (DSP) is a poorly characterized disease of
growing concern associated most frequently with the ingestion of mussels and
scallops. It is characterized primarily by generally mild gastrointestinal
disorders (nausea, vomition, diarrhea, abdominal pain, chills, headache,
fever); it has not been reported to be fatal.[7]

Neurotoxic shellfish poisoning (NSP) also is a poorly characterized toxicosis (in terms of frequency of occurrence, degree of severity, etc.). It
has been associated with ingestion of toxic shellfish along the Florida coast
and the Gulf of Mexico in the U.S., and results in both gastrointestinal and
neurological disorders (tingling and numbness of lips, tongue and throat,
muscular aches, dizziness, reversal of the sensations of hot and cold, diarrhea, vomiting). Red tides responsible for this toxicosis are easily discernible, result in massive fish kills, and are frequently associated with
respiratory problems after disruption of the organism and aerosolization of
the toxin(s).[9]

Amnesic shellfish poisoning (ASP) first came to the attention of public
health authorities in 1987 when 156 cases of acute intoxication occurred as a
result of ingestion of cultured blue mussels (*Mytilus edulis* L.) harvested
off Prince Edward Island, Nova Scotia; 22 individuals were hospitalized with
3 elderly patients eventually succumbing to the toxicosis.[11,39] The toxicosis
is characterized by the onset of gastrointestinal disorders within 24 hours
(vomiting, diarrhea, abdominal pain), and within 48 hours, neurological disorders (confusion, memory loss, disorientation, seizures, coma). The toxicosis appears to be particularly serious in elderly patients, inducing symptoms reminiscent of Alzheimer's disease.

The occurrence of seafood-associated human toxicoses in the United States
is well-documented. The CDC, in the course of their foodborne disease surveillance activities, have developed statistics on confirmed foodborne disease
outbreaks in the United States (Table 1), grouping such cases into bacterial,
chemical parasitic and viral categories.[4] Such statistics confirm the
general view that from a human health standpoint, bacterial contamination of
foods is by far the most serious problem. On the other hand, it is interesting to note that of the chemical contaminants identified in foodborne disease
outbreaks, "ciguatoxin" accounts for over half of the confirmed outbreaks.

Table 1. Confirmed foodborne disease outbreaks,
United States (1978-1982).[4]

Etiologic Agent	Number	Percent
Bacterial	696	68.4
Chemical	237	23.3
Ciguatoxin	75	7.4
Heavy Metals	10	1.0
MSG	5	0.5
Mushroom	17	1.7
PSP	10	1.0
Scombrotoxin	73	7.2
Other	47	4.6
Parasitic	36	3.5
Viral	50	4.9
Confirmed Total	1017	100.0

PSP is associated with relatively few outbreaks, most likely because of the
strong control programs exerted in the United States to prevent exposure to
toxic shellfish. Such statistics are not a good indicator of incidence of
NSP and DSP primarily because most cases of such toxicoses go unreported.

There are published reports indicating that the true incidence of human
illness resulting fron ingestion of seafood is much greater than is reflected
in the above statistics (Table 2). For example, estimates of the yearly
incidence of ciguatera alone worldwide is ca. 50,000 cases.[3] The WHO estimates a fatality rate of 0.1-12% in episodes of ciguatera. Locally, in tropical areas in particular, the incidence of poisoning may be quite high, e.g.
Vernoux estimated that 0.3-1% of the population of the French West Indies
each year is affected. Considering that in many areas of the Caribbean seafood is a staple food, and that fish in these areas are frequently toxic,

Table 2. Phycotoxicoses - Worldwide

Toxicosis	Cases (Date)	Location	Reference
Ciguatera	50,000/yr	Worldwide	3
		United States	23
	27/1000*/yr	Virgin Islands	35
	0.5/1000*/yr	Dade Co., FLA	61
	60/1000*/yr	Gambier Islands	10
PSP	2500 (to 1984)	Worldwide	68
	222 (1880-1970)	Canada	36
	>100 (1976)	Western Europe	16
	187 (1987)	Guatemala	51
DSP	>1300 (1976-1982)	Japan	68
	>5000 (1981)	Spain	7
NSP	5 (1970-1974)	United States	68
ASP	156 (1987)	Canada	11,39

the magnitude of the public health problem becomes clear. For example, Olsen[43] reported that "in St. Thomas, nearly 50 of 84 species of fish in the commercial catch and 56% of the total landings bear some risk of intoxication if eaten."

The incidence of documented cases of PSP, NSP, DSP, and ASP is much less than that for ciguatera. This is undoubtedly a reflection of our inability to accurately measure the true incidence because of: (a) misdiagnoses; (b) poor reporting of poisoning cases; and (c) the application of effective control measures to the harvest of shellfish. Of these toxicoses, the most serious appears to be PSP where the extreme potency of the toxins has in the past resulted in an unusually high mortality rate. That PSP can be a serious public health problem to this day was demonstrated vividly in a recent Guatemalan outbreak of 187 cases of PSP, resulting from ingestion of a clam soup, whereby 26 deaths occurred.[51] A recent publication describing the tumor-promoting activity of the DSP toxins (okadaic acid and pectenotoxin-1) on mouse skin is also of some concern; whether such compounds are tumor promoters also of the digestive system remains to be seen.[18]

Origins of the Toxins

In Table 3 the principal organisms involved in the genesis of the toxins are listed. Undoubtedly, as time goes on additional algal species will be identified as toxigenic. Examination of this list leads to the following observations: (a) In the case of ciguatera, the major toxins isolated have yet to be identified structurally. Of considerable interest is the recent find by Dr. Dickey and his colleagues that cultures of *P. concavum* produce okadaic acid, implicating this acid (one of the DSP toxins) in ciguatera.[15] (b) With respect to PSP, one is struck by the large number of compounds (20) contributing to this toxicosis. Of special interest is the finding that *Aphanizomenon flos-aquae*, a blue-green alga, is capable of elaborating saxitoxin and some of its derivatives,[13,26] and the finding that a *Moraxella* sp., an intracellular bacterium associated with *P. tamarensis*, also produces saxitoxin.[28] This raises the interesting possibility that the PSP-producing dinoflagellates" may, in the end, be only vectors and not the originators, of the toxins.

Chemistry

Ciguatera: Apparently ciguatera is caused by several toxins, the total number at present unknown. Examinations of toxic fish have revealed the presence of two lipid soluble toxins (ciguatoxin and scaritoxin) and a water-soluble toxin (maitotoxin). The relative proportions of the individual toxins varies with the source; e.g. extracts from the parrot fish contained mainly ciguatoxin and scaritoxin, whereas those from the surgeonfish contained mainly ciguatoxin and maitotoxin ("maito" being the Tahitian name for surgeon fish).[71] In general, herbivorous fish seem to contain more maitotoxin than ciguatoxin, whereas in carnivorous fish ciguatoxin is predominant.[9]

Although the literature is replete with references to research into the isolation, purificatioan, and characterization of the toxins responsible for ciguatera, we still do not know the chemical structures of these compounds. This of course is partly due to the difficulty in isolation, purification, and crystallization of an extremely toxic material present in a biological matrix at extremely low concentrations even in the most toxic fish (e.g. a moray eel liver contains ca. 1 μg ciguatoxin). Even so, by 1967 Prof. Scheuer had obtained a highly purified sample of ciguatoxin.[54] To date the information available (Table 4) leads one to conclude that these compounds are high molecular weight polyethers, the molecular weight of ciguatoxin often being referred to as slightly over 1000 and that of maitotoxin about 3400. To make matters more difficult, Nukina and co-workers[41] have

Table 3. Phycotoxins - Genesis

Toxicosis	Organism Implicated	Toxins Produced	Reference
Ciguatera	*Gambierdiscus toxicus*	Ciguatoxin	67
	Prorocentrum concavum,	Maitotoxin	67
	lima, mexicanum	Scaritoxin	67
	Amphidinium spp.	Okadaic acid	15
	Ostreopsis siamensis, lenticularia, ovata		
PSP	*Gonyaulax excavata catenella, tamarensis acatenella*	Saxitoxins (STX) Gonyautoxins (GTX)	56
	Protogonyaulax catenella, bahamense		56
	Pyrodinium bahamense		56
	Cochlodinium	Zn complexes of GTX	44
	Gymnodinium catenatum	GTX 5,6 and C 1-4	56
	Jania sp.	GTX 1-3	56
	Aphanizomenon flos-aquae	STX, N-STX, GTX 1-3,5	26
	Moraxella sp.	STX	
DSP	*Halichondria okadai, melandocia*	Okadaic acid	62
	Dinophysis fortii, acuminata,	Dinophysis toxins (DPT)	7
	acuta, norvegica	Pectenotoxins	37
	Prorocentrum lima, minimum	Yessotoxin	38
NSP	*Ptychodiscus brevis (Gymnodinium breve)*	Brevetoxins 1-8	7
ASP	*Nitzshia pungens Chondria armata*	Domoic acid	42

suggested, and provided evidence, that ciguatoxin exists as two interchangeable forms.

PSP: Paralytic shellfish poisoning is caused by a group of closely related compounds containing the very uncommon (for naturally occurring compounds) tricyclic diguanidine moiety (Fig. 1). As was adumbrated earlier, these compounds originate with many algal species from the genus *Gonyaulax*, and are now grouped under the genus *Alexandrium*.[12,60]

The shellfish exhibit marked differences in their ability to assimilate the toxin and to eliminate/metabolize it. For example, the Bay mussel (*Mya edulis*) concentrates the toxins mainly in the digestive gland and retains the toxin for about two weeks; it accumulates and releases the toxin rapidly. The soft shell clam (*Mya arenaria*) accumulates and releases the toxin slowly; the toxin is concentrated in the digestive glands in the summer months, and in the gills during the autumn and winter months. The Alaskan butter clam accumulates the toxin primarily in the siphon and only slowly releases the toxin.[9]

Although PSP has been known for hundreds of years it was not until 1957 that Schantz and co-workers isolated pure toxin from the California mussel and the Alaskan butter clam as a highly hygroscopic, colorless powder. The

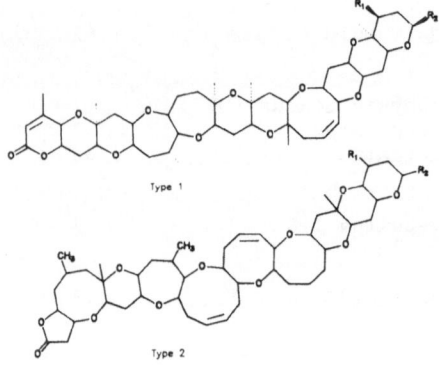

PARALYTIC SHELLFISH POISONS (PSP's)

			Carbamate	Sulfamate	Decarbamoyl
R_1	R_2	R_3	R_4=CONH$_2$	R_4=CONHSO$_3$	R_4=H
H	H	H	Stx	B1 (Gtx5)	dc Stx
H	H	OSO$_3$	Gtx2	C1 (epi-Gtx8)	dc Gtx2
H	OSO$_3$	H	Gtx3	C2 (Gtx8)	dc Gtx3
OH	H	H	neo-Stx	B2 (Gtx6)	dc neo-Stx
OH	H	OSO$_3$	Gtx1	C3	dc Gtx1
OH	OSO$_3$	H	Gtx4	C4	dc Gtx4

DIARRHETIC SHELLFISH POISONS

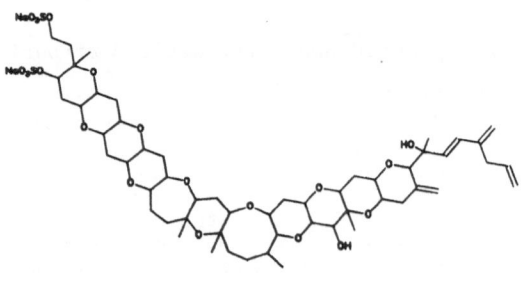

	R_1	R_2
Okadaic Acid	H	H
Dinophysis Toxin-1	H	Me
Dinophysis Toxin-3	Acyl	Me

Type 1

Type 2

NEUROTOXIC SHELLFISH POISONS

Brevetoxins

DIARRHETIC SHELLFISH POISONS

	R
Pectenotoxin 1	CH$_2$OH
2	CH$_3$
3	CHO
6	COOH

AMNESIC SHELLFISH POISON

DOMOIC ACID

C$_{15}$H$_{21}$NO$_6$

DIARRHETIC SHELLFISH POISONS

Yessotoxin

Fig. 1. Seafood Toxins

Table 4. Seafood Toxins: Chemical and Physical Properties

Toxin	Mol. Wt.	Formula	Physical Properties	Ref.
Ciguatoxin	1112	$C_{59}H_{85}NO_{19}$		24
Maitotoxin	3424	$C_{160}H_{259}O_{70}S_2N$	Colorless, amorphous α^{D21} 16(0.36 MeOH/H$_2$O) UV^{MeOH} 229 (10500)	71
Saxitoxin	372	$C_{10}H_{19}Cl_2N_7O_4$	pKa 11.5, 8.1	56
Neosaxitoxin	388	$C_{10}H_{19}Cl_2N_7O_5$	pKa 8.65, 6.75	56
Okadaic Acid	804	$C_{44}H_{68}O_{13}$	MP 164.5-165°C α^{D25} 23.9 (0.088, CHCl$_3$)	27
Brevetoxin B	894	$C_{50}H_{70}O_{14}$	MP 270°C (dec.) UV^{MeOH} 208 (16000) IR^{KBr} 1725, 1685	40
Domoic Acid	311	$C_{15}H_{21}NO_6$	MP 213°C (dec.) Colorless Needles α^{D25} -111 (0.2, H$_2$O) UV^{HOH} 242 (4.42)	42

chemical structure was finally settled in 1975.[53] Further research revealed that PSP was caused not by a single toxin, but by a group of toxins, all
derivatives of saxitoxin (Fig. 1).[60] These include the five hydroxy/hy-
droxysulfate derivatives (N-STX, GTX 1-4), the six N-fulfamoyl toxins (the
so-called "cryptic" toxins), the six decarbamoyl toxins, and most recently
the two Zn-bound PSP toxins (Zn-GTX 1,4). These classes of compounds are
easily interconverted; e.g. the sulfamate toxins are easily hydrolyzed to the
corresponding carbamates and these, either chemically or enzymatically, to
the decarbamoylated derivatives. For example, B1 is converted quantitatively into saxitoxin and B2 into neosaxitoxin by heating 5 min. at 100° in
0.1 M HCl. Conversely the decarbamoyl saxitoxin (dcSTX) is readily converted into saxitoxin by treatment with chlorosulfonylisocyanate in formic acid
at 0°C. Neosaxitoxin itself is readily converted into saxitoxin by reduction with zinc/acetic acid. Both the biosynthetic pathway and a total synthesis of saxitoxin have been accomplished and published.[56]

   DSP: Diarrheic shellfish poisoning has been frequently observed in Japan
and northern Europe,[29] and is apparently caused by a group of polyether type
compounds (Fig. 1);[7] these may be grouped in an acidic fraction, including
okadaic acid, a compound originally isolated from a sponge (*Halichondria
okadai*);[62] dinophysis toxin-1, the 35-methyl derivative of okadaic acid and
dinophysis toxin-2, the 7-0-acyl derivative, and a neutral fraction composed
of some polyether lactones, pectenotoxins 1-3, 6,[37] and a large polyether,
yessotoxin,[38] reminiscent of the brevetoxins.

   Some of the chemical and physical properties of okadaic acid are shown in
Table 4. It is a fairly high molecular weight (804), crystalline compound,
with an astonishing number (17) of asymmetric centers. The free carboxylic
acid functions are easily derivatized using the anthryl diazomethane (ADAM)
reagent.[32] A total synthesis of okadaic acid has been published.[27]

   NSP: The structures of the toxins thought to be responsible for NSP[55,7]
are shown in Fig. 1. Again the toxins are complex polycyclic polyethers of

relatively high molecular weight. Some of the physical data on brevetoxin B as a representative of this type of compound are shown on Table 4.[34]

ASP: ASP is caused by an unusual amino acid, domoic acid (Fig. 1).[11,39] Some of the physical properties of this material are shown in Table 4.[42]

Of course, the key to the control of the problem of contamination of seafood by these toxins is the availability of good analytical methodology for their detection. Table 5 summarizes the current status of the available methodology. In each case the mouse bioassay has been used to advantage as a qualitative or semiquantitative assay procedure. In each case the disadvantages of such an assay (i.e. poor quantitation, large CVs, objections to the use of test animals, etc.) has encouraged the development of alternate analytical

Table 5.   Seafood Toxins: Analytical Methodology

Toxin	Assay Procedure		Comments	Ref.
Ciguatoxin	Bioassay:	Mouse Mosquito Cat, Mongoose Guinea pig ileum	Non-specific	67
	Stick EIA		Separates toxic from non-toxic fish	24
	HPLC/FL			57
Maitotoxin	HPTLC		Red color with H_2SO_4	
Saxitoxin & Dervs.	Bioassay:	Mouse Fly	Det. Lt. 32-58 μg/100g	66 52
	HPLC/Post-column ox. TLC Cellulose acetate electrophoresis		Det. Lt. ca. 20 μg/100g	59 45
	RIA			69
	Indirect ELISA		Det. Lt. 50-100 ppb	14
	ELISA		Det. Lt. 20 ppb	50
Okadaic acid	Bioassay:	Rat, Mouse Suckling mouse		22
	GLC (TMS ethers) TLC HPLC/UV			
	HPLC/Fl (ADAM)		Det. Lt. 100-200 μg/100g	32
	ELISA		Det. Lt. 1 μg/100g	33
	ELISA		Det. Lt. 10 ppb	64
Brevetoxins	Bioassay:	Mouse		7
	TLC			
	HPLC			48
	ELISA			8
Domoic acid	Bioassay:	Mouse		
	HPLC/DAD		Det. Lt. 75 μg/100g	49
	HPLC/UV		30 μg/100g	49

procedures. In the case of the PSP toxins the mouse bioassay usually under-timates toxin concentration due to the insensitivity of the organism to some of the toxins (e.g. the "cryptic" toxins).[21] Unlike PSP, the dose-survival time of the DSP toxins in the mouse assay fluctuates considerabaly and fatty acids interfere with the analysis giving false positive results (this is a particularly serious problem in scallops in late spring - early summer). This prompted development of the suckling mouse bioassay in which one measures the fluid accumulation ratio (FAR), i.e. the ratio of the weight of the intestine/weight of the remaining body, 4 hours after treatment with a scallop extract.[22] Interestingly, studies using this assay showed that there was no difference in toxicity between cooked and uncooked mussels.[22]

Of the analytical chemical methods developed thus far, HPLC-based methods appear to show the most promise and greatest potential for replacing the mouse bioassay. For the saxitoxins the use of HPLC with post-column oxidation (H_2O_2/OH- , IO_4- , tBUOO-) of saxitoxins to fluorescent derivatives shows the most promise for quantitation of the individual toxins.[59] For okadaic acid, derivatization with 9-anthryldiazomethane to form a highly fluorescent product appears to give good results.[32] For domoic acid, direct UV detection of the extended conjugated system gives a very sensitive assay.[49] Of course, in all cases the need for dependable primary standards is the key to good results and is currently the most vexing problem.

Several immunoassays have been developed that show great promise, for use in control programs. For ciguatera the "Hokama stick test" has been applied successfully in Hawaii to control exposure to ciguatoxic fish.[24] Interestingly the immunogen has also been found to cross react with the brevetoxins, presumably due to structural similarities between ciguatoxin and the brevetoxins, presumably due to structural similarities between ciguatoxin and the brevetoxins. UBE Industries, Ltd.[64] of Japan has now marketed an ELISA test kit for detection of the DSP toxins. The kit has an advertised detection limit of 10 ppb with a detection range of 10-300 ppb. The monoclonal antibodies used are specific for okadaic acid and dinophysis toxins 1 and 2, with little cross reactivity with the pectenotoxins and yessotoxin.

Comparative Toxicology and Human Consequences

In Table 6 a comparison is made of the mouse (i.p.) toxicity of the seafood toxins with some well-known toxins. One can readily conclude that these toxins must be handled with respect. Ciguatoxin and maitotoxin are extremely toxic naturally occurring substances. Yet ciguatera is rarely lethal, presumably due to the low concentration levels normally found in edible fish flesh (1-10 ppb). Incidence of death is <0.5%; in cases where death occurred, the victim has consumed those parts of the fish in which ciguatoxin is known to concentrate (liver, viscera, organs, roe). There is some evidence that certain individuals are more susceptible to the toxins than others, and that multiple exposure results in increased sensitivity to the toxins.[25]

The saxitoxins, although slightly less toxic than ciguatoxin, are much more dangerous to humans, because they normally occur at higher concentrations. By far the grestest number of recorded fatalities due to ingestion of seafood have been caused by the saxitoxins. In humans, sensitivity to the saxitoxins is so variable (partly because toxin mixtures are being consumed) that estimates of the human lethal dose varies from 500-1200 μg/100 kg;[68] of course, recovery is near certain if respiratory support is provided in time. In 1944 in Canada the blue mussels were so highly contaminated that 100 g meats would have been sufficient to kill 175,000 mice;[36] this would be equivalent to 2800 μg saxitoxin/100 g meat (1 mouse unit = 0.16 μg). Such contamination levels would be lethal to humans ingesting as little as 20 g meats.

Table 6. Comparative Toxicology

Toxin	LD50* (μg/kg)	LD100* (μg/kg)	Reference
Botulinum A Toxin	2.5 X 10^{-5}		67
Maitotoxin	0.13		67
Palytoxin	0.15		67
Cobra toxin	0.3		67
Ciguatoxin	0.45		67
Saxitoxin	3		67
Tetrodotoxin	8		67
Aflatoxin	9		68
Brevetoxin A	95		67
Yessotoxin		100	70
DTX-1		160	70
Okadaic acid	192		67
DTX-3		500	70
Curare	200		67
Brevetoxin B	500		67
Domoic acid		600	70
Sodium cyanide	10,000		

*Mouse, i.p.

Control

The economic incentive for control of human exposure to these toxins is immense. Some estimates presented at the 3rd International Conference on Toxic Dinoflagellates[3] were that the cost of hospitalization and lost work-time due to ciguatera each year was ca. $2.7 million in Canada and $17.9 million in the United States (based upon 300 cases in Canada and 2000 cases in the United States). To this cost must be added the cost of regulatory control programs (in Canada this amounted to ca. $1 million per year for shellfish) and the cost of this control program on the seafood industry. Nevertheless these societal costs must be borne to insure the safety of seafood and generate the consumer confidence essential in a growth industry.

Table 7 summarizes the control programs currently in effect to insure seafood safety. These control programs have been particularly effective in the control of shellfish toxicity. Some of these programs are based upon monitoring the dinoflagellate concentrations in the shellfish growing areas; others monitor, using the mouse bioassay, the toxin levels in the shellfish themselves. In the case of ciguateric fish, only local ordinances are currently in force; for example, in Florida the sale of barracuda weighing over 4 lbs. is prohibited. As we learn more about the toxins involved and develop methodology for detecting/quantitating these toxins, this situation will undoubtedly change.

CONCLUSIONS

The fact that in this chapter only the commonly (up to this point) observed poisonings due to the ingestion of seafood have been addressed, of course excluding decomposition-related illnesses, should not be construed to mean that other algal-derived toxins are not of concern. Quite to the contrary, our experiences to date with ciguatera and the shellfish-related toxicoses opens a Pandora's box of concern for contamination of seafood with toxins of algal/bacterial origin. Seafood-related illnesses other than those described (*vide supra*) have been reported; for example, Prof. Yasumoto

Table 7. Control of Human Exposure to Seafood Toxins

Country	Control Actions	References
United States	PSP - 80 μg/100g meats - harvest prohibited Ciguatera - Local laws NSP - Local laws	7
Canada	PSP - 80 μg/100g meats - harvest prohibited ASP - 20 μg/100g meats - harvest prohibited	3
Japan	PSP - 4MU/g edible meat (64 μg/100g) - harvest/ sale prohibited DSP - 0.05 MU/g edible meat - harvest/sale prohibited - 200 cells *D. fortii*/l - closure	6
Norway	PSP - 80 μg/100g mussel meat - distribution banned DSP - >5-7 MU/100g mussel meat - distribution banned	72
Sweden	PSP - 80 μg/100g - harvest prohibited DSP - 60 μg Okadaic acid eq./100g - closure	19
France	DSP - 200 cells *D. acuminata*/l - closure	30

recently described a case of human intoxication and death due to ingestion of toxic crabs; the toxin was isolated and identified as palytoxin.[1] Bacterial toxins, such as tetrodotoxin, and its derivatives are also troublesome toxicants of concern in seafood.[58,63]

REFERENCES

1. Alcala AC, Alcala LC, Garth JS, Yasumura D, Yasumoto T, 1988, Human fatality due to ingestion of the crab *Demania reynaudii* that contained a palytoxin-like toxin, *Toxicon*, 26:105-107.
2. Anderson DM, Private Communication, 1989.
3. Anderson DM, White AW, Baden DG, eds., 1985, "Toxic Dinoflagellates", Elsevier Science Publishing Co., Inc., New York.
4. Anon, 1985, Foodborne disease outbreaks annual summary 1982, Centers for Disease Control, Atlanta, GA.
5. Anon, 1984, Environmental Health Criteria 37: Aquatic (Marine and Freshwater) Biotoxins, World Health Organization, Geneva, Switzerland.
6. Asakawa M, 1988, Surveillance for PSP- and DSP-infested bivalves in Hokkaido, Japan, 7th International Symposium on Mycotoxins and Phycotoxins, IUPAC, Tokyo, Japan.
7. Baden DG, 1988, Public health problems of red tides, *In*: "Marine Toxins and Venoms," Tu AT, ed., Marcel Dekker, Inc., New York.
8. Baden DG, Mende TJ, Walling J, Schultz DR, 1984, Specific antibodies directed against toxins of *Ptychodiscus brevis* (Florida's red tide dinoflagellate), *Toxicon*, 22:783-789.
9. Baden DG, 1983, Marine food-borne dinoflagellate toxins, *Int'l Rev of Cytology*, 82:99-150.
10. Bagnis R, Celerier P, Cruchet P, Pascal H, Legrand A, 1988, Human intoxication with ciguateric phycotoxins in French Polynesia: Incidence and epidemiological features in 1987, 7th International Symposium on

Mycotoxins and Phycotoxins, IUPAC, Tokyo, Japan.

11. Bird C, 1988, Identification of the toxic agent responsible for the Prince Edward Island contaminated mussel incident: A summary of work conducted at the Atlantic Research Laboratory, Atlantic Research Laboratory Technical Report #56, NRRC No. 29083.

12. Blanco J, Campos MJ, 1988, The effect of water conditioned by a PSP-producing dinoflagellate on the growth of four algal species used as food for invertebrates, *Aquaculture*, 68:289-298.

13. Carmichael WW, 1988, Toxins of freshwater algae, *In*: "Marine Toxins and Venoms," Tu AT, ed., Marcel Dekker, Inc., New York.

14. Chu F, Fan T, 1985, Indirect ELISA for saxitoxin in shellfish, *J Assoc Off Anal Chem.*, 68:13-16.

15. Dickey RW, Bobzin S, Faulkner DJ, Bencsath FA, Andrzejewski D, 1989, The identification of okadaic acid from a Caribbean dinoflagellate *Prorocentrum concavum*, *Toxicon*, submitted for publication.

16. Fraga S, Sanchez FJ, 1985, Toxic and potentially toxic dinoflagellates found in *Galician rias* (NW Spain), *In*: "Toxic Dinoflagellates," Anderson DM, White AW, Baden DG, eds., Elsevier Science Publishing Co., New York.

17. Frenette C, MacLean, JD, Gyorkos TW, 1988, A large common-source outbreak of ciguatera fish poisoning, *J Infect Dis.*, 158:1128-1131.

18. Fujiki H, Suganuma M, Suguri H, Yoshizawa S, Takagi K, Uda N, Wakamatsu K, Yamada K, Murata M, Yasumoto T, Sugimura T, 1988, Diarrhetic shellfish toxin, dinophysistoxin-1, is a potent tumor promotor on mouse skin, *Jpn J Cancer Res.*, 79:1089-1093.

19. Hageltorn M, 1988, Algal toxin contaminating shellfish and toxicity monitoring in Sweden, 7th International Symposium on Mycotoxins and Phycotoxins, IUPAC, Tokyo, Japan.

20. Hall S, Page S, 1989, Paralytic shellfish poisoning in Guatemala, *In*: "Marine Toxins: Studies of Origin, Structure and Molecular Pharmacology," Hall S, ed., American Chemical Society, Washington, DC.

21. Hall S, Reichardt PB, 1984, Cryptic paralytic shellfish toxins, *In*: "Seafood Toxins," Ragelis EP, ed., ACS Symposium Series 262, American Chemical Society, Washington, DC.

22. Hamano Y, Kinoshita Y, Yasumoto T, 1985, Suckling mice assay for diarrhetic shellfish poisoning, *In*: "Toxic Dinoflagellates," Anderson, DM, White AW, Baden DG, eds., Elsevier Science Publishing Co., New York.

23. Higerd T, 1983, Ciguatera seafood poisoning: A circumtropical fisheries problem, *In*: "Natural Toxins and Human Pathogens in the Marine Environment," Colwell R, ed., Maryland Seagrant Publi.

24. Hokama Y, 1988, Ciguatera fish poisoning, *J Clin Lab Anal.*, 2:44-50.

25. Hokama Y, Miyahara JT, 1986, Ciguatera poisoning: Clinical and immunological aspects, *J Toxicol-Toxin Rev.*, 5:25-53.

26. Ikawa M, Auger K, Mosley SP, Sasner JJ, Noguchi T, Hashimoto K, 1985, Toxin profiles of the blue-green alga *Aphanizomenon flos-aquae*, *In*: "Toxic Dinoflagellates," Anderson DM, White AW, Baden DG, eds., Elsevier Science Publishing Co., New York.

27. Isobe I, Ichikawa Y, Bai D, Masaki H, Goto T, 1987, Synthesis of a marine polyether toxin, okadaic acid (4) -- Total synthesis, *Tet.*, 43:4767-4776.

28. Kodama M, Ogata T, Sato S, 1988, Bacterial production of saxitoxin, *Agric. Biol. Chem.*, 52:1075-1077.

29. Kumagai M, Yanagi T, Murata M, Yasumoto T, Kat M, Lassus P, Rodriguez-Vazquez JA, 1986, Okadaic acid as the causative toxin of diarrhetic shellfish poisoning in Europe, *Agric Biol Chem.*, 50:2853-2857.

30. Lassus P, Bardouil M, Truquet I, Truquet P, LeBaut C, Pierre MJ, 1985, *Dinophysis acuminata* distribution and toxicity along the southern Brittany coast (France): Correlation with hydrological parameters, *In*: "Toxic Dinoflagellates," Anderson DM, White AW, Baden DG, eds., Elsevier Science Publishing Co., New York.

31. Lee MS, Repeta DJ, Nakanishi K, 1986, Biosynthetic origins and assignments of ^{13}C NMR peaks of brevetoxin B, *J Amer Chem Soc.*, 108:7855-7856.

32. Lee JS, Yanagi T, Kenma R, Yasumoto T, 1987, Fluorometric determination of diarrhetic shellfish toxins by high performance liquid chromatography, *Agric Biol Chem.*, 51:877-881.
33. Levine L, Fujiki H, Yamada K, Ojika M, Gjika HB, Van Vunakis H, 1988, Production of antibodies and development of a radioimmunoassay for okadaic acid, *Toxicon*, 26:1123-1128.
34. Lin Y, Risk M, Ray SM, Van Engen D, Clardy J, Golik J, James JC, Nakanishi K, 1981, Isolation and structure of brevetoxin B from the "red tide" dinoflagellate *Ptychodiscus brevis (Gymnodinium breve)*, *J Amer Chem Soc.*, 103:6773-6775.
35. McMillan JP, Granade HR, Hoffman P, 1980, Ciguatera fish poisoning in the United States Virgin Islands: Preliminary studies, *J Coll Virgin Is.*, 6:84.
36. Medcof JC, 1985, Life and death with *Gonyaulax*: An historical perspective, *In*: "Toxic Dinoflagellates," Anderson DM, White AW, Baden DG, eds., Elsevier Science Publishing Co., New York.
37. Murata M, Sano M, Iwashita T, Naoki H, Yasumoto T, 1986, The structure of pectenotoxin-3, a new constituent of diarrhetic shellfish toxins, *Agric Biol. Chem.*, 50:2693-2695.
38. Murata M, Kumagai M, Lee JS, Yasumoto T, 1987, Isolation and structure of yessotoxin, a novel polyether compound implicated in diarrhetic shellfish poisoning, *Tet. Let.*, 28:5869-5872.
39. National Research Council of Canada, 1988, Solving the toxic mussel problem, *Can Chem News*, 40:15-17.
40. Nakanishi K, 1985, The chemistry of brevetoxins: A review, *Toxicon*, 23:473-479.
41. Nukina M, Koyanagi LM, Scheuer PJ, 1984, Two interchangeable forms of ciguatoxin, *Toxicon*, 22:169-176.
42. Ohfune Y, Tomita M. 1982, Total synthesis of (-)-domoic acid. A revision of the original structure, *J Amer Chem Soc.*, 104:3511-3513.
43. Oksen D, Nellis D, Wood R, 1984, Ciguratera in the eastern Caribbean, *Marine Fisheries Review*, 46:13-18.
44. Onoue Y, Nazawa K, 1988, Zinc-bound PSP toxins separated from *Cochlidinium* red tide, 7th International Symposium on Mycotoxins and Phycotoxins, Tokyo, Japan.
45. Oshima Y, Hasegawa M, Yasumoto T, Hallegraeff G, Blackburn S, 1987, Dinoflagellate *Gymnodinium catenatum* as the source of paralytic shellfish toxins in Tasmanian shellfish, *Toxicon*, 10:1105-1111.
46. Palafox NA, Jain LG, Pinano AZ, Gulik TM, Willians RK, Schatz IJ, 1988, Successful treatment of ciguatera fish poisoning with intravenous mannitol, *JAMA*, 259:2740-2742.
47. Pawlak J, Tempesta MS, Golik J, Zagorski MG, Lee MS, Nakanishi K, Iwashita T, Gross ML, Tomer KB, 1987, Structure of brevetoxin A as constructed from NMR and MS data, *J Amer Chem Soc.*, 109:1144-1150.
48. Pierce RH, Brown RC, Kucklick JR, 1985, Analysis of *Ptychodiscus brevis* toxins by reverse phase HPLC, *In*: "Toxic Dinoflagellates," Anderson DM, White AW, Baden DG, eds., Elsevier Science Publishing Co., New York.
49. Quilliam M, Sim P, McCulloch A, McInnes A, 1988, Determination of domoic acid in shellfish tissue by high performance liquid chromatography, *Atlantic Res Lab Tech Report*, 55 (NRCC 29015).
50. Renz V, Terplan G, 1988, Ein enzymimmunologischer nachweis von saxitoxin, *Arch Lebensm.*, 39:25-33.
51. Rodrigue DC, Tauxe RV, Blake P, Etzel RA, Kilbourne EM, Hall S, DeCampos M, Velasquez OH, 1989, Outbreak of lethal paralytic shellfish poisoning in Guatemala, Lancet, Submitted for publication.
52. Ross MR, Siger A, Abbott BC, 1985, The house fly: An acceptable subject for paralytic shellfish toxin bioassay, *In*: "Toxic Dinoflagellates," Anderson DM, White AW, Baden DG, eds., Elsevier Science Publishing Co., New York.

53. Schantz EJ, Ghazarossian VE, Schnoes HK, Strong FM, Springer JP, Pezzanite JD, Clardy J, 1975, The structure of saxitoxin, *J Amer Chem Soc.*, 97:1238-1239.

54. Scheuer PJ, Takahashi W, Tsutsumi J, Yoshida T, 1967, Ciguatoxin: Isolation and chemical nature, *Science*, 155:1267-1268.

55. Shimizu Y, Bando H, Chou H, Van Duyne G, Clardy JC, 1986, Absolute configuration of brevetoxins, *J Chem Soc.*, ___:1656-1658.

56. Shimzu Y, 1988, The chemistry of paralytic shellfish toxins, *In*: "Marine Toxins and Venoms," Tu AT, ed., Marcel Dekker, Inc., New York.

57. Sick LV, Hansen DC, Babinchak JA, Higerd TB, 1986, An HPLC-fluorescence method for identifying a toxic fraction extracted from the marine dinoflagellate *Gambierdiscus toxicus*, *Marine Fisheries Rev.*, 48:29-34.

58. Simidu U, Noguchi T, Hwang D, Shida Y, Hashimoto K, 1987, Marine bacteria which produce tetrodotoxin, *Appl Env Micro.*, 53:1714-1715.

59. Sullivan JJ, Wekell MM, Hall S, 1988, Detection of paralytic shellfish toxins, *In*: "Marine Toxins and Venoms," Tu AT, ed., Marcel Dekker, Inc., New York.

60. Taylor FJR, 1985, The taxonomy and relationships of red tide dino-flagellates, *In*: "Toxic Dinoflagellates," Anderson DM, White AW, Baden DG, eds., Elsevier Science Publishing Co., New York.

61. Taylor SL, 1988, Marine toxins of microbial origin, *Food Tech.*, March:94-98.

62. Tachibana K, Scheuer PJ, Tsukitani Y, Kikuchi H, Van Engen D, Clardy J, Gopichand Y, Schmitz F, 1981, Okadaic acid, a cytotoxic polyether from two marine sponges of the genus *Halichondria*, *J Amer Chem Soc.*, 103:2469-2472.

63. Tibballs J, 1988, Severe tetrodotoxic fish poisoning, *Anaesthesia Int Care*, 16:215-217.

64. UBE Industries Ltd., Diagnostic Development Group, ARK Mori Bldg., 12-32, Akasaki 1-chome, Minato-ku, Tokyo 107, Japan.

65. Vernoux J, 1988, Ciguatera fish poisoning: Epidemiology, toxicology and prevention of the illness on Saint-Barthelemy Island, French West Indies, *Oceanologica Acta*, 11:37-48.

66. Williams S, ed., 1984, "Official Methods of Analysis," 14th Edition, Association of Official Analytical Chemists, Arlington, VA.

67. Withers NW, 1988, Ciguatera fish toxins and poisoning, *In*: "Marine Toxins and Venoms," Tu AT, ed., Marcel Dekker, Inc., New York.

68. World Health Organization, International Program on Chemical Safety, 1979, Environmental Health Criteria 11 -- Mycotoxins, Geneva, Switzerland.

69. Yang GC, Imagire SJ, Yasei P, Ragelis EP, Park DL, Page SW, Carlson RE, Guire PE, 1987, Radioimmunoassay of paralytic shellfish toxins in clams and mussels, *Bull Environ Contam Toxicol.*, 39:264-271.

70. Yasumoto T, 1985, Recent progress in the chemistry of dinoflagellate toxins, *In*: "Toxic Dinoflagellates," Anderson DM, White AW, Baden DG, eds., Elsevier Science Publishing Co., New York.

71. Yasumoto T, Murata M, Yokoyama M, 1989, Studies on the chemical structure of maitotoxin, *In*: "Marine Toxins: Studies of Origin, Structure and Molecular Pharmacology," Hall S, ed., American Chemical Society, Washington, DC.

72. Yndestad M, 1988, Regulations concerning mycotoxins and phycotoxins in Norway, 7th International Symposium on Mycotoxins and Phycotoxins, IUPAC, Tokyo, Japan.

THE POLYETHER BREVETOXINS AND SITE FIVE OF THE VOLTAGE-SENSITIVE SODIUM

CHANNEL

Daniel G. Baden

University of Miami
Rosenstiel School of Marine and Atmospheric Science
Division of Biology and Living Resources
4600 Rickenbacker Causeway
Miami, Florida 33149

and

School of Medicine
Department of Biochemistry and Molecular Biology
P.O. Box 016129
Miami, Florida 33101

ABSTRACT

Florida red tide brevetoxins interact with a specific site associated with the voltage-sensitive sodium channel (VSSC) in a number of selected species. It is this same site which is involved in the molecular action of the marine polyether ciguatoxin. The site is located in a hydrophobic portion of the channel, the portion which is involved in activation/inactivation of ion flux across excitable membranes. Several different types of derivatized brevetoxins are available for investigation of sodium channel topography. Four tritiated brevetoxin probes, PbTx-3, PbTx-9, PbTx-7, and PbTx-10 are available for specific binding studies. In all cases, measurement of specific binding affinity and capacity parallels *in vivo* and *in vitro* potency. Dissociation constants are in the nanomolar concentration range, and binding maxima approximate 6-8 pmoles/mg protein. Brevetoxin photoaffinity probes bind in a specific covalent fashion to a protein component of approximately 260 kDa from synaptosomes, suggesting an association with the α-subunit of sodium channel. Bound toxin has been detected following polyacrylamide gel electrophoresis of sodium dodecylsulfate solubilized rat brain synaptosomes using: (1) goat antibrevetoxin-rabbit anti-goat-peroxidase sandwich immunoassays following Western blotting; (2) goat antibrevetoxin-peroxidase direct immunoassay; or (3) liquid scintillation detection of tritium-labeled brevetoxin photoaffinity probe-sodium channel conjugates.

TOXINS

Natural

We routinely isolate six brevetoxins from laboratory cultures of *Ptychodiscus brevis*, all based on the two polyether backbones as described

Microbial Toxins in Foods and Feeds, Edited by A.E. Pohland *et al.*,
Plenum Press, New York, 1990

by Shimizu et al.[1] In logarithmic cells, the two predominant toxins are PbTx-1 and PbTx-2 (Fig. 1). In stationary cells, approximately the same relative amounts of PbTx-1 and PbTx-2 are present on a per cell basis, but now in addition PbTx-3, PbTx-5, PbTx-6 (based on the backbone present in PbTx-2), and PbTx-7 (based on the backbone present in PbTx-1) appear.

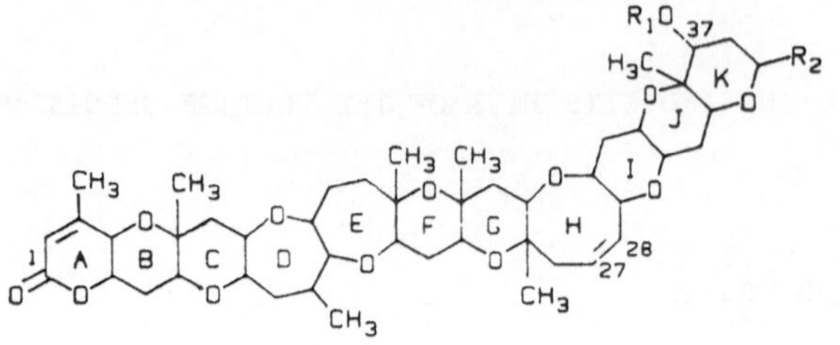

Type-1

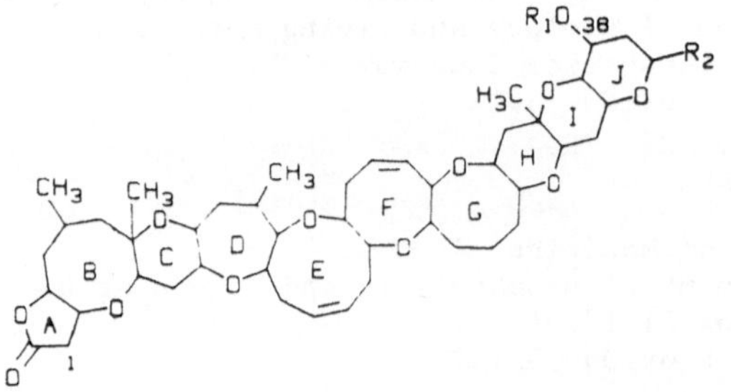

Type-2

Toxin	Type	R₁	R₂
PbTx-1	2	H	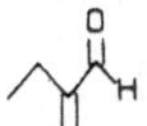
PbTx-2	1	H	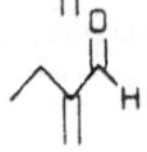
PbTx-3	1	H	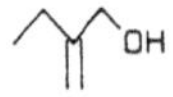
PbTx-5	1	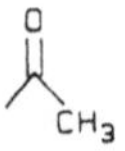	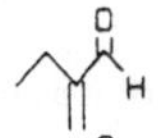
PbTx-6	1	H	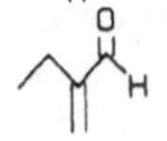 27, 28 epoxide
PbTx-7	2	H	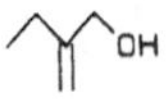
PbTx-8	1	H	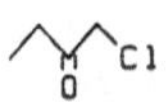
PbTx-9	1	H	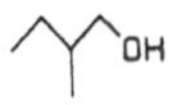
PbTx-10	2	H	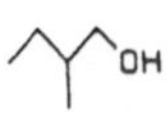

Fig. 1. The brevetoxins are comprised of two types, with substituent derivatization as indicated. Type-1 toxins are, as a rule, less potent than are Type-2 toxins.

Two additional synthetic toxins, PbTx-9 and PbTx-10, are available by chemical reduction.[2] Synthetic tritiated PbTx-3 and unlabeled PbTx-3 were prepared by chemical reduction of PbTx-2 using sodium borotritiide or sodium borohydride, respectively. Toxin PbTx-7 was produced by identical chemical reduction of PbTx-1 using borohydride. HPLC-purified radioactive PbTx-3 and PbTx-7 had a specific activity of 10-15 Ci/mmole, or one-fourth the specific activity of the chemical reductant. HPLC-purified radioactive PbTx-9 and PbTx-10 had a specific activity of 20-30 Ci/mmole, or one-half the specific activity of the chemical reductant. Labeled type-1 toxin is stable for 4-6 months, repurification by HPLC being performed as necessary. Labeled type-2 toxin is stable for a matter of days and is used as soon after preparation as possible.

MOLECULAR PHARMACOLOGY

We have previously shown that tritiated PbTx-3 binds to "high affinity" site 5 associated with voltage-sensitive sodium channels, and have determined a K_D of 2.9 nM and a B_{max} of approximately 7 picomoles/mg synaptosomal protein (Fig. 2) for this site.[3] We also demonstrated that tritiated PbTx-3 could be displaced in a specific manner from its binding site by either natural or synthetic brevetoxins. Our initial observation was that displacement efficiency was linked in a positive fashion with potency in animals. Specific displacement curves correlated well with the potency of each individual purified toxin. Differential lipid solubility of each of the natural brevetoxins made it imperative to include Emulphor EL-620 in all experimental tubes.

In addition to developing displacement curves for the six toxins (n=2), we had sufficient toxin material for PbTx-1,-2,-3, and -7 to calculate K_is. These are summarized for several species in Table 1.[4]

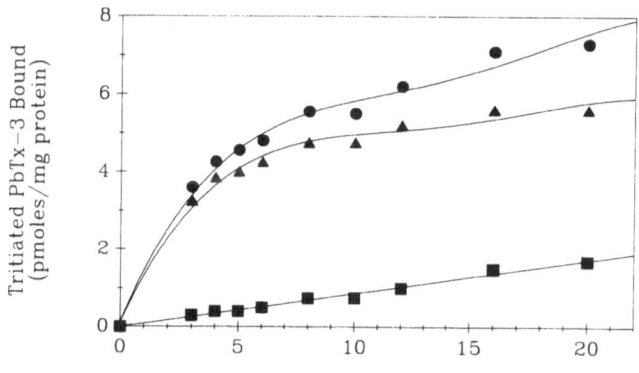

Fig. 2. Tritiated PbTx-3 (14.33 Ci/mmole) at increasing concentrations was incubated together with synaptosomes prepared from rat brain. Following 1 hr incubations in the presence (squares) and absence (circles) of 10 μM unlabeled PbTx-3, synaptosomes were centrifuged and washed to determine total and nonspecific binding respectively. Specific binding (triangles) was determined by a subtraction of nonspecific binding from total binding. Half-maximal binding was observed at 2.6 nM tritiated toxin, and the maximum number of specific sites was calculated at 6.9 pmoles/mg synaptosome protein.

Table 1.   Inhibition constants for derivative brevetoxins derived
           from the Cheng-Prusoff[8] equation.

Toxin	K_i (nM)		
	Turtle	Fish	Rat
PbTx-1	0.39	10.10	0.72
PbTx-2	1.34	23.57	3.51
PbTx-3	1.96	37.04	2.47
PbTx-5	- - - -	- - - -	2.68
PbTx-6	- - - -	- - - -	6.60
PbTx-7	- - - -	- - - -	0.85

Table 2.   Comparison of dissociation constant (K_d) and binding
           maximum (B_{max}) in fish, turtles, and rats.[*]

Species	K_d (nM)	B_{max} (pMol/mg Protein)	Temp. Optimum (°C)	Specific Binding at K_d
Fish	6.1	1.40	23	80%
Turtle	1.5	2.25	4	80%
Rat	2.6	6.80	4	90%

[*]   Mean values for K_d and B_{max}, n = 9,4,6 for fish, turtles, and rats
      respectively.

Species specificity of binding

    Synaptosomes from rats,[3] turtles,[5] or fish[6] were prepared to examine
the binding characteristics of each with respect to brevetoxins.   Table 2
outlines the results of the comparison, and illustrates that any of the
three systems examined bind brevetoxins in a reproducible manner with
approximately equal efficacy.

Specific binding of four tritiated brevetoxins

    A preliminary comparison of specific binding of tritiated PbTx-3,
PbTx-7, PbTx-9, and PbTx-10 in rat brain synaptosomes indicates an equivalent B_{max} and a progression of K_d values which parallel the relative potencies of the labeled brevetoxins.   This is a further indication to us that
binding affinity is the conservative requirement in the potency of the
brevetoxins (Table 3), and further, that we may be able to utilize the
toxins which are of higher specific activity for more detailed receptor
characterization.[2]

    Our evidence indicates that, at a K_d concentration of tritiated
PbTx-3, the $t_{1/2}$ for on- and off-rates approximate 1-2 minutes.   A closer
approximation cannot be derived utilizing present protocols.   There is no
membrane potential dependence of brevetoxin binding to the high affinity,
low capacity binding site known as Site 5, K_d = 2.6 (intact), 2.9 (lysed),

Table 3. Comparison of K_d and B_{max} for four different tritiated brevetoxin probes in rat brain synaptosomes.

Toxin	K_d (nM)	B_{max} (pmoles/mg protein)
PbTx-3	2.13	6.99
PbTx-9	8.76	6.75
PbTx-7	1.91	6.38
PbTx-10	1.56	6.46

3.3 (depolarized) and B_{max}= 6.01 (intact), 5.83 (lysed) and 5.75 pmoles/mg protein (depolarized).

Regardless of the organism used for synaptosomal preparations, it is apparent to us that the topographic characteristics of the brevetoxin binding site on the VSSC are comparable. Using brevetoxins PbTx-1-3, and PbTx-5-7, K_i data for specific displacement of tritiated PbTx-3 shows comparable data in each case (Table 1). The more hydrophobic type-2 brevetoxins are most efficacious in their ability to compete for site 5 binding.[4]

Classes of brevetoxin binding sites

Two separate brevetoxin binding sites have been discovered in rat brain synaptosomes. The brevetoxins bind with an affinity constant which is consistently in the 1-5 nM concentration range, in good agreement with affinity data for other potent marine toxins like saxitoxin. In addition, the binding maximum in synaptosomes is also in good agreement with data for Site 1 toxins, which are known to bind to channels with a 1:1 stoichiometry. However, the allosteric modulation of sodium channel binding by other natural toxins by brevetoxins occurs at brevetoxin concentrations much higher, ca. 20-100 nM. These data are inconsistent with high affinity, low capacity binding.

Converse to this allosteric modulation which occurs at higher brevetoxin concentrations, is the finding that membrane depolarization, [22]Na influx and *competitive* displacement of tritiated brevetoxin binding by unlabeled competitors, is dose dependent in the same concentration ranges observed for the high affinity binding site.[7] Thus, the allosteric modulation at other sodium channel binding sites appears to arise from brevetoxin interaction with a lower affinity, high capacity binding site.

Using classical Rosenthal analysis, we have been able to distinguish two separate specific brevetoxin sites (Fig. 3).

The two site hypothesis is supported by brevetoxin inhibition constant data and double reciprocal competition plots, which indicate a deviation from competitive type patterns to non-competitive type patterns at higher competitor brevetoxin concentrations. The noncompetitive displacement appears to be specific in nature, and is not likely due to changes in membrane fluidity. But, certainly more investigation is required before concrete conclusions can be offered.

Biochemical identification of the brevetoxin binding site

Using the tritiated p-azidobenzoyl-ester (C-42-linked) of PbTx-3, we have succeeded in demonstrating that brevetoxins interact with the α subunit of the voltage-sensitive sodium channel. In the dark, the complete brevetoxin photoaffinity probe exhibits virtually identical K_d and B_{max} values for unaltered PbTx-3. Incubation of 2 nM tritiated p-azidobenzoyl-PbTx-3 (14.33 Ci/mmole) with rat brain synaptosomes for one hour, followed by 10 minute irradiation at 254 nm results in the covalent modification of site 5 with the radioactive brevetoxin photoaffinity probe. Subsequent solubilization of the site with sodium dodecylsulfate and electrophoresis on SDS-polyacrylamide gels against protein standards, yields a tritiated fraction at 230-260 kDa which tests positive for protein by either silver or Coomassie blue staining (Fig. 4). Incubation of tritiated p-azidobenzoyl-PbTx-3 in the presence of excess unlabeled p-azidobenzoyl-PbTx-3 prevents the specific interaction of the labeled photoaffinity probe with site 5, and no radioactivity is associated with the high molecular weight fraction under these conditions. Following Western blotting, the same protein fraction reacts positive using a sandwich immunoassay procedure involving goat antibrevetoxin antibodies followed by rabbit anti-goat-peroxidase linked antibodies. Our preliminary experiments indicate that PbTx-3-peroxidase conjugates also react in a positive fashion with the subunit following Western blotting.

ACKNOWLEDGEMENTS

This work was supported in part by the US Army Medical Research and Development Command under contract numbers DAMD17-85-C-5171, DAMD17-97-C-7001 and DAMD17-88-C-8148. Opinions, interpretations, conclusions, and recommendations are those of the author and are not necessarily endorsed by the US Army. Portions of the work reviewed herein constitutes research conducted at the University of Miami for the degree of Master of Marine Science (Richard A. Edwards) and the Ph.D. in Biochemistry and Molecular Biology (Vera L. Trainer) and shall be reported fully elsewhere.

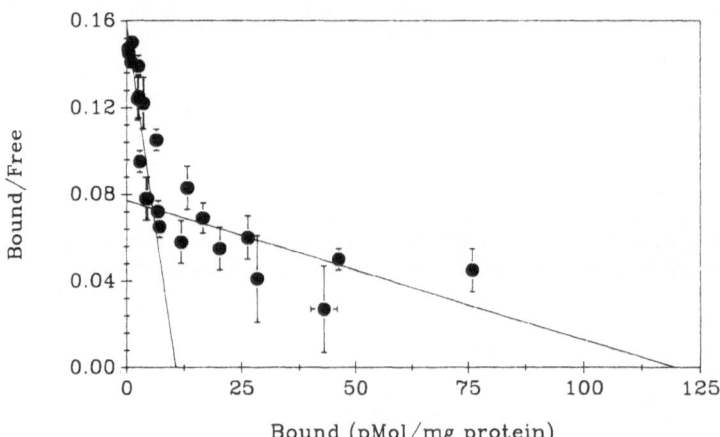

Fig. 3. Employing the protocol utilized in Fig. 2, synaptosomes were exposed to higher concentrations of PbTx-3 approaching 200 nM and classical Rosenthal plots were constructed. Two different binding sites, classified as high affinity low capacity (site 5) and low affinity high capacity (uncertain localization), were observed. High affinity site: K_d= 2.6-2.9 nM; B_{max}= 5.6-7.0 pmoles/mg protein. Low affinity site: K_d= 79-125 nM; B_{max}= 63.7-180 pmoles/mg protein.

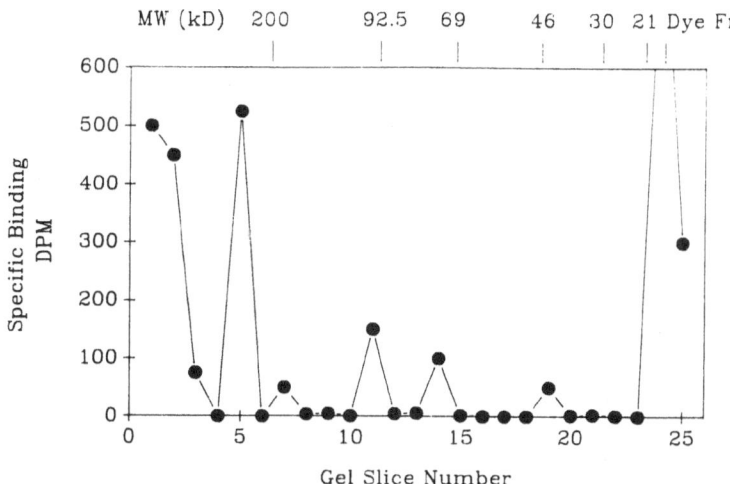

Fig. 4. Identification of the high affinity protein fraction specifically labeled by 2.0 nM tritiated PbTx-3 photoaffinity probes.
SDS-polyacrylamide mini-slab gel electrophoresis of sodium
dodecyl sulfate solubilized tritiated photoaffinity brevetoxin
PbTx-3, followed by slicing the gel and quantifying tritiated
PbTx-3 photoaffinity probe bound, yielded the profile shown
above. Molecular weight standards run on the same gel indicate that the majority of specific tritium binding comigrates
with a protein fraction of approximately 240-260 kDa. Western
blotting of the gel, followed by anti-brevetoxin-peroxidase
visualization or by goat antibrevetoxin-rabbit antigoat-
peroxidase sandwich assay visualization, identifies the same
protein fraction.

REFERENCES

1. Shimizu Y, Chou HN, Bando H, VanDuyne G, Clardy J, 1986, Structure of
    brevetoxin-a (GB-1), the most potent toxin in the Florida red tide
    organism *Gymnodinium breve (Ptychodiscus brevis)*, *JACS*, 108:514-515.
2. Baden DG, Mende TJ, Trainer VL, 1989, Derivatized brevetoxins and
    their use as quantitative tools in detection, *7th IUPAC Symposium on
    Mycotoxins and Phycotoxins*, Tokyo, Japan, in press.
3. Poli MA, Mende TJ, Baden DG, 1986, Brevetoxins, unique activators of
    voltage-sensitive sodium channels, bind to specific sites in rat brain
    synaptosomes, *Mol Pharm.*, 30:129-135.
4. Baden DG, Mende TJ, Szmant AM, Trainer VL, Edwards RA, Roszell LE,
    1988, Brevetoxin binding: molecular pharmacology versus immunoassay,
    Toxicon., 26:97-104.
5. Edwards RA, 1988, The response of sodium channels in turtle brain
    synaptosomes to conditions present during oxygen deprivation, M.S.
    Thesis, University of Miami.
6. Stuart AM, Baden DG, 1988, Florida red tide brevetoxins and binding
    in fish brain synaptosomes, *Aq Toxicol.*, 13:271-280.
7. Poli MA, 1985, Characterization of the binding of the *Ptychodiscus
    brevis* neurotoxin T17 to the sodium channels in rat brain synaptosomes,
    Ph.D. Dissertation, University of Miami.
8. Cheng YS, Prusoff, WH, 1973, Relationship between the inhibition con-
    stant (Ki) and the concentration of inhibitor which causes 50 percent
    inhibition (IC50) of an enzyme reaction, *Biochem Pharmacol.*, 22:3099-
    3112.

CELLULAR MECHANISMS OF ACTION FOR FRESHWATER CYANOBACTERIA (BLUE-GREEN ALGAE) TOXINS

Wayne W. Carmichael, Nik A. Mahmood[1] and
Edward G. Hyde

Department of Biological Sciences
Wright State University
Dayton, Ohio 45435

and

The National Toxicology Program[1]
NIEHS
Research Triangle Park
Durham, North Carolina 27709

ABSTRACT

Acute lethal toxicity from cyanobacteria is caused by ingestion of toxic cells or toxins from certain freshwater/brackish water species of *Anabaena*, *Aphanizomenon*, *Microcystis*, *Nodularia*, and *Oscillatoria*. The toxins include a related family of hepatotoxic cyclic hepta and pentapeptides, produced by strains of *Anabaena*, *Microcystis*, *Nodularia*, and *Oscillatoria*, termed microcystins or cyanoginosins, that contain both D and L amino acids plus two novel amino acids. These toxins appear to be taken up into the cell by bile acid carriers and exert their cell disaggregating effect by altering the cell's cytoskeletal system. Species and strains of *Anabaena* produce at least two neurotoxins, termed anatoxins: one a depolarizing neuromuscular blocking agent, the other an irreversible anticholinesterase. Strains of *Aphanizomenon flos-aquae* produce saxitoxin and neosaxitoxin the primary toxins in cases of paralytic shellfish poisoning (PSP). These various toxins have intraperitoneal mouse LD_{50} values of 10-500 μg/kg. Current research indicates that a third group of toxins with contact irritant properties are produced by some freshwater cyanobacteria. Indirect evidence indicates that one or all of these toxins are responsible for certain cases of human gastroenteritis and dermatitis from municipal and recreational water supplies.

INTRODUCTION

Reports of toxic algae in the freshwater environment are almost exclusively due to members of the division Cyanophyta, commonly called blue-green algae or cyanobacteria. Although cyanobacteria are found in almost any environment ranging from hot springs to Antarctic soils, known toxic members are mostly planktonic. Published accounts of field poisonings by cyanobacteria have been known since the late 19th century. These reports describe sickness and death of livestock, pets, and wildlife following

Microbial Toxins in Foods and Feeds, Edited by A.E. Pohland *et al.,*
Plenum Press, New York, 1990

ingestion of water containing toxic algal cells or the toxin released by the aging cells. Recent reviews of these poisonings and the toxins of freshwater cyanobacteria are given by Carmichael,[19,21,22] Codd and Bell,[32] and Gorham and Carmichael.[45]

Toxins produced by cyanobacteria are in two general groups. The first, which is emphasized in this report, includes those toxins responsible for acute lethal poisonings. About 12 genera have been implicated in producing these toxins but only *Anabaena*, *Aphanizomenon*, *Microcystis*, *Nodularia*, and *Oscillatoria* have had toxins isolated, and investigated chemically and toxicologically. The second group includes a number of secondary chemicals that are not highly lethal to animals but instead show more selective bioactivity. These bioactive chemicals (Table 1) include scytophycins produced by certain strains of *Scytonema pseudohofmanni*. *Scytonema pseudohofmanni* strain BC-1-2, isolated from a forested area on the island of Oahu, Hawaii, was found to produce two lipophilic toxins termed scytophycin A (MW 821) and B (MW 819). Scytophycin B is moderately toxic to mice by the intraperitoneal route (LD_{50j} = 650 μg/kg). Scytophycins show a very strong cytotoxic activity, however, when tested against cell cultures, such as KB human epidermoid carcinoma and NIH/3T3 mouse fibroblast cell lines (1 and 0.65 ng/ml, respectively). These toxins are also moderately active against intraperitoneally implanted cells from P388 lymphocytic leukemia and Lewis lung carcinoma.[68] *Scytonema hofmanni* UTEX 1581 has been shown to produce the chlorine-containing diaryl-lactone called cyanobacterin. This compound has been shown to have anticyanobacterial activity and has been proposed as a possible algicide against cyanobacteria.[44,46]

A cytotoxic alkaloid has been isolated from the filamentous species *Hapalosiphon fontinalis* strain V-3-1. This isolate was made from soil samples collected in the Marshall Islands in 1981. This strain produces the lipophilic compound hapalindole A, that has a broad range of antialgal and antimycotic activity.[67] *Oscillatoria acutissima* strain B-1, isolated from a freshwater pond in Oahu, was found to produce two novel macrolide compounds termed acutiphycin and 20, 21-didehydroacutiphycin. These macrolides show cytotoxicity (KB and NIH/3T3) and antitumor activity (Murine Lewis lung carcinoma).[5]

Other toxins that show low lethal toxicity to laboratory test animals include lipopolysaccharide endotoxin produced as part of the cell wall by all cyanobacteria[75] and certain toxins of some cyanobacteria suspected of causing contact irritation in recreational water supplies.[10,32]

Economic losses due to water-based diseases from freshwater cyanobacteria toxins have only been reported from the first group and are the result of contact with or consumption of water containing toxin and/or toxic cells. These toxins are all water soluble and temperature stable. They are either released by the cyanobacterial cell or loosely bound so that changes in cell permeability or age allow their release into the environment. Known occurrences of toxic cyanobacteria in water supplies include Canada (four provinces), Europe (12 countries), United States (20 states), USSR (Ukraine), Australia, India, Bangladesh, South Africa, Israel, Japan, New Zealand, Argentina, Chile, Thailand and the Peoples Republic of China[25,29,88] (Fig. 1). The economic impact from toxic freshwater cyanobacteria include the costs incurred from deaths of domestic animals; allergic and gastrointestinal problems after human contact with waterblooms; lost income from recreational areas; and increased expense for the detection and removal of taste, odor, and toxins. The remainder of this report discusses the acute lethal hepato- and neurotoxins produced by freshwater cyanobacteria and their cellular/molecular mechanism of action (Table 1).

Table 1. Toxins of freshwater cyanobacteria.

Species, strain, and source	Toxin term	Structure	LD$_{50}$ μg/kg IP, mouse
NEUROTOXINS			
Anabaena flos-aquae	Anatoxin-A	Secondary amine alkaloid MW 165	200
Strain NRC-44-1 (Canada, Saskatchewan)			
Strain NRC-525-17 (Canada, Saskatchewan)	Anatoxin-A(S)	Unknown	50
Aphanizomenon flos-aquae	Aphantoxin (neosaxitoxin)	Purine alkaloid MW 315 (neoSTX) MW 299 (STX)	10
Strain NH-1 & NH-5 (U.S., New Hampshire)	Aphantoxin II (saxitoxin)		
HEPATOTOXINS			
Anabaena flos-aquae	Microcystins[a]	Heptapeptides MW 994	50
Strain S-23-g-1 (Canada, Saskatchewan)			
Microcystis aeruginosa	Cyanoginosins[a]	Heptapeptides MW 909-1044	50
Strain WR-70 (=UV-010) (South Africa, Transvaal)			
(Waterbloom, Australia, New South Wales)	Cyanoginosin	Heptapeptide MW 1035	50
(Waterbloom, U.S., Wisconsin)	Microcystin	Heptapeptide MW 994	50
Strain NRC-1(SS-17) (Canada, Ontario)	Microcystin	Heptapeptide MW 994	50
Strain 7820 (Scotland, Loch Balgaves)	Microcystin	Heptapeptide MW 994 MW 994	50
(Waterbloom, Norway, Lake Akersvatn)	Microcystin	Heptapeptide MW 994	50
Microcystis aeruginosa	Microcystin	Heptapeptide	50
Strain M-228 (Japan, Tokyo)		MW 994 MW 1044	

Table 1. Toxins of freshwater cyanobacteria. (Continued)

Species, strain, and source	Toxin term	Structure	LD$_{50}$ μg/kg IP, mouse
Microcystis aeruginosa	Cyanogenosin[a]	Heptapeptide MW 1039	not reported
Microcystis viridis	Cyanoviridin[a]	Heptapeptide MW 1039	not reported
Nodularia spumigena	Nodularin	Pentapeptide MW 824	30-50
Oscillatoria agardhii var. *isothrix* (Waterbloom, Norway, Lake Froylandsvatn)	Microcystins	Heptapeptides MW 1009	300-500
Oscillatoria agardhii var.	Microcystins (Waterbloom, Norway, Lake Kolbotnvatn)	Heptapeptides MW 1023	500-1000
CYTOTOXINS			
Scytonema pseudohofmanni Strain BC-1-2 (U.S., Hawaii)	Scytophycin A & B	Methylformamide A=MW 821; B=MW 819	650 (scyto-phycin B)
Scytonema hofmanni Strain UTEX-1581 (U.S., Texas)	Cyanobacterin	Chlorinated diaryllactone	not reported
Hapalosiphon fontinalis Strain V-3-1 (Marshall Islands)	Hapalindole A	Substituted indole alkaloid	not reported
Oscillatoria acutissima Strain B-1 (U.S., Hawaii)	Acutiphycin	Macrolide	not reported

[a] See text for explanation of terminology.

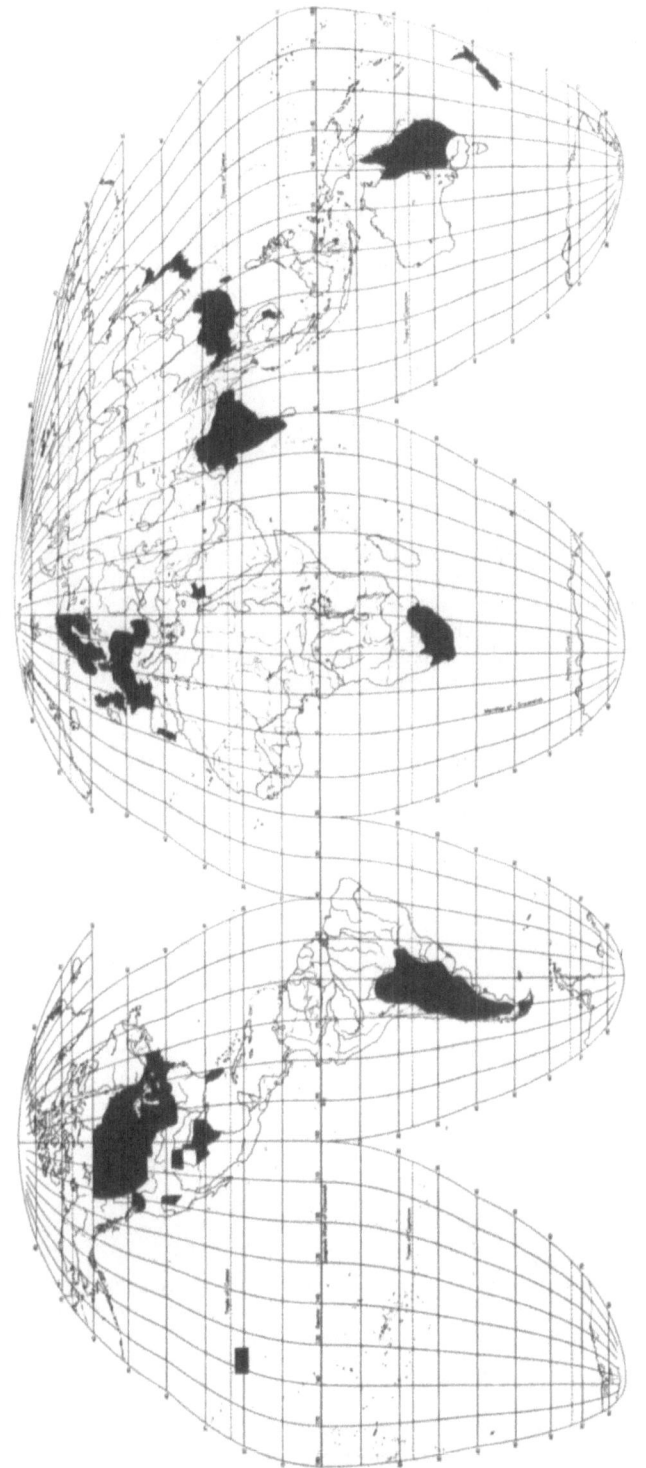

Fig. 1. Map showing countries and areas (dark) where toxic cyanobacteria waterblooms have been detected.

Anatoxins: Neurotoxins produced by filamentous *Anabaena flos-aquae* are called anatoxins (ANTXS).[25] Currently two anatoxins, from differ- ent strains of *A. flos-aquae*, have been isolated and at least partially characterized. ANTX-A from strain (single filament isolate) NRC-44-1 is the first toxin from a freshwater cyanobacteria to be chemically defined. It is the secondary amine, 2-acetyl-9-azabicyclo (4-2-1) non-2-ene[35,46] molecular weight 165 daltons (Fig. 2). It has been synthesized through a ring expansion of cocaine,[17,18] from iminium salts,[6,52,73] from nitrone,[95,96] from 4-cycloheptenone or tetrabromotricyclooctane[34] by construction of the azabicyclo ring, from 9-methyl-9-azabicyclo [3.3.1] nonan-1.ol[98] and by starting with 9-methyl-9-aza [4.2.1] nonan-2-one.[56]

ANTX-A is a potent, postsynaptic, depolarizing, neuromuscular blocking agent that affects both nicotinic and muscarinic acetylcholine (ACH) recep- tors.[2,27,90,91] Signs of poisoning in field reports for wild and domestic animals include staggering, muscle fasciculations, gasping, convulsions, and opisthotonos (birds). Death by respiratory arrest occurs within minutes to a few hours depending on species, dosage, and prior food consumption. The intraperitoneal LD_{50} (IP) in a mouse for purified toxin is about 200 μg/ kg body weight, with a survival time of 4-7 minutes. This means that animals need only to ingest a few milliliters to a few liters of the toxic surface bloom to receive a lethal dose.[23,24,28] Detection of ANTX-A while still primarily by mouse bioassay is being supplemented by three analytical detection methods. These methods are based on high performance liquid chromatography (HPLC),[3,99] gas chromatography-mass spectrometry (GC-MS),[89] and gas chromatography-electron capture detection (GC-ECD).[92]

All known occurrences of ANTX-A production have been from Canada or the United States. More recently, ANTX-A has been detected in *Anabaena flos-aquae* blooms from Japan (Watanabe, personal communication); Norway (Skulberg, personal communication) and Finland (Sivonen, personal communication).

anatoxin - a hydrochloride

R = H; saxitoxin dihydrochloride
R = OH; neosaxitoxin dihydrochloride

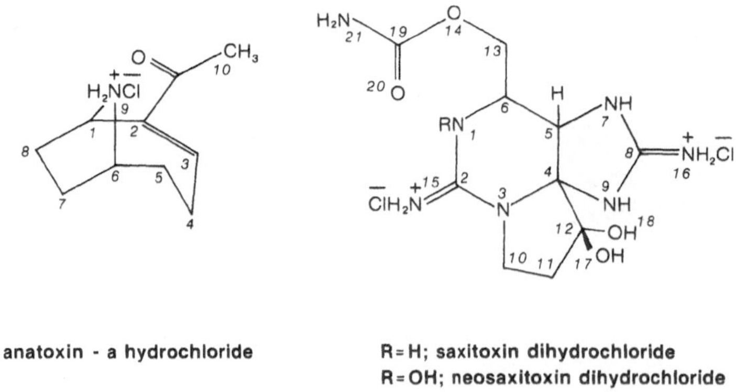

Fig. 2. (Left) Anatoxin-a (ANTX-A) hydrochloride. Produced by the freshwater filamentous cyanobacterium *Anabaena flos-aquae* NRC-44-1.

(Right) Aphantoxin-I (neosaxitoxin) and Aphantoxin-Il (saxitoxin) produced by certain strains of the filamentous cyanobacterium *Aphanizomenon flos-aquae*.

A neurotoxin more recently isolated and under current study is referred to as anatoxin-a(s) [ANTX-A(S)]. The primary source of this toxin is *A. flos-aquae* strain NRC-525-17 isolated in 1965 from Buffalo Pound Lake in Saskatchewan, Canada. ANTX-A(S) is physiologically and chemically different from ANTX-A. It produces opisthotonos in chicks, as does ANTX-A, but also causes viscous salivation [the reason for the (S) label] and lachrymation in mice, chromodacryorrhea (bloody tears) in rats, urinary incontinence, muscular weakness, fasciculation, convulsion and defecation prior to death by respiratory arrest. Also observed is a dose-dependent fasciculation of limbs for 1-2 minutes after death. ANTX-A(S) was purified, from *A. flos-aquae* cells, by column chromatography and (HPLC),[26,58] but its structure is still being worked on. ANTXA(S) is acid stable, unstable in basic conditions, has very low ultraviolet (uv) absorbance, gives a positive alkaloid test, and has a molecular weight estimated by gel exclusion chromatography of about 250 daltons.

The mouse IP LD_{50} for ANTX-A(S) is approximately 30 μg/kg, seven times more lethally toxic than ANTX-A. At the LD_{50} the survival time for mice is 10-30 minutes. Mahmood and Carmichael[62] concluded from the signs of poisoning, i.e., salivation and muscle twitch potentiation, that the cholinergic system was the primary target of ANTX-A(S). No serum cholinesterase activity was observed in rats dosed with 350 and 600 μg/kg of ANTX-A(S) suggesting an anticholinesterase mechanism. ANTX-A(S) also sensitized the frog rectus abdominus muscle and chick biventer cervicis muscle to exogenous acetylcholine, but had no effect on the action of ANTX-A (a depolaring agent) on frog rectus abdominus. Further work with ANTX-A(S)[63] has confirmed and extended the conclusion that this toxin is an anticholinesterase agent. ANTX-A(S) was shown to inhibit *in vitro* electric eel acetylcholinesterase (AChE, EC 3.1.1.7) (Fig. 3) and horse serum butyrylcholinesterase (BUChE, EC 3.1.1.8) in a time and concentration dependent fashion by a mechanism similar to the organophosphate anticholinesterases as illustrated in scheme (I).

(I)

$$EOH+IX \quad \overset{k_1}{\underset{k_{-1}}{\rightleftharpoons}} \quad \underset{HX}{EOH(IX)} \quad \overset{k_2}{\rightarrow} \quad \underset{H_2O}{EOI} \quad \overset{k_3}{\rightarrow} \quad EOH+HIO$$

In this scheme, EOH is the enzyme, IX is the inhibitor (either a carbamate or an organophosphate). EOH(IX) is analogous to the Michaelis Menton complex seen with the substrate reaction. EOI is the acyl-enzyme intermediate for carbamates or a phosphoro-enzyme intermediate for the organophosphates. The equilibrium constant for this reaction (K_d) is defined as k_{-1}/k_1 and the phosphorylation or carbamylation constant is defined as k_2. In this study,[63] ANTX-A(S) was found to be more specific for AChE than BUChE. The double reciprocal and Dixon plot of the inhibition of electric eel AChE indicated that the toxin is a non-competitive inhibitor (V_{max} decreases, K_m remains unchanged). Because non-competitive inhibitors can be reversible or irreversible a plot of V_{max} vs. the total amount of enzyme was used to distinguish which of the two, ANTX-A(S) resembled. In this type of plot the lines for control and non-competitive inhibitor will pass through the origin, while that of the irreversible inhibitor parallels the control. Figure 4 shows that ANTX-A(S) is an irreversible inhibitor of electric eel AChE. Further comparison of the kinetic result of ANTX-A(S) inhibition of AChE with diisopropyl-fluorophosphate (DFP; an irreversible cholinesterase inhibitor) indicated that inhibition follows the generalized scheme for irreversible inhibitors.[63]

Work is continuing on the molecular mechanism of AChE/ANTX-A(S) binding. This work focuses on two areas: the first is comparison of the

kinetic inhibition from different sources of AChE by ANTXA(S) and then
comparing the kinetic constants against that found with DFP.

The second set of investigations is examining the nature of the interaction of ANTX-A(S) and AChE. Binding of substrate and the irreversible
inhibitors of AChE involves attachment at two sites on the AChE molecule.[64,77] These two sites are 5A apart and comprise the active site of
the enzyme. The first is known as the anionic site which contains one or
more negatively charged residues surrounded by regions of hydrophobicity.[97]
This area binds the cationic portion of the molecule, and for uncharged
molecules the hydrophobic regions act as an anchoring point.[47] The second
site is the esteratic site where the ester linkage of the substrate or
inhibitor is cleaved by a serine residue made highly reactive by a "charge-
relay" system made up of a histidine and an aspartate or tyrosine residue.

There are two ways of producing inhibition of anticholinesterase
agents. The first termed irreversible inhibition occurs when a carbamate
or organophosphate reacts with AChE. The serine residue can become carbamylated with the subsequent hydrolysis of the carbamyl group. This process
requires only minutes for reactivation of the enzyme. In the second process, the serine becomes phosphorylated and it requires hours to days for
hydrolysis back to active enzyme. The second type of inhibition termed
reversible includes the true reversible anticholinesterases (i.e., tetra-
methylammonium, TMA) that block only the anionic site and are displaced by
increasing substrate concentration. AChE can also be inhibited by high

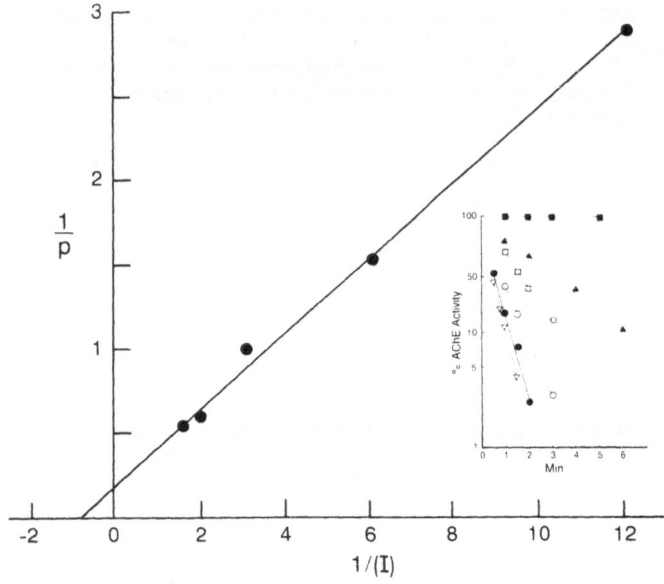

Fig. 3.   Inhibition of eel AChE by ANTX-A(S) - the 'secondary plot'. P,
          the first-order rate constant which was the rate of inhibition
          at that ANTX-A(S) concentration obtained from the 'primary
          plot' (insert). The intercept on the 1/P axis is $1/k_2$ and
          the intercept on the 1/[I] axis is $-1/K_d$. *Figure insert:*
          Progressive irreversible inhibition of eel AChE by ANTX-A(S).
          The inactivation followed first-order kinetics. ANTX-A(S) con-
          centrations, μg/ml: (▲) - 0.083; (□) - 0.166; (○) -
          0.331; (●) - 0.497; (▽) - 0.599; (■) - Control. Each point
          represents the mean of 3 or 4 determinations.

substrate concentrations where the cationic head of one molecule and the ester linkage of another molecule bind concurrently at the active site and substrate hydrolysis cannot occur resulting in inhibition.[64,94] In our studies to date, and reported here, we have used ACh; physostigmine and TMA to investigate the binding of ANTX-A(S) to AChE.

In the first group of studies, involving kinetic inhibition studies, comparisons of the equilibrium (K_d), phosphorylation (k_2) and inhibition constant (k_i) for the inhibition of electric eel and human erythrocyte AChE by ANTX-A(S) and DFP were done (Table 2). ANTX-A(S) has a higher affinity for human erythrocyte AChE $(K_d=0.25\ \mu M)$ than electric eel AChE $(K_d=3.67\ \mu M)$ (Table II). ANTX-A(S) also shows greater affinity for AChE than DFP $(K_d=300\ \mu M)$. And finally the biomolecular rate constant, k_i, which indicates the overall rate of reaction, shows AChE is more sensitive toward inhibition by ANTX-A(S) $(k_i=5.4\times10^5\ M^{-1}min^{-1})$ and DFP $(k_i=3.4\times10^{4}/M^{-1}min^{-1})$.

These studies add information on the comparative activity of ANTX-A(S) and other irreversible AChE inhibitors but do not show the site of inhibition.

The second group of studies was designed to investigate the site of inhibition. The results of these protection studies are shown in Figs. 5 and 6. By using ACh, one immediately determines if ANTX-A(S) is an active site directed inhibitor. ANTX-A(S) competes with ACh for the active site but its preference for either the anionic or esteratic subsite cannot be determined using ACh alone. TMA, a reversible anionic site blocker, protection shows a similar dose-dependence. Physostigmine at $10^{-5}M$ protects approximately 50% of the enzyme from ANTX-A(S) inhibition (Fig. 7).

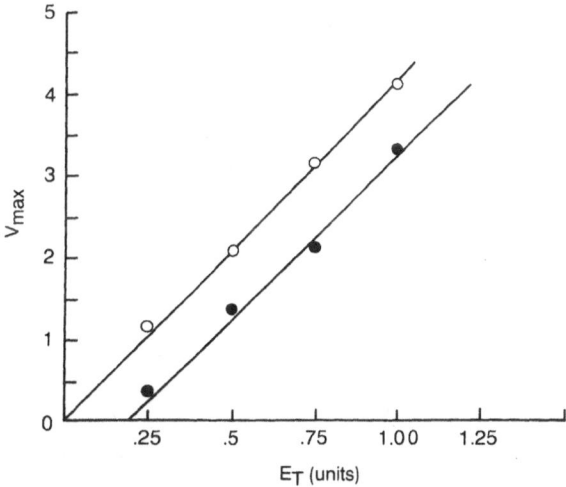

Fig. 4.   Plot of V_{max} against total enzyme $[E_T]$ showing the irreversible inhibition of electric eel acetylcholinesterase (AChE) by ANTX-A(S). The enzymes were incubated with 0.32 $\mu g/ml$ ANTX-A(S) for 1.0 min and acetylthiocholine (final concentrations 2.5, 4.7, 6.3 and 7.8 x $10^{-4}M$) was added. V_{max} was determined from the double reciprocal plots (not shown). O = control; ● = ANTX-A(S). (Reproduced with permission from Ref. 42. Copyright 1987 Pergamon Press).

561

Table 2. Affinity equilibrium (K_d) phosphorylation rate (k_2) and bimolecular rate (k_i) constants for the inhibition of AChE by ANTX-A(S) and DFP.

	AChE					
	Electric Eel			Human Erythrocytes		
	K_d*+ (μM)	k_2* (min^{-1})	k_i (M^{-1}min^{-1})	K_d*+ (μM)	k_2* (min^{-1})	k_i (M^{-1}min^{-1})
Toxin						
ANTX-A(S)	3.67	2.0	5.4×10^5	0.25	21.40	8.6×10^7
DFP	300	10	3.4×10	-	-	-

* K_d and k_2 were determined from the secondary plots (Fig. 3). K_d is the reciprocal of the abscissa intercept and k_2 the reciprocal of the ordinate intercept.
+ An estimated MW of 267 for ANTX-A(S) was used to calculate K_d.

Comparing TMA and physostigmine at 10^{-5}M indicates that physostigmine is the superior protectant and the main ANTX-A(S) attack site is the esteratic site, with the anionic site anchoring the molecule in the active site. In addition to these laboratory studies on ANTX-A(S), field poisonings by ANTX-A(S) have now been confirmed in at least one case.[59] The poisonings involved the death of five dogs, eight pups and two calves that ingested quantities of *A. flos-aquae*, contaning ANTX-A(S), in Richmond Lake, South Dakota, in late summer 1985.

In summary, ANTX-A(S) uses the two site attachment mechanism analogous to substrate and does not inhibit AChE in the manner of the reversible anticholinesterases.

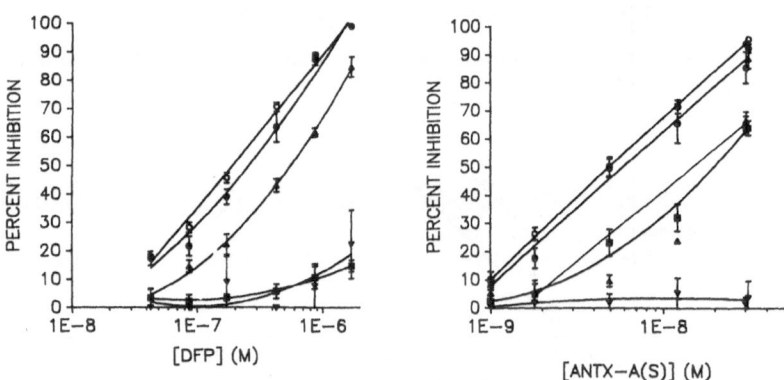

Fig. 5. (Left) Inhibition of eel AChE by DFP in the absence and presence of acetylcholine. DFP and buffer or DFP and ACh were coincubated for 2 minutes before percent inhibition (100 percent activity) was determined. Symbols: ($\bigcirc$) - DFP control; ($\bullet$) - DFP and 1 μM ACh; ($\blacktriangle$) - DFP and 10 μM ACh; ($\blacksquare$) - DFP and 100 μM ACh; ($\blacktriangledown$) - DFP and 10 mM ACh. Each point is the mean ± SEM of at least 3 experiments.

(Right) Inhibition of eel ACh by ANTX-A(S) in the absence and presence of acetylcholine.

Aphantoxins: Occurrence of neurotoxins (aphantoxins) in the freshwater filamentous cyanobacterium *Aphanizomenon flos-aquae* was first demonstrated by Sawyer et al.[85] All aphantoxins (APHTXS) studied to date have come from waterblooms and laboratory strains of nonfasciculate (non-flake forming) *Aph. flos-aquae* that occurred in lakes and ponds of New Hampshire from 1966 through 1980. Toxic cells and extracts of *Aph. flos-aquae* were shown to be toxic to mice, fish, and waterfleas *(Daphnia catawba)* by Jakim and Gentile.[49] Chromatographic and pharmacological evidence established that APHTXS consist mainly of two neurotoxic alkaloids that strongly resemble saxitoxin (STX) and neosaxitoxin (neoSTX), the two primary toxins of red tide paralytic shellfish poisoning (PSP).[84] The bloom material and toxic strain used in studies before 1980 came from collections made between 1960 and 1970. The more recent work on APHTXS has used two strains (NH-1 and NH-5) isolated by Carmichael in 1980 from a small pond near Durham, New Hampshire.[20,48] These APHTXS, as well as neoSTX and STX, are fast-acting neurotoxins that inhibit nerve conduction by blocking sodium channels without affecting permeability to potassium, the transmembrane resting potential, or membrane resistance.[1] Mahmood and Carmichael,[61] using the NH-5 strain showed that batch-cultured cells have a mouse IP LD_{50} of about 5 mg/kg. Each gram of lyophilized cells yields about 1.3 mg aphantoxin I (neosaxitoxin) and 0.1 mg aphantoxin II (Saxitoxin) (Fig. 2). Also detected are three labile neurotoxins that are not similar to any of the known paralytic shellfish poisons.

Shimizu et al.[87] studied the biosynthesis of the STX analog neoSTX using *Aph. flos-aquae* NH-1. They were able to confirm its presence in strain NH-1 and to explain the biosynthetic pathway for this important group of secondary chemicals.

Hepatotoxins: Low-molecular-weight peptide toxins that affect the liver have been the predominant toxins involved in cases of animal poisonings due to cyanobacterial toxins.[21,45,86] After almost 25 years of structure analysis on toxic peptides of the colonial bloom-forming cyanobacterium *Microcystis aeruginosa*, Botes et al.[12-14] and Santikarn et al.[83] provided structure details on one of four toxins (designated toxin BE-4) produced by

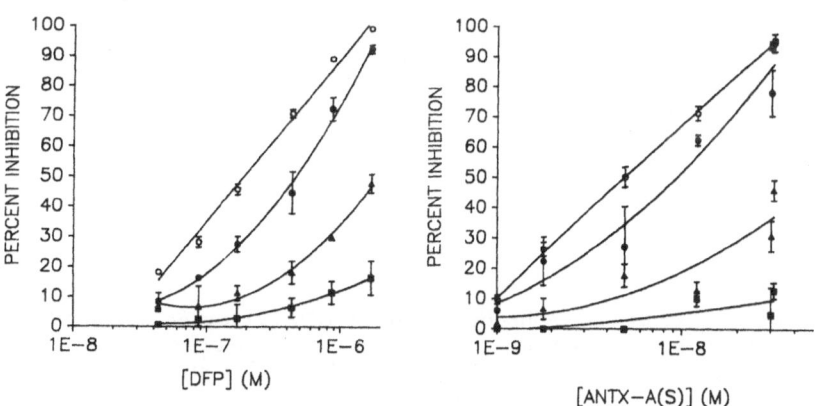

Fig. 6. (Left) Inhibition of eel AChE by DFP in the absence and presence of tetramethyl ammonium iodide. Incubation conditions are the same as in Fig. 5. Symbols: (○) - DFP control; (●) - 11 μM TMA; (▲) - 110 μM TMA; (■) - 550 μM TMA.

(Right) Inhibition of eel AChE by ANTX-A(S) in the absence and presence of tetramethyl ammonium iodide. Conditions and symbols are the same as in Fig. 5.

the South African *M. aeruginosa* strain WR70 (=UV-010). They concluded that it was monocyclic and contained three D-amino acids--alanine, erythro-β-methylaspartic acid, and glutamic acid, two L-amino acids--leucine and alanine--plus two unusual amino acids. These were *N*-methyldehydroalanine (Medha) and a nonpolar side chain of 20 carbon atoms that turned out to be a novel β-amino acid; 3-amino-o-methoxy-2,6,8-trimethyl-10-phenyldeca-4,6-dienoic acid (ADDA). Based on fast atom bombardment mass spectrometry (FABMS) and nuclear magnetic resonance (NMR) studies, BE-4 toxin is now known to be a cyclic heptapeptide having a molecular weight of 909 daltons (Fig. 10).

Instead of calling the BE-4 toxin microcystin, as previous *Microcystis* toxins were called[51,69,74] and using alphabetical or numerical suffixes to indicate chromatographic elution order or structural differences, Botes[12] proposed the generically derived designation cyanoginosin (CYGSN). This prefix, which indicates that cyanobacterial origin, is followed by a two-letter suffix that indicates the identity and sequence of the two L-amino acids relative to the N-Me-dehydroalanyl-D-alanine bond. Thus, toxin BE-4 was renamed cyanoginosin-LA because leucine and alanine are the L-amino acids. These L-amino acids were shown by Eloff et al.[39] and Botes et al.[15] to vary between strains and to account for structural differences for different toxic fractions within a single strain. Botes et al.[15] showed that the other three toxins of strain WR-70 all had the same three D-amino acids and two novel amino acids (Medha and ADDA). The L-amino acids were leucine-arginine (CYGSN-LR/MCYST-LR), tyrosine-arginine (CYGSN-YR/MCYST-YR), and tyrosine-alanine (CYGSN-YA/MCYST-YA). They were also able to show that the hepatotoxin isolated by Elleman et al.[37] from waterbloom material collected in Malpas Dam, New South Wales, Australia, contained the five characteristic amino acids plus the L-amino acid variance tyrosine-methionine (therefore, it is termed CYGSN-YM/MCYST-YM).

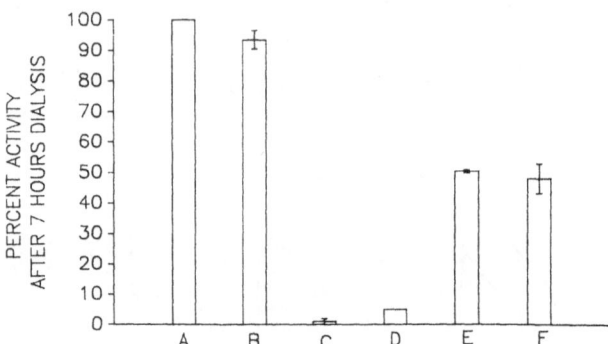

Fig. 7. Inhibition of eel AChE by DFP or ANTX-A(S) after carbamylation by 10^{-5}M physostigmine. All samples were dialyzed against 100 mM sodium phosphate buffer, pH 8 for 7 hours before activity was determined. A - enzyme treated with buffer for 30 minutes then dialyzed; B - enzyme treated with 10^{-5}M physostigmine for 30 minutes then dialyzed: C - enzyme treated with buffer for 30 minutes then treated with 2 μM DFP for two minutes then dialyzed; D - enzyme treated with buffer for 30 minutes then 0.03 μM ANTX-A(S) for two minutes then dialyzed; E - same as in C but treated for 30 minutes with 10^{-5}M physostigmine prior to DFP; F - same as in D but treated for 30 minutes with 10^{-5}M physostigmine prior to ANTX-A(S). The results are means $\pm$ SEM of six experiments.

Microcystin (MCYST) is the term given to the fast death factor (FDF) produced by *M. aeruginosa* strain NRC-1 and its daughter strain NRC-1 (SS-17).[11,51] An absolute structure for the toxin of strain NRC-1 (SS-17) is not yet available but is known to be a peptide (MW 994) containing the variant amino acids leucine and arginine. Krishnamurthy et al.[53,54] have shown that the toxin isolated from a waterbloom of *M. aeruginosa* collected in Lake Akersvatn, Norway[9] has a structure similar to that of MYCST from NRC-1 (SS-17) and CYGSN-LR (Fig. 8). This toxin has also been found to be the main toxin produced by the Scottish strain of *M. aeruginosa* PCC-7820 and a Canadian *A. flos-aquae* strain S-23-g-1. The identification of a peptide toxin from *A. flos-aquae* S-23-g-1 provides the first evidence that these hepatotoxins are produced by filamentous as well as coccoid cyanobacteria. *A. flos-aquae* S-23-g-1 and toxic *M. aeruginosa* from a waterbloom in Wisconsin also produced a second cyclic heptapeptide hepatotoxin, which has been found to have six of the same amino acids, that is, leucine-arginine, but has aspartic acid instead of β-methylaspartic acid.[54]

The filamentous genus *Oscillatoria* has also been shown to produce a hepatotoxin.[40,70] From waterblooms of *O. agardhii* and *O. agardhii* var. isothrix, two similar cyclic heptapeptides have been isolated. Both toxins have the variant amino acids arginine-arginine and aspartic acid instead of β-methylaspartic acid. *O. agardhii* var. isothrix also has dehydroalanine instead of methyldehydroalanine.[53] More recently *M. virdis*[55] and *M. aeruginosa*[71] have been shown to produce the cyclic heptapeptide with an arginine-arginine "L" amino acid variant.

Nodularia spumigena has also been shown to produce a peptide with hepatotoxic activity. The more recent reports come from Australia,[65] the German Democratic Republic,[50] Denmark,[57] Sweden,[36] Finland[41,72] and New Zealand.[30] Recently structure information on *Nodularia* toxin has been presented by Rinehart et al.[76] for waterbloom material collected from Lake Forsythe, New Zealand in 1984; by Eriksson et al.[41] from water bloom material collected in the Baltic Sea in 1986, Runnegar et al.[81] for a field isolate from the Peel Inlet, Perth, Australia, and Carmichael et al.[30] for a field isolate from Lake Ellesmere, New Zealand. Structure work by these researchers all indicate that the peptide is smaller than the heptapeptide toxins. Rinehart's work shows that the toxin is a pentapeptide with a similar structure

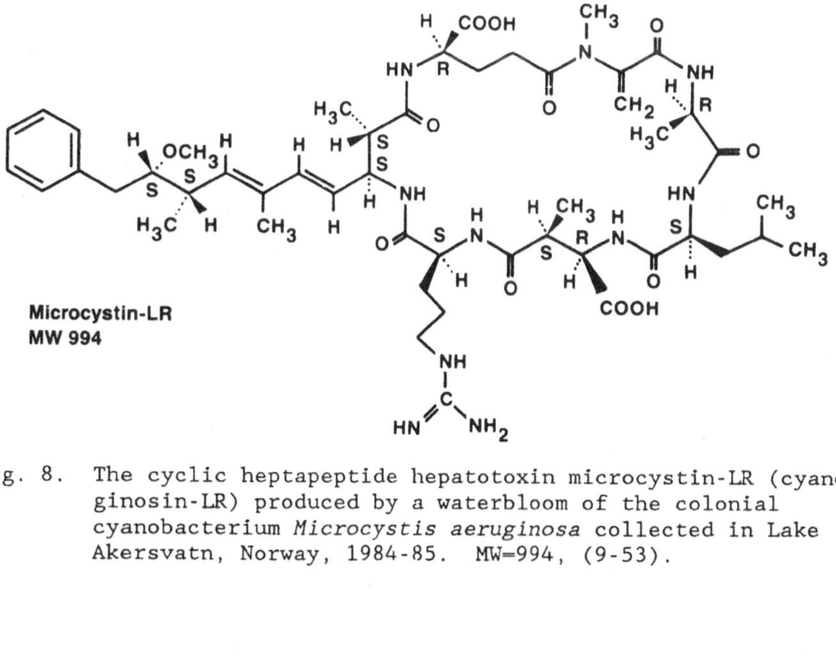

Microcystin-LR
MW 994

Fig. 8. The cyclic heptapeptide hepatotoxin microcystin-LR (cyanoginosin-LR) produced by a waterbloom of the colonial cyanobacterium *Microcystis aeruginosa* collected in Lake Akersvatn, Norway, 1984-85. MW=994, (9-53).

to the heptapeptides and containing β-methylaspartic acid, glutamic acid, arginine, dehydrobutyrine and ADDA (M.W. 824) (Fig. 9).

The hepatotoxins have been called Fast-Death Factor,[11] microcystin,[51] cyanoginosin,[12] cyanoviridin[55] and cyanogenosin[71] (apparently a misspelling of cyanoginosin). Because of the different terms used to refer to similar compounds it has been proposed[31] that the term microcystin (MCYST) plus the suffix "XY" (designating the variant L-amino acids) be used as the basis for naming all the existing and future monocyclic heptapeptide toxins of cyanobacteria. Cyclic peptides with fewer or greater than seven peptides or peptide linked components (i.e., the pentapeptide from *Nodularia* is termed nodularin [NODLN]) should be named according to the genus name from which they are originally isolated or by their chemical composition relative to the existing microcystins. The hepatotoxic heptapeptide general structure thus becomes:

```
          1      2            3              4   5   6           7
Cyclo (-D-Ala-L-"X"-D-erythro-β-methylAsp-L-"Y"-ADDA-D-Glu-N-Methyldehy-
droAla) (Fig. 10).
```

    X = Leucine (L), Arginine (R), Tryosine (Y)
    Y = Arginine (R), Alanine (A), Methionine (M)

    "XY" combinations for heptapeptide toxins currently defined:
    LR; LA; YA; YM; YR; RR

    ADDA=3-amino-9-methoxy-2,6,8-trimethyl-10-phenyldeca-4,6-dienoic acid

Mode of Action for Microcystins

The liver has always been reported as the organ that showed the greatest degree of histopathological change when animals are poisoned by these cyclic peptides. The molecular basis of action for these cyclic peptides is not yet understood but the cause of death from toxin and toxic cells administered to laboratory mice and rats is at least partially known and is concluded to be hypovolemic shock caused by interstitial hemorrhage into the liver.[93] This work with small animal models is currently being extended to larger animals to study the uptake, distribution, and metabolism of the toxins.[7] From studies using ^{125}I-labeled MCYST-YM (CYGSN-YM), the liver is the organ for both accumulation and excretion.[42,82] Brooks and

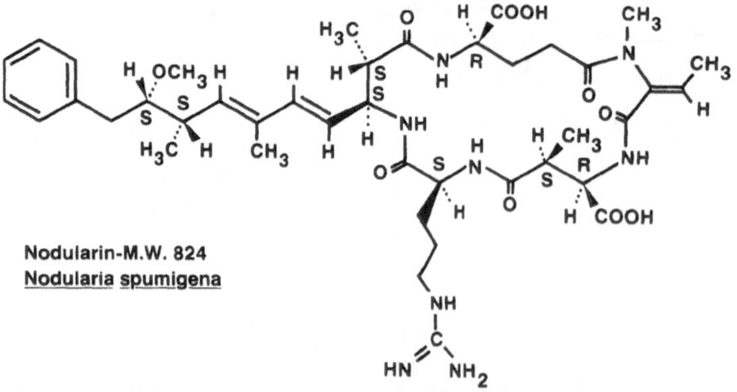

Nodularin-M.W. 824
Nodularia spumigena

Fig. 9. The cyclic pentapeptide hepatotoxin nodularin produced by a New Zealand water bloom and strain of the filamentous brackish water cyanobacterium *Nodularia spumigena*.

Codd,[16] using ^{14}C labeled MCYST-LR, showed that seventy percent of the labeled toxin was localized in the mouse liver 1 minute following intraperitoneal injection of the toxin.

Studies at both the light and electron microscopic (EM) level of time course histopathological changes in mouse liver show rapid and extensive centrilobular necrosis of the liver with loss of characteristic architecture of the hepatic cords. Sinusoid endothelial cells and then hepatocytes show extensive fragmentation and vesiculation of cell membranes.[43,78] Using microcystin-LR from *M. aeruginosa* strain PCC7820, Dabholkar and Carmichael[33] found that at both lethal and sublethal toxin levels hepatocytes show progressive intracellular changes beginning at about 10 min postinjection. The most common response to lethal and sublethal injections is vesiculation of rough endoplasmic reticulum (RER), swollen mitochondria, and degranulation (partial or total loss of ribosomes from vesicles). The vesicles appear to form from dilated parts of RER by fragmentation or separation. Affected hepatocytes remain intact and do not lyse. Use of the isolated perfused rat liver to study the pathology of these toxins shows similar results to the *in vivo* work. Berg et al.[8] used three structurally different cyclic heptapeptide hepatotoxins (MCYST-LR; desmethyl 3-MCYST-RR and didesmethyl 3,7-MCYST-RR). All three toxins had a similar effect on the perfused liver system although both "RR" toxins required higher concentrations (5-7x) to produce their effect. This was consistent with the lower toxicity of the "RR" toxins, which was about 500 and 1000 μg/kg IP mouse compared to 50 μg/kg for MCYST-LR.

In vitro studies on isolated cells including hepatocytes, erythrocytes, fibroblasts and alveolar cells continue to demonstrate the specificity of

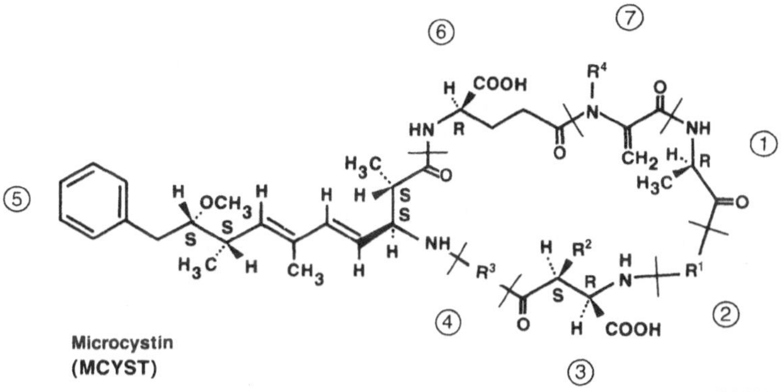

Microcystin
(MCYST)

M.W.

MCYST-LA: R^1 = Leu; R^2 = CH_3; R^3 = Ala; R^4 = CH_3 909

MCYST-YA: R^1 = Tyr; R^2 = CH_3; R^3 = Ala; R^4 = CH_3 959

MCYST-LR: R^1 = Leu; R^2 = CH_3; R^3 = Arg; R^4 = CH_3 994

desmethyl 3- MCYST-LR: R^1 = Leu; R^2 = H; R^3 = Arg; R^4 = CH_3 980

MCYST-YM: R^1 = Tyr; R^2 = CH_3; R^3 = Met; R^4 = CH_3 1035

MCYST-RR: R^1 = Arg; R^2 = CH_3; R^3 = Arg; R^4 = CH_3 1037

desmethyl 3- MCYST-RR: R^1 = Arg; R^2 = H; R^3 = Arg; R^4 = CH_3 1023

desmethyl 3,7- MCYST-RR: R^1 = Arg; R^2 = H; R^3 = Arg; R^4 = H   1009

MCYST-YR: R^1 = Tyr; R^2 = CH_3; R^3 = Arg; R^4 = CH_3 1044

Fig. 10. Structure of six microcystins varying only in L-amino acids and three microcystins with desmethyl portions of amino acids 3 and 7.

567

action that these toxins have for liver cells.[42,81] This has led Aune and Berg[4] to use isolated rat hepatocytes as a screen for detecting hepatotoxic waterblooms of cyanobacteria.

The cellular/molecular mechanism of action for these cyclic peptide toxins is now an area of active research in several laboratories. These peptides cause striking ultrastructural changes in isolated hepato- cytes[79] including a decrease in the polymerization of actin. This effect on the cells' cytoskeletal system continues to be investigated and recent work indirectly supports the idea that these toxins interact with the cells' cytoskeletal system.[39,42] Why there is a specificity of these toxins for liver cells is not clear although it has been suggested that the bile uptake system may be at least partly responsible for penetration of the toxin into the cell.[8]

SUMMARY

Acute poisoning of humans by freshwater cyanobacteria as occurs with paralytic shellfish poisoning, while reported, has never been confirmed. Humans are probably just as susceptible as pets, livestock or wildlife but people naturally avoid contact with heavy waterblooms of cyanobacteria. In addition, there are no known vectors, like shellfish, to concentrate toxins from cyanobacteria into the human food chain. Susceptibility of humans to cyanobacteria toxins is supported mostly by indirect evidence. In many of these cases, however, if a more thorough epidemiological study had been possible these cases probably would have shown direct evidence for toxicity.

In addition to acute lethal poisonings, episodes of dermatitis and/or irritation from contact with freshwater cyanobacteria are occurring with increasing frequency; partly because more eutrophic waters are being used for recreational purposes. Investigations in the USA,[10] Canada,[29] Scotland,[32] and Norway[88] have shown that dermatotoxic blooms may be dominated by *Anabaena*, *Aphanizomenon*, *Gloeotrichia* and *Oscillatoria*, respectively.

Toxic cyanobacteria can produce neurotoxic, hepatotoxic and dermato- toxic compounds that are a direct threat to animal and human water supplies. This threat increases as water bodies become more eutrophic, thus support- ing higher production of toxic and nontoxic cyanobacteria. The presence of these potent natural toxins pose an increasing threat to the maintenance of quality water supplies for agriculture, municipal and recreational use.

ACKNOWLEDGEMENTS

Work with European toxic cyanobacteria was partially supported by a NATO collaborative research grant between W. W. Carmichael and G. A. Codd, University of Dundee, Scotland, and O. M. Skulberg, Norwegian Water Research Institute, Oslo, Norway. Toxin structure work on European and North American peptide toxins is supported in part by U.S. AMRDC contract DAMD17-87-C-7019 to W. W. Carmichael. Portions of the work represent part of the Ph.D. dissertation research of N. A. Mahmood and E. G. Hyde. Their work was supported in part by fellowship support from the Biomedical Ph.D. Program, Wright State University.

REFERENCES

1. Adelman WJ, Jr, Fohlmeister JF, Sasner JJ, Jr, Ikawa M, 1982, Sodium channels blocked by aphantoxin obtained from the blue-green alga,

568

Aphanizomenon flos-aquae, Toxicon., 20:513-516.

2. Aronstam RS, Witkop B, 1981, Anatoxin-a interactions with cholinergic synaptic molecules, *Proc Natl Acad Sci.*, 78:4639-4643.

3. Astrachan NB, Archer BG, 1981, Simplified monitoring of anatoxin-a by reverse-phase high performance liquid chromatography and the sub-acute effects of anatoxin-a in rats, *In:* "The Water Environment: Algal Toxins and Health," Carmichael WW, ed., Plenum Press, New York City.

4. Aune T, Berg K, 1986, Use of freshly prepared rat hepatocytes to study toxicity of blooms of the blue-green algae *Microcystis aeruginosa* and *Oscillatoria agardhii, J Toxicol Envir Hlth.*, 19:325-336.

5. Barchi JJ, Moore RE, Patterson GML, 1984, Acutiphycin and 20,21-didehydroacutiphycin, new antineoplastic agents from the cyanophyta *Oscillatoria acutissima, J Amer Chem Soc.*, 106:8194-8197.

6. Bates HA, Rapaport H, 1979, Synthesis of anatoxin-a via intramolecular cyclization of iminium salts, *J Amer Chem Soc.*, 101:1259-1265.

7. Beasley VR, Lovell R, Cook W, Lundeen G, Holmes K, Hooser S, Haschek-Hock W, Carmichael WW, 1986, Findings in preliminary pathophysiology studies in hepatotoxic and neurotoxic blue-green algae using swine and laboratory animals, *ACS 8th Rocky Mtn Region Meet.*, June 8-12, 1986, Denver.

8. Berg K, Wyman J, Carmichael WW, Dabholkar AS, 1988, Isolated rat liver perfusion studies with cyclic heptapeptide toxins of *Microcystis* and *Oscillatoria* (freshwater cyanobacteria), *Toxicon.*, 26:827-837.

9. Berg K, Carmichael WW, Skulberg OM, Benestad C, Underdal B, 1987, Investigation of toxic waterbloom of *Microcystis aeruginosa* (cyanophyceae) in Lake Akersvatn, Norway, *Hydrobiol.*, 144:97-103.

10. Billings WH, 1981, Water-associated human illness in northeast Pennsylvania and its suspected association with blue-green algae blooms, *In:* "The Water Environment: Algal Toxins and Health," Carmichael WW, ed., Plenum Press, New York City.

11. Bishop CT, Anet EFLJ, Gorham PR, 1959, Isolation and identification of the fast-death factor in *Microcystis aeruginosa* NRC-1, *Can J Biochem Physiol.*, 37:453-471.

12. Botes DP, 1986, Cyanoginosins--isolation and structure, *In:* "Mycotoxins and Phycotoxins, Bioactive Molecules," Steyn PS, Vleggar R, eds., Elsevier, Amsterdam.

13. Botes DP, Kruger H, Viljoen CC, 1982, Isolation and characterization of four toxins from the blue-green alga, *Microcystis aeruginosa, Toxicon*, 20:945-954.

14. Botes DP, Tuinman AA, Wessels PL, Viljoen CC, Kruger H, Williams DH, Santikarn S, Smith RJ, Hammond SJ, 1984, The structure of cyanoginosin-LA, a cyclic heptapeptide toxin from the cyanobacterium *Microcystis aeruginosa, J Chem Soc Perkin Trans.*, 1:2311-2318.

15. Botes DP, Wessels PL, Kruger H, Runnegar MTC, Santikarn S, Smith RJ, Barna JCJ, Williams DH, 1985, Structural studies on cyanoginosins -LR, -YR, -YA, and -YM, peptide toxins from *Microcystis aeruginosa, J Chem Soc Perkin Trans.*, 1:2747-2748.

16. Brooks WP, Codd GA, 1987, Distribution of *Microcystis aeruginosa* peptide toxin and interactions with hepatic microsomes in mice, *Pharm Tox.*, 60:187-191.

17. Campbell HF, Edwards OE, Elder JW, Kolt RJ, 1979, Total synthesis of DL-anatoxin-A and DL-isoanatoxin-a, *Pol J Chem.*, 53:27-37.

18. Campbell HF, Edwards OE, Kolt R, 1977, Synthesis of noranatoxin-A and anatoxin-A, *Can J Chem.*, 55:1372-1379.

19. Carmichael WW, 1981, Freshwater blue-green algae (cyanobacteria) toxins - a review, *In:* "The Water Environment: Algal Toxins and Health," Carmichael WW, ed., Plenum Press, New York City.

20. Carmichael WW, 1982, Chemical and toxicological studies of the toxic freshwater cyanobacteria *Microcystis aeruginosa, Anabaena flos-aquae* and *Aphanizomenon flos-aquae, S Afr J Sci.*, 78:367-372.

21. Carmichael WW, 1986, Algal toxins, *In*: "Advances in Botanical Research," Callow JA, ed., Academic Press, London.
22. Carmichael WW, 1988, Toxins of freshwater algae, *In*: "Handbook of Natural Toxins; V.3 Marine Toxins and Venoms," Tu AT, ed., Marcel Dekker, New York City.
23. Carmichael WW, Biggs DF, 1978, Muscle sensitivity differences in two avian species to anatoxin-a produced by the freshwater cyanophyte *Anabaena flos-aquae* NRC-44-1, *Can J Zool.*, 56:510-512.
24. Carmichael WW, Gorham PR, 1977, Factors influencing the toxicity and animal susceptibility of *Anabaena flos-aquae* (cyanophyta) blooms, *J Phycol.*, 13:97-101.
25. Carmichael WW, Gorham PR, 1978, Anatoxins from clones of *Anabaena flos-aquae* isolated from lakes of western Canada, *Mitt Int Verein Limnol.*, 21:285-295.
26. Carmichael WW, Mahmood NA, 1984, Toxins from freshwater cyanobacteria (blue-green algae), *In*: "Seafood Toxins," Ragelis EP, ed., Am Chem Soc Symp, Series 262, Washington DC.
27. Carmichael WW, Biggs DF, Peterson MA, 1979, Pharmacology of anatoxin-a, produced by the freshwater cyanophyte *Anabaena flos-aquae* NRC-44-1, *Toxicon*, 17:229-236.
28. Carmichael WW, Gorham PR, Biggs DF, 1977, Two laboratory case studies on the oral toxicity to calves of the freshwater cyanophyte (blue-green alga) *Anabaena flos-aquae* NRC-44-1, *Can Vet J.*, 18:71-75.
29. Carmichael WW, Jones CLA, Mahmood NA, Theiss WC, 1985, Algal toxins and water-based diseases, *In*: "Critical Reviews in Environmental Control, Vol. 15," Straub CP, ed., Chemical Rubber Co. Press, Florida.
30. Carmichael WW, Eschedor JT, Patterson GML, Moore RE, 1988, Toxicity and partial structure for a hepatotoxic peptide produced by *Nodularia spumigena* Mertens emend. strain L575 (cyanobacteria) from New Zealand, *Appl Environ Microbiol.*, 54:2257-2263.
31. Carmichael W, Beasley V, Bunner DL, Eloff JN, Falconer I, Gorham PR, Harada K-I, Yu M-J, Krishnamurthy T, Moore RE, Rinehart K, Runnegar M, Skulberg OM, Watanabe M, 1988, Naming of cyclic heptapeptide toxins of cyanobacteria (blue-green algae), *Toxicon*, Letter to the Editor, 26: 971-973.
32. Codd GA, Bell SG, 1985, Eutrophication and toxic cyanobacteria in freshwater, *Water Pollution Control*, 84:225-232.
33. Dabholkar AS, Carmichael WW, 1987, Ultrastructural changes in the mouse liver induced by hepatotoxin from the freshwater cyanobacterium *Microcystis aeruginosa* strain 7820, *Toxicon.*, 25:285-292.
34. Danheiser RL, Morin JM, Jr, Salaski, EJ, 1985, Efficient total synthesis of (±)-anatoxin a, *J Amer Chem Soc.*, 107:8066-8073.
35. Devlin JP, Edwards OE, Gorham PR, Hunter NR, Pike RK, Stavric B, 1977, Anatoxin-a, a toxic alkaloid from *Anabaena flos-aquae* NRC-44h, *Can J Chem.*, 55:1367-1371.
36. Edler L, Ferno S, Lind MG, Lundberg R, Nilsson PO, 1985, Mortality of dogs associated with a bloom of the cyanobacterium *Nodularia spumigena* in the Baltic Sea, *Ophelia*, 24:103-109.
37. Elleman TC, Falconer IR, Jackson ARB, Runnegar MT, 1978, Isolation, characterization and pathology of the toxin from a *Microcystis aeruginosa* (= *Anacystis cynea* bloom), *Aust J Biol Sci.*, 31:209-218.
38. Eloff JN, 1987, Isolation, chemical characterization and toxonomic implications of the toxins in *Microcystis aeruginosa* isolates, Ph.D. Thesis, Univ. of the Orange Free State, Republic of South Africa (8 chapters).
39. Eriksson J, Hagerstr H, Isomaa B, 1987, Cell selective cyto-toxicity of a peptide toxin from cyanobacterium (technical note), *Biochem Biophys Acta*, 930:304-310.
40. Eriksson JE, Meriluoto JAO, Kujari HP, Skulberg OM, 1988, A comparison of toxins isolated from the cyanobacteria *Oscillatoria agardhii* and *Microcystis aeruginosa*, *Comp Biochem Physiol.*, 89c:207-210.

41. Eriksson JE, Meriluoto JAO, Kujari HP, Osterlund K, Fagerlund K, Hallbom L, 1988, Preliminary characterization of a toxin isolated from the cyanobacterium *Nodularia spumigena*, *Toxicon*, 26:161-166.
42. Falconer IR, Runnegar MTC, 1987, Effects of the peptide toxin from *Microcystis aeruginosa* on intracellular calcium, pH and membrane integrity in mammalian cells, *Chem Biol Interactions*, 63:215-225.
43. Foxall TL, Sasner JJ, Jr, 1981, Effects of a hepatic toxin from the cyanophyte *Microcystis aeruginosa*, In: "The Water Environment: Algal Toxins and Health," Carmichael WW, ed., Plenum Press, New York City.
44. Gleason FK, Paulson JL, 1984, Site of action of the natural algicide, cyanobacterin, in the blue-green alga *Synechococcus* sp., *Arch Microbiol.*, 138:273-277.
45. Gorham PR, Carmichael WW, 1988, Hazards of freshwater blue-greens (cyanobacteria), Ch 16 In: "Algae and Human Affairs," Lembi CA, ed., Cambridge University Press, Cambridge.
46. Huber CS, 1972, The crystal structure and absolute configuration of 2,9- diacetyl-o-azabicyclo (4,2,1) non-2,3-3ene, *Acta Crystallograph.*, B28:2577-2582.
47. Husan FB, Cohen SG, Cohen JB, 1980, Hydrolysis by acetylcholinesterase: apparent molal volumes and trimethyl and methyl subsites, *J Biol Chem.*, 255:3895-3904.
48. Ikawa M, Wegener K, Foxall TL, Sasner JJ, Jr, 1982, Comparison of the toxins of the blue-green alga *Aphanizomenon flos-aquae* with the *Gonyaulax* toxins, *Toxicon*, 20:747-752.
49. Jackim E, Gentile JH, 1968, Toxins of blue-green alga: similarity to saxitoxin, *Science*, 162:915-916.
50. Kalbe L, Thiess D, 1964, Entenmassensterben durch *Nodularia*-wasserblute am Kleinen Jusmunden godden auf Ragen, *Arch Exp Vet Med.*, 18:535-539.
51. Konst H, McKercher PD, Gorham PR, Robertson A, Howell J, 1965, Symptoms and pathology produced by toxic *Microcystis aeruginosa* NRC-1 in laboratory and domestic animals, *Can J Comp Med Vet Sci.*, 29:221-228.
52. Koskinen MP, Rapoport H, 1985, Synthetic and conformational studies on anatoxin-a: a potent acetylcholine agonist, *J Med Chem.*, 28:1301-1309.
53. Krishnamurthy T, Carmichael WW, Sarver EW, 1986, Investigations of freshwater cyanobacteria (blue-green algae) toxic peptides. I. Isolation-purification and characterization of peptides from *Microcystis aeruginosa* and *Anabaena flos-aquae*, *Toxicon*, 24:865-873.
54. Krishnamurthy T, Szafraniec L, Sarver EW, Hunt DF, Shanbanowitz J, Carmichael WW, Missler S, Skulberg O, Codd G, 1986, Amino acid analysis of freshwater blue-green algal toxic peptides by fast atom bombardment tandem mass spectrometric technique, In: "Proc 34th Ann Conf on Mass Spectrometry and Allied Topics," Cincinnati.
55. Kusumi T, Ooi T, Watanabe M, Kakisawa H, 1987, Structure of cyanoviridin RR, a toxin from the cyanobacterium (blue-green alga) *Microcystis viridis*, In: "Proc of the 29th Symposium on Chemistry of Natural Products, Sapporo, Japan.
56. Lindgren B, Stjernlof P, Trogen L, 1987, Synthesis of anatoxin-a, a constituent of blue-green freshwater algae, *Acta Chem Scand.*, B41: 180-183.
57. Lindstrom T, Eriksson J, 1985, Problemalger och fiskdod i alandska vattentakter, *Ymparisto ja Terveys*, 16:41-44.
58. Mahmood NA, 1985, Neurotoxins from the cyanobacteria *Anabaena flos-aquae* NRC 525-17 and *Aphanizomenon flos-aquae* NH-5, Ph.D. Thesis, Wright State University, Dayton.
59. Mahmood NA, Carmichael WW, Pfahler D, 1988, Anticholinesterase poisonings in dogs from a cyanobacterial (blue-green algae) bloom dominated by *Anabaena flos-aquae*, *Amer J Vet Res.*, 49:500-503.
60. Mahmood NA, Carmichael WW, 1987, Anatoxin-a(s), an anticholinesterase from cyanobacterium *Anabaena flos-aquae* NRC-525-17, *Toxicon*, 25: 1221-1227.

61. Mahmood NA, Carmichael WW, 1986, Paralytic shellfish poisons produced by the freshwater cyanobacterium *Aphanizomenon flos-aquae* NH-5, *Toxicon*, 24:175-186.

62. Mahmood NA, Carmichael WW, 1986, The pharmacology of anatoxin-a(s), a neurotoxin produced by the freshwater cyanobacterium *Anabaena flos-aquae* NRC 525-17, *Toxicon*, 24:425-434.

63. Mahmood NA, Carmichael WW, 1987, Anatoxin-a(s), an anticholinesterase from the cyanobacterium *Anabaena flos-aquae* NRC-525-17, *Toxicon*, 25:1221-1227.

64. Main AR, 1980, Cholinesterase inhibitors, *In*: "Introduction to Biochemical Toxicology," Hodgson E, Guthrie F, eds., Elsevier, New York.

65. Main DC, Berry PH, Peet RL, Robertson JP, 1977, Sheep mortalities associated with the blue-green alga *Nodularia spumigena*, *Aust Vet J.*, 53:578-581.

66. Mason CP, Edwards KR, Carlson RE, Pignatello J, Gleason FK, Wood JM, 1982, Isolation of chlorine-containing antibiotic from the freshwater cyanobacterium *Scytonema hofmanni*, *Science*, 215:400-402.

67. Moore RE, Chenk C, Patterson GML, 1984, Hapalindoles: New alkaloids from the blue-green alga *Hapalosiphon fontinalis*, *J Amer Chem Soc.*, 106:6456-6457.

68. Moore RE, Patterson GML, Mynderse JS, Barchi J, Jr, Norton TR, Furusawa E, Furusawa S, 1986, Toxins from cyanophytes belonging to the Scytonemataceae, *Pure & Appl Chem.*, 58:263-271.

69. Murthy JR, Capindale JB, 1970, A new isolation and structure for the endotoxin from *Microcystis aeruginosa* NRC-1, *Can J Biochem.*, 48:508-510

70. Ostensvik O, Skulberg OM, Soli NE, 1981, Toxicity studies with blue-green algae from Norwegian inland waters, *In*: "The Water Environment: Algal Toxins and Health," Carmichael WW, ed., Plenum Press, New York.

71. Painuly P, Perez R, Fukai T, Shimizu Y, 1988, The structure of a cyclic peptide toxin, cyanogenosin-RR from *Microcystis aeruginosa*, *Tetrahedr. Let.*, 29:11-14.

72. Perrson PE, Sivonen K, Keto J, Kononen K, Niemi M, Viljamaa H, 1984, Potentially toxic blue-green algae (cyanobacteria) in Finnish natural waters, *Aqua Fenn.*, 14:147-154.

73. Petersen JS, Toteberg-Kaulen S, Rapoport H, 1984, Synthesis of (±)-w-Aza[x.y.1] bicycloalkanes by an intramolecular Mannich reaction, *J Org Chem.*, 49:2948-2953.

74. Rabin P, Darbre A, 1975, An improved extraction procedure for the endotoxin from *Microcystis aeruginosa* NRC-1, *Biochem Soc Trans.*, 3, 428-430.

75. Raziuddin S, Siegelman HW, Tornabene TG, 1983, Lipopolysaccharides of the cyanobacterium *Microcystis aeruginosa*, *Eur J Biochem.*, 137:333-336.

76. Rinehart KL, Harada KI, Namikoshi M, Chen C, Harvis C, Munro MHG, Bhent JW, Mulligan PE, Beasley VR, Dahlem AM, Carmichael WW, Nodularin, microcystin, and the configuration of Adda, *J Amer Chem Soc.*, in press.

77. Rosenberry T, 1975, Acetylcholinesterase, *In*: "Advances in Enzymology," Meister A, ed., John Wiley and Sons, New York.

78. Runnegar MTC, Falconer IR, 1981, Isolation, characterization and pathology of the toxin from the blue-green alga *Microcystis aeruginosa*, *In*: "The Water Environment: Algal Toxins and Health," Carmichael WW, ed., Plenum Press, New York.

79. Runnegar MTC, Falconer IR, 1986, Effect of toxin from the cyanobacterium *Microcystis aeruginosa* on ultrastructural morphology and actin polymerization in isolated hepatocytes, *Toxicon*, 24:109-115.

80. Runnegar MTC, Jackson ARB, Falconer IA, 1988, Toxicity of the cyanobacteria *Nodularia spumigena* Mertens, *Toxicon*, 26:143-151.

81. Runnegar MTC, Andrews J, Gerdes RG, Falconer IR, 1987, Injury to hepatocytes induced by a peptide toxin from the cyanobacterium *Microcystis aeruginosa*, *Toxicon*, 25:1235-1239.

82. Runnegar MTC, Falconer IR, Buckley T, Jackson ARB, 1986, Lethal potency and tissue distribution of ^{125}I-labelled toxic peptides from the blue-green alga *Microcystis aeruginosa, Toxicon*, 24:506-509.

83. Santikarn S, Williams DH, Smith RJ, Hammond SJ, Botes DP, Tuinman A, Wessels PL, Viljoen CC, Kruger H, 1983, A partial structure for the toxin BE-4 from the blue-green algae, *Microcystis aeruginosa, J Chem Soc Chem Commun.*, 12:652-654.

84. Sasner JJ, Jr, Ikawa M, Foxall TL, 1984, Studies on *Aphanizomenon* and *Microcystis* toxins, *In*: "Seafood Toxins," Ragelis EP, ed., Amer Chem Soc Symposium, Series 262, Washington, D.C.

85. Sawyer PJ, Gentile JH, Sasner JJ, Jr, 1968, Demonstration of a toxin from *Aphanizomenon flos-aquae* (L.) Ralf, *Can J Microbiol.*, 14:1199-1204.

86. Schwimmer M, Schwimmer D, 1968, Medical aspects on phycology, *In*: "Algae, Man and the Environment," Jackson DF, ed., Syracuse University Press, New York.

87. Shimizu Y, Norte M, Hori A, Genenah A, Kobayushi M, 1984, Biosynthesis of saxitoxin analogues: the unexpected pathway, *J Amer Chem Soc.*, 106: 6433-6434.

88. Skulberg OM, Codd GA, Carmichael WW, 1984, Toxic blue-green algal blooms in Europe: a growing problem, *Ambio.*, 13:244-247.

89. Smith RA, Lewis D, 1987, A rapid analysis of water for anatoxin-a, the unstable toxic alkaloid from *Anabaena flos-aquae*, the stable non-toxic alkaloids left after bioreduction and a related amine which may be nature's precursor to anatoxin-a, *Vet Hum Tox.*, 29:153-154.

90. Spivak CE, Waters J, Witkop B, Albuquerque EX, 1983, Potencies and channel properties induced by semirigid agonists at frog nicotinic acetylcholine receptors, *Mol Pharm.*, 23:337-343.

91. Spivak CE, Witkop B, Albuquerque EX, 1980, Anatoxin-a: a novel, potent agonist at the nocotinic receptor, *Mol Pharmacol.*, 18:384-394.

92. Stevens DK, Krieger RI, 1988, Analysis of anatoxin-a by GC/ECD, *J Analyt Toxicol.*, 12:126-131.

93. Theiss WC, Carmichael WW, Wyman J, Bruner R, 1988, Blood pressure and hepatocellular effects of the cyclic heptapeptide toxin produced by *Microcystis aeruginosa* strain PCC-7820, *Toxicon.*, 26:603-613.

94. Tomlinson G, Mutusi B, McLennan I, 1980, Modulation of acetylcholinesterase activity by peripheral site ligands, *Mol Pharm.*, 18:33-39.

95. Tufariello JJ, Meckler H, Senaratne KPA, 1984, Synthesis of anatoxin-a: very fast death factor, *J Amer Chem Soc.*, 106:7979-7980.

96. Tufariello JJ, Meckler H, Senaratne KPA, 1985, The use of nitrones in the synthesis of anatoxin-a, very fast death factor, *Tetrahedr.*, 41: 3447-3453.

97. Wilson IB, Quan C, 1958, Acetylcholinesterase studies on molecular complementariness, *Arch Biochem Biophys.*, 73:131-143.

98. Wiseman JR, Lee SY, 1986, Synthesis of anatoxin-a, *J Org Chem.*, 51: 2485-2487.

99. Wong SH, Hindin E, 1982, Detecting an algal toxin by high-pressure liquid chromatography, *Amer Water Works Assoc J.*, 74:528-529.

NEW ASPECTS OF TETRODOTOXIN

Kanehisa Hashimoto, Tamao Noguchi and Shugo Watabe

Laboratory of Marine Biochemistry
Faculty of Agriculture
University of Tokyo
Yayoi 1-1-1
Bunkyo, Tokyo 113, Japan

INTRODUCTION

In Japan, people have been accustomed to eating pufferfish, sporadically resulting in poisoning or even death. Japanese scientists continued
to study the causative toxin, tetrodotoxin (TTX), and Yokoo[31] for the first
time isolated and crystallized this toxin from pufferfish ovaries. This
event stimulated studies on TTX, leading to elucidation of its structure in
1964. TTX was long believed to occur exclusively in pufferfish. Since
1964, however, this toxin also has been detected in many other vertebrates
and invertebrates. Some intestinal bacteria of TTX-bearing animals were
demonstrated to produce TTX. This suggested that TTX accumulated in those
animals could pass along the food web starting from such TTX-producing
bacteria. Furthermore, attempts have also been made to prepare a monoclonal
antibody to TTX. This review deals with some new aspects of TTX.

Distribution of TTX in nature

In 1964, Mosher et al.[12] detected TTX in the California newt, which
was the first TTX-containing organism other than pufferfish. Since then,
the toxin has been detected in a tropical goby,[16] Costa Rican frogs,[3]
the blue-ringed octopus,[24] and several species of carnivorous gastropods
such as trumpet shell,[14] ivory shell,[20] and frog shell.[19] In addition,
some species of starfish on which these gastropods prefer to feed also contained TTX.[21] Toxic crabs,[22,30] flatworms,[2,11] a horseshoe crab,[6] ribbon-
worms[10] and arrowworms[27] were also added to the list of TTX-bearing animals
(Table 1).

TTX productivity of intestinal and other bacteria from TTX-bearing animals

Whether the origin of TTX is endogenous or exogenous in pufferfish has
been controversial for a long time. Recently, it was found that most of the
cultured pufferfish were nontoxic[23] but easily became toxic when fed a TTX-
containing diet.[9] Furthermore, shell debris of TTX-containing small gastropods were detected rather frequently in the intestinal contents of pufferfish.[7] Starfish arms were also found in the intestines of the trumpet shell,
another TTX bearer. Subsequent studies established that the starfish, which
the trumpet shell prefers, also contains TTX.[21]

Microbial Toxins in Foods and Feeds, Edited by A.E. Pohland *et al.,*
Plenum Press, New York, 1990

Table 1. Distribution of Tetrodotoxin and Related Substances in Animals.

Animals		Parts
1. Platyhelminthes:		
Turbellaria;	Flatworms	
	Planocera sp.	Whole body
2. Nemertinea:	Ribbonworms	
	Lineus fuscoviridis	Whole body
	Tubulanus punctatus	Whole body
	Cephalothrix linearis	Whole body
3. Mollusca:		
Gastropoda;	*Charonia sauliae*	Digestive gland
	Babylonia japonica	Digestive gland
	Tutufa lissostoma	Digestive gland
	Zeuxis siquijorensis	Digestive gland
	Niotha clathrata	Digestive gland
	Cymatium echo	Digestive gland
	Pugilina ternotona	Digestive gland
Cephalopoda;	*Octopus maculosus*	Posterior salivary gland
4. Annelida:		
Polychaeta;	*Pseudopotamilla acelata*	Whole body
5. Arthropoda:	*Atergatis floridus*	Whole body
	Zosimus aeneus	Whole body
	Carcinoscorpius rotundicauda	Egg
6. Chaetognatha:	Arrowworms	
	Parasagitta sp.	Head
	Flaccisagirtta sp.	Head
7. Echinodermata:	Starfishes	
	Astropecten polyacanthus	Whole body
	A. latespinosus	Whole body
	A. scoparius	Whole body
8. Vertebrata:		
Pisces;	*Fugu* sp.	Skin, liver, ovary
	Gobius criniger	Skin, viscera, gonad
Amphibia;	*Tarichia* sp.	Skin, egg, ovary, muscle, blood
	Cynops sp.	Skin, egg, ovary, muscle, blood
	Triturus sp.	Skin, egg, ovary, muscle, blood
	Atelopus sp.	Skin

All these observations suggest that TTX-bearers become toxic through the food chain in which TTX is transferred from lower to higher strata animals. This, along with the phylogenetically irregular occurrence of TTX (Table 1), suggested that some microorganism(s) could be the true producer(s) of this toxin. This was demonstrated by the finding of TTX-producing bacteria as described below.

A xanthid crab, *Atergatis floridus*, usually contains large amounts of paralytic shellfish poison (PSP), causing frequent poisoning incidents in the tropics. This species also inhabits the Pacific coast of Japan, but contains TTX as the major toxin, in place of PSP.[22]

We collected several specimens of this crab from Shimoda, Shizuoka Prefecture, and isolated the dominant strains of bacteria from their intestinal contents. After cultivating in 1% NaCl-1% Phytone peptone medium at 25°C for 10 days, cells were harvested, ultrasonicated in 0.1% acetic acid, and examined for toxicity by the mouse assay. A portion of the cells was used for identification according to Bergey's.[5] Toxin was partially purified from the lysate by ultrafiltration and Bio-Gel P-2 column chromatography. The "TTX fraction" thus prepared was subjected to HPLC and GC-MS analyses. The results showed that a *Vibrio fischeri*-like bacterium produced TTX and anhydroTTX.[18,26]

The dominant bacteria were also isolated from intestinal contents as well as several tissues of other TTX-bearing animals such as the puffer *Fugu vermicularis vermicularis*,[17] the starfish *Astropecten polyacanthus*,[13] the blue-ringed octopus *Octopus maculosus*,[1] the flatworm *Planocera reticulata*,[11] and the horseshoe crab *Carcinoscorpius rotundicauda*.[6]

Table 2 shows the analytical results obtained by examining the metabolites produced by the dominant intestinal bacteria of the starfish.[13] As far as strains No. 2-4 are concerned, essentially the same results were obtained with the culture broth. A total of 18 dominant strains were isolated, and classified into two genera, *Vibrio* (14 strains) and *Staphylococcus* (4 strains). The 14 *Vibrio* strains were composed of four of *V. alginolyticus*, six of *V. damsela*, and four unidentified strains. Instrumental analyses demonstrated that the majority of the strains examined produced TTX and/or related substances. In addition to the four strains of *V. alginolyticus*, some strains of *V. damsela* and *Staphylococcus* were recognized to be TTX producers.

Strain No. 7 (*V. alginolyticus*), which showed TTX production most typically, was cultivated in several 500-ml flasks containing a 1% NaCl-1% Phytone peptone medium at 25°C. At 24 h-intervals, cells were harvested and extracted as described above, and assayed for lethal potency. As much as 213 MU per flask rather suddenly appeared in the 72 h-culture[13] (Table 3). These results suggested that *Vibrio* and *Staphylococcus* bacteria, especially *V. alginolyticus*, are involved in toxification of the starfish *A. polyacanthus*.

A total of 33 strains of aerobic and facultatively anaerobic bacteria were isolated from the puffer.[17] Among them, 26 strains were identified as *Vibrio* spp., divided into seven groups (I-VIII, excluding VI) on the basis of several characters.

TTX fractions which were separated from bacterial cells and/or culture broths of these strains were subjected to instrumental analyses. The data obtained for cell extracts are collectively shown in Table 4. All strains classified into Group I, which showed a toxicity of 3 MU, were identified as *V. alginolyticus*.

The toxic crab species *A. floridus* inhabiting a limited area (Kojima) of Ishigaki Island, Okinawa Prefecture, rather surprisingly contains TTX, not not PSP, as the major toxin.[15] Four dominant strains were isolated from the intestinal contents of the *A. floridus* specimens, classified into *Vibrio* groups I, III, V and VIII.

Table 5 shows the results obtained for these four strains. An appreciable TTX production (30 MU per flask) by the group VIII strain was noted. HPLC and GC-MS data obtained for the TTX fraction are shown in Figs. 1-3.

Table 2. Instrumental analyses for TTXs in TTX fraction from *Vibrio* and *Staphylococcus* strains isolated from a starfish *A. polyacanthus*.

Serial No. of Strain	Species	HPLC[*]	UV[*]	GC-MS[*]
1	*Vibrio alginolyticus*	-	±	+
2	*Vibrio sp.*	-	±	+
3	*V. alginolyticus*	+	+	+
4	*Vibrio sp.*	-	-	+
5	*V. damsela*	+	-	+
6	*V. damsela*	-	-	-
7	*V. alginolyticus*	+	+	+
8	*V. alginolyticus*	±	-	+
9	*V. damsela*	-	-	+
10	*V. damsela*	-	-	±
11	*V. damsela*	-	-	-
12	*Vibrio sp.*		-	±
13	*V. damsela*		-	-
14	*Vibrio sp.*		-	±
15	*Staphylococcus*		-	-
16	*Staphylococcu*		-	-
17	*Staphylococcus*		-	+
18	*Staphylococcus*		-	+

[*] +, Detected; ±, judgement difficult; -, not detected.

Table 3. Changes in lethal potency of *V. alginolyticus* cells from *A. polyacanthus* when cultured in a 500-ml flask for 24, 48 and 72 hours.

Time of culture (h)	Lethal potency (MU/flask)
24	-[*]
48	-
72	213

[*] Less than 2 MU.

The TTX fraction gave rise to two peaks in HPLC, a large one corresponding to TTX and a small one corresponding to anhydrotetrodotoxin (Anh-TTX) (Fig. 1). When degraded with alkali, trimethylsilylated and subjected to GC-MS, the TTX fraction exhibited mass fragment ions at m/z 407 (molecular peak), 392 (base peak) and 376 all of which are specific to the trimethylsilylated C_9-base derived from authentic TTX (Figs. 2 and 3).

Table 4. Instrumental analyses for TTXs, along with lethal potency of the extracts of six *Vibrio* groups isolated from puffer intestine.

Group No.	Lethal potency[1] (MU)	HPLC[2] TTX	HPLC[2] Anh-TTX	GC-MS[2]
I	3	+	+	+
II	-	±	+	+
III	-	±	±	+
IV	-	±	±	+
V	-	±	-	±
VII	-	±	±	-

[1] 3; A total of 3 MU of TTX was detected in *Vibrio* cells harvested after culture for 7-days in a 500-ml flask.
-; Less than 2 MU.
[2] +; Detected; ±; judgement difficult; -; not detected.

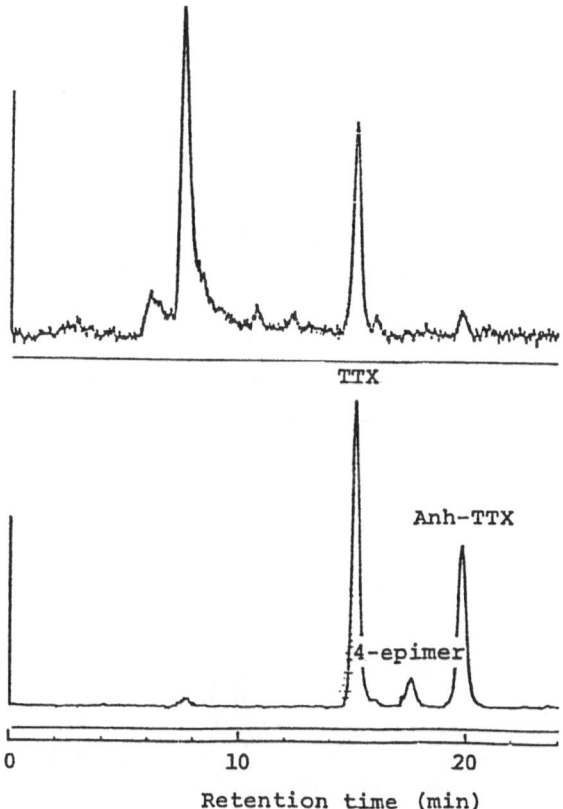

TTX

Anh-TTX

4-epimer

0   10   20

Retention time (min)

Fig. 1. HPLC of TTX fraction from *Vibrio* Group VIII isolated from a "Kojima specimen" (upper) and of authentic TTXs (lower).

Table 5. Analyses for TTX and related substances in the extracts of several *Vibrio* groups of bacteria, which were isolated from intestines of *A. floridus* specimens collected from Kojima in Ishigaki Island, Okinawa.

Vibrio group	Lethal[*1] potency (MU)	HPLC[*2]		GC-MS[*2]
		TTX	Anh-TTX	
I	-	+	-	
III	-	+	-	
V	-	±	-	
VIII	30	+	+	+

[*1,2] Refer to the footnotes in Table 4.

The results obtained for 16 dominant strains from several tissues of the octopus[1] are summarized in Table 6. Two strains (No. 1 and 2) from the arm and another two (No. 9 and 12) from the intestines, produced TTX, as demonstrated by HPLC and GC-MS analyses. However, two strains (No. 14 and

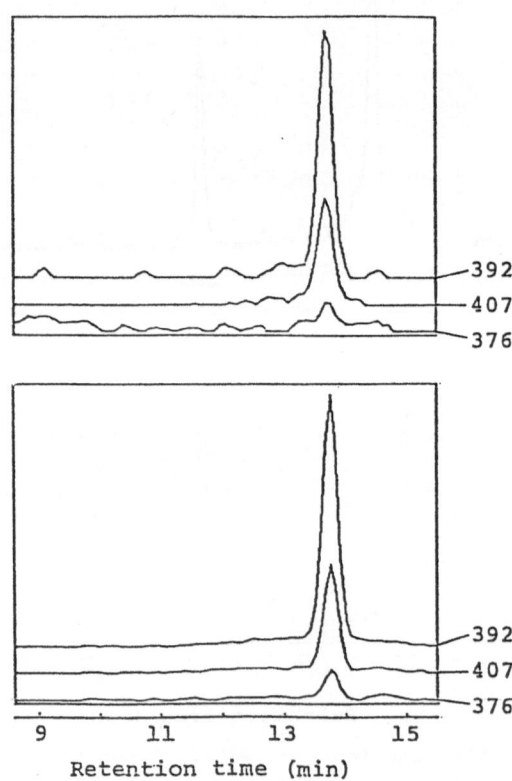

Retention time (min)

Fig. 2. Selected ion-monitored chromatograms of the trimethylsilylated derivative from authentic TTXs (lower) and of the corresponding derivative from TTX fraction of *Vibrio* Group VIII isolated from a "Kojima specimen" (upper).

Table 6. Instrumental analyses for TTXs, along with lethal potency,
in TTX fraction from bacterial strains isolated from arms,
intestines and salivary glands of the octopus *O. maculosus*.

Serial No. of Strain	Source	Lethal[*1] Potency (MU/Flask)	HPLC[*2] TTX	HPLC[*2] Anh-TTX	UV[*2]	GC-MS[*2]
1	Arms	-	-	+	+	+
2	"	-	-	+	+	+
3	"	-	-	+	+	±
4	"	-	-	-	+	+
5	"	-	-	±	-	-
6	"	-	-	±	+	±
7	"	-	-	±	+	+
8	Intestines	-	-	±	-	+
9	"	-	-	+	+	+
10	"	-	-	±	+	-
11	"	-	-	+	+	±
12	"	-	-	+	+	+
13	"	-	-	±	+	+
14	Salivary glands	3	±	±	+	+
15	"	-	-	-	-	±
16	"	5	±	-	+	+

[*1,2] Refer to the footnotes in Tables 2 and 4.

16) out of the three isolated from the salivary glands produced toxin, as
confirmed by the mouse assay as well as instrumental analyses.   Those
strains were identified as *Bacillus* sp. and *Pseudomonas* sp., respectively.
In this connection, Yasumoto et al.[29] isolated a *Pseudomonas* sp. from
a red calcareous alga *Jania* sp. which produced TTX and a related substance.

Thirteen of the sixty dominant strains isolated from the flatworm
showed more or less TTX productivity,[10] as summarized in Table 7.   Four
strains (No. 2, 3, 4 and 10) gave positive results in the mouse assay.

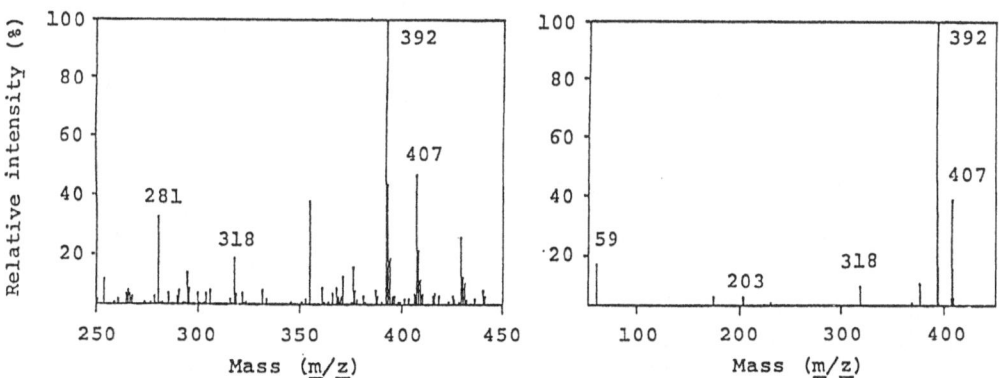

Fig. 3. Mass spectra of the trimethylsilylated derivative from
authentic TTXs (right) and of the corresponding derivative
from TTX fraction of *Vibrio* Group VIII isolated from a
"Kojima specimen" (left).

Table 7.   Instrumental analyses for TTXs, along with lethal potency, in TTX fraction from bacteria isolated from the flatworm *P. multitentaculata* intestines.

Serial No. of Strain	Lethal Potency[1] (MU/Flask)	HPLC[2] TTX	Anh-TTX	GC-MS[2]
1	-	+	+	-
2	> 2	+	+	-
3	>10	+	+	+
4	>10	+	±	+
5	-	+	-	+
6	-	±	-	±
7	-	±	-	+
8	-	+	-	+
9	-	+	-	+
10	2	±	-	+
11	-	±	-	+
12	-	+	±	+
13	-	±	±	+

[1,2]   Refer to the footnotes in Table 4.

Thirteen of the 16 dominant strains isolated from the intestines of the horseshoe crab were recognized to be TTX-producers by HPLC and GC-MS analyses[6] (Table 8).

TTX productivity of marine bacteria

To confirm the involvement of marine bacteria in TTX production, screening experiments with some typical marine bacterial strains were carried out.[25] The bacteria used were from culture collections, mostly the American Type Culture Collection (Rockville, MD, U.S.A.) and the National Collection of Marine Bacteria (Aberdeen, Scotland), and included 15 strains of the family Vibrionaceae, 5 strains of the genus *Alteromonas*, and 1 strain of *Escherichia coli*. The strains were incubated in a seawater medium. For three strains, *Aeromonas hydrophyla*, *Aeromonas salmonicida*, and *E. coli*, the medium was also prepared with distilled water. After incubation, cells were harvested by centrifugation. TTX fractions obtained from bacaterial cells by essentially the same technique as described above, were subjected to HPLC and GC-MS analyses.[25] The data obtained are summarized in Table 9.

Among the 15 strains of the family Vibrionaceae, 10 strains including those of *V. alginolyticus*, *V. parahaemolyticus*, and *V. anguillarum*, along with *Photobacterium phosphoreum*, clearly showed the ability to produce anh-TTX. The cultured cells of *V. alginolyticus* ATCC 17749 killed five mice.

In this respect, Kogure et al.[4] reported that TTX was accumulated at a detectable level in sediment samples from 2 coastal stations and one deep sea station in the western Pacific, and suggested that the bacteria inhabiting the sediment may be responsible for the TTX accumulation.

Mechanism involved in toxification of TTX-bearers

All these studies clearly demonstrated that there are many TTX-produc-

Table 8. Instrumental analyses for TTXs in TTX fraction from
bacterial strains isolated from the intestnes of
horseshoe crab *C. rotundicauda*.

Serial No. of Strain	Source	HPLC[*] TTX	HPLC[*] Anh-TTX	UV[*]	GC-MS[*]
1	Intestines	-	+	+	+
2	"	-	+	+	+
3	"	-	+	+	+
4	"	-	+	+	+
5	"	-	±	+	+
6	"	-	±	+	+
7	"	+	+	+	+
8	"	-	+	+	+
9	"	-	±	+	+
10	"	-	±	+	+
11	"	-	±	+	+
12	"	-	+	+	+
13	"	-	±	+	+

[*] Refer to the footnotes in Tables 2 and 4.

other tissues of TTX-bearing animals. It was also shown that many strains
of standard marine bacteria including *V. alginolyticus* are endowed with more
or less TTX-producing ability.

This, along with various findings so far obtained, seems to indicate
the involvement of the following mechanism in the toxification of TTX-bearing animals: First, some TTX-producing marine bacteria enter and inhabit
the intestines of invertebrates of lower-strata in the food chain. These
bacteria produce TTX and/or related substances, which are accumulated in the
hosts and then transferred to organisms of middle strata through predation.
The physiological relationship between those bacteria and lower-strata
invertebrates is still unclear. Finally, carnivorous animals, whether
invertebrates or vertebrates as represented by pufferfish, feed on these
toxic lower strata invertebrate, accumulating TTX and/or relatd substances
efficiently. A portion of the toxin may come directly from TTX-producing
bacteria inhabiting the intestines, or even from the sediment.

Anti-tetrodotoxin antibody

TTX is officially determined by the mouse assay method in Japan. TTX
can also be assayed by HPLC and GC-MS as described above.

Attempts have been made to prepare a monoclonal antibody to TTX,
without much success. Recently, we have obtained somewhat promising
results.[28] Tetrodonic acid, a relatively nontoxic derivative of TTX, was
conjugated with bovine serum albumin and injected intraperitoneally into
BALB/c mice. After several injections, spleen cells were isolated, fused
with myeloma cells X63-Ag8-6.5.3 and cloned by the limiting dilution method.
The monoclonal antibody was produced in ascites fluid in the mouse by the
cloned cell. The reactivity of the antibody examined by ELISA correlated
with TTX concentrations from 0.03 to 100 μg/well (Fig. 4).

Toxic livers of *F. vermicularis vermicularis* and nontoxic livers of *F.
rubripes rubripes* were extracted with an equal volume of phosphate-buffered

Table 9. Analyses of tetrodotoxin and anhydrotetrodotoxin in bacterial cells.

Bacterial Strain	Toxin detected by:[a]		
	HPLC		GC-MS
	TTX	Anh-TTX	
Vibrio alginolyticus ATCC 17749	-	+	+
V. alginolyticus NCMB 1903	-	+	+
V. anguillarum NCMB 829	-	+	+
V. anguillarum NCMB 1291	-	+	+
V. costicola (V. costicolus) NCMB 701	-	+	+
V. fischeri NCMB 1281	-	±	-
V. fischeri (Photobacterium fischeri) NCMB 1381	-	±	±
V. harveyi (Aeromonas harveyi) NCMB 2	-	±	+
V. marinus Ps 207	-	±	±
V. parahaemolyticus NCMB 1902	-	+	+
V. parahaemolyticus ATCC 17802	-	+	+
Photobacterium phosphoreum NCMB 844	±	+	+
Aeromonas hydrophila NCMB 89	-	+	-
A. hydrophila NCMB 89[b]	-	±	-
A. salmonicida ATCC 14174	±	+	+
A. salmonicida ATCC 14174[b]	±	±	+
Plesiomonas shigelloides ATCC 14029	±	+	+
Escherichia coli IAM 1268	-	-	-
E. coli IAM 1268[b]	-	±	±
Alteromonas communis IAM 12914	-	±	±
A. haloplanktis IAM 12918	-	-	-
A. nigrifaciens IAM 13010	±	±	±
A. undina IAM 12922	-	-	-
A. vaga IAM 12923	±	-	±

[a] HPLC, High-performance liquid chromatography; GC-MS, gas chromatography-mass spectrometry, TTX, Tetrodotoxin; Anh-TTX, anhydrotetrodotoxin, +, Clearly detected; ±, difficult to detect; -, not detected.

[b] Cultivated in a freshwater medium.

saline (pH 7.2), and used in some TTX-antibody reactions. In the nontoxic puffer liver extract, a close relationship was obtained between reactivity and TTX in a range from 0.35 to 6 µg/well (Fig. 5). For some unknown reason, the reactivity was inversely related to the concentration of TTX below 0.35 µg/well. Reactivity of the monoclonal antibody to TTX was examined in the toxic puffer liver extract. As shown in Fig. 6, the ELISA reaction progressively increased with increasing toxicity from 40 to 152 MU/g liver, which corresponded to 0.2-0.76 µg TTX/50 µl.

In our study, the reactivity of the monoclonal antibody from the mouse was rather low and actually antibodies did not neutralize the toxicity of TTX which was administered intraperitoneally into mice. Further attempts are now in progress to produce a monoclonal antibody with stronger reactivity.

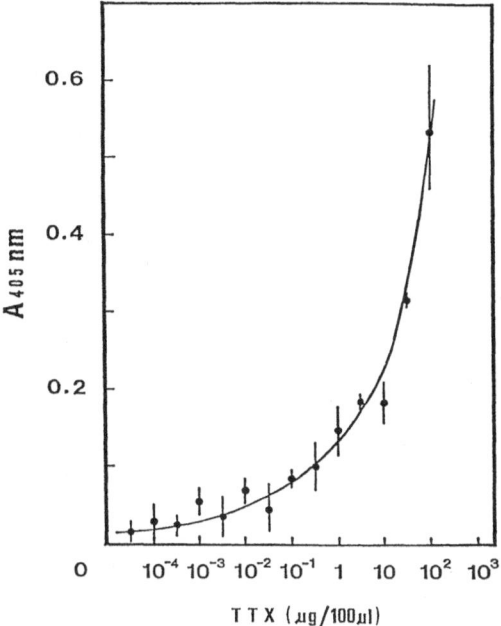

Fig. 4. Reactivity of monoclonal antibody from D12E8 cell line with TTX alone.

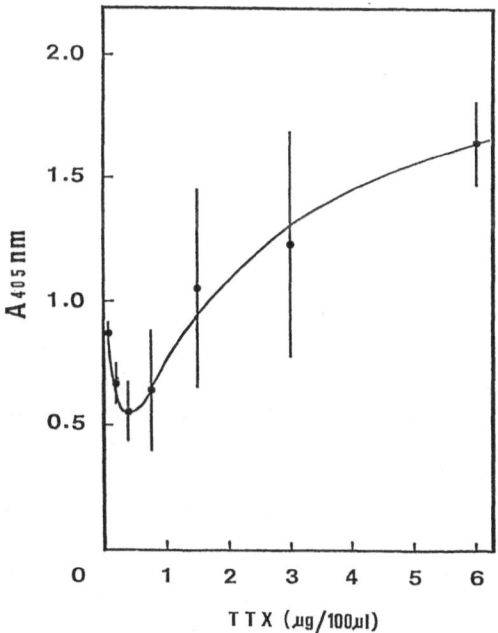

Fig. 5. Reactivity of monoclonal antibody from D12E8 cell line with TTX in the presence of the non-toxic puffer liver extract.

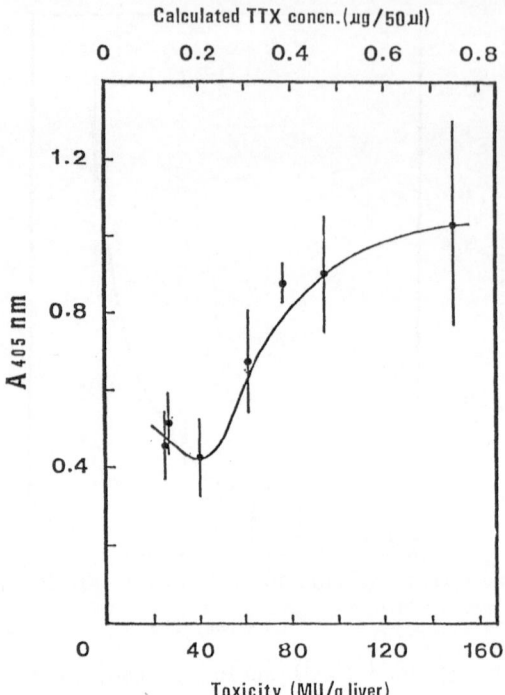

Fig. 6.   Reactivity of monoclonal antibody from D12E8 cell line with
          the toxic puffer liver extract.

REFERENCES

1.   Hwang DF, Arakawa O, Saito T, Noguchi T, Simidu U, Tsukamoto K, Shida
     Y, Hashimoto K, 1988, Tetrodotoxin-producing bacteria from the blue-
     ringed octopus, Octopus maculosus, Mar Biol. (in press).
2.   Jeon JK, Miyazawa K, Noguchi T, Narita H, Ito K, Hashimoto K, 1986,
     Occurrence of paralytic toxicity in marine flatworms, Nippon Suisan
     Gakkaishi, 52:1065-1069.
3.   Kim YH, Brown GB, Mosher HS, Fuhrman FA, 1975, Tetrodotoxin:
     Occurrence in atelopid frogs of Costa Rica, Science, 189:1151-1152.
4.   Kogure K, Do HK, Thuesen EV, Nanba K, Ohwada K, Simidu U, 1988,
     Accumulation of tetrodotoxin in marine sediment, Mar Ecol Prog Ser.,
     45:303-305 pp.
5.   Krieg NR, Holt JG, 1984, Bergey's Manual of Systematic Bacteriology,
     Williams & Wilkins, Baltimore, 1:518-538.
6.   Kungsuwan A, Nagashima Y, Noguchi T, Shida Y, Suvapeepan S,
     Suwansakornkul P, Hashimoto K, 1987, Tetrodotoxin in the horseshoe
     crab Carcinoscorpius rotundicauda inhabiting Thailand, Nippon Suisan
     Gakkaishi, 53:261-266.
7.   Kungsuwan A, Noguchi T, Arakawa O, Simidu U, Tsukamoto K, Shida Y,
     Hashimoto K, 1988, Tetrodotoxin-producing bacteria from the horseshoe
     crab Carcinoscorpius rotundicauda, Nippon Suisan Gakkaishi (in press).
8.   Kuwabara R, Kanoh S, Noguchi T, Hashimoto K, 1983, Composition of
     stomach contents in a puffer Fugu pardalis, Abstr Ann Mtg., (Spring),
     Japan Soc Sci Fish., Tokyo: p. 221.
9.   Matsui T, Hamada S, Konosu S, 1981, Difference in accumulation of
     pufferfish toxin and crystalline tetrodotoxin in the pufferfish,
     Nippon Suisan Gakkaishi, 47:535-537.

10. Miyazawa K, 1988, Distribution of tetrodotoxin in arthropods, flat-worms, tapeworms, etc., *In*: "Recent Advances in Tetrodotoxin Research," Hashimoto K, ed., (Sym Ser No 70), Koseisha-Koseikaku, Tokyo.

11. Miyazawa K, Jeon JK, Maruyama J, Noguchi T, Ito K, Hashimoto K, 1986, Occurrence of tetrodotoxin in the flatworm *Planocera multitentaculata*, *Toxicon.*, 24:645-650.

12. Mosher HS, Fuhrman FA, Buchwald HD, Fischer HG, 1964, Tarichatoxin-tetrodotoxin: a potent neurotoxin, *Science*, 144:1100-1110.

13. Narita H, Matsubara S, Miwa N, Akahane S, Murakami M, Goto T, Nara M, Noguchi T, Saito T, Shida Y, Hashimoto K, 1987, *Vibrio alginolyticus*, a TTX-producing bacterium isolated from the starfish *Astropecten polyacanthus*, *Nippon Suisan Gakkaishi*, 53:617-621.

14. Narita H, Noguchi T, Maruyama J, Ueda Y, Hashimoto K, Watanabe Y, Hida K, 1981, Occurrence of tetrodotoxin in a trumpet shell, "boshubora" *Charonia sauliae*, *Nippon Suisan Gakkaishi*, 47:935-941.

15. Noguchi T, Arakawa O, Daigo K, Hashimoto K, 1986, Local differences in toxin com position of a xanthid crab *Aterqatis floridus* inhabiting Ishigaki Island, Okinawa, *Toxicon.*, 24: 705-712.

16. Noguchi T, Hashimoto Y, 1973, Isolation of tetrodotoxin from a goby *Gobius criniger*, *Toxicon.*, 11:305-307.

17. Noguchi T, Hwang DF, Arakawa O, Sugita H, Deguchi Y, Shida Y, Hashimoto K, 1987, *Vibrio alginolyticus*, a tetrodotoxin-producing bacterium, in the intestines of fish *Fugu vermicularis vermicularis*, *Mar Biol.*, 94:625-630.

18. Noguchi T, Jeon JK, Arakawa O, Sugita H, Deguchi Y, Hashimoto K, 1986, Occurrence of tetrodotoxin and anhydrotetrodotoxin in *Vibrio* sp. isolated from the intestines of a xanthid crab, *Atergatis floridus*, *J Biochem.*, 99:311-314.

19. Noguchi T, Maruyama J, Narita H, Hashimoto K, 1984, Occurrence of tetrodotoxin in the gastropod mollusk *Tutufa lissostoma* (frog shell), *Toxicon.*, 22:219-226.

20. Noguchi T, Maruyama J, Ueda Y, Hashimoto K, Harada T, 1981, Occurrence of tetrodotoxin in the Japanese ivory shell *Babylonia japonica*, *Nippon Suisan Gakkaishi*, 47:901-913.

21. Noguchi T, Narita H, Maruyama J, Hashimoto K, 1982, Tetrodotoxin in the starfish *Astropecten polyacanthus*, in association with toxifica-tion of a trumpet shell, "boshubora" *Charonia sauliae*, *Nippon Suisan Gakkaishi*, 48:1173-1177.

22. Noguchi T, Uzu A, Koyama K, Hashimoto K, 1983, Occurrence of tetrodo-toxin as the major toxin in a xanthid crab *Atergatis floridus*, *Nippon Suisan Gakkaishi*, 49:1887-1892.

23. Saito T, Maruyama J, Kanoh S, Jeon JK, Noguchi T, Harada T, Murata O, Hashimoto K, 1984, Toxicity of the cultured pufferfish *Fugu rubripes rubripes*, along with their resistibility against tetrodotoxin (Japanese), *Nippon Suisan Gakkaishi*, 50:1573-1576.

24. Sheumack DD, Howden MEH, Spence I, Quinn RJ, 1978, Maculotoxin: a neurotoxin from the venom glands of the octopus *Hapalochlaena maculosa* identified as tetrodotoxin, *Science*, 199:188-189.

25. Simidu U, Noguchi T, Hwang DF, Shida Y, Hashimoto K, 1987, Marine bacteria which produce tetrodotoxin, *Appl Environ Microbiol.*, 53:714-1715.

26. Sugita H, Ueda R, Noguchi T, Arakawa O, Hashimoto K, Deguchi Y, 1987, Identification of a tetrodotoxin-producing bacterium isolated from the xanthid crab *Atergatis floridus*, *Nippon Suisan Gakkaishi*, 53:693.

27. Thuesen EV, Kogure K, Hashimoto K, Nemoto T, 1988, Poison arrowworms: a tetrodotoxin venom in the marine phylum Chaetognatha, *J Exp Mar Biol Ecol.*, 116:249-256.

28. Watabe S, Sato Y, Nakaya M, Hashimoto K, Enomoto A, Kaminogawa S, Yamauchi K, 1988, Monoclonal antibody raised against tetrodonic acid, a derivative of tetrodotoxin, *Toxicon.*, (in press).

29. Yasumoto T, Yasumura D, Yotsu M, Michishita T, Endo A, Kotaki Y, 1986, Bacterial production of tetrodotoxin and anhydrotetrodotoxin, *Agric Biol Chem.*, 50:793-795.

30. Yasumura D, Oshima Y, Yasumoto T, Alcala AC, Alcala LC, 1986, Tetrodotoxin and paralytic shellfish toxins in Philippine crabs, *Agric Biol Chem.*, 50:593-598.

31. Yokoo A, 1950, Chemical studies on tetrodotoxin Rept. III. Isolation of spheroidine, *J Chem Soc Japan*, 71:591-592.

POLYETHER TOXINS IMPLICATED IN CIGUATERA AND SEAFOOD POISONING

Michio Murata and Takeshi Yasumoto

Department of Food Chemistry
Faculty of Agriculture
Tohoku University
Tsutsumidori Amamiya
Sendai 980 Japan

Naturally occurring polyether compounds are secondary metabolites presumably derived from polyketide biosynthetic pathways. Organisms of *Streptomycetales* were the first reported biogenetic origin of this class of compounds. Some of the compounds have been used as antibiotics for domestic animals. Recently, a variety of novel polyethers has been reported from marine biota, and dinoflagellates have been established as another important source of the compounds. Many of the polyethers are implicated in seafood poisonings; ciguatoxin and maitotoxin in ciguatera, okadaic acid and pectenotoxins in the diarrhetic shellfish poisoning (DSP), and brevetoxins in neurotoxic shellfish poisoning. This paper reviews recent progress in the chemistry of polyether toxins implicated in seafood poisonings with emphasis on ciguatera and DSP.

CIGUATOXIN (CTX)

The principle toxin responsible for ciguatera has been presumed to be ciguatoxin. The epiphytic dinoflagellate, *Gambierdiscus toxicus*, was identified as the origin of CTX.[23] While a significant amount of CTX or its congener was detected in the wild algae, the organism seems to lose the CTX-producing activity when cultured.

The structural study of CTX has been handicapped by the extreme difficulty in obtaining enough materials for NMR measurements or crystallization. Since the success of Scheuer's group in isolating CTX from the viscera of moray eels, mostly *Gymnothorax javanicus*,[12,15] some chemical and pharmacological features of the toxin have been disclosed. The lethal potency of CTX against mice (0.45 μg/kg)[15] is comparable with that of palytoxin. The most important pharmacological aspect of CTX is that it increases permeability of Na^+ through sodium channels.[11] Nukina et al. reported that a high resolution FAB mass spectrum suggested the molecular weight to be 1111.7 ± 0.3, which corresponded to a formula of $C_{59}H_{85}NO_{19}$.[5] The ^{1}H NMR spectrum showed 25-30 oxymethines, most of which are presumably assignable on the basis of their chemical shifts to juncture carbons of the ether rings.[15] These characteristics as well as the highly unsaturated nature of the molecule imply the presence of many ether rings. The outline of the structure probably resembles those of the brevetoxins[5,13] or yessotoxin.[7]

Microbial Toxins in Foods and Feeds, Edited by A.E. Pohland *et al.*,
Plenum Press, New York, 1990

MAITOTOXIN (MTX)

The toxin was discovered as one of the ciguatera causing toxins from surgeonfish *Ctenochaetus striatus* and was named after the Tahitian name of the fish 'matio'.[21] Unlike CTX, MTX can be obtained from the cultured cells of *G. toxicus*. MTX has a mouse lethal potency of 0.13 ug/kg which makes the toxin the most potent among nonproteinaceous toxins reported.[26]

Since our isolation of pure MTX from cultured *G. toxicus*, a wide range of studies have been done on its biological action. Following the discovery by Ohizumi's group that MTX acts as a specific agonist to calcium channels,[17] MTX has been widely used as a research probe to investigate biochemical reactions in cells initiated by Ca^{2+} influx. In contrast to the expanding biochemical and pharmacological studies, the structural studies have been lagging because of the small amount of sample available and the high molecular weight of the toxin.

Maitotoxin was obtained as an amorphous solid; $[\alpha]^{21}D+16.8°$ (c 0.36, MeOH-H_2O 1:1); UV max (MeOH-H_2O 1:1) 230 nm (ϵ 9600).[26] The presence of a nitrogen atom was suggested by the positive reaction to Dragendorff's reagent. The molecular weight, measured by negative FAB mass spectrometry (JEOL, HX-110), was suggested to be 3424.5 0.5 as a disodium salt (exact mass mode). IR bands at 1250 and 1280 cm^{-1}, and inorganic ion determination done on hydrolysate indicated the presence of two sulfate esters in the mole- cule. Judging from ^1H and ^{13}C NMR spectra (Fig. 1 and 2), the toxin has 21 methyls, ca. 38 methylenes, ca. 83 methines and 17 quaternary carbons. $^{13}13C$ NMR signals corresponding to carbonyl or ketal/acetal were not observed, suggesting that the toxin has no repeating units such as amino acids or sugars. On the basis of the spectral data, we have deduced a possible molecular formula of maitotoxin to be $C_{157-165}H_{259-267}NS_2O_{68-72}$. This composi- tion suggests that the toxin has 20-29 ether rings and 29-43 hydroxyls. These molecular characteristics remind us of palytoxin,[6,19] which has a linear skeleton of 115 carbons bearing 10 ether rings, one amino, and 41 hydroxyl groups. Yet, maitotoxin possesses an even larger molecular weight and more ether rings but fewer double bonds than those of palytoxin. These features may pose greater difficulty in elucidating the structure of maitotoxin.

DIARRHETIC SHELLFISH TOXINS

Diarrhetic shellfish poisoning (DST) is caused by ingestion of mussels or scallops[24] contaminated with toxins produced by dinoflagellates, usually members of the genus *Dinophysis*.[25] Although the acute symptoms of DST in humans are milder than those of paralytic shellfish poisoning, the long duration of the infestation periods and wide geographical occurrence of the dinoflagellates pose serious problems to both human health and the fishery industries.

As to the pathological effects of DSP toxins, Terao et al. examined ultrastructural changes induced by dinophysistoxin-1 (DTX1) and pecteno- toxin-1 (PTX1).[18] DTX1 produced severe mucosal injuries in the small intes- tine, and PTX1 caused hepatotoxicity by the formation of non-fatty vacuoles in hepatocytes. Moreover, Fujiki's group recently reported that okadaic acid, which was the dominant toxin in European mussels, had a potent tumor promoting activity.[14] Attention should be paid to chronic toxicity of these toxins for humans caused by continuous consumption of shellfish contaminated with subintoxication levels of the toxin.

Since the discovery of the poisoning, the authors have continued the structural studies of the causative toxins and isolated 11 toxic constitu- ents, and divided them into three groups on the basis of skeletal structural differences. The first group comprises okadaic acid (Fig. 3, 1) and its congeners (2, 3) and is responsible for the gastrointestinal disorders in human.[1,20] Okadaic acid was first isolated from the sponge, *Halichondria okadai*[16] and was identified as the toxic principle of European DSP.[3] Dinophysistoxin-1 (DTX1, 2) was obtained from Japanese mussels and was named after its elaborator, *Dinophysis fortii*. The structure was identified as 35-methylokadaic acid on the basis of spectroscopic comparisons with okadaic acid.[9] One of the dominant toxins in Japanese scallops was less polar than okadaic acid and was named dinophysistoxin-3 (DTX3, 3).[22] The toxin was shown to be a mixture of 7-O-acyl DTX1 with a variety of a fatty acid esters ranging from a saturated one like palmitoyl to a highly unsaturated one like docosahexaenoyl.[20,22]

The second group is a series of polyether lactons named pectenotoxin-1, -2, -3 and -6 (PTXs, Fig. 4, 4, 5, 6, 7), which have been isolated from scallops from Mutsu Bay, the northern edge of Honshu Island, Japan.[22] PTX4, PTX5 and PTX7 have been isolated from the same shellfish but their structures are not known. The known PTXs are a series of congeners with substituents having various degrees of oxidation at C43 from methyl (PTX2) to carboxylic acid (PTX6).[8,22] Most recently, *D. fortii* was identified as a producer of PTX.[4] Only PTX2 was detected in the dinoflagellates whereas the other homologs coexisted in the scallops when the algae were collected. This finding suggests that oxidation at C42 takes place in shellfish.

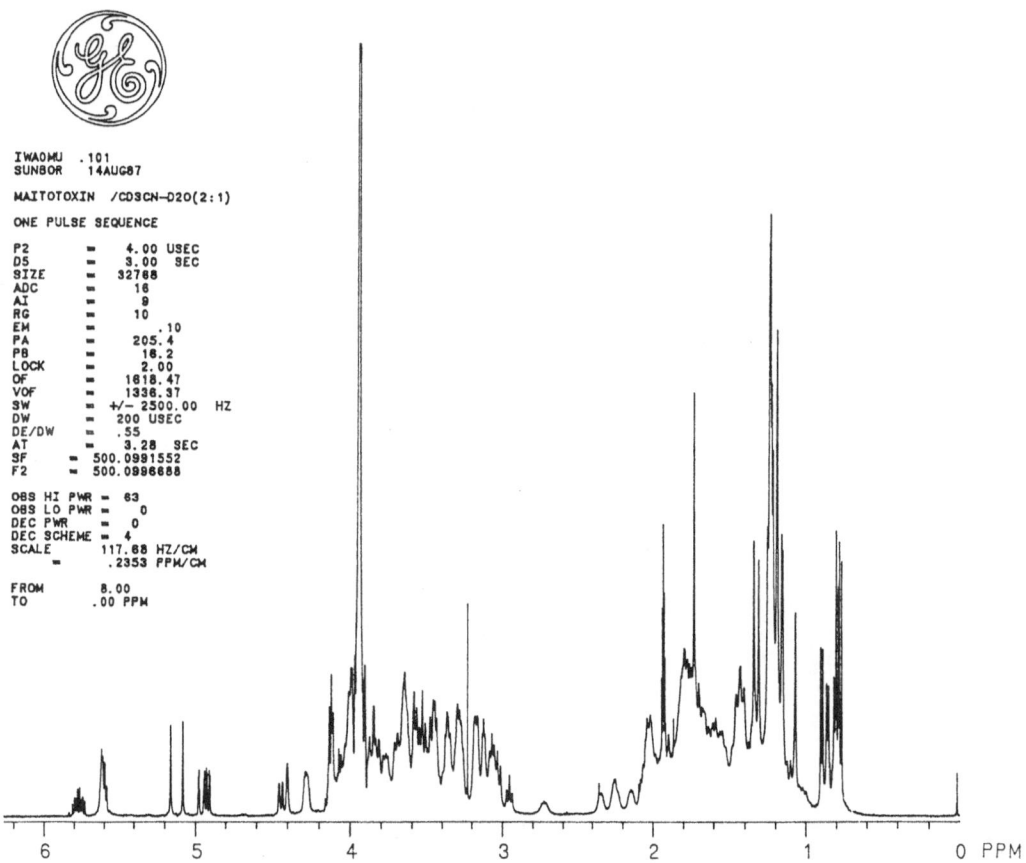

Fig. 1. ^{1}H NMR Spectrum of Maitotoxin in CD_3CN-D_2O (1:1) Measured by GN-500 (General Electric, 500 MHz).

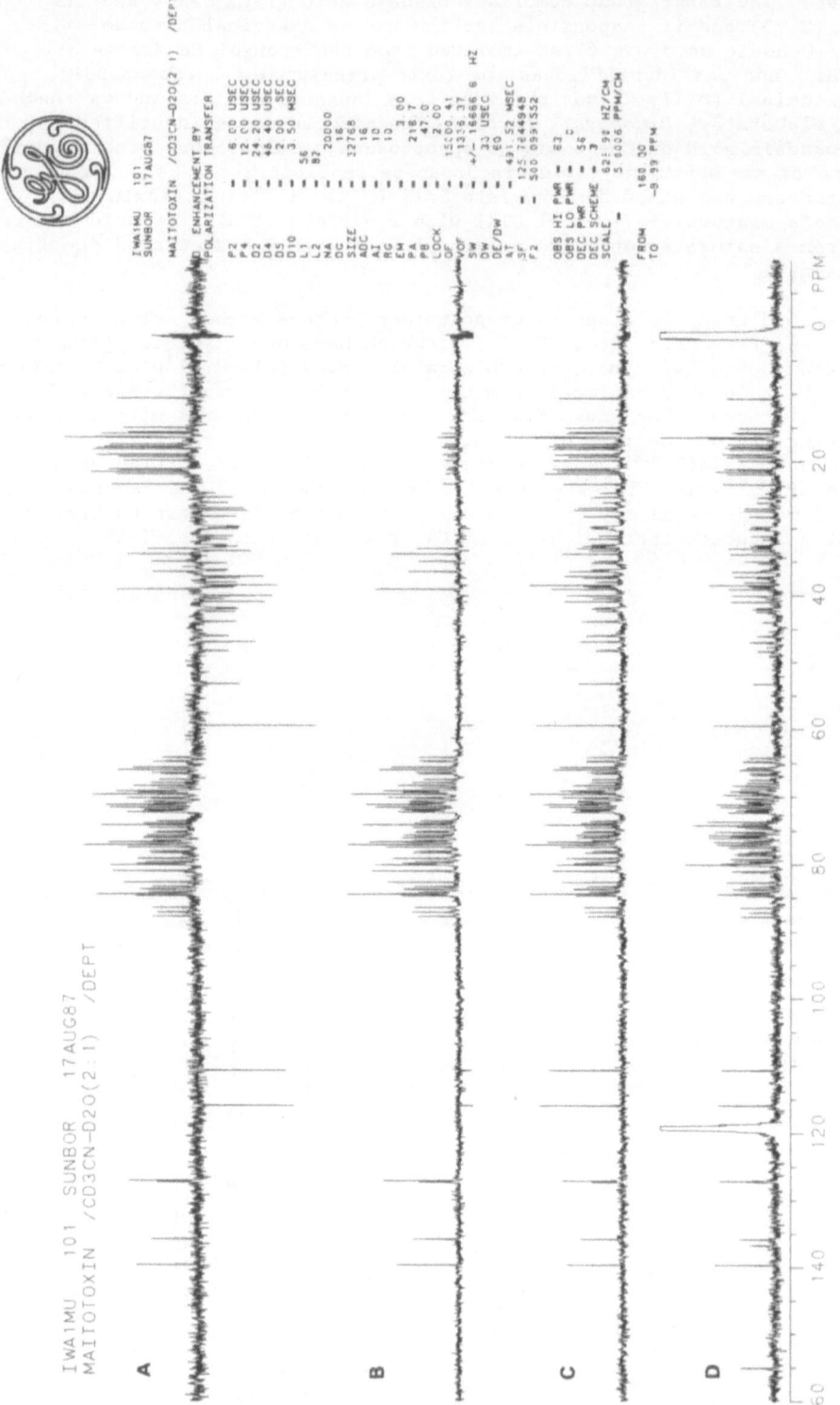

Fig. 2. ^{13}C NMR Spectra of Maitotoxin in CD_3CN-D_2O (1:1) by Broad Band Decoupling (D) and DEPT Sequence (A, B, C). (A) CH_3 and CH_2 give positive peaks, while CH negative. (B) Only CH appears. (C) No quaternary carbons appear.

The newest member of the DSP toxins is yessotoxin (YTX, Fig. 5). It was previously named DTX2 and later given the present name because of its completely different structural features compared to those of DTX1. The structure partly resembles the brevetoxins, which are known to be the causative agents in massive fish kills frequently occurring along the Florida coast.[5,13] Elucidation of the structure of YTX was accomplished mainly by means of advanced NMR techniques.[7] The relative stereochemistry of the fused ether rings was assigned by analyzing the $^3J_{CH}$ and NOE's of the ether linkages (Fig. 5).[2] The transfusion manner of the ether rings (rings A, D, F, H, I and J) was readily determined by NOE's measured in phase-sensitive NOESY spectra. Significant NOE's were observed on the protons or the methyls on the ring juncture carbons, which took the 2,6-diaxial configuration in the tetrahydropyran system. As to rings B and C, overlapping signals of oxymethines prevented NOE analyses. Nevertheless, the signal

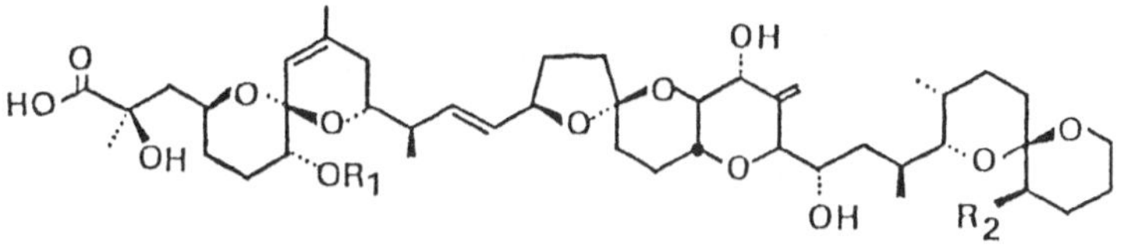

1 OA: $R_1 = H$ $R_2 = H$

2 DTX1: $R_1 = H$ $R_2 = CH_3$

3 DTX3: $R_1 =$ [structure] $R_2 = CH_3$

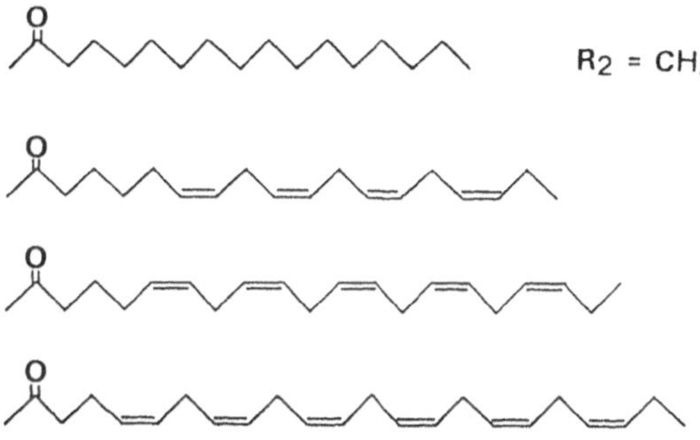

Fig. 3.   Structures of okadaic acid and dinophysistoxin-1 and -3.

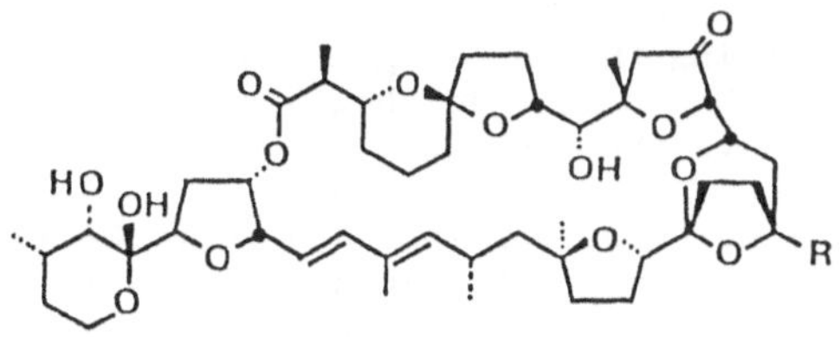

4 PTX1: R= CH_2OH
5 PTX2: CH_3
6 PTX3:   CHO
7 PTX6:   COOH

Fig. 4.   Structures of pectenotoxins.

593

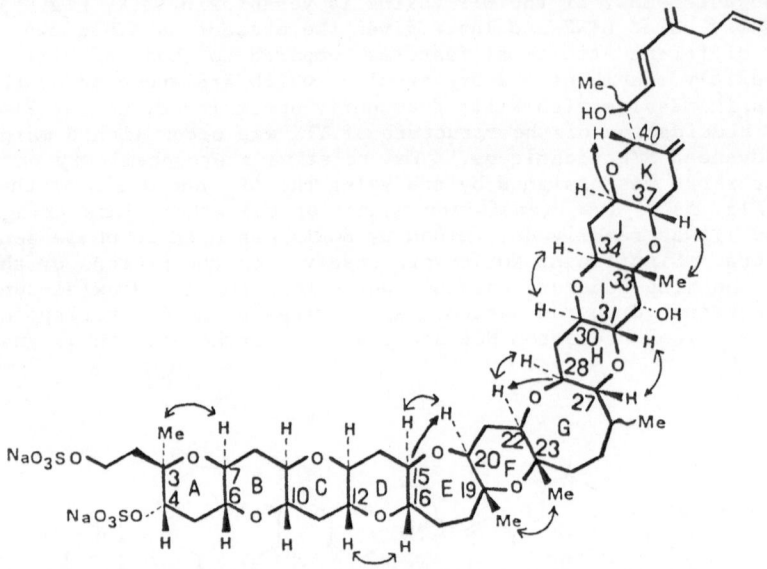

Fig. 5.   Probable relative stereochemistry of yessotoxin deduced from
          NOE's and CH long range couplings.

          '———→' denote the correlations between ^{13}C (tail) and
          ^{1}H (head) observed by the COLOC sequence at 7 Hz.
          '←——→' indicate pairs of the protons giving cross peaks by
          phase-sensitive NOESY.   All the 2D spectra were measured in
          CD_3OD on an AM 500 (500 MHz, Bruker) spectrometer.

widths of H_2-8 and H_2-11 were determined by the cross peaks of the 1H-1H
COSY and were shown to coincide with those of the methylenes on rings D, H
and J.   Thus, these two rings presumably have the same stereochemistry as
ring D.   Three bond J_{CH}'s across ether linkages observed on C15/H-20, C28/
H-22, and C36/H-40 by the COLOC (10 Hz, COlleration of LOng range Coupling),
indicated the presence of seven, eight and six membered rings (rings E, G
and K).   In the six membered rings, except ring K, a dihedral angle formed
with the oxycarbon of the ether rings (C3) and the proton on the other side
of the ether carbon (H-7) are ca. 66° so that $^{3}J_{CH}$ is too small to be
detected in the COLOC sequence.   On the other hand, the transfused seven and
eight membered ring (rings E and G) give the dihedral angle of ca. 0°, that
is large enough for detecting the couplings.   The coupling between C36 and
H-40 observed in the sequence suggested that a side chain of YTX is substituted axially on C40 to give the dihedral angle 180° between H-40 and C36.

     Molecular mechanics calculations indicate that the ring G tends to
take the crown form rather than the boat-chair form with energy difference
of 1.4 Kcal.   The gap, however, might be too small to fix the conformation.
The flexibility of the molecule due to the conformational interchange at
ring G may play an important roll in manifesting the toxicity as was
reported for brevetoxin-A.[13]   The pathological effects and the biogenic
origin of YTX are unknown at the present time.

REFERENCES

1. Hamano Y, Kinoshita Y, Yasumoto T, 1986, Enteropathogenicity of diarrhetic shellfish toxins in intestinal models, *J Food Hyg Soc Jpn.*, 27:375-379.
2. Kumagai M, Murata M, Lee JS, Yasumoto T, 1987, Symposium paper of 29th symposium on the chemistry of natural products, Sapporo, pp. 600-607.
3. Kumagai M, Yanagi T, Murata M, Yasumoto T, Kat M, Lassus R, Rodoriguez-Vazquez JA, Okadaic acid as the causative toxin of diarrhetic shellfish poisoning in Europe, *Agric Biol Chem.*, 50:2853-2857.
4. Lee JS, personal communication.
5. Lin YY, Risk M, Ray SM, VanEngen D, Clardy J, Golik J, James JC, Nakanishi K, 1981, Isolation and structure of brevetoxin B from "red tide" dinoflagellate *Ptychodiscus brevis (Gymnodinium breve)*, *J Amer Chem Soc.*, 103:6773-6774.
6. Moore RE, Bartolini G, Barchi J, Bother-By AA, Dodok J, Ford J, 1982, Absolute stereochemistry of palytoxin, *J Amer Chem Soc.*, 104:3776-3778.
7. Murata M, Kumagai M, Lee JS, Yasumoto T, 1987, Isolation and structure of yessotoxin, a novel polyether compound implicated in diarrhetic shellfish poisoning, *Tetrahedron Lett.*, 28:5869-5872.
8. Murata M, Sano M, Iwashita T, Naoki H, Yasumoto T, 1986, The structure of pectenotoxin-3, a new constituent of diarrhetic shellfish toxins, *Agric Biol Chem.*, 50:2693-2695.
9. Murata M, Shimatani M, Sugitani H, Oshima Y, Yasumoto T, 1982, Isolation and structural elucidation of the causative toxin of the diarrhetic shellfish poisoning, *Nippon Suisan Gakkaishi*, 48:549-552.
10. Nukina M, Tachibana K and Scheuer JP, 1987, Abstract paper of Annual Meeting of *Agric Chem Soc Jpn.*, Tokyo, pp. 512.
11. Ohizumi Y, Ishida T, Shibata S, 1982, Mode of the ciguatoxin-induced supersensitivity in the guinea-pig vas deferens, *J Pharmacol Exp Ther.*, 221, 748-752.
12. Scheuer PJ, Takahashi W, Tsutsumi J, Yoshida T, 1967, Ciguatoxin: isolation and chemical nature, *Science*, 155:1267.
13. Shimizu Y, Chou HN, Bando H, Duyne GV, Clardy JC, 1986, Structure of brevetoxin A (GB-1 toxin), the most potent toxin in the Florida red tide organism *Gymnodinium breve (Ptychodiscus brevis)*, *J Amer Chem Soc.*, 108:514-515.
14. Suganuma M, Fujiki H, Suguri H, Yoshizawa S, Hirota M, Nakayasu M, Ojika M, Wakamatsu K, Yamada K, Sugimura T, 1988, Okadaic acid: An additional non-phorbol-12-tetradecanoate-13-acetate-type tumor promoter, *Proc Natl Acad Sci USA*, 85:1768-71.
15. Tachibana K, 1980, Structural Studies on Marine Toxins, PhD Thesis, University of Hawaii, Honolulu.
16. Tachibana K, Scheuer JP, Tsukitani Y, Kikuchi H, Engen DV, Clardy J, Gopichand, Y, Schmitz FJ, 1981, Okadaic acid, a cytotoxic polyether from two marine sponges of the genus *Halichondria*, *J Amer Chem Soc.*, 103:2469-2471.
17. Takahashi M, Ohizumi Y, Yasumoto T, 1982, Maitotoxin, a Ca^{2+} channel activator candidate, *J Biol Chem.*, 257:7287-7289.
18. Terao K, Ito E, Yanagi T, Yasumoto T, 1986, Histophathological studies on experimental marine toxin poisoning. I. Ultrastructural changes in the small intestine and liver of suckling mice induced by dinophysis-toxin-1 and pectenotoxin-1, *Toxicon*, 24:1141-1151.
19. Uemura D, Ueda K, Hirata Y, Naoki H, Iwashita T, 1981, Further studies on palytoxin. II. Structure of palytoxin, *Tetrahedron Lett.*, 22:2781-2784.
20. Yanagi T, Murata M, Torigoe K, Yasumoto T, 1989, Biological activities of semisynthetic analogs of dinophysistoxin-3, the major diarrhetic shellfish toxin, *Agric Biol Chem.*, in press.

21. Yasumoto T, Bagnis R, Vernoux JP, 1976, Toxicity study on surgeon-fishes-II. Properties of the principal water-soluble toxin. *Nippon Suisan Gakkaishi*, 42:359-365.

22. Yasumoto T, Murata M, Oshima Y, Sano, M, Matsumoto GK, Clardy J, 1985, Diarrhetic shellfish toxins, *Tetrahedron*, 41:1019-1025.

23. Yasumoto T, Nakajima I, Bagnis R, Adachi R, 1977, Findings of a dino-flagellate as a likely culprit of ciguatera, *Nippon Suisan Gakkaishi*, 48:1021-26.

24. Yasumoto T, Oshima Y, Yamaguchi M, 1978, Occurrence of a new type of shellfish poisoning in the Tohoku district, *Nippon Suisan Gakkaishi*, 44:1249-1255.

25. Yasumoto T, Oshima Y, Sugawara W, Fukuyo Y, Oguri H, Igarashi T, Fujita N, 1980, Identification of *Dinophysis forii* as a causative organism of diarrhetic shellfish poisoning, *Nippon Suisan Gakkaishi*, 46:1405-1411.

26. Yokoyama A, Murata M, Oshima Y, Iwashita T, Yasumoto T, 1988, Some chemical properties of maitotoxin, a putative calcium channel agonist isolated from a marine dinoflagellate, *J Biochem.*, 104:184-187.

THE CELLULAR MECHANISM OF ACTION OF MAITOTOXIN AND CIGUATOXIN

Yasushi Ohizumi

Mitsubishi Kasei Institute of Life Sciences
11 Minamiooya
Machida, Tokyo 194, Japan

ABSTRACT

To clarify the cellular mechanism of action of maitotoxin and ciguatoxin, the effects of both the toxins were studied using isolated cardiac myocytes and a pheochromocytoma cell line (PC12). Maitotoxin caused an increase in norepinephrine release from PC12 cells and Ca^{2+} influx into them, and induced a Ca^{2+}-dependent arrhythmogenic effect on cardiac myocytes. Maitotoxin and ciguatoxin caused an increase in the degree and rate of contraction of myocytes. Pharmacological, electrophysiological and morphological data suggest that maitotoxin elevates the Ca permeability of the cell membrane to increase the intracellular Ca^{2+} concentration, and thus causes pharmacological actions. Also suggested is that CTX activates muscle Na channels by modifying the voltage-dependence of channel activation to increase Na inward currents, and thus causes excitatory effects.

INTRODUCTION

Marine toxins such as tetrodotoxin, saxitoxin and sea anemone toxins have been extensively studied because of their action on specific channel sites. Ciguatera is well known to be a seafood poisoning caused by ingestion of a variety of poisonous reef fishes. Maitotoxin (MTX), a water soluble substance, is the most potent marine toxin known with the minimum lethal dose of 0.17 μg/kg (i.p.) in mice. We have shown for the first time that MTX caused a Ca^{2+}-dependent excitatory effect in a rat pheochromocytoma cell line, PC12,[29,30] smooth muscle[20,22,23] and cardiac muscle.[7-9,11,12] MTX has been demonstrated to modify Ca channel functions[3,10,12,17,33] and phosphoinositide metabolism.[1,5]

Ciguatoxin (CTX), the principal toxin of ciguatera seafood poisoning, causes a depolarization of the cell membrane[2,14,25,26] and a contraction of smooth muscle.[13,19,21] Furthermore, CTX causes an inotropic action.[14,16,24,28] Recently, it has been shown that CTX and brevetoxins bind to common receptor sites of Na channels.

Microbial Toxins in Foods and Feeds, Edited by A.E. Pohland *et al.,*
Plenum Press, New York, 1990

METHODS

Cell culture

PC12 cells were maintained in Dulbecco's modified Eagle's medium and were subcultured on polylysine-coated Falcon dish at a density of 1.5 x 10^6 cells/dish two days before experimentation.[29]

Preparation of single myocytes

Rat or guinea-pig isolated myocardial cells were prepared by enzymatic digestion as reported previously.[9,10]

Mechanical responses

Cardiac cells were perfused with the medium on a glass slide with a pair of platinum wire electrodes and stimulated rectangular pulses (3 ms, 15 V/cm, 2 Hz). The beating activity of isolated myocytes was observed under a phase contrast microscope and was recorded with a video recording system or a high-speed movie camera as described previously.[9]

Assay of ^{3}H-norepinephrine (NE) release

The ^{3}H-NE release from PC12 cells was measured by the method previously described.[29]

Assay of ^{45}Ca uptake

The ^{45}Ca uptake into PC12 cells was determined using the procedure previously reported.[29]

Measurement of intracellular Ca^{2+} concentration

The intracellular free Ca^{2+} concentration of isolated cardiac cells was determined according to the method previously described.[7]

Patch-clamp experiments

Isolated myocytes were voltage clamped using the whole-cell patch-clamp technique and single-channel currents were recorded by means of patch-clamp technique in the cell-attached configuration.[10]

Ultrastructure

The guinea-pig isolated atria were immersed in fixative solution, and then washed with phosphate buffer. The material was postfixed in osmic acid (1%) and was further processed by standard techniques for electron microscopy.[8]

RESULTS

Mechanical response

The time course of contraction-relaxation cycles in a single beating action of atrial cells was observed with a high-speed movie camera. MTX and CTX (3-10 ng/ml) induced an increase in the degree and rate of longitudinal contraction of the myocytes.[9,28] The image of the cells was recorded with a video recording system. Both the toxins caused an abnormal movement of the myocytes and then the percentage of damaged round cells increased.[9,28] In the case of MTX, a delay was observed in the change of normally beating myocytes into cells showing arrhythmic movements in the presence of verapamil.

^{3}H-NE release

^{3}H-NE was markedly released from PC12 cells into external medium after treatment with MTX (1-100 ng/ml).[29,30] The MTX-induced release of ^{3}H-NE was abolished in Ca^{2+}-free medium, but not affected by tetrodotoxin (TTX) (Table 1). MTX still caused a marked release of ^{3}H-NE even in the absence of Na^+.

^{45}Ca uptake

After an exposure of PC12 cells to MTX, a marked increase in ^{45}Ca uptake was observed. MTX is still able to cause ^{45}Ca uptake in the Na^+-free medium. As shown in Table 1, ^{45}Ca influx induced by MTX or K^+ was inhibited or abolished by treatment with verapamil, Mn^{2+} or tetracaine.[29]

Intracellular Ca^{2+} concentration

In the isolated cardiac myocytes MTX (1 ng/ml) caused a marked increase in the Ca^{2+} concentration from 121.5 ± 9 nM (control) to 380 ± 23 nM.[7]

Whole-cell patch-clamp experiments

MTX caused a steadily flowing current in the cells superfused with Na^+-free, K^+-free Tyrode solution and a gradual inward shift of the holding current level at -80 mV.[10] This current was enhanced by epinephrine and abolished by Ca^{2+}. The I-V relation of MTX-activated current was almost linear and reversal potential was about -23 mV (Fig. 1). In Cs^+-Tyrode solution, CTX (6-20 ng/ml) shifted the peak Na current-voltage curve by 10 to 40 mV in the negative direction, indicating a shift of the voltage-dependence of channel activation toward more hyperpolarized potentials.[28] As shown in Fig. 2 Na inward currents were evoked by depolarizing pulses and then were completely inactivated within 20 ms. In the presence of CTX (20 ng/ml), a small sustained inward current was observed even after the depolarizing pulse.

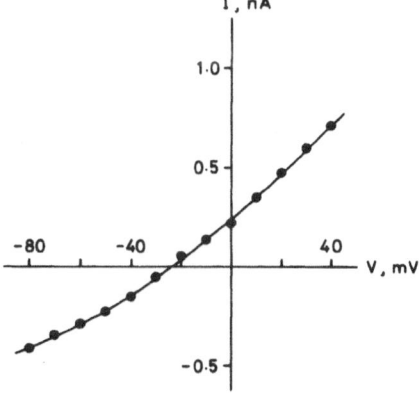

Fig. 1. Effects of MTX (1 ng/ml) on the current-voltage relationship of isolated cardiac cells. Currents were elicited by 200 ms depolarizing pulses from a holding potential (-80 mV) to various potentials. Ordinate scale, steady-state current levels measured at the end of each pulse. Abscissa scale, membrane potential during the pulse.

Table 1. Effects of several agents on the norepinephrine release and Ca^{2+} influx induced by MTX, K^+ and A23187 in PC12 cells.

| Addition | Percentage[a] of response to | | | | |
| | [³H]Norepinephrine | | | ⁴⁵Ca influx | |
	Maitotoxin	K^+	A23187	Maitotoxin	K^+
	10 ng/ml	51.4 mM	20 μM	10 ng/ml	51.4 mM
None	100	100	100	100	100
Tetrodotoxin (1 μM)	90	ND	ND	ND	ND[b]
Verapamil (30 μM)	0	7	118	53	0
(300 μM)	4	ND	130	5	0
Mn^{2+} (5 mM)	3	1	0	6	2
Tetracaine (1 mM)	0	0	105	0	1

[a] Values are expressed as a percent of the amount in the respective control cultures.
[b] ND, not determined.

Cell-attached patch-clamp experiments

MTX added to the pipette solution induced single channel openings with novel properties.[10] The gating showed little voltage-dependence. The open-time histograms of inward currents occurring at each burst indicated that the distribution of opening time was essentially single exponential for MTX activated channels. The mean open time of the channels (10.4 ms) was about 10 times longer than that of voltage-dependent Ca channels.

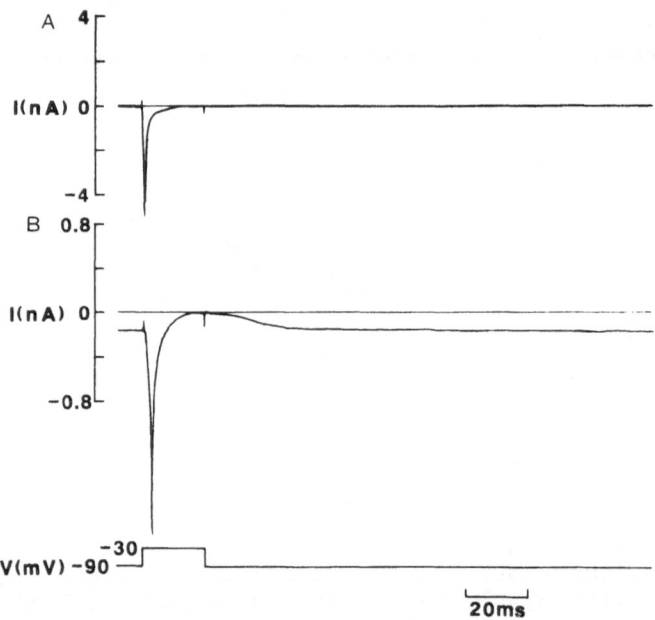

Fig. 2. Na inward currents in the presence (B) or absence (A) of CTX (20 ng/ml) in the isolated cardiac cells.

Ultrastructure

After treatment with MTX (5-30 ng/ml) cardiac cells showed heavily swollen mitochondria with disorganized cristae.[8] The chromatin of the nucleus was markedly aggregated peripherally. The sarcoplasm had no glycogen granules. The sarcomeres were contracted severely and broken segments of myofibrils were occasionally seen. However, no morphological changes were seen in atria incubated in a Ca^{2+}-free medium containing MTX (30 ng/ml).[8]

DISCUSSION

Because we showed for the first time that MTX induced Ca^{2+} dependent pharmacological actions,[29,30] MTX has drawn the attention of numerous investigators because of its characteristic activity.[4,6,15,18,27,31,32] In PC12 cells MTX caused a profound increase in Ca^{2+} uptake and the Ca^{2+}-dependent release of ^{3}H-NE.[29,30] In cardiac myocytes MTX induced an increase in the degree of contraction and subsequent arrhythmogenic action. The cytoplasmic free Ca^{2+} concentration of myocytes markedly increased in the presence of MTX. Electron microscopic observations revealed that MTX induced severe pathomorphological changes in the atria. These pharmacological and toxicological actions of MTX were inhibited or abolished by Ca antagonists or Ca^{2+}-free medium. Furthermore, the data obtained from patchclamp experiments suggest that MTX activates Ca channels with novel properties such as voltage-independent channel gating and very long open time. An attractive interpretation of all these observations is that MTX activate a new class of the voltage-independent Ca channel to increase the Ca^{2+} permeability of the cell membrane, which may account for the mechanism of Ca^{2+} dependent pharmacological actions of MTX.

In single cardiac cells CTX caused an arrhythmogenic action preceded by a marked increase in the amplitude of longitudinal contractions. Wholecell clamp experiments showed that CTX shifts the voltage-dependence of Na channel activation and inactivation to more negative potentials. These results suggest that CTX increases the membrane permeability of Na^{+} to increase Ca^{2+} availability in the cardiac cell and thus caused the cardiotonic effects. Furthermore, CTX caused a sustained inward current after the depolarizing pulse, which may contribute a small membrane depolarization.

ACKNOWLEDGEMENT

I wish to thank Professor T. Yasumoto of Tohoku University for generously supplying maitotoxin and ciguatoxin, and my coworkers, Professor R. Ochi of Juntendo University, Professor K. Momose of Showa University, Dr. M. Kobayashi, Mr. S. Kondo and Miss A. Seino of this institute. I am grateful to Ms. A. Muroyama for skillful assistance and Ms. Y. Murakami for typing the manuscript.

REFERENCES

1. Berta P, Sladeczek F, Derancourt J, Durand M, Travo P, Haiech J, 1986, Maitotoxin stimulates the formation of inositol phosphates in rat aortic myocytes, *FEBS Lett.*, 197:349-352.
2. Bidard JN, Vijverberg HPM, Frelin C, Chungue E, Legrand AM, Bagnis R, Lazdunski M, 1984, Ciguatoxin is a novel type of Na^{+} channel toxin. *J Biol Chem.*, 259:8353-8357.
3. Freedman SB, Miller RJ, Miller DM, Tindall DR, 1984, Interactions of maitotoxin with voltage-sensitive calcium channels in cultured neuronal cells, *Proc Nat Acad Sci USA*, 81:4582-458. .

4. Gomi S, Chaen S, Sugi H, 1984, The mode of action of maitotoxin on the membrane systems of frog skeletal muscle fibers, *Proc Jpn Acad.*, 60: 28-31.
5. Gusovsky F, Daly JW, Yasumoto T, Rojas E, 1988, Differential effects of maitotoxin on ATP secretion and on phosphoinositide breakdown in rat pheochromocytoma cells. *FEBS Lett.*, 233:139-142.
6. Kim YI, Login IS, Yasumoto T, 1985, Maitotoxin activates quantal trans-mitter release at the neuromuscular junction: evidence for elevated intraterminal Ca^{2+} in the motor nerve terminal, *Brain Res.*, 346:357-362.
7. Kobayashi M, Goshima K, Ochi R, Ohizumi Y, 1987, Arrhythmogenic action of maitotoxin in guinea-pig and rat cardiac muscle, *Eur J Pharmacol.*, 142:1-8.
8. Kobayashi M, Kondo S, Yasumoto T, Ohizumi Y, 1986, Cardiotoxic effects of maitotoxin, a principal toxin of seafood poisoning, on guinea pig and rat cardiac muscle, *J Pharmac Exp Ther.*, 238:1077-1083.
9. Kobayashi M, Miyakoda G, Nakamura T, Ohizumi Y, 1985a, Ca-dependent arrhythmogenic effects of maitotoxin, the most potent marine toxin known, on isolated rat cardiac muscle cells, *Eur J Pharmac.*, 111: 121-123.
10. Kobayashi M, Ochi R, Ohizumi Y, 1987, Maitotoxin-activated single calcium channels in guinea-pig cardiac cells, *Br J Pharmacol.*, 92: 665-671.
11. Kobayashi M, Ohizumi Y, Yasumoto T, 1985b, The mechanism of action of maitotoxin in relation to Ca^{2+} movements in guinea pig and rat cardiac muscles, *Br J Pharmac.*, 86:385-391.
12. Legrand AM, Bagnis R, 1984, Effects of highly purified maitotoxin extracted from dinoflagellate *Gambierdiscus toxicus* on action potential of isolated rat heart, *J Mol Cell Cardiol.*, 16:663-666.
13. Lewis RJ, Endean R, 1984, Mode of action of ciguatoxin from the Spanish mackerel, *Scomberomorus commersoni*, on the guinea-pig ileum and vas deferens, *J Pharmacol Exp Ther.*, 228:756-760.
14. Lewis RJ, Endean R, 1986, Direct and indirect effects of ciguatoxin on the guinea-pig atria and papillary muscles, *Naunyn-Schmiedeberg's Arch Pharmacol.*, 334:313-322.
15. Login IS, Judd AM, Cronin MJ, Koike K, Schettini G, Yasumoto T, Macleod RM, 1985, The effects of maitotoxin on $^{45}Ca^{2+}$ flux and hormone release in GH_3 rat pituitary cells, *Endocrinol.*, 116:622-627.
16. Miyahara JT, Akau CK, Yasumoto T, 1979, Effects of ciguatoxin and maitotoxin on the isolated guinea pig atria, *Res Comm Chem Path Pharmac.*, 25:177-180.
17. Miyamoto T, Ohizumi Y, Washio H, Yasumoto T, 1984, Potent excitatory effect of maitotoxin on Ca channels in the insect skeletal muscle, *Pflugers Arch.*, 400:439-441.
18. Niki I, Tamagawa T, Niki H, Niki A, Koide T, Sakamoto N, 1986, Stimu-lation of insulin release by maitotoxin, an activator of voltage-dependent calcium channels, *Biomed Res.*, 7:107-112.
19. Ohizumi Y, Ishida Y, Shibata S, 1982, Mode of ciguatoxin-induced supersensitivity in the guinea-pig vas deferens, *J Pharmacol Exp Ther.*, 221:748-752.
20. Ohizumi Y, Kajiwara A, Yasumoto T, 1983, Excitatory effects of the most potent marine toxin, maitotoxin, on the guinea-pig vas deferens, *J Pharmac Exp Ther.*, 227:199-204.
21. Ohizumi Y, Shibata S, Tachibana K, 1981, Mode of the excitatory and inhibitory actions of ciguatoxin in the guinea-pig vas deferens, *J Pharmacol Exp Ther.*, 217:475-480.
22. Ohizumi Y, Yasumoto T, 1983, Contraction and increase in tissue calcium content induced by maitotoxin, the most potent known marine toxin, in intestinal smooth muscle, *Br J Pharmacol.*, 79:3-5.
23. Ohizumi Y, Yasumoto T, 1983, Contractile response of the rabbit aorta to maitotoxin, the most potent marine toxin, *J Physiol.*, 337:711-721.

24. Ohshika H, 1971, Marine toxins from the pacific-IX, Some effects of ciguatoxin on isolated mammalian atria, *Toxicon*, 9:337-343.
25. Rayner MD, 1972, Mode of action of ciguatoxin, *Fed Proc.*, 31:1139-1145.
26. Rayner MD, Kosaki TI, 1970, Ciguatoxin: Effects on Na fluxes in frog muscle, *Fed Proc.*, 29:548.
27. Schettini G, Koike K, Login IS, Judd AM, Cronin MJ, Yasumoto T, Macleod, RM, 1984, Maitotoxin stimulates hormonal release and calcium flux in rat anterior pituitary cells *in vitro*, *Am J Physiol.*, 247: E520-E525.
28. Seino A, Kobayashi M, Momose K, Yasumoto T, Ohizumi Y, 1988, The mode of inotropic action of ciguatoxin on guineapig cardiac muscle, *Br J Pharmacol.*, in press.
29. Takahashi M, Ohizumi Y, Yasumoto T, 1982, Maitotoxin, a Ca^{2+} channel activator candidate, *J Biol Chem.*, 257:7287-7289.
30. Takahashi M, Tatsumi M, Ohizumi Y, Yasumoto T, 1983, Ca^{2+} channel activating function of maitotoxin, the most potent marine toxin known, in clonal rat pheochromocytoma cells, *J Biol Chem.*, 258: 10944-10949.
31. Terao K, Ito E, Sakamaki Y, Igarashi K, Yokoyama A, Yasumoto T, 1988, Histopathological studies of experimental marine toxin poisoning, II. The acute effects of maitotoxin on the stomach, heart and lymphoid tissue in mice and rats, *Toxicon*, 26:395-402.
32. Ueda H, Tamura S, Fukushima N, Takagi H, 1986, Pertussis toxin (IAP) enhances maitotoxin (a putative Ca^{2+} channel agonist)-induced Ca^{2+} entry into synaptosomes, *Eur J Pharmac.*, 122:379-380.
33. Yoshii M, Tsunoo A, Kuroda Y, Wu CH, Narahashi T, 1987, Maitotoxin-induced membrane current in neuroblastoma cells, *Brain Res.*, 424: 119-125.

RECENT PROGRESS IN THE CHEMISTRY AND TOXICOLOGY OF PARALYTIC SHELLFISH

POISONING

Y. Shimizu and N. K. Gulavita

Department of Pharmacognosy and Environmental
Health Sciences
College of Pharmacy
The University of Rhode Island
Kingston, Rhode Island 02881

INTRODUCTION

There has been tremendous progress in research involving the chemistry,
biochemistry and toxicology of paralytic shellfish poisons (PSP) in the past
15 years.[11] A number of new toxins have been isolated from various biological sources (Fig. 1), and the list of alleged progenitors also expanded
to many organisms including prokaryotes such as cyanobacteria and gram-negative bacteria. In this paper, we would like to report some of the most
recent work conducted in our laboratory.

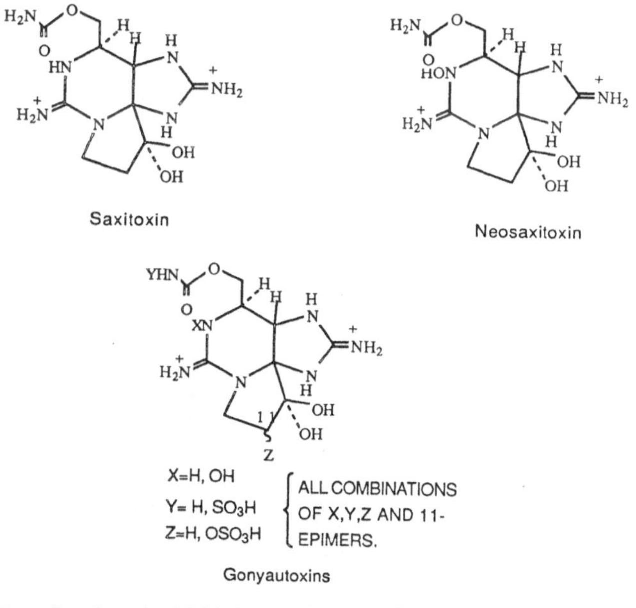

Saxitoxin

Neosaxitoxin

X=H, OH
Y= H, SO$_3$H
Z=H, OSO$_3$H

ALL COMBINATIONS
OF X,Y,Z AND 11-
EPIMERS.

Gonyautoxins

Fig. 1. Paralytic shellfish toxins isolated from various sources.

Microbial Toxins in Foods and Feeds, Edited by A.E. Pohland *et al.,*
Plenum Press, New York, 1990

We have recently determined all building blocks of the toxin molecule and elucidated some intricate mechanisms involved in the biosynthetic pathway by feeding various ^{14}C, ^{13}C, ^{2}H, and ^{15}N-labeled precursors.[12,16] The results are summarized in Fig. 2.

First, the major carbon skeleton is formed by the Claisen-type condensation of acetate on arginine. After concomitant decarboxylation, the amino group is converted to a guanidino group. The origin of the added amidine moiety was proved to be also arginine in an earlier experiment.[12]

After cyclization to an imidazole ring, the next step is speculated to be the formation of the pyrimidine ring. The side chain carbon moiety is introduced from methionine via S-adenosylmethionine, followed by a hydride ion migration and deprotonation to form an exocyclic methylene intermediate. Epoxidation, opening to an aldehyde, and reduction lead to a carbinol intermediate, which is then carbamylated. The carbamyl group was also shown to

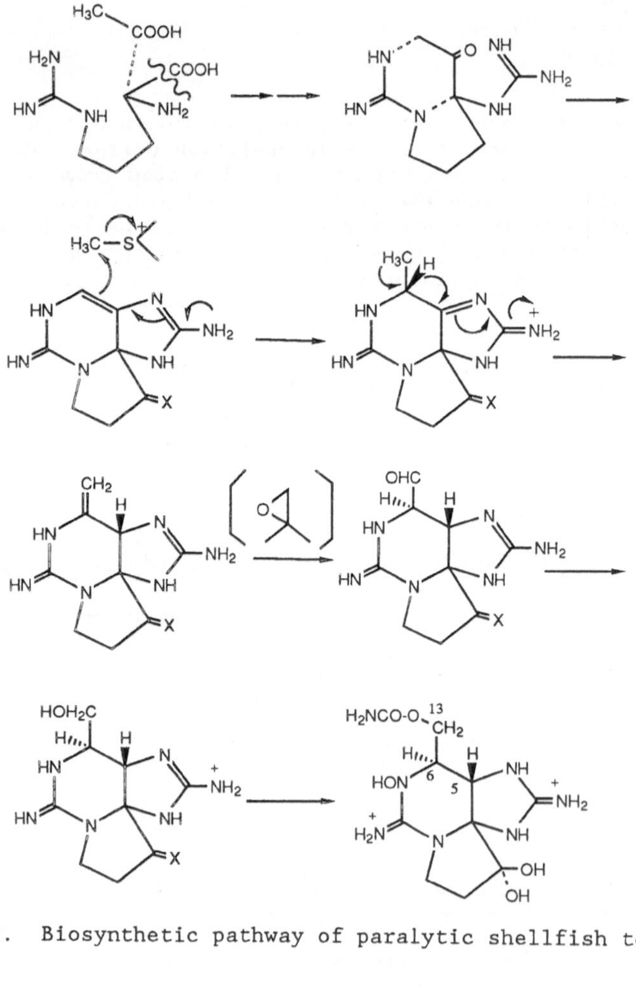

Fig. 2. Biosynthetic pathway of paralytic shellfish toxins.

be derived from arginine via an ornithine-urea cycle. It is important to note that the whole biosynthetic process involves heavy participation of arginine molecules.

Molecular structures of saxitoxin and neosaxitoxin around physiological pH

Saxitoxin and tetrodotoxin are important tools in pharmacological research as highly selective sodium channel blockers.[2] The discovery of new toxins has provided a unique opportunity to further study the mechanism of action and topology of sodium channels.[3,4,6,17,18,19] Revised models for the binding of the toxin to the channels have been proposed. Such models can conveniently explain the rather suprising fact that the sulfocarbamyl toxins have drastically diminished toxicity.[15]

In 1977, our group first isolated neosaxitoxin (neoSTX) from Alaskan butter clams, *Saxidomus giganteus*.[8] The compound was later discovered in most PSP-containing samples and was also found to be a basic structure of many toxins. The action of neoSTX on sodium channels has been studied by Kao and Stricharz's groups.[3,4,17,18,19] In the meantime, recent developments in electrophysiology have enabled the measurement of a single-channel current, pointing to the presence of subtypes of sodium channels in the tissues, where the channels were once thought to be homogeneous. For example, bull-frog skeletal muscles were found to possess saxitoxin-sensitive, but neo-saxitoxin-insensitive, sodium channels.[5] The finding seems to be highly significant in considering the multifaceted control of neuro-muscular systems Therefore, we decided to undertake the investigation of the molecular structures of neoSTX in solutions with respect to those of saxitoxin.

The only difference in the structures of saxitoxin and neosaxitoxin is the presence of an N-hydroxyl group on N-1 of the pyrimidine guanidinium group, which is generally considered to be the non-participating part of the molecule. The essential moieties for binding are the imidazoline guanidinium and two hydroxyl groups of the hydrated ketone.[3,12]

The molecular structures of saxitoxin solutions around physiological pH have been extensively studied by the authors' group[13] and others.[7,10] The toxin exists in an equilibrium of three molecular structures: a) deprotonated hydrate form, b) monoprotonated hydrate form, and c) monoprotonated keto-form (Fig. 3). Since the diprotonated form and the keto-form are considered to be inactive or less active, the finding is in fairly good agreement with the results of pH-dependent activity or binding studies. Previously, we also reported data on the protonation and deprotonation of the neoSTX molecule,[14] but more precise information about the solution structures of neoSTX was deemed urgent for interpretation of the binding or electrophysiological experiments.

NeoSTX was dissolved in D_2O and examined by 300-MHz NMR. The pH was adjusted by DCl and NaOD solutions using a micro combination glass electrode (Ingold). The results are summarized in Fig. 4. The chemical shifts of all protons attached to 6-C, 10-C, and 13-C are unchanged up to pH 5. However, at pH 5, they start to shift to the higher field, indicating electronic changes due to deprotonation of a guanidinium group. This change is especially distinct with the 10-beta and 10-alpha protons. Also, a steep drop in chemical shift was observed with 6-H at pH above 6. There is another large up-field shift at pH 7 ~ 8, which is associated with the second deprotonation. Examination of the spectra showed the appearance of signals due to the keto-form above pH 7.21. The H-6 signal was the most distinct signal of the keto-form in a mixture of the keto-form and the hydrated-form.

The approximate proportions of the keto-form to the total molecular species are 15% at pH 8.12, 30% at pH 8.57, and 50 % at pH 9.04. These values are much higher than those observed with STX (e.g., 36% at pH 9.8), which coincide with the observation that neoSTX undergoes more facile exchange of 11-H by enolization of the keto group than STX. There is no visible change in spin-spin coupling constants at different pH, suggesting that no major conformational change takes place in the equilibrium. Here, an important question is how can the above data be translated into the real solution structures at physiological conditions?

The dissociation of weak acids including protonated amines is known to be subject to considerable isotope effects by replacing H with D. The effect for -N$^+$-D -- -N + D$^+$ with respect to hydrogen is approximately:

$$pK = pK(in\ D_2O) - pK(in\ H_2O) = \sim 0.50.^9$$

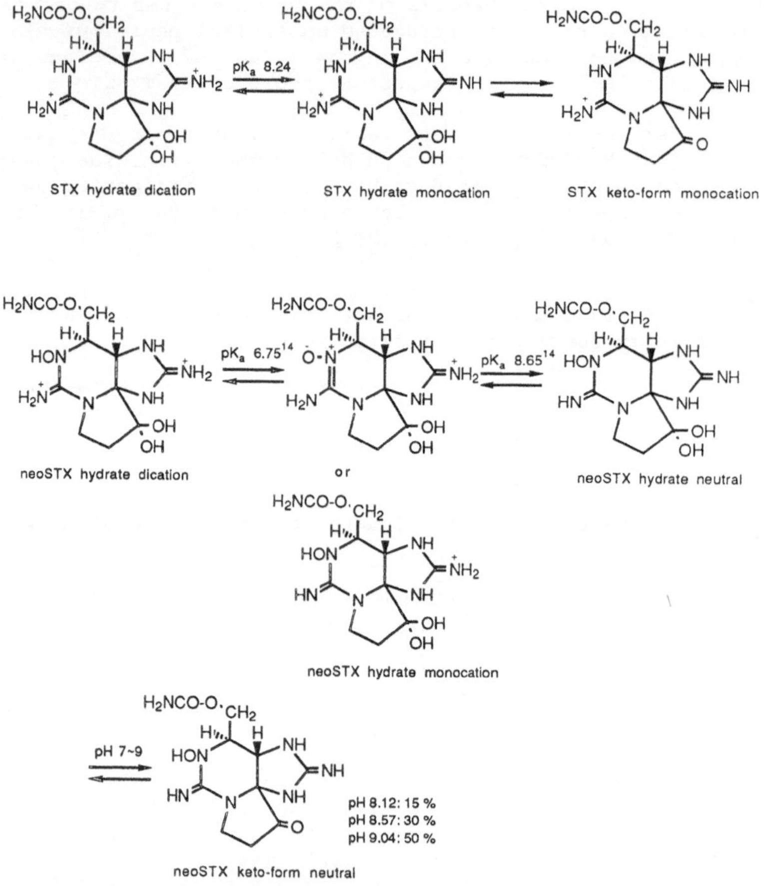

Fig. 3.   Molecular structures of saxitoxin (STX) and neosaxitoxin (neoSTX) in solution.

This means that the deprotonation is considerably restricted in D_2O. However, an ordinary glass electrode calibrated with normal standard buffer solutions requires a large correction for pD measurement, which is roughly:

$$pD = pH \text{ (operational)} + 0.45.\text{[1]}$$

Therefore, the observed pH in D_2O should be very close to the real values for the compound in H_2O.

At present, it is difficult to judge how these observed structural changes can be precisely correlated to the different responses of saxitoxin and neosaxitoxin to certain receptors.

ACKNOWLEDGEMENT

This work is supported by NIH grants, GM 28754 and GM 24425, which are greatly appreciated.

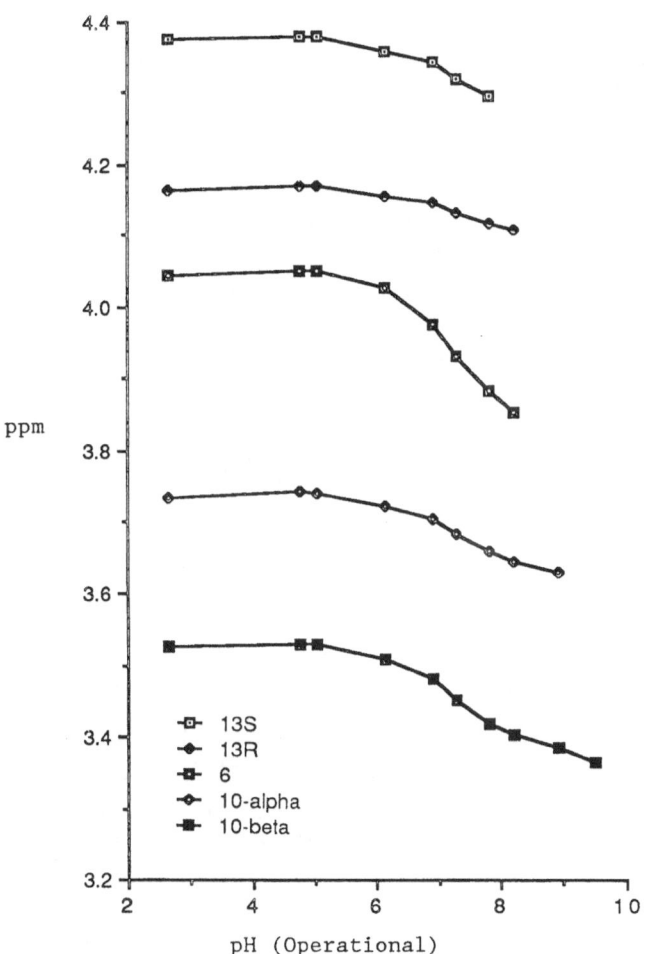

Fig. 4.   pH-Dependent chemical shift changes of neosaxitoxin (neoSTX) in D_2O.

REFERENCES

1. Covington AK, Paabo M, Robinson RA, Bates RG, 1968, Use of the glass electrode in deuterium oxide and the relation between the standardized pD (pa_D) scale and the operational pH in heavy water. *Anal Chem.*, 40:700-706.
2. Hille B, 1974, The receptor for tetrodotoxin and saxitoxin: A structural hypothesis, *Biophys J.*, 15:615-619.
3. Kao CY, 1983, New perspectives on the interaction of tetrodotoxin and saxitoxin with excitable membranes, *Toxicon*, Suppl. 3:211-219.
4. Kao CY, Walker SE, 1982, Active groups of saxitoxin and tetrodotoxin as deduced from actions of saxitoxin analogues on frog muscle and squid axon, *J Physiol.*, 323:619-637.
5. Moczydlowski E, Private Communication.
6. Moczydlowski E, Hall S, Garber SS, Strichartz GS, Miller C, 1984, Voltage-dependent blockade of muscle Na^+ channels by guanidinium toxins. Effects of toxin charge, *J Gen Physiol.*, 84:687-704.
7. Niccolai N, Schnoes HK, Gibbons WH, 1980, Study of the stereochemistry, relaxation mechanisms, and internal motions of natural products utilizing proton relaxation parameters: Solution and crystal structures of saxitoxin, *J Amer Chem Soc.*, 80:1513-1517.
8. Oshima Y, Buckley LJ, Alam M, Shimizu Y, 1977, Heterogeneity of paralytic shellfish poisons. Three new toxins from cultured *Gonyaulax tamarensis* cells, *Mya arenaria*, and *Saxidonus giganteus*, *Comp Biochem Physiol.*, 57c:31-34.
9. Robinson RA, Paabo M, Bates RG, 1969, Deuterium isotope effect on the dissociation of weak acids in water and deuterium oxide, *J Res NBS (Phys and Chem)*, 73A:299-308.
10. Rogers RS, Rapoport H, 1980, The pKa's of saxitoxin, *J Amer Chem Soc.*, 102:7335-7339.
11. Shimizu, Y, Paralytic Shellfish Poisons, 1984, *In*: "Progress in the chemistry of organic natural products", Herz W, Grisebach H, Kirby GW, eds., Springer-Verlag, New York, pp 235-264.
12. Shimizu Y, 1982, Recent Progress in Marine toxin research, *Pure & Appl Chem.*, 54:1973-1980.
13. Shimizu Y, Hsu CP, Genenah A, 1981, Structure of saxitoxin in solution and stereochemistry of dihydrosaxitoxin, *J Amer Chem Soc.*, 103:605-609.
14. Shimizu Y, Hsu CP, Fallon WE, Oshima Y, Miura I, Nakanishi K, 1978, The structure of neosaxitoxin, *Amer Chem Soc.*, 100:6791-6793.
15. Shimizu Y, Kobayashi M, Genenah A, Oshima Y, 1984, Isolation of side-chain sulfated saxitoxin analogs - their significance in interpretation of the mechanism of action, *Tetrahedron*, 40:539-544.
16. Shimizu Y, Norte M, Hori A, Genenah A, Kobayashi M, 1984, Biosynthesis of saxitoxin analogues: The unexpected pathway, *J Am Chem Soc.*, 106:6433-6434.
17. Strichartz G, 1981, Relative potencies of several derivatives of saxitoxin: Electrophysiological and toxin-binding studies, *Biophys J.*, 33:209a.
18. Strichartz G, 1982, Structure of the saxitoxin binding site at sodium channels in nerve membranes. Exchange of tritium from bound toxin molecules. *Mol Pharmacol.*, 21:343-350.
19. Strichartz G, 1984, Structural determinants of the affinity of saxitoxin for neuronal sodium channels, electrophysiological studies on frog peripheral nerve, *J Gen Physiol.*, 84:281-305.

CONTRIBUTORS